Springer Proceedings in Mathematics

Volume 17

For further volumes:
http://www.springer.com/series/8806

Springer Proceedings in Mathematics

The book series will feature volumes of selected contributions from workshops and conferences in all areas of current research activity in mathematics. Besides an overall evaluation, at the hands of the publisher, of the interest, scientific quality, and timeliness of each proposal, every individual contribution will be refereed to standards comparable to those of leading mathematics journals. It is hoped that this series will thus propose to the research community well-edited and authoritative reports on newest developments in the most interesting and promising areas of mathematical research today.

Christian Bär • Joachim Lohkamp
Matthias Schwarz
Editors

Global Differential Geometry

Springer

Editors

Christian Bär
Universität Potsdam
Institut für Mathematik
Postfach 60 15 53
14415 Potsdam Brandenburg
Germany
baer@math.uni-potsdam.de

Matthias Schwarz
Universität Leipzig
Fakultät für Mathematik und Informatik
Johannisgasse 26
04103 Leipzig Sachsen
Germany
mschwarz@math.uni-leipzig.de

Joachim Lohkamp
Universität Münster
Mathematisches Institut
Einsteinstraße 62
48149 Münster
Germany
j.lohkamp@uni-muenster.de

ISSN 2190-5614
ISBN 978-3-642-22841-4 e-ISBN 978-3-642-22842-1
DOI 10.1007/978-3-642-22842-1
Springer Heidelberg Dordrecht London New York

Library of Congress Control Number: 2011943494

Mathematical Subject Classification (2010): 53CXX; 53DXX; 58JXX

Printed on acid-free paper

Springer is part of Springer Science+Business Media (www.springer.com)

Preface

This collection of research papers provides an overview over some of the progress that has been made in major areas in Differential Geometry in the past few years. It is centred around the scientific activities within the Priority Programme "Global Differential Geometry" supported by the German Research Foundation – Deutsche Forschungsgemeinschaft (DFG) – from 2003 until 2009. This Priority Programme, and hence the present volume, covers the following areas as well as their mutual connections:

- Global Riemannian Geometry
- Geometric Analysis
- Symplectic Geometry

In particular this volume offers the following topics:

The contributions to Global Riemannian Geometry include existence and obstruction results for metrics with particular properties, such as metrics under particular curvature and/or holonomy constraints, or metrics of low-dimensional geometries. Interesting aspects of geometric limits are also included. Some papers discuss asymptotic geometries, Euclidean buildings or singular spaces.

One of the topics in Geometric Analysis is the spectral geometry of elliptic operators on Riemannian manifolds, including their applications in differential topology. Another one is the geometry and analysis of Lorentzian manifolds, as well as classical and quantum fields on Lorentzian manifolds. Progress on mean curvature flow and scalar curvature constraints are also discussed.

Finally, the Symplectic Geometry section considers new aspects of Floer Homology and Contact Structures on odd-dimensional manifolds.

We hope this panoramic collection of papers will be helpful and inspiring.

Potsdam — *Christian Bär*
Münster — *Joachim Lohkamp*
Leipzig — *Matthias Schwarz*

Contents

Part I
Riemannian Geometry

Holonomy Groups and Algebras

Lorenz J. Schwachhöfer

1 Introduction

An affine connection is one of the basic objects of interest in differential geometry. It provides a simple and invariant way of transferring information from one point of a connected manifold M to another and, not surprisingly, enjoys lots of applications in many branches of mathematics, physics and mechanics. Among the most informative characteristics of an affine connection is its holonomy group which is defined as the subgroup $Hol_p(M) \subset \mathrm{Aut}(T_pM)$ consisting of all automorphisms of the tangent space T_pM at $p \in M$ induced by parallel translations along p-based loops.

The notion of *holonomy* first arose in classical mechanics at the end of the 19th century. It was Heinrich Hertz who used the terms *"holonomic" and "non-holonomic" constraints* in his magnum opus *Die Prinzipien der Mechanik, in neuen Zusammenhängen dargestellt* ("The principles of mechanics presented in a new form") which appeared one year after his death in 1895. For a more detailed exposition of the early origins of the holonomy problem, see also [21].

The notion of holonomy in the mathematical context seems to have appeared for the first time in the work of E.Cartan [30, 31, 33]. He considered the Levi-Civita connection of a Riemannian manifold M, so that the holonomy group is contained in the orthogonal group. He showed that in this case, the holonomy group is always connected if M is simply connected. Moreover, he observed that $Hol_p(M)$ and $Hol_q(M)$ are conjugate via parallel translation along any path from p to q, hence the holonomy group $Hol(M) \subset Gl(n, \mathbb{R})$ is well defined up to conjugation.

L. J. Schwachhöfer (✉)
Fakultät für Mathematik, Technische Universität Dortmund, Vogelpothsweg 87, 44221 Dortmund, Germany
e-mail: lschwach@math.uni-dortmund.de

C. Bär et al. (eds.), *Global Differential Geometry*, Springer Proceedings in Mathematics 17, DOI 10.1007/978-3-642-22842-1_1,

Cartan's interest in holonomy groups was due to his observation that for a *Riemannian symmetric space*, the holonomy group and the isotropy group coincide up to connected components, as long as the symmetric space contains no Euclidean factor. This insight he used to classify Riemannian symmetric spaces [32].

In the 1950s, the concept of holonomy groups was treated more thoroughly. In 1952, Borel and Lichnerowicz [16] proved that the holonomy group of a Riemannian manifold is always a Lie subgroup, possibly with infinitely many components. In the same year, de Rham [37] proved what is nowadays called the *de Rham Splitting theorem.* Namely, if the holonomy of a Riemannian manifold is reducible, then the metric must be a local product metric; if the manifold is in addition complete and simply connected, then it must be a Riemannian product globally. In 1954, Ambrose and Singer proved a result relating the Lie algebra of the holonomy group and the curvature map of the connection [2].

A further milestone was reached by M.Berger in his doctoral thesis [9]. Based on the theorem of Ambrose and Singer, he established necessary conditions for a Lie algebra $\mathfrak{g} \subset \mathrm{End}(V)$ to be the Lie algebra of the holonomy group of a torsion free connection, and used it to classify all irreducible non-symmetric holonomy algebras of Riemannian metrics, i.e., such that $\mathfrak{g} \subset \mathfrak{so}(n)$. This list is remarkably short. In fact, it is included in (and almost coincides with) the list of connected linear groups acting transitively on the unit sphere. This fact was proven later directly by J.Simons [66] in an algebraic way. Recently, C.Olmos gave a beautiful simple argument showing this transitivity using elementary arguments from submanifold theory only [59].

Together with his list of possible *Riemannian* holonomy groups, Berger also gave a list of possible irreducible holonomy groups of *pseudo-Riemannian manifolds*, i.e., manifolds with a non-degenerate metric which is not necessarily positive definite. Furthermore, in 1957 he generalized Cartan's classification of Riemannian symmetric spaces to the isotropy irreducible ones [10].

In the beginning, it was not clear at all if the entries on Berger's list occur as the holonomy group of a Riemannian manifold. In fact, it took several decades until the last remaining cases were shown to occur by Bryant [16]. As it turns out, the geometry of manifolds with special holonomy groups are of utmost importance in many areas of differential geometry, algebraic geometry and mathematical physics, in particular in string theory. It would lead too far to explain all of these here, but rather we refer the reader to [11] for an overview of the geometric significance of these holonomies.

In 1998, S.Merkulov and this author classified all *irreducible* holonomy groups of torsion free connections [69]. In the course of this classification, some new holonomies were discovered which are *symplectic*, i.e., they are defined on a symplectic manifold such that the symplectic form is parallel. The first such symplectic example was found by Bryant [17]; later, in [34, 35] an infinite family of such connections was given. These symplectic holonomies share some striking rigidity properties which later were explained on a more conceptual level by M.Cahen and this author [26], linking them to parabolic contact geometry.

In this article, we shall put the main emphasis on the investigation of connections on principal bundles as all other connections can be deduced from these. This allows us to prove most of the basic results in greater generality than they were originally stated and proven. Thus, Sect. 2 is devoted to the collection of the basic definitions and statements, where in most cases, sketches of the proofs are provided. In Sect. 3, we shall collect the known classification results where we do not say much about the proofs, and finally, in Sect. 4 we shall describe the link of special symplectic connections with parabolic contact geometry.

2 Basic Definitions and Results

2.1 Connections on Principal Bundles

Let $\pi : P \to M$ be a (right)-principal G-bundle, where M is a connected manifold and G is a Lie group with Lie algebra $\mathfrak{g}$. A *principal connection on P* may be defined as a $\mathfrak{g}$- valued one-form $\omega \in \Omega^1(P) \otimes \mathfrak{g}$ such that:

1. ω is G-equivariant, i.e., $r^*_{g^{-1}}(\omega) = Ad_g \circ \omega$ for all $g \in G$,
2. $\omega(\xi^*) = \xi$ for all $\xi \in \mathfrak{g}$, where $\xi^*_p := \frac{\mathrm{d}}{\mathrm{d}t}|_{t=0}(p \cdot \exp(t\xi))$ denotes the action field corresponding to ξ.

Here, $r_g : P \to P$ denotes the right action of G. Alternatively, we may define a principal connection to be a G-invariant splitting of the tangent bundle

$$TP = \mathscr{H} \oplus \mathscr{V}, \text{ where } \mathscr{V}_p = \ker(d\pi)_p = span(\{\xi^*_p \mid \xi \in \mathfrak{g}\}) \text{ for all } p \in P. \quad (1)$$

In this case, $\mathscr{H}$ and $\mathscr{V}$ are called the *vertical* and *horizontal space*, respectively.

To see that these two definitions are indeed equivalent, note that for a given connection one-form $\omega \in \Omega^1(P) \otimes \mathfrak{g}$, we may define $\mathscr{H} := \ker(\omega)$; conversely, given the splitting (1), we define ω by $\omega|_{\mathscr{H}} \equiv 0$ and $\omega(\xi^*) = \xi$ for all $\xi \in \mathfrak{g}$; it is straightforward to verify that this establishes indeed a one-to-one correspondence.

The *curvature form* of a principal connection is defined as

$$\Omega := \mathrm{d}\omega + \frac{1}{2}[\omega, \omega] \in \Omega^2(P) \otimes \mathfrak{g}. \quad (2)$$

For its exterior derivative we get

$$\mathrm{d}\Omega + [\omega, \Omega] = 0. \quad (3)$$

By the *Maurer-Cartan equations*, it follows from (2) that

$$\xi^* \lrcorner \Omega = 0 \text{ for all } \xi \in \mathfrak{g}, \text{ and } \mathrm{d}r^*_g(\Omega) = Ad_g \circ \Omega. \quad (4)$$

A (piecewise smooth) curve $c : [a,b] \to P$ is called *horizontal* if $c'(t) \in \mathscr{H}_{c(t)}$ for all $t \in [a,b]$. Evidently, for every curve $\underline{c} : [a,b] \to M$ and $p \in \pi^{-1}(\underline{c}(a))$, there is a unique horizontal curve $c^p : [a,b] \to P$, called *horizontal lift of c*, with $\underline{c} = \pi \circ c^p$ and $c^p(a) = p$. Since by the G-equivariance of $\mathscr{H}$ we have $c^{p \cdot g} = r_g \circ c^p$, the correspondence

$$\Pi_{\underline{c}} : \pi^{-1}(\underline{c}(a)) \longrightarrow \pi^{-1}(\underline{c}(b)), \qquad p \longmapsto c^p(b)$$

is G-equivariant and is called *parallel translation along* $\underline{c}$. The *holonomy at* $p \in P$ is then defined as

$$\begin{aligned} Hol_p := \{g \in G \mid p \cdot g = \Pi_{\underline{c}}(p) \text{ for } \underline{c} : [a,b] \to M \text{ with } \underline{c}(a) = \underline{c}(b) \\ = \pi(p)\} \subset G. \end{aligned} \tag{5}$$

Evidently, $Hol_p \subset G$ is a subgroup as we can concatenate and invert loops. Also, the G- equivariance of $\mathscr{H}$ implies that

$$Hol_{p \cdot g} = g^{-1}\, Hol_p\, g. \tag{6}$$

Moreover, if we pick any path $\underline{c} : [a,b] \to M$ then, again by concatenating paths, we obtain for $p \in \pi^{-1}(\underline{c}(a))$

$$Hol_{\Pi_{\underline{c}}(p)} = Hol_p. \tag{7}$$

Thus, by (6) and (7) it follows that the holonomy group $Hol \cong Hol_p \subset G$ is well defined up to conjugation in G, independent of the choice of $p \in P$.

We define the equivalence relation $\sim$ on P by saying that

$$p \sim q \text{ if } p \text{ and } q \text{ can be joined by a horizontal path.} \tag{8}$$

Then definition (5) can be equivalently formulated as

$$Hol_p := \{g \in G \mid p \cdot g \sim p\}. \tag{9}$$

Theorem 2.1. (Ambrose-Singer-Holonomy Theorem [2]). *Let $\pi : P \to M$ be a principal G-bundle with a connection $\omega \in \Omega^1(P) \otimes \mathfrak{g}$ and the corresponding horizontal distribution $\mathscr{H} \subset TP$.*

1. The smallest involutive distribution on P which contains $\mathscr{H}$ is the distribution

$$\hat{\mathscr{H}}_p := \mathscr{H}_p \oplus \{\xi_p^* \mid \xi \in \mathfrak{hol}_p\},$$

where $\mathfrak{hol}_p \subset \mathfrak{g}$ is the Lie subalgebra generated by

$$\mathfrak{hol}_p = \langle\{\Omega(d\Pi_{\underline{c}}(v), d\Pi_{\underline{c}}(w)) \mid v, w \in T_pP, \underline{c} : [a,b] \\ \to M \textit{ any path with } \underline{c}(a) = \pi(p)\}\rangle. \tag{10}$$

2. *The identity component of* $(Hol_p)_0 \subset G$ *is a (possibly non-regular) Lie subgroup with Lie algebra* $\mathfrak{hol}_p$.

Proof. Observe first that the dimension of the right hand side of (10) is independent of $p \in P$. Indeed, from the definition, $\hat{\mathscr{H}}_{q\cdot g} = dr_g(\hat{\mathscr{H}}_q)$, so that this dimension is independent of the point in the fiber of P; moreover, if $p \sim q$ and $\underline{c} : [a,b] \to M$ is a path with horizontal lift joining p and q, then it follows from the very definition that $\hat{\mathscr{H}}_q \cap \mathscr{V}_q = d\Pi_{\underline{c}}(\hat{\mathscr{H}}_p \cap \mathscr{V}_p)$, and $d\Pi_{\underline{c}}$ is an isomorphism.

To see that $\hat{\mathscr{H}}$ is involutive, let $\underline{X}, \underline{Y} \in \mathscr{X}(M)$ be vector fields and $X, Y \in \mathscr{X}(P)$ be their horizontal lifts. Note that the flows Φ^t_X and $\Phi^t_{\underline{X}}$ relate as

$$\Phi^t_X = \Pi_{\underline{c}^t_X}, \quad \text{where } \underline{c}^t_X : [0,t] \to M \text{ is a trajectory of } \underline{X}.$$

Therefore, if we let $\hat{\mathscr{V}}_p := \{\xi^*_p \mid \xi \in \mathfrak{hol}_p\}$, then the definition of $\mathfrak{hol}_p$ implies that $\Phi^t_X(\hat{\mathscr{V}}_p) = \hat{\mathscr{V}}_q$, where $q = \Phi^t_X(p)$ and thus, $[X, \hat{\mathscr{V}}_p] \subset \hat{\mathscr{V}}_p$ for all horizontal vector fields X, i.e., $[\mathscr{H}, \hat{\mathscr{V}}] \subset \hat{\mathscr{H}}$.

Next, by (2), $[X,Y] = -\xi^*_{\Omega(X,Y)}$ mod $\mathscr{H}$ for all horizontal vector fields X, Y so that $[\mathscr{H}, \mathscr{H}] \subset \hat{\mathscr{H}}$; finally, $[\hat{\mathscr{V}}, \hat{\mathscr{V}}] \subset \hat{\mathscr{V}}$ as $\mathfrak{hol}_p$ is a Lie algebra by definition.

Thus, $\hat{\mathscr{H}} \subset P$ is an involutive distribution. Conversely, the above arguments show that any involutive distribution containing $\mathscr{H}$ also contains $\hat{\mathscr{H}}$, so that $\hat{\mathscr{H}}$ is minimal as asserted.

Let $P_0 \subset P$ be a maximal leaf of $\hat{\mathscr{H}}$, let $p_0 \in P_0$ and let

$$H := \{g \in G \mid p_0 \cdot g \in P_0\} \subset G.$$

Since $\mathscr{H}$ and hence $\hat{\mathscr{H}}$ is G-invariant, it follows that $H \subset G$ is a subgroup. In fact, $H \subset G$ is a (possibly non-regular) Lie subgroup since $H \cong P_0 \cap \pi^{-1}(\pi(p_0))$. In fact, the restriction $\pi : P_0 \to M$ is a principal H-bundle.

Standard arguments now show that P_0 is indeed a single equivalence class w.r.t. $\sim$, so that $H = Hol_{p_0}$ is a Lie subgroup of G with Lie algebra $\mathfrak{hol}_p$. See e.g. [5] for details. □

Definition 2.2. Let $P \to M$ be a principal G-bundle, and let $H \subset G$ be a (possibly non-regular) Lie subgroup of G. We call a (possibly non-regular) submanifold $P' \subset P$ an H- *reduction of* P if the restriction $\pi : P' \to M$ is a principal H-bundle.

In particular, a maximal leaf $P_0 \subset P$ of the distribution $\hat{\mathscr{H}}$ from Theorem 2.1 is called a *holonomy reduction of* P which is therefore a reduction with structure group $Hol \subset G$. We denote the restriction of ω, Ω and $\mathscr{H}$ to P_0 by the same symbols.

By the G-equivariance of the distributions $\mathscr{H}$ and $\hat{\mathscr{H}}$ it follows that if $P_0, P_0' \subset P$ are two holonomy reductions then $P_0' = r_g(P_0)$ for some $g \in G$. That is, the holonomy reduction $P_0 \subset P$ is unique up to the right G-action, and this allows to speak of *the* holonomy reduction.

The connection ω is called *locally flat* if $\Omega = 0$. By the above definitions, $\Omega = 0$ if and only if the horizontal distribution $\mathscr{H}$ from (1) is involutive, hence the holonomy reduction $P_0 \to M$ is a regular covering with deck group Hol. It follows that the pull-back of this covering, $(\pi|_{P_0})^*(P) = P_0 \times G$, is the trivial bundle, and ω is simply the pull back of the *Maurer-Cartan form* on G under projection onto the second factor.

This idea can be generalized as follows.

Proposition 2.3. *Let $\pi : P \to M$ be a principal G-bundle with a connection with holonomy group $Hol \subset G$, and let $Hol^0 \subset Hol$ denote the identity component. Then there is a regular covering $p : \tilde{M} \to M$ with deck group $\Gamma := Hol/Hol^0$ such that the pull-back bundle $p^*(P) \to \tilde{M}$ with the connection $p^*(\omega)$ has holonomy group Hol^0.*

In particular, if M is simply connected, then the holonomy group of any connection on $P \to M$ is connected.

Proof. Let $P_0 \subset P$ be a holonomy reduction, and let $\tilde{M} := P_0/Hol^0$. Then the induced map $p : \tilde{M} \to M$ is a principal Γ-bundle, and since Γ is discrete, it follows that p is a regular covering. Thus, we have the commutative diagram of principal bundles

$$\begin{array}{ccc} P_0 & & \\ {\scriptstyle Hol^0}\downarrow & \searrow{\scriptstyle Hol} & \\ \tilde{M} & \xrightarrow{\ \Gamma\ } & M \end{array}$$

with the indicated structure groups, and the distribution $\mathscr{H}$ on P_0 induces a connection on each of the principal bundles indicated by the vertical arrows. It follows now that $P_0 \to \tilde{M}$ is the holonomy reduction of $p^*(P_0) \to \tilde{M}$. □

The regular covering $p : \tilde{M} \to M$ yields a short exact sequence

$$0 \longrightarrow \pi_1(\tilde{M}) \xrightarrow{\ p_*\ } \pi_1(M) \xrightarrow{\ m\ } \Gamma \longrightarrow 0,$$

and the map $m : \pi_1(M) \to \Gamma = Hol/Hol^0$ is called the *monodromy map*. It can be interpreted geometrically as follows. The parallel translation along a contractible loop in M always lies in Hol^0 since it can be joined to the identity by the parallel translations along a family of paths which define a homotopy to the trivial loop. Thus, the parallel translation along any loop, regarded mod Hol^0, only depends on the represented homotopy class, and this yields the monodromy map.

We finish this section by mentioning the following result.

Theorem 2.4. *[46] Let $P \to M$ be a principal G bundle, let $H \subset G$ be a (possibly non-regular) Lie subgroup. Moreover, let $P_0 \subset P$ be a connected (possibly non-embedded) H-reduction of P. Then there is a connection on P such that P_0 is the holonomy reduction of this connection.*

In particular, there is a connection on P with holonomy group H if and only if P admits a connected H-reduction $P_0 \subset P$.

Proof. Any connection on P_0 can be extended to a connection on P in a unique way, using the G-equivariance of the connection form. Thus, the problem reduces to showing that the principal H-bundle $P_0 \to M$ has a connection whose holonomy equals all of H.

If we pick a "generic" (i.e., maximally non-integrable) horizontal distribution in the neighborhood of some $p \in P_0$, then $\{\Omega(v, w) \mid v, w \in \mathscr{H}_p\} = \mathfrak{h}$. Thus, by Ambrose-Singer holonomy Theorem 2.1, the holonomy reduction has the same dimension as P_0, and since P_0 is connected, it is the holonomy reduction, showing that H is the holonomy group. □

If $H \subset G$ is a regular subgroup, then the existence of an H-reduction is equivalent to the existence of a global section of the G/H-fiber bundle $P/H \to M$. That is, the existence of a connection with prescribed holonomy is merely a topological property.

2.2 Connections on Vector Bundles

Let $P \to M$ be a principal G-bundle, and let $\rho : G \to \mathrm{Aut}(V)$ be a representation on a finite dimensional (real or complex) vector space. Then the *associated vector bundle* is the bundle

$$E := P \times_G V \longrightarrow M,$$

where $P \times_G V$ is the quotient of $P \times V$ by the free G-action $g \star (p, v) := (p \cdot g^{-1}, \rho(g)v)$. Evidently, the fibers of E are isomorphic to V. In fact, every vector bundle $E \to M$ can be described (non-uniquely) in this way: we fix a (real or complex) vector space V isomorphic to the fibers of E, and let

$$P_E := \{u_x : E_x \to V \text{ a linear isomorphism, where } x \in M\}$$

with the obvious projection to M. This is called the *full frame bundle of E*. The structure group of P_E is $\mathrm{Aut}(V)$ which acts by composition from the right, and it is straightforward to verify that $P_E \to M$ becomes a principal $\mathrm{Aut}(V)$-bundle, and $E = P_E \times_{\mathrm{Aut}(V)} V$ with the natural action of $\mathrm{Aut}(V)$ on V.

In general, if $E = P \times_G V$ is such a vector bundle and ω is a connection on P, then the splitting (1) of TP induces a splitting

$$T_{(p,v)}(P \times V) = \mathscr{H}_p \oplus \mathscr{V}_p \oplus V,$$

and since $\mathscr{H}$ is invariant under the diagonal action of G, it descends to a distribution $\mathscr{H}_E \subset TE$ on $E = P \times_G V$, which is transversal to the fibers of $E \to M$. Since $\ker(d\pi) = E$ as a bundle in a canonical way, the connection ω on P induces a bundle splitting

$$TE = \mathscr{H}_E \oplus E. \tag{11}$$

Thus, we have an induced projection $TE \to E$, and this defines a *covariant derivative* on E, i.e., a map

$$\nabla : \Gamma^\infty(M, E) \longrightarrow \Omega^1(M) \otimes \Gamma^\infty(M, E) \qquad \text{as} \qquad \nabla\sigma := (d\sigma)_E.$$

Let $\underline{c} : [a,b] \to M$ be a (piecewise smooth) path, pick a horizontal lift $c : [a,b] \to P$ and some $v_0 \in V$. We let $v : [a,b] \to E$ be defined as $v(t) := (c(t), v_0)/G \in P \times_G V = E$. Then v is parallel along $\underline{c}$, i.e., $\nabla_{\underline{c}'(t)} v(t) = 0$. Thus, as in the case of a connection on a principal bundle, we have the notion of parallel translation

$$P^E_{\underline{c}} : E_{\underline{c}(a)} \longrightarrow E_{\underline{c}(b)}$$

which is a linear isomorphism. Thus, the definition of the *holonomy group of* ∇ is given analogously as

$$\begin{aligned} Hol_x(E \to M, \nabla) := \big\{ P^E_{\underline{c}} \mid \underline{c} : [a,b] \to M \text{ a path with } \underline{c}(a) \\ = \underline{c}(b) = x \big\} \subset \mathrm{Aut}(E_x). \end{aligned}$$

If $c : [a,b] \to P$ is a horizontal lift of some loop, then $c(b) = c(a) \cdot g$ for some $g \in G$, and hence, $v(b) = g \cdot v(a)$ with $v(t) \in E_{\underline{c}(t)}$ as above. Therefore, we have the following

Proposition 2.5. *Let $P \to M$ be a principal G-bundle and let $E := P \times_G V$ be an associated principal bundle w.r.t. some representation $\rho : G \to Aut(V)$. Let $\omega \in \Omega^1(P) \otimes \mathfrak{g}$ be a connection on P and let $\nabla : \Gamma^\infty(M, E) \longrightarrow \Omega^1(M) \otimes \Gamma^\infty(M, E)$ be the induced covariant derivative on E.*

Then for $p \in P$ and $x := \pi(p) \in M$ we have $Hol_x(E \to M, \nabla) \cong \rho(Hol_p) \subset Aut(V)$. In particular, if $P_0 \subset P$ is the holonomy reduction of ω, then $E = P_0 \times_{Hol} V$.

Therefore, connections on vector bundles and their holonomies can be described in terms of the holonomy on an associated principle bundle.

2.3 The Spencer Complex

We shall briefly summarize the construction of the Spencer complex for a Lie subalgebra $\mathfrak{g} \subset \mathrm{End}(V)$. For a more detailed exposition, we refer the interested reader to [18, 44, 58].

Let V be a finite dimensional vector space over $\mathbb{R}$ or $\mathbb{C}$. We let $A^{p,q}(V) := \odot^p V^* \otimes \Lambda^q V^*$. This space can be thought of as the space of q-forms on V with values in the space of homogeneous polynomials on V of degree p. Exterior differentiation thus yields a map $\delta : A^{p,q}(V) \to A^{p-1,q+1}(V)$, which turns $A^{*,*}(V) = \bigoplus_{p,q\geq 0} A^{p,q}(V)$ into a bigraded complex. Likewise, $\bigoplus_{p,q\geq 0}(V \otimes A^{p,q}(V))$ becomes a bigraded complex by the maps $\delta_V := Id_V \otimes \delta$.

Let $\mathfrak{g} \subset \mathrm{End}(V) \cong V^* \otimes V$ be a subalgebra. The *k-th prolongation of* $\mathfrak{g}$, denoted by $\mathfrak{g}^{(k)}$ for an integer k, is defined inductively by the formulae $\mathfrak{g}^{(-1)} = V$, $\mathfrak{g}^{(0)} = \mathfrak{g}$, and

$$\mathfrak{g}^{(k)} = \delta_V^{-1}(\mathfrak{g}^{(k-1)} \otimes V^*).$$

That is,

$$\mathfrak{g}^{(k)} = (\mathfrak{g} \otimes \odot^k V^*) \cap (V \otimes \odot^{k+1} V^*),$$

where we use exterior differentiation $\delta : \odot^{k+1} V^* \to V^* \otimes \odot^k V^*$ to regard both $\mathfrak{g} \otimes \odot^k V^*$ and $V \otimes \odot^{k+1} V^*$ as subspaces of $V \otimes V^* \otimes \odot^k V^*$. Alternatively, we can define $\mathfrak{g}^{(k)}$ inductively by $\mathfrak{g}^{(-1)} = V$, $\mathfrak{g}^{(0)} = \mathfrak{g}$ and the exact sequence

$$0 \longrightarrow \mathfrak{g}^{(k)} \longrightarrow \mathfrak{g}^{(k-1)} \otimes V^* \longrightarrow \mathfrak{g}^{(k-2)} \otimes \Lambda^2 V^*. \tag{12}$$

For example,

$$\mathfrak{g}^{(1)} = \{\alpha \in V^* \otimes \mathfrak{g} \mid \alpha(x)y = \alpha(y)x \text{ for all } x, y \in V\}.$$

Furthermore, we define the *Spencer complex of* $\mathfrak{g}$ to be $(C^{p,q}(\mathfrak{g}), \delta)$ with

$$C^{p,q}(\mathfrak{g}) = \mathfrak{g}^{(p-1)} \otimes \Lambda^q(V^*) \subset V \otimes \odot^p V^* \otimes \Lambda^q V^* = V \otimes A^{p,q}(V).$$

It is not hard to see that $\delta(C^{p,q}(\mathfrak{g})) \subset C^{p-1,q+1}(\mathfrak{g})$, and thus, $(C^{p,q}(\mathfrak{g}), \delta)$ is indeed a complex where we denote the boundary maps by

$$\delta_{\mathfrak{g}}^{p,q} : C^{p,q}(\mathfrak{g}) \longrightarrow C^{p-1,q+1}(\mathfrak{g}). \tag{13}$$

Its cohomology groups $H^{p,q}(\mathfrak{g})$ are called the *Spencer cohomology groups of* $\mathfrak{g}$. The lower corner of this bigraded complex takes the form

$$\begin{array}{ccccccccc}
\mathfrak{g}^{(2)} & & \mathfrak{g}^{(2)} \otimes V^* & & \cdots & & & & \\
& \searrow & & \searrow & & & & & \\
\mathfrak{g}^{(1)} & & \mathfrak{g}^{(1)} \otimes V^* & & \mathfrak{g}^{(1)} \otimes \Lambda^2 V^* & & \cdots & & \\
& \searrow & & \searrow & & \searrow & & & \\
\mathfrak{g} & & \mathfrak{g} \otimes V^* & & \mathfrak{g} \otimes \Lambda^2 V^* & & \mathfrak{g} \otimes \Lambda^3 V^* & & \cdots \\
& \searrow & & \searrow & & \searrow & & \searrow & \\
V & & V \otimes V^* & & V \otimes \Lambda^2 V^* & & V \otimes \Lambda^3 V^* & & \cdots
\end{array}$$

Table 1 List of irreducible complex matrix Lie groups G with $\mathfrak{g}^{(1)} \neq 0$

	Group G	Representation V	$\mathfrak{g}^{(1)}$	$\mathfrak{g}^{(2)}$	$H^{1,2}(\mathfrak{g})$
1	$\mathrm{SL}(n,\mathbb{C})$	$\mathbb{C}^n, n \geq 2$	$(V \otimes \odot^2 V^*)_0$	$(V \otimes \odot^3 V^*)_0$	$\odot^2 V^*$
2	$\mathrm{GL}(n,\mathbb{C})$	$\mathbb{C}^n, n \geq 1$	$V \otimes \odot^2 V^*$	$V \otimes \odot^3 V^*$	0
3	$\mathrm{GL}(n,\mathbb{C})$	$\odot^2\mathbb{C}^n, n \geq 2$	V^*	0	0
4	$\mathrm{GL}(n,\mathbb{C})$	$\Lambda^2\mathbb{C}^n, n \geq 5$	V^*	0	0
5	$\mathrm{GL}(m,\mathbb{C}) \cdot \mathrm{GL}(n,\mathbb{C})$	$\mathbb{C}^m \otimes \mathbb{C}^n, m,n \geq 2$	V^*	0	0
6	$\mathrm{Sp}(n,\mathbb{C})$	$\mathbb{C}^{2n}, n \geq 2$	$\odot^3 V^*$	$\odot^4 V^*$	0
7	$\mathbb{C}^* \cdot \mathrm{Sp}(n,\mathbb{C})$	$\mathbb{C}^{2n}, n \geq 2$	$\odot^3 V^*$	$\odot^4 V^*$	0
8	$\mathrm{CO}(n,\mathbb{C})$	$\mathbb{C}^n, n \geq 3$	V^*	0	$\mathscr{W}$[1]
9	$\mathbb{C}^* \cdot \mathrm{Spin}(10,\mathbb{C})$	$\mathbb{C}^{16}$	V^*	0	0
10	$\mathbb{C}^* \cdot \mathrm{E}_6^{\mathbb{C}}$	$\mathbb{C}^{27}$	V^*	0	0

[1] $\mathscr{W}$ denotes the space of *formal Weyl curvatures* (see e.g. [11]).

It is worth pointing out that all of these spaces are $\mathfrak{g}$-modules in an obvious way, and that all maps are $\mathfrak{g}$-equivariant. Thus, the Spencer cohomology groups are $\mathfrak{g}$-modules as well. Also, we define $K(\mathfrak{g}) := \ker \delta_{\mathfrak{g}}^{1,2}$, so that we have the exact sequence

$$0 \longrightarrow \mathfrak{g}^{(2)} \longrightarrow \mathfrak{g}^{(1)} \otimes V^* \longrightarrow K(\mathfrak{g}) \longrightarrow H^{1,2}(\mathfrak{g}) \longrightarrow 0, \tag{14}$$

where the map in the middle is given by $R_{\alpha\otimes\phi}(x,y) = \phi(x)\alpha(y) - \phi(y)\alpha(x)$ for $\alpha \otimes \phi \in \mathfrak{g}^{(1)} \otimes V^*$.

If we assume that $\mathfrak{g} \subset \mathrm{End}(V)$ acts *irreducibly*, then there are only very few possibilities for which $\mathfrak{g}^{(1)} \neq 0$. These subalgebras have been classified by Cartan [29] and Kobayashi and Nagano [53]. The result is listed in Table 1 for *complex* Lie algebras. The Spencer cohomologies $H^{1,2}(\mathfrak{g})$ of these Lie algebras are well-known. (See e.g. [18] and [69] who use considerably different techniques for the calculations).

2.4 G-Structures and Intrinsic Torsion

Consider the tangent bundle $TM \to M$ of a connected manifold M. We define the *(total) coframe bundle*

$$\mathfrak{F}_V := \{u_x : T_xM \to V \text{ a linear isomorphism}\} \longrightarrow M,$$

where $\dim V = \dim M$ as in Sect. 2.2. Thus, $\mathfrak{F}_V \to M$ is an $\mathrm{Aut}(V)$- principal bundle. On $\mathfrak{F}_V$, we define the *tautological one-form*

$$\theta \in \Omega^1(\mathfrak{F}_V) \otimes V, \qquad \theta_u(v) := u(d\pi(v)) \text{ for } v \in T_u(\mathfrak{F}_V), \tag{15}$$

where $\pi : TM \to M$ denotes the canonical projection. We have the equivariance condition

$$r_g^*(\theta) = g \cdot \theta. \tag{16}$$

Let $G \subset \mathrm{Aut}(V)$ be a (possibly non-regular) Lie subgroup. A *G-structure on M* is, by definition, a reduction of $\mathfrak{F}_V$ with structure group G, i.e., it is a (possibly non-regular) submanifold $F \subset \mathfrak{F}_V$ such that we have the commuting diagram

$$\begin{array}{ccc} F & \overset{\iota}{\hookrightarrow} & \mathfrak{F}_V \\ & {\scriptstyle G}\searrow & \downarrow {\scriptstyle \mathrm{Aut}(V)} \\ & & M \end{array}$$

Note that if $G \subset \mathrm{Aut}(V)$ is a regular subgroup, then these reductions are in a one-to-one correspondence to sections of the $\mathrm{Aut}(V)/G$-bundle $\mathfrak{F}_V/G \to M$. Also, $\iota^*(\theta)$ is called the *tautological one form of F*, and we shall denote it by θ instead of $\iota^*(\theta)$. In fact, the existence of such a form $\theta \in \Omega^1(F) \otimes V$ characterizes G-structures on M as the next result shows.

Proposition 2.6. *Let $\pi : P \to M$ be a principal G-bundle and let V be a vector space of the same dimension as M. If there exists one form $\underline{\theta} \in \Omega^1(P) \otimes V$ with* $\ker(\underline{\theta}) = \ker(d\pi)$ *and a faithful representation $\rho : G \to Aut(V)$ such that $r_g^*(\underline{\theta}) = \rho(g) \cdot \underline{\theta}$ for all $g \in G$, then there is a G-invariant immersion $\iota : P \hookrightarrow \mathfrak{F}_V$ such that*

$$\begin{array}{ccc} P & \overset{\iota}{\hookrightarrow} & \mathfrak{F}_V \\ & {\scriptstyle G}\searrow & \downarrow {\scriptstyle Aut(V)} \\ & & M \end{array}$$

commutes and $\underline{\theta} = \iota^*(\theta)$, where θ is the tautological one form on $\mathfrak{F}_V$. In particular, $\iota(P) \subset \mathfrak{F}_V$ is a G-structure on M with tautological form $\underline{\theta}$.

Proof. Since for $p \in P$ we have $\ker \mathrm{d}\pi_p = \ker \underline{\theta}_p$, it follows that there is a unique isomorphism $\iota_p : T_{\pi(p)}M \to V$ such that $\underline{\theta}_p = \mathrm{d}\pi_p \circ \iota_p$. Thus, $\iota_p \in \mathfrak{F}_V$, so that we get a smooth map $\iota : P \to \mathfrak{F}_V$. The equivariance of $\underline{\theta}$ implies that ι is G-equivariant, hence $\iota(P) \subset \mathfrak{F}_V$ is a G-structure, and the fact that $\underline{\theta} = \iota^*(\theta)$ follows immediately from definition (15). □

Note that $\mathrm{End}(V)$ is the Lie algebra of $\mathrm{Aut}(V)$, hence any connection on $\mathfrak{F}_V$ is a one form $\omega \in \Omega^1(\mathfrak{F}_V) \otimes \mathrm{End}(V)$. Its *torsion* is defined as

$$\Theta := \mathrm{d}\theta + \omega \wedge \theta \in \Omega^2(\mathfrak{F}_V) \otimes V \tag{17}$$

whose derivative yields

$$\Omega \wedge \theta = d\Theta + \omega \wedge \Theta. \tag{18}$$

Then (17) implies the conditions

$$\xi^* \lrcorner \Theta = 0 \ \text{ for all } \xi \in \mathfrak{g}, \text{ and } dr_g^*(\Theta) = g \cdot \Theta. \tag{19}$$

Therefore, there is a Aut(V)-equivariant map $Tor : \mathfrak{F}_V \to Hom(\Lambda^2 V, V)$ such that

$$\Theta = Tor(\theta \wedge \theta).$$

The equivariance of Tor implies that its derivative takes the form

$$dTor + \omega \cdot Tor = \nabla_\theta Tor,$$

where the multiplication on the left hand side refers to the action of $\mathfrak{g} \subset \text{End}(V)$ on $\Lambda^2 V^* \otimes V$, and where $\nabla_\theta Tor \in \Omega^1(\mathfrak{F}_V) \otimes (V^* \otimes \Lambda^2 V^* \otimes V)$. Analogously, by (4), it follows that there is a Aut(V)-equivariant map $R : \mathfrak{F}_V \to Hom(\Lambda^2 V, \text{End}(V))$ such that

$$\Omega = R(\theta \wedge \theta), \tag{20}$$

and (18) implies the *first Bianchi identity*

$$\sum_{cycl.} R(v_1, v_2)v_3 = \sum_{cycl.} (\nabla_{v_1} Tor)(v_2, v_3) + Tor(Tor(v_1, v_2), v_3). \tag{21}$$

Again, by the Aut(V)-equivariance of R, taking the derivative of (20) yields

$$dR + \omega \cdot R = \nabla_\theta R, \tag{22}$$

where the multiplication on the left hand side refers to the action of $\mathfrak{g} \subset \text{End}(V)$ on $\Lambda^2 V^2 \otimes \text{End}(V)$, and where $\nabla_\theta R \in \Omega^1(\mathfrak{F}_V) \otimes (V^* \otimes \Lambda^2 V^* \otimes \text{End}(V))$. In fact, (3) implies that for all $v_1, v_2, v_3 \in V$ we have the *second Bianchi identity*

$$\sum_{cycl.} (\nabla_{v_1} R)(v_2, v_3) + R(Tor(v_1, v_2), v_3) = 0. \tag{23}$$

As $Hom(\Lambda^2 TM, TM) = \mathfrak{F}_V \times_{\text{Aut}(V)} Hom(\Lambda^2 V, V)$ and $Hom(\Lambda^2 TM, \text{End}(TM)) = \mathfrak{F}_V \times_{\text{Aut}(V)} Hom(\Lambda^2 V, \text{End}(V))$ the equivarianve of Tor and R implies that they induce sections which by abuse of notation we denote by the same symbols, namely

$$Tor \in \Gamma^\infty(Hom(\Lambda^2 TM, TM)) \ \text{ and } \ R \in \Gamma^\infty(Hom(\Lambda^2 TM, \text{End}(TM)),$$

These sections are also called the *torsion* and the *curvature of the connection*, respectively. In terms of the covariant derivative on TM corresponding to ω they are given by the formulas

$$Tor(X,Y) = \nabla_X Y - \nabla_Y X - [X,Y]$$

and

$$R(X,Y)Z = \nabla_X \nabla_Y Z - \nabla_Y \nabla_X Z - \nabla_{[X,Y]} Z.$$

If ω_F is a connection one form on the G-bundle $F \to M$, then by Aut(V)-equivariant continuation, there is a unique connection one form $\omega \in \Omega^1(\mathfrak{F}_V) \otimes \mathrm{End}(V)$ such that $\omega_F = \iota^*(\omega)$. The converse is not true; in fact, given a connection one form $\omega \in \Omega^1(\mathfrak{F}_V) \otimes \mathrm{End}(V)$, its restriction $\iota^*(\omega)$ is a connection one form on F if and only if F contains a holonomy reduction of ω.

Let us now consider two connections ω_F and ω'_F on a G structure $F \hookrightarrow \mathfrak{F}_V$. By definition, $\omega_F(\xi^*) = \omega'_F(\xi^*) = \xi$ for all $\xi \in \mathfrak{g}$, hence there is a G- equivariant map $\alpha : F \to Hom(V, \mathfrak{g})$ such that

$$\omega'_F = \omega_F + \alpha \circ \theta.$$

This means that for the torsion two forms of ω_F and ω'_F we have

$$\Theta'_F - \Theta_F = (\alpha \circ \theta) \wedge \theta = \delta^{1,1}_{\mathfrak{g}}(\alpha)(\theta \wedge \theta)$$

with the map $\delta^{1,1}_{\mathfrak{g}} : \mathfrak{g} \otimes V^* \to V \otimes \Lambda^2 V$ from the Spencer complex (13). That is,

$$Tor'_F = Tor_F + \delta^{1,1}_{\mathfrak{g}}(\alpha).$$

This immediately implies the following

Theorem 2.7. *Let $F \hookrightarrow \mathfrak{F}_V$ be a G-structure for some (possibly non-regular) Lie subgroup $G \subset Aut(V)$ with Lie algebra $\mathfrak{g} \subset End(V)$. Then the following hold:*

1. *Let ω_F be a connection on F, let $Tor : F \to \Lambda^2 V^* \otimes V$ be its torsion and $[Tor] : F \to H^{1,1}(\mathfrak{g})$ be the element represented by Tor in the Spencer cohomology group. Then $[Tor]$ is independent of the choice of connection on F, and is thus called the* intrinsic torsion *of F.*
2. *There exists a torsion free connection on F, i.e., a connection with $\Theta \equiv 0$, if and only if the intrinsic torsion of F vanishes.*
3. *If $\mathfrak{g}^{(1)} = 0$, then the torsion of a connection on F uniquely determines the connection.*

Let us now assume that ω is a *torsion free* connection on the G-structure $F \hookrightarrow \mathfrak{F}_V$ on M. Then from (21) it follows that $\sum_{cycl.} R(v_1, v_2)v_3 = 0$ which means that the image of the curvature map $R : \mathfrak{F}_V \to Hom(\Lambda^2 V, \mathrm{End}(V))$ is contained in

$K(\mathfrak{g})$, the kernel of the map $\delta^{1,2}_{\mathfrak{g}}$ of the Spencer complex as defined in (14). This kernel may be describes as

$$K(\mathfrak{g}) = \ker \delta^{1,2}_{\mathfrak{g}} = Hom(\Lambda^2 V, \mathfrak{g}) \cap (\ker \delta^{1,2})$$
$$= \left\{ R \in Hom(\Lambda^2 V, \mathfrak{g}) \mid \sum_{cycl.} R(v_1, v_2)v_3 = 0 \right\},$$

and is called the *space of formal curvatures of* $\mathfrak{g} \subset End(V)$. Furthermore, (23) implies that for a torsion free connection we have

$$\nabla R : F \longrightarrow K^1(\mathfrak{g}) := Hom(V, K(\mathfrak{g})) \cap \ker(V^* \otimes \Lambda^2 V^* \otimes \mathfrak{g} \longrightarrow \Lambda^3 V^* \otimes \mathfrak{g})$$
$$= \left\{ \phi \in Hom(V, K(\mathfrak{g})) \mid \sum_{cycl.} \phi(v_1)(v_2, v_3) = 0 \right\}.$$

These two identities for torsion free connections have some remarkable consequences. For a Lie subalgebra $\mathfrak{g} \subset \mathrm{End}(V)$ we define the ideal

$$\underline{\mathfrak{g}} := \{R(v, w) \mid R \in K(\mathfrak{g}),\ v, w \in V\} \lhd \mathfrak{g}$$

If $\omega \in \Omega^1(P) \otimes \mathfrak{g}$ is a torsion free connection on some G-structure $P \hookrightarrow \mathfrak{F}_V$ on some manifold M, then the linear maps

$$\Lambda^2 T_p P \longrightarrow \mathfrak{hol}_p, \qquad v \wedge w \longmapsto \Omega(d\Pi_{\underline{c}}(v), d\Pi_{\underline{c}}(w)),$$

where $\underline{c} : [a, b] \rightarrow M$ is any path with $\underline{c}(a) = \pi(p)$, satisfy the first Bianchi identity and hence are elements of $K(\mathfrak{hol}_p)$. Therefore, $\Omega(d\Pi_{\underline{c}}(v), d\Pi_{\underline{c}}(w)) \in \underline{\mathfrak{hol}}_p \lhd \mathfrak{hol}_p$. On the other hand, the Ambrose-Singer Holonomy Theorem 2.1 immediately implies that $\mathfrak{hol}_p$ is generated by $\Omega(d\Pi_{\underline{c}}(v), d\Pi_{\underline{c}}(w))$, so that we must have

$$\underline{\mathfrak{hol}}_p = \mathfrak{hol}_p.$$

Furthermore, if $K^1(\mathfrak{g}) = 0$, then evidently $\nabla R \equiv 0$, and a torsion free connection with this property is called *locally symmetric*. Thus, we can deduce the following

Theorem 2.8. (Berger's criteria [9]). *Let $\mathfrak{g} \subset End(V)$ be a Lie subalgebra, and define $K(\mathfrak{g})$, $K^1(\mathfrak{g})$ and $\underline{\mathfrak{g}} \lhd \mathfrak{g}$ as above.*

1. *If $\mathfrak{g}$ is the Lie algebra of the holonomy group of a torsion free connection on some manifold, then $\underline{\mathfrak{g}} = \mathfrak{g}$.*
2. *If $K^1(\mathfrak{g}) = 0$, then any torsion free connection on some manifold whose holonomy Lie algebra is contained in $\mathfrak{g}$ must be locally symmetric.*

Lie algebras $\mathfrak{g} \subset \mathrm{End}(V)$ satisfying $\underline{\mathfrak{g}} = \mathfrak{g}$ are called *Berger subalgebras*. Moreover, a Berger algebra is called *symmetric* if $K^1(\mathfrak{g}) = 0$ and *non-symmetric* otherwise. Thus, Theorem 2.8 says that the Lie algebra of the holonomy group of a torsion free connection must be a Berger algebra, and if this Berger algebra is symmetric, then any torsion free connection with this holonomy must be locally symmetric.

We shall describe locally symmetric connections in more detail in the following section.

2.5 Symmetric Connections

Let $H \subset G$ be a closed Lie subgroup with Lie algebra $\mathfrak{h} \subset \mathfrak{g}$, so that $M := G/H$ is a homogeneous space. Furthermore, assume that the Lie algebra $\mathfrak{g}$ of G admits an Ad_H- invariant decomposition

$$\mathfrak{g} = \mathfrak{h} \oplus \mathfrak{m}. \tag{24}$$

The Ad_H-invariance implies that also

$$[\mathfrak{h}, \mathfrak{m}] \subset \mathfrak{m}. \tag{25}$$

Let $\mu \in \Omega^1(G) \otimes \mathfrak{g}$ be the left invariant Maurer-Cartan form of G and decompose it according to (24) as

$$\mu = \omega + \theta, \text{ where } \omega \in \Omega^1(G) \otimes \mathfrak{h} \text{ and } \theta \in \Omega^1(G) \otimes \mathfrak{m}.$$

Since μ is left invariant, it satisfies the equivariance condition $r^*_{g^{-1}}(\mu) = Ad_g \circ \mu$, hence the corresponding equivariance holds for ω and θ as well. Moreover, the action fields ξ^* of the *right* H-action on G are *left* invariant, hence $\omega(\xi^*) = \mu(\xi^*) = \xi$ for all $\xi \in \mathfrak{h}$. It follows that ω is a connection one form on $G \to M$, and by Proposition 2.6, we may regard G as an H-structure on M and θ as the tautological form of this H-structure. The Maurer-Cartan equation

$$d\mu + \frac{1}{2}[\mu, \mu] = 0 \tag{26}$$

implies for the torsion and the curvature of this connection

$$\Theta = -\frac{1}{2}[\theta, \theta]_{\mathfrak{m}} \text{ and } \Omega = -\frac{1}{2}[\theta, \theta]_{\mathfrak{h}}.$$

Thus, ω is *torsion free* if and only if $[\mathfrak{m}, \mathfrak{m}] \subset \mathfrak{h}$ which together with (25) implies that the involution

$$d\sigma : \mathfrak{g} \longrightarrow \mathfrak{g}, \qquad d\sigma|_{\mathfrak{h}} = Id_{\mathfrak{h}}, \qquad d\sigma|_{\mathfrak{m}} = -Id_{\mathfrak{m}}$$

is a Lie algebra isomorphism and hence – after passing to an appropriate covering of G – is the differential of a Lie group involution

$$\sigma : G \to G, \qquad \sigma^2 = Id_G.$$

This process can be reverted, and we obtain the following result.

Proposition 2.9. *Let $\sigma : G \to G$ be an involution, i.e., an isomorphism with $\sigma^2 = Id_G$, and let $\mathfrak{g} = \mathfrak{h} \oplus \mathfrak{m}$ be the eigenspace decomposition of $\mathfrak{g}$ w.r.t. the differential $d\sigma : \mathfrak{g} \to \mathfrak{g}$. Let $H \subset G$ be any σ-invariant subgroup with Lie algebra $\mathfrak{h}$. Then the following hold:*

1. *The principal H-bundle $G \to G/H =: M$ is an H-structure on M and carries a canonical G- invariant torsion free connection induced by the splitting $\mathfrak{g} = \mathfrak{h} \oplus \mathfrak{m}$ from (24).*
2. *For each $p \in G/H$, there is a connection preserving involution $\sigma_p : G/H \to G/H$ such that p is an isolated fixed point of σ_p.*

Thus, G/H is an (affine) symmetric space. Moreover, the Lie algebra of the holonomy group of the canonical connection is $[\mathfrak{m}, \mathfrak{m}] \lhd \mathfrak{h}$.

Indeed, for $p = gH$ and $q = g'H \in G/H$, the involution σ_p is defined as

$$\sigma_p(q) := g\sigma(g')g^{-1}H.$$

The statement on the holonomy algebra follows once again from the Ambrose-Singer-Holonomy Theorem 2.1.

From (22) and the definition of the symmetric connection it follows immediately that $\nabla R \equiv 0$. In fact, this equation characterizes symmetric connections, at least locally.

Proposition 2.10. *Let $P \to M$ be an H-structure with a torsion free connection ω such that $\nabla R \equiv 0$. Then the connection is* locally symmetric, *i.e., after replacing M, P and H by appropriate covers, there are local diffeomorphism $\imath : P \to G$ and $\underline{\imath} : M \to G/H$, where G is a Lie group containing H such that G/H is a symmetric space, and so that the diagram*

$$\begin{array}{ccc} P & \xrightarrow{\ \imath\ } & G \\ {\scriptstyle H}\downarrow & & \downarrow \\ M & \xrightarrow{\ \underline{\imath}\ } & G/H \end{array}$$

commutes. Moreover, if $\mu = \theta_{\mathfrak{g}} + \omega_{\mathfrak{g}}$ is the decomposition of the Maurer-Cartan form of G which induces the symmetric connection, then $\theta = \imath^*(\theta_{\mathfrak{g}})$ and $\omega = \imath^*(\omega_{\mathfrak{g}})$. That is, $\underline{\imath} : M \to G/H$ is a connection preserving local diffeomorphism.

Proof. We define a Lie algebra structure on $\mathfrak{g} := \mathfrak{h} \oplus V$ by the conditions that:

1. $\mathfrak{h} \subset \mathfrak{g}$ is a Lie subalgebra.
2. $[\mathfrak{h}, V] \subset V$, and $ad_h|_V \in \mathrm{End}(V)$ is given by the embedding $\mathfrak{h} \subset \mathrm{End}(V)$ for $h \in \mathfrak{h}$.
3. $[V, V] \subset \mathfrak{h}$ is given by requiring the identity $\omega(v_1, v_2) = -[\theta(v_1), \theta(v_2)]$ for all $v_1, v_2 \in T_pP$.

It is straightforward to verify that the condition $\nabla R \equiv 0$ implies that the definition of $[V, V]$ is independent of the choice of $p \in P$, and that this is indeed a Lie algebra structure on $\mathfrak{g} = \mathfrak{h} \oplus V$.

Thus, we may define the form $\mu := \omega + \theta \in \Omega^1(P) \otimes \mathfrak{g}$, and the structure equations (2) and (17) with $\Theta = 0$ imply that μ satisfies the Maurer- Cartan equation (26). Therefore, by Cartan's theorem, after replacing P by an appropriate cover, there is a local diffeomorphism $\imath : P \to G$ with $\mu = \imath^*(\mu_G)$, where G is a Lie group with Lie algebra $\mathfrak{g}$, and μ_G is the Maurer-Cartan form on G.

The involution $d\sigma : \mathfrak{g} \to \mathfrak{g}$ which has $\mathfrak{h}$ and V as its $(+1)$ and (-1)-eigenspace, respectively, integrates to an involution $\sigma : G \to G$, and since we assume H to be connected, it follows that $G/\imath(H)$ is a symmetric space, and after identifying H and $\imath(H)$, it follows that $\imath : P \to G$ induces a connection preserving map $\underline{\imath} : M \to G/H$. □

2.6 Splitting Theorems

Let $P_i \to M_i$ be principal G_i-bundles for $i = 1, 2$. The *product bundle* $P \to M$ is the principal $(G_1 \times G_2)$-bundle $P := P_1 \times P_2$ with $M := M_1 \times M_2$.

If $\omega_i \in \Omega^1(P_i) \otimes \mathfrak{g}_i$ is a connection on P_i, then $\omega := p_1^*(\omega_1) + p_2^*(\omega_2) \in \Omega^1(P) \otimes (\mathfrak{g}_1 \oplus \mathfrak{g}_2)$ with the canonical projections $p_i : P \to P_i$ is again a connection one form on P and is called the *product connection of* ω_1 *and* ω_2 *on* P. It follows that the curvature $\Omega \in \Omega^2(P) \otimes (\mathfrak{g}_1 \oplus \mathfrak{g}_2)$ of the product connection is given by

$$\Omega = p_1^*(\Omega_1) + p_2^*(\Omega_2).$$

For the equivalence relation $\sim$ from (8) it now follows from the definition that

$$(p_1, p_2) \sim (q_1, q_2) \text{ in } P \text{ if and only if } p_i \sim q_i \text{ in } P_i \text{ for } i = 1, 2.$$

Thus, since by the proof of the Ambrose-Singer Holonomy Theorem 2.1 the holonomy reduction of $P \to M$ is a single equivalence class w.r.t. $\sim$, it follows that a holonomy reduction $P_0 \subset P$ is of the form $P_0 = P_0^1 \times P_0^2$, where $P_0^i \subset P_i$ are holonomy reductions. In particular, for the holonomy group we have

$$Hol_{(p_1,p_2)}(P) = Hol_{p_1}(P_1) \times Hol_{p_2}(P_2) \subset G_1 \times G_2.$$

Similarly, if $P_i \hookrightarrow \mathfrak{F}_{V_i}$ is a G_i-structure on M_i and $\theta_i \in \Omega^1(P_i) \otimes V_i$ is the corresponding tautological form for $i = 1, 2$, then $\theta := p_1^*(\theta_1) + p_2^*(\theta_2) \in \Omega^1(P) \otimes (V_1 \oplus V_2)$ satisfies the hypotheses of Proposition 2.6, hence there is an induced $(G_1 \times G_2)$-structure $P \hookrightarrow \mathfrak{F}_{V_1 \oplus V_2}$ on M. This structure is called the *product structure of* $P_1 \hookrightarrow \mathfrak{F}_{V_1}$ *and* $P_2 \hookrightarrow \mathfrak{F}_{V_2}$.

In the following, we shall derive some conditions which imply that a connection is a product connection. For this, let us consider a G-structure $P \subset \mathfrak{F}_V \to M$. If there is a G- invariant decomposition

$$V = V_1 \oplus \ldots \oplus V_r \tag{27}$$

with $r \geq 2$ and $V_j \neq 0$ for all j, then we call $G \subset \mathrm{Aut}(V)$ *decomposable*, otherwise, $G \subset \mathrm{Aut}(V)$ is called *indecomposable.*

Theorem 2.11. *Let M be a manifold with a torsion free connection, and suppose that the holonomy group Hol_p is connected and decomposable, i.e., there is a Hol_p-invariant splitting (27) with $r \geq 2$ and $V_j \neq 0$ and $V \cong T_pM$. Let $H_i \subset Aut(V_i)$, $i = 1, \ldots, r$, be the restriction of the action on Hol_p on V_i, and let $\mathfrak{h}_i \subset End(V_i)$ be its Lie algebra.*

Suppose further that the first prolongations of $\mathfrak{h}_i$ vanish, i.e., $\mathfrak{h}_i^{(1)} = 0$ for $i = 1, \ldots, r$. Then the following hold.

1. *The connection is locally a product connection, i.e., each $x \in M$ has a neighborhood $U = U_1 \times \ldots \times U_r$ such that the restriction of the connection to $U \subset M$ is a product of connections on U_i.*
2. *$Hol_p = H_1 \times \ldots \times H_r$, where H_i acts trivially on V_j for $i \neq j$.*

Proof. Let $\pi : P \to M$ be a holonomy reduction and let $\mathscr{H}$ be the horizontal distribution on P which we decompose as

$$\mathscr{H} = \mathscr{H}_1 \oplus \ldots \oplus \mathscr{H}_r, \quad \text{where } \mathscr{H}_i := \mathscr{H} \cap \theta^{-1}(V_i).$$

The G-equivariance implies that there are distributions $\mathscr{D}_i \subset TM$ such that

$$TM = \mathscr{D}_1 \oplus \ldots \oplus \mathscr{D}_r, \quad \text{and } d\pi : \mathscr{H}_i \to \mathscr{D}_i \text{ is a pointwise linear isomorphism.}$$

We assert that for the space of formal curvatures, we have

$$K(\mathfrak{h}_1 \oplus \cdots \oplus \mathfrak{h}_r) = K(\mathfrak{h}_1) \oplus \ldots \oplus K(\mathfrak{h}_r). \tag{28}$$

The inclusion $\supset$ is evident. For the converse, consider an element of $R \in K(\mathfrak{h}_1 \oplus \cdots \oplus \mathfrak{h}_r)$. Since $R(V_i, V_i)|_{V_i} \in K(\mathfrak{h}_i)$, we may assume w.l.o.g. that $R(V_i, V_i)|_{V_i} = 0$. Then the first Bianchi identity for R implies

$$0 = \underbrace{R(x_i, y_j)z_k}_{\in V_k} + \underbrace{R(y_j, z_k)x_i}_{\in V_i} + \underbrace{R(z_k, x_i)y_j}_{\in V_j},$$

where the subscripts refer to the decomposition (27). Thus, if i, j, k are pairwise distinct, then all terms vanish, so that we have $R(V_i, V_j)V_k = 0$ in this case. Next, if $i = j \neq k$, then

$$R(x_i, y_i)z_k = 0 \text{ and } R(z_k, y_i)x_i - R(z_k, x_i)y_i = 0. \tag{29}$$

The first equation in (29) implies that $R(V_i, V_i)V_k = 0$ for $k \neq i$, and therefore, $R(V_i, V_i) = 0$. The second equation in (29) implies that for fixed $z_k \in V_k$, the map $V_i \to \mathfrak{h}_i$, $x_i \mapsto R(z_k, x_i)|_{V_i}$ is an element of $\mathfrak{h}_i^{(1)}$. Since by hypothesis $\mathfrak{h}_i^{(1)} = 0$, it follows that $R(z_k, x_i)|_{V_i} = 0$. Thus, $R(V_k, V_i) = 0$ for $k \neq i$ and hence, $R = 0$ which shows (28).

This together with Theorem 2.8 implies that $\mathfrak{hol}_p \subset \mathfrak{h}_1 \oplus \ldots \oplus \mathfrak{h}_r$, and since the revers inclusion is obvious, we have equality. Thus, we also have the asserted equality for the holonomy groups as we assume these to be connected.

Let us decompose $\theta := \theta_1 + \ldots + \theta_r$, $\omega = \omega_1 + \ldots + \omega_r$ and $\Omega = \Omega_1 + \ldots + \Omega_r$ with $\theta_i \in \Omega^1(P) \otimes V_i$ and $\omega_i, \Omega_i \in \Omega^*(P) \otimes \mathfrak{h}_i$, and we define the distributions

$$\hat{\mathscr{H}_i} := \mathscr{H}_i \oplus \{\xi^* \mid \xi \in \mathfrak{h}_i\}.$$

Then $\hat{\mathscr{H}_i} \lrcorner \theta_j = 0$, $\hat{\mathscr{H}_i} \lrcorner \omega_j = 0$ and $\hat{\mathscr{H}_i} \lrcorner \Omega_j = 0$ for $i \neq j$. Thus $\hat{\mathscr{H}_i}$ is integrable, and

$$d\theta_i + \omega_i \wedge \theta_i = 0 \text{ and } \Omega_i = d\omega_i + \frac{1}{2}[\omega_i, \omega_i] \text{ for all } i.$$

If we let $P_i \subset P$ be an integral leaf of $\hat{\mathscr{H}_i}$, then $U_i := \pi(P_i) \subset M$ is an integral leaf of $\mathscr{D}_i \subset TM$, and the restriction $\pi : P_i \to U_i$ is a principal H_i- bundle. Indeed, Proposition 2.6 implies that θ_i may be regarded as the tautological form of an H_i-structure $P_i \hookrightarrow \mathfrak{F}_{V_i}$ on U_i, and ω_i is a torsion free connection with holonomy H_i on P_i.

Thus, each $x \in M$ has a neighborhood $U = U_1 \times \ldots \times U_r$ where the U_i are integral leafs of $\mathscr{D}_i$, and θ induces the product structure on U with structure group $H_1 \times \ldots \times H_r$, and ω is the product connection of the ω_i as asserted. □

We would like to point out that in general, the decomposability of the holonomy group alone does *not* imply connection to be a direct sum connection in general. See [54, p.290] for an example. That is, the condition $\mathfrak{h}_i^{(1)} = 0$ is essential in our argument.

As a special application of the splitting theorem, note that the prolongations of the orthogonal Lie algebras vanish, i.e., $\mathfrak{so}(p,q)^{(1)} = 0$. Thus, Theorem 2.11 implies the following results.

Theorem 2.12. **(de Rham-Wu Splitting Theorem [37,70])**. *Let (M, g) be a (pseudo-) Riemannian manifold, and suppose that the holonomy group of its Levi- Civita connection is decomposable. Then locally, (M, g) is isometric to a product metric*

$(\mathbb{R}^{k_1}, g_1) \times \ldots \times (\mathbb{R}^{k_r}, g_r)$ *with* $k_j = \dim V_j$, *and* $Hol_p^0(M) = H_1 \times \ldots \times H_r$ *with* $H_j \subset O(V_j, g_j)$.

This was first shown by de Rham in the Riemannian case and later generalized by Wu to the pseudo-Riemannian case. Note that in the Riemannian case, decomposability and irreducibility are equivalent.

In fact, there is also a *global* version of these splitting theorems which we shall not prove here. It relies on the *Cartan-Ambrose-Hicks-theorem* which explains the behavior of the curvature tensor under parallel translation. For a proof, see e.g. [5].

Theorem 2.13. (Global de Rham-Wu Splitting Theorem [37, 71]). *Let* (M, g) *be a geodesically complete simply connected (pseudo-)Riemannian manifold, and suppose that the holonomy group of its Levi-Civita connection is decomposable. Then* $(M, g) = (M_1, g_1) \times \ldots \times (M_r, g_r)$ *is the Riemannian product of complete (pseudo-)Riemannian manifolds* (M_i, g_i).

By virtue of this theorem, it is generally natural to assume the holonomy to be indecomposable as we may regard (local) product connections as "trivial" compositions.

3 Classification Results

In this section, we address the question which subgroups $G \subset \mathrm{Aut}(V)$ can occur as the holonomy of a given principal bundle. By Theorem 2.4, this reduces to deciding for which subgroups of the structure group there is a reduction. In particular, locally *any* connected subgroup $G \subset \mathrm{Aut}(V)$ can be realized as a holonomy group.

However, if we deal with *torsion free* connections on some G-structure, then the answer is far more difficult. By Theorem 2.8, for a subgroup $G \subset \mathrm{Aut}(V)$ to be the holonomy group of a torsion free connection, it is necessary that its Lie algebra is a *Berger algebra*. Therefore, the problem of classifying all holonomy groups of *torsion free* connections has as an algebraic subproblem the classification of all Berger algebras $\mathfrak{g} \subset \mathrm{End}(V)$.

Surprisingly, to the authors knowledge, there is no instance known of a Berger algebra which *cannot* be the Lie algebra of such a holonomy group. However, since there is no complete classification of Berger algebras, it is not clear if this is the case in general.

We shall now collect some classification results for certain subclasses of holonomy groups and algebras.

3.1 Irreducible Symmetric Spaces

Recall from Sect. 2.5 that a symmetric space is a manifold with an affine connection (M, ∇) such that for each $x \in M$ there is a connection preserving involution

$\sigma_x : M \to M$ which has $x \in M$ as an isolated fixed point. Let G be the *transvection group of* M, i.e., the identity component of the group generated by all σ_x, $x \in M$. Then G is a Lie group which acts transitively on M. Hence we can write $M = G/H$ for some closed subgroup $H \subset G$, and we call this symmetric space *irreducible* if the isotropy representation of H on $T_{x_0}M$ is irreducible, where $x_0 := eH \in M$. It follows that there is an involution

$$\sigma_0 := Ad_{\sigma_{x_0}} : G \longrightarrow G,$$

and $H \subset G$ is σ_0-invariant, and its Lie algebra $\mathfrak{h} \subset \mathfrak{g}$ is the $(+1)$- eigenspace of the differential $d\sigma_0 : \mathfrak{g} \to \mathfrak{g}$. Thus, Proposition 2.9 applies to G/H, and in fact the connection defined there coincides with the given connection on M.

Theorem 3.1. *[32] Let (M, ∇) be an irreducible symmetric space with transvection group G, and let $\mathfrak{g}$ denote its Lie algebra. Then*

1. *G is semi-simple.*
2. *If $\mathfrak{g} = \mathfrak{h} \oplus \mathfrak{m}$ is the symmetric decomposition, then $[\mathfrak{m}, \mathfrak{m}] = \mathfrak{h}$.*
3. *The holonomy of the connection satisfies $H^0 \subset Hol \subset H$, i.e., up to connected components, the isotropy group and the holonomy group coincide.*

We should remark here that Cartan proved Theorem 3.1 only in the case of irreducible *Riemannian* symmetric spaces. However, his proof can be adapted to the general case immediately; see e.g. [47].

In fact, Cartan even succeeded in providing a classification of *simply connected irreducible Riemannian symmetric spaces* [32], i.e., those symmetric spaces whose holonomy group is contained in the orthogonal group. In this case, irreducibility and indecomposability are equivalent.

Later, M.Berger [10] gave a classification of simply connected affine symmetric spaces with irreducible holonomy. These are all pseudo-Riemannian since the Killing form of $\mathfrak{g}$ induces a non-degenerate inner product on $\mathfrak{m}$ which induces a pseudo-Riemannian metric.

In general, the classification of affine symmetric spaces is far from complete. In the case of *pseudo-Riemannian symmetric spaces*, this classification was established for metrics of signature $(1, n)$ by M.Cahen and N.Wallach [27], and recently by I.Kath and M.Olbrich for signatures $(2, n)$ [52]. They also give a general construction method for such spaces of arbitrary signature.

3.2 Holonomy of Riemannian Manifolds

Let (M, g) be a Riemannian manifold. In this case, the holonomy group is contained in the orthogonal group $O(n)$ or, equivalently, its identity component is compact. Moreover, indecomposability is equivalent to irreducibility of the group. As we mentioned in Sect. 3.1, the Riemannian symmetric spaces are classified by Cartan

[32]. Thus, Theorems 2.8 and 2.12 imply that it suffices to classify all irreducible non-symmetric Berger algebras which are contained in $O(n)$. This has been achieved by Berger [9] where the below classification table was established.

Another important question is the determination of *parallel spinors*, i.e., parallel sections of the spinor bundle of a spin manifold M. If we assume that the holonomy of M is connected (e.g. if M is simply connected, cf. Proposition 2.3), then the space of parallel spinors corresponds to the subspace of the spinor representation on which the holonomy algebra $\mathfrak{hol}(M) \subset \mathfrak{so}(n) \cong \mathfrak{spin}(n)$ acts trivially. These spaces have been described by M.Wang [43] for all entries in Berger's list, and we add the dimension of the space of parallel spinors for each of the holonomies in question.

It was noted immediately that this list is contained in the list of subgroups of the orthogonal group which act transitively on the unit sphere. This fact was later proven directly by J.Simons [66] in an algebraic way. Recently, C.Olmos gave a beautiful simple argument showing this transitivity using submanifold theory only [59].

As it turns out, *all* of the groups in Table 2 do occur as holonomy of Riemannian connections:

1. $SO(n)$ is the reduced holonomy of a "generic" Riemannian manifold.
2. If $Hol \subset U(m)$, then the metric g is called *Kähler*. Kähler metrics form a natural class of complex manifolds, and the "generic" Kähler manifold has holonomy equal to $U(m)$.
3. If $Hol \subset SU(m)$ then the metric is called a *Calabi-Yau metric*. Since $SU(m) \subset U(m)$, each Calabi-Yau metric is necessarily Kähler. In fact, a Kähler metric with connected holonomy group is Calabi-Yau if and only if its Ricci curvature vanishes.

   The first examples of *complete* Calabi-Yau metrics were given by E.Calabi [28]. Later, S.T.Yau's solution to the Calabi conjecture [72] showed that a compact Kähler manifold with trivial canonical line bundle or, equivalently, with vanishing first Chern class admits a unique Calabi-Yau metric whose Kähler form represents the same cohomology class as the Kähler form of the original Kähler metric. For explicit examples, we refer to the books by A. Besse [11], S. Salamon [61] and D. Joyce [50].

Table 2 Classification of connected irreducible non-symmetric Holonomies contained in SO(n)

n	H	Associated geometry	Dim. of space of parallel spinors
$n \geq 2$	SO(n)	Generic Riemannian manifold	0
$2m \geq 4$	U(m)	Generic Kähler manifold	0
$2m \geq 4$	SU(m)	Special Kähler manifold	2
$4m \geq 8$	Sp(m) · Sp(1)	Quaternionic Kähler manifold	0
$4m \geq 8$	Sp(m)	Hyper-Kähler manifold	$m+1$
7	G_2	Exceptional holonomy	1
8	Spin(7)	Exceptional holonomy	1

4. Metrics with $Hol = Sp(m) \cdot Sp(1)$ are called *quaternionic Kähler*, although this terminology is somewhat misleading: quaternionic Kähler manifolds are *not* Kähler, as $Sp(1) \cdot Sp(m)$ is not contained in $U(m)$. Quaternionic Kähler manifolds are always Einstein, but not Ricci flat.

   Homogeneous quaternionic-Kähler manifolds were classified by D.Aleksevskii and V. Cortés [1, 36]. For more details on the theory of these manifolds, see [42, 46, 61, 62]. It is worth pointing out that there are so far no known examples of closed quaternionic Kähler manifolds with positive scalar curvature other than quaternionic projective space.
5. Metrics with $Hol \subset Sp(m)$ are called *hyper-Kähler*. These metrics are Kähler as $Sp(m) \subset SU(2m)$. In fact, hyper-Kähler metrics admit a whole two-sphere worth of Kähler structures which induce the quaternionic structure. First explicit examples were found by Calabi [28]. Compact examples were constructed using Yau's proof of the Calabi conjecture, see [6] for details.
6. The holonomy groups G_2 and $Spin(7)$ are called *exceptional holonomies* as they only occur in dimension 7 and 8, respectively. The existence of metrics with exceptional holonomy was shown locally by R.Bryant [16]. Complete examples were given by Bryant and Salamon [8], and compact examples were given by Joyce [48, 49] . See also [50] for a more detailed exposition.

Note that in general, the holonomy group of a Riemannian manifold may be non-compact as it may have infinitely many connected components. For example, let $M := \mathbb{R}^3/\Gamma$, where $\mathbb{R}^3$ is equipped with the standard flat connection and Γ is the cyclic group generated by an affine map whose linear part is rotation around an axis with irrational rotation angle and whose translation part is in direction of this axis. Then the holonomy of M is a cyclic group whose closure is $SO(2) \subset O(3)$. Thus, the holonomy group is non-compact. However, note that M is non-compact either.

It remained an open question for a long time whether the holonomy of a *compact* Riemannian manifold is necessarily compact (where the connection considered here is of course the Levi-Civita connection). In fact, it was conjectured in [11] that this is the case.

However, as it turns out, the answer is negative. In fact, B.Wilking [69] constructed examples of compact Riemannian manifolds with noncompact holonomy. He also showed that any such manifold must be finitely covered by a torus bundle over a compact manifold, where the dimension of the torus fiber is at least four.

3.3 Holonomy Groups of Pseudo-Riemannian Manifolds

3.3.1 Irreducible Holonomy Groups

In [9], Berger also classified all connected irreducible Berger groups which are subgroups of $SO(p, q)$ which are therefore candidates for the holonomy group of a pseudo-Riemanian manifold with a metric of signature (p,q). There were some minor omissions and errata on his list which were corrected by Bryant [18]. As in the case of Riemannian holonomies, one can obtain the dimension of the space of parallel spinors in each of these cases which has been worked out by H.Baum and I.Kath [5]. Summarizing all these results, we obtain Table 3 below:

We should also point out that all of these Berger groups do occur as the holonomy of pseudo-Riemannian manifolds, as has been shown by Bryant [18].

3.3.2 Indecomposable Lorentzian Holonomy Groups

A *Lorentzian manifold* is a pseudo-Riemannian manifold of signature $(n,1)$. According to Berger's classification in Table 3, there is no proper irreducible subgroup of $SO_0(n,1)$ which can occur as the holonomy group of a Lorentzian manifold. In fact, there is no proper irreducible subgroup of $SO_0(n,1)$ at all - this fact has been shown in a purely geometric manner by Di Scala and Olmos [38].

Therefore, we may assume that the holonomy representation is indecomposable but not irreducible. This implies that there must be a one-dimensional Hol-invariant subspace $\mathbb{R}\xi \subset \mathbb{R}^{n,1}$ with $\xi \neq 0$ such that $\langle \xi, \xi \rangle = 0$. Let $\Xi := (\mathbb{R}\xi)^\perp/(\mathbb{R}\xi)$ which is well defined as $\mathbb{R}\xi \subset (\mathbb{R}\xi)^\perp$. Since Hol leaves $\mathbb{R}\xi$ and hence its orthogonal complement invariant, it follows that there is an induced action of Hol on Ξ which preserves the induced positive definite inner product. (Note that this action may fail to be irreducible).

Based on work of L.Bérard-Bergery and A. Ikemakhen [7], the following classification result was established by T.Leistner.

Table 3 Classification of connected irreducible non-symmetric Holonomies contained in $SO_0(r,s)$

$n = r + s$	H	Associated geometry	Dim. of space of parallel spinors
$p+q \geq 2$	$SO_0(p,q)$	Generic	0
$2p \geq 4$	$SO(p,\mathbb{C})$	Generic complex	0
$2(p+q) \geq 4$	$U(p,q)$	Pseudo-Kähler	0
$2(p+q) \geq 4$	$SU(p,q)$	Special pseudo-Kähler	2
$4(p+q) \geq 8$	$Sp(p,q)$	Pseudo-hyper-Kähler	$p+q+1$
$4(p+q) \geq 8$	$Sp(p,q) \cdot Sp(1)$	Pseudo-quaternionic Kähler	0
$4p \geq 8$	$Sp(p,\mathbb{R}) \cdot SL(2,\mathbb{R})$		0
$8p \geq 16$	$Sp(p,\mathbb{C}) \cdot SL(2,\mathbb{C})$		0
7	G_2		1
7	G_2'		1
14	$G_2^{\mathbb{C}}$		2
8	$Spin(7)$		1
8	$Spin(4,3)$		1
16	$Spin(7,\mathbb{C})$		1

Theorem 3.2. *[55] Let $H \subset SO_0(n,1)$ be a connected indecomposable, but not irreducible subgroup, and let $\hat{H} \subset SO(\Xi)$ be the image of the induced representation described above. Then the following are equivalent:*

1. *H is a Berger group,*
2. *$\hat{H}$ is a Berger group, i.e., it is either the isotropy group of an irreducible Riemannian symmetric space, or is one of the entries of Table 2, or the direct product of such groups.*

Moreover, if a Lorentzian Spin-manifold admits a parallel spinor, then $H = \hat{H} \ltimes \mathbb{R}^n$, and the dimension of the space of parallel spinors coincides with this dimension for a Riemannian Spin- manifold with holonomy $\hat{H}$.

Furthermore, each indecomposable Berger group which is contained in $SO_0(n,1)$ does occur as the holonomy group of a Lorentzian manifold [40, 55].

3.3.3 Indecomposable Holonomy Groups of Pseudo-Riemannian Manifolds of Signature (p,q) with $p,q \geq 2$

In the non-Lorentzian case, there are a number of results which we shall not describe here in more detail. We already mentioned the partial classification of symmetric spaces [52]; another striking result is the classification of *Kählerian* holonomies of complex signature $(1,n)$ (hence of real signature $(2,2n)$) by Galaev [41]. Further results on signature $(2,n)$ may be found e.g. in [39,45], and for split signature (n,n) e.g. in [8].

3.4 Special Symplectic Holonomy Groups

A *symplectic connection* is a torsion free connection on a symplectic manifold (M,ω) such that ω is parallel or, equivalently, $Hol \subset Sp(n,\mathbb{R})$. We say that this connection has *special symplectic holonomy* if Hol acts absolutely irreducibly on the tangent space or, equivalently, if Hol acts irreducibly and does not preserve any complex structure.

First special symplectic holonomies were given by Bryant [17] and by Q.-S.Chi, S.Merkulov and the author [34, 35]. Finally, these holonomies were classified by Merkulov and the author ([69], see also [65]), and the possible holonomies are listed in Table 4.

As it turns out, all of these holonomies share striking rigidity properties which we shall explain in more detail in Sect. 4.

Table 4 Special symplectic holonomy groups

Group H	Representation space	Group H	Representation space
$\mathrm{SL}(2,\mathbb{R})$	$\mathbb{R}^4 \simeq \odot^3\mathbb{R}^2$	E_7^5	$\mathbb{R}^{56}$
$\mathrm{SL}(2,\mathbb{C})$	$\mathbb{C}^4 \simeq \odot^3\mathbb{C}^2$	E_7^7	$\mathbb{R}^{56}$
$\mathrm{SL}(2,\mathbb{R})\cdot\mathrm{SO}(p,q)$	$\mathbb{R}^{2(p+q)} \simeq \mathbb{R}^2\otimes\mathbb{R}^{p+q},\ p+q\geq 3$	$\mathrm{E}_7^{\mathbb{C}}$	$\mathbb{C}^{56}$
$\mathrm{SL}(2,\mathbb{C})\cdot\mathrm{SO}(n,\mathbb{C})$	$\mathbb{C}^{2n} \simeq \mathbb{C}^2\otimes\mathbb{C}^n,\ n\geq 3$	$\mathrm{Spin}(2,10)$	$\mathbb{R}^{32}$
$\mathrm{Sp}(1)\cdot\mathrm{SO}(n,\mathbb{H})$	$\mathbb{H}^n \simeq \mathbb{R}^{4n},\ n\geq 2$	$\mathrm{Spin}(6,6)$	$\mathbb{R}^{32}$
$\mathrm{SL}(6,\mathbb{R})$	$\mathbb{R}^{20} \simeq \Lambda^3\mathbb{R}^6$	$\mathrm{Spin}(6,\mathbb{H})$	$\mathbb{R}^{32}$
$\mathrm{SU}(1,5)$	$\mathbb{R}^{20} \subset \Lambda^3\mathbb{C}^6$	$\mathrm{Spin}(12,\mathbb{C})$	$\mathbb{C}^{32}$
$\mathrm{SU}(3,3)$	$\mathbb{R}^{20} \subset \Lambda^3\mathbb{C}^6$	$\mathrm{Sp}(3,\mathbb{R})$	$\mathbb{R}^{14} \subset \Lambda^3\mathbb{R}^6$
$\mathrm{SL}(6,\mathbb{C})$	$\mathbb{C}^{20} \simeq \Lambda^3\mathbb{C}^6$	$\mathrm{Sp}(3,\mathbb{C})$	$\mathbb{C}^{14} \subset \Lambda^3\mathbb{C}^6$

3.5 *Irreducible Holonomy Groups*

Holonomy groups which are irreducible, but of none of the above types, i.e., which preserve neither a (pseudo-)Riemannian nor a symplectic structure, have been investigated already by Berger [9] and later by Bryant [18, 19]. A complete classification was obtained by Merkulov and the author ([69], see also [65]). We shall not deal much with the geometric content of these holonomies here, but rather conclude this survey with the classification table (Table 5), referring the interested reader to the cited references.

4 Special Symplectic Connections

In this section, we shall deal with the notion of a *special symplectic connection* in the sense of [26]. First of all, if (M,ω) is a symplectic manifold, then a *symplectic connection* is defined to be a torsion free connection such that ω is parallel or, equivalently, such that the holonomy of the connection is contained in the symplectic group

$$\mathrm{Sp}(V,\omega) := \{g \in \mathrm{Aut}(V) \mid \omega(gx,gy) = \omega(x,y) \text{ for all } x,y \in V\},$$

whose Lie algebra is given as

$$\mathfrak{sp}(V,\omega) := \{h \in \mathrm{End}(V) \mid \omega(hx,y) + \omega(x,hy) = 0 \text{ for all } x,y \in V\}.$$

A *special symplectic connection* is defined as a symplectic connection on a manifold of dimension at least 4 which belongs to one of the following classes.

1. **Bochner-Kähler and Bochner-bi-Lagrangian connections**
   If the symplectic form is the Kähler form of a (pseudo-)Kähler metric, then its curvature decomposes into the Ricci curvature and the Bochner curvature [13].

Table 5 List of non-Riemannian, non-symplectic holonomy groups

Group H	Representation space V	Restrictions remarks
$T_{\mathbb{R}} \cdot \mathrm{SL}(n,\mathbb{C})$	$\{A \in M_n(\mathbb{C}) \mid A = A^*\}$	$n \geq 3$
$T_{\mathbb{R}} \cdot \mathrm{SL}(n,\mathbb{R}) \cdot \mathrm{SL}(m,\mathbb{R})$	$\mathbb{R}^n \otimes \mathbb{R}^m$	$n \geq m \geq 2, nm \neq 4$
$T_{\mathbb{R}} \cdot \mathrm{SL}(n,\mathbb{H}) \cdot \mathrm{SL}(m,\mathbb{H})$	$\mathbb{H}^n \otimes_{\mathbb{H}} \mathbb{H}^m$	$n \geq m \geq 1, nm \neq 1$
$T_{\mathbb{C}} \cdot \mathrm{SL}(n,\mathbb{C}) \cdot \mathrm{SL}(m,\mathbb{C})$	$\mathbb{C}^n \otimes \mathbb{C}^m$	$n \geq m \geq 2, nm \neq 4$
$T_{\mathbb{R}} \cdot \mathrm{SL}(n,\mathbb{R})$	$\mathbb{R}^n$	$n \geq 2$
$T_{\mathbb{R}} \cdot \mathrm{SL}(n,\mathbb{H})$	$\mathbb{H}^n$	$n \geq 1$
$T_{\mathbb{C}} \cdot \mathrm{SL}(n,\mathbb{R})$	$\mathbb{C}^n$	$n \geq 2$
$T_{\mathbb{C}} \cdot \mathrm{SL}(n,\mathbb{C})$	$\mathbb{C}^n$	$n \geq 2$
$\mathrm{U}(p,q)$	$\mathbb{C}^{p+q}$	$p+q \geq 2$
$\mathrm{SU}(p,q)$	$\mathbb{C}^{p+q}$	$p+q \geq 2,\ pq \neq 1$
$T_{\mathbb{C}} \cdot \mathrm{SU}(p,q)$	$\mathbb{C}^2$	$p+q = 2$
$T_{\mathbb{R}} \cdot \mathrm{SL}(n,\mathbb{R})$	$\Lambda^2\mathbb{R}^n$	$n \geq 5$
$T_{\mathbb{C}} \cdot \mathrm{SL}(n,\mathbb{C})$	$\Lambda^2\mathbb{C}^n$	$n \geq 5$
$T_{\mathbb{R}} \cdot \mathrm{SL}(n,\mathbb{H})$	$\{A \in M_n(\mathbb{H}) \mid A = A^*\}$	$n \geq 3$
$T_{\mathbb{R}} \cdot \mathrm{SL}(n,\mathbb{R})$	$\odot^2\mathbb{R}^n$	$n \geq 3$
$T_{\mathbb{C}} \cdot \mathrm{SL}(n,\mathbb{C})$	$\odot^2\mathbb{C}^n$	$n \geq 3$
$T_{\mathbb{R}} \cdot \mathrm{SL}(n,\mathbb{H})$	$\{A \in M_n(\mathbb{H}) \mid A = -A^*\}$	$n \geq 2$
$T_{\mathbb{R}} \cdot \mathrm{SO}(p,q)$	$\mathbb{R}^{p+q}$	$p+q \geq 3$
$T_{\mathbb{C}} \cdot \mathrm{SO}(n,\mathbb{C})$	$\mathbb{C}^n$	$n \geq 3$
$T_{\mathbb{R}} \cdot \mathrm{Spin}(5,5)$	$\Delta^+_{(5,5)}$	
$T_{\mathbb{R}} \cdot \mathrm{Spin}(1,9)$	$\Delta^+_{(1,9)}$	
$T_{\mathbb{C}} \cdot \mathrm{Spin}(10,\mathbb{C})$	$(\Delta^+_{10})^{\mathbb{C}}$	
$T_{\mathbb{R}} \cdot \mathrm{E}_6^1$	$\mathbb{R}^{27}$	
$T_{\mathbb{R}} \cdot \mathrm{E}_6^4$	$\mathbb{R}^{27}$	
$T_{\mathbb{C}} \cdot \mathrm{E}_6^{\mathbb{C}}$	$\mathbb{C}^{27}$	
$\mathrm{SL}(2,\mathbb{R}) \cdot \mathrm{SO}(p,q)$	$\mathbb{R}^2 \otimes \mathbb{R}^{p+q}$	$p+q \geq 3$
$\mathrm{Sp}(1) \cdot \mathrm{SO}(n,\mathbb{H})$	$\mathbb{H}^n$	$n \geq 2$
$\mathrm{Sp}(n,\mathbb{R})$	$\mathbb{R}^{2n}$	$n \geq 2$
$\mathbb{R}^* \cdot \mathrm{Sp}(2,\mathbb{R})$	$\mathbb{R}^4$	
$\mathrm{Sp}(p,q)$	$\mathbb{H}^{p+q}$	$p+q \geq 2$
$T_{\mathbb{R}} \cdot \mathrm{SL}(2,\mathbb{R})$	$\odot^3\mathbb{R}^2$	
$T_{\mathbb{C}} \cdot \mathrm{SL}(2,\mathbb{C})$	$\odot^3\mathbb{C}^2$	

Notations: $T_{\mathbb{F}}$ denotes any connected subgroup of $\mathbb{F}^*$
$\odot^p V$ denotes the symmetric tensors of V of degree p

If the latter vanishes, then (the Levi-Civita connection of) this metric is called Bochner-Kähler.

Similarly, if the manifold is equipped with a bi-Lagrangian structure, i.e., two complementary Lagrangian distributions, then the curvature of a symplectic connection for which both distributions are parallel decomposes into the Ricci curvature and the Bochner curvature. Such a connection is called Bochner-bi-Lagrangian if its Bochner curvature vanishes.

For results on Bochner-Kähler and Bochner-bi-Lagrangian connections, see [20] and [51] and the references cited therein.

2. **Connections of Ricci type**

   Under the action of the symplectic group, the curvature of a symplectic connection decomposes into two irreducible summands, namely the Ricci curvature and a Ricci flat component. If the latter component vanishes, then the connection is said to be of Ricci type.

   Connections of Ricci type are critical points of a certain functional on the moduli space of symplectic connections [15]. Furthermore, the canonical almost complex structure on the twistor space induced by a symplectic connection is integrable if and only if the connection is of Ricci type [9, 67]. For further properties see also [3, 23–25].

3. **Connections with special symplectic holonomy**

   A symplectic connection is said to have *special symplectic holonomy* if its holonomy is contained in a proper absolutely irreducible subgroup of the symplectic group. Thus, these are connections whose holonomy is listed in Table 4 on page 28.

   The special symplectic holonomies have been classified in [69] and further investigated in [17, 34, 38, 63, 65].

We can consider all of these conditions also in the complex case, i.e., for complex manifolds of complex dimension at least 4 with a holomorphic symplectic form and a holomorphic connection.

At first, it may seem unmotivated to collect all these structures in one definition, but we shall provide ample justification for doing so. Indeed, there is a beautiful link between special symplectic connections and parabolic contact geometry.

For this, consider a simple Lie group G with Lie algebra $\mathfrak{g}$. We say that $\mathfrak{g}$ is 2-gradable, if $\mathfrak{g}$ contains the root space of a long root. This is equivalent to saying that there is a decomposition as a graded vector space

$$\mathfrak{g} = \mathfrak{g}^{-2} \oplus \mathfrak{g}^{-1} \oplus \mathfrak{g}^{0} \oplus \mathfrak{g}^{1} \oplus \mathfrak{g}^{2}, \quad \text{and} \quad [\mathfrak{g}^{i}, \mathfrak{g}^{j}] \subset \mathfrak{g}^{i+j}, \tag{30}$$

with $\dim \mathfrak{g}^{\pm 2} = 1$. Indeed, there is a (unique) element $H_{\alpha_0} \in [\mathfrak{g}^{-2}, \mathfrak{g}^{2}] \subset \mathfrak{g}^{0}$ such that $\mathfrak{g}^{i}$ is the eigenspace of $ad(H_{\alpha_0})$ with eigenvalue $i = -2, \ldots, 2$, and any non-zero element of $\mathfrak{g}^{\pm 2}$ is a long root vector.

Denote by $\mathfrak{p} := \mathfrak{g}^{0} \oplus \mathfrak{g}^{1} \oplus \mathfrak{g}^{2} \leq \mathfrak{g}$ and let $P \subset$ G be the corresponding connected Lie subgroup. It follows that the homogeneous space $\mathscr{C} :=$ G$/P$ carries a canonical G-invariant contact structure which is determined by the Ad_P-invariant distribution $\mathfrak{g}^{-1} \bmod \mathfrak{p} \subset \mathfrak{g}/\mathfrak{p} \cong T\mathscr{C}$. In fact, we may regard $\mathscr{C}$ as the projectivisation of the adjoint orbit of a maximal root vector. That is, we view $\mathscr{C} \subset \mathbb{P}^o(\mathfrak{g})$ where $\mathbb{P}^o(V)$ denotes the set of oriented lines through 0 of a vector space V, so that $\mathbb{P}^o(V)$ is the sphere if V is real, and $\mathbb{P}^o(V)$ is a complex projective space if V is complex.

Each $a \in \mathfrak{g}$ induces an action field a^* on $\mathscr{C}$ with flow $\mathrm{T}_a := \exp(\mathbb{R}a) \subset \mathrm{G}$ which hence preserves the contact structure on $\mathscr{C}$. Let $\mathscr{C}_a \subset \mathscr{C}$ be the open subset on which a^* is transversal to the contact distribution. There is a unique contact form $\alpha \in \Omega^1(\mathscr{C}_a)$ such that $\alpha(a^*) \equiv 1$. That is, a^* is a Reeb vector field of the contact form α.

We can cover $\mathscr{C}_a$ by open sets U such that the local quotient $M_U := \mathrm{T}_a^{loc}\backslash U$, i.e., the quotient of U by a sufficiently small neighbourhood of the identity in T_a, is a manifold. Then M_U inherits a canonical symplectic structure $\omega \in \Omega^2(M_U)$ such that $\pi^*(\omega) = d\alpha$ for the canonical projection $\pi : U \to M_U$.

It is now our aim to construct a connection on M_U which is "naturally" associated to the given structure. For this, we let $\mathrm{G}_0 \subset \mathrm{G}$ be the connected subgroup with Lie algebra $\mathfrak{g}^0 \leq \mathfrak{g}$. Since $\mathfrak{g}^0 \leq \mathfrak{p}$ and hence $\mathrm{G}_0 \subset P$, it follows that we have a fibration

$$P/\mathrm{G}_0 \longrightarrow \mathrm{G}/\mathrm{G}_0 \longrightarrow \mathscr{C} = \mathrm{G}/P. \tag{31}$$

In fact, we may interpret $\mathrm{G}/\mathrm{G}_0 := \{(\alpha, v) \in T_p^*\mathscr{C} \times T_p\mathscr{C} \mid p \in \mathscr{C}, \alpha(\mathscr{D}_p) = 0, \alpha(v) = 1\}$, where $\mathscr{D} \subset T\mathscr{C}$ denotes the contact distribution. Thus, given $a \in \mathfrak{g}$, then for each $p \in \mathscr{C}_a$ we may regard the pair (α_p, a_p^*) from above as a point in G/G_0, i.e., we have a canonical embedding $\imath : \mathscr{C}_a \hookrightarrow \mathrm{G}/\mathrm{G}_0$.

Let $\Gamma_a := \pi^{-1}(\imath(\mathscr{C}_a)) \subset \mathrm{G}$ where $\pi : \mathrm{G} \to \mathrm{G}/\mathrm{G}_0$ is the canonical projection. Then the restriction $\pi : \Gamma_a \to \imath(\mathscr{C}_a) \cong \mathscr{C}_a$ becomes a principal G_0-bundle.

Consider the *Maurer-Cartan form* $\mu := g^{-1}dg \in \Omega^1(\mathrm{G}) \otimes \mathfrak{g}$ which we decompose according to (30) as $\mu = \sum_{i=-2}^{2} \mu_i$ with $\mu_i \in \Omega^1(\mathrm{G}) \otimes \mathfrak{g}^i$. Then we can show the following.

Proposition 4.1. *[26] Let $a \in \mathfrak{g}$ be such that $\mathscr{C}_a \subset \mathscr{C}$ is non-empty, define the action field $a^* \in \mathfrak{X}(\mathscr{C})$ and the principal G_0-bundle $\pi : \Gamma_a \to \mathscr{C}_a$ with $\Gamma_a \subset \mathrm{G}$ from above. Then we have the following.*

1. *The restriction of the components $\mu_0 + \mu_{-1} + \mu_{-2}$ of the Maurer-Cartan form to Γ_a yields a pointwise linear isomorphism $T\Gamma_a \to \mathfrak{g}^0 \oplus \mathfrak{g}^{-1} \oplus \mathfrak{g}^{-2}$.*
2. *There is a linear map $R : \mathfrak{g}^0 \to \Lambda^2(\mathfrak{g}^1)^* \otimes \mathfrak{g}^0$ and a smooth function $\rho : \Gamma_a \to \mathfrak{g}^0$ with the following property. If we define the differential forms $\kappa \in \Omega^1(\Gamma_a)$, $\theta \in \Omega^1(\Gamma_a) \otimes \mathfrak{g}^1$ and $\eta \in \Omega^1(\Gamma_a) \otimes \mathfrak{g}^0$ by the equation*

$$\mu_0 + \mu_{-1} + \mu_{-2} = -2\kappa \left(\frac{1}{2} e_{-2} + \rho \right) + \theta + \eta$$

for a fixed element $0 \neq e_{-2} \in \mathfrak{g}^{-2}$, then the following equations hold:

$$d\kappa = \frac{1}{2} \langle e_{-2}, [\theta, \theta] \rangle, \tag{32}$$

where $\langle\ ,\ \rangle$ refers to the Killing form of $\mathfrak{g}$, *and*

$$\begin{aligned} d\theta + \eta \wedge \theta &= 0, \\ d\eta + \tfrac{1}{2}[\eta, \eta] &= R_\rho(\theta \wedge \theta). \end{aligned} \tag{33}$$

Since the Maurer-Cartan form and hence κ, θ and η are invariant under the left action of the subgroup $\mathrm{T}_a \subset \mathrm{G}$, we immediately get the following

Corollary 4.2. *[26] On $T_a \backslash \Gamma_a$, there is a coframing $\eta + \theta \in \Omega^1(T_a \backslash \Gamma_a) \otimes (\mathfrak{g}^0 \oplus \mathfrak{g}^1)$ satisfying the structure equations (33) for a suitable function $\rho : T_a \backslash \Gamma_a \to \mathfrak{g}^0$.*

Thus, we could, in principle, regard θ and η as the tautological and the connection 1-form, respectively, of a connection on the principal bundle $\mathrm{T}_a \backslash \Gamma_a \to \mathrm{T}_a \backslash \Gamma_a / \mathrm{G}_0$ whose curvature is represented by R_ρ. However, $\mathrm{T}_a \backslash \Gamma_a / \mathrm{G}_0 \cong \mathrm{T}_a \backslash \mathscr{C}_a$ will in general be neither Hausdorff nor locally Euclidean, so the notion of a principal bundle cannot be defined globally.

The way out of this difficulty is to consider *local* quotients only, i.e., we restrict to sufficiently small open subsets $U \subset \mathscr{C}_a$ for which the local quotient $T_a^{loc} \backslash U$ *is* a manifold. Clearly, $\mathscr{C}_a$ can be covered by such open cells.

Moreover, if we describe explicitly the curvature endomorphisms R_ρ for $\rho \in \mathfrak{g}^0$, then one can show that – depending on the choice of the 2-gradable simple Lie algebra $\mathfrak{g}$ – the connections constructed above satisfy one of the conditions for a special symplectic connection mentioned before.

More precisely, we have the following

Theorem 4.3. *[26] Let $\mathfrak{g}$ be a simple 2-gradable Lie algebra with* $\dim \mathfrak{g} \geq 14$, *and let $\mathscr{C} \subset \mathbb{P}^o(\mathfrak{g})$ be the projectivisation of the adjoint orbit of a maximal root vector. Let $a \in \mathfrak{g}$ be such that $\mathscr{C}_a \subset \mathscr{C}$ is non-empty, and let $T_a = \exp(\mathbb{R}a) \subset G$. If for an open subset $U \subset \mathscr{C}_a$ the local quotient $M_U = T_a^{loc} \backslash U$ is a manifold, then M_U carries a special symplectic connection.*

The dimension restriction on $\mathfrak{g}$ guarantees that $\dim M_U \geq 4$ and rules out the Lie algebras of type A_1, A_2 and B_2.

The type of special symplectic connection on M_U is determined by the Lie algebra $\mathfrak{g}$. In fact, there is a one-to-one correspondence between the various conditions for special symplectic connections and simple 2-gradable Lie algebras. More specifically, if the Lie algebra $\mathfrak{g}$ is of type A_n, then the connections in Theorem 4.3 are Bochner–Kähler of signature (p, q) if $\mathfrak{g} = \mathfrak{su}(p+1, q+1)$ or Bochner-bi-Lagrangian if $\mathfrak{g} = \mathfrak{sl}(n, \mathbb{R})$; if $\mathfrak{g}$ is of type C_n, then $\mathfrak{g} = \mathfrak{sp}(n, \mathbb{R})$ and these connections are of Ricci-type; if $\mathfrak{g}$ is a 2-gradable Lie algebra of one of the remaining types, then the holonomy of M_U is contained in one of the special symplectic holonomy groups in Table 4 on page 28. Also, for two elements $a, a' \in \mathfrak{g}$ for which $\mathscr{C}_a, \mathscr{C}_{a'} \subset \mathscr{C}$ are non-empty, the corresponding connections from Theorem 4.3 are equivalent if and only if a' is G-conjugate to a positive multiple of a.

If $\mathrm{T}_a \cong S^1$ then $\mathrm{T}_a\backslash\mathscr{C}_a$ is an orbifold which carries a special symplectic orbifold connection by Theorem 4.3. Hence it may be viewed as the "standard orbifold model" for (the adjoint orbit of) $a \in \mathfrak{g}$. For example, in the case of positive definite Bochner–Kähler metrics, we have $\mathscr{C} \cong S^{2n+1}$, and for connections of Ricci-type, we have $\mathscr{C} \cong \mathbb{RP}^{2n+1}$. Thus, in both cases the orbifolds $\mathrm{T}_a\backslash\mathscr{C}$ are weighted projective spaces if $\mathrm{T}_a \cong S^1$, hence the standard orbifold models $\mathrm{T}_a\backslash\mathscr{C}_a \subset \mathrm{T}_a\backslash\mathscr{C}$ are open subsets of weighted projective spaces.

Surprisingly, the connections from Theorem 4.3 exhaust *all* special symplectic connections, at least locally. Namely we have the following

Theorem 4.4. *[26] Let (M, ω) be a symplectic manifold with a special symplectic connection of class C^4, and let $\mathfrak{g}$ be the Lie algebra associated to the special symplectic condition as above.*

1. *Then there is a principal $\hat{T}$-bundle $\hat{M} \to M$, where $\hat{T}$ is a one dimensional Lie group which is not necessarily connected, and this bundle carries a principal connection with curvature ω.*
2. *Let $T \subset \hat{T}$ be the identity component. Then there is an $a \in \mathfrak{g}$ such that $T \cong T_a \subset G$, and a T_a-equivariant local diffeomorphism $\hat{\imath} : \hat{M} \to \mathscr{C}_a$ which for each sufficiently small open subset $V \subset \hat{M}$ induces a connection preserving diffeomorphism $\iota : T^{loc}\backslash V \to T_a^{loc}\backslash U = M_U$, where $U := \hat{\imath}(V) \subset \mathscr{C}_a$ and M_U carries the connection from Theorem 4.3.*

The situation in Theorem 4.4 can be illustrated by the following commutative diagram, where the vertical maps are quotients by the indicated Lie groups, and $\mathrm{T}\backslash\hat{M} \to M$ is a regular covering.

$$\begin{array}{ccccc} & & \hat{M} & \xrightarrow{\hat{\imath}} & \mathscr{C}_a \\ & \swarrow^{\hat{\mathrm{T}}} & \downarrow{\scriptstyle \mathrm{T}} & & \downarrow{\scriptstyle \mathrm{T}_a} \\ M & \longleftarrow & \mathrm{T}\backslash\hat{M} & \xrightarrow{\iota} & \mathrm{T}_a\backslash\mathscr{C}_a \end{array} \tag{34}$$

In fact, one might be tempted to summarize Theorems 4.3 and 4.4 by saying that for each $a \in \mathfrak{g}$, the quotient $\mathrm{T}_a\backslash\mathscr{C}_a$ carries a canonical special symplectic connection, and the map $\iota : \mathrm{T}\backslash\hat{M} \to \mathrm{T}_a\backslash\mathscr{C}_a$ is a connection preserving local diffeomorphism. If $\mathrm{T}_a\backslash\mathscr{C}_a$ is a manifold or an orbifold, then this is indeed correct. In general, however, $\mathrm{T}_a\backslash\mathscr{C}_a$ may be neither Hausdorff nor locally Euclidean, hence one has to formulate these results more carefully.

As consequences, we obtain the following

Corollary 4.5. *All special symplectic connections of C^4-regularity are analytic, and the local moduli space of these connections is finite dimensional, in the sense that the germ of the connection at one point up to 3rd order determines the* *connection entirely. In fact, the generic special symplectic connection associated to the Lie algebra* $\mathfrak{g}$ *depends on* $(rk(\mathfrak{g}) - 1)$ *parameters.*

Moreover, the Lie algebra $\mathfrak{s}$ *of* affine vector fields, *i.e., vector fields on* M *whose flow preserves the connection, is isomorphic to* $\mathfrak{z}(a)/(\mathbb{R}a)$ *with* $a \in \mathfrak{g}$ *from Theorem 4.4, where* $\mathfrak{z}(a) = \{x \in \mathfrak{g} \mid [x, a] = 0\}$. *In particular,* $\dim \mathfrak{s} \geq rk(\mathfrak{g}) - 1$ *with equality implying that* $\mathfrak{s}$ *is abelian.*

When counting the parameters in the above corollary, we regard homothetic special symplectic connections as equal, i.e., (M, ω, ∇) is considered equivalent to $(M, e^{t_0}\omega, \nabla)$ for all $t_0 \in \mathbb{R}$ or $\mathbb{C}$.

We can generalize Theorem 4.4 and Corollary 4.5 easily to orbifolds. Indeed, if M is an orbifold with a special symplectic connection, then we can write $M = \hat{\mathrm{T}}\backslash\hat{M}$ where $\hat{M}$ is a manifold and $\hat{\mathrm{T}}$ is a one dimensional Lie group acting properly and locally freely on $\hat{M}$, and there is a local diffeomorphism $\hat{\imath} : \hat{M} \to \mathscr{C}_a$ with the properties stated in Theorem 4.4.

There is a remarkable similarity between the cones $\mathscr{C}_i \subset \mathfrak{g}_i$, $i = 1, 2$, for the simple Lie algebras $\mathfrak{g}_1 := \mathfrak{su}(n+1, 1)$ and $\mathfrak{g}_2 := \mathfrak{sp}(n, \mathbb{R})$. Namely, $\mathscr{C}_1 = S^{2n+1}$ with the standard CR-structure, and $\mathfrak{g}_1$ is the Lie algebra of the group $\mathrm{SU}(n+1, 1)$ of CR-isomorphisms of S^{2n+1} [51]. On the other hand, $\mathscr{C}_2 = \mathbb{RP}^{2n+1}$, regarded as the lines in $\mathbb{R}^{2n+2}$ with the projectivised action of $\mathfrak{sp}(n+1, \mathbb{R})$ on $\mathbb{R}^{2n+2}$. Thus, $\mathscr{C}_1$ is the universal cover of $\mathscr{C}_2$, so that the local quotients $\mathrm{T}_a\backslash\mathscr{C}_a$ are related. In fact, we have the following result.

Proposition 4.6. *[60] Consider the action of the 2-gradable Lie algebras* $\mathfrak{g}_1 := \mathfrak{su}(n+1, 1)$ *and* $\mathfrak{g}_2 := \mathfrak{sp}(n+1, \mathbb{R})$ *on the projectivised orbits* $\mathscr{C}_1$ *and* $\mathscr{C}_2$*, respectively. Then the following are equivalent.*

1. *For* $a_i \in \mathfrak{g}_i$ *the actions of* $T_{a_i} \subset G_i$ *on* $\mathscr{C}_i$ *are conjugate for* $i = 1, 2$,
2. $a_i \in \mathfrak{u}(n+1)$ *where* $\mathfrak{u}(n+1) \subset \mathfrak{g}_i$ *for* $i = 1, 2$ *via the two standard embeddings.*

This together with the preceding results yields the following

Theorem 4.7. *[60]*

1. *Let* (M, ω, ∇) *be a symplectic manifold with a connection of Ricci type, and suppose that the corresponding element* $a \in \mathfrak{sp}(n+1, \mathbb{R})$ *from Theorem 4.3 is conjugate to an element of* $\mathfrak{u}(n+1) \subset \mathfrak{sp}(n+1, \mathbb{R})$. *Then* M *carries a canonical Bochner–Kähler metric whose Kähler form is given by* ω.
2. *Conversely, let* (M, ω, J) *be a Bochner-Kähler metric such that the element* $a \in \mathfrak{su}(n+1, 1)$ *from Theorem 4.4 is conjugate to an element of* $\mathfrak{u}(n+1) \subset \mathfrak{su}(n+1, 1)$. *Then* (M, ω) *carries a canonical connection of Ricci-type.*

Note that in [20], Bochner–Kähler metrics have been locally classified. In this terminology, the Bochner–Kähler metrics in the above theorem are called *Bochner–Kähler metrics of type I*. For more details, we also refer the reader to [12].

References

1. Alekseevskii, D.V.: Classification of quaternionic spaces with a transitive solvable group of motions. Math. USSR Izv. **9**, 297–339 (1975)
2. Ambrose, W., Singer, I.M.: A Theorem on holonomy, Trans. Amer. Math. Soc. **75**, 428–443 (1953)
3. Baguis, F., Cahen, M.:, A construction of symplectic connections through reduction. Let. Math. Phys **57**, 149–160 (2001)
4. Baum, H.: Eichfeldtheiorie. Springer-Verlag Berlin Heidelberg (2009)
5. Baum, H., Kath, I.: Parallel spinors and holonomy groups on pseudo-Riemannian spin-manifolds, Ann. Glob. Anal. Geom. **17**(1), 1–17 (1999)
6. Beauville, A.: Variétés Kählériennes dont la première classe de Chern est nulle. Jour. Diff. Geom. **18**, 755–782 (1983)
7. Bérard-Bergery, L., Ikemakhen, A.: On the holonomy of Lorentzian manifolds. Geometry in mathematical physics and related topics (Los Angeles, CA, 1990), Proc. Sympos. Pure Math., AMS, **54**(2), 27–40 (1993)
8. Bérard-Bergery, L., Ikemakhen, A.:, *Sur l'holonomie des variétés pseudo- riemaniennes de signature* (n,n). Bull. Soc. Math. France **125**(1), 93–114 (1997)
9. Berger, M.: Sur les groupes d'holonomie des variétés à connexion affine et des variétés Riemanniennes. Bull. Soc. Math. France **83**, 279–330 (1955)
10. Berger, M.: Les espaces symétriques noncompacts. Ann. Sci. Ecol. Norm. Sup. **74**, 85–177 (1957)
11. Besse, A.L.: Einstein Manifolds, Ergebnisse der Mathematik und ihrer Grenzgebiete, 3. Folge, Band 10, Springer-Verlag, Berlin, New York (1987)
12. Bieliavsky, P., Cahen, M., Gutt, S., Rawnsley, J., Schwachhöfer, L.J.: Symplectic Connections, International Journal of Geometric Methods in Modern Physics, **3**(3), 375–420 (2006)
13. Bochner, S.: Curvature and Betti numbers, II, Ann. Math. **50**, 77–93 (1949)
14. Borel, A., Lichnerowicz, A.: Groupes d'holonomie des variétés riemanniennes. C.R. Acad. Sci. Paris **234**, 1835–1837 (1952)
15. Bourgeois, F., Cahen, M.: A variational principle for symplectic connections. J. Geom. Phys **30**, 233–265 (1999)
16. Bryant, R.:, Metrics with exceptional holonomy. Ann. Math. **126**, 525–576 (1987)
17. Bryant, R.: Two exotic holonomies in dimension four, path geometries, and twistor theory, Proc. Symp. in Pure Math. **53**, 33–88 (1991)
18. Bryant, R.: Classical, exceptional, and exotic holonomies: A status report. Actes de la Table Ronde de Géométrie Différentielle en l'Honneur de Marcel Berger, Collection SMF Séminaires and Congrès 1 (Soc. Math. de France) 93–166 (1996).
19. Bryant, R.: Recent Advances in the Theory of Holonomy, Séminaire Bourbaki, exposé no. 861, Paris: Société Mathématique de France, Astérisque. **266**, 351–374 (2000)
20. Bryant, R.: Bochner-Kähler metrics, J. Amer. Math. Soc. **14**, 623–715 (2001)
21. Bryant, R.: Geometry of manifolds with Special Holonomy: "100 years of Holonomy" Cont. Math. **395**, 29–38 (2006)
22. Bryant, R., Salamon, S.: On the construction of some complete metrics with exceptional holonomy, Duke. Math. Jour. **58**, 829–850 (1989)
23. Cahen, M., Gutt, S., Rawnsley, J.: Symmetric symplectic spaces with Ricci-type curvature. In: Dito, G., Sternheimer, D. (eds.) Conférence Moshé Flato 1999, Vol. II, Math. Phys. Stud. **22**, 81–91 (2000)
24. Cahen, M., Gutt, S., Horowitz, J., Rawnsley, J.: Homogeneous symplectic manifolds with Ricci-type curvature. J. Geom. Phys. **38**, 140–151 (2001)
25. Cahen, M., Gutt, S., Schwachhöfer, L.J.: Construction of Ricci-type connections by reduction and induction, In: "The Breadth of Symplectic and Poisson Geometry", Festschrift in Honour of Alan Weinstein, pp. 41–58. Birkhäuser (2004)

26. Cahen, M., Schwachhöfer, L.J.: Special Symplectic Connections. J. Diff. Geom. **83**(2), 229–271 (2009)
27. Cahen, M., Wallach, N.: Lorentzian symmetric spaces. Bull. Amer. Math. Soc. **76**, 585–591 (1970)
28. Calabi, E.: Métriques kähleriennes et fibrés holomorphes, Ann. Ec. Norm. Sup. **12**, 269–294 (1979)
29. Cartan, É.: Les groupes de transformations continus, infinis, simples. Ann. Éc. Norm. **26**, 93–161 (1909)
30. Cartan, E.: Sur les variétés à connexion affine et la théorie de la relativité généralisée I & II, Ann. Sci. Ecol. Norm. Sup. **40**, 325–412 (1923); et **41**, 1–25 (1924) ou Oeuvres complètes, tome III, 659–746 et 799–824
31. Cartan, E.: La géométrie des espaces de Riemann, Mémorial des Sciences Mathématiques. vol. 5, Paris, Gauthier-Villars (1925)
32. Cartan, E.:, Sur une classe remarquable d'espaces de Riemann, Bull. Soc. Math. France **54**, 214–264 (1926); **55**, 114–134 (1927) ou Oeuvres complètes, tome I, vol. 2, 587–659
33. Cartan, E.: Les groupes d'holonomie des espaces généralisés, Acta. Math. **48**, 1–42 (1926); ou Oeuvres complètes, tome III, vol. 2, 997–1038
34. Chi, Q.-S., Merkulov, S.A., Schwachhöfer, L.J.: On the existence of infinite series of exotic holonomies. Inv. Math. **126**, 391–411 (1996)
35. Chi, Q.-S., Merkulov, S.A., Schwachhöfer, L.J.: Exotic holonomies $E_7^{(a)}$. Int. Jour. Math. **8**, 583–594 (1997)
36. Cortés, V.: Alekseevskian spaces. Diff. Geom. Appl. **6**, 129–168 (1996)
37. de Rham, G.: Sur la réductibilité d'un espace de Riemann. Comm. Math. Helv. **26**, 328–344 (1952)
38. Di Scala, A., Olmos, C.: The geometry of homogeneous submanifolds in hyperbolic space, Math. Zeit. **237**(1), 199–209 (2001)
39. Galaev, K.: Remark on holonomy groups of pseudo-Riemannian manifolds of signature $(2, n+2)$, arXive: math.DG/0406397 (2004)
40. Galaev, A.: Metrics that realize all Lorentzian holonomy algebras. Int. J. Geom. Methods Mod. Phys. **3**(5–6), 1025–1045 (2006)
41. Galaev, K.: Holonomy groups and special geometric structures of pseudo-K ählerian manifolds of index 2, arXive: math.DG/0612392 (2006)
42. Galicki, K.: A generalization of the momentum mapping construction for quaternionic Kähler manifolds. Comm. Math. Phys. **108**, 117–138 (1987)
43. Galicki, K., Lawson, H.B.: Quaternionic reduction and quaternionic orbifolds. Math. Ann. **282**, 1–21 (1988)
44. Guillemin, V.: The integrability problem for G-structures. Trans. Amer. Math. Soc. **116**, 544–560 (1965)
45. Ikemakhen, A.: Sur l'holonomie des variétés pseudo-riemaniennes de signature $(2, 2+n)$. Publ. Math. **43**(1), 55–84 (1999)
46. Hano, J., Ozeki, H.: On the holonomy groups of linear connections. Nagoya Math. Jour. **10**, 97–100 (1956)
47. Helgason, S.: Differential Geometry, Lie Groups, and Symmetric Spaces. Graduate Studies in Mathematics, Vol. 34, AMS, Providence, RI (2001)
48. Joyce, D.: Compact 8-manifolds with holonomy G_2: I & II. Jour. Diff. Geo. **43**, 291–375 (1996)
49. Joyce, D.: Compact Riemannian 7-manifolds with holonomy $Spin(7)$. Inv. Math. **123**, 507–552 (1996)
50. Joyce, D.: Compact manifolds with Special holonomy. Oxford Mathem. Monographs, Oxford Sci. Pub. (2000)
51. Kamishima, Y.: Uniformization of Kähler manifolds with vanishing Bochner tensor. Acta Math. **172**, 229–308 (1994)
52. Kath, I., Olbrich, M.: On the structure of pseudo-Riemannian symmetric spaces. Transform. Groups **14**(4), 847–885 (2009)

53. Kobayashi, S., Nagano, K.: On filtered Lie algebras and geometric structures II J. Math. Mech. **14**, 513–521 (1965)
54. Kobayashi, S., Nomizu, K.: Foundations of Differential Geometry, I, Interscience publishers (1963)
55. Leistner, T.: On the classification of Lorentzian holonomy groups. Jour. Diff. Geo. **76**(3), 423–484 (2007)
56. Merkulov, S.A., Schwachhöfer, L.J.: Classification of irreducible holonomies of torsion free affine connections. Ann. Math. **150**, 77–149 (1999); Addendum: Classification of irreducible holonomies of torsion-free affine connections, Ann. Math. **150**, 1177–1179 (1999)
57. O'Brian, N.R., Rawnsley, J. Twistor spaces. Ann. Glob. Anal. Geom **3**, 29–58 (1985)
58. Ochiai, T.: Geometry associated with semi-simple flat homogeneous spaces. Trans. Amer. Math. Soc. **152**, 159–193 (1970)
59. Olmos, C.: A geometric proof of the Berger holonomy theorem. Ann. Math. **161**(1), 579–588 (2005)
60. Panák, M., Schwachhöfer, L.: Bochner-Kähler metrics and connections of Ricci type. Proceedings of the 10th International Conference on Differential Geometry and Its Applications 2007, Differential geometry and its applications, 339–352 (2008)
61. Salamon, S.: Riemannian geometry and holonomy groups, Pitman Research Notes in Mathematics, no. 201, Longman Scientific & Technical, Essex (1989)
62. Salamon, S.: Quaternionic Kähler manifolds. Inv. Math. **67**, 143–171 (1982)
63. Schwachhöfer, L.J.: On the classification of holonomy representations. Habilitationsschrift, Universität Leipzig (1998)
64. Schwachhöfer, L.J.: Homogeneous connections with special symplectic holonomy. Math. Zeit. **238**, 655–688 (2001)
65. Schwachhöfer, L.J.: Connections with irreducible holonomy representations. Adv. Math. **160**, 1–80 (2001)
66. Simons, J.: On transitivity of holonomy systems. Ann. Math. **76**, 213–234 (1962)
67. Vaisman, I.: Variations on the theme of twistor spaces. Balkan J. Geom. Appl. **3**, 135–156 (1998)
68. Wang, M.: Parallel spinors and parallel forms. Ann. Global Anal. Geom. **7**(1), 59–68 (1989)
69. Wilking, B.: On compact Riemannian manifolds with noncompact holonomy groups. J. Diff. Geom. **52**(2), 223–257 (1999)
70. Wu, H.: On the de Rham decomposition theorem. Illinois. J. Math. **8**, 291–311 (1964)
71. Wu, H.: Holonomy groups of indefinite metrics. Pac. J. Math. **20**(2), 351–392 (1967)
72. Yau, S.T.: On the Ricci curvature of a compact Kähler mainfold and the complex Monge-Ampère equation I. Com. Pure and Appl. Math **31**, 339–411 (1978)

Entropies, Volumes, and Einstein Metrics

D. Kotschick

Abstract We survey the definitions and some important properties of several asymptotic invariants of smooth manifolds, and discuss some open questions related to them. We prove that the (non-)vanishing of the minimal volume is a differentiable property, which is not invariant under homeomorphisms. We also formulate an obstruction to the existence of Einstein metrics on four-manifolds involving the volume entropy. This generalizes both the Gromov–Hitchin–Thorpe inequality proved in [Kotschick, On the Gromov–Hitchin–Thorpe inequality, C. R. Acad. Sci. Paris 326 (1998), 727–731], and Sambusetti's obstruction [Sambusetti, An obstruction to the existence of Einstein metrics on 4-manifolds, Math. Ann. 311 (1998), 533–547].

1 Introduction

In his seminal paper on bounded cohomology [15], Gromov introduced both the simplicial volume and the minimal volume of manifolds. While the definition of the simplicial volume belongs to quantitative algebraic topology, that of the minimal volume belongs to asymptotic Riemannian geometry, in that one takes an infimum of a certain quantity over the space of all Riemannian metrics on a fixed manifold. The simplicial volume provides a lower bound for the minimal volume, via the relations of these quantities to another asymptotic invariant, the minimal volume entropy of a manifold. This invariant is an asymptotic one in two ways: first one considers the asymptotics at infinity of a quantity on a non-compact manifold, and then one takes an infimum over all metrics, in the same way as for the definition of the minimal volume.

D. Kotschick (✉)
Mathematisches Institut, LMU München, Theresienstr. 39, 80333 München, Germany
e-mail: dieter@member.ams.org

C. Bär et al. (eds.), *Global Differential Geometry*, Springer Proceedings in Mathematics 17, DOI 10.1007/978-3-642-22842-1_2,

In Sect. 2 of this paper we survey several asymptotic invariants, including the three mentioned above. The invariants we discuss are related to each other by an interesting chain of inequalities that we explain, with the simplicial volume at the bottom, and the minimal volume at the top. We shall also discuss the nature of these invariants. Several of the asymptotic Riemannian invariants actually turn out to be rather crude topological invariants, by the recent work of Brunnbauer [11], who completed and generalized earlier work of Babenko [1–3]. This is not so for the minimal volume, since Bessières [7] has given examples of high-dimensional manifolds that are homeomorphic but have different minimal volumes, both of them positive. In Sect. 5 of this paper we exhibit another way in which the minimal volume fails to be invariant under homeomorphisms: there are pairs of homeomorphic four-manifolds for which the minimal volume is zero for one, and is positive for the other. As the vanishing of the minimal volume implies the vanishing of all real characteristic numbers, such examples cannot be simply connected. In fact, our examples are parallelizable and exhibit for the first time the existence of exotic smooth structures on parallelizable closed four-manifolds.

In Sect. 3 of this paper we discuss two further asymptotic invariants that can be used to bound the volume entropy, and, therefore, the minimal volume, from below. These invariants, based on spectral and isoperimetric quantities, are more complicated than the invariants defined in Sect. 2, and it is as yet unclear whether Brunnbauer's results [11] about topological invariance can be adapted for them.

In Sect. 4 we give an application of the discussion of asymptotic invariants to (non-)existence of Einstein metrics on four-manifolds. We prove that a closed oriented four-manifold X admitting an Einstein metric must satisfy the inequality

$$\chi(X) \geq \frac{3}{2}|\sigma(X)| + \frac{1}{108\pi^2}\lambda(X)^4,$$

where $\chi(X)$ denotes the Euler characteristic, $\sigma(X)$ the signature, and $\lambda(X)$ the volume entropy. Using a recent rigidity result for the entropy due to Ledrappier and Wang [26] we characterize the case of equality.

Whenever one has a positive lower bound for the volume entropy, the above inequality implies a particular improvement of the classical Hitchin–Thorpe inequality [18]

$$\chi(X) \geq \frac{3}{2}|\sigma(X)|.$$

Two specific cases of lower bounds for the entropy which we discuss in Sect. 4 are the following. First, from Sect. 2, the spherical volume and the simplicial volume give lower bounds for the entropy. In the case of the simplicial volume this leads to

$$\chi(X) \geq \frac{3}{2}|\sigma(X)| + \frac{1}{162\pi^2}||X||,$$

which improves on our result in [22]. Second, using the lower bound for the entropy arising from the existence of maps to locally symmetric spaces of rank one proved by Besson–Courtois–Gallot [9], we deduce the results of Sambusetti [35]. Thus, our obstruction to the existence of Einstein metrics subsumes all the known obstructions which are homotopy invariant. See [21,23] for obstructions which are not homotopy invariant.

2 Entropies and Volumes

Let M be a connected closed oriented manifold of dimension n. In this section we discuss the following chain of inequalities between its topological invariants:

$$\frac{n^{n/2}}{n!}||M|| \leq 2^n n^{n/2} T(M) \leq \lambda(M)^n \leq h(M)^n \leq (n-1)^n \mathrm{MinVol}(M). \qquad (1)$$

This chain appears in [33], without $T(M)$, and with an unspecified constant in front of $||M||$. The inequalities involving $T(M)$ are from [8], where $T(M)$ was defined.

We now explain the terms in (1), and the sources for the different inequalities.

2.1 *Simplicial Volume*

The simplicial volume was introduced by Gromov [15]. Let $c = \Sigma_i r_i \sigma_i$ be a chain with real coefficients r_i, where $\sigma_i \colon \Delta^k \to X$ are singular k–simplices in M. One defines the norm of c to be

$$||c|| = \Sigma_i |r_i|.$$

If $\alpha \in H_k(M, \mathbb{R})$, set

$$||\alpha|| = inf\, \{||c|| \mid c \text{ a cycle representing } \alpha\},$$

where the infimum is taken over all cycles representing α. This is a seminorm on the homology.

The simplicial volume is defined to be the norm of the fundamental class, and is denoted

$$||M|| = ||[M]||.$$

Gromov [15] has shown that the simplicial volume of M is determined by the image of its fundamental class under the classifying map $f \colon M \to B\pi_1(M)$ of the universal covering.

This motivates the following definition.

Definition 1. An invariant I of closed oriented manifolds M is said to be homologically invariant if it depends only on $f_*[M] \in H_*(B\pi_1(M))$.

Homological invariance implies that I is homotopy-invariant, and that it is a bordism invariant of $[M, f] \in \Omega_n(B\pi_1(M))$.

2.2 Spherical Volume

The spherical volume was introduced by Besson–Courtois–Gallot [8].

Let $\tilde{M}$ be the universal covering of M. Fix a positive $\pi_1(M)$-invariant measure $\tilde{\mu}$ on $\tilde{M}$ that is absolutely continuous with respect to Lebesgue measure. Denote by (S^∞, can) the unit sphere in the Hilbert space $L^2(\tilde{M}, \tilde{\mu})$ endowed with the metric induced by the scalar product. For every $\pi_1(M)$-equivariant immersion $\Phi\colon \tilde{M} \to S^\infty$ we have an induced metric $\Phi^*(can)$ which descends to M, and we write $vol(\Phi)$ for $Vol(M, \Phi^*(can))$.

The spherical volume of M is defined to be

$$T(M) = inf\,\{vol(\Phi) \mid \Phi \text{ a } \pi_1(M) - \text{equivariant immersion } \tilde{M} \to S^\infty\}.$$

This does not depend on the choice of measure $\tilde{\mu}$, see [8].

Brunnbauer [11] has proved that the spherical volume is homologically invariant in the sense of Definition 1. In particular it is homotopy-invariant.

Besson, Courtois and Gallot proved the first inequality in (1) as Theorem 3.16 in [8].

2.3 Volume Entropy

For a Riemannian metric g on M consider the lift $\tilde{g}$ to the universal covering $\tilde{M}$. For an arbitrary basepoint $p \in \tilde{M}$ consider the limit

$$\lambda(M, g) = \lim_{R\to\infty} \frac{log Vol(B(p, R))}{R},$$

where $B(p, R)$ is the ball of radius R around p in $\tilde{M}$ with respect to $\tilde{g}$, and the volume is taken with respect to $\tilde{g}$ as well. After earlier work by Efremovich, Shvarts, Milnor [29] and others, Manning [27] showed that the limit exists and is independent of p. It follows from [29] that $\lambda(M, g) > 0$ if and only if $\pi_1(M)$ has exponential growth.

We call $\lambda(M, g)$ the volume entropy of the metric g, and define the volume entropy of M to be

$$\lambda(M) = inf\,\{\lambda(M, g) \mid g \in Met(M) \text{ with } Vol(M, g) = 1\}.$$

This sometimes vanishes even when $\lambda(M, g) > 0$ for every g. The normalization of the total volume is necessary because of the scaling properties of $\lambda(M, g)$.

Any metric on M can be scaled so that $Ric_g \geq -\frac{1}{n-1}g$. By the Bishop volume comparison theorem, cf. [10], this implies $\lambda(M, g) \leq 1$. Setting

$$\hat{g} = \frac{1}{Vol(M, g)^{2/n}} g$$

we have $Vol(M, \hat{g}) = 1$ and $\lambda(M, \hat{g}) = Vol(M, g)^{1/n}\lambda(M, g) \leq Vol(M, g)^{1/n}$, implying

$$\lambda(M) = \inf\{\lambda(M, \hat{g})\} \leq \inf\{Vol(M, g)^{1/n} \mid g \in Met(M) \text{ s.t. } Ric_g \geq -\frac{1}{n-1}g\}. \tag{2}$$

This is not an equality because the Bishop estimate is not sharp, except for metrics of constant sectional curvature.

Babenko has shown that the volume entropy $\lambda(M)$ is homotopy invariant [1], and is also an invariant of the bordism class $[M, f] \in \Omega_n(B\pi_1(M))$, cf. [3]. These results were sharpened by Brunnbauer [11], who proved the homological invariance of $\lambda(M)$. Although these results show that $\lambda(M)$ is a rather crude invariant which is understandable in many situations, there are still many open questions about this invariant. For example, its behavior under taking finite coverings is not understood. We do not even know whether the vanishing of λ on a finite covering of M implies the vanishing on M itself. This question came up in [19].

The second inequality in (1) follows from Theorem 3.8 of Besson–Courtois–Gallot [8].

2.4 Topological Entropy

For a Riemannian metric g on M consider the topological entropy $h(M, g)$ of its geodesic flow as a dynamical system on the unit sphere bundle, cf. [27]. The topological entropy of M is defined to be

$$h(M) = inf\,\{h(M, g) \mid g \in Met(M) \text{ with } Vol(M, g) = 1\}.$$

Here again the normalization of the total volume is necessary because of scaling properties.

It seems to be unkown what exactly the minimal topological entropy depends on, e.g. the homotopy type, the homeomorphism type, or the diffeomorphism type. It is not clear whether this is a subtle invariant, or a crude one like the minimal volume entropy and the spherical volume.

Manning [27] proved $\lambda(M, g) \leq h(M, g)$ for all closed M. Taking the infimum over all metrics with normalized volume yields the third inequality in (1).

2.5 Minimal Volume

The minimal volume was introduced by Gromov [15]. It is defined by

$$\mathrm{MinVol}(M) = inf\{Vol(M,g) \mid g \in Met(M) \text{ with } |K_g| \leq 1\},$$

where K_g denotes the sectional curvature of g.

It is known that the minimal volume is a very sensitive invariant of M, which depends on the smooth structure in an essential way. Bessières [7] has given examples of pairs of high-dimensional manifolds which are homeomorphic, but have different positive minimal volumes. In Sect. 5 we shall use Seiberg–Witten theory to show that the vanishing of the minimal volume is not invariant under homeomorphisms. This represents a more dramatic failure of topological invariance than Bèssières's result, since it shows that there are homeomorphic manifolds such that one collapses with bounded sectional curvature, and the other one does not.

Manning [28] proved that for a closed Riemannian manifold with sectional curvature bounded by $|K_g| \leq k$ the topological entropy is bounded by

$$h(M,g) \leq (n-1)\sqrt{k}.$$

By rescaling, this implies the last of the inequalities in (1), cf. [31] p. 129.

3 Isoperimetric Constants and Minimal Eigenvalues

One of Gromov's original motivations for introducing the simplicial volume was to obtain lower bounds for the minimal volume. However, the lower bounds given by the simplicial volume via (1) are usually rather weak. In particular, there seems to be no known example of a manifold with $T(M) > 0$ for which the inequality

$$\frac{n^{n/2}}{n!}||M|| \leq 2^n n^{n/2} T(M) \tag{3}$$

is sharp. All the other inequalities in (1) are sharp for surfaces of genus $g \geq 2$, where

$$4(g-1) = ||\Sigma_g|| < 8T(\Sigma_g) = \lambda(\Sigma_g)^2 = h(\Sigma_g)^2 = \mathrm{MinVol}(\Sigma_g) = 4\pi(g-1).$$

One approach to understanding and quantifying the failure of (3) to be sharp is to consider intermediate invariants which may interpolate between the simplicial and the spherical volumes. In fact, a candidate for such an intermediate invariant, based on the minimal eigenvalue of the Laplacian on the universal covering, occurs in the work of Besson–Courtois–Gallot [8]. We now elaborate on this to discuss the following alternative to (1):

$$I(M)^n \leq 2^n \Lambda_0(M)^{n/2} \leq 2^n n^{n/2} T(M) \leq \lambda(M)^n \leq h(M)^n, \qquad (4)$$

where we can also continue on the right with the same minimal volume term as in (1). Only the first two terms and the first two inequalities on the left need any explanation, as the rest has been explained already in the previous section.

3.1 Minimal Eigenvalue

Given a closed Riemannian manifold (M, g), we consider the Riemannian universal cover $(\tilde{M}, \tilde{g})$, and define

$$\lambda_0(\tilde{M}, \tilde{g}) = \inf_{f \in C_0^\infty} \frac{\int_{\tilde{M}} f \cdot \Delta f \, d vol_{\tilde{g}}}{\int_{\tilde{M}} f^2 d vol_{\tilde{g}}},$$

where C_0^∞ denotes the smooth compactly supported functions on $\tilde{M}$. Extending the Laplacian to L^2-functions, $\lambda_0(\tilde{M}, \tilde{g})$ is the greatest lower bound for its spectrum. It is tempting now to define the minimal eigenvalue of M as

$$\inf\{\lambda_0(\tilde{M}, \tilde{g}) \mid g \in Met(M) \text{ with } Vol(M, g) = 1\},$$

which would be the naive generalization of the definition of the minimal volume entropy. However, this definition always gives zero, cf. [36]. The correct definition of an invariant of M derived from the minimal eigenvalue is the following. First, we define an invariant of a conformal class $[g]$ by setting

$$\Lambda_0(M, [g]) = \sup\{\lambda_0(\tilde{M}, \tilde{g}') \mid g' \in [g] \text{ with } Vol(M, g') = 1\},$$

and then define

$$\Lambda_0(M) = \inf\{\Lambda_0(M, [g]) \mid [g] \in Conf(M)\}$$

as an infimum over conformal classes.

This is a meaningful invariant, which for surfaces of genus $g \geq 2$ can be shown to equal $\pi(g-1)$, cf. [36]. The work of Besson–Courtois–Gallot [8] implies the second inequality in (4) in all dimensions n. It is sharp for surfaces.

3.2 Isoperimetric Constant

Cheeger's isoperimetric constant of (M, g) is defined to be

$$i(\tilde{M}, \tilde{g}) = \inf_N \frac{Vol_{n-1}(\partial N)}{Vol_n(N)},$$

where $N \subset \tilde{M}$ ranges over all compact connected subsets with smooth boundary, say. Again one would naively define the isoperimetric constant of M to be

$$\inf\{i(\tilde{M}, \tilde{g}) \mid g \in Met(M) \text{ with } Vol(M, g) = 1\},$$

but this always gives zero, cf. [36]. The definition has to be modified in the same way as for the minimal eigenvalue. First we define an invariant of a conformal class $[g]$ by setting

$$I(M, [g]) = \sup\{i(\tilde{M}, \tilde{g}') \mid g' \in [g] \text{ with } Vol(M, g') = 1\},$$

and then we take the infimum over conformal classes:

$$I(M) = \inf\{I(M, [g]) \mid [g] \in Conf(M)\}.$$

Cheeger's inequality $i(\tilde{M}, \tilde{g})^2 \leq 4\lambda_0(M)$, see [13], implies the first inequality in (4).

For both the minimal eigenvalue $\Lambda_0(M)$ and for the isoperimetric constant $I(M)$ it is not immediately clear what they depend on. One could try to use Brunnbauer's axiomatic approach [11] to prove that these are homological invariants of manifolds. However, because of the complicated definitions, using a supremum within each conformal class before taking the infimum over conformal classes, it is hard to verify the required axioms for these invariants. Because of this difficulty we do not yet know how subtle these invariants are.

The chain (4) shows that one can use the minimal eigenvalue and the isoperimetric constant instead of the simplicial volume to obtain lower bounds for the minimal volume entropy and the minimal volume.

A very important outstanding problem is to decide whether there is an upper bound for the simplicial volume in terms of the isoperimetric constant. More precisely, one would like to know whether the inequality

$$\frac{n^{n/2}}{n!}||M|| \leq I(M)^n$$

holds. If this were the case, then we could insert (4) into (1), and we could hope to measure the gap between the simplicial volume and the spherical volume. Potentially this could lead to an improvement of (1), by improving the constant in front of the simplicial volume to close the gap between the simplicial and spherical volumes in (1).

4 Einstein Metrics on Four-Manifolds

We now prove the following constraint on the topology of four-manifolds admitting Einstein metrics.

Theorem 1. *Let X be a closed oriented Einstein 4–manifold. Then*

$$\chi(X) \geq \frac{3}{2}|\sigma(X)| + \frac{1}{108\pi^2}\lambda(X)^4, \tag{5}$$

where $\chi(X)$ denotes the Euler characteristic, $\sigma(X)$ the signature, and $\lambda(X)$ the volume entropy.

Equality in (5) occurs if and only if every Einstein metric on X is flat, is non-flat locally Calabi–Yau, or is of constant negative sectional curvature.

Remark 1. Note that the right-hand side of (5) is an invariant of the cobordism class $[X, f] \in \Omega_4(B\pi_1(X))$, where $f: X \to B\pi_1(X)$ is the classifying map of the universal covering. For the signature this is due to Thom, and for the volume entropy it was proved by Babenko[1] in [3]. Rationally, one has the well-known isomorphism

$$\Omega_4(B\pi_1(X)) \otimes \mathbb{Q} = H_4(B\pi_1(X); \Omega_0(\star) \otimes \mathbb{Q}) \oplus H_0(B\pi_1(X); \Omega_4(\star) \otimes \mathbb{Q}).$$

The second summand is responsible for the signature term (=first Pontryagin number) and the first one for the entropy term in (5). Indeed, Brunnbauer's homological invariance result for the entropy [11] shows that $\lambda(X)$ depends only on $f_*[X] \in H_4(B\pi_1(X))$.

Proof of Theorem 1. By the Gauss–Bonnet theorem the Euler characteristic of a closed oriented Riemannian 4–manifold (X, g) is

$$\chi(X) = \frac{1}{8\pi^2}\int_X \frac{1}{24}s_g^2 + |W|^2 - |Ric_0|^2 d\,\mathrm{vol}_g,$$

where s_g is the scalar curvature, W is the Weyl tensor, and Ric_0 is the trace-less Ricci tensor of g. For an Einstein metric this reduces to

$$\chi(X) = \frac{1}{8\pi^2}\int_X \frac{1}{24}s_g^2 + |W|^2 d\,\mathrm{vol}_g. \tag{6}$$

Thus $\chi(X) \geq 0$, with equality if and only if every Einstein metric on X is flat.

If the scalar curvature of an Einstein metric is positive, then $\pi_1(X)$ is finite by Myers's theorem and the volume entropy vanishes, so that (5) reduces to the Hitchin–Thorpe inequality obtained by comparing (6) with the Chern–Weil formula

$$\sigma(X) = \frac{1}{12\pi^2}\int_X |W_+|^2 - |W_-|^2 d\,\mathrm{vol}_g. \tag{7}$$

If the scalar curvature is zero, then the Cheeger–Gromoll splitting theorem implies that either the fundamental group is finite, or the Euler characteristic

---

[1]His paper assumes that manifolds are of dimension ≥ 5, but that is not important here; cf. [11].

vanishes. In the latter case the Einstein metric is flat and the volume growth is only polynomial, so that in both cases the volume entropy vanishes and we are done as before.

If the scalar curvature is negative, we scale the metric so that $Ric_g = -\frac{1}{3}g$. Using the Chern–Weil formula (7) the second term in (6) is

$$\frac{1}{8\pi^2}\int_X |W|^2 d\,\mathrm{vol}_g \geq \frac{1}{8\pi^2}\left|\int_X |W_+|^2 - |W_-|^2 d\,\mathrm{vol}_g\right| = \frac{3}{2}|\sigma(X)|. \tag{8}$$

The first term in (6) is

$$\frac{1}{8\pi^2}\frac{1}{24}\left(-\frac{4}{3}\right)^2 Vol(X,g) = \frac{1}{108\pi^2}Vol(X,g).$$

Now using the Bishop estimate to bound the volume entropy $\lambda(X,g)$ from above by 1 (see (2)) we find

$$\frac{1}{8\pi^2}\frac{1}{24}\left(-\frac{4}{3}\right)^2 Vol(X,g) \geq \frac{1}{108\pi^2}Vol(X,g)\lambda(X,g)^4 \geq \frac{1}{108\pi^2}\lambda(X)^4. \tag{9}$$

This completes the proof of (5). Equality cannot hold in the case of positive scalar curvature because in this case we threw away the scalar curvature term in (6). In the case of zero scalar curvature the entropy vanishes and the discussion of equality reduces to the corresponding discussion for the Hitchin–Thorpe inequality, see Hitchin [18]. The conclusion is that every Einstein metric is flat or (up to choosing the orientation suitably) locally Calabi–Yau but non-flat. For negative scalar curvature equality in (5) implies equality in (9), so that $\lambda(X) > 0$ and every Einstein metric on X has to be entropy-minimizing. Moreover, the Einstein metric must have entropy $\lambda(X,g) = 1$, and so by the recent rigidity theorem of Ledrappier and Wang [26] it has constant negative sectional curvature.[2] Thus these are the only candidates for equality. They do indeed give equality because they are conformally flat and therefore (8) is an equality for them, and they are entropy-minimizing by the celebrated result of Besson–Courtois–Gallot [9].

Remark 2. The three cases giving rise to equality in (5) correspond to the vanishing of all three terms in (5) for flat manifolds, to the vanishing of $\lambda(X)$ only for Calabi–Yaus, and to the vanishing of $\sigma(X)$ only in the hyperbolic case. After earlier, unpublished, work of Calabi, Charlap–Sah, and Levine, the closed orientable flat four-manifolds were classified by Hillman [17] and by Wagner [37], who showed that there are 27 distinct ones. By Bieberbach's theorems, all these manifolds are finite quotients of T^4. In the locally Calabi–Yau case, Hitchin [18] showed that the

[2]Our scaling normalization for the lower Ricci curvature bound is different from the one used in [26], so entropy $= 1$ in our case corresponds to entropy $= n - 1$ in [26].

only possible manifolds are the $K3$ surface and its quotients by certain free actions of $\mathbb{Z}_2$ and of $\mathbb{Z}_2 \times \mathbb{Z}_2$. In the hyperbolic case there are of course infinitely many manifolds.

Combining (4) with Theorem 1 we obtain:

Corollary 1. *Let X be a closed oriented Einstein 4–manifold. Then we have the following lower bounds for the Euler characteristic of X:*

$$\chi(X) \geq \frac{3}{2}|\sigma(X)| + \frac{64}{27\pi^2}T(X), \tag{10}$$

$$\chi(X) \geq \frac{3}{2}|\sigma(X)| + \frac{4}{27\pi^2}\Lambda_0(X)^2, \tag{11}$$

$$\chi(X) \geq \frac{3}{2}|\sigma(X)| + \frac{1}{108\pi^2}I(X)^4. \tag{12}$$

In all cases, equality occurs if and only if every Einstein metric on X is flat, is non-flat locally Calabi–Yau, or is of constant negative sectional curvature.

The point of the last statement is that equality in one of these estimates implies equality in (5).

Combining (1) with Theorem 1 we obtain the following improvement of the Gromov–Hitchin–Thorpe inequality proved in [22]:

Corollary 2. *Let X be a closed oriented Einstein 4–manifold. Then*

$$\chi(X) \geq \frac{3}{2}|\sigma(X)| + \frac{1}{162\pi^2}||X||. \tag{13}$$

Note that this is not sharp in any interesting cases with non-vanishing simplicial volume term, because the lower bound for the entropy in terms of the simplicial volume is not optimal.

Theorem 1 together with the work of Besson–Courtois–Gallot [9] implies the following:

Corollary 3. (Sambusetti [35]). *Let X be a closed oriented Einstein 4-manifold.*

If X admits a map of non-zero degree d to a real hyperbolic manifold Y, then

$$\chi(X) \geq \frac{3}{2}|\sigma(X)| + |d| \cdot \chi(Y), \tag{14}$$

with equality only if X is also real hyperbolic.

If X admits a map of non-zero degree d to a complex hyperbolic manifold Y, then

$$\chi(X) > \frac{3}{2}|\sigma(X)| + |d| \cdot \frac{32}{81}\chi(Y). \tag{15}$$

Proof of Theorem 1. To prove (14) we combine (5) with the inequality

$$\lambda(X)^4 \geq d \cdot \lambda(Y)^4 = d \cdot 3^4 Vol(Y, g_0)$$

from [9]. Here g_0 is the hyperbolic metric on Y normalized so that $K = -1$. With this normalization the Gauss–Bonnet formula (6) gives $Vol(Y, g_0) = \frac{4\pi^2}{3}\chi(Y)$ so that $3^4 Vol(Y, g_0) = 108\pi^2\chi(Y)$.

In the case of equality X must itself be hyperbolic, by the equality case of Theorem 1.

For maps to complex hyperbolic manifolds one uses the corresponding statement from [9]. Again $\lambda(X)^4 \geq d \cdot \lambda(Y)^4$ and the hyperbolic metric on Y is entropy-minimizing. Thus one only has to check the proportionality factors between the fourth power of the entropy and the Euler characteristic, and the weak form of (15) follows. By Theorem 1 equality cannot occur in this case.

It would be interesting to know whether (5) remains true if we replace the volume entropy λ by the topological entropy h, cf. (1). The results of Paternain and Petean [32,33] are very suggestive in this regard. While the volume entropy can only be positive for manifolds with fundamental groups of exponential growth, the topological entropy may be positive even for simply connected manifolds. Unlike the volume entropy, the topological entropy is not known to be homotopy invariant. Notice that we can definitely not replace λ^4 by a positive multiple of MinVol, because the $K3$ surface satisfies $\chi(K3) = \frac{3}{2}|\sigma(K3)|$, and has positive MinVol as its Euler characteristic and signature are non-zero. We will show in the next section that the non-vanishing of the minimal volume depends in an essential way on the smooth structure. That the existence of an Einstein metric depends on the smooth structure was first shown in [21]. Recently Brunnbauer, Ishida and Suárez-Serrato [12] have constructed some interesting examples showing that the smooth obstructions to the existence of Einstein metrics are completely independent of those provided by Theorem 1.

To end this section, we briefly discuss the history of Theorem 1. The Hitchin–Thorpe inequality $\chi(X) \geq \frac{3}{2}|\sigma(X)|$ was, for a long time, the only known obstruction to the existence of Einstein metrics on four-manifolds. This changed when Gromov [15] gave a lower bound for the Euler number by a multiple of the simplicial volume. For many years after that, the Hitchin–Thorpe inequality and the "Gromov obstruction" were treated as separate, unrelated obstructions; see for example the discussion in [6, 9, 35]. In 1997, trying to understand Gromov's argument, I found that the signature term of the Hitchin–Thorpe inequality and the simplicial volume term of Gromov's inequality can actually be combined, to obtain an inequality like (13); see [22]. That inequality still has the flaw that it is not sharp in any interesting cases, because the simplicial volume term is too weak. Thinking about the chain (1), I found the sharp Theorem 1 in 2004, wrote it down in [24], and also explained that it subsumes all the known homotopy-invariant obstructions to Einstein metrics. That paper has remained unpublished, because various referees claimed that Theorem 1 was not interesting, or that it was the same as the result of [22], or that it was known to the authors of [9, 35]. What the referees did not notice, and I myself only noticed recently, is that the discussion of the limiting case in [24] was actually incomplete. The only way I know how to characterize the limiting case, is through the recent rigidity result of Ledrappier and Wang [26], as used in the proof of Theorem 1 given above. In particular there seems to be no way of obtaining the desired conclusion from [9, 35].

5 Minimal Volumes and Smooth Structures

In this section we show that vanishing of the minimal volume is a property of the smooth structure, which is not invariant under homeomorphisms. The proof below actually shows that one can change the smooth structure of a manifold with a smooth free circle action so that for the new smooth structure any smooth circle action must have fixed points.

Theorem 2. *For every $k \geq 0$ the manifold $X_k = k(S^2 \times S^2)\#(1+k)(S^1 \times S^3)$ with its standard smooth structure has zero minimal volume.*

If k is odd and large enough, then there are infinitely many pairwise non-diffeomorphic smooth manifolds Y_k homeomorphic to X_k, all of which have strictly positive minimal volume.

Proof of Theorem 1. Note that $X_0 = S^1 \times S^3$ has obvious free circle actions, and therefore collapses with bounded sectional curvature. To see that all X_k have vanishing minimal volume it suffices to construct fixed-point-free circle actions on them.

The product $S^2 \times S^2$ has a diagonal effective circle action which on each factor is rotation around the north-south axis. It has four fixed pints, and the linearization of the action induces one orientation at two of the fixed points, and the other orientation at the remaining two. The induced action on the boundary of an S^1-invariant small ball around each of the fixed points is the Hopf action on S^3. By taking equivariant connected sums at fixed points, pairing fixed points at which the linearizations give opposite orientations, we obtain effective circle actions with $2+2k$ fixed points on the connected sum $k(S^2 \times S^2)$ for every $k \geq 1$. Now we have $1+k$ fixed points at which the linearization induces one orientation, and $1+k$ at which it induces the other orientation. Then making equivariant self-connected sums at pairs of fixed points with linearizations inducing opposite orientations we finally obtain a free circle action on $X_k = k(S^2 \times S^2)\#(1+k)(S^1 \times S^3)$.

If k is odd and large enough, then there are symplectic manifolds Z_k homeomorphic (but not diffeomorphic) to $k(S^2 \times S^2)$, see for example [15]. By the construction given in [15], we may assume that Z_k contains the Gompf nucleus of an elliptic surface. By performing logarithmic transformations inside this nucleus, we can vary the smooth structures on the Z_k in such a way that the number of Seiberg–Witten basic classes with numerical Seiberg–Witten invariant $= \pm 1$ becomes arbitrarily large, cf. Theorem 8.7 of [14] and Example 3.5 of [23].

Consider $Y_k = Z_k\#(1+k)(S^1 \times S^3)$. This is clearly homeomorphic to X_k. Although the numerical Seiberg–Witten invariants of Y_k must vanish, cf. [20, 25], we claim that each of the basic classes with numerical Seiberg–Witten invariant $= \pm 1$ on Z_k gives rise to a monopole class on Y_k, that is the characteristic class of a Spin^c-structure for which the monopole equations have a solution for every Riemannian metric on Y_k. There are two ways to see this. One can extract our claim from the connected sum formula [4] for the stable cohomotopy refinement of Seiberg–Witten invariants introduced by Bauer and Furuta [5], cf. [Furuta, Private communication (2003)]. Alternatively, one uses the invariant defined by the homology class of the moduli space of solutions to the monopole equations, as in [20]. This means that the first homology of the manifold is used, and here this is enough to obtain a non-vanishing invariant. Using this invariant, our claim follows from Proposition 2.2 of Ozsváth–Szabó [30].

As Y_k has non-torsion monopole classes c with $c^2 = 2\chi(Z_k) + 3\sigma(Z_k) = 4 + 4k > 0$, the bound

$$c^2 \leq \frac{1}{32\pi^2} \int_{Y_k} s_g^2 d\,\mathrm{vol}_g,$$

where s_g is the scalar curvature of any Riemannian metric g, shows that Y_k cannot collapse with bounded scalar curvature, cf. [23]. *A fortiori* it cannot collapse with bounded sectional curvature, and so its minimal volume is strictly positive.

The monopole classes we constructed on Y_k are all generic monopole classes in the sense of [23]. By Lemma 2.4 of *loc. cit.* each manifold has at most finitely many such classes. As we can change the smooth structure to make the number of generic monopole classes arbitrarily large, we have infinitely many distinct smooth structures we can choose for Y_k.

The above proof also gives the following:

Corollary 4. *There are pairs of homeomorphic closed manifolds such that one collapses with bounded sectional curvature and the other one cannot collapse even with bounded scalar curvature.*

Remark 3. That connected sums of manifolds with vanishing minimal volumes may have non-vanishing minimal volumes is immediate by looking at connected sums of tori. The manifolds X_k discussed above have the property that their minimal volumes vanish, although they are connected sums of manifolds with non-vanishing minimal volumes. Thus the minimal volume, and even its (non-)vanishing, does not behave in a straightforward manner under connected sums.

This remark was motivated by the recent paper [34] of Paternain and Petean. After Theorem 2 appeared on the arXiv in [24], these authors remarked on the complicated behaviour of the minimal volume under connected sums based on some 6-dimensional examples, see Remark 3.1 in [34].

Acknowledgements This work is part of the project *Asymptotic invariants of manifolds*, supported by the *Schwerpunktprogramm Globale Differentialgeometrie* of the *Deutsche Forschungsgemeinschaft*.

References

1. Babenko, I.K.: Asymptotic invariants of smooth manifolds. Izv. Ross. Akad. Nauk. Ser. Mat. **56**, 707–751 (1992) (Russian) Engl. transl. in Russian Acad. Sci. Izv. Math. **41**, 1–38 (1993)
2. Babenko, I.K.: Asymptotic volumes and simply connected surgeries of smooth manifolds. Izv. Ross. Akad. Nauk. Ser. Mat. **58**, 218–221 (1994) (Russian); Engl. transl. in Russian Acad. Sci. Izv. Math. **44**, 427–430 (1995)
3. Babenko, I.K.: Extremal problems of geometry, surgery on manifolds, and problems in group theory. Izv. Ross. Akad. Nauk. Ser. Mat. **59**, 97–108 (1995) (Russian); Engl. translation in Izv. Math. **59**, 321–332 (1995)
4. Bauer, S.: A stable cohomotopy refinement of Seiberg–Witten invariants: II. Invent. Math. **155**, 21–40 (2004)
5. Bauer, S., Furuta, M.: A stable cohomotopy refinement of Seiberg–Witten invariants: I. Invent. Math. **155**, 1–19 (2004)
6. Besse, A.L.: Einstein Manifolds. Springer, Berlin (1987)
7. Bessières, L.: Un théorème de rigidité différentielle. Comment. Math. Helv. **73**, 443–479 (1998)
8. Besson, G., Courtois, G., Gallot, S.: Volume et entropie minimale des espaces localement symétriques. Invent. Math. **103**, 417–445 (1991)
9. Besson, G., Courtois, G., Gallot, S.: Entropies et rigidités des espaces localement symétriques de courbure strictement négative. Geom. Func. Anal. (GAFA) **5**, 731–799 (1995)
10. Bishop, R.L., Crittenden, R.J.: Geometry of Manifolds. Academic, New York (1964); reprinted by AMS Chelsea Publishing, Providence 2001
11. Brunnbauer, M.: Homological invariance for asymptotic invariants and systolic inequalities. Geom. Func. Anal. (GAFA) **18**, 1087–1117 (2008)
12. Brunnbauer, M., Ishida, M., Suárez-Serrato, P.: An essential relation between Einstein metrics, volume entropy, and exotic smooth structures. Math. Res. Lett. **16**, 503–514 (2009)
13. Cheeger, J.: A lower bound for the smallest eigenvalue of the Laplacian. In: Problems in Analysis (Papers dedicated to Salomon Bochner, 1969), pp. 195–199. Princeton University Press, Princeton, NJ (1970)
14. Fintushel, R., Stern, R.J.: Rational blowdowns of smooth 4–manifolds. J. Diff. Geom. **46**, 181–235 (1997)
15. Gromov, M.: Volume and bounded cohomology. Publ. Math. I.H.E.S. **56**, 5–99 (1982)
16. Hanke, B., Kotschick, D., Wehrheim, J.: Dissolving four-manifolds and positive scalar curvature. Math. Zeit. **245**, 545–555 (2003)
17. Hillman, J.A.: Flat 4-manifold groups. New Zealand J. Math. **24**, 29–40 (1995)
18. Hitchin, N.J.: Compact four-dimensional Einstein manifolds. J. Diff. Geom. **9**, 435–441 (1974)
19. Kedra, J., Kotschick, D., Morita, S.: Crossed flux homomorphisms and vanishing theorems for flux groups. Geom. Func. Anal. (GAFA) **16**, 1246–1273 (2006)
20. Kotschick, D.: The Seiberg-Witten invariants of symplectic four-manifolds. Séminaire Bourbaki, 48ème année, 1995–96, no. 812. Astérisque **241**, 195–220 (1997)
21. Kotschick, D.: Einstein metrics and smooth structures. Geom. Topology **2**, 1–10 (1998)
22. Kotschick, D.: On the Gromov–Hitchin–Thorpe inequality. C. R. Acad. Sci. Paris **326**, 727–731 (1998)
23. Kotschick, D.: Monopole classes and Einstein metrics. Intern. Math. Res. Notices **2004**(12), 593–609 (2004)
24. Kotschick, D.: Entropies, volumes, and Einstein metrics. Preprint arXiv:math/0410215 v1 [math.DG] 8 Oct 2004
25. Kotschick, D., Morgan, J.W., Taubes, C.H.: Four-manifolds without symplectic structures but with non-trivial Seiberg–Witten invariants. Math. Res. Lett. **2**, 119–124 (1995)
26. Ledrappier, F., Wang, X.: An integral formula for the volume entropy with applications to rigidity. J. Diff. Geom. **85**, 461–477 (2010)
27. Manning, A.: Topological entropy for geodesic flows. Ann. Math. **110**, 567–573 (1979)

28. Manning, A.: More topological entropy for geodesic flows. In: Dynamical Systems and Turbulence, Springer LNM vol. 898, pp. 234–249. Springer, Berlin (1981)
29. Milnor, J.: A note on curvature and the fundamental group. J. Diff. Geom. **2**, 1–7 (1968)
30. Ozsváth, P., Szabó, Z.: Higher type adjunction inequalities in Seiberg–Witten theory. J. Diff. Geom. **55**, 385–440 (2000)
31. Paternain, G.P.: Geodesic Flows. Progress in Mathematics, vol. 180. Birkäuser, Basel (1999)
32. Paternain, G.P., Petean, J.: Einstein manifolds of non-negative sectional curvature and entropy. Math. Res. Lett. **7**, 503–515 (2000)
33. Paternain, G.P., Petean, J.: Minimal entropy and collapsing with curvature bounded from below. Invent. Math. **151**, 415–450 (2003)
34. Paternain, G.P., Petean, J.: Collapsing manifolds obtained by Kummer-type constructions. Trans. Am. Math. Soc. **361**, 4077–4090 (2009)
35. Sambusetti, A.: An obstruction to the existence of Einstein metrics on 4-manifolds. Math. Ann. **311**, 533–547 (1998)
36. Schmidt, R.: Spectral geometry, asymptotic invariants, and geometric group theory. Diplomarbeit München (2009)
37. Wagner, M.: Über die Klassifikation flacher Riemannscher Mannigfaltigkeiten. Diplomarbeit Basel (1997)

Kac-Moody Geometry

Walter Freyn

Abstract The geometry of symmetric spaces, polar actions, isoparametric submanifolds and spherical buildings is governed by spherical Weyl groups and simple Lie groups. The most natural generalization of semisimple Lie groups are affine Kac-Moody groups as they mirror their structure theory and have good explicitly known representations as groups of operators. In this article we describe the infinite dimensional differential geometry associated to affine Kac-Moody groups: Kac-Moody symmetric spaces, isoparametric submanifolds in Hilbert space, polar actions on Hilbert spaces and universal geometric twin buildings.

1 What is It All About

In this article we describe the infinite dimensional differential geometry governed by affine Kac-Moody groups. The theory of Kac-Moody algebras and Kac-Moody groups emerged around 1960 independently in the works of Kac [29], Moody [40], Kantor [31] and Verma (unpublished). From a formal, algebraic point of view, Kac-Moody algebras can be understood as realizations of generalized Cartan matrices [30]. Hence Kac-Moody algebras appear as a natural generalization of simple Lie algebras. Furthermore it was pointed out that there is an explicit description of affine Kac-Moody algebras in terms of extensions of loop algebras. This point of view links the theory of affine Kac-Moody algebras to the theory of simple Lie algebras in another very elementary geometric way. A completion of the loop algebras with respect to various norms opens the way to the use of functional analytic methods [46].

W. Freyn (✉)
Universität Münster, Institut für Mathematik, Einsteinstrasse 62, 48149 Münster, Germany
e-mail: walter.freyn@uni-münster.de

C. Bär et al. (eds.), *Global Differential Geometry*, Springer Proceedings in Mathematics 17, DOI 10.1007/978-3-642-22842-1_3,

The next milestone was the discovery of a close link between Kac-Moody algebras and infinite dimensional differential geometry around 1990 [26]. Chuu-Lian Terng proved that certain isoparametric submanifolds in Hilbert spaces and polar actions on Hilbert spaces can be described using completions of algebraic Kac-Moody algebras [50]. In the parallel finite dimensional theory, the isometry groups are semisimple Lie groups. Hence this points again to the close similarity between semisimple Lie groups and Kac-Moody groups. During the same time Jacques Tits developed the idea of twin buildings and twin BN-pairs which are associated to algebraic Kac-Moody groups much the same way as buildings and BN-pairs are associated to simple Lie groups [53]. All those developments hint to the idea, that the rich subject of finite dimensional geometric structures whose symmetries are described by Lie groups should have an infinite dimensional counterpart, whose symmetries are affine Kac-Moody groups [22].

The most important among those objects are the following four classes:

- Symmetric spaces,
- Polar actions,
- Buildings over the fields $\mathbb{R}$ or $\mathbb{C}$,
- Isoparametric submanifolds.

In the first two examples, homogeneity is intrinsic, in the second two examples homogeneity is a priori an additional assumption. However, homogeneity can be proven under the assumption of sufficiently high rank. The geometry of those examples reflects the two most important decompositions of the Lie groups: the structure of symmetric spaces is a consequence of the Iwasawa decomposition, the structure of the building is a consequence of the system of parabolic subgroups and connected to the Bruhat decomposition. Let us note, that similar decompositions exist also for affine Kac-Moody groups.

In the investigation of those classes of objects one encounters two decisive structural elements: First, they all have a class of special "flat" subspaces (resp. subcomplexes) equipped with the action of a spherical reflection group. Subgroups of this reflection group fix special subspaces (resp. lower dimensional cells). Those reflection groups correspond to the Weyl groups of simple Lie algebras. Second, those flat subspaces are pieced together such that they meet in spheres around this set of special subspaces.

Evidently we want a similar feature for our infinite dimensional theory. Taking into account that Weyl groups of affine Kac-Moody algebras are affine Weyl groups, that is reflection groups on spaces of signature $(+ + + + + + + +0)$, we expect the reflection groups appearing in Kac-Moody geometry to be of this type. Unfortunately the metric $(+ + + + + + + +0)$ is degenerate. To get a nondegenerate metric, as we wish it for the construction of pseudo-Riemannian manifolds (and hence symmetric spaces), we have to add an additional dimension and choose a metric, that is defined by a pairing between the last coordinate and our new one. Hence the resulting spaces will be Lorentzian.

This philosophy gives us the following picture:

There is a class of Lorentz symmetric spaces whose group of transvections is an affine Kac-Moody group. The classification of irreducible Kac-Moody symmetric spaces is analogous to the one of irreducible finite dimensional symmetric spaces. Their isotropy representations induce polar actions on the Lie algebra. Because of the additional dimension we expect the essential part to be 1-codimensional. Principal orbits are isoparametric submanifolds. There is a class of buildings, whose chambers correspond to the points of the isoparametric submanifolds. As the system of parabolic subgroups consists of two "opposite" conjugacy classes, the building will consist of two parts; as the Bruhat decompositions do not exist on the whole Kac-Moody group, each of those parts is highly disconnected.

By work done during the last 20 years this picture is now well established. In this survey we describe the main objects and connections between them, focusing on more recent work.

In Sect. 2 we start by the investigation of spherical and affine Coxeter groups and describe the connections between them [1, 11]. Then we describe the finite dimensional theory (Sect. 3) [1,4,27]. After that, we turn to the infinite dimensional theory and introduce affine Kac-Moody algebras (Sect. 4.1) and their Kac-Moody groups (Sect. 4.2) [14, 30, 46]. Having described the symmetry groups, we turn to the geometry governed by them: In Sect. 5 we introduce Kac-Moody symmetric spaces [13, 14]. Then we turn to infinite dimensional polar actions (Sect. 6) [51], and isoparametric submanifolds (Sect. 7) [50]. Finally, in Sect. 8, we investigate universal geometric twin buildings [14]. The last section, Sect. 9, is devoted to a description of open problems and directions for future research.

2 Reflection Groups

Let V be a vector space equipped with a not necessarily positive definite inner product $\langle\ ,\ \rangle$. An (affine) hyperplane H in V uniquely determines an (affine) reflection $s_H : V \longrightarrow V$. A reflection group W is a group of reflections, such that W equipped with the discrete topology acts properly discontiuous on V. Hence it is defined by a set of hyperplanes $\mathcal{H} = \{H_i, i \in I\}$ such that $s_j(H_i) \subset \mathcal{H}\ \forall i, j$. Each reflection group is a discrete subgroup of the group $O(\langle\ ,\rangle)$.

The structure of the resulting groups depends now on the metric $\langle\ ,\ \rangle$ on V.

The most common case is the one such that $\langle\ ,\ \rangle$ is positive definite, i.e. without loss of generality the standard Euclidean metric. The resulting reflection group is a discrete subgroup of the orthogonal group $O(\langle\ ,\ \rangle)$. As the unit sphere of $\langle\ ,\ \rangle$ is compact and W is supposed to act properly discontinuous, we deduce that W is a finite group.

Define a group W to be a Coxeter group if it admits a presentation of the following type:

$$W := \left\langle s_1, \dots, s_n, n \in \mathbb{N} \middle| s_i^2 = 1, (s_i s_k)^{m_{ik}} = 1, i, k = 1, \dots, n \text{ and } m_{ik} \in \mathbb{N} \cup \infty \right\rangle.$$

Finite reflection groups are exactly finite Coxeter groups. Furthermore any finite Coxeter group admits a realization as a subgroup of the orthogonal group of a positive definite metric [6].

$K := \{1, \ldots, n\}$ is called the indexing set. If $K' \subset K$, we use the notation $W_{K'} \subset W \cong W_K$ for the sub-Coxeter group generated by K'. The matrix $M = (m_{ik})_{i,k \in K}$ is called the Coxeter matrix. A Coxeter system is a pair (W, S) consisting of a Coxeter group W and a set of generators S such that $\text{ord}(s) = 2$ for all $s \in S$.

Any group element $w \in W$ is a word in the generators $s_i \in S$. We define the length $l(w)$ of an element $w \in W$ to be the length of the shortest word representing w. The specific word – and thus the length $l(w)$ – depends on the specified set S of generators. Nevertheless many global properties are preserved by a change of generators – see [11].

Simple finite Coxeter groups have a complete classification:

1. Type A_n is the symmetric group in n-letters.
2. Type C_n resp. B_n is the group of signed permutations of n elements.
3. Type D_n is the Weyl group of the orthogonal groups $SO(2n, \mathbb{C})$.
4. Type G_2, F_4, E_6, E_7, E_8 are the Weyl groups of the Lie groups of the same names.
5. Type H_3 is the symmetry group of the 3-dimensional Dodecahedron and the Icosahedron, H_4 is the symmetry group of a regular 120-sided solid in 4 space whose 3-dimensional faces are dodecahedral.
6. $I_n, n = 5$ or $n > 7$ is the dihedral group of order $2n$. We have the equivalences $I_1 = \mathbb{Z}_2, I_2 = A_1 \times A_1, I_3 = A_2, I_4 = C_2, I_6 = G_2$.

Call a Coxeter group "crystallographic" if it stabilizes a lattice. The crystallographic Coxeter groups are types $A_n, C_n, D_n, E_6, E_7, E_8, F_4, G_2$. Crystallographic Coxeter groups are exactly those that appear as Weyl groups of the root systems of finite simple Lie algebras [6].

Closely related to finite Coxeter groups are affine Weyl groups. They are discrete subgroups of the group of affine transformations of a vector space. This group is a semidirect product of $O(\langle\ ,\ \rangle)$ with the group of translations of V. Hence $W_{\text{aff}} = L \rtimes W$ with W a finite Coxeter group and $L \cong \mathbb{Z}^k$ a lattice in V. In particular W is crystallographic. There is an easy way to linearize affine transformations by embedding the affine n-space V^n into an affine subspace i.e. the one defined by $x_{n+1} = 1$ of an $n+1$-dimensional vector space V^{n+1}. Then the group of affine transformations on V^n embeds into the general linear group $GL(V^{n+1})$ [1]. V^{n+1} carries a degenerate metric of signature $(n, 0)$. Hence we can interpret affine Weyl group as linear reflection groups in a vector space with a degenerate metric. In this form they appear as Weyl groups of affine Kac-Moody algebras. To get a non-degenerate metric on the Kac-Moody algebra we will add a further direction, to pair it with the degenerate direction. Hence the resulting $n+1$-dimensional vector space will carry a Lorentz structure [30].

The next natural case are reflection groups in a space with a Lorentz metric. They appear as Weyl groups of hyperbolic Kac-Moody algebras [47]. While hyperbolic

Kac-Moody algebras have applications in physics and in mathematics for example in M-theory and supergravity, there is up to now no differential geometry developed admitting hyperbolic Kac-Moody groups as symmetry groups.

3 The Finite Dimensional Blueprint

This section introduces the finite dimensional geometry whose symmetries are governed by semisimple Lie groups. Useful references are [1, 4, 27, 44].

3.1 Semisimple Lie Groups

We define a Lie algebra to be semisimple if it has no Abelian ideals. It is simple if it is not 1-dimensional and has no nontrivial ideal. A Lie group is (semi-)simple, if its Lie algebra is (semi-)simple.

Classical examples are $SL(n,\mathbb{C}) := \{X \in \text{Mat}^{n\times n}(\mathbb{C})|\det(X) = 1\}$ with Lie algebra $\mathfrak{sl}(n,\mathbb{C}) := \{X \in \text{Mat}^{n\times n}(\mathbb{C})|\text{trace}(X) = 0\}$, the orthogonal groups $SO(n,\mathbb{C}) := \{X \in \text{Mat}^{n\times n}(\mathbb{C})|XX^t = \text{Id}\}$. Every complex semisimple Lie algebra (resp. Lie group) is the direct product of complex simple Lie algebras (resp. groups). Hence we focus our study on simple Lie algebras and Lie groups.

We call a real simple Lie algebra $\mathfrak{g}$ a real form of the complex simple Lie algebra $\mathfrak{g}_\mathbb{C}$ if its complexification is isomorphic to $\mathfrak{g}_\mathbb{C}$. A simple Lie algebra or Lie group has various real forms: Real forms of $SL(n,\mathbb{C})$ are among others $SU(n) := \{g \in SL(n,\mathbb{C})|g\overline{g}^t = \text{Id}\}$ and $SL(n,\mathbb{R}) := \{X \in \text{Mat}^{n\times n}(\mathbb{R})|\det(X) = 1\}$

Real forms are in bijection with conjugate linear involutions: Fixing a real form, complex conjugation along this real form defines the involution. Starting with a conjugate linear involution, the fixed point set is a real form.

Among those real forms there is a unique distinguished compact one:

Theorem 3.1. *Each complex simple Lie group has up to conjugation a unique compact real form. The same is true for simple Lie algebras.*

For later reference we note an important decompositions theorem of simple Lie groups:

Theorem 3.2 (Iwasawa decomposition). *A simple Lie group G has a decomposition $G = KAN$, where K is maximal compact, A is Abelian and N is nilpotent.*

Now we need to relate simple Lie algebras to spherical reflection groups:

Let G be a compact simple Lie group. A torus T is a maximal Abelian subgroup of G. In a matrix representation a torus is a subgroup of simultaneously diagonalizable elements – a maximal torus is a maximal subgroup with this property. For example a torus in $SU(n)$ consists of all diagonal matrices $T := \text{diag}(a_1,\dots,a_n|a_i \in \mathbb{C}, |a_i| = 1, a_1\cdot\dots\cdot a_n = 1)$. It is known that all maximal tori are conjugate and that each element in G is contained in at least one maximal torus. Choose an arbitrary torus T. The Weyl group is defined to be $W = N/T$ where N is the normalizer of T. In the case of $SU(n)$ the Weyl group is the group of permutations of the n-elements $a_1, \dots, a_n$ - hence it is the symmetric group in n letters. The Weyl group is automatically a finite reflection group. Via the exponential map tori correspond to Abelian subalgebras.

For complex simple Lie groups a similar procedure is possible, but the situation is a little more complicated as not every element lies in a torus. We focus on the Lie algebra.

Definition 3.1 (Cartan subalgebra). A Cartan subalgebra of $\mathfrak{g}$ is a subalgebra $\mathfrak{h}$ of $\mathfrak{g}$ such that

1. $\mathfrak{h}$ is maximal Abelian in $\mathfrak{g}$
2. For each $h \in \mathfrak{h}$ the endomorphism $\mathrm{ad}(h) : \mathfrak{g} \longrightarrow \mathfrak{g}$ is semisimple.

While all Cartan subalgebras are conjugate, it is no longer true, that they cover $\mathfrak{g}$.

Via the adjoint action of a Cartan subalgebra on the Lie algebra one defines the root system which is a refinement of the Weyl group action. The structure of this root system can be encoded into a matrix, called Cartan matrix.

Definition 3.2 (Cartan matrix). A Cartan matrix $A^{n \times n}$ is a square matrix with integer coefficients such that

1. $a_{ii} = 2$ and $a_{i \neq j} \leq 0$,
2. $a_{ij} = 0 \Leftrightarrow a_{ji} = 0$,
3. There is a vector $v > 0$ (component wise) such that $Av > 0$ (component wise).

Example 3.1 (2×2-Cartan matrices). There are – up to equivalence – four different 2-dimensional Cartan matrices:

$$\begin{pmatrix} 2 & 0 \\ 0 & 2 \end{pmatrix}, \begin{pmatrix} 2 & -1 \\ -1 & 2 \end{pmatrix}, \begin{pmatrix} 2 & -1 \\ -2 & 2 \end{pmatrix}, \begin{pmatrix} 2 & -1 \\ -3 & 2 \end{pmatrix}.$$

They correspond to the Weyl groups of types $A_1 \times A_1, A_2, B_2, G_2$.

Definition 3.3. A Cartan matrix $A^{n \times n}$ is called decomposable iff $\{1, 2, \dots, n\}$ has a decomposition in two non-empty sets N_1 and N_2 such that $a_{ij} = 0$ for $i \in N_1$ and $j \in N_2$. Otherwise it is called indecomposable.

A complete list of indecomposable Cartan matrices consists of A_n, $B_{n,n \geq 2}$, $C_{n,n \geq 3}$, $D_{n,n \geq 4}$, E_6, E_7, E_8, F_4, G_2 [6].

Conversely, starting with a Cartan matrix A one can construct a Lie algebra $\mathfrak{g}(A)$ called its realization:

Definition 3.4 (Realization). Let $A^{n \times n}$ be a Cartan matrix. The realization of A, denoted $\mathfrak{g}(A)$, is the algebra

$$\mathfrak{g}(A^{n \times n}) = \langle e_i, f_i, h_i, i = 1, \dots, n | \mathrm{R}_1, \dots, \mathrm{R}_6 \rangle,$$

where

$$\begin{aligned}
\mathrm{R}_1 &: \quad [h_i, h_j] = 0, \\
\mathrm{R}_2 &: \quad [e_i, f_j] = h_i \delta_{ij}, \\
\mathrm{R}_3 &: \quad [h_i, e_j] = a_{ji} e_j, \\
\mathrm{R}_4 &: \quad [h_i, f_j] = -a_{ji} f_j, \\
\mathrm{R}_5 &: \quad (\mathrm{ad} e_i)^{1-a_{ji}}(e_j) = 0 \ (i \neq j), \\
\mathrm{R}_6 &: \quad (\mathrm{ad} f_i)^{1-a_{ji}}(f_j) = 0 \ (i \neq j).
\end{aligned}$$

In consequence there is a bijection between indecomposable Cartan matrices and complex simple Lie algebras; hence the classification of Cartan matrices yields also a complete list of simple Lie algebras. If a Cartan matrix $A^{(n+m)\times(n+m)}$ is decomposable into the direct sum of two Cartan matrices $A^{n\times n}$ and $A^{m\times m}$ then the same decomposition holds for the realizations: It is a direct product of (semi-)simple Lie algebras. This decomposition into direct factors is furthermore reflected in the structure of the geometric objects associated to those Lie algebras.

A further important structure property of simple Lie groups is the BN-pairs structure:

Definition 3.5 (Borel subgroup, parabolic subgroup). Let $G_{\mathbb{C}}$ be a complex simple Lie group. A Borel subgroup B is a maximal solvable subgroup. A subgroup $P \subset G_{\mathbb{C}}$ is called parabolic iff it contains a Borel subgroup.

Example 3.2. – The standard Borel subgroup in $SL(n,\mathbb{C})$ is the group of upper triangular matrices. All Borel subgroups are conjugate.

– A standard parabolic subgroup in $SL(n,\mathbb{C})$ is an upper block-triangular matrix, that is an upper triangular matrix having blocks on its diagonal. There are several conjugacy classes of parabolic subgroups, corresponding to the various block-triangular matrices.

The BN-pair structure formalises the way those groups are assembled to yield a simple Lie group:

Definition 3.6 (BN-pair). Let $G_{\mathbb{C}}$ be a complex simple Lie group. A set (B, N, W, S) is a BN-pair for G iff:

1. $G = \langle B, N \rangle$. Moreover $T = B \cap N \lhd N$ and $W = N/T$.
2. $s^2 = 1 \ \forall s \in S$ and $W = \langle S \rangle$ and (W, S) is a Coxeter system.
3. Let $C(w) := BwB$. Then $C(s)C(w) \subseteq C(w) \cup C(sw) \ \forall s \in S$ and $w \in W$.
4. $\forall s \in S : sBs \nsubseteq B$.

Theorem 3.3 (BN-pairs and Bruhat decomposition). *Every complex simple Lie group G has a unique BN-pair structure. Let W be the Weyl group of G. Then*

$$G = \coprod_{w \in W} C(w).$$

Proof. See [9], Sect. 30. □

The Bruhat decomposition encodes the structure of the Tits buildings – compare Sect. 3.4.

3.2 Symmetric Spaces

Definition 3.7. A (pseudo-)Riemannian symmetric space M is a pseudo-Riemannian manifold M such that for each $m \in M$ there is an isometry $\sigma_m : M \longrightarrow M$ such that $\sigma_m(m) = m$ and $d\sigma_m|_{T_m M} = -Id$.

Direct consequences of the definition are that symmetric spaces are geodesically complete homogeneous spaces. Let $I(M)$ denote the isometry group of M and $I(M)_m$ the isotropy subgroup of the point $m \in M$ then $M = I(M)/I(M)_m$. Let g_m denote the metric on $T_m M$. Clearly $I(M)_m \subset O(g_m)$. Hence for a Riemannian symmetric space, we find that $I(M)_m$ is a closed subgroup of a compact orthogonal group, hence compact.

We formalize those concepts

Definition 3.8 (Symmetric pair). Let G be a connected Lie group, H a closed subgroup. The pair (G, H) is called a symmetric pair if there exists an involutive analytic automorphism $\sigma : G \longrightarrow G$ such that $(H_\sigma)_0 \subset H \subset H_\sigma$. Here H_σ denotes the fixed points of σ and $(H_\sigma)_0$ its identity component. If $Ad_G(H)$ is compact, it is said to be Riemannian symmetric.

Each symmetric space defines a symmetric pair. Conversely, each symmetric pair describes a symmetric space [27].

Definition 3.9 (OSLA). An orthogonal symmetric Lie algebra is a pair $\mathfrak{g}, s$ such that

1. $\mathfrak{g}$ is a Lie algebra over $\mathbb{R}$,
2. s is an involutive automorphism of $\mathfrak{g}$,
3. The set of fixed points of s, denoted $\mathfrak{k}$, is a compactly embedded subalgebra.

Clearly each Riemannian symmetric pair defines an OSLA. The converse is true up to coverings.

Hence to give a classification of Riemannian symmetric spaces, we just have to classify OSLA's.

We focus now our attention to the Riemannian case: The most important result is the following:

Theorem 3.4. *Let M be an irreducible Riemannian symmetric space. Then either its isometry group is semisimple or $M = \mathbb{R}$.*

In the non-Riemannian case this is no longer true. While the pseudo-Riemannian symmetric spaces with semisimple isometry group are completely classified [17], recent results of Ines Kath and Martin Olbricht [32, 33] show, that for pseudo-Riemannian symmetric spaces with a non-semisimple isometry group, a classification needs a classification of solvable Lie algebras, which is out of reach.

There are two classes of irreducible Riemannian symmetric spaces with semisimple isometry group: Spaces of compact type and spaces of noncompact type.

Definition 3.10. A Riemannian symmetric space is of compact type iff its isometry group is a compact semisimple Lie group. It is called of noncompact type if its isometry group is a noncompact semisimple Lie group.

Theorem 3.5. *Let* $(\mathfrak{g}, s)$ *be an orthogonal symmetric Lie algebra and* (L, U) *a Riemannian symmetric pair associated to* $(\mathfrak{g}, s)$*:*

1. *If* (L, U) *is of the compact type, then* L/U *has sectional curvature* ≥ 0.
2. *If* (L, U) *is of the noncompact type, then* L/U *has sectional curvature* ≤ 0.
3. *If* (L, U) *is of the Euclidean type, then* L/U *has sectional curvature* $= 0$.

The Cartan-Hadamard theorem tells us that symmetric spaces of noncompact type are diffeomorphic to a vector space. Hence for every orthogonal symmetric Lie algebra of noncompact type, there is exactly one symmetric space of noncompact type. In contrast the topology of symmetric spaces of compact type is more complicated. As the fundamental group need not be trivial, for one orthogonal symmetric Lie algebra of the compact type there may be different symmetric spaces. There is always a simply connected one, which is the universal cover of all the others.

Symmetric spaces of compact type and noncompact type appear in duality: For every simply connected, irreducible symmetric space of the compact type, there is exactly one of the noncompact type and vice versa.

Example 3.3. Take $(\mathfrak{l}, \mathfrak{u}) := (\mathfrak{so}(n+1), \mathfrak{so}(n))$. A Riemannian symmetric pair associated to $(\mathfrak{l}, \mathfrak{u})$ is $(SO(n+1), SO(n))$. The corresponding symmetric space is isomorphic to the quotient L/U, hence is a sphere. Another symmetric space associated to $(\mathfrak{so}(n+1), \mathfrak{so}(n))$ is the projective space $\mathbb{R}P(n)$. The noncompact dual symmetric space is the hyperbolic space $\mathbb{H}^n = SO(n,1)/SO(n)$.

Besides $\mathbb{R}^n$, there are four classes of Riemannian symmetric spaces, two classes of spaces of compact type and two classes of spaces of noncompact type

1. **Type I** consists of coset spaces G/K, where G is a compact simple Lie group and K is a compact subgroup satisfying $Fix(\sigma)_0 \subseteq K \subseteq Fix(\sigma)$ for some involution σ. In this case $(L, U) = (G, K)$
2. **Type II** consists of compact simple Lie groups G equipped with their bi-invariant metric. $(L, U) = (G \times G, \Delta)$, where $\Delta = \{(x, x) \in G \times G\}$ is the diagonal subgroup.
3. **Type III** consists of spaces G/K where G is a noncompact, real simple Lie group and K a maximal compact subgroup. $(L, U) = (G, K)$.
4. **Type IV** consists of spaces $G_{\mathbb{C}}/G$ where $G_{\mathbb{C}}$ is a complex simple Lie group and G a compact real form. $(L, U) = (G_{\mathbb{C}}, G)$.

Types **I** and **III** are in duality as are types **II** and **IV**.

3.3 Polar Representations and Isoparametric Submanifolds

Definition 3.11 (Polar action). Let M be a Riemannian manifold. An isometric action $G : M \longrightarrow M$ is called polar if there exists a complete, embedded, closed submanifold $\Sigma \subset M$, that meets each orbit orthogonally.

Definition 3.12 (Polar representation). A polar representation is a polar action on an Euclidean vector space, acting by linear transformations.

Example 3.4 (Adjoint representation). Let G be a compact simple Lie group, $\mathfrak{g}$ its Lie algebra. The adjoint representation is polar. Sections are the Cartan subalgebras.

Example 3.5 (s-representation). Let (G, K) be a Riemannian symmetric pair, $M = G/K$ the corresponding symmetric space and $m = eK \in M$ (hence $K = I(M)_m$). The action of K on M induces an action on $T_m M$. Let $\mathfrak{g} = \mathfrak{k} \oplus \mathfrak{p}$ be the corresponding Cartan decomposition (i.e. $\mathfrak{k} = Lie(K)$. Then $\mathfrak{p} \cong T_m M$. Using this isomorphism, we get an action of K on $\mathfrak{p}$. This action, called the s-representation of M, is polar. Each Abelian subalgebra in $\mathfrak{p}$ is a section.

Theorem 3.6 (Dadok). *Every polar representation on $\mathbb{R}^n$ is orbit equivalent to an s-representation.*

The orbits of polar actions have an interesting geometric structure: We define

Definition 3.13 (Isoparametric submanifold). A submanifold $S \subset V$ is called isoparametric if S has a flat normal bundle and constant principal curvatures.

Theorem 3.7. *Principal orbits of a polar representation is an isoparametric submanifold.*

This result can be proven by a direct verification. Conversely we have:

Theorem 3.8 (Thorbergsson). *Each full irreducible isoparametric submanifold of $\mathbb{R}^n$ of rank at least three is an orbit of an s-representation.*

The original proof by Gudlaugur Thorbergsson [52] proceeds by associating a Tits building to any isoparametric submanifold, relying on the fact that spherical buildings of rank at least three are classified [52]. A second proof of Carlos Olmos proceeds by showing homogeneity [4].

Hence most isoparametric submanifolds are homogeneous. Nevertheless, there are some examples in codimension 2 which are non-homogeneous. A complete classification is still missing.

3.4 Spherical Buildings

3.4.1 Foundations

There are several equivalent definitions of a building [1]. For us the most convenient one is the W-metric one:

Definition 3.14 (Building – W-metric definition). Let (W, S) be a Coxeter system. A building of type (W, S) is a pair $(\mathcal{C}, \delta)$ consisting of a nonempty set $\mathcal{C}$ whose elements are called chambers together with a map $\delta : \mathcal{C} \times \mathcal{C} \longrightarrow W$, called the Weyl distance function, such that for all $C, D \in \mathcal{C}$ the following conditions hold:

1. $\delta(C, D) = 1$ iff $C = D$.
2. If $\delta(C, D) = w$ and $C' \in \mathcal{C}$ satisfies $\delta(C', C) = s \in S$ then $\delta(C', D) = sw$ or w. If in addition $l(sw) = l(w) + 1$ then $\delta(C', D) = sw$.
3. If $\delta(C, D) = w$ then for any $s \in S$ there is a chamber $C' \subset \mathcal{C}$, such that $\delta(C', C) = s$ and $\delta(C', D) = sw$.

Example 3.6. Let $G_{\mathbb{C}}$ be a complex simple Lie group. Set $\mathcal{C} = G_{\mathbb{C}}/B$ and define $\delta(f, g) = w$ iff $f^{-1}g \subset BwB$ in the Bruhat decomposition. Then $(\mathcal{C}, \delta)$ is a spherical building.

Hence chambers are in bijection to Borel subgroups B. This definition is equivalent to the standard definition of a building as a simplicial complex – for a proof see [1]:

Definition 3.15 (Building – chamber complex definition). A building $\mathfrak{B}$ is a thick chamber complex Σ together with a set $\mathcal{A}$ of thin chamber complexes $A \in \mathcal{A}$, called apartments, satisfying the following axioms:

1. For every pair of simplices $x, y \in \Sigma$ there is an apartment $A_{x,y} \subset \mathcal{A}$ containing both of them.
2. Let A and A' be apartments, x a simplex and C a chamber such that $\{x, C\} \subset A \cap A'$. Then there is a chamber complex isomorphism $\varphi : A \longrightarrow A'$ fixing x and C pointwise.

In this definition the finite reflection groups appear via their action on the apartments: To each Coxeter System (W, S) one can associate a simplicial complex C of dimension $|S| - 1$, called the Coxeter complex of type (W, S).

We start with (W, S): Let $\mathcal{S}$ denote the power set of S and define a partial order relation on $\mathcal{S}$ by $S' < S'' \in \mathcal{S}$ iff $(S')^c \subset (S'')^c$ as subsets of S. Here $(S')^c$ denotes the complement of S' in S. Now construct a simplicial complex $\Sigma(S)$ associated to $\mathcal{S}$ by identifying a set $S' \subsetneq \mathcal{S}$ with a simplex $\sigma(S')$ of dimension $|S'^c| - 1$ and defining the boundary relations of $\Sigma(S)$ via the partial order on $\mathcal{S}$: $\sigma(S')$ is in the boundary of $\sigma(S'')$ iff $S' < S''$. In this simplicial complex $\emptyset = (S)^c$ corresponds to a simplex of maximal dimension and the $|S|$ sets consisting of the single elements $s_i \in S$ correspond to faces.

The simplicial complex $\Sigma(W, S)$ consists of all W-translates of elements in $\Sigma(S)$. Its elements $\sigma(w, S')$ correspond to pairs consisting of an element $w \in W$ and an element $S' \subset \mathcal{S}$ subject to the equivalent relation $\sigma(w_1, S_1') \simeq \sigma(w_1, S_1')$ iff $S_1' = S_2'$ and $w_1 \cdot \langle S_1' \rangle = w_2 \cdot \langle S_2' \rangle$. It carries a natural W-action that is transitive on simplices of maximal dimension (called chambers). For a simplex $\sigma(w, \{s_i, i \in S'\})$, the stabilizer subgroup is $wW_{S'}w^{-1}$ as elements in $W_{S'}$ stabilize $\{s_i, i \in S'\}$.

As a simplicial complex the Coxeter complex is independent of the choice of S.

The action of elements $s_i \in S$ on $\Sigma(W, S)$ can be interpreted geometrically as reflections at the faces $\sigma(e, s_i)$ of $\sigma(e, \emptyset)$, where e denotes the identity element of W. For further details we refer to [11].

A chamber complex is a simplicial complex satisfying two taming properties: first it is required that every simplex is contained in the boundary of a simplex of maximal dimension and second that for every pair of simplices x and y there exists a sequence of simplices of maximal dimension $S := \{z_1, \dots, z_n\}$ such that $x \in z_1$, $y \in z_n$ and $z_i \cap z_{i+1}$ contains a codimension 1 simplex. Simplices of maximal dimension are called chambers; simplices of codimension 1 are called walls. A sequence S is called a gallery connecting x and y.

A chamber complex will be called thin if every wall is a face of exactly 2 chambers. It will be called thick if every wall is a face of at least 3 chambers.

A chamber complex map $\varphi : A \longrightarrow A'$ is a map of simplicial complexes, mapping k-simplices onto k-simplices and respecting the face relation. It is a chamber complex isomorphism iff it is bijective.

3.4.2 Buildings and Polar Actions

Let G be a real simple Lie group of compact type and $\mathfrak{g}$ its Lie algebra. To understand how buildings fit into our picture, we define a G-equivariant embedding into the Lie algebra.

Start with the adjoint action:

$$\varphi : G \times \mathfrak{g} \longrightarrow \mathfrak{g}, \qquad (g, X) \mapsto gXg^{-1}.$$

We want first to restrict the domain of definition to a fundamental domain for this action. As $\mathfrak{g}$ is covered by conjugate maximal Abelian subalgebras $\mathfrak{t}$, the map

$$\varphi : G \times \mathfrak{t} \longrightarrow \mathfrak{g}, \qquad (g, X) \mapsto gXg^{-1}$$

is surjective. As $T := \exp \mathfrak{t}$ acts trivially on $\mathfrak{t}$,

$$\varphi : G/T \times \mathfrak{t} \longrightarrow \mathfrak{g}, \qquad (gT, X) \mapsto gXg^{-1}$$

is well defined and surjective. Let $\Delta \subset \mathfrak{t}$ be a fundamental domain for the action of the Weyl group $W := N(T)/T \subset G/T$ on $\mathfrak{t}$. Then the map

$$\varphi : G/T \times \Delta_G \longrightarrow \mathfrak{g}, \qquad (gT, X) \mapsto gXg^{-1}$$

is again surjective.

For regular elements, i.e. all those in the interior Δ, this map is injective [8]. For all elements Y in the boundary, the stabilizer subgroups are generated by the reflections $s_i \subset W$ fixing Y. For an element $X \subset \Delta_{\mathfrak{g}}$ in the intersection of the faces $\Delta_{i_1}, \dots, \Delta_{i_k}, i \in I$ the stabilizer is $W_I = \langle S_I \rangle$.

As the adjoint action preserves the Cartan-Killing form $B(X,Y)$, it is an isometry with respect to the induced metric $\langle X,Y\rangle = -B(X,Y)$. Hence it preserves every sphere $S^n_R \subset \mathfrak{g}$ of fixed radius R. Thus we can define an Ad-invariant subspace $\mathfrak{g}_R := \mathfrak{g} \cap S^m_R \subset \mathfrak{g}$. A fundamental domain for the Adjoint action on $\mathfrak{g}_R$ is the set $\Delta_{(\mathfrak{g},R)} := \Delta_{\mathfrak{g}} \cap S^n_R$.

Accordingly we get a surjective map:

$$\mathrm{Ad}(G/T) : \Delta_{(\mathfrak{g},R)} \mapsto \mathfrak{g}_R, \qquad (g,X) \mapsto gXg^{-1}.$$

Now construct a simplicial complex like that: For each element $X \in \Delta_{(\mathfrak{g},R)}$ there is a subgroup $W_X \subset W$ stabilizing X. If X is in the interior, then $W_X = \{e\}$. If it is in the boundary, W_X is the group $W_I \subset W$ generated by the reflections s_i that fix X. Thus we can replace $\Delta_{(\mathfrak{g},R)}$ by the complex $\Sigma(S)$. It has exactly $n :=$ rank $\mathfrak{g}$ faces $f_1,\dots,f_n$. Let $S = \{s_i\}, i = 1,\dots,n$ where s_i denotes the reflection at f_i. (W,S) is a Coxeter system. The cells in $\Delta_{(\mathfrak{g},R)}$ correspond bijectively to subsets $S' \subset S$.

The W-translates $\Sigma(S)$ tessellate $\mathfrak{t}_R := \mathfrak{t} \cap S^n_R$; thus the tessellation of $\mathfrak{t}_R$ corresponds to the thin Coxeter complex $\Sigma(W,S)$. The G/T-translates tessellate the whole sphere $\mathfrak{g}_R$. We get a simplicial complex

$$\mathfrak{B}_G := (G/T \times \Delta)/\sim,$$

By the Iwasawa decomposition we get $B := T \times A \times N$ and $G/T := G_{\mathbb{C}}/B$. Hence:

$$\mathfrak{B}_G := (G_{\mathbb{C}}/B \times \Delta)/\sim$$

Our simplicial complex is the building for G and we get an embedding of the building $\mathfrak{B}_G := (G/T \times \Delta)/\sim$ into the sphere S_R.

Moreover let $h \in G$ act on the building $\mathfrak{B}_G$ by left multiplication. Then the following diagram commutes:

$$\begin{array}{ccc} \mathfrak{B}_G & \xrightarrow{\ h\ } & \mathfrak{B}_G \\ {\scriptstyle i}\downarrow & & \downarrow{\scriptstyle i} \\ \mathfrak{g}_R & \xrightarrow{\ \mathrm{Ad}(h)\ } & \mathfrak{g}_R \end{array}$$

The description $\mathfrak{B}_G := (G_{\mathbb{C}}/B \times \overline{\Delta})/\sim$ shows that $G_{\mathbb{C}}$ acts from the left on the building $\mathfrak{B}_G = (G_{\mathbb{C}}/B \times \Delta)/\sim$.

In this complex description chambers correspond exactly to Borel subgroups, hence we find:

Theorem 3.9. *Let $G_{\mathbb{C}}$ be a simple Lie group of type X_l and $\mathfrak{B}$ the building of the same type.*

1. *The chambers of $\mathfrak{B}$ correspond to the Borel subgroups of $G_{\mathbb{C}}$, simplices of $\mathfrak{B}$ correspond to parabolic subgroups. The correspondence can be realized by associating to every simplex $c \in \mathfrak{B}$ its stabilizer subgroup P_c in $G_{\mathbb{C}}$ with respect to the left action, described above.*
2. *A simplex c is in the boundary of a cell d iff $P_d \subset P_c$.*

Remark 3.1. It is an important observation that in a simple complex Lie group all Borel subgroups are conjugate. In building theoretical language this translates to the fact that the building is a connected simplicial complex.

The association of buildings to symmetric spaces can be done in the same way as we did it for Lie groups, with the adjoint representation replaced by the isotropy representation [12, 27, 39]. A good survey for applications of buildings in finite dimensional geometry is [28].

4 Kac-Moody Groups and Their Lie Algebras

4.1 Geometric Affine Kac-Moody Algebras

The classical references for (algebraic) Kac-Moody algebras are the books [30] and [41]. We will encounter algebraic Kac-Moody algebras and various analytic completions. In some cases of ambiguity we use the denomination algebraic Kac-Moody algebra for the classical Kac-Moody algebras.

Definition 4.1 (affine Cartan matrix). An affine Cartan matrix $A^{n\times n}$ is a square matrix with integer coefficients, such that

1. $a_{ii} = 2$ and $a_{i\neq j} \leq 0$.
2. $a_{ij} = 0 \Leftrightarrow a_{ji} = 0$.
3. There is a vector $v > 0$ (component wise) such that $Av = 0$.

Example 4.1 (2×2-affine Cartan matrices). There are – up to equivalence – two different 2-dimensional affine Cartan matrices:

$$\begin{pmatrix} 2 & -2 \\ -2 & 2 \end{pmatrix}, \begin{pmatrix} 2 & -1 \\ -4 & 2 \end{pmatrix}.$$

They correspond to the non-twisted algebra $\tilde{A}_1$ and the twisted algebra $\tilde{A}'_1$.

1. The indecomposable non-twisted affine Cartan matrices are

$$\widetilde{A}_n, \widetilde{B}_n, \widetilde{C}_n, \widetilde{D}_n, \widetilde{E}_6, \widetilde{E}_7, \widetilde{E}_8, \widetilde{F}_4, \widetilde{G}_2.$$

Every non-twisted affine Cartan matrix $\widetilde{X}_l$ can be constructed from a (finite) Cartan matrix X_l by the addition of a further line and column. The denomination as "non-twisted" points to the explicit construction as loop algebras.

2. The indecomposable twisted affine Cartan matrices are

$$\widetilde{A}'_1, \widetilde{C}'_l, \widetilde{B}^t_l, \widetilde{C}^t_l, \widetilde{F}^t_4, \widetilde{G}^t_2.$$

The Kac-Moody algebras associated to them can be constructed as fixed point algebras of certain automorphisms σ of a non-twisted Kac-Moody algebra X. This construction suggests an alternative notation describing a twisted Kac-Moody algebra by the order of σ and the type of X. This yields the following equivalences:

$\widetilde{A}'_1$	${}^2\widetilde{A}_2$
$\widetilde{C}'_l$	${}^2\widetilde{A}_{2l}, l \geq 2$
$\widetilde{B}^t_l$	${}^2\widetilde{A}_{2l-1}, l \geq 3$
$\widetilde{C}^t_l$	${}^2\widetilde{D}_{l+1}, l \geq 2$
$\widetilde{F}^t_4$	${}^2\widetilde{E}_6$
$\widetilde{G}^t_2$	${}^3\widetilde{D}_4$

As in the finite dimensional case to every affine Cartan matrices A one can associate a realizations $\mathfrak{g}(A)$. Those correspond exactly to the affine Kac-Moody algebras. Fortunately besides this abstract approach using generators and relations there is a very concrete second description for affine Kac-Moody algebras, namely the loop algebra approach. To describe the loop algebra approach to Kac-Moody algebras we follow the terminology of the article [24].

Let $\mathfrak{g}$ be a finite dimensional reductive Lie algebra over the field $\mathbb{F} = \mathbb{R}$ or $\mathbb{C}$. Hence $\mathfrak{g}$ is a direct product of a semisimple Lie algebra $\mathfrak{g}_s$ with an Abelian Lie algebra $\mathfrak{g}_a$. Let furthermore $\sigma \in \mathrm{Aut}(\mathfrak{g}_s)$ denote an automorphism of finite order of $\mathfrak{g}_s$ such that $\sigma|_{\mathfrak{g}_a} = Id$. If $\mathfrak{g}_s$ is a Lie algebra over $\mathbb{R}$ we suppose it to be of compact type.

$$L(\mathfrak{g},\sigma) := \{f : \mathbb{R} \longrightarrow \mathfrak{g} \,|\, f(t+2\pi) = \sigma f(t), f \text{ satisfies some regularity condition}\}.$$

We use the notation $L(\mathfrak{g},\sigma)$ to describe in a unified way algebraic constructions that apply to various explicit realizations of loop algebras satisfying sundry regularity conditions – i.e. smooth, real analytic, (after complexification) holomorphic or algebraic loops. If we discuss loop algebras of a fixed regularity we use other precise notations: $M\mathfrak{g}$ for holomorphic loops on $\mathbb{C}^*$, $L_{alg}\mathfrak{g}$ for algebraic, $A_n\mathfrak{g}$ for holomorphic loops on the annulus $A_n = \{z \in \mathbb{C} | e^{-n} \leq |z| \leq e^n\}$.

Definition 4.2 (Geometric affine Kac-Moody algebra). The geometric affine Kac-Moody algebra associated to a pair $(\mathfrak{g},\sigma)$ is the algebra:

$$\widehat{L}(\mathfrak{g},\sigma) := L(\mathfrak{g},\sigma) \oplus \mathbb{F}c \oplus \mathbb{F}d,$$

equipped with the lie bracket defined by:

$$[d, f] := f'; [c, c] = [c, d] = [c, f] = [d, d] = 0;$$
$$[f, g] := [f, g]_0 + \omega(f, g)c.$$

Here $f \in L(\mathfrak{g}, \sigma)$ and ω is a certain antisymmetric 2-form on $M\mathfrak{g}$, satisfying the cocycle condition.

Let us remark that in contrast to the usual Kac-Moody theory, $\mathfrak{g}$ has not to be simple, but may be reductive, i.e. a product of semisimple Lie algebra with an Abelian one. The algebra $\widetilde{L}(G, \sigma) := L(\mathfrak{g}, \sigma) \oplus \mathbb{F}c$ is called the derived algebra.

We give some further definitions:

Definition 4.3. A real form of a complex geometric affine Kac-Moody algebra $\widehat{L}(\mathfrak{g}_{\mathbb{C}}, \sigma)$ is the fixed point set of a conjugate linear involution.

Involutions of a geometric affine Kac-Moody algebra restrict to involutions of irreducible factors of the loop algebra. Hence the invariant subalgebras are direct products of invariant subalgebras in those factors together with the appropriate torus extension.

Definition 4.4 (compact real affine Kac-Moody algebra). A compact real form of a complex affine Kac-Moody algebra $\widehat{L}(\mathfrak{g}_{\mathbb{C}}, \sigma)$ is a real form which is isomorphic to $\widehat{L}(\mathfrak{g}_{\mathbb{R}}, \sigma)$ where $\mathfrak{g}_{\mathbb{R}}$ is a compact real form of $\mathfrak{g}_{\mathbb{C}}$.

Remark 4.1. A semisimple Lie algebra is called of "compact type" iff it integrates to a compact semisimple Lie group. The infinite dimensional generalization of compact Lie groups are loop groups of compact Lie groups and their Kac-Moody groups, constructed as extensions of those loop groups (see Sect. 4.2). Thus the denomination is justified by the fact that "compact" affine Kac-Moody algebras integrate to "compact" Kac-Moody groups.

To define a loop group of the compact type we use an infinite dimensional version of the Cartan-Killing form:

Definition 4.5 (Cartan-Killing form). The Cartan-Killing form of a loop algebra $L(\mathfrak{g}_{\mathbb{C}}, \sigma)$ is defined by

$$B_{(\mathfrak{g}_{\mathbb{C}},\sigma)}(f, g) = \int_0^{2\pi} B(f(t), g(t))\, dt.$$

Definition 4.6 (compact loop algebra). A loop algebra of compact type is a subalgebra of $L(\mathfrak{g}_{\mathbb{C}}, \sigma)$ such that its Cartan-Killing form is negative definite.

Lemma 4.1. *Let $\mathfrak{g}_{\mathbb{R}}$ be a compact semisimple Lie algebra. Then the loop algebra $L(\mathfrak{g}_{\mathbb{R}}, \sigma)$ is of compact type.*

Proof. The Cartan-Killing form on $\mathfrak{g}_{\mathbb{R}}$ is negative definite. Hence $B_{(\mathfrak{g}_{\mathbb{C}},\sigma)}(f, g)$ is negative definite. □

To find noncompact real forms we need the following result of Ernst Heintze and Christian Groß (Corollary 7.7. of [24]):

Theorem 4.1. *Let $\mathcal{G}$ be an irreducible complex geometric affine Kac-Moody algebra i.e. $\mathcal{G} = \widehat{L}(\mathfrak{g}, \sigma)$ with $\mathfrak{g}$ simple, $\mathcal{U}$ a real form of compact type. The conjugacy classes of real forms of noncompact type of $\mathcal{G}$ are in bijection with the conjugacy classes of involutions on $\mathcal{U}$. The correspondence is given by $\mathcal{U} = \mathcal{K} \oplus \mathcal{P} \mapsto \mathcal{K} \oplus i\mathcal{P}$ where $\mathcal{K}$ and $\mathcal{P}$ are the ± 1-eigenspaces of the involution.*

Thus to find noncompact real forms we have to study automorphism of order 2 of a geometric affine Kac-Moody algebra of the compact type. From now on we restrict to involutions $\widehat{\varphi}$ of type 2, that is $\widehat{\varphi}(c) = -c$.

A careful examination of the construction of a geometric affine Kac-Moody algebra of a non simple Lie algebra allows, to extend this result to the broader class of geometric affine Kac-Moody algebras [14]:

Theorem 4.2. *Let $\mathcal{G}$ be a complex geometric affine Kac-Moody algebra, $\mathcal{U}$ a real form of compact type. The conjugacy classes of real forms of noncompact type of $\mathcal{G}$ are in bijection with the conjugacy classes of involutions on $\mathcal{U}$. The correspondence is given by $\mathcal{U} = \mathcal{K} \oplus \mathcal{P} \mapsto \mathcal{K} \oplus i\mathcal{P}$ where $\mathcal{K}$ and $\mathcal{P}$ are the ± 1-eigenspaces of the involution. Furthermore every real form is either of compact type or of noncompact type. A mixed type is not possible.*

Lemma 4.2. *Let $\mathfrak{g}$ be semisimple and $\widehat{L}(\mathfrak{g}, \sigma)_D$ be a real form of the noncompact type. Let $\widehat{L}(\mathfrak{g}, \sigma)_D = \mathcal{K} \oplus \mathcal{P}$ be a Cartan decomposition. The Cartan Killing form is negative definite on $\mathcal{K}$ and positive definite on $\mathcal{P}$*

Proof. Suppose first σ is the identity. Let φ be an automorphism. Then without loss of generality $\varphi(f) = \varphi_0(f(-t))$ [24]. Let $\mathfrak{g} = \mathfrak{k} \oplus \mathfrak{p}$ be the decomposition of $\mathfrak{g}$ into the ± 1-eigenspaces of φ_0. Then $f \in \text{Fix}(\varphi)$ iff its Taylor expansion satisfies

$$\sum_n a_n e^{int} = \sum \varphi_0(a_{-n}) e^{int}.$$

Let $a_n = k_n \oplus p_n$ be the decomposition of a_n into the ± 1 eigenspaces with respect to φ_0. Hence

$$f(t) = \sum_n k_n \cos(nt) + \sum_n p_n \sin(nt).$$

Then using bilinearity and the fact that $\{\cos(nt), \sin(nt)\}$ are orthonormal we can calculate $B_{\mathfrak{g}}$:

$$B_{\mathfrak{g}} = \int_0^{2\pi} \sum_n \cos^2(nt) B(k_n, k_n) - \int_0^{2\pi} \sum_n \sin^2(nt) B(p_n, p_n).$$

Hence $B_{\mathfrak{g}}$ is negative definite on $\mathrm{Fix}(\varphi)$. Analogously one calculates that it is positive definite on the -1-eigenspace of φ. If $\sigma \neq Id$ then one gets the same result by embedding $L(\mathfrak{g}, \sigma)$ into an algebra $L(\mathfrak{h}, \mathrm{id})$ which is always possible [30]. □

Lemma 4.3. *Let $\mathfrak{g}$ be Abelian. The Cartan-Killing form of $L(\mathfrak{g})$ is trivial.*

Proof. Direct calculation. □

Now we can define OSAKAs, the Kac-Moody analogue of orthogonal symmetric Lie algebras [14]:

Definition 4.7 (Orthogonal symmetric Kac-Moody algebra). An orthogonal symmetric affine Kac-Moody algebra (OSAKA) is a pair $\left(\widehat{L}(\mathfrak{g}, \sigma), \widehat{\rho}\right)$ such that

1. $\widehat{L}(\mathfrak{g}, \sigma)$ is a real form of an affine geometric Kac-Moody algebra,
2. $\widehat{\rho}$ is an involutive automorphism of $\widehat{L}(\mathfrak{g}, \sigma)$ of the second kind,
3. $\mathrm{Fix}(\widehat{\rho})$ is a compact real form.

Following Helgason, we define 3 types of OSAKAs:

Definition 4.8 (Types of OSAKAs). Let $\left(\widehat{L}(\mathfrak{g}, \sigma), \widehat{\rho}\right)$ be an OSAKA. Let $\widehat{L}(\mathfrak{g}, \sigma) = \mathcal{K} \oplus \mathcal{P}$ be the decomposition of $\widehat{L}(\mathfrak{g}, \sigma)$ into the eigenspaces of $\widehat{L}(\mathfrak{g}, \sigma)$ of eigenvalue $+1$ resp. -1.

1. If $\widehat{L}(\mathfrak{g}, \sigma)$ is a compact real affine Kac-Moody algebra, it is said to be of the compact type.
2. If $\widehat{L}(\mathfrak{g}, \sigma)$ is a noncompact real affine Kac-Moody algebra, $\widehat{L}(\mathfrak{g}, \sigma) = \mathcal{U} \oplus \mathcal{P}$ is a Cartan decomposition of $\widehat{L}(\mathfrak{g}, \sigma)$.
3. If $L(\mathfrak{g}, \sigma)$ is Abelian it is said to be of Euclidean type.

Definition 4.9 (irreducible OSAKA). An OSAKA $\left(\widehat{L}(\mathfrak{g}, \sigma), \widehat{\rho}\right)$ is called irreducible iff its derived algebra has no non-trivial derived Kac-Moody subalgebra invariant under $\widehat{\rho}$.

Thus we can describe the different classes of irreducible OSAKAs of compact type.

1. The first class consists of pairs consisting of compact real forms $\widehat{M\mathfrak{g}}$, where $\mathfrak{g}$ is a simple Lie algebra together with an involution of the second kind. Complete classifications are available [23]. This are irreducible factors of type I. They correspond to Kac-Moody symmetric spaces of type I.
2. Let $\mathfrak{g}_{\mathbb{R}}$ be a simple real Lie algebra of the compact type. The second class consists of pairs of an affine Kac-Moody algebra $\widehat{M(\mathfrak{g}_{\mathbb{R}} \times \mathfrak{g}_{\mathbb{R}})}$ together with an involution of the second kind, interchanging the factors. Those algebras correspond to

Kac-Moody symmetric spaces that are compact Kac-Moody groups equipped with their Ad-invariant metrics (type II).

Dualizing OSAKAs of the compact type, we get the OSAKAs of the noncompact type.

1. Let $\mathfrak{g}_{\mathbb{C}}$ be a complex semisimple Lie algebra, and $\widehat{L}(\mathfrak{g}_{\mathbb{C}}, \sigma)$ the associated affine Kac-Moody algebra. This class consists of real forms of the noncompact type that are described as fixed point sets of involutions of type 2 together with a special involution, called Cartan involution. This is the unique involution on $\mathcal{G}$, such that the decomposition into its ± 1 eigenspaces $\mathcal{K}$ and $\mathcal{P}$ yields: $\mathcal{K} \oplus i\mathcal{P}$ is a real form of compact type of $\widehat{L}(\mathfrak{g}_{\mathbb{C}}, \sigma)$. Those orthogonal symmetric Lie algebras correspond to Kac-Moody symmetric spaces of type III.
2. Let $\mathfrak{g}_{\mathbb{C}}$ be a complex semisimple Lie algebra. The fourth class consists of $\widehat{L}(\mathfrak{g}_{\mathbb{C}}, \sigma)$ with the involution given by the complex conjugation $\widehat{\rho_0}$ with respect to a compact real form, i.e. $\widehat{L}(\mathfrak{g}_{\mathbb{R}}, \sigma)$. Those algebras correspond to Kac-Moody symmetric spaces of type IV.

The derived algebras of the last class of OSAKAs – the ones of Euclidean type – are Heisenberg algebras [46]. The maximal subgroup of compact type is trivial.

4.2 Affine Kac-Moody Groups

There are several different approaches to affine Kac-Moody groups. The usual algebraic approach follows the definition of algebraic groups via a functor. Tits defines a group functor from the category of rings into the category of groups, whose evaluation on the category of fields yields Kac-Moody groups. Various completions of the groups defined this way are possible. Nevertheless, restricting to affine Kac-Moody groups, there is a second much more down to earth approach which consists in the definition of Kac-Moody groups as special extensions of loop groups - for those two and other approaches compare the article [53]. This second approach for affine Kac-Moody groups relies on a curious identification: Let G denote an affine algebraic group scheme and study the group $G(k[t,t^{-1}])$, where k is a field. Either, defining a torus extension of $G(k[t,t^{-1}])$ we get a Kac-Moody group over the field k or tensoring with the quotient field $k(t)$ of $k[t,t^{-1}]$ we get an algebraic group over $k(t)$. This hints to a close connection between algebraic groups and affine algebraic Kac-Moody groups, with a loop group as the intermediate object. Now "analytic" completions can be defined just by completing the ring $k[t,t^{-1}]$ with respect to some norm.

In this section we focus on the loop group approach to Kac-Moody groups. Our presentation follows the book [46].

We use again the regularity-independent notation $L(G_{\mathbb{C}}, \sigma)$ for the complex loop group and $L(G, \sigma)$ for its real form of compact type. To define groups of polynomial maps, we use the fact, that every compact Lie group is isomorphic to a subgroup of some unitary group. Hence we can identify it with a matrix group. Similarly, the complexification can be identified with a subgroup of some general linear group [46].

Kac-Moody groups are constructed in two steps.

1. The first step consists in the construction of an S^1-bundle in the real case (resp. a $\mathbb{C}^*$-bundle in the complex case) over $L(G,\sigma)$ that corresponds via the exponential map to the derived algebra. The fiber corresponds to the central term $\mathbb{R}c$ (resp. $\mathbb{C}c$) of the Kac-Moody algebra.
2. In the second step we construct a semidirect product with S^1 (resp. $\mathbb{C}^*$). This corresponds via the exponential map to the $\mathbb{R}d$- (resp. $\mathbb{C}d$-) term

Study first the extension of $L(G,\sigma)$ with the short exact sequence:

$$1 \longrightarrow S^1 \longrightarrow X \longrightarrow L(G,\sigma) \longrightarrow 1.$$

There are various groups X that fit into this sequence. We need to find a group $\widetilde{L}(G,\sigma)$ such that its tangential Lie algebra at $e \in \widetilde{L}(G,\sigma)$ is isomorphic to $\widetilde{L}(\mathfrak{g},\sigma)$.

As described in [46] this S^1-bundle is best represented by triples (g,p,z) where g is an element in the loop group, p a path connecting the identity to g and $z \in S^1$ (respective $\mathbb{C}^*$) subject to the relation of equivalence: $(g_1,p_1,z_1) \sim (g_2,p_2,z_2)$ iff $g_1 = g_2$ and $z_1 = C_\omega(p_2 * p_1^{-1})z_2$. Here $C_\omega(p_2 * p_1^{-1}) = e^{\int_{S(p_2 * p_1^{-1})}\omega}$ where $S(p_2 * p_1^{-1})$ is a surface bounded by the closed curve $p_2 * p_1^{-1}$ and ω denotes the 2-form used to define the central extension of $L(\mathfrak{g},\sigma)$. The term $z_1 = C_\omega(p_2 * p_1^{-1})z_2$ defines a twist of the bundle. The law of composition is defined by

$$(g_1,p_1,z_1)\cdot(g_2,p_2,z_2) = (g_1g_2, p_1 * g_1(p_2), z_1z_2).$$

If G is simply connected and ω integral (which is the case in our situation), it can be shown that this object is a well defined group [46], Theorem 4.4.1. If G is not simply connected, the situation is a little more complicated: Let $G = H/Z$ where H is a simply connected Lie group and $Z = \pi_1(G)$. Let $(LG)_0$ denote the identity component of LG. We can describe the extension using the short exact sequence [46], Sect. 4.6. :

$$1 \longrightarrow S^1 \longrightarrow \widetilde{LH}/Z \longrightarrow (LG)_0 \longrightarrow 1$$

In case of complex loop groups, the S^1-bundle is complexified to a $\mathbb{C}^*$-bundle.

The second much easier extension yields now Kac-Moody groups:

Definition 4.10 (Kac-Moody group).

1. Let G be a compact real Lie group. The compact real Kac-Moody group $\widehat{L}(G,\sigma)$ is the semidirect product of S^1 with the S^1-bundle $\widetilde{L}(G,\sigma)$.

2. Let $G_{\mathbb{C}}$ be a complex simple Lie group. The complex Kac-Moody group $\widehat{L}(G_{\mathbb{C}},\sigma)$ is the semidirect product of $\mathbb{C}^*$ with the $\mathbb{C}^*$-bundle $\widetilde{L}(G_{\mathbb{C}},\sigma)$. It is the complexification of $\widehat{L}(G,\sigma)$.

The action of the semidirect S^1-(resp. $\mathbb{C}^*$-) factor is in both cases given by a shift of the argument: $\mathbb{C}^* \ni w : MG \to MG : f(z) \mapsto f(z \cdot w)$.

5 Kac-Moody Symmetric Spaces

The existence of Kac-Moody symmetric spaces was conjectured by Chuu-Lian Terng 1995 in her article [50], nevertheless she explains that a rigorous definition faces serious difficulties. In his thesis Bogdan Popescu investigates possible constructions and is able to define Kac-Moody symmetric spaces of the compact type; nevertheless his approach fails for spaces of the noncompact type [45].

While we have freedom in choosing which regularity to use for the construction of many objects in Kac-Moody geometry, the regularity of Kac-Moody symmetric spaces is completely fixed by the algebraic operations required: we described that the semidirect factor acts on the loop group by a shift of the argument. As in the complex case this factor is isomorphic to $\mathbb{C}^*$, we need holomorphic loops on $\mathbb{C}^*$ for the complexification.

The realizations, we use for Kac-Moody symmetric spaces are the non-twisted groups $MG := \{f : \mathbb{C}^* \longrightarrow G_{\mathbb{C}} | f \text{ is holomorphic}\}$ and the twisted groups $MG^{\sigma} := \{f \in MG | \ \sigma \circ f(z) = f(\omega z)\}$. The real form of compact type is defined by $MG_{\mathbb{R}} := \{f \in MG_{\mathbb{C}} | f(S^1) \subset G_{\mathbb{R}}\}$.

Theorem 5.1. *$MG_{\mathbb{C}}$ and $MG^{\sigma}_{\mathbb{C}}$ are tame Fréchet manifolds. The same is true for compact real forms and all quotients that appear in the definition of Kac-Moody symmetric spaces.*

The idea of the proof is to use logarithmic derivatives. The concept of logarithmic derivatives for regular Lie groups is developed in the book [36], chapters 38 and 40. Furthermore it is used by Karl-Hermann Neeb to prove regularity results for locally convex Lie groups [42]. Recall the definitions of a tame Fréchet space [21]:

Definition 5.1 (Fréchet space). A Fréchet vector space is a locally convex topological vector space which is complete, Hausdorff and metrizable.

Definition 5.2 (Grading). Let F be a Fréchet space. A grading on F is a collection of seminorms $\{\| \ \|_n, n \in \mathbb{N}_0\}$ that define the topology and satisfy

$$\|f\|_0 \leq \|f\|_1 \leq \|f\|_2 \leq \|f\|_3 \leq \dots .$$

Definition 5.3 (Tame equivalence of gradings). Let F be a graded Fréchet space, $r, b \in \mathbb{N}$ and $C(n), n \in \mathbb{N}$ a sequence with values in $\mathbb{R}^+$. The two gradings $\{\| \ \|_n\}$ and $\{\widetilde{\| \ \|}_n\}$ are called $(r, b, C(n))$-equivalent iff

$$\|f\|_n \leq C(n)\|\widetilde{f}\|_{n+r} \text{ and } \|\widetilde{f}\|_n \leq C(n)\|f\|_{n+r} \text{ for all } n \geq b.$$

They are called tame equivalent iff they are $(r, b, C(n))$-equivalent for some $(r, b, C(n))$.

Example 5.1. Let B be a Banach space with norm $\| \ \|_B$. Denote by $\Sigma(B)$ the space of all exponentially decreasing sequences $\{f_k\}$, $k \in \mathbb{N}_0$ of elements of B. On this space, we can define various different gradings; among them are the following:

$$\|f\|_{l_1^n} := \sum_{k=0}^{\infty} e^{nk}\|f_k\|_B$$

$$\|f\|_{l_\infty^n} := \sup_{k \in \mathbb{N}_0} e^{nk}\|f_k\|_B$$

Definition 5.4 (Tame map). A linear map $\varphi : F \longrightarrow G$ is called $(r, b, C(n))$-tame if it satisfies the inequality

$$\|\varphi(f)\|_n \leq C(n)\|f\|_{n+r}.$$

φ is called tame iff it is $(r, b, C(n))$-tame for some $(r, b, C(n))$.

Definition 5.5 (Tame isomorphism). A map $\varphi : F \longrightarrow G$ is called a tame isomorphism iff it is a linear isomorphism and φ and φ^{-1} are tame maps.

Definition 5.6 (Tame direct summand). F is a tame direct summand of G iff there exist tame linear maps $\varphi : F \longrightarrow G$ and $\psi : G \longrightarrow F$ such that $\psi \circ \varphi : F \longrightarrow F$ is the identity.

Definition 5.7 (Tame space). F is tame iff there is a Banach space B such that F is a tame direct summand of $\Sigma(B)$.

Theorem 5.2. *The space $F := Hol(\mathbb{C}^*, \mathbb{C}^n)$ is a tame Fréchet space.*

The quite technical proof can be found in [14]. As a consequence $M\mathfrak{g}$ is a tame Lie algebra (For a Lie algebra to be tame, we require additionally, that the adjoint action of each element is tame).

Proof. Proof of theorem 5.1 [14]: We concentrate on the special case MG.

Start with an embedding

$$\begin{aligned} \varphi : MG_{\mathbb{C}} &\hookrightarrow \quad \Omega^1(\mathbb{C}^*, \mathfrak{g}_{\mathbb{C}}) \times G_{\mathbb{C}} \\ f \quad &\mapsto (\delta(f) = f^{-1}df, f(1)) \end{aligned}$$

Let π_1 and π_2 denote the projections:

$$\begin{aligned} \pi_1 &: \Omega^1(\mathbb{C}^*, \mathfrak{g}_{\mathbb{C}}) \times G_{\mathbb{C}} \mapsto \Omega^1(\mathbb{C}^*, \mathfrak{g}_{\mathbb{C}}) \\ \pi_2 &: \Omega^1(\mathbb{C}^*, \mathfrak{g}_{\mathbb{C}}) \times G_{\mathbb{C}} \mapsto \quad G_{\mathbb{C}} \end{aligned}$$

Charts for $MG_{\mathbb{C}}$ will be products of charts for $\pi_1 \circ \varphi(MG_{\mathbb{C}})$ and $\pi_2 \circ \varphi(MG_{\mathbb{C}})$.

- $\pi_2 \circ \varphi$ is surjective; hence to describe the second factor, we can choose charts for G. Via the exponential mapping, we use charts in $\mathfrak{g}_{\mathbb{C}}$. To describe the norms, we use for $\| \;\|_n$ on this factor the Euclidean metric.
- The first factor is more difficult to deal with as $\pi_1 \circ \varphi$ is not surjective. While every form $\alpha \in \Omega(\mathbb{C}^*, \mathfrak{g}_{\mathbb{C}})$ is locally integrable, the monodromy may prevent global integrability. A form $\alpha \in \Omega^1(\mathbb{C}^*, \mathfrak{g}_{\mathbb{C}})$ is in the image of $\pi_1 \circ \varphi$ iff its monodromy vanishes, that is iff

$$e^{\int_{\mathbb{S}^1} \alpha} = e \in G_{\mathbb{C}}.$$

This is equivalent to the condition $\int_{\mathbb{S}^1} \alpha = a_{-1}(\alpha) \subset \frac{1}{2\pi i} \exp^{-1}(e)$ where $a_{-1}(\alpha)$ denotes the (-1)-coefficient of the Laurent development of $\alpha = f(z)dz$. So we can describe $\Im(\pi_1 \circ \varphi)$ as the inverse image of the monodromy map of $e \in G_{\mathbb{C}}$.

Thus we have to show that this inverse image is a tame Fréchet manifold. To this end, we use composition with a chart $\psi : U \longrightarrow V$ for $e \in U \subset G$ with values in $G_{\mathbb{C}}$. This gives us a tame map $\Omega(\mathbb{C}^*, \mathfrak{g}_{\mathbb{C}}) \longrightarrow \mathfrak{g}_{\mathbb{C}}$. The proof is completed by showing that this map is regular and the proof that inverse images of regular maps are tame Fréchet submanifolds [14]. This proves that $\pi_1 \circ \varphi$ is a tame Fréchet submanifold.

Thus $MG_{\mathbb{C}}$ is a product of a tame Fréchet manifold with a Lie group and hence a tame Fréchet manifold. This completes the proof of theorem 5.1. □

A similar result for Kac-Moody groups, is a direct consequence of a result of Bogdan Popescu, stating that fiber bundles whose fiber is a Banach space over tame Fréchet manifolds are tame [45].

Kac-Moody symmetric spaces are tame Fréchet manifolds. Those manifolds have tangential spaces, that can be equipped with weak metrics. We suppose furthermore the existence of a Levi-Civita connection. Then one can prove, that the usual algebraic identities – i.e. the Bianchi identities – hold [21].

Other concepts well known from the finite dimensional case can be generalized too:

A nice example is Kulkarni's theorem about the sectional curvature on Lorentz manifolds: Define as usual the sectional curvature by $K_f(g,h) = \frac{\langle R(f)\{g,h,g\},h\rangle}{|g\wedge h|^2}$ if $|g \wedge h|^2 \neq 0$. Then

Theorem 5.3 (generalized Kulkarni-type). *Let M be a tame Fréchet Lorentz manifold with Levi-Civita connection. Then the following conditions are equivalent:*

- *$K_f(g,h)$ is constant.*
- *$a \leq K_f(g,h)$ or $K_f(g,h) \leq b$ for some $a, b \in \mathbb{R}$.*
- *$a \leq K_f(g,h) \leq b$ on all definite planes for some $a \leq b \in \mathbb{R}$.*
- *$a \leq K_f(g,h) \leq b$ on all indefinite planes for some $a \leq b \in \mathbb{R}$.*

For the finite dimensional proof of this theorem and more generally finite dimensional Lorentz geometry [43].

Let us now turn to Kac-Moody symmetric spaces themselves:

Definition 5.8 (Kac-Moody symmetric space). An (affine) Kac-Moody symmetric space M is a tame Fréchet Lorentz symmetric space such that its isometry group $I(M)$ contains a transitive subgroup isomorphic to an affine geometric Kac-Moody group H.

We can distinguish Kac-Moody symmetric spaces of the Euclidean, the compact and the noncompact type, corresponding to the respective types of Riemannian symmetric spaces:

We have the following results:

Theorem 5.4 (affine Kac-Moody symmetric spaces of the "compact" type). *Both the Kac-Moody group* $\widehat{MG}^{\sigma}_{\mathbb{R}}$ *equipped with its* Ad*-invariant metric, and the quotient space* $X = \widehat{MG}^{\sigma}_{\mathbb{R}}/Fix(\rho_*)$ *equipped with their* Ad*-invariant metric are tame Fréchet symmetric spaces of the "compact" type with respect to their natural* Ad*-invariant metric. Their curvatures satisfy*

$$\langle R(X,Y)X,Y\rangle \geq 0.$$

Theorem 5.5 (affine Kac-Moody symmetric spaces of the "noncompact" type). *Both quotient spaces* $X = \widehat{MG}^{\sigma}_{\mathbb{C}}/\widehat{MG}^{\sigma}_{\mathbb{R}}$ *and* $X = H/Fix(\rho_*)$*, where* H *is a noncompact real form of* $\widehat{MG}^{\sigma}_{\mathbb{C}}$ *equipped with their* Ad*-invariant metric, are tame Fréchet symmetric spaces of the "noncompact" type. Their curvatures satisfy*

$$\langle R(X,Y)X,Y\rangle \leq 0.$$

Furthermore Kac-Moody symmetric spaces of the noncompact type are diffeomorphic to a vector space.

Define the notion of duality as for finite dimensional Riemannian symmetric spaces.

Theorem 5.6 (Duality). *Affine Kac-Moody symmetric spaces of the compact type are dual to Kac-Moody symmetric spaces of the noncompact type and vice versa.*

A complete classification of Kac-Moody symmetric spaces following the lines of the classification of finite dimensional Riemannian symmetric spaces follows from the correspondence between simply connected Kac-Moody symmetric spaces and OSAKAs [14].

Kac-Moody symmetric spaces have several conjugacy classes of flats. For our purposes the most important class are those of finite type. A flat is called of finite type iff it is finite dimensional. A flat is called of exponential type iff it lies in the image of the exponential map and it is called maximal iff it is not contained in another flat. Adapting a result of Bogdan Popescu [45] to our setting, and generalizing it to symmetric spaces of the noncompact type, we find:

Theorem 5.7. *All maximal flats of finite exponential type are conjugate.*

The isotropy representations of Kac-Moody symmetric spaces correspond exactly to the s-representations for involutions on affine Kac-Moody algebras that are studied in [20] and induce hence polar actions on Fréchet- resp. Hilbert spaces. This is the subject of the next section.

6 Polar Actions

Let $P(G, H) := \{g \in \hat{G} = H^1([0, 1], G) \mid (g(0), g(1)) \in H \subset G \times G\}$ denote the space of all H^1-Sobolev path in G whose endpoints are in H and let $V = H^0([0, 1], \mathfrak{g})$ denote the space of all H^0-Sobolev path in V.

We quote the following theorem of [51]:

Theorem 6.1. *Suppose the action of H on G is hyperpolar. Let A be a torus section through e and let $\mathfrak{a}$ denote its Lie algebra. Then*

1. *The $P(G, H)$-action on V is polar and the space $\widehat{\mathfrak{a}} = \{\widehat{a} | a \in \mathfrak{a}\}$ is a section, where $\widehat{a} : [0, 1] \longrightarrow \mathfrak{g}$ denotes the constant map with value a.*
2. *Let $N(\mathfrak{a})$ be the normalizer of $\mathfrak{a}$ in $P(G, H)$, $Z(\mathfrak{a})$ the centralizer. The quotient $N(\mathfrak{a})/Z(\mathfrak{a})$ of the normalizer of $\mathfrak{a}$ is an affine Weyl group.*

Similar to finite dimensional polar representations, that are induced by the isotropy representations of symmetric spaces, the classical examples of polar actions on Hilbert space come from isotropy representations of Kac-Moody symmetric spaces:

Theorem 6.2. *Let $\widehat{L}(G, \sigma)$ be a Kac-Moody group of H^1-loops and let $\rho : \widehat{L}(G, \sigma) \longrightarrow \widehat{L}(G, \sigma)$ be an involution of the second kind i.e. an involution such that on the Lie algebra level $d\rho(c) = -c$ and $d\rho(d) = -d$. Let $L(\mathfrak{g}, \sigma) = \mathcal{K} \oplus \mathcal{P}$ be the decomposition of $L(\mathfrak{g}, \sigma)$ into the ± 1-eigenspaces of $d\rho$. Then the restriction of the Adjoint action of Fix(ρ) $\subset \widehat{L}(G, \sigma)$ to the subspace $\{x \in \mathcal{P} | r_d = 1$ and $|x| = -1\}$ where r_d denotes the coefficient of d, is polar.*

For the definition of involutions of the second kind and a classification of involutions see [23] and [24].

Proof. [20] □

We describe the four classes of $P(G, H)$-actions more explicitly [26]:

1. The diagonal subgroup $H := \{(x, x) \in G \times G\}$. Elements of $P(G, H)$ are closed loops. The polar actions defined by this group are induced by the isotropy representations of non-twisted Kac-Moody symmetric spaces of type II resp. type IV.
2. The twisted diagonal subgroup $H := \{(x, \sigma(x)) \in G \times G\}$, where σ is a diagram automorphism. The associated polar actions are induced by the isotropy representations of twisted Kac-Moody symmetric spaces of types II resp. IV.

3. Let $H := K \times K \subset G \times G$. The polar actions defined by these groups are induced by the isotropy representations of non-twisted Kac-Moody symmetric spaces of type I resp. type III.
4. Let $H = K_{\sigma_1} \times K_{\sigma_2}$ where K_{σ_i} are fixed point groups of suitable involutions. The polar actions defined by these groups are induced by the isotropy representations of twisted Kac-Moody symmetric spaces of type I resp. type III.

7 Isoparametric Submanifolds

Definition 7.1 (PF-Submanifold). An immersed finite codimensional submanifold $M \subset V$ is proper Fredholm (PF) if the restriction of the end point map to a disk normal bundle of M on any radius r is proper Fredholm.

Definition 7.2. An immersed PF submanifold $f : M \longrightarrow V$ of a Hilbert space V is called isoparametric if

1. $codim(M)$ is finite.
2. $\nu(M)$ is globally flat.
3. for any parallel normal field v on M, the shape operators $A_{v(x)}$ and $A_{v(y)}$ are orthogonally equivalent for all x, y in M

The two main references are [50] and [44]. One can associate to isoparametric submanifolds in Hilbert spaces affine Weyl groups, that describe – as in the finite dimensional case – the positions of the curvature spheres. Similar to the finite dimensional case, most examples for isoparametric submanifolds in Hilbert space arise from polar actions:

Theorem 7.1. *A principal orbit of a polar action on a Hilbert space is isoparametric.*

A partial converse also exists:

Theorem 7.2 (Heintze-Liu). *A (complete, connected, irreducible, full) isoparametric submanifold of a Hilbert space V with codimension $\neq 1$ ($\neq 2$ if $\dim V < \infty$) is a principal orbit of a polar action.*

For the proof see [25]. In principle a proof along the lines of Thorbergsson should be possible also.

8 Universal Geometric Twin Buildings

For Kac-Moody symmetric spaces universal twin buildings take over the role played by spherical buildings for finite dimensional Riemannian symmetric spaces.

For algebraic loop groups (resp. affine Kac-Moody groups) there appear two constructions of buildings in the literature:

- If one replaces semisimple Lie groups (or more generally reductive linear algebraic groups) by algebraic Kac-Moody groups, it turns out that algebraic twin buildings take over the role played by spherical buildings for Lie groups.

  As in the Lie group situation we want an equivalence between Borel subgroups of the Kac-Moody group and chambers of the building. Due to the fact that Kac-Moody groups have two conjugacy classes of Borel subgroups the associated "building" breaks up into two pieces: Such a a twin building consists of a pair $\mathfrak{B}^+ \cup \mathfrak{B}^-$ of buildings that are "twinned": The twinning can be described in different ways: From the point of view of apartments the twinning is most easily defined by the introduction of a system of twin apartments, that is subcomplexes $\mathcal{A}^+ \cup \mathcal{A}^- \subset \mathfrak{B}^+ \cup \mathfrak{B}^-$, consisting of two apartments $\mathcal{A}^+$ and $\mathcal{A}^-$, such that $\mathcal{A}^+$ is contained in $\mathfrak{B}^+$ and $\mathcal{A}^-$ is contained in $\mathfrak{B}^-$. Imposing some axioms similar to those used for spherical buildings, many features known from apartments in spherical buildings generalize to the system of twin apartments [1].
- For groups of algebraic loops there is a theory of affine buildings (but not for twin buildings) developed from the point of view of loop groups [39].

To associate twin buildings to Kac-Moody symmetric spaces the main problem is again to unify algebraic and analytic aspects of the theory: twin buildings associated to affine Kac-Moody groups consist of pairs of Euclidean buildings. These are purely algebraic constructions and thus they correspond only to the subgroup of algebraic loops. Written in an algebraic notion these groups are of the form $G(\mathbb{C}[z, z^{-1}])$, that is: groups of polynomial loops in z and z^{-1}. Their affine Weyl groups act transitively on the chambers of the apartments whereas the groups $G(\mathbb{C}[z, z^{-1}])$ act transitively on the apartments.

Unfortunately a straightforward process of completion – even in only one direction z or z^{-1} – destroys the twinned structure. The classically used remedy can be found in Shrawan Kumar's book [37]: A group that is completed only in one direction – (let's say in the direction of z: hence the Laurent polynomials in z and z^{-1} are replaced by holomorphic functions with finite principal part) – acts on the part of the twin building corresponding to this direction (i.e. $\mathfrak{B}^+$). For our purposes however it is not enough to restrict the theory to one half of the twin building: On the one hand we need a completion that is symmetric in z and z^{-1} in order to be able to define the involutions of the second kind which are needed for Kac-Moody symmetric spaces. On the other hand our long-term objective is a proof of Mostow rigidity and there are good rigidity results only for twin buildings themselves, but not for their affine parts separately.

Our solution in [14] is to define on the level of the building a "completion" of the two parts of the twin building that corresponds to the completed groups. For different completions of the loop group (i.e. H^1, C^∞, analytic or holomorphic loops), the associated completions of the building are different. The process of completing algebraic twin buildings leads to a simplicial complex $\mathfrak{B} = \mathfrak{B}^+ \cup \mathfrak{B}^-$, whose (uncountable many) positive and negative connected components are affine buildings; we call these objects "universal geometric buildings".

8.1 Universal BN-Pairs

Definition 8.1 (Universal geometric BN-pair for $\widehat{L}(G,\sigma)$). Let $\widehat{L}(G,\sigma)$ be an affine Kac-Moody group. Data $(\widehat{B}^+, \widehat{B}^-, N, W, S)$ is a twin BN-pair for $\widehat{L}(G,\sigma)$ iff there are subgroups $\widehat{L}(G,\sigma)^+$ and $\widehat{L}(G,\sigma)^-$ of $\widehat{L}(G,\sigma)$ such that $\widehat{L}(G,\sigma) = \langle \widehat{L}(G,\sigma)^-, \widehat{L}(G,\sigma)^+ \rangle$ subject to the following axioms:

1. $(\widehat{B}^+, N, W, S)$ is a BN-pair for $\widehat{L}(G,\sigma)^+$ (called B^+N),
2. $(\widehat{B}^-, N, W, S)$ is a BN-pair for $\widehat{L}(G,\sigma)^-$ (called B^-N),
3. $(\widehat{B}^+ \cap \widehat{L}(G,\sigma)^-, \widehat{B}^- \cap \widehat{L}(G,\sigma)^+, N, W, S)$ is a twin BN-pair for $\widehat{L}(G,\sigma)^+ \cap \widehat{L}(G,\sigma)^-$.

The subgroups $\widehat{L}(G,\sigma)^+$ and $\widehat{L}(G,\sigma)^-$of $\widehat{L}(G,\sigma)$ depend on the choice of $\widehat{B}^+$ and $\widehat{B}^-$. A choice of a different subgroup $\widehat{B}^{+'}$ (resp. $\widehat{B}^{-'}$) gives the same subgroup $\widehat{L}(G,\sigma)^+$ (resp. $\widehat{L}(G,\sigma)^-$) of $\widehat{L}(G,\sigma)$ if $\widehat{B}^{+'} \subset \widehat{L}(G,\sigma)^+$ (resp. if $\widehat{B}^{-'} \subset \widehat{L}(G,\sigma)^-$)). For all positive (resp. negative) Borel subgroups the positive (resp. negative) subgroups $\widehat{L}(G,\sigma)^+$ resp $\widehat{L}(G,\sigma)^-$ are conjugate. Hence without loss of generality we can think of $\widehat{B}^\pm$ to be the standard positive (resp negative) affine Borel subgroup. The groups $\widehat{L}(G,\sigma)^\pm$ – called the standard positive (resp. negative) subgroups are then characterized by the condition that 0 (resp. ∞) is a pole for all elements.

Remark 8.1. For an algebraic Kac-Moody group a generalized BN-pair coincides with a BN-pair. Hence we get $\widehat{L}(G,\sigma)^+ = \widehat{L}(G,\sigma)^- = \widehat{L}(G,\sigma)$.

We use the equivalent definition for the loop groups $L(G,\sigma)$.

Lemma 8.1. *1. The groups $L(G,\sigma)^+$ (resp. $L(G,\sigma)^-$) have a positive (resp. negative) Bruhat decomposition and a Bruhat twin decompositions.*
2. The group $L(G,\sigma)$ has a Bruhat twin decomposition but no Bruhat decomposition.

Proof. The Bruhat decompositions in the first part follow by definition, the Bruhat twin decomposition by restriction and the second part. The second part is a restatement of the decomposition results in chapter 8 of [46]. □

Compare also the similar decomposition results stated in [53].

Theorem 8.1 (Bruhat decomposition). *Let $\widehat{L}(G,\sigma)$ be an affine Kac-Moody group with affine Weyl group W_{aff}. Let furthermore $\widehat{B}^\pm$ denote a positive (resp. negative) Borel group. There are decompositions*

$$\widehat{L}(G,\sigma)^+ = \coprod_{w\in W_{aff}} \widehat{B}^+ w \widehat{B}^+ \quad \textit{and} \quad \widehat{L}(G,\sigma)^- = \coprod_{w\in W_{aff}} \widehat{B}^- w \widehat{B}^- .$$

Proof. This is a consequence of lemma 8.1. □

Theorem 8.2 (Bruhat twin decomposition). *Let $\widehat{L}(G,\sigma)$ be an affine algebraic Kac-Moody group with affine Weyl group W_{aff}. Let furthermore $\widehat{B}^{\pm}$ denote a positive and its opposite negative Borel group. There are two decompositions*

$$\widehat{L}(G,\sigma) = \coprod_{w \in W_{aff}} \widehat{B}^{\pm} w \widehat{B}^{\mp}.$$

Remark 8.2. Note that the Bruhat twin decomposition is defined on the whole group $\widehat{L}(G,\sigma)$. This translates into the fact that any two chambers in $\mathfrak{B}^+$ resp. $\mathfrak{B}^-$ have a well-defined Weyl codistance. In contrast Bruhat decomposition are only defined for subgroups. This translates into the fact that there are positive (resp. negative) chambers without a well-defined Weyl distance.

Example 8.1. Kumar studies Kac-Moody groups and algebras that are completed "in one direction". In the setting of affine Kac-Moody groups of holomorphic loops this means: holomorphic functions with finite principal part. There is an associated twin BN-pair; the positive Borel subgroups are completed affine Borel subgroups while the negative ones are the algebraic affine Borel subgroups. Thus for a universal geometric twin BN-pair we have to use: $\widehat{L}(G,\sigma)^+ = \widehat{L}(G,\sigma)$ and $\widehat{L}(G,\sigma)^- = \widehat{L_{alg}G^{\sigma}}$ [37].

Lemma 8.2. *The intersection $\widehat{L}(G,\sigma)^0$ of $\widehat{L}(G,\sigma)^+$ with $\widehat{L}(G,\sigma)^-$ is isomorphic to the group of algebraic loops*

$$\widehat{L}(G,\sigma)^0 \simeq \widehat{L_{alg}G^{\sigma}}$$

Proof. $\widehat{L_{alg}G^{\sigma}}$ is the maximal subgroup of $\widehat{L}(G,\sigma)$ having both Bruhat decompositions. □

8.2 Universal Geometric Twin Buildings

We now define a universal geometric twin building using the W-metric approach:

Definition 8.2 (Universal geometric twin building). Let $\widehat{L}(G,\sigma)$ be an affine Kac-Moody group with a universal geometric BN-pair. Define $\mathcal{C}^+ := \widehat{L}(G,\sigma)/B^+$ and $\mathcal{C}^- := \widehat{L}(G,\sigma)/B^-$.

1. The distance $\delta^{\epsilon} : \mathcal{C}^{\epsilon} \times \mathcal{C}^{\epsilon} \longrightarrow W$, $\epsilon \in \{+,-\}$ is defined via the Bruhat decompositions: $\delta^{\epsilon}(gB^{\epsilon}, fB^{\epsilon}) = w$ iff $g^{-1}f$ is in the w-class of the Bruhat decomposition of $\widehat{L}(G,\sigma)^{\epsilon}$. Otherwise it is ∞.
2. The codistance $\delta^* : \mathcal{C}^+ \times \mathcal{C}^- \cup \mathcal{C}^- \times \mathcal{C}^+ \longrightarrow W$ is defined by $\delta^*(gB^-, fB^+) = w$ (resp. $\delta^*(gB^+, fB^-) = w$) iff $g^{-1}f$ is in the w-class of the corresponding Bruhat twin decomposition of $\widehat{L}(G,\sigma)$.

The elements of $\mathcal{C}^{\pm}$ are called the positive (resp. negative) chambers of the universal geometric twin building. The building is denoted by $\mathfrak{B} = \mathfrak{B}^{+} \cup \mathfrak{B}^{-}$. One can define a simplicial complex realization in the usual way. We define connected components in $\mathfrak{B}^{\pm}$ in the following way: Two elements $\{c_1, c_2\} \in \mathfrak{B}^{\pm}$ are in the same connected component iff $\delta^{\pm}(c_1, c_2) \in W_{\text{aff}}$. This is an equivalence relation. Denote the set of connected components by $\pi_0(\mathfrak{B})$ resp. $\pi_0(\mathfrak{B}^{\pm})$.

Remark 8.3. Let $L(G,\sigma)$ be an algebraic affine Kac-Moody group. Then the universal geometric twin building coincides with the algebraic twin building.

Lemma 8.3 (Properties of a universal geometric twin building).

1. *The connected components of $\mathfrak{B}^{\epsilon}$ are buildings of type (W, S),*
2. *Each pair consisting of one connected component in $\mathfrak{B}^{+}$ and one in $\mathfrak{B}^{-}$ is an algebraic twin building of type (W, S),*
3. *The connected components of $\mathfrak{B}^{\epsilon}$ are indexed by elements in $\widehat{L}(G,\sigma)/\widehat{L}(G,\sigma)^{\epsilon}$,*
4. *The action of $\widehat{L}(G,\sigma)$ on $\mathfrak{B}$ by left multiplication is isometric,*
5. *The Borel subgroups are exactly the stabilizers of the chambers, parabolic subgroups are the stabilizers of simplices,*
6. *$\widehat{L}(G,\sigma)^{\epsilon}$ acts on the identity component Δ_0^{ϵ} by isometries,*
7. *Let $\Delta_0^{+} \cup \Delta_0^{-}$ be a twin building in $\mathfrak{B}$. There is a subgroup isomorphic to the group of algebraic loops, that acts transitively on $\Delta_0^{+} \cup \Delta_0^{-}$.*

Proof. We sketch the proofs of some parts. For additional details [16]

1. The relation defined by finite codistance is clearly symmetric and self-reflexive, transitivity follows by calculation. One has to check the axioms for a building.
2. Each pair consisting of a connected component in $\mathfrak{B}^{+}$ and one in $\mathfrak{B}^{-}$ fulfills the axioms of an algebraic twin building. As the Bruhat decomposition is defined on $\widehat{L}(G,\sigma)$, the codistance is defined between arbitrary chambers in $\mathfrak{B}^{\epsilon}$ resp. $\mathfrak{B}^{-\epsilon}$.
3. $\widehat{L}(G,\sigma)$ has a decomposition into subsets of the form $\widehat{L}(G,\sigma)^{\epsilon}$. Those subsets are indexed with elements in $\widehat{L}(G,\sigma)/\widehat{L}(G,\sigma)^{\epsilon}$. The class corresponding to the neutral element is $\widehat{L}(G,\sigma)^{\epsilon} \subset \widehat{L}(G,\sigma)$. Thus it corresponds to a connected component and a building of type (W, S). The result follows via translation by elements in $\widehat{L}(G,\sigma)/\widehat{L}(G,\sigma)^{\epsilon}$: a connected component of $\mathfrak{B}^{\epsilon}$ containing fB^{ϵ} consists of all elements $f\widehat{L}(G,\sigma)^{\epsilon}B^{\epsilon}$ as $\delta(fhB^{\epsilon}, fh'B^{\epsilon}) = w((fh)^{-1}fh) = w(h^{-1}f^{-1}fh') = w(h^{-1}h') \in W$ as $h, h' \in \widehat{L}(G,\sigma)^{\epsilon}$.
4. G acts isometrically on a twin building if the action on both parts preserves the distances and the codistance [1], 6.3.1; the result follows by direct calculation.
5. The chamber corresponding to fB^{ϵ} is stabilized by the Borel subgroup $B_f^{\epsilon} := fB^{\epsilon}f^{-1}$. The converse follows as each Borel subgroup is conjugate to a standard one. Analogous for the parabolic subgroups. □

Theorem 8.3 (Embedding of twin buildings). *Denote by $H_{l,r}$ the intersection of the sphere of radius l of a real affine Kac-Moody algebra $\widehat{M\mathfrak{g}}$ with the horospheres $r_d = \pm r$, where r_d is the coefficient of d in the Kac-Moody algebra. There is a*

2-parameter family $(l,r) \in \mathbb{R}^+ \times \mathbb{R}^+$ of $\widehat{MG}$-equivariant immersion of $\mathfrak{B}^+ \cup \mathfrak{B}^-$ into $\widehat{M\mathfrak{g}}$. It is defined by the identification of $H_{l,r}$ with $\mathfrak{B}$. The two complexes $\mathfrak{B}^+$ and $\mathfrak{B}^-$ are immersed into the two sheets of $H_{l,r}$.

As in the finite dimensional situation, one can extend this result to the isotropy representations of Kac-Moody symmetric spaces. This gives embeddings of the universal geometric twin building into the tangential spaces. A consequence is the following

Corollary 8.1. *Points of the isoparametric submanifolds, corresponding to Kac-Moody symmetric spaces, are in bijection to chambers in one half of the universal twin building*

This shows in particular that, from a geometric point of view, universal geometric twin buildings are the correct generalization of spherical buildings for Kac-Moody groups.

8.3 Linear Representations for Universal Geometric Buildings

The matrix representations for the classical Lie groups give rise to linear representations for the associated buildings: For example the building of type A_{n-1} over $\mathbb{C}$ corresponds to the flag complex of subspaces in the vector space $V^n := \mathbb{C}^n$. Buildings for the other classical groups correspond to complexes of "special" subspaces [17].

In this section we describe the infinite dimensional flag representation for buildings of type $\widetilde{A}_{n-1}$. As linear representations for the loop groups, we use the operator representations studied in [46], Chap. 6. Subspaces are elements in suitable Grassmannians [46] (Chap. 7). Originally those Grassmannians were introduced by Mikio Sato in the context of integrable systems [49].

Let $H^n = L^2(S^1, \mathbb{C}^n)$ denote the separable Hilbert space of square summable functions on S^1 with values in $\mathbb{C}^n$. Let $H = H^{++} \oplus H^0 \oplus H^{--}$ be a polarization (for example the one induced by the action of $-i\frac{d}{d\theta}$). Set $H^+ = H^{++} \oplus H^0$ and $H^- = H^0 \oplus H^{--}$.

Following [46] Definition 7.1, the positive Grassmannian is defined as follows:

Definition 8.3 (Grassmannian). The positive Grassmannian $Gr^+(H)$ is the set of all closed subspaces W of H such that

1. The orthogonal projection $pr_+ : W \longrightarrow H^+$ is a Fredholm operator,
2. The orthogonal projection $pr_{--} : W \longrightarrow H^{--}$ is a Hilbert-Schmidt operator.

We define the virtual dimension of W to be $\nu(W) = \dim(\ker pr_+) - \dim(\operatorname{coker} pr_+)$.

There are various subgrassmannians corresponding to sundry regularity conditions. The most important examples are:

Definition 8.4 (positive algebraic Grassmannian). The positive algebraic Grassmannian $Gr_0^+(H)$ consists of subspaces $W \subset Gr^+(H)$ such that $z^k H^+ \subset W \subset z^{-k} H^+$.

Using the explicit description $H = L^2(S^1, \mathbb{C}^n)$, $Gr_0^+(H)$ consists exactly of the elements $W \in Gr(H)$ such that the images of $pr_{--} : W \longrightarrow H^{--}$ and $pr_+ : W^\perp \longrightarrow H^+$ are polynomials [46] and [18].

There are various other Grassmannians, i.e. the rational Grassmannian $Gr_1(H)$, $Gr_\omega(H)$ and the smooth Grassmannian $Gr_\infty(H)$ [46], for the definition of the tame Fréchet Grassmannian G_t and the H^1-Grassmannian [14], corresponding to different regularity conditions.

Definition 8.5 (reduced Grassmannian). The reduced positive Grassmannian $Gr^{n,+}(H^n)$ consists of subspaces $W \subset Gr^+(H^n)$ such that $G(W) \subset W$ (or explicitly $zW \subset W$).

The definition of the other types of reduced Grassmannians, especially reduced algebraic Grassmannians, reduced H^1- and reduced tame Grassmannians is self explaining.

The following theorem [46], Theorem 8.3.2 – shows this to be the correct notion to work well with the action of loop groups.

Theorem 8.4. *The group $L_{\frac{1}{2}} U_n$ of $\frac{1}{2}$-Sobolev loops acts transitively on $Gr^{n,+}$ and the isotropy group of H^+ is the group U_n of constant loops.*

This theorem yields the equivalences $\Omega_{\frac{1}{2}} U_n = Gr^n(H)$ and $\Omega_{alg} U_n = Gr_0^n$. Similar statements hold for $Gr_1^{n,+}(H)$, $Gr_\omega^{n,+}(H)$, $Gr_\infty^{n,+}(H)$ and $Gr_t^{n,+}(H)$; for $Gr_1^{n,+}(H)$, $Gr_\omega^{n,+}(H)$, $Gr_\infty^{n,+}(H)$ a proof can be found in [46]; this proof adapts to the case of $Gr_t^{n,+}(H)$ straight forwardly.

The next step is the definition of the flag varieties:

As subspaces in a flag satisfy $zW_k = W_{k+n} \subset W_k$, all subspaces of periodic flags are taken from $Gr^{n,+}$.

Let $\{e_1, \ldots, e_n\}$ be a basis of $V^n \simeq \mathbb{C}^n$ and $V_i := \text{span}\langle e_{i+1}, \ldots, e_n\rangle$. We define the positive normal flag to be the flag $\{W_{k'}\}_{k' \in \mathbb{Z}}$ such that for $k' = kn + l$, $k \in \mathbb{Z}, l \in \{0, \ldots, n-1\}$ we have $W_{k'} = W_{kn+l} := z^k W_l$ and W_l consists of all functions whose negative Fourier coefficients vanish and whose 0-coefficient is in V_i.

To define the manifolds of partial periodic flags the virtual dimension is used. For this purpose let $\mathcal{K} \subset \mathbb{Z}$ be a subset such that with $k \in \mathcal{K}$ also $k + nl \in \mathcal{K}$, $\forall l \in \mathbb{Z}$.

Furthermore set $m_{\mathcal{K}} := \#\{\mathcal{K} \cap \{1, \ldots, n\}\}$. Denote those $m_{\mathcal{K}}$-numbers $k_1, \ldots, k_{m_{\mathcal{K}}}$.

Definition 8.6 (positive periodic flag manifold). The positive periodic flag manifold $Fl_{\mathcal{K}}^{n,+}$ consists of all flags $\{W_k\}, k \in \mathbb{Z}$ in H^n such that

1. $W_k \subset \mathrm{Gr}^+ H$,
2. $W_{k+1} \subset W_k \forall k$,
3. $zW_k = W_{k+n}$.
4. For every flag $\{W_k\}_{k\in\mathbb{Z}}$ the map $\nu : (\{W_k\}_{k\in\mathbb{Z}}) \longrightarrow \mathbb{Z}$ mapping every subspace W_k to its virtual dimension is a surjection onto $\mathcal{K}$.

A flag is full iff $\mathcal{K} = \mathbb{Z}$. It is trivial iff $m_{\mathcal{K}} = 1$. Trivial flags are in bijection with elements of $Gr^{n,+}(H)$ under the identification $Gr^{n,+}(H) \ni W_0 \leftrightarrow \{z^k W_0\}_{k\in\mathbb{Z}} \in Fl^{n,+}_{\nu(W_0)+n\mathbb{Z}}$. A completely symmetric theory can be developed for H^-.

Theorem 8.5. *The flag complex of all periodic flags is a universal geometric twin building $\mathfrak{B}$. Positive flags correspond to simplices in $\mathfrak{B}^+$, negative ones to simplices in $\mathfrak{B}^-$.*

Proof. This is the main result of chapter 6 of [14]. □

We give some further hints: One starts by defining apartments via frames, i.e. a set of 1-dimensional subspaces, that span V. Then one proves two lemmas showing that those apartments satisfy the axioms in the simplicial complex definition of a building. Using those, one shows, that each connected component is a building. Then one checks using twin apartments, that $\mathfrak{B} = \mathfrak{B}^+ \cup \mathfrak{B}^-$ is a universal twin building (cf. [3] for the concept of twin apartments).

Definition 8.7 (frames and affine Weyl group). A frame is a sequence of subspaces $\{U_k\}_{k\in\mathbb{Z}} \subset H^n$ such that $U_{k+n} = zU_k$ and $H^n = \bigcup U_k$. We call a permutation $\pi : \{U_k\}_{k\in\mathbb{Z}} \longrightarrow \{U_k\}_{k\in\mathbb{Z}}$ admissible if $\pi(U_{k+n}) = \pi(U_k) + n$. The affine Weyl group W_{aff} is defined to be the group of admissible permutations of $\{U_k\}_{k\in\mathbb{Z}}$.

Definition 8.8 (apartment). Let $\{W_k\}$ be a full flag, $\{\leq W_k\}$ the set of all partial flags in $\{W_k\}$, hence in simplicial complex language, the boundary of $\{W_k\}$ and $\{U_k\}_{k\in\mathbb{Z}}$ a frame. The apartment $\mathcal{A}(\{W_k\}, U_k)$ consists of all flags that are transformations of flags in $\{\leq W_k\}$ by admissible permutations of $\{\{U_k\}_{k\in\mathbb{Z}}\}$.

Lemma 8.4. *For a pair of two flags $\{W_k\}$ and $\{W'_{k'}\}$ there exists an apartment containing both of them iff they are compatible in the sense that for all elements $W_k \in \{W_k\}$ there are elements $W'_{i'}, W'_{l'} \in \{W'_{k'}\}$ such that $W'_{i'} \subset W_k \subset W'_{l'}$ and vice versa. Compatibility defines an equivalence relation on the space of flags.*

Proof. As we have seen the complex of all flags is a chamber complex. Hence without loss of generality we can assume that $\{W_k\}$ and $\{W'_k\}$ are two maximal compatible flags. For each $k \in \mathbb{Z}$, we define the set $\pi(k) := \{m | \exists v \in (W_m \backslash W_{m+1}) \cap (W'_k \backslash W'_{k+1})\}$. We have to show that $|\pi(k)| = 1$ for all k. So for $i \in \{0, \ldots, n-1\}$ we choose vectors $v_i \subset \pi(i)$ and put $U_i = \operatorname{span}\langle v_i \rangle$. Furthermore for $i' = ln + i$ set $U_{i'} = U_{ln+i} = z^l U_i$.

The proof now consists of several steps:

- $\{U_k\}$ is a periodic frame. As the flags $\{W_k\}$ and $\{W'_k\}$ are periodic, $v \in (W_m \backslash W_{m+1}) \cap (W'_k \backslash W'_{k+1})$ is equivalent to $z^l v \in (W_{m+ln} \backslash W_{m+1+ln}) \cap (W'_{k+ln} \backslash W'_{k+1+ln})$.
- $W_k = W_{k+1} \oplus U_k$, $W_m = W_{m+1} \oplus U_k$ for all k and $m \subset \pi(k)$. So the apartment associated to $\{U_k\}$ contains $\{W_k\}$ and $\{W'_k\}$.
- $\pi(k+n) = \pi(k) + n$ follows from the periodicity of $\{W_k\}$ and $\{W'_k\}$.
- So we are left with showing that π is a permutation, that is $|\pi(k)| = 1\ \forall k$. The compatibility condition gives

$$z^{l+1}\{W_k\} = \{W_{k+(l+1)n}\} \subset z\{W'_k\} = \{W'_{k+n}\} \subset \{W'_{k+1}\} \subset \{W'_k\} \subset z^{-l}\{W_k\}.$$

So $W'_k \backslash W'_{k+1} \subset W_{k-ln}$. This shows that there are numbers m such that the set $(W_m \cap W'_k \backslash W'_{k+1})$ is nonempty. On the other hand $W_{k+(l+1)n} \subset W'_{k+1}$ shows that the set of those m is bounded from above. So there is for every k a maximal m such that $(W_m \cap (W'_k \backslash W'_{k+1})$ is nonempty. But then $(W_{m+1} \cap (W'_k \backslash W'_{k+1}))$ is empty. So $(W_m \backslash W'_{m+1}) \cap (W'_k \backslash W'_{k+1})$ is nonempty. So $\pi(k)$ is nonempty for all k. Symmetrically also $\pi^{-1}(m)$ is nonempty for all m. We use now the periodicity condition: $\pi(n+k) = \pi(k) + n$. As each set $\pi(k)$ is nonempty, for every $l \in \{0, \ldots, n-1\}$ there is k, such that $\pi(k) = l (\mathrm{mod}\, n)$.

This means that $\pi(k + n\mathbb{Z}) = l + n\mathbb{Z}$ as for each l, $l + n\mathbb{Z}$ is in the image. The pigeon hole principle asserts that $|\pi(k)| = 1(\mathrm{mod\ n})$. Then the periodicity shows that $|\pi(k)| = 1$. Hence π is a permutation and thus an element of W_{aff}.

Furthermore by direct calculation one finds:

Lemma 8.5 (apartments are isomorphic). *For every pair of apartments $\mathcal{A}$ and $\mathcal{A}'$ there is an isomorphism $\varphi : \mathcal{A} \longrightarrow \mathcal{A}'$. If $\{W_k\}, \{W'_{k'}\} \subset \mathcal{A} \cap \mathcal{A}'$ such that $\{W_k\}$ is full, one can choose φ in a way that it fixes $\{W_k\}$ and $\{W'_{k'}\}$ pointwise.*

Those two lemmas yield the theorem:

Theorem 8.6 (Tits building). *The simplicial complex associated to each equivalence class of compatible flags is an affine Tits building of type $\widetilde{A}_{n-1}$.*

Corollary 8.2. *The simplicial complex of positive (partial) flags in $Gr_0^{n,+}$ is an algebraic affine Tits building.*

For the special case of algebraic subspaces i.e. elements of the algebraic Grassmannian this construction coincides with the well-known lattice description [2, 17, 35, 39]. Similar constructions are possible for the other classical groups [14] for a sketch.

9 Conclusion and Outlook

As we have shown Kac-Moody geometry has reached a mature state. The classical objects whose symmetries are described by semisimple Lie groups – symmetric spaces, polar actions, isoparametric submanifolds, buildings – have infinite dimensional counterparts whose symmetries are defined by analytic affine Kac-Moody groups. Furthermore important results known from the finite dimensional theory that describe the connections between those objects, generalize to the infinite dimensional setting. The crucial point of the theory is a control of the functional analytic framework that permits a generalization of the algebraic operations. Hence while differential geometric in spirit, the field incorporates various concepts of infinite dimensional Lie theory, functional analysis and algebra.

There are still many important open problems.

1. **Mostow rigidity:** Study quotients of Kac-Moody symmetric spaces of the noncompact type. We conjecture the existence of a Mostow-type theorem for Kac-Moody symmetric spaces of the noncompact type, if the rank r of each irreducible factor satisfies $r \geq 4$; in the finite dimensional situation the main ingredients for the proof of Mostow rigidity are the spherical buildings which are associated to the universal covers $\widetilde{M}$ and $\widetilde{M}'$ of two homotopy equivalent locally symmetric spaces $M = \widetilde{M}/\Gamma$ and $M' = \widetilde{M}'/\Gamma'$ (suppose the rank r of each de Rham factor satisfies $\text{rank}(M) \geq 2$). To prove Mostow rigidity one has to show that a homotopy equivalence of the quotients lifts to a quasi isometry of the universal covers and induces a building isomorphism. This step is done via a description of the boundary at infinity. By rigidity results of Jacques Tits this building isomorphism is known to introduce a group isomorphism which in turn leads to an isometry of the quotients.

   Hence to prove a generalization of Mostow rigidity to quotients of Kac-Moody symmetric spaces of the noncompact type along these lines, one needs a description of the boundary of Kac-Moody symmetric spaces of the non compact type. In view of their Lorentz structure those spaces are not CAT(0). Consequently a direct generalization of the finite dimensional ideas to construct a boundary seems difficult. Nevertheless as for each point in the symmetric space a boundary can be defined, a complete definition using the action of the Kac-Moody groups seems within reach. We note that a generalization or adaption of the methods developed in [10] and in [19] might lead to a proof of Mostow rigidity based on local-global methods. By work of Andreas Mars this is possible for algebraic Kac-Moody groups in case of rings with sufficiently many units [38].
2. **Holonomy:** Study the holonomy of infinite dimensional Lorentz manifolds. More precisely: Is there a generalization of the Berger holonomy theorem to Kac-Moody symmetric spaces? In the finite dimensional Riemannian case, Berger's holonomy theorem tells us the following: Let M be a simply connected manifold with irreducible holonomy. Then either the holonomy acts transitively on the tangent sphere or M is symmetric. Kac-Moody symmetric spaces are

Lorentz manifolds and have a distinguished, parallel lightlike vector field, corresponding to c. In general relativity similar finite dimensional objects are known as Brinkmann waves [7]. Hence it seems that Kac-Moody symmetric spaces should be understood as a kind of infinite dimensional Brinkmann wave. The conjectured holonomy theorem states then that the holonomy of an infinite dimensional Brinkmann wave is either transitive on the horosphere around c or the space is a Kac-Moody symmetric space.

3. **Characterization:** Is there some intrinsic characterization of Kac-Moody symmetric spaces? Probably the existence of sufficiently many finite dimensional flats and the closely related Fredholm property of the induced polar actions will be part of this characterization. From an algebraic point of view, it would be very interesting, to investigate if there are symmetric spaces associated to Lorentzian Kac-Moody algebras - nevertheless, the absence of good explicit realizations seems to be a serious impediment. From a functional analytic point of view, a generalization to other classes of maps, i.e. holomorphic maps on Riemann surfaces or symplectic maps would be interesting. From the finite dimensional blueprint the development of an infinite dimensional version of the theory of non-reductive pseudo-Riemannian symmetric spaces as developed in the finite dimensional case by Ines Kath and Martin Olbricht in [33] seems promising. In view of the characterization of Kac-Moody symmetric spaces as infinite dimensional Brinkmann waves, we would like to understand, how the existence of this special direction is related to other structure properties. It is hence natural, to ask, if there is a class of infinite dimensional Lorentz symmetric spaces without such a distinguished direction?
4. **"Semi-Riemannian" Kac-Moody Symmetric Spaces:** We argued that the Kac-Moody symmetric spaces, we developed correspond to Riemannian symmetric spaces. Marcel Berger studied reductive symmetric spaces of arbitrary index [17]. The Kac-Moody analogue should exist but is not constructed nor is a classification available. Remark that a classification can be given without a construction of the spaces just by considering involutions of Kac-Moody algebras [24].
5. **Kac-Moody Symmetric Spaces as Moduli Space:** Recent results by Shimpey Kobayashi and Josef Dorfmeister show that the moduli spaces of different classes of integrable surfaces can be understood as real forms of loop groups of $\mathfrak{sl}(2,\mathbb{C})$ [34]. We conjecture that Kac-Moody symmetric spaces can be interpreted as Moduli spaces of special classes of submanifolds in more general situations.

References

1. Abramenko, P., Brown, K.: Buildings, vol. 248 of Graduate Texts in Mathematics. Springer, New York (2008)
2. Abramenko, P., Nebe, G.: Lattice chain models for affine buildings of classical type. Mathematische Annalen **322**, 537–562 (2002)

3. Abramenko, P., Ronan, M.: A characterization of twin buildings by twin apartments. Geometriae Dedicata **73**, 1–9 (1998)
4. Berndt, J., Console, S., Olmos, C.: Submanifolds and holonomy, vol. 434 of Research Notes in Mathematics. Chapman&Hall, Boca Raton (2003)
5. Berger, M.: Les espaces symétriques non compacts. Ann. Sci Ecole Norm. Sup. **74**, 85–177 (1957)
6. Bourbaki, N.: Lie groups and Lie algebras, chapters 4–6. Elements of Mathematics. Springer, New York (2002)
7. Brinkmann, H.-W.: Einstein spaces which are mapped conformally on each other. Mathematische Annalen **94**, 119–145 (1925)
8. Bröcker, T., tom Dieck, T.: Representations of compact Lie groups, vol. 98 of Graduate Texts in Mathematics. Springer, New York (1998)
9. Bump, D.: Lie groups, vol. 225 of Graduate Texts in Mathematics. Springer, New York (2004)
10. Caprace, P.-E.: Abstract homomorphisms of split Kac-Moody groups. Memories of the AMS **198**(924) (2009)
11. Davis, M.W.: The Geometry and Topology of Coxeter Groups. Princeton University Press, Princeton (2008)
12. Eberlein, P.B.: Geometry of nonpositively curved manifolds. Chicago Lectures in Mathematics, Chicago (1996)
13. Freyn, W.: A general theory of affine Kac-Moody symmetric spaces. Kongressberichte der Süddeutschen Geometrietagung **32**, 4–18 (2007)
14. Freyn, W.: Kac-Moody symmetric spaces and universal twin buildings. PhD thesis, Universität Augsburg (2009)
15. Freyn, W.: Functional analytic methods for cities. (submitted, 2011)
16. Freyn, W.: Kac-Moody groups, analytic regularity conditions and cities (submitted, 2010)
17. Garrett, P.: Buildings and Classical Groups. Chapman Hall, London (1997)
18. Gohberg, I., Goldberg, S., Kaashoek, M.: Basic Classes of Linear Operators. Birkhäuser, Basel (2003)
19. Gramlich, R., Mars, A.: Isomorphisms of unitary forms of Kac-Moody groups over finite fields. J. Algebra **322**, 554–561 (2009)
20. Groß, C.: s-Representations for involutions on affine Kac-Moody algebras are polar. Manuscripta Math. **103**, 339–350 (2000)
21. Hamilton, R.S.: The inverse function theorem of Nash and Moser. Bull. Am. Math. Soc. (N.S.) **7**(1), 65–222 (1982)
22. Heintze, E.: Towards symmetric spaces of affine Kac-Moody type. Int. J. Geom. Methods Mod. Phys. **3**(5–6), 881–898 (2006)
23. Heintze, E.: Real forms and finite order automorphisms of affine Kac-Moody algebras – an outline of a new approach. Opus Bayern. http://www.opus-bayern.de/uni-augsburg/volltexte/2008/763:15 (2008)
24. Heintze, E., Groß, C.: Involutions, finite order automorphisms and real forms of affine Kac-Moody algebras (2009)
25. Heintze, E., Liu, X.: Homogeneity of infinite-dimensional isoparametric submanifolds. Ann. Math. (2), **149**(1), 149–181 (1999)
26. Heintze, E., Palais, R.S., Terng, C.-L., Thorbergsson, G.: Hyperpolar actions on symmetric spaces, pp. 214–245. Conf. Proc. Lecture Notes Geom. Topology, IV. Int. Press, Cambridge, MA (1995)
27. Helgason, S.: Differential geometry, Lie groups, and symmetric spaces, vol. 34 of Graduate Studies in Mathematics. American Mathematical Society, Providence, RI (2001) Corrected reprint of the 1978 original
28. Ji, L.: Buildings and their applications in geometry and topology. Asian J. Math. **10**, 11–80 (2006)
29. Kac, V.G.: Simple irreducible graded Lie algebras of finite growth (russian). Math. USSR-Izvestiya **2**, 1271–1311 (1968)

30. Kac, V.G.: Infinite-dimensional Lie algebras, 3rd edn. Cambridge University Press, Cambridge (1990)
31. Kantor, I.L.: Graded Lie algebras (russian). Trudy Sem. Vektor. Tenzor. Anal. **15**, 227–266 (1970)
32. Kath, I., Olbrich, M.: On the structure of pseudo-Riemannian symmetric spaces. Transformation groups, **14**(4), 847–885 (2009)
33. Kath, I., Olbrich, M.: Recent developments in pseudo-Riemannian geometry, ESI Lect. in math. Phzsics, pp. 1–52, European mathematical society, 2008
34. Kobayashi, S.: Real forms of complex surfaces of constant mean curvature. Transactions of the AMS, **363**(4), 1765–1788 (2011)
35. Kramer, L.: Loop groups and twin buildings. Geometriae Dedicata **92**, 145–178 (2002)
36. Kriegl, A., Michor, P.W.: The convenient setting of global analysis, vol. 53 of Mathematical Surveys and Monographs. American Mathematical Society, Providence, RI (1997)
37. Kumar, S.: Kac-Moody groups, their flag varieties and representation theory, vol. 204 of Progress in Mathematics. Birkhäuser, Boston, MA (2002)
38. Mars, A.: Rigidity. On the topology and geometry of Kac-Moody groups, PhD-thesis, TU Darmstadt, 2011
39. Mitchell, S.A.: Quillen's theorem on buildings and the loops on a symmetric space. L'Enseignement Mathematique **34**, 123–166 (1988)
40. Moody, R.V.: Euclidean Lie algebras. Canad. J. Math. **21**, 1432–1454 (1969)
41. Moody, R.V., Pianzola, A.: Lie algebras with triangular decomposition. Wiley, New York (1995)
42. Neeb, K.-H.: Towards a Lie theory of locally convex groups. Jpn. J. Math. **1**(2), 291–468 (2006)
43. O'Neill, B.: Semi-Riemannian geometry, vol. 103 of Pure and Applied Mathematics. Academic Press [Harcourt Brace Jovanovich Publishers], New York (1983). With applications to relativity
44. Palais, R., Terng, C.-L.: Critical point theory and submanifold geometry, vol. 1353 of Lecture Notes in Mathematics. Springer, Berlin (1988)
45. Popescu, B.: Infinite dimensional symmetric spaces. Thesis. University of Augsburg, Augsburg (2005)
46. Pressley, A., Segal, G.: Loop groups. Oxford Mathematical Monographs. The Clarendon Press Oxford University Press, New York (1986)
47. Ray, U.: Automorphic Forms and Lie Superalgebras. Springer, New York (2006)
48. Rémy, B.: Groupes de Kac-Moody déployés et presque déployés. Société mathematiques de France, Paris (2002)
49. Sato, M., Sato, Y.: Soliton equations as dynamical systems on infinite dimensional Grassmann manifolds, vol. 81 of North-Holland Math. Stud, pp. 259–271. North-Holland, Amsterdam (1983)
50. Terng, C.-L.: Proper Fredholm submanifolds of Hilbert space. J. Diff. Geom. **29**(1), 9–47 (1989)
51. Terng, C.-L.: Polar actions on Hilbert spaces, J. Geom. Anal. **5**(1), 129–150 (1995)
52. Thorbergsson, G.: Isoparametric foliations and their buildings. Ann. Math. **133**, 429–446 (1991)
53. Tits, J.: Groups and group functors attached to Kac-Moody data, vol. 1111 of Lecture notes in mathematics, pp. 193–233. Springer, Berlin (1984)

Collapsing and Almost Nonnegative Curvature

Wilderich Tuschmann

1 Introduction

Almost nonnegatively curved manifolds are charming spaces for at least two reasons: From a classical point of view, they are natural generalizations of almost flat as well as nonnegatively and positively curved manifolds, and the study of all of the latter has a long tradition in Riemannian geometry. Secondly, almost nonnegatively curved manifolds are precisely the spaces which can be collapsed to a point under a fixed lower bound on sectional curvature, so that in degenerations and convergence of metrics under lower curvature bounds they play the same fundamental role that almost flat manifolds do in Cheeger–Fukaya–Gromov's theory of collapse with curvature bounded in absolute value.

The aim of this paper is to report on several results and questions on almost nonnegatively curved manifolds from both perspectives. After an introduction to basic results in collapsing and convergence under a lower bound on sectional curvature we will discuss constructions of and obstructions to almost nonnegative curvature, then proceed to manifolds with almost nonnegative operator and close with a section on several open questions in these fields. Many important related subjects, like convergence and collapsing with both-sided bounds on sectional or Ricci curvature, or manifolds with lower bounds on (or even constant) Ricci curvature and their degenerations, will (almost) not be treated, and I apologize in advance for also leaving numerous other important topics out.

W. Tuschmann (✉)
Fakultät für Mathematik Karlsruher Institut für Technologie (KIT) Kaiserstraße 89-93 D-76133 Karlsruhe, Germany
e-mail: tuschmann@kit.edu

C. Bär et al. (eds.), *Global Differential Geometry*, Springer Proceedings in Mathematics 17, DOI 10.1007/978-3-642-22842-1_4,

2 Convergence and Collapsing Under a Lower Curvature Bound

Let us first recall the notion of Gromov–Hausdorff distance, introduced by Gromov around 1980 as an abstract version of the classical Hausdorff distance (compare [23]). For nonempty closed and bounded subsets A and B of a metric space X the Hausdorff distance between A and B in X, $d_H(A, B) = d_H^X(A, B)$, is defined as the infimum of all positive real numbers $\varepsilon > 0$ such that the open ε-neighborhood of A in X contains B and vice versa. The Hausdorff distance then defines a metric on the set of all closed and bounded nonempty subsets of X. If now X and Y are two compact metric spaces, their *Gromov–Hausdorff distance* $d_{GH}(X, Y)$ is the infimum of all numbers $d_H^Z(f(X), g(Y))$, where Z ranges over all metric spaces and f and g over all isometric embeddings $f : X \to Z$ and $g : Y \to Z$. The resulting number is clearly independent of any ambient space; for practical purposes, however, it is sometimes better to define the Gromov–Hausdorff distance of X and Y equivalently by the infimum of all Hausdorff distances $d_H^Z(X, Y)$, where Z is the disjoint union of X and Y and where the infimum is now taken over all metrics on Z which extend the metrics on X and Y.

The Gromov–Hausdorff distance defines a metric on the set of all compact metric spaces, considered up to isometry, and therefore gives rise to the notion of *Gromov–Hausdorff convergence* of sequences of compact metric spaces. There is in fact also a useful notion of GH-convergence for noncompact metric spaces, at least for locally compact complete length ones in which the distance between points is given by the infimum of the lengths of all curves joining the points: A sequence $(X_n, x_n)_{n \in \mathbb{N}}$ of such spaces with given basepoints $x_n \in X_n$ is said to converge to a pointed metric space (Y, y) in the pointed Gromov–Hausdorff sense if for all $r > 0$ the closed r-balls $B_r(x_n)$ Gromov–Hausdorff converge to the closed r-ball $B_r(y)$ in Y. Under this so-called *pointed Gromov–Hausdorff convergence* (of locally compact complete length spaces) the collection of all such spaces, considered up to isometries that preserve basepoints, is moreover complete. For further, e.g., equivariant notions of (pointed) Gromov–Hausdorff convergence, as well as many examples, see besides [23] for instance also Fukaya's comprehensive article [15].

In applying his extension of Bishop's volume comparison theorem, Gromov arrived at his celebrated precompactness result ([20], compare [23]):

Theorem 2.1 (Gromov's Precompactness Theorem). *For any $m \in \mathbb{N}$ and real numbers κ and D the class of closed Riemannian m-manifolds with Ricci curvature $Ric \geq (m-1)\kappa$ and diameter $diam \leqslant D$ is precompact in the Gromov–Hausdorff topology, and the class of pointed complete Riemannian m-manifolds with $Ric \geq (m-1)\kappa$ is precompact with respect to the pointed Gromov–Hausdorff topology.*

Thus, any sequence of Riemannian m-manifolds satisfying the assumptions of the precompactness theorem will have a Gromov–Hausdorff convergent subsequence and, most important, this subsequence will always have a limit! This limit is, moreover, a compact (respectively, locally compact and complete) length space of Hausdorff dimension at most m, and if its dimension is strictly less than m, one says that the sequence *collapses* to it.

If one now restricts attention to complete pointed Riemannian m-manifolds with a lower bound on *sectional* curvature, the Gromov–Hausdorff closure of this class is given by very special length spaces, also called Alexandrov spaces. Here and in what follows an *Alexandrov space* will denote a complete length space of finite Hausdorff dimension which has curvature *curv* bounded from below in comparison sense (for the precise definition see, e.g., [15]). For a Riemannian manifold the property *curv* $\geq \kappa$ is just equivalent to saying that its sectional curvature is bounded from below by κ, and for general length spaces it means that small geodesic triangles in the space are not thinner than their corresponding triangles in the two-dimensional simply connected space form of constant Gauss curvature κ.

Alexandrov spaces have always integer Hausdorff dimension which also equals their topological dimension, and their general theory has been initiated by Burago, Gromov and Perelman in the articles [7, 36]. For more on Alexandrov spaces and the geometry and analysis that can be done on them see, in particular, Petrunin's article [37] and the forthcoming book [1].

Of special importance in convergence and collapsing under a lower bound on sectional curvature are Perelman's stability and Yamaguchi's fibration theorem. A pointed version of the former (see [36] as well as [30] for a detailed proof) can be stated as follows.

Theorem 2.2 (Perelman's Stability Theorem). *Assume that (M_n, p_n) is a Gromov–Hausdorff convergent sequence of pointed complete Riemannian m-manifolds with sectional curvature $\geq \kappa$ which converges to an Alexandrov space (Y, p) of the same dimension and let C be a compact subset of Y. Then for all large n there exist compact subsets C_n of M_n and homeomorphisms $h_n : C \to C_n$ which are almost isometric in the sense that for all $x, y \in C$ one has that $|d(x, y) - d_n(h_n(x), h_n(y))| \to 0$ as $n \to \infty$. If, in particular, all M_n have also uniformly bounded diameters, then for large n they are thus all homeomorphic to Y.*

Yamaguchi's fibration theorem ([46], compare also [7] and [47] for more general versions as well as [13] for bounded curvature collapse) deals with the collapsing case and may be stated as follows:

Theorem 2.3 (Yamaguchi's Fibration Theorem). *Suppose that (M_n, p_n) is a Gromov–Hausdorff convergent sequence of pointed complete Riemannian m-manifolds with sectional curvature $\geq \kappa$ which collapses to a Riemannian manifold (Y, p) of lower dimension, and let C be a compact subset of Y. Then for all large n there exist compact subsets C_n of M_n and mappings π_n from C_n to C such that all $\pi_n : C_n \to C$ are smooth fibrations over C whose fibers are almost nonnegatively curved in the following generalized sense.*

Definition 2.4. A closed smooth manifold M is called almost nonnegatively curved in the generalized sense if for some nonnegative integer k there exists a sequence of complete Riemannian metrics g_n on $M \times \mathbb{R}^k$ and points $p_n \in M \times \mathbb{R}^k$ such that:

1. The sectional curvatures of the metric balls of radius n around p_n satisfy

$$\sec(B_n(p_n)) \geq -1/n.$$

2. For $n \to \infty$ the pointed Riemannian manifolds $((M \times \mathbb{R}^k, g_n), p_n)$ converge in the pointed Gromov–Hausdorff distance to flat Euclidean space $(\mathbb{R}^k, 0)$.
3. The regular fibers over 0 are diffeomorphic to M for all large n.

If $k = 0$, this definition reduces to the standard one. It is, however, an open problem whether all manifolds which are almost nonnegatively curved in the generalized sense are almost nonnegatively curved in the standard one.

Yamaguchi's result shows in particular that almost nonnegatively curved manifolds are crucial for understanding degenerations of metrics under a lower curvature bound, as, to give a prominent example, this is for instance the case in Perelman's work on geometrization.

3 Constructions of Almost Nonnegative Curvature

Almost nonnegatively curved manifolds were introduced by Gromov in the late seventies (see [18, 19]). As was already alluded to in the introduction, a closed smooth manifold is said to be almost nonnegatively curved if it can Gromov–Hausdorff converge to a single point under a lower curvature bound. In more classical terms, this means the following:

Definition 3.1. A closed smooth manifold M is called almost nonnegatively curved if for all $\varepsilon > 0$ there exists a Riemannian metric g_ε on M whose curvature and diameter satisfy the scaling-invariant inequality

$$\sec(M, g_\varepsilon) \cdot (diam(M, g_\varepsilon))^2 > -\varepsilon.$$

Almost nonnegatively curved manifolds thus generalize almost flat as well as nonnegatively curved manifolds, and their class is in fact strictly bigger:

Example 3.2. Let N be the three-dimensional Heisenberg group, i.e., the nilpotent group of upper triangular real 3×3-matrices with ones on the diagonal, and let Γ be its lattice $\Gamma := N \cap GL(3, \mathbb{Z})$. Then $M := S^2 \times N/\Gamma$ is almost nonnegatively curved, but neither almost flat nor nonnegatively curved.

Due to the following constructions, there are much more almost nonnegatively curved manifolds known today than nonnegatively curved ones. One large source of examples stems from the following result of Fukaya and Yamaguchi [16]:

Theorem 3.3. *Let $F \to M \to N$ be a smooth fiber bundle with compact Lie structure group G and almost nonnegatively curved base N whose fiber F admits a G-invariant metric of nonnegative curvature. Then the total space M has almost nonnegative curvature.*

Example 3.4. The total space of any (linear) sphere bundle over a sphere or any torus bundle over a torus admits almost nonnegative curvature.

The second main source of examples of almost nonnegatively curved manifolds stems from *cohomogeneity one manifolds*, i.e., smooth manifolds M on which a compact Lie group G acts smoothly with codimension one principal orbits. Here one has (see [42, 43]):

Theorem 3.5. *Any closed cohomogeneity one manifold is almost nonnegatively curved.*

Remark 3.6. Grove and Ziller showed that any cohomogeneity one manifold (M, G) with nonregular orbits of codimension at most two admits a nonnegatively curved metric which is invariant under the cohomogeneity one action of G (see [26]), and, moreover, that any closed cohomogeneity one manifold admits an invariant metric of nonnegative Ricci curvature as well as an invariant metric of positive Ricci curvature if its fundamental group is finite (compare [27]).

All ε-almost nonnegatively curved metrics that implicitly occur in Theorem 3.5 can also be chosen to be *invariant* under the cohomogeneity one action and will, moreover, simultaneously also have nonnegative Ricci curvature and positive Ricci curvature if $\pi_1(M)$ is finite, see [43]. Moreover, in this respect Theorem 3.5 is in fact also optimal, since there are closed cohomogeneity one manifolds which do not admit any invariant metric of nonnegative sectional curvature (see [25]).

Theorem 3.5 yields, as we shall explain now, also many new examples of manifolds with almost nonnegative sectional curvature. Notice first that particularly interesting examples of closed cohomogeneity one manifolds are given by the odd-dimensional Brieskorn manifolds W_d^{2n-1} (see [5, 28, 35]). Given integers $n \geq 2$ and $d \geq 1$, the manifolds W_d^{2n-1} are the $(2n-1)$-dimensional real algebraic submanifolds of $\mathbb{C}^{n+1}$ defined by the equations

$$z_0^d + z_1^2 + \cdots + z_n^2 = 0 \quad \text{and} \quad |z_0|^2 + |z_1|^2 + \cdots + |z_n|^2 = 1.$$

All manifolds W_d^{2n-1} are invariant under the standard linear action of $O(n)$ on the $(z_1, \ldots, z_n)$-coordinates, and the action of S^1 via the diagonal matrices of the form $diag(e^{2i\theta}, e^{di\theta}, \ldots, e^{di\theta})$. The resulting action of the product group $G := O(n) \times S^1$ has cohomogeneity one (see [29]) with two nonprincipal orbits of codimensions 2 and $n-1$, respectively.

The topology of the Brieskorn manifolds W_d^{2n-1} is quite well understood. Notice as special cases that W_1^{2n-1} is equivariantly diffeomorphic to S^{2n-1} with the linear action of $G = O(n) \times S^1 \subset U(n)$. Moreover, W_2^{2n-1} is equivariantly diffeomorphic to the Stiefel manifold $V_{n+1,2} = O(n+1)/O(n-1)$ of orthonormal two-frames in $\mathbb{R}^{n+1}$, with the action of G given by the standard inclusion $G \subset O(n+1) \times O(2) = O(n+1) \times Norm_{O(n+1)} O(n-1)/O(n-1)$. Also, for $n = 2$ the manifold W_d^3 is diffeomorphic to the lens space $S^3/\mathbb{Z}_d$, whereas for $n \geq 3$ all manifolds W_d^{2n-1} are simply connected.

If now $n \geq 3$ and $d \geq 1$ are odd, then W_d^{2n-1} is a homotopy sphere. Indeed, if $d \equiv \pm 1 \bmod 8$ then W_d^{2n-1} is diffeomorphic to the standard $(2n-1)$-sphere, while for $d \equiv \pm 3 \bmod 8$, W_d^{2n-1} is diffeomorphic to the *Kervaire sphere* K^{2n-1}. The Kervaire spheres are generators of the groups of homotopy spheres which bound parallelizable manifolds, and all Kervaire spheres can be realized as Brieskorn manifolds W_d^{2n-1} [4, 28]. The sphere K^{2n-1} is exotic, i.e., homeomorphic but not diffeomorphic to the standard sphere, if $n+1$ is not a power of 2 [6]. On the other hand, if $n = 2m$ is even, then W_d^{4m-1} is a rational homology sphere whose only nontrivial cohomology groups are given by $H^0(W_d^{4m-1}) \cong H^{4m-1}(W_d^{4m-1}) \cong \mathbb{Z}$ and $H^{2m}(W_d^{4m-1}) \cong \mathbb{Z}_d$ ([4], p.275). All in all this yields the following:

Corollary 3.7. *For $n \geq 2$ and $d \geq 1$ all Brieskorn manifolds W_d^{2n-1}, hence all Kervaire spheres and infinite families of rational homology spheres in each dimension $4m-1$ for $m \geq 2$ admit almost nonnegative curvature. In particular, there is an infinite number of dimensions in which there exist almost nonnegatively curved exotic spheres.*

There are also interesting quotients of the Brieskorn manifolds by finite cyclic groups to which Theorem 3.5 can be directly applied. This yields, for example, for each integer $k \geq 1$ at least 4^k oriented diffeomorphism types of almost nonnegatively curved homotopy $\mathbb{RP}^{4k+1}$ (compare [43]).

However, one can also leave the cohomogeneity one case and study free actions of positive-dimensional Lie groups on the Brieskorn manifolds W_d^{2n-1}: If $n = 2m$ is even, there is a free circle action on W_d^{4m-1} which is given by the action of the circle subgroup $S^1 = Z(U(m)) \subset O(2m)$ where Z denotes the center. Likewise, if we assume that $n = 4m$, then the subgroup $Sp(1) \subset O(4m)$, viewed as scalar multiplication of the unit quaternions on $\mathbb{R}^{4m} \cong \mathbb{H}^m$, acts also freely on W_d^{8m-1}. Combining these observations with the curvature nondecreasing property of Riemannian submersions and some topology (for details see again [43]), one obtains a multitude of rational cohomology projective spaces with almost nonnegative curvature:

Corollary 3.8. *For every integer $m \geq 2$ there are infinite families $(N_d^{4m-2})_{d \geq 1}$ and $(\tilde{N}_d^{8m-4})_{d \geq 1}$ of almost nonnegatively curved and mutually not homotopy equivalent simply connected manifolds of dimension $4m-2$ and $8m-4$ with the rational cohomology ring of $\mathbb{CP}^{2m-1}$ and $\mathbb{HP}^{2m-1}$, respectively.*

It is hard to tell how big the class of almost nonnegatively curved manifolds actually is. However, the following result from [31] reduces, in a certain sense, the classification of manifolds with almost nonnegative curvature to the simply connected case.

Theorem 3.9. *Let M be an almost nonnegatively curved manifold. Then a finite cover $\tilde{M}$ of M is the total space of a fiber bundle*

$$F \to \tilde{M} \to N$$

over a nilmanifold N *with a simply connected fiber* F *which is almost nonnegatively curved in the generalized sense.*

Remark 3.10. Let us point out that Theorem 3.9 *does not extend* to manifolds with almost nonnegative Ricci curvature. It was conjectured in [16] that a finite cover of an almost nonnegatively Ricci curved manifold M fibers over a nilmanifold with a fiber which has nonnegative Ricci curvature and whose fundamental group is finite, but this conjecture was later refuted by Anderson (see [2]).

4 Obstructions to Almost Nonnegative Curvature

After having seen that that there is indeed a huge number of almost nonnegatively curved manifolds out there, in this section we describe, in chronological order, the main topological obstructions to almost nonnegative curvature that are known today.

Gromov proved in [18, 20, 21] the following.

Theorem 4.1. *If* $M = M^m$ *is an almost nonnegatively curved* m*-manifold, then the following holds:*

1. *The minimal number of generators of the fundamental group* $\pi_1(M)$ *of* M *can be estimated by a constant* $C_1(m)$ *depending only on* m.
2. *The first real Betti number of* M *satisfies* $b_1(M;\mathbb{R}) \leqslant m$ *(and this holds, actually, also for manifolds with almost nonnegative Ricci curvature).*
3. *The sum of the Betti numbers of* M *with respect to any field of coefficients does not exceed some uniform constant* $C_2 = C_2(m)$.

Example 4.2. The manifold $S^2 \times S^2$ has nonnegative curvature, but the manifolds M_k^4 given by the connected sum of k copies of $S^2 \times S^2$ do not admit almost nonnegative curvature as soon as their total Betti number is bigger than $C_2(4)$.

By work of Gromov and Gallot (see [17, 22]) one further has:

Theorem 4.3. *The* $\hat{A}$*-genus of a closed spin manifold* X *of almost nonnegative Ricci curvature is bounded by* $\hat{A}(X) \leqslant 2^{\dim(X)/2}$.

Work of Schoen and LeBrun on scalar curvature functionals (see [33, 41]) yields yet another obstruction:

Theorem 4.4. *If a closed manifold has negative Yamabe constant, then it cannot volume collapse with scalar curvature bounded from below. In particular, no such manifold can be almost nonnegatively curved.*

Example 4.5. Complex algebraic surfaces of general type have negative Yamabe constant (see [32]) and thus do not admit almost nonnegative curvature.

Along with his fibration theorem, in [46] Yamaguchi also showed:

Theorem 4.6. *If $M = M^m$ is an almost nonnegatively curved m-manifold, then a finite cover of M fibers over a flat $b_1(M;\mathbb{R})$-dimensional torus, and M^m is diffeomorphic to a torus if $b_1(M;\mathbb{R}) = m$.*

Example 4.7. Closed manifolds with infinite fundamental group and nonvanishing Euler characteristic do not admit almost nonnegative curvature.

Remark 4.8. Important research on manifolds of almost nonnegative Ricci curvature and Gromov–Hausdorff convergence under lower bounds on Ricci curvature has in particular been conducted by Colding and Cheeger–Colding (compare here especially the articles [9–12, 14]). As regards the second statement in Theorem 4.6, in [10] Cheeger and Colding showed that an almost nonnegatively Ricci curved manifold with maximal first Betti number is diffeomorphic to a torus.

In [16] Fukaya and Yamaguchi studied in detail the fundamental groups of almost nonnegatively curved manifolds and proved the following version of Gromov's conjecture from [20] that manifolds with almost nonnegative Ricci curvature have almost nilpotent fundamental groups:

Theorem 4.9. *Let $M = M^m$ be an almost nonnegatively curved m-manifold. Then the fundamental group $\pi_1(M)$ is almost nilpotent, i.e., contains a nilpotent subgroup of finite index. Moreover, $\pi_1(M)$ is $C(m)$-solvable in the sense that it contains a solvable subgroup of index at most $C(m)$.*

Theorem 4.9 was sharpened by Kapovitch, Petrunin and the present author in [31] as follows:

Theorem 4.10. *Let $M = M^m$ be an almost nonnegatively curved m-manifold. Then $\pi_1(M)$ is $C(m)$-nilpotent, i.e., $\pi_1(M)$ contains a nilpotent subgroup of index at most $C(m)$.*

Example 4.11. For any $C > 0$ there exist prime numbers $p > q > C$ and a finite group G_{pq} of order pq which is solvable but not nilpotent. In particular, G_{pq} does not contain any nilpotent subgroup of index less than or equal to C.

Whereas none of the results mentioned so far excludes G_{pq} from being the fundamental group of some almost nonnegatively curved m-manifold, Theorem 4.10 shows that for $C > C(m)$ none of the groups G_{pq} can be realized as the fundamental group of such a manifold.

Remark 4.12. Kapovitch and Wilking have recently given a proof of Theorem 4.10 for manifolds of almost nonnegative Ricci curvature (and, moreover, also a proof that the fundamental groups of closed m-manifolds with a lower bound on Ricci curvature and an upper bound on diameter have a presentation with a universally bounded number of generators and relations). The preprint is now available on the arxiv as arXiv1105:5955v2.

The last result we would like to mention in this section concerns a relation between almost nonnegative curvature and the actions of the fundamental group on the higher homotopy groups.

Recall that an action by automorphisms of a group G on an abelian group V is called nilpotent if V admits a finite sequence of G-invariant subgroups $V = V_0 \supset V_1 \supset \ldots \supset V_k = 0$ such that the induced action of G on V_i / V_{i+1} is trivial for any i. A connected CW-complex X is called *nilpotent* if $\pi_1(X)$ is a nilpotent group that operates nilpotently on $\pi_k(X)$ for every $k \geq 2$. One then has, also from [31], the following:

Theorem 4.13. *Let M be an almost nonnegatively curved manifold. Then a finite cover of M is a nilpotent space.*

Example 4.14. Let h: $S^3 \times S^3 \to S^3 \times S^3$ be the diffeomorphism defined by $h : (x, y) \mapsto (xy, yxy)$. The induced map h_* on $\pi_3(S^3 \times S^3)$ is given by a matrix A_h whose eigenvalues are different from one in absolute value. Let M be the mapping cylinder of h. Then M has the structure of a fiber bundle $S^3 \times S^3 \to M \to S^1$, and the action of $\pi_1(M) \cong \mathbb{Z}$ on $\pi_3(M) \cong \mathbb{Z}^2$ is generated by A_h. In particular, M is not a nilpotent space and hence, by Theorem 4.13, it does not admit almost nonnegative curvature.

5 Almost Nonnegative Curvature Operator

If one does not impose conditions on the fundamental group, it is a well-known fact that starting from dimension three the classes of closed positively respectively nonnegatively respectively almost nonnegatively curved manifolds are strictly contained in each other. Whether, however, within the category of closed *simply connected* manifolds the classes of almost nonnegatively curved and nonnegatively curved or even positively curved spaces do or do not coincide, is a completely open problem.

On the other hand, if one replaces sectional curvature by the curvature *operator* on tangent bivectors, then the respective classifications (see [3] as well as the survey [34] with the further references given there) of manifolds with positive resp. nonnegative curvature operator show that here distinctions can be made. In fact, starting from dimension $m \geq 4$, the class of closed simply connected m-manifolds with nonnegative curvature operator is always strictly bigger than the corresponding positive curvature operator one (which by [3] consists just of the standard m-sphere).

How about these problems when studying closed manifolds with almost nonnegative curvature operator?

First of all, all almost flat manifolds fall inside this class, and being almost flat is in fact equivalent to having almost flat curvature operator. Thus, in particular, from dimension three on there are many closed manifolds with almost nonnegative curvature operator which do not admit nonnegative curvature operator (nor just nonnegative sectional curvature).

Let us now, however, look at this problem for simply connected manifolds. In contrast to the sectional curvature case here an answer is known. Namely, from recent work of Sebastian [44] as well as from [45] one can infer the following:

Theorem 5.1. *There exist closed simply connected Riemannian manifolds with almost nonnegative curvature operator which do not admit any metric with nonnegative curvature operator.*

6 Conjectures and Questions

We conclude this article with a number of open questions and conjectures.

Recall first that in rational homotopy theory a *simply connected* topological space S is called *rationally elliptic* if it is homotopy equivalent to a finite CW-complex and if $\dim \pi_*(S, \mathbb{Q}) < \infty$.

A famous conjecture attributed to R. Bott (see [24]) states that simply connected closed nonnegatively curved manifolds are rationally elliptic. This conjecture was extended by Grove to almost nonnegatively curved manifolds (see [8]), and it seems natural to include in it almost nonnegatively curved manifolds in the generalized sense as well.

In order to also cover here manifolds with infinite fundamental groups, one may, following B. Totaro, employ the following generalized definition of rationally elliptic spaces:

A connected topological space S is *rationally elliptic* if it is homotopy equivalent to a finite CW complex, it has a finite covering which is a nilpotent space, and its universal covering is rationally elliptic in the ordinary sense.

With this definition one can extend Bott's original conjecture as well as Grove's version of it to manifolds which are not simply connected as follows (compare [31]):

Conjecture 6.1. Any manifold which is almost nonnegatively curved in the generalized sense is rationally elliptic.

Notice that Theorem 4.13 implies that Conjecture 6.1 is true in full generality if and only if it is true for simply connected manifolds.

One interesting potential way to approach Conjecture 6.1 is given as follows. It has been shown by Paternain and Petean (see [38]) that nilpotent closed manifolds which admit Riemannian metrics with zero topological entropy have rationally elliptic universal coverings. Together with Theorem 4.13, this means that Conjecture 6.1 would follow from a positive answer to the following question:

Question 6.2. Do manifolds with almost nonnegative curvature in the generalized sense always admit a Riemannian metric with zero topological entropy?

Let us now turn to some open conjectures which concern the fundamental groups of (almost) nonnegatively and positively curved spaces. It is a classical fact that nonnegatively curved manifolds have almost abelian ones, and Fukaya and Yamaguchi conjectured here the following (see [16]):

Conjecture 6.3. The fundamental group of an nonnegatively curved m-manifold is $C(m)$-abelian.

In this regard in [31] the following two conjectures were posed:

Conjecture 6.4. There exists $C = C(m)$ such that if M^m is almost nonnegatively curved, then there is a nilpotent subgroup $N \subset \pi_1(M)$ of index $\leqslant C$ whose torsion is contained in its center (or, at least, whose torsion is abelian).

Conjecture 6.5. If M^m is almost nonnegatively curved, then the action of $\pi_1(M)$ on $\pi_2(M)$ is almost trivial (or maybe even $C(m)$-trivial), i.e., there exists a finite index subgroup of $\pi_1(M)$ (or, respectively, a subgroup of index $\leqslant C(m)$) which acts trivially on $\pi_2(M)$.

Since the fundamental group of a closed positively curved m-manifold is finite, Conjecture 6.4 would immediately imply that such a fundamental group has to be $C(m)$-abelian.

As was pointed out by Wilking, if true, Conjecture 6.4 would imply a positive answer to Conjecture 6.5, and Conjectures 6.4 and 6.5 are also related to the following conjecture of Rong (see [39, 40]) which has been proved in [40] under the additional assumption of a uniform upper curvature bound:

Conjecture 6.6. Positively curved m-manifolds have $C(m)$-cyclic fundamental groups.

Here is one last more "$C(m)$ question". It arises naturally from Theorem 4.13:

Question 6.7. Is it true that almost nonnegatively curved m-manifolds M^m are $C(m)$-nilpotent spaces?

And, of course, one would also like to know the answer to the following questions.

Question 6.8. Is it true that manifolds which are almost nonnegatively curved in the generalized sense are almost nonnegatively curved in the standard sense?

Question 6.9. Does Theorem 4.13 generalize to manifolds with almost nonnegative Ricci curvature?

Finally, in turning once more to manifolds with almost nonnegative curvature operator, in view of Theorem 3.9 it is reasonable to ask:

Question 6.10. Let M be manifold with almost nonnegative curvature operator. Is it true that a finite cover $\tilde{M}$ of M is the total space of a fiber bundle

$$F \to \tilde{M} \to N$$

over a nilmanifold N with a simply connected fiber F which admits almost nonnegative curvature operator?

Question 6.11. Are there any closed simply connected manifolds with almost nonnegative sectional curvature which do not admit almost nonnegative curvature operator?

References

1. Alexander, S., Kapovitch, V., Petrunin, A.: Alexandrov Geometry. Monograph. In preparation.
2. Anderson, M.T.: Hausdorff perturbations of Ricci-flat manifolds and the splitting theorem. Duke Math. J. **68**(1):67–82 (1992)
3. Böhm, C., Wilking, B.: Manifolds with positive curvature operators are space forms. Ann. Math. (2) **167**(3), 1079–1097 (2008)
4. Bredon, G.: Introduction to Compact Transformation Groups. Academic Press, New York (1972)
5. Brieskorn, E.: Beispiele zur Differentialtopologie von Singularitäten. Invent. Math. **2**, 1–14 (1966)
6. Browder, W.: The Kervaire invariant of framed manifolds and its generalization. Ann. Math. (2) **90**, 157–186 (1969)
7. Burago, Y., Gromov, M., Perelman, G.: A. D. Alexandrov's spaces with curvatures bounded from below I. Uspechi Mat. Nauk **47**, 3–51 (1992)
8. Chang, S.-Y.A., Yang, P.C., Grove, K., Wolfson, J.G.: Conformal., Riemannian and Lagrangian geometry. In: Freire A. (eds.) Papers from the 30th John H. Barrett Memorial Lectures delivered at the University of Tennessee, Knoxville, TN, May 2000. University Lecture Series, vol. 27, pp. x+85. American Mathematical Society, Providence (2002). ISBN: 0-8218-3210-753-06, MR1911593 (2003c:53001)
9. Cheeger, J., Colding, T.: Lower bounds on Ricci curvature and the almost rigidity of warped products. Ann. Math. **144**, 189–237 (1996)
10. Cheeger, J., Colding, T.: On the structure of spaces with Ricci curvature bounded below. I. J. Differ. Geom. **46**, 406–480 (1997)
11. Cheeger, J., Colding, T.: On the structure of spaces with Ricci curvature bounded below. II. J. Differ. Geom. **54**(1), 13–35 (2000)
12. Cheeger, J., Colding, T.: On the structure of spaces with Ricci curvature bounded below. III. J. Differ. Geom. **54**(1), 37–74 (2000)
13. Cheeger, J., Fukaya, K., Gromov, M.: Nilpotent structures and invariant metrics on collapsed manifolds. J. Am. Math. Soc. **5**(2), 327–372 (1992)
14. Colding, T.: Ricci curvature and volume convergence. Ann. Math. (2) **145**(3), 477–501 (1997)
15. Fukaya, K.: Metric Riemannian geometry. In: Handbook of Differential Geometry, vol. II, pp. 189–313. Elsevier, Amsterdam (2006)
16. Fukaya, K., Yamaguchi, T.: The fundamental groups of almost nonnegatively curved manifolds. Ann. Math. (2) **136**(2), 253–333 (1992)
17. Gallot, S.: Inégalités isopérimétriques, courbure de Ricci et invariants géométriques. II. C. R. Acad. Sci. Paris Sér. I Math. **296**(8), 365–368 (1983)
18. Gromov, M.: Almost flat manifolds. J. Differ. Geom. **13**, 231–241 (1978)
19. Gromov, M.: Synthetic geometry in Riemannian manifolds. In: Proceedings of the International Congress of Mathematicians, Helsinki, pp. 415–419 (1978), Acad. Sci. Fennica, Helsinki, (1980)
20. Gromov, M.: Structures métriques pour les variétés riemanniennes. In: Lafontaine, J., Pansu, P. (eds.) Textes Mathématiques, vol. 1, CEDIC, Paris (1981)

21. Gromov, M.: Curvature, diameter and Betti numbers. Comment. Math. Helv. **56**, 179–195 (1981)
22. Gromov, M.: Volume and bounded cohomology. Inst. Hautes Études Sci. Publ. Math. **56**, 5–99 (1982)
23. Gromov, M.: Metric Structures for Riemannian and Non-Riemannian Spaces. Progress in Mathematics, vol. 152. Birkhäuser, Boston (1999)
24. Grove, K., Halperin, S.: Contributions of rational homotopy theory to global problems in geometry. Inst. Hautes Études Sci. Publ. Math. **56**, 171–178 (1982)
25. Grove, K., Verdiani, L., Wilking, B., Ziller, W.: Non-negative curvature obstructions in cohomogeneity one and the Kervaire spheres. Ann. Sc. Norm. Super. Pisa Cl. Sci. (5) **5**(2), 159–170 (2006)
26. Grove, K., Ziller, W.: Curvature and symmetry of Milnor spheres. Ann. Math. (2) **152**, 331–36 (2000)
27. Grove, K., Ziller, W.: Cohomogeneity one manifolds with positive Ricci curvature. Invent. Math. **149**, 619–646 (2002)
28. Hirzebruch, F., Mayer, K.: $O(n)$-Mannigfaltigkeiten, exotische Sphären und Singularitäten. Lecture Notes in Mathematics, vol. 57, Springer, Berlin-New York (1968)
29. Hsiang, W.-C., Hsiang, W.-Y.: On compact subgroups of the diffeomorphism groups of Kervaire spheres. Ann. Math. (2) **85**, 359–369 (1967)
30. Kapovitch, V.: Perelman's stability theorem. In: Cheeger, J., Grove, K. (eds.) Metric and comparison geometry. Surveys in Differential Geometry, vol. 11, pp. xii+347. International Press, Somerville (2007). ISBN: 978-1-57146-117-9
31. Kapovitch, V., Petrunin, A., Tuschmann, W.: Nilpotency, almost nonnegative curvature, and the gradient flow on Alexandrov spaces. Ann. Math. (2) **171**(1), 343–373 (2010)
32. LeBrun, C.: Kodaira dimension and the Yamabe problem. Comm. Anal. Geom. **7**(1), 133–156 (1999)
33. LeBrun, C.: Ricci curvature, minimal volumes, and Seiberg–Witten theory. Invent. Math. **145**(2), 279–316 (2001)
34. Mercuri, F., Noronha, M.H.: On the topology of complete Riemannian manifolds with nonnegative curvature operator. Rend. Sem. Fac. Sci. Univ. Cagliari **63**(2), 149–171(1993)
35. Milnor, J.: Singular Points of Complex Hypersurfaces. Annals of Mathematics Studies, No. 61. Princeton University Press, Princeton, N.J.; University of Tokyo Press, Tokyo (1968)
36. Perelman, G.: Alexandrov spaces with curvatures bounded from below II. Preprint (1991)
37. Petrunin, A.: Semiconcave functions in Alexandrov's geometry. In: Cheeger, J., Grove, K. (eds.) Metric and comparison geometry. Surveys in Differential Geometry, vol. 11, pp. xii+347. International Press, Somerville (2007). ISBN: 978-1-57146-117-9
38. Paternain, G., Petean, J.: Zero entropy and bounded topology. Comment. Math. Helv. **81**(2), 287–304 (2006)
39. Rong, X.: On the fundamental groups of manifolds of positive sectional curvature. Ann. Math., II. Ser. **143**(2), 397–411 (1996)
40. Rong, X.: The almost cyclicity of the fundamental groups of positively curved manifolds. Invent. Math. **126**(1), 47–64 (1996)
41. Schoen, R.M.: Variational theory for the total scalar curvature functional for Riemannian metrics and related topics. In: Giaquinta, M. (ed.) Topics in calculus of variations. Lectures given at the Second 1987 C.I.M.E. Session held in Montecatini Terme, July 20-28, 1987. Lecture Notes in Mathematics, 1365, pp. x+196. Springer, Berlin (1989). ISBN: 3-540-50727-2
42. Schwachhöfer, L. Tuschmann, W.: Almost Nonnegative Curvature and Cohomogeneity One. Max Planck Institute for Mathematics in the Sciences Preprint Series, Preprint no. 62/2001, http://www.mis.mpg.de (2001)
43. Schwachhöfer, L.J., Tuschmann,W.: Metrics of positive Ricci curvature on quotient spaces. Math. Ann. **330**(1), 59–91 (2004)

44. Sebastian, D.: Konstruktionen von Mannigfaltigkeiten mit fast nichtnegativem Krümmungsoperator. Dissertationsschrift, Karlsruher Institut für Technologie (2011)
45. Sebastian, D., Tuschmann, W.: Manifolds with almost nonnegative curvature operator. In preparation.
46. Yamaguchi, T.: Collapsing and pinching under a lower curvature bound. Ann. Math. **133**, 317–357 (1991)
47. Yamaguchi, T.: A convergence theorem in the geometry of Alexandrov spaces. In: Actes de la Table Ronde de Géométrie Differentielle, Séminaires & Congrès, vol. 1, pp. 601–642. Soc. Math. France, Paris, (1996)

Algebraic Integral Geometry

Andreas Bernig

Abstract A survey on recent developments in (algebraic) integral geometry is given. The main focus lies on algebraic structures on the space of translation invariant valuations and applications in integral geometry.

1 Algebraic Integral Geometry

Algebraic integral geometry is a relatively modern part of integral geometry. It aims at proving geometric formulas (kinematic formulas, Crofton formulas, Brunn-Minkowski-type inequalities etc.) by taking a structural viewpoint and employing various algebraic techniques, including abstract algebra, Lie algebras and groups, finite- and infinite-dimensional representations, classical invariant theory, Gröbner bases, cohomology theories, algebraic geometry and so on.

The situation can be roughly compared to symbolic integration. In order to integrate a given (say sufficiently elementary) function, there is no need to know anything about the definition of the integral. It suffices to know a certain number of integration rules, like partial integration and the substitution rule. In algebraic integral geometry, the corresponding rules for computing geometric integrals are worked out. The fundamental theorem of algebraic integral geometry is one of these rules.

The main object of the theory is the *space of all translation invariant valuations*. Here the emphasis is on *space*, since in general not a single valuation but the set of all valuations is studied. Roughly speaking, this space is a graded commutative algebra satisfying Poincaré duality and Hard Lefschetz theorem. Moreover, there is

A. Bernig (✉)
Institut für Mathematik, Goethe-Universität Frankfurt, Robert-Mayer-Str. 10, 60054 Frankfurt, Germany
e-mail: bernig@math.uni-frankfurt.de

C. Bär et al. (eds.), *Global Differential Geometry*, Springer Proceedings in Mathematics 17, DOI 10.1007/978-3-642-22842-1_5,

a Fourier transform and a convolution product on this algebra and all these algebraic structures reflect geometric properties and formulas.

Among the most important contributions to algebraic integral geometry are Nijenhuis observation (6) (who moreover speculated about a possible algebraic structure explaining it), the theorems by Hadwiger (Theorem 2.3), P. McMullen (Theorem 3.1), Klain (Sect. 3.2) and Schneider (see Sect. 3.3). A spectacular breakthrough was achieved by Alesker in 2001, who proved the *McMullen conjecture* (in fact a much stronger version of it, see Sect. 3.4) and subsequently introduced many of the algebraic structures mentioned above.

The structure of the present paper is as follows.

After a short reminder of some classical integral-geometric formulas in Sect. 2, we will explain the new algebraic tools in Sect. 3.

The transition between algebra and geometry is done in Sect. 4, where the theoretical background for integral geometry of subgroups of $\mathrm{SO}(n)$ is given.

In Sect. 5, this program is carried out in a special and important case, namely for the group $G = \mathrm{U}(n)$, yielding *hermitian integral geometry*. Section 6 gives an overview of integral geometry for other groups and three important open problems are stated in Sect. 7.

The reader is invited to read J. Fu's survey [28] which has some non-empty intersection with the present paper.

2 Classical Integral-Geometric Formulas

Let us fix some notations for the rest of the paper. The n-dimensional unit ball is denoted by B. Let ω_n be its volume. The *flag coefficients* are defined by

$$\begin{bmatrix} n \\ k \end{bmatrix} := \binom{n}{k} \frac{\omega_n}{\omega_k \omega_{n-k}}.$$

For an odd number $2m+1$, we set

$$(2m+1)!! = 1 \cdot 3 \cdot 5 \cdot \ldots \cdot (2m+1)$$

and use the convention $(-1)!! = 1$.

If V is a finite-dimensional Euclidean vector space and G is any subgroup of $\mathrm{SO}(V)$, we let $\bar{G}$ be the group generated by G and translations. This group has a canonical measure, which is the product of the Haar probability measure on G and the Lebesgue measure on the translation group. In particular, the measure of the set $\{\bar{g} \in \bar{G} : \bar{g}(x) \in K\}$, $x \in V$ equals the volume of K for every compact convex set K.

We will use the following terminology: *subspaces* will always be *linear subspaces*, while *planes* will always be *affine planes*. The *Grassmann manifold* of all

k-dimensional subspaces in V is denoted by $\mathrm{Gr}_k V$. The *affine Grassmann manifold* of all k-planes is denoted by $\overline{\mathrm{Gr}_k} V$.

2.1 Valuations

Throughout this paper, V denotes a finite-dimensional vector space. The space of non-empty compact convex subsets in V is denoted by $\mathcal{K}(V)$. With respect to *Minkowski addition*

$$K + L = \{x + y | x \in K, y \in L\},$$

$\mathcal{K}(V)$ is a semigroup. This space has a natural topology, called the *Hausdorff-topology* which is defined as follows:

$$d_H(K, L) := \inf_{r \geq 0} \{K \subset L + rB, L \subset K + rB\}, \quad K, L \in \mathcal{K}(V).$$

Here B is the unit ball for some euclidean scalar product. The metric d_H depends on the choice of this scalar product, but the induced topology does not.

Definition 2.1. Let A be a semigroup. A functional $\mu : \mathcal{K}(V) \to A$ is called a valuation if

$$\mu(K \cup L) + \mu(K \cap L) = \mu(K) + \mu(L)$$

whenever $K, L, K \cup L \in \mathcal{K}(V)$.

The case of $A = \mathbb{R}$ (or $A = \mathbb{C}$) is the most important one. Everyone is familiar with at least two examples of real-valued valuations. The first one is the constant valuation $\mu(K) = 1$ for all $K \in \mathcal{K}(V)$. This valuation is called *Euler characteristic* and denoted by χ. The name needs some explanation: in fact there is a canonical way to extend this valuation to finite unions of compact convex sets, and this extension equals the Euler characteristic with respect to Borel-Moore homology.

The second familiar example of a valuation is the volume, which we denote by vol. Note that this valuation depends on the choice of a Euclidean metric on V (or at least on the choice of a Lebesgue measure).

A particularly important class of valuations is that of continuous, translation invariant valuations. The valuation μ is called translation invariant if $\mu(K + t) = \mu(K)$ for all $t \in V$. Euler characteristic and volume clearly have this property. We will see later on that all continuous, translation invariant valuations arise in some way from these two basic ones.

Definition 2.2. The space of complex-valued, continuous, translation invariant valuations is denoted by Val. If G is a subgroup of $\mathrm{GL}(V)$, then

$$\mathrm{Val}^G(V) := \{\mu \in \mathrm{Val} \,|\, \mu(gK) = \mu(K) \quad \forall g \in G\}.$$

Before studying the space Val, let us give some other important examples of valuations, which do not belong to Val.

First of all, non-continuous valuations (on the space of polytopes), the *Dehn functionals*, played a central role in Dehn's solution of Hilbert's 3rd problem. Another famous non-continuous example is the *affine surface area*, which is semi-continuous (see [44] and [43] and the references therein for more information).

Since $\mathcal{K}(V)$, endowed with the Minkowski addition, is a semigroup, we may take $A = \mathcal{K}(V)$ in Definition 2.1. It is easy to see that $\mu(K) = K$ defines a valuation. More interesting examples are the *intersection body operator* (defined on a subset of $\mathcal{K}(V)$) and the *projection body operator*. See [2,41,42,55] for more information.

If $A = Sym^* V$, the space of symmetric tensors over V, then A-valued valuations are called *tensor valuations*. Their study has been initiated by McMullen [47] and Alesker [4]. Recently, remarkable progress in the study of kinematic formulas for tensor valuations was made by Hug, Schneider and R. Schuster [35, 36]. One may hope and expect that some algebraic tools will be useful in simplifying their formulas.

2.2 Intrinsic Volumes

Let V be a Euclidean vector space of dimension n. At the heart of integral geometry are the *intrinsic volumes* $\mu_0, \dots, \mu_n$. We give four equivalent definitions.

First of all, we may use *projections* onto lower-dimensional subspaces. For $0 \leq k \leq n$, the group $\mathrm{SO}(V)$ acts transitively on the Grassmannian $\mathrm{Gr}_k(V)$ of k-dimensional subspaces in V. We endow this manifold with the unique invariant probability measure dL. Then

$$\mu_k(K) := \begin{bmatrix} n \\ k \end{bmatrix} \int_{\mathrm{Gr}_k(V)} \mathrm{vol}_k(\pi_L K)dL \tag{1}$$

defines an element $\mu_k \in \mathrm{Val}^{\mathrm{SO}(V)}$. Here vol_k denotes the k-dimensional Lebesgue measure on the subspace L and $\pi_L K$ is the orthogonal projection of K onto L. This formula (and some more general versions) is called *Kubota formula*.

For the second definition, we use *intersections* instead of projections. We let $\overline{\mathrm{Gr}_k}(V)$ be the k-dimensional affine Grassmannian on which we use the unique $\overline{\mathrm{SO}(V)}$-invariant measure dE such that the measure of planes intersecting the unit ball equals ω_{n-k}. Then we set

$$\mu_k(K) := \begin{bmatrix} n \\ k \end{bmatrix} \int_{\overline{\mathrm{Gr}}_{n-k}(V)} \chi(K \cap E)dE. \tag{2}$$

The equivalence of this definition with the previous one is an elementary exercise. Formula (2) is called *Crofton formula*. More general Crofton formulas play an important role in algebraic integral geometry, see Sect. 3.5.

The third description of the μ_k is rather a characterization than a definition. Looking at the μ_k defined as above, one sees that

1. μ_k is a continuous, translation invariant and $\mathrm{SO}(V)$-invariant valuation,
2. μ_k is of degree k, i.e. $\mu_k(tK) = t^k \mu_k(K)$ for all $t \geq 0$ and
3. The restriction of μ_k to a k-plane equals the k-dimensional Lebesgue measure on that plane.
4. μ_k is even, i.e. $\mu_k(-K) = \mu_k(K)$.

In fact, the μ_k are uniquely characterized by these properties, as we will see in Sect. 3.2.

Some of the μ_k are well-known: μ_0 is the Euler characteristic χ which was mentioned in the introduction. μ_n is the usual Lebesgue measure, μ_{n-1} is half of the surface area and μ_1 is a constant times the *mean width*.

The intrinsic volumes may also be defined by the *Steiner formula*. For $t \geq 0$, let $K + tB$ be the t-tube around K. Then $\mathrm{vol}(K + tB)$ turns out to be a polynomial in t given by

$$\mathrm{vol}(K + tB) = \sum_{k=0}^{n} \mu_{n-k}(K)\omega_k t^k. \tag{3}$$

Taking $K = B$, we easily get

$$\mu_k(B) = \binom{n}{k} \frac{\omega_n}{\omega_{n-k}}. \tag{4}$$

2.3 Kinematic Formulas

The most important formulas of integral geometry are the *kinematic formulas*:

$$\int_{\mathrm{SO}(V)} \mu_i(K \cap \bar{g}L) d\bar{g} = \sum_{k,l} c_{k,l}^i \mu_k(K)\mu_l(L), \tag{5}$$

where

$$c_{k,l}^i = \begin{cases} \begin{bmatrix} n+i \\ i \end{bmatrix} \begin{bmatrix} n+i \\ k \end{bmatrix}^{-1} & k+l = n+i \\ 0 & k+l \neq n+i. \end{cases}$$

It may be checked that the constants on the right hand side are correct by plugging in balls of different radii (template method, see below).

Looking at the formula, one makes the following observations. Since

$$\begin{bmatrix} n+i \\ k \end{bmatrix} = \begin{bmatrix} n+i \\ n+i-k \end{bmatrix},$$

the coefficients on the right hand side are symmetric, i.e. $c_{k,l}^i = c_{l,k}^i$. This reflects of course the fact that changing the role of K and L in the integral on the left hand side of the formula does not change its value. Next, we observe that the *total degree* $n+i$ at the right hand side is the degree i on the left hand side plus the dimension of the ambient space. Another symmetry property for the coefficients comes from Fubini's theorem:

$$\int_{\overline{SO(V)}} \int_{\overline{SO(V)}} \mu_i(K \cap \bar{g}L \cap \bar{h}M) d\bar{g} d\bar{h} = \int_{\overline{SO(V)}} \int_{\overline{SO(V)}} \mu_i(K \cap \bar{g}L \cap \bar{h}M) d\bar{h} d\bar{g}.$$

This translates to

$$\sum_r c_{r,m}^i c_{k,l}^r = \sum_r c_{r,l}^i c_{k,m}^r.$$

Nijenhuis [49] made a less obvious observation: Renormalizing

$$\tilde{\mu}_k := \frac{\pi^n k! \omega_k}{\pi^k n! \omega_n} \mu_k,$$

the kinematic formula (5) becomes

$$\int_{\overline{SO(V)}} \tilde{\mu}_i(K \cap \bar{g}L) d\bar{g} = \sum_{k+l=n+i} \tilde{\mu}_k(K) \tilde{\mu}_l(L). \tag{6}$$

Hence all coefficients on the right hand side become 1.

At first glance, this may seem to be trivial, since we may change the constants on the right hand side to whatever we want by rescaling the μ_k. However, a closer look reveals that we only have $n+1$ free parameters (one for the scaling of each μ_k), but $\binom{n+2}{2}$ non-zero coefficients on the right hand side. Nijenhuis speculated that there exists some algebraic structure explaining this strange fact ("...the suggestion of an underlying algebra with the c's as structure constants was inevitable" [49]). It turns out that this is indeed the case, as we will see below.

An array of *additive kinematic formulas* arises if we replace intersection by Minkowski addition:

$$\int_{SO(V)} \mu_i(K + gL) dg = \begin{bmatrix} 2n-i \\ n-i \end{bmatrix} \sum_{k+l=i} \begin{bmatrix} 2n-i \\ n-k \end{bmatrix}^{-1} \mu_k(K) \mu_l(L). \tag{7}$$

In this case, there is an analogous statement as in Nijenhuis' observation: after renormalizing

$$\tilde{\mu}_k := \frac{(n-k)! \omega_{n-k}}{n! \omega_n} \mu_k, k = 0, \ldots, n,$$

the additive kinematic formula (7) reads

$$\int_{SO(V)} \tilde{\mu}_i(K + gL) dg = \sum_{k+l=i} \tilde{\mu}_k(K) \tilde{\mu}_l(L).$$

We will see later an explanation of this fact too. It will also turn out that the usual and the additive kinematic formula are *dual to each other* in a precise sense and one can be derived from the other.

2.4 Hadwiger's Theorem

We have already seen that $\mu_k \in \mathrm{Val}^{SO(V)}$. Hadwiger's theorem states conversely that *all* valuations in $\mathrm{Val}^{SO(V)}$ are obtained by linear combinations of intrinsic volumes.

Theorem 2.3. *The vector space* $\mathrm{Val}^{SO(V)}$ *of continuous, translation invariant,* $SO(V)$*-invariant valuations on a Euclidean vector space* V *of dimension* n *has the basis*

$$\mu_0, \mu_1, \ldots, \mu_n.$$

An elementary and nice proof may be found in [40].

Hadwiger's theorem is quite powerful. It leads to all formulas which we have stated before. Indeed, let us look for instance at the additive kinematic formula (7). For each fixed body L, the left hand side of this formula is a valuation in K. It is easy to prove that this valuation belongs to $\mathrm{Val}^{SO(V)}$, hence it may be written in the form $\sum_{k=0}^n d_k(L)\mu_k(K)$. Next, fixing K, one easily gets that d_k is also an element of $\mathrm{Val}^{SO(V)}$ for each fixed k, hence $d_k(L) = \sum_{l=0}^n d_{kl}\mu_l(L)$ with complex numbers d_{kl}. We thus know that

$$\int_{SO(V)} \mu_i(K + gL)dg = \sum_{k,l=0}^n d_{kl}^i \mu_k(K)\mu_l(L)$$

for some fixed constants d_{kl}^i. There is a nice trick to determine these constants, which is called the *template method*. We plug in on both sides of the equation special convex bodies K and L for which we may compute the integral on the left hand side and the intrinsic volumes on the right hand side to obtain a system of linear equations on the d_{kl}^i. Solving this system yields the d_{kl}^i. More precisely, let us take $K = rB, L = sB$ (where B is as always the unit ball). The left hand side equals $\mu_i((r+s)B) = (r+s)^i \binom{n}{i}\frac{\omega_n}{\omega_{n-i}}$. The right hand side equals

$$\sum_{k,l} d_{k,l}^i r^k \binom{n}{k}\frac{\omega_n}{\omega_{n-k}} s^l \binom{n}{l}\frac{\omega_n}{\omega_{n-l}}.$$

Comparing the coefficients of $r^j s^{i-j}$ on both sides gives us

$$\binom{i}{j}\binom{n}{i}\frac{\omega_n}{\omega_{n-i}} = d_{j,i-j}^i \binom{n}{j}\frac{\omega_n}{\omega_{n-j}} \binom{n}{i-j}\frac{\omega_n}{\omega_{n-i+j}},$$

which is (7).

2.5 General Hadwiger Theorem

The last theorem from classical integral geometry which we want to mention is the *general Hadwiger theorem*. It applies to more general valuations, but we will state (and prove) it only in the special case of translation invariant valuations.

Theorem 2.4. *Let* $\phi \in \mathrm{Val}$. *Then*

$$\int_{\overline{\mathrm{SO}(V)}} \phi(K \cap \bar{g}L) d\bar{g} = \sum_{l=0}^{n} c_l(K)\mu_l(L),$$

where

$$c_l(K) := \int_{\overline{\mathrm{Gr}}_{n-l}} \phi(K \cap E) dE.$$

The theorem can be proved using Hadwiger's characterization theorem and a limit argument. We will give another, more conceptual proof, in Sect. 4.3.

3 Algebraic Structures on Valuations

The main object of algebraic integral geometry is the space $\mathrm{Val} = \mathrm{Val}(V)$ of continuous, translation invariant valuations on an n-dimensional vector space V. This space has a surprisingly rich algebraic structure which we are going to describe in this section.

3.1 McMullen's Decomposition

A valuation μ is *of degree* k if $\mu(tK) = t^k \mu(K)$ for all $t \geq 0$ and all K. It is *even* if $\mu(-K) = \mu(K)$ and *odd* if $\mu(-K) = -\mu(K)$. The corresponding subspaces of Val are denoted by $\mathrm{Val}_k^+, \mathrm{Val}_k^-$.

McMullen [45] proved the following decomposition:

Theorem 3.1.

$$\mathrm{Val} = \bigoplus_{\substack{k=0,\dots,n \\ \epsilon=\pm}} \mathrm{Val}_k^\epsilon . \tag{8}$$

In particular, the degree of a valuation is an integer between 0 and the dimension of the ambient space. We refer to (8) as the *McMullen grading*.

McMullen's theorem allows us to introduce a Banach space structure on Val by setting

$$\|\mu\| := \sup \{|\mu(K)| : K \in \mathcal{K}(V), K \subset B\}.$$

Then (Val, $\|\cdot\|$) is a Banach space. Choosing another scalar product on V gives an equivalent norm. Hence we get a uniquely defined *Banach space structure* on Val.

3.2 Klain Embedding

We now suppose that we have a fixed Euclidean structure on V. This is not strictly necessary but simplifies the exposition.

Klain found a nice way to describe *even* continuous, translation invariant valuations. In [39], he first characterized the volume as the only valuation (up to a multiplicative constant) which is continuous, translation invariant, even and *simple* (i.e. vanishing on lower-dimensional sets).

Klain's characterization of the volume implies that given $\mu \in \mathrm{Val}_k^+$ and a k-dimensional subspace E, the restriction $\mu|_E$ is a multiple $\mathrm{Kl}_\mu(E)$ of the k-dimensional volume on E. Indeed, this follows once we know that $\mu|_E$ is simple. If $F \subset E$ is a subspace of minimal dimension such that $\mu|_F \neq 0$, then $\mu|_F$ is simple and hence a multiple of the volume on F. Since μ is of degree k, this is only possible if $E = F$.

The continuous function

$$\mathrm{Kl}_\mu : \mathrm{Gr}_k(V) \to \mathbb{C} \tag{9}$$

is called the *Klain function* of μ. The induced map

$$\mathrm{Kl} : \mathrm{Val}_k^+ \hookrightarrow C(\mathrm{Gr}_k V)$$

is called the *Klain embedding*. To see that this map is indeed injective, we suppose that $\mathrm{Kl}_\mu = 0$ for some $\mu \in \mathrm{Val}_k^+$. Then the restriction of μ to any $(k+1)$-dimensional subspace F is simple, hence a multiple of the $(k+1)$-dimensional volume on F. Since μ is k-homogeneous, this is only possible if $\mu|_F = 0$. Iterating this procedure, we see that the restriction to any subspace of V (including V itself) vanishes, hence $\mu = 0$.

3.3 Schneider Embedding

The counterpart of Klain's embedding theorem for *odd* valuations was given by Schneider. He showed in [54] that an odd, simple, continuous, translation invariant valuation μ can be written as

$$\mu(K) = \int_{S(V)} f(v) dS_{n-1}(K, v),$$

where $S_{n-1}(K, \cdot)$ is the $(n-1)$-th surface area measure of K [53] and f is an odd function on the unit sphere in V (which will be denoted by $S(V)$). In particular, μ is of degree $n-1$.

The function f is unique up to linear functions. Equivalently, f is unique under the additional condition

$$\int_{S(V)} v f(v) dv = 0. \tag{10}$$

Similarly as in the even case, this implies a description of odd valuations of a given degree. Namely, suppose $\mu \in \mathrm{Val}_k^-$. Then μ vanishes on k-dimensional sets, hence the restriction $\mu|_E$ to a $(k+1)$-dimensional subspace E is simple and can be described by an odd function on the unit sphere of E satisfying the condition (10) with V replaced by E. To put these functions into one object, one can use the *partial flag manifold* $\mathrm{Flag}_{k+1,1}$ consisting of pairs (E, L), where $E \in \mathrm{Gr}_{k+1}(V)$ and L is an oriented line in E. Then the *Schneider function* is an odd function on $\mathrm{Flag}_{k+1,1}$ (i.e. a function that changes sign if (E, L) is replaced by $(E, -L)$). The space of continuous, odd functions on $\mathrm{Flag}_{k+1,1}$ is denoted by $C^{odd}(\mathrm{Flag}_{k+1,1}\, V)$.

The valuation μ is uniquely determined by its Schneider function, as follows easily by induction on the dimension. Hence we get an embedding (the *Schneider embedding*)

$$S : \mathrm{Val}_k^- \hookrightarrow C^{odd}(\mathrm{Flag}_{k+1,1}\, V).$$

3.4 Irreducibility Theorem and Smooth Valuations

Let V be an n-dimensional vector space. Without fixing a Euclidean structure on V, we still have the general linear group $\mathrm{GL}(V)$ acting on V and on $\mathcal{K}(V)$. This action induces an action on Val by

$$g\mu(K) := \mu(g^{-1}K),$$

which preserves degree and parity of a valuation.

Theorem 3.2. (Alesker's irreducibility theorem) *The spaces* $\mathrm{Val}_k^\epsilon, k = 0, \ldots, n, \epsilon = \pm$ *are irreducible* $\mathrm{GL}(V)$*-representations.*

We remind the reader that these spaces are in general infinite-dimensional Banach spaces and that in this context, *irreducible* means that they do not admit any non-trivial, invariant, closed subspaces.

One way to understand the statement of the theorem is as follows. Start with a non-zero valuation $\mu \in \mathrm{Val}_k^\epsilon$ and consider its orbit under the group $\mathrm{GL}(V)$, i.e. the set of all $g\mu$. Then the space of linear combinations of such valuations are dense in Val_k^ϵ, which means that every valuation in Val_k^ϵ may be approximated by these special ones.

The proof of Theorem 3.2 is contained in [5]. It uses the Klain-Schneider embedding as well as heavy machinery from representation theory.

Alesker's irreducibility theorem is of fundamental importance in algebraic integral geometry. Let us explain the reason for this.

If we give some construction of translation invariant valuations, which does not use any extra structure (like Euclidean metric), then we obtain a $\mathrm{GL}(V)$-invariant subspace of Val. By Alesker's irreducibility theorem, its intersection with any of the spaces Val_k^ϵ is either trivial or dense. From this, one obtains several characterization theorems for translation invariant valuations.

We will see three main examples for this construction. The first is a positive answer to a conjecture by McMullen [46].

Corollary 3.3. *Valuations of the form* $K \mapsto \mathrm{vol}(K+A)$*, where* vol *is any Lebesgue measure on* V *and* A *is a fixed convex body, span a dense subspace of* Val.

The proof follows from the trivial observation that valuations of the form $K \mapsto \mathrm{vol}(K+A)$ span a $\mathrm{GL}(V)$-invariant subspace of Val and that its intersection with each Val_k^ϵ is non-trivial.

For the second example, we need the notion of *conormal cycle* of a compact convex set. We suppose that V is oriented and let $S^*V = V \times S(V^*)$ be the spherical cotangent bundle of V, defined as follows. For $p \in V$, let T_p^*V be the dual of the tangent space at p. On the space $T_p^*V \setminus \{0\}$, there is an equivalence relation given by $\xi \sim \xi'$ if and only if $\xi = \lambda \xi'$ for some real $\lambda > 0$. The equivalence class of ξ is denoted by $[\xi]$.

The space S^*V consists of the pairs $(p,[\xi])$, where $p \in V, \xi \in T_p^*V \setminus \{0\}$. An element $(p,[\xi]) \in S^*V$ can be thought of as a pair (p,E), where $E = p + \ker \xi$ is an oriented affine hyperplane in V containing p.

The conormal cycle $N(K)$ of $K \in \mathcal{K}(V)$ is an oriented $(n-1)$-dimensional Lipschitz submanifold in S^*V. It is given by the set of all (p,E) such that $p \in \partial K$ and E is an oriented support plane of K at p.

If V is a Euclidean vector space, then we can identify S^*V with the sphere bundle SV. The image of the conormal cycle under this identification is the *normal cycle* of K.

Let ω be a translation invariant $(n-1)$-form on S^*V and ϕ be a translation invariant n-form on V. Then the valuation

$$K \mapsto \int_{N(K)} \omega + \int_K \phi \tag{11}$$

is a continuous, translation invariant valuation. A valuation in Val of this form is called *smooth* and Val^{sm} denotes the corresponding subspace. By Alesker's irreducibility theorem, Val^{sm} is a dense subspace of Val.

The representation (11) opens the door to another central fact of algebraic integral geometry: *Smooth valuations can be extended to a large class of compact non-convex sets*. Indeed, many compact sets $X \subset V$ admit a normal cycle $N(X)$ and (11) may be used to define $\mu(X)$. Examples of such sets are polyconvex sets (i.e. finite unions of convex sets), sets with positive reach, in particular smooth submanifolds (possibly with boundary or corner), compact sets which are definable in some o-minimal structure (see [59] for o-minimal structures and [20, 30] for the normal cycle of a definable set), in particular compact subanalytic or semialgebraic sets.

Smooth valuations are natural from the viewpoint of representation theory. Alesker's original definition of a smooth valuation in [6] was the following.

Theorem 3.4. (Alesker, [10]) *A valuation $\mu \in \mathrm{Val}$ is smooth if and only if the map*

$$\mathrm{GL}(V) \to \mathrm{Val}$$
$$g \mapsto g\mu$$

is smooth as a map from a Lie group to an infinite-dimensional Banach space.

The proof uses the *Casselman-Wallach theorem* from representation theory. Furthermore, there is a natural way to endow Val^{sm} with a Fréchet space topology which is finer than the induced topology.

The third application of Alesker's irreducibility theorem concerns *Crofton formulas* for even, homogeneous valuations. If m is a (signed) translation invariant measure on the affine Grassmannian manifold $\overline{\mathrm{Gr}}_{n-k}(V)$, then the valuation

$$\mu(K) := \int_{\overline{\mathrm{Gr}}_{n-k}(V)} \chi(K \cap E)dm(E) \tag{12}$$

is an element of Val_k^+. The signed measure m is called *Crofton measure* of μ. Since this construction is $\mathrm{GL}(V)$-invariant, it follows that the space of even, k-homogeneous valuations admitting such a Crofton measure is dense in Val_k^+. If we restrict to smooth Crofton measures (i.e. measures which are given by integration over some smooth top-dimensional translation invariant form on $\overline{\mathrm{Gr}}_{n-k}(V)$), then this subspace is precisely $\mathrm{Val}_k^{+,sm}$, i.e. the space of *smooth*, even, k-homogeneous valuations. For this last statement, the Casselman-Wallach theorem is used again, compare [16].

3.5 *Product*

One of the milestones of algebraic integral geometry is the introduction of a product structure on the space Val^{sm} by Alesker [9]. To define it, Alesker used his solution of McMullen's conjecture (Corollary 3.3). If A_1, A_2 are smooth convex bodies with positive curvature, the valuations

$$\phi(K) = \mathrm{vol}_n(K + A_1), \psi(K) = \mathrm{vol}_n(K + A_2) \tag{13}$$

are smooth and the Alesker product is defined by

$$\phi \cdot \psi(K) = \mathrm{vol}_{2n}(\Delta K + A_1 \times A_2), \tag{14}$$

where $\Delta : V \to V \times V$ is the diagonal embedding.

Alesker proved that this definition extends uniquely to a linear and continuous product

$$\mathrm{Val}^{sm} \times \mathrm{Val}^{sm} \to \mathrm{Val}^{sm}$$
$$(\phi, \psi) \mapsto \phi \cdot \psi.$$

If ϕ and ψ are given as in (13), then

$$\begin{aligned}
\phi \cdot \psi(K) &= \mathrm{vol}_{2n}(\Delta K + A_1 \times A_2) \\
&= \int_V \int_V 1_{\Delta K + A_1 \times A_2}(x, y) dx dy \\
&= \int_V \mathrm{vol}_n((y - A_2) \cap K + A_1) dy \\
&= \int_V \phi((y - A_1) \cap K) dy. \qquad (15)
\end{aligned}$$

This last expression extends to arbitrary $\phi \in \mathrm{Val}^{sm}$ by linearity.

In the case of even smooth valuations, there is another description of the product based on general Crofton formulas. We have seen that if $\phi \in \mathrm{Val}_k^{+,sm}$, there is a smooth, translation invariant measure m_ϕ on the space of $(n-k)$-planes in V such that

$$\phi(K) = \int_{\overline{\mathrm{Gr}}_{n-k}(V)} \chi(K \cap E) dm_\phi(E).$$

For ψ as in (13), applying (15) and Fubini's theorem gives us

$$\begin{aligned}
\phi \cdot \psi(K) &= \int_V \int_{\overline{\mathrm{Gr}}_{n-k}(V)} \chi((y - A_2) \cap K \cap E) dm_\phi(E) dy \\
&= \int_{\overline{\mathrm{Gr}}_{n-k}(V)} \int_V \chi((y - A_2) \cap K \cap E) dy dm_\phi(E) \\
&= \int_{\overline{\mathrm{Gr}}_{n-k}(V)} \mathrm{vol}_n(K \cap E + A_2) dm_\phi(E) \\
&= \int_{\overline{\mathrm{Gr}}_{n-k}(V)} \psi(K \cap E) dm_\phi(E). \qquad (16)
\end{aligned}$$

Again, this equation holds true for all $\psi \in \mathrm{Val}^{sm}$. In particular, we get that $\chi \cdot \psi = \psi$, i.e. *the Euler characteristic is the unit with respect to the Alesker product.*

From (16) and (2) we see that the coefficient $c_k(K)$ in the general Hadwiger theorem 2.4 is given by

$$c_k(K) = \begin{bmatrix} n \\ k \end{bmatrix}^{-1} \phi \cdot \mu_k(K), \tag{17}$$

which is already half of the "algebraic" proof of the general Hadwiger theorem.

3.6 Alesker-Poincaré Duality

The Alesker product satisfies a remarkable Poincaré duality which is in fact a central ingredient in the algebraic approach to kinematic formulas. By Klain's theorem, the space Val_n (where n is the dimension of V) is generated by any Lebesgue measure. Fixing a Euclidean structure on V, we thus get an isomorphism $\mathrm{Val}_n \cong \mathbb{C}$. Given two smooth valuations ϕ, ψ, let $\langle \phi, \psi \rangle \in \mathbb{C}$ be the image of the n-homogeneous component of $\phi \cdot \psi$ under this isomorphism.

Alesker proved that the pairing

$$\begin{aligned} \mathrm{Val}^{sm} \times \mathrm{Val}^{sm} &\to \mathbb{C} \\ (\phi, \psi) &\mapsto \langle \phi, \psi \rangle \end{aligned}$$

is *perfect*, which means that the induced map

$$\mathrm{PD} : \mathrm{Val}^{sm} \to \mathrm{Val}^{sm,*}$$

is injective and has dense image [9]. Roughly speaking, the space Val^{sm} is *self-dual*.

3.7 Alesker-Fourier Transform

There is another remarkable duality on the space of translation invariant valuations, which shares many formal properties with the Fourier transform of functions. It was introduced by Alesker in the even case in [6] and in the odd case in [14].

In invariant terms, the Alesker-Fourier transform is a map

$$\wedge : \mathrm{Val}^{sm} \to \mathrm{Val}^{sm}(V^*) \otimes \mathrm{Dens}(V^*),$$

where $\mathrm{Dens}(V^*) = \Lambda^n V \otimes \mathbb{C}$ denotes the 1-dimensional space of complex-valued Lebesgue measures on V^*. Given a scalar product on V, we will identify $\mathrm{Val}^{sm}(V^*) \otimes \mathrm{Dens}(V^*)$ with $\mathrm{Val}^{sm}(V)$.

The definition in the even case is easy to write down. If $\mu \in \mathrm{Val}_k^{sm,+}$ has Klain function $\mathrm{Kl}_\mu \in C^\infty(\mathrm{Gr}_k)$, then $\hat{\mu} \in \mathrm{Val}_{n-k}^{sm,+}$ is defined by the condition $\mathrm{Kl}_{\hat{\mu}}(E) = \mathrm{Kl}_\mu(E^\perp)$. Alesker showed that $\hat{\mu}$ indeed exists. One way to see this is to note that (12) can be rewritten in the form

$$\mu(K) = \int_{\mathrm{Gr}_k V} \mathrm{vol}(\pi_L K) dm(L), \tag{18}$$

where m is a (signed) smooth measure on $\mathrm{Gr}_k V$. Taking $m^\perp$ to be the image of m under the map $L \mapsto L^\perp$, we can construct $\hat{\mu}$ by setting

$$\hat{\mu}(K) = \int_{\mathrm{Gr}_{n-k} V} \mathrm{vol}(\pi_L K) dm^\perp(L).$$

The definition in the odd case is much more involved and we refer to the original paper [14] for the details. One of the main points of the construction is an odd version of a Crofton formula. If $\phi \in \mathrm{Val}_k^{-,sm}$, then one can write ϕ (non-uniquely) in the form

$$\phi(K) = \int_{\mathrm{Gr}_{k+1} V} \psi_L(\pi_L K) dL,$$

where $\psi_L \in \mathrm{Val}_k^{-,sm}(L)$ depends smoothly on L and π_L is the orthogonal projection onto L. The reader should compare this formula with (18).

Using this formula, Alesker defined the Fourier transform on odd valuations in an inductive way and showed that the result does not depend on several choices (like the choice of the ψ_L).

The Alesker-Fourier transform satisfies a Plancherel-type formula:

$$\hat{\hat{\mu}}(K) = \mu(-K). \tag{19}$$

3.8 Convolution

Given a product and a Fourier transform, it is natural to consider the *convolution* product on Val^{sm} in such a way that the Fourier transform is an algebra isomorphism between $(\mathrm{Val}^{sm}, \cdot)$ and $(\mathrm{Val}^{sm}, *)$, i.e.

$$\widehat{\phi \cdot \psi} = \hat{\phi} * \hat{\psi}, \quad \phi, \psi \in \mathrm{Val}^{sm}. \tag{20}$$

It was shown in [24] that such a convolution product exists, and that there is the following equivalent definition, which is similar to (14). Suppose

$$\phi(K) = \mathrm{vol}(K + A_1), \psi(K) = \mathrm{vol}(K + A_2),$$

where A_1, A_2 are smooth convex bodies with positive curvature. Then

$$(\phi * \psi)(K) := \mathrm{vol}(K + A_1 + A_2), \tag{21}$$

and the so-defined convolution extends to a unique linear and continuous product on Val^{sm}. As the product, $*$ is commutative and associative. The volume is the unit in $(\mathrm{Val}^{sm}, *)$. The degree of $\phi * \psi$ is the sum of the degrees of ϕ and ψ minus the dimension n of V.

Note that the definition (21) was given before the Alesker-Fourier transform was extended to odd valuations. Equation (20) in the odd case was established in [14].

Let us make an important remark here. Since we use some volume in the definition (21) of the convolution, it is not independent of the choice of a Euclidean scalar product on V. Without any choices, $*$ is defined on the twisted space $\mathrm{Val}^{sm}(V^*) \otimes \mathrm{Dens}(V)$, where $\mathrm{Dens}(V) \cong \Lambda^n V^* \otimes \mathbb{C}$ denotes the 1-dimensional space of complex-valued Lebesgue measures on V .

On the other hand, the product definition (14) does not depend on any Euclidean structure, taking vol_{2n} to be the product measure. The coordinate free version of the Alesker-Fourier transform is an isomorphism

$$\mathrm{Val}^{sm}(V) \to \mathrm{Val}^{sm}(V^*) \otimes \mathrm{Dens}(V)$$

and with these modifications (20) is independent of a choice of Euclidean structure.

Equation (21) implies another nice property of the convolution. Namely, if ϕ, ψ are mixed volumes (see [53] for the definition and properties of mixed volumes) then their convolution product is again a mixed volume. More precisely, if $k + l \geq n$ and $A_1, \ldots, A_{n-k}, B_1, \ldots, B_{n-l}$ are convex bodies with smooth boundary and positive curvature, then the convolution product of the mixed volumes

$$\phi(K) := V(K[k], A_1, \ldots, A_{n-k})$$
$$\psi(K) := V(K[l], B_1, \ldots, B_{n-l})$$

is the mixed volume

$$\phi * \psi(K) = \binom{k+l}{k}^{-1} \binom{k+l}{n} V(K[k+l-n], A_1, \ldots, A_{n-k}, B_1, \ldots, B_{n-l}).$$

3.9 *Hard Lefschetz Theorems*

The well-known *Hard Lefschetz theorem* from complex algebraic geometry states that iterates of the Lefschetz operator (multiplication by the symplectic form) realize the Poincaré-isomorphisms in the cohomology of Kähler manifolds. See [37] for more information.

In algebraic integral geometry, there is a similar theorem (in fact two versions of it). The Lefschetz operator is replaced by the multiplication with the first intrinsic volume μ_1 (we fix some Euclidean structure here). The corresponding operator is denoted by

$$L : \mathrm{Val}_*^{sm} \to \mathrm{Val}_{*+1}^{sm}.$$

Intertwining with the Alesker-Fourier transform, we get an operator

$$\Lambda : \mathrm{Val}_*^{sm} \to \mathrm{Val}_{*-1}^{sm}, \Lambda\phi = 2\widehat{L\hat{\phi}} = 2\mu_{n-1} * \phi.$$

Explicitly, this operator is given by

$$\Lambda\mu(K) = \left.\frac{d}{dt}\right|_{t=0} \mu(K + tB),$$

where B is the unit ball and $K + tB$ is the parallel set of radius t around K.

From the Steiner formula (3) and the trivial fact $\hat{\mu}_k = \mu_{n-k}$ one gets

$$L\mu_k = \frac{(k+1)\omega_{k+1}}{2\omega_k}\mu_{k+1}$$

$$\Lambda\mu_k = \frac{(n-k+1)\omega_{n-k+1}}{\omega_{n-k}}\mu_{k-1}.$$

Theorem 3.5. *Let V be an n-dimensional Euclidean vector space.*

1. *For $k \leq \frac{n}{2}$, the map*
$$L^{n-2k} : \mathrm{Val}_k^{sm} \to \mathrm{Val}_{n-k}^{sm}$$
is an isomorphism.
2. *For $k \geq \frac{n}{2}$, the map*
$$\Lambda^{2k-n} : \mathrm{Val}_k^{sm} \to \mathrm{Val}_{n-k}^{sm}$$
is an isomorphism.

Corollary 3.6. *The multiplication operator*

$$L : \mathrm{Val}_k^{sm} \to \mathrm{Val}_{k+1}^{sm}$$

is injective if $2k+1 \leq n$ and surjective if $2k+1 \geq n$. The derivation operator

$$\Lambda : \mathrm{Val}_k^{sm} \to \mathrm{Val}_{k-1}^{sm}$$

is injective if $2k-1 \geq n$ and surjective if $2k-1 \leq n$.

This corollary tells us that it is enough to understand valuations in the middle degree and that all other valuations are found by applying a simple operator to a valuation of middle degree. This is particularly useful when studying G-invariant valuations. The corollary also tells us that, roughly speaking, most valuations concentrate close to the middle degree.

Several authors have contributed to the proof of Theorem 3.5. Building on previous work with Bernstein [16], Alesker first proved both versions of the Hard Lefschetz theorem in the even case [6, 7]. The second version was extended to odd valuations in [23]. The proof used the Laplacian acting on differential forms on the sphere and some results from complex geometry (Kähler identities). Next, it was shown in [24] that in the even case, both versions of the Hard Lefschetz theorem are in fact equivalent via the Alesker-Fourier transform (which was at that time defined only for even valuations). Finally, Alesker extended in [14] the Fourier transform to odd valuations and derived the first version of the Hard Lefschetz theorem in the odd case from the second one.

4 Applications in Integral Geometry

4.1 Abstract Hadwiger-Type Theorem

We have sketched in the first section how the kinematic formulas and Crofton formulas can be easily proved with Hadwiger's theorem. A similar argument will give analogous (although more complicated) formulas for all subgroups G of $\mathrm{SO}(n)$ such that $\dim \mathrm{Val}^G < \infty$.

The next theorem was formulated by Alesker [12].

Theorem 4.1. *A compact subgroup G of $\mathrm{SO}(n), n \geq 2$ satisfies $\dim \mathrm{Val}^G < \infty$ if and only if G acts transitively on the unit sphere. In this case, every G-invariant, translation invariant and continuous valuation is smooth.*

Let us give the idea of the proof (taken from [31]). First of all, remember that smooth, translation invariant valuations are dense in the space of all translation invariant valuations. A smooth valuation is given by integration over the conormal cycle of some translation invariant differential form ω. If the valuation is G-invariant, then we may assume (by averaging over the group) that ω is also G-invariant. If G acts transitively on the unit sphere, then a G-invariant, translation invariant differential form on the sphere bundle is uniquely determined by its value at any given point. Hence the space of all such forms is finite-dimensional.

Now take any continuous, translation invariant, G-invariant valuation μ. We may approximate it by a sequence of smooth, translation invariant valuations. Averaging these valuations with respect to the Haar measure on G, we may in fact approximate μ by a sequence of smooth G-invariant, translation invariant valuations. But this space is finite-dimensional, hence closed. Therefore μ itself belongs to this finite-dimensional space. In particular,

$$\mathrm{Val}^G \subset \mathrm{Val}^{sm}.$$

Let us now prove the inverse implication. The Klain embedding (9) in the case $k = 1$ induces an isomorphism

$$\mathrm{Kl} : \mathrm{Val}_1^+ \cong C^\infty(\mathrm{Gr}_1 V).$$

This follows from the fact that the *cosine transform* is an isomorphism on even smooth functions on the unit sphere [40]. Since μ is G-invariant if and only if Kl_μ is G-invariant, we have

$$\mathrm{Kl} : \mathrm{Val}_1^{G,+} \cong C^\infty(\mathrm{Gr}_1 V)^G.$$

If G does not act transitively on the sphere, then the space of smooth G-invariant functions on the projective space $\mathrm{Gr}_1 V = \mathbb{P}V$ is infinite-dimensional. Therefore $\mathrm{Val}_1^{G,+}$ is also infinite-dimensional, which implies that Val^G is infinite-dimensional.

The classification of connected compact Lie groups G acting transitively on some sphere is a topological problem which was solved by Montgomery-Samelson [48] and Borel [27]. There are six infinite lists

$$\mathrm{SO}(n), \mathrm{U}(n), \mathrm{SU}(n), \mathrm{Sp}(n), \mathrm{Sp}(n) \cdot \mathrm{U}(1), \mathrm{Sp}(n) \cdot \mathrm{Sp}(1) \tag{22}$$

and three exceptional groups

$$\mathrm{G}_2, \mathrm{Spin}(7), \mathrm{Spin}(9). \tag{23}$$

These groups are important in differential geometry and topology, since the holonomy group of an irreducible non-symmetric Riemannian manifold is always from this list and each group from this list except $\mathrm{Sp}(n) \cdot \mathrm{U}(1)$ and $\mathrm{Spin}(9)$ does appear as the holonomy group of such a manifold.

There are various natural inclusions among these groups:

$$\mathrm{U}(n), \mathrm{SU}(n) \subset \mathrm{SO}(2n), \quad \mathrm{Sp}(n), \mathrm{Sp}(n) \cdot \mathrm{U}(1), \mathrm{Sp}(n) \cdot \mathrm{Sp}(1) \subset \mathrm{SO}(4n),$$

$$\mathrm{SU}(4) \subset \mathrm{Spin}(7), \mathrm{G}_2 \subset \mathrm{SO}(7), \mathrm{SU}(3) \subset \mathrm{G}_2 \subset \mathrm{Spin}(7) \subset \mathrm{SO}(8), \mathrm{Spin}(9) \subset \mathrm{SO}(16).$$

The last two inclusions are the spin representations. We refer to [26] for more information on holonomy groups.

4.2 The Kinematic Coproduct

The first thing we need to do in order to relate the kinematic formulas to the algebraic structures introduced in the previous section is to give a more abstract description of these formulas.

Let V be a Euclidean vector space and let G be a subgroup of $\mathrm{SO}(V)$ which acts transitively on the unit sphere. We have seen that in this case, the space Val^G is finite-dimensional and consists only of smooth valuations.

If $\phi_1, \dots, \phi_m$ is a basis of Val^G, then by the same trick as in Sect. 2.4 we obtain kinematic formulas

$$\int_{\bar{G}} \phi_i(K \cap \bar{g}L) d\bar{g} = \sum_{k,l=1}^{m} c_{k,l}^i \phi_k(K) \phi_l(L). \tag{24}$$

There is a very nice and clever way to encode these formulas in a purely algebraic way. For this, Fu [32] defined the kinematic operator

$$k_G : \mathrm{Val}^G \to \mathrm{Val}^G \otimes \mathrm{Val}^G$$

$$\phi_i \mapsto \sum_{k,l=1}^{m} c_{k,l}^i \phi_k \otimes \phi_l.$$

This map is in fact a *cocommutative, coassociative coproduct* on Val^G. Let us remind the reader of the definition of a coproduct. Loosely speaking, we write down the corresponding usual property (commutativity or associativity) in terms of a commuting diagram and reverse all arrows to obtain the co-property.

For instance, cocommutativity means that the following diagram commutes:

$$\begin{array}{ccc} \mathrm{Val}^G & \xrightarrow{k_G} & \mathrm{Val}^G \otimes \mathrm{Val}^G \\ {\scriptstyle id}\downarrow & & \downarrow{\scriptstyle \iota} \\ \mathrm{Val}^G & \xrightarrow{k_G} & \mathrm{Val}^G \otimes \mathrm{Val}^G . \end{array}$$

Here ι is the map that interchanges the factors of $\mathrm{Val}^G \otimes \mathrm{Val}^G$.

In more concrete terms, this says that the coefficients in the kinematic formula (24) satisfy $c_{k,l}^i = c_{l,k}^i$, which expresses the symmetry of the formula (in K and L) as in Sect. 2.3.

The coassociativity property is the commutativity of the following diagram:

$$\begin{array}{ccc} \mathrm{Val}^G & \xrightarrow{(k_G \otimes id) \circ k_G} & \mathrm{Val}^G \otimes \mathrm{Val}^G \otimes \mathrm{Val}^G \\ {\scriptstyle id}\downarrow & & \downarrow{\scriptstyle id \otimes id \otimes id} \\ \mathrm{Val}^G & \xrightarrow{(id \otimes k_G) \circ k_G} & \mathrm{Val}^G \otimes \mathrm{Val}^G \otimes \mathrm{Val}^G . \end{array}$$

This property is equivalent to the formula

$$\sum_r c_{r,m}^i c_{k,l}^r = \sum_r c_{r,l}^i c_{k,m}^r,$$

and this comes just from Fubini's theorem, compare Sect. 2.3.

In a similar vein, there are additive kinematic formulas for G:

$$\int_G \phi_i(K + gL)d\bar{g} = \sum_{k,l=1}^{m} d^i_{k,l}\phi_k(K)\phi_l(L). \tag{25}$$

which can be encoded by the cocommutative, coassociative coproduct

$$\begin{aligned} a_G : \mathrm{Val}^G &\to \mathrm{Val}^G \otimes \mathrm{Val}^G \\ \phi_i &\mapsto \sum_{k,l=1}^{m} d^i_{k,l}\phi_k \otimes \phi_l. \end{aligned} \tag{26}$$

4.3 Fundamental Theorem of Algebraic Integral Geometry

The *fundamental theorem of algebraic integral geometry* relates the kinematic coproduct and the product structure and is the basis for a fuller understanding of the kinematic formulas (24).

Theorem 4.2. *Let G be a group acting transitively on the unit sphere,* m_G : $\mathrm{Val}^G \otimes \mathrm{Val}^G \to \mathrm{Val}^G$ *the restriction of the Alesker product to* Val^G; $\mathrm{PD}_G : \mathrm{Val}^G \to \mathrm{Val}^{G*}$ *the restriction of the Alesker-Poincaré duality to* Val^G *and* k_G *the kinematic coproduct. Then the following diagram commutes*

$$\begin{array}{ccc} \mathrm{Val}^G & \xrightarrow{k_G} & \mathrm{Val}^G \otimes \mathrm{Val}^G \\ \Big\downarrow{\scriptstyle \mathrm{PD}_G} & & \Big\downarrow{\scriptstyle \mathrm{PD}_G \otimes \mathrm{PD}_G} \\ \mathrm{Val}^{G*} & \xrightarrow{m_G^*} & \mathrm{Val}^{G*} \otimes \mathrm{Val}^{G*}. \end{array}$$

This theorem, based on a basic version which we discuss below, was proven in [24].

Let us work out the most important case, namely the principal kinematic formula $k_G(\chi)$. First note that $\mathrm{Val}^G \otimes \mathrm{Val}^G = \mathrm{Hom}(\mathrm{Val}^{G*}, \mathrm{Val}^G)$. We may thus regard $k_G(\chi)$ as a map from Val^{G*} to Val^G. Recall that PD_G is a map from Val^G to Val^{G*}.

Given $K \in \mathcal{K}(V)$, let $\tau_K \in \mathrm{Val}^{G*}$ be defined by $\tau_K(\phi) := \phi(K)$. Then the τ_K span Val^{G*} and we get

$$k_G(\chi)(\tau_K)(\cdot) = \int_{\bar{G}} \chi(K \cap \bar{g}\cdot)d\bar{g} \in \mathrm{Val}^G$$

and hence for any $\phi \in \mathrm{Val}^G$ by (16)

$$\phi \cdot k_G(\chi)(\tau_K)(\cdot) = \int_{\bar{G}} \phi(K \cap \bar{g}\cdot) d\bar{g} \in \mathrm{Val}^G .$$

We plug in a ball B_R of radius R into this equation. If R is large, the measure of all $\bar{g}$ with $K \subset \bar{g} B_R$ is approximately $\mathrm{vol}(B_R)$, while the measure of all $\bar{g}$ with $K \cap \bar{g} B_R \neq \emptyset, K$ is $o(R^n)$. It follows that the n-th homogeneous component of $\phi \cdot k_G(\chi)(\tau_K)$ is given by $\phi(K)$ vol. In other words,

$$\mathrm{PD}_G(k_G(\chi)(\tau_K))(\phi) = \phi(K) = \tau_K(\phi),$$

which implies that

$$\mathrm{PD}_G \circ k_G(\chi) = Id.$$

Hence PD_G *and* $k_G(\chi)$ *are inverse to each other*, and this statement is equivalent to the fact that

$$(\mathrm{PD}_G \otimes \mathrm{PD}_G) \circ k_G(\chi) = m_G^* \circ \mathrm{PD}_G(\chi),$$

which follows from Theorem 4.2. See also [24, 32] for more details.

The fundamental theorem of algebraic integral geometry says roughly that the knowledge of k_G is the same as the knowledge of m_G. It may be used in two ways. If we know k_G, then we may first compute $\mathrm{PD}_G := k_G(\chi)^{-1}$ and, using the above diagram, we may compute the whole product structure. Conversely, knowing the product, we can compute PD_G and hence k_G. This is how the theorem will be used in the sequel.

Nevertheless, in concrete situations, things turn out to be not so easy, since in order to compute m_G^* we have to invert some potentially huge matrix which might be a challenge. We will come back to this point when we discuss the hermitian case.

A consequence from the fundamental theorem of algebraic integral geometry is

$$k_G(\phi \cdot \psi) = (\phi \otimes \chi) \cdot k_G(\psi) = (\chi \otimes \psi) \cdot k_G(\phi), \quad \phi, \psi \in \mathrm{Val}^G . \tag{27}$$

We give a proof of the more general statement

$$\int_{\bar{G}} \phi \cdot \psi(K \cap \bar{g} L) d\bar{g} = ((\phi \otimes \chi) \cdot k_G(\psi))(K, L), \tag{28}$$

where ψ is supposed to be smooth and translation invariant, but not necessarily G-invariant.

By linearity and density, it is enough to assume that ϕ has the form $\phi(K) = \mathrm{vol}(K + A)$ for some smooth convex body A with positive curvature. Then

$$\phi \cdot \psi(K \cap \bar{g} L) = \int_V \psi((x - A) \cap K \cap \bar{g} L) dx$$

by (15) and hence

$$\begin{aligned}\int_{\bar{G}} \phi \cdot \psi(K \cap \bar{g}L) d\bar{g} &= \int_{\bar{G}} \int_V \psi((x-A) \cap K \cap \bar{g}L) dx d\bar{g} \\ &= \int_V \int_{\bar{G}} \psi((x-A) \cap K \cap \bar{g}L) d\bar{g} dx \\ &= \int_V k_G(\psi)((x-A) \cap K, L) dx \\ &= (\phi \otimes \chi) \cdot k_G(\psi)(K, L).\end{aligned}$$

In the special case $G = \mathrm{SO}(n)$, $\psi = \chi$, Equation (28) and the principal kinematic formula (5) imply the general Hadwiger theorem 2.4:

$$\begin{aligned}\int_{\overline{\mathrm{SO}(n)}} \phi(K \cap \bar{g}L) d\bar{g} &= ((\phi \otimes \chi) \cdot k_{\mathrm{SO}(n)}(\chi))(K, L) \\ &= \sum_{k=0}^{n} \begin{bmatrix} n \\ k \end{bmatrix}^{-1} (\phi \cdot \mu_k)(K) \mu_{n-k}(L) \\ &= \sum_{k=0}^{n} c_k(K) \mu_{n-k}(L),\end{aligned}$$

where

$$c_k(K) = \begin{bmatrix} n \\ k \end{bmatrix}^{-1} (\phi \cdot \mu_k)(K) = \int_{\overline{\mathrm{Gr}}_{n-k}(V)} \phi(K \cap E) dE$$

by (17).

In the same situation, Nijenhuis' observation becomes evident. We set $t := \frac{2}{\pi}\mu_1 \in \mathrm{Val}_1^{\mathrm{SO}(n)}$. By the Hard Lefschetz theorem 3.5 we must have $t^n = c \operatorname{vol}_n$ for some constant $c \neq 0$. Therefore

$$\mathrm{Val}^{\mathrm{SO}(n)} = \mathbb{C}[t]/(t^{n+1}).$$

Then we have $\mathrm{PD}(t^i) = c(t^{n-i})^*$, where $\{(t^k)^*, k = 0, \dots, n\}$ is the dual basis to the basis $\{t^k, k = 0, \dots, n\}$ of $\mathrm{Val}^{\mathrm{SO}(n)}$. From Theorem 4.2 it follows that

$$k_G(t^i) = \frac{1}{c} \sum_{k+l=n+i} t^k \otimes t^l.$$

Setting $\tilde{\mu}_k = \frac{1}{c} t^k$ thus gives us

$$k_G(\tilde{\mu}_i) = \sum_{k+l=n+i} \tilde{\mu}_k \otimes \tilde{\mu}_l.$$

In fact, it is easily computed (see Sect. 3.9 or [25]) that

$$t^k = \frac{k!\omega_k}{\pi^k}\mu_k,$$

hence $c = \frac{n!\omega_n}{\pi^n}$ and

$$\tilde{\mu}_k = \frac{\pi^n k!\omega^k}{\pi^k n!\omega_n}\mu_k.$$

4.4 Additive Formulas

There is a similar statement relating the convolution product to the additive kinematic formulas (25). It was proved in [24] under the assumption $\mathrm{Val}^G \subset \mathrm{Val}^+$, which turns out to be always the case [18].

Theorem 4.3. *Let G be a group acting transitively on the unit sphere. Let a_G be the additive kinematic coproduct, see (26). Let $c_G : \mathrm{Val}^G \otimes \mathrm{Val}^G \to \mathrm{Val}^G$ be the restriction of the convolution to Val^G. Then the following diagram commutes*

$$\begin{array}{ccc} \mathrm{Val}^G & \xrightarrow{a_G} & \mathrm{Val}^G \otimes \mathrm{Val}^G \\ {\scriptstyle \mathrm{PD}_G}\downarrow & & \downarrow{\scriptstyle \mathrm{PD}_G \otimes \mathrm{PD}_G} \\ \mathrm{Val}^{G*} & \xrightarrow{c_G^*} & \mathrm{Val}^{G*} \otimes \mathrm{Val}^{G*}. \end{array}$$

Corollary 4.4. *Kinematic formulas* (24) *and additive kinematic formulas* (7) *are related by the formula*

$$a_G = (\wedge \otimes \wedge) \circ k_G \circ \wedge. \tag{29}$$

Explicitly, this means that if the kinematic formulas are given by

$$\int_{\bar{G}} \phi_i(K \cap \bar{g}L) d\bar{g} = \sum_{k,l=1}^{m} c_{k,l}^i \phi_k(K)\phi_l(L),$$

in some basis $\phi_1, \ldots, \phi_m$ of Val^G, then the additive kinematic formulas in the Fourier-dual basis $\hat{\phi}_1, \ldots, \hat{\phi}_m$ are given by

$$\int_G \hat{\phi}_i(K + gL) dg = \sum_{k,l=1}^{m} c_{k,l}^i \hat{\phi}_k(K)\hat{\phi}_l(L),$$

with the same constants.

This corollary gives a precise meaning to the fact which we have mentioned in Sect. 2.3: *Kinematic formulas and additive kinematic formulas are dual to each other.*

This explains also the observation from Sect. 2.3: since in some basis of $\mathrm{Val}^{\mathrm{SO}(n)}$ all coefficients of $k_{\mathrm{SO}(n)}$ are 1, the same holds true for $a_{\mathrm{SO}(n)}$ in the Fourier-dual basis.

5 The Hermitian Case

In his 1976 book on integral geometry [52], Santaló wrote that *Integral geometry on complex spaces has not been sufficiently developed and probably deserves further study.*

In the previous two sections, we have described the theoretical framework relating algebraic structures on valuations and integral-geometric formulas. The aim of this section is to show how this works in practice for the first non-classical example from list (22), namely the group $G = \mathrm{U}(n)$.

We let $V \cong \mathbb{C}^n$ be a complex vector space of (complex) dimension n, endowed with a hermitian inner product H. Recall that H is

1. Conjugate linear in the first component and linear in the second component, i.e.

$$H(\lambda v, \mu w) = \bar{\lambda} H(v, w)\mu, \quad v, w \in V, \lambda, \mu \in \mathbb{C},$$

2. Conjugate symmetric, i.e. $H(w, v) = \overline{H(v, w)}$ and
3. Positive definite, i.e. $H(v, v) > 0$ for $v \neq 0$.

The subgroup of $\mathrm{GL}(V, \mathbb{C})$ fixing H is the unitary group $\mathrm{U}(n)$.

The real part of H is a real inner product on V, while the imaginary part of H is a symplectic form Ω on V. In particular, $\mathrm{U}(n)$ is a subgroup of $\mathrm{SO}(2n)$.

Before going into details, let us remark that $-1 \in \mathrm{U}(n)$, hence all unitarily invariant valuations are even.

5.1 $\mathrm{Val}^{\mathrm{U}(n)}$ *as a Vector Space*

The abstract Hadwiger Theorem 4.1 tells us that $\dim \mathrm{Val}^{\mathrm{U}(n)} < \infty$, but it says nothing about the actual value of this dimension. Alesker showed in [5] that

$$\dim \mathrm{Val}_k^{\mathrm{U}(n)} = \min\left\{\left\lfloor \frac{k}{2} \right\rfloor, \left\lfloor \frac{2n-k}{2} \right\rfloor\right\} + 1. \tag{30}$$

Note that these dimensions have the typical behavior predicted by the Hard Lefschetz Theorem 3.5: they are increasing for degrees smaller than half the (real) dimension and decreasing for degrees bigger than half the (real) dimension.

There are various ways of proving this formula. Alesker's original proof used representation theoretical methods to decompose the space of even valuations on an $2n$-dimensional vector space as a direct sum of irreducible $\mathrm{SO}(2n)$-modules. Since it is known which irreducible $\mathrm{SO}(2n)$-modules contain a $\mathrm{U}(n)$-invariant vector, the above formula follows easily.

A second possible proof goes as follows. Since $\mathrm{Val}^{\mathrm{U}(n)} \subset \mathrm{Val}^{sm}$, we can represent each unitarily invariant valuation by a pair (ω, ϕ) of differential forms as in (11). Since we may average over the group, we may actually take ω, ϕ to be $\mathrm{U}(n)$-invariant too. But the $\mathrm{U}(n)$-invariant, translation invariant smooth forms on the sphere bundle SV can be explicitly described. This was carried out by Park [50] using the *first fundamental theorem* for the group $\mathrm{U}(n)$. Different pairs (ω, ϕ) may induce the same valuation. Fortunately, one can characterize the kernel of the normal cycle map in terms of a certain second-order differential operator, called *Rumin operator* which was introduced by Rumin in [51]. This works even in the much more general setting of *valuations on manifolds*, see [23]. The Rumin operator of the unitarily and translation invariant forms on SV was (somehow implicitly) computed in [25]. These computations imply (30).

A third way to prove (30) is sketched in [19]. It uses the fact that Val_k^G and some spaces of G-invariant differential forms on the unit sphere bundle SV fit into an exact sequence.

Knowing the dimension of $\mathrm{Val}^{\mathrm{U}(n)}$, the next question is to find a basis. Alesker gave in fact two of them, which are dual to each other with respect to the Alesker-Fourier transform. The idea is to mimic the definition of the intrinsic volumes in (1) and (2) and using complex Grassmannians instead of real ones. Using intersections with complex planes, Alesker defined

$$U_{k,p}(K) := \int_{\overline{\mathrm{Gr}}_{n-p}^{\mathbb{C}}} \mu_{k-2p}(K \cap \bar{E})\, d\bar{E}.$$

The $U_{k,p}$, as p ranges over $0, 1, \ldots, \min\left\{\left\lfloor \frac{k}{2} \right\rfloor, \left\lfloor \frac{2n-k}{2} \right\rfloor\right\}$, constitute a basis of $\mathrm{Val}_k^{\mathrm{U}(n)}$.

Fu renormalized these valuations by setting

$$t := \frac{2}{\pi}\mu_1 = \frac{2}{\pi}U_{1,0} \in \mathrm{Val}_1^{\mathrm{U}(n)}$$

$$s := nU_{2,1} \in \mathrm{Val}_2^{\mathrm{U}(n)}$$

which implies that

$$s^p t^{k-2p} = \frac{(k-2p)!n!\omega_{k-2p}}{(n-p)!\pi^{k-2p}} U_{k,p}.$$

The second basis given by Alesker uses projections onto complex subspaces instead of intersections:

$$C_{k,q}(K) := \int_{\mathrm{Gr}_q^{\mathbb{C}}} \mu_k(\pi_E(K))\, dE.$$

As q ranges over all values from $n-\min\left\{\left\lfloor \frac{k}{2} \right\rfloor, \left\lfloor \frac{2n-k}{2} \right\rfloor\right\}$ to n, the $C_{k,q}$ constitute a basis of $\mathrm{Val}_k^{\mathrm{U}(n)}$. Up to a normalizing constant, the Fourier transform of $U_{k,p}$ is $C_{2n-k,n-p}$.

5.2 $\mathrm{Val}^{\mathrm{U}(n)}$ *as an Algebra*

The monomials $s^p t^{k-2p}$, with $0 \leq k \leq 2n$ and $0 \leq p \leq \min\left\{\left\lfloor \frac{k}{2} \right\rfloor, \left\lfloor \frac{2n-k}{2} \right\rfloor\right\}$, constitute a basis of $\mathrm{Val}^{\mathrm{U}(n)}$. We therefore speak of the *monomial basis* or the *ts-basis* of $\mathrm{Val}^{\mathrm{U}(n)}$.

We have a graded algebra epimorphism

$$\mathbb{C}[t,s] \twoheadrightarrow \mathrm{Val}^{\mathrm{U}(n)},$$

where t, s on the left hand side are formal variables of degree 1 resp. 2 (in the following, the distinction between variables and actual valuations will not be made, which is quite in the spirit of algebraic integral geometry). The kernel of this map is an ideal I_n in $\mathbb{C}[t,s]$, which, by Hilbert basis theorem, must be generated by finitely many polynomials.

There is a relatively easy way to compute these polynomials, which was given by Fu [29].

First, one deduces from (30) that I_n is generated by two polynomials f_{n+1} and f_{n+2} of total degree $n+1$ and $n+2$ respectively.

Next, by Alesker-Poincaré duality, in order to show that some polynomial f of total degree d in t and s is zero, it is enough to show that $f \cdot s^p t^{2n-d-2p} = 0$ for all p. Since $\mathrm{Val}_{2n}^{\mathrm{U}(n)}$ is spanned by the Lebesgue measure, this amounts to some combinatorial identity among the coefficients of f once we know how to evaluate the monomials $s^p t^{2n-2p}$ on a unit ball. Using the *transfer principle*, which relates valuations on $\mathbb{C}^n$ and on $\mathbb{CP}^n$, [34], one can compute these values. The final result (which was first proved by Fu in [33] using another method) is as follows:

Theorem 5.1. *There is an isomorphism between graded algebras*

$$\mathrm{Val}^{\mathrm{U}(n)} \cong \mathbb{C}[t,s]/(f_{n+1}, f_{n+2}),$$

where

$$\log(1+t+s) = f_1 + f_2 + f_3 + \cdots = t + \left(s - \frac{t^2}{2}\right) + \left(-st + \frac{t^3}{3}\right) + \cdots$$

is the expansion in (weighted) homogeneous polynomials.

As explained in Sect. 4.3, from the product structure, we can compute $\mathrm{PD}_{\mathrm{U}(n)}$ and $m_{\mathrm{U}(n)}$ and therefore $k_{\mathrm{U}(n)}$. Theorem 5.1 thus implies the knowledge of the kinematic formulas in the ts-basis.

Working this out in higher dimensions is rather cumbersome, because some huge matrix has to be inverted. Also, one would like to have not only the value of the coefficients in the kinematic formulas (24), but some closed forms. They seem to be hard to obtain from Theorem 5.1. Another missing point is the knowledge of the kinematic formula in another basis of $\mathrm{Val}^{\mathrm{U}(n)}$, for instance in the C-basis.

5.3 Hermitian Intrinsic Volumes and Tasaki Valuations

It seems difficult to describe the value of a basis element of the ts-basis on, say a polytope or a submanifold (since all unitarily invariant valuations are smooth, they may be canonically extended to submanifolds with boundary or corners, see Sect. 3.4). Therefore we introduce another, more geometric basis. This mimics the third characterization of the intrinsic volumes in Sect. 2.2.

Recall that a real subspace E of V is called *isotropic* if the restriction of the symplectic form to E vanishes. Then the dimension of E does not exceed n, and an isotropic subspace of dimension n is called *Lagrangian*. We call E of type (k,q) if E can be written as the orthogonal sum of a complex subspace of (complex) dimension q and an isotropic subspace of dimension $k-2q$. Then $k-q \leq n$.

Theorem 5.2. *There is a unique valuation $\mu_{k,q} \in \mathrm{Val}_k^{\mathrm{U}(n)}$ whose Klain function evaluated at a subspace of type (k,q') equals $\delta_{qq'}$. Moreover,*

$$\hat{\mu}_{k,q} = \mu_{2n-k,n-k+q}.$$

The idea of the construction of $\mu_{k,q}$ is as follows. We know from the discussion in Sect. 5.1 that every unitarily invariant valuation of degree $k < 2n$ is given by integration over the normal cycle of some translation invariant, unitarily invariant differential form on SV. Park [50] showed that the algebra of these forms is generated by three 1-forms and four 2-forms, and integrating a suitable product of these basic forms over the normal cycle yields the valuation $\mu_{k,q}$.

Since the $\mu_{k,q}$ with $\max(0,k-n) \leq q \leq \lfloor \frac{k}{2} \rfloor$ are linearly independent, it follows from (30) that they form a basis of $\mathrm{Val}_k^{\mathrm{U}(n)}$.

Finally, the statement on the Fourier transform boils down to the fact that the orthogonal complement of a subspace of type (k,q) is of type $(2n-k, n-k+q)$, which is easy to prove.

A version of these valuations was considered by Tasaki. He showed that the orbits of the $\mathrm{U}(n)$-action on $\mathrm{Gr}_k(V)$ are characterized by $\left\lfloor \frac{\min\{k,2n-k\}}{2} \right\rfloor$ *Kähler angles*. We use a slight modification of his construction. Let $p := \lfloor \frac{k}{2} \rfloor$. Given a k-dimensional subspace $E \subset V$, the restriction of the symplectic form Ω of V to E can be written as

$$\Omega|_E = \sum_{i=1}^{p} \cos\theta_i \alpha_{2i-1} \wedge \alpha_{2i},$$

where $\alpha_1, \ldots, \alpha_k$ is dual to an orthonormal basis of E and $0 \leq \theta_1 \leq \ldots \leq \theta_p \leq \frac{\pi}{2}$. The p-tuple $(\theta_1, \ldots, \theta_p)$ is called *multiple Kähler angle* of E.

For instance, a subspace is isotropic if all its Kähler angles are $\frac{\pi}{2}$, while it is complex if all Kähler angles are 0. More generally, a subspace is of type (k,q) if q of its Kähler angles are 0 and the remaining $p-q$ Kähler angles are $\frac{\pi}{2}$. Tasaki [56] showed that two k-dimensional subspaces belong to the same $\mathrm{U}(n)$-orbit if and only if their multiple Kähler angles agree.

The *Tasaki valuations* $\tau_{k,q} \in \mathrm{Val}^{\mathrm{U}(n)}, 0 \leq q \leq p$ are defined by their Klain function:

$$\mathrm{Kl}_{\tau_{k,q}}(E) = \sigma_q(\cos^2\theta_1(E), \ldots, \cos^2\theta_p(E)), \tag{31}$$

where σ_q is the the qth elementary symmetric function.

It is of course elementary to compute the relations between the Tasaki valuations and the hermitian intrinsic volumes:

$$\tau_{k,q} = \sum_{i=q}^{\lfloor k/2 \rfloor} \binom{i}{q} \mu_{k,i}, \quad \mu_{k,q} = \sum_{i=q}^{\lfloor k/2 \rfloor} (-1)^{i-q} \binom{i}{q} \tau_{k,i}. \tag{32}$$

If M is a compact k-dimensional manifold, then the canonical extension of $\tau_{k,q}$ to M is given by

$$\int_M \sigma_q(\cos^2 \Theta(T_x M))\, dx.$$

Using such expressions, Tasaki [57] formulated general *Poincaré formulas*, which are special instances of the principal kinematic formula $k_{\mathrm{U}(n)}(\chi)$, with K, L replaced by compact submanifolds of complementary dimension.

5.4 Kinematic Formulas

Let us now explain, in an informal style, how the hermitian intrinsic volumes may be used to compute the relations between the different bases (U-basis and C-basis), and to compute the kinematic formulas.

One can easily compute the derivation operator Λ (compare Sect. 3.9) on the hermitian intrinsic volumes. This comes from the fact that the hermitian intrinsic volumes are given by integration over the normal cycle of certain differential forms. The operator Λ corresponds to a certain Lie derivative on the level of forms which is easy to compute.

Since we also know the Alesker-Fourier transform of $\mu_{k,q}$, we can compute $L\mu_{k,q}$ (which is multiplication by t, up to a factor). Now a crucial (and somehow mysterious) observation is that (some renormalizations of) L and Λ and some degree counting operator define a representation of the Lie algebra $\mathfrak{sl}_2$ on $\mathrm{Val}^{\mathrm{U}(n)}$. In the general translation invariant setting Val, this is not the case.

The next observation is that $\mu_{n,0}$ (which is also known as *Kazarnovskii's pseudovolume [38]*) is a multiple of the polynomial f_n from Theorem 5.1. This follows from the fact that the kernel of the restriction map $\mathrm{Val}_n^{\mathrm{U}(n)} \to \mathrm{Val}_n^{\mathrm{U}(n-1)}$ is 1-dimensional and contains $\mu_{n,0}$ and f_n.

With some more tricks, one can compute the scaling factor and compute the relations between the hermitian intrinsic volumes and the ts-basis. The result can be most easily expressed in terms of the Tasaki valuations:

$$\tau_{k,q} = \frac{\pi^k}{\omega_k (k-2q)!(2q)!} t^{k-2q}(4s-t^2)^q. \tag{33}$$

Since $\mathrm{Val}^{\mathrm{U}(n)}$ is a finite-dimensional $\mathfrak{sl}_2$-representation, it admits a canonical decomposition (*Lefschetz decomposition*). The corresponding basis is called the *primitive basis* and is defined for all $0 \leq r \leq \frac{\min(k,2n-k)}{2}$ by

$$\pi_{k,r} = (-1)^r (2n-4r+1)!! \sum_{i=0}^{r} (-1)^i \frac{(k-2i)!}{(2r-2i)!} \frac{(2r-2i-1)!!}{(2n-2r-2i+1)!!} \tau_{k,i}. \tag{34}$$

This new basis is quite helpful for computational purposes, since in this basis, the matrix describing the Alesker-Poincaré-duality is anti-diagonal and we can easily compute its inverse (which is what we have to do in order to compute $k_{\mathrm{U}(n)}(\chi)$, see Theorem 4.2).

As a result, the principal kinematic formula $k_{\mathrm{U}(n)}(\chi)$ in terms of the primitive basis was established in [25].

Theorem 5.3. *Set* $p := \min\left\{\lfloor \frac{k}{2} \rfloor, \lfloor \frac{2n-k}{2} \rfloor\right\}$.

$$k_{U(n)}(\chi) = \frac{1}{\pi^n} \sum_{k=0}^{2n} \omega_k \omega_{2n-k} \sum_{r=0}^{p} \frac{(n-r)!}{8^r (2n-4r)!} \frac{(2n-2r+1)!!}{(2n-4r+1)!!} \binom{n}{2r}^{-1} \pi_{k,r} \otimes \pi_{2n-k,r}. \tag{35}$$

Using (34) and (32), we may restate this formula in terms of the Tasaki basis or in terms of intrinsic volumes, but the corresponding formulas are rather difficult.

In conclusion, the vector space structure as well as the algebra structure on $\mathrm{Val}^{\mathrm{U}(n)}$ in terms of the different bases are now rather well understood.

5.5 *Positive and Monotone Cone*

A valuation is called *positive* if $\mu(K) \geq 0$ for all K. It is easy to see that an $\mathrm{SO}(n)$-invariant valuation $\sum c_k \mu_k$ is positive if and only if all c_k are positive. In fact, μ evaluated at a k-dimensional disk of radius r behaves like $c_k r^k + o(r^k)$. On the other hand, it clearly follows from (1) or from (2) that each μ_k is positive. Moreover, the μ_k and each positive linear combination μ of them is *monotone*, i.e. $\mu(K) \leq \mu(L)$ if $K \subset L$. Hence in the classical setting, the cones of positive and monotone invariant valuations coincide.

The situation in the $\mathrm{U}(n)$-case is more involved. A similar argument as above shows that a valuation $\mu = \sum_{k,q} c_{k,q}\mu_{k,q}$ can only be positive if the Klain function of each homogeneous component is positive, hence $c_{k,q} \geq 0$. That the $\mu_{k,q}$ are indeed positive does not follow immediately from their definition. But it can be shown (using the fact that the $\mu_{k,q}$ are *constant coefficient valuations*) that $\mu_{k,q}$ evaluated at a polytope is positive from which the positivity of $\mu_{k,q}$ follows by continuity.

What about the monotone cone? One way to construct an invariant monotone valuation is to use a positive invariant Crofton measure. It is not hard to see that the cone of all invariant valuations admitting a positive Crofton measure is dual to the positive cone with respect to the scalar product $\langle \phi, \psi \rangle := \mathrm{PD}(\phi)(\hat{\psi})$.

But there are more monotone valuations. The idea to test monotonicity of a smooth, translation invariant valuation μ is to use a variation of a smooth convex body and to describe the first variation $\delta\mu$ of μ as a *curvature measure*, which is a signed measure concentrated on the boundary of K.

The main observation is that μ is monotone if and only if the corresponding curvature measure is positive, and that this happens if and only if some *infinitesimal valuations* associated to $\delta\mu$ are positive.

In the $\mathrm{U}(n)$-case, Park [50] has written down a list of equivariant curvature measures. The first variation map δ may be computed in terms of the hermitian intrinsic volumes and Park's curvature measures.

Since the positive cone in $\mathrm{Val}^{\mathrm{U}(n)}$ is known (see above), we can thus determine the monotone cone too. The result is that a valuation

$$\mu = \sum_{k,q} c_{k,q}\mu_{k,q}$$

is monotone if and only if

$$(k-2q)c_{k,q} \geq (k-2q-1)c_{k,q+1}, \quad \max\{0, k-n\} \leq q \leq \left\lfloor \frac{k-1}{2} \right\rfloor$$

and

$$(n+q-k+1)c_{k,q} \leq (n+q-k+3/2)c_{k,q+1}, \quad \max\{0, k-n-1\} \leq q \leq \left\lfloor \frac{k-2}{2} \right\rfloor.$$

From this description we see that $\mu \in \mathrm{Val}^{\mathrm{U}(n)}$ is monotone if and only if each homogeneous component of μ is monotone. This is a general fact [25]: *A translation invariant continuous valuation is monotone if and only if each homogeneous component is monotone*. This answers a question of P. McMullen [45]. The corresponding statement with *monotone* replaced by *positive* seems to be unknown.

We can draw some more consequences of the above result. The cone of monotone invariant valuations is a polyhedral cone. It is not closed under any of the algebraic constructions from Sect. 3. Let us give some examples (which are extremal rays of the polyhedral cone of monotone invariant valuations).

The valuation

$$\mu := \mu_{4,1} + \frac{2}{3}\mu_{4,2} \in \mathrm{Val}_4^{\mathrm{U}(3)}$$

is monotone, but its Fourier transform

$$\hat{\mu} = \mu_{2,0} + \frac{2}{3}\mu_{2,1} \in \mathrm{Val}_2^{\mathrm{U}(3)}$$

is not monotone (the second inequality with $q = 0$ is violated).

Consider

$$\mu := \mu_{4,0} + \frac{6}{7}\mu_{4,1} + \frac{12}{7}\mu_{4,2} \in \mathrm{Val}_4^{\mathrm{U}(6)}$$
$$\phi := \mu_{4,0} + \frac{4}{3}\mu_{4,1} + \frac{32}{27}\mu_{4,2} \in \mathrm{Val}_4^{\mathrm{U}(6)}.$$

Then μ, ϕ are monotone valuations. From the technique described in Sect. 5.4, one obtains that

$$\mu \cdot \phi = \frac{1002}{81}\mu_{8,2} + \frac{2552}{189}\mu_{8,3} + \frac{6112}{567}\mu_{8,4} \in \mathrm{Val}_8^{\mathrm{U}(6)},$$

which is not monotone (the second inequality with $q = 3$ is violated).

Similarly, the invariant valuations

$$\mu := \mu_{4,0} + \frac{2}{3}\mu_{4,1} + \frac{4}{3}\mu_{4,2} \in \mathrm{Val}_4^{\mathrm{U}(4)}$$

$$\phi := \mu_{6,2} + \frac{2}{3}\mu_{6,3} \in \mathrm{Val}_6^{\mathrm{U}(4)}.$$

are monotone, but their convolution product

$$\mu * \phi = 4\mu_{2,0} + \frac{8}{3}\mu_{2,1} \in \mathrm{Val}_2^{\mathrm{U}(4)},$$

is not monotone (the second inequality with $q = 0$ is violated).

This can be used to show that a monotone version of McMullen's conjecture does not hold true. Taking linear combinations of valuations of the form $K \mapsto \mathrm{vol}(K+A)$ with *positive* coefficients clearly yields monotone valuations and one would expect that every monotone valuation is the limit of such positive linear combinations. But this would imply that the monotone cone is closed under convolution, which is not the case.

6 Other Group Actions

Let G be any compact connected Lie group acting transitively on the unit sphere. We have seen that Val^G is a finite-dimensional algebra, and that there are kinematic and additive G-kinematic formulas. Groups with this property are listed in (22) and (23). The classical case $G = \mathrm{SO}(n)$ was sketched in Sect. 2, while the case $G = \mathrm{U}(n)$ was the subject of Sect. 5.

In this section, we will explain what is known for other G.

6.1 Special Unitary Group

The difference between the integral geometry of $\mathrm{U}(n)$ and that of $\mathrm{SU}(n)$ is not large. Naturally enough, it comes from the complex determinant.

Let V be a hermitian vector space of dimension n, and let $\mathrm{SU}(V) \cong \mathrm{SU}(n)$ be the special unitary group acting on V.

For $k \neq n$, two k-dimensional subspaces are in the same $\mathrm{SU}(n)$-orbit if and only if they are in the same $\mathrm{U}(n)$-orbit. Klain's theorem thus implies that if $\mu \in \mathrm{Val}_k^+(V)$ is even and $\mathrm{SU}(n)$-invariant, then it is already $\mathrm{U}(n)$-invariant.

As it turns out, all $\mathrm{SU}(n)$-invariant valuations are even. This is not trivial if $n \equiv 1 \mod 2$, since in this case $-1 \notin \mathrm{SU}(n)$.

In the middle degree however, things are different. Given an n-dimensional subspace W in a complex n-dimensional vector space, one defines

$$\Theta(W) := \det(w_1, \dots, w_n),$$

where $w_1, \dots, w_n$ is an orthonormal basis of W. Since another choice of basis $w_1, \dots, w_n$ will affect Θ by the factor ± 1 (depending on the orientations), this invariant is a well-defined element of $\mathbb{C}/\{\pm 1\}$. If the restriction of the symplectic form of V on W is not degenerated (which can only happen if n is even), there is a natural choice of orientation of W and $\Theta(W)$ is well-defined in $\mathbb{C}$.

Two $\mathrm{U}(n)$-equivalent n-dimensional subspaces in V belong to the same $\mathrm{SU}(n)$-orbit if and only if their Θ-invariants agree. Using this, one can show that

$$\dim \mathrm{Val}_k^{\mathrm{SU}(n)} = \begin{cases} \dim \mathrm{Val}_k^{\mathrm{U}(n)} & k \neq n \\ \dim \mathrm{Val}_k^{\mathrm{U}(n)} + 4 & k = n, n \equiv 0 \mod 2 \\ \dim \mathrm{Val}_k^{\mathrm{U}(n)} + 2 & k = n, n \equiv 1 \mod 2. \end{cases}$$

The Klain functions of the new valuations may be explicitly described in terms of Tasaki angles and the Θ-invariant. The algebra structure and the kinematic formulas for $\mathrm{SU}(n)$ are variations from the $\mathrm{U}(n)$-case, see [21].

6.2 Exceptional Groups

The group $\mathrm{Spin}(9)$ is the universal (two-fold) cover of $\mathrm{SO}(9)$. It can be explicitly described in a number of ways, for instance using Clifford algebras or using octonions. It acts on a 16-dimensional space $\mathbb{R}^{16}$ which may be interpreted as an octonionic plane $\mathbb{O}^2$. The group $\mathrm{Spin}(7)$ acts on $\mathbb{R}^8$, which is an octonionic line. The group of automorphisms of $\mathbb{O}$ is called G_2, it acts on the space of purely octonionic elements, which is $\mathbb{R}^7$.

Let v be a point of the corresponding unit sphere. The stabilizers of $\mathrm{Spin}(9), \mathrm{Spin}(7)$ and G_2 are given by $\mathrm{Spin}(7), \mathrm{G}_2$ and $S\mathrm{U}(3)$. The action of G_2 on T_vS^7 and that of $\mathrm{SU}(3)$ on T_vS^6 are again transitive on the corresponding unit spheres, which is not the case for the action of $\mathrm{Spin}(7)$ on T_vS^{15}. This makes it rather easy to describe the integral geometry of G_2 and $\mathrm{Spin}(7)$, but for $\mathrm{Spin}(9)$ other methods will be necessary.

Let us first consider G_2. The stabilizer is $\mathrm{SU}(3)$ acting on $W := T_vS^6$. Any G_2-invariant valuation μ may be restricted to a $\mathrm{SU}(3)$-invariant valuation on W. The restriction of μ to W vanishes if and only if μ is simple, since G_2 acts transitively on 6-dimensional subspaces. But simple valuations are of degree 7 (in the even case) or 6 (in the odd case). Hence, if μ is of degree $k \leq 5$, $\mu|_W = 0$ if and only if $\mu = 0$, and therefore $\dim \mathrm{Val}_k^{\mathrm{G}_2} \leq \dim \mathrm{Val}_k^{\mathrm{SU}(3)}$ for $0 \leq k \leq 5$. Using furthermore the symmetry induced by the Hard Lefschetz theorem, we obtain that $\dim \mathrm{Val}_k^{\mathrm{G}_2} = 1$ for $k \neq 3, 4$ and that $\dim \mathrm{Val}_3^{\mathrm{G}_2} = \dim \mathrm{Val}_4^{\mathrm{G}_2}$ is either 1 or 2.

Now we repeat the argument with $\mathrm{Spin}(7)$ instead of G_2 and G_2 instead of $\mathrm{SU}(3)$ to obtain that $\dim \mathrm{Val}_k^{\mathrm{Spin}(7)} = 1$ for $k \neq 4$ and that $\dim \mathrm{Val}_4^{\mathrm{Spin}(7)}$ equals 1 or 2.

It remains to decide whether $\dim \mathrm{Val}_4^{\mathrm{Spin}(7)}$ equals 1 or 2. Since Spin(7) contains SU(4) as a subgroup, it is easy to find a Spin(7)-invariant, not SO(8)-invariant element of degree 4 in $\mathrm{Val}^{\mathrm{SU}(4)}$. Going back, we see that this implies $\dim \mathrm{Val}_3^{\mathrm{G}_2} = \dim \mathrm{Val}_4^{\mathrm{G}_2} = 2$, hence we get the following table:

k	$\dim \mathrm{Val}_k^{\mathrm{G}_2}$	$\dim \mathrm{Val}_k^{\mathrm{Spin}(7)}$
0	1	1
1	1	1
2	1	1
3	2	1
4	2	2
5	1	1
6	1	1
7	1	1
8	–	1

The new valuations in these spaces may be explicitly described. Since there are relatively few of them valuations, it is an easy task to describe the product structures. There are isomorphisms of graded algebras

$$\mathrm{Val}^{\mathrm{G}_2} \cong \mathbb{C}[t,u]/(t^2u, u^2+4t^6)$$
$$\mathrm{Val}^{\mathrm{Spin}(7)} \cong \mathbb{C}[t,v]/(v^2-t^8, vt),$$

where u is of degree 3 and v of degree 4.

From these isomorphisms and the fundamental theorem of algebraic integral geometry, one can derive kinematic formulas and additive formulas for G_2 and Spin(7). We refer to [18] for details.

6.3 Symplectic Groups

The integral geometry of the remaining three sequences $\mathrm{Sp}(n), \mathrm{Sp}(n)\cdot \mathrm{U}(1), \mathrm{Sp}(n)\cdot \mathrm{Sp}(1)$ in the list (22) seems to be quite difficult. The case $n=1$ is already contained in the SU(n)-theory, since $\mathrm{Sp}(1) \cong \mathrm{SU}(2)$ (see also [8, 22]). But even for $n=2$, things are mysterious. From a combinatorial formula in [19], one gets the following dimensions

k	0	1	2	3	4	5	6	7	8
$\dim \mathrm{Val}_k^{\mathrm{Sp}(2)}$	1	1	7	13	29	13	7	1	1
$\dim \mathrm{Val}_k^{\mathrm{Sp}(2)\cdot \mathrm{U}(1)}$	1	1	3	5	9	5	3	1	1
$\dim \mathrm{Val}_k^{\mathrm{Sp}(2)\cdot \mathrm{Sp}(1)}$	1	1	2	3	5	3	2	1	1

It is not known how to describe these valuations geometrically. For general n, there is a combinatorial formula using Young diagrams and Schur functions to compute the dimensions of the spaces $\mathrm{Val}_k^{\mathrm{Sp}(n)}, \mathrm{Val}_k^{\mathrm{Sp}(n)\cdot \mathrm{U}(1)}, \mathrm{Val}_k^{\mathrm{Sp}(n)\cdot \mathrm{Sp}(1)}$. The behavior of these numbers is rather irregular. For large n (in fact $n \geq k$ is enough), these dimensions stabilize to some value $\dim \mathrm{Val}_k^{\mathrm{Sp}(\infty)}$ (resp. $\dim \mathrm{Val}_k^{\mathrm{Sp}(\infty)\cdot \mathrm{U}(1)}$, $\dim \mathrm{Val}_k^{\mathrm{Sp}(\infty)\cdot \mathrm{Sp}(1)}$). These asymptotic values can be explicitly computed, their Poincaré series is given by

$$\sum_{k=0}^{\infty} \dim \mathrm{Val}_k^{\mathrm{Sp}(\infty)} x^k = \frac{x^4-3x^3+6x^2-3x+1}{(1-x)^7(1+x)^3}$$

$$\sum_{k=0}^{\infty} \dim \mathrm{Val}_k^{\mathrm{Sp}(\infty)\cdot \mathrm{U}(1)} x^k = \frac{x^6-2x^5+2x^4+2x^2-2x+1}{(x^2+1)(x^2+x+1)(1+x)^2(1-x)^6}$$

$$\sum_{k=0}^{\infty} \dim \mathrm{Val}_k^{\mathrm{Sp}(\infty)\cdot \mathrm{Sp}(1)} x^k = \frac{x^5+2x^4+x^3+1}{(x^2+1)(x^2+x+1)(1+x)^2(1-x)^4}.$$

This is another hint that quaternionic integral geometry is difficult, since it follows from these expressions that none of the algebras $\mathrm{Val}^{\mathrm{Sp}(\infty)}, \mathrm{Val}^{\mathrm{Sp}(\infty)\cdot \mathrm{U}(1)}$, $\mathrm{Val}^{\mathrm{Sp}(\infty)\cdot \mathrm{Sp}(1)}$ is a freely generated algebra (in contrast to the $\mathrm{U}(n)$-case, where $\mathrm{Val}^{\mathrm{U}(\infty)} \cong \mathbb{C}[t,s]$). It is not even known whether these algebras are finitely generated.

7 Some Open Problems

Let us describe three main problems whose solutions will probably stimulate further progress in algebraic integral geometry.

1. In the Euclidean setting, there are more elaborate versions of the kinematic formulas, the *local* kinematic formulas [53]. They apply to *curvature measures*, which are local versions of the intrinsic volumes. Each intrinsic volume is related to exactly one curvature measure. In the hermitian case, the invariant curvature measures were described in [25]. It is known that there are local kinematic formulas [30]. However, the computation of the coefficients in such a formula is a challenge, since the algebraic machinery from Sect. 3 only applies to valuations and not to curvature measures.
2. We have described in detail the theory of valuations on an affine space. The theory of *valuations on manifolds* was recently worked out, mainly by Alesker [10–13, 15, 17, 22, 23]. This gives the appropriate framework to study integral geometry of projective and hyperbolic spaces. It turns out that on compact rank one symmetric spaces (CROSS), the space of (smooth) invariant valuations is

finite-dimensional and that a version of the fundamental theorem of algebraic integral geometry holds true [15]. To work out the algebraic structure of the space of valuations on a CROSS is a challenge. In the case of $\mathbb{CP}^n$, Abardia [1] and Abardia-Gallego-Solanes [3] studied various Crofton- and Chern-Gauss-Bonnet-type formulas. There also exists a (rather mysterious) conjecture by J. Fu [58] concerning the algebra structure of the space of invariant valuations on $\mathbb{CP}^n$.

3. The intrinsic volumes satisfy a number of important inequalities, like the isoperimetric inequality and the Brunn-Minkowski inequality. What is the corresponding statement in the hermitian case? A special case of this general question is the following. Let $\mu := a\mu_{2,0} + b\mu_{2,1} \in \mathrm{Val}_2^{\mathrm{U}(2)}$ be a positive valuation. What is the minimum of $\mu(K)$ as K ranges over all compact convex bodies of volumes 1? By a version of the isoperimetric inequality, the minimum in the case $a = b$ is achieved by a ball, but the case $a \neq b$ is open.

Acknowledgements I was happy to profit from many discussions with and talks by Semyon Alesker and Joseph Fu on algebraic integral geometry. The present text is strongly influenced by their ideas and I am grateful to them. The terms *Algebraic integral geometry* and *Fundamental theorem of algebraic integral geometry* were invented by Fu. I also thank Gautier Berck, Franz Schuster and Christoph Thäle for numerous useful remarks on this text.

References

1. Abardia, J.: Geometria integral en espais de curvatura holomorfa constant. PhD-Thesis, Universitat Autònoma de Barcelona (2009)
2. Abardia, J., Bernig, A.: Projection bodies in complex vector spaces. Adv. Math. **227**(2), 830–846 (2011)
3. Abardia, J., Gallego, E., Solanes, G.: Gauss-Bonnet theorem and Crofton type formulas in complex space forms. Israel J. Math., arXiv:0904.0336 (to appear)
4. Alesker, S.: Description of continuous isometry covariant valuations on convex sets. Geom. Dedicata **74**(3), 241–248 (1999)
5. Alesker S.: Description of translation invariant valuations on convex sets with solution of P. McMullen's conjecture. Geom. Funct. Anal. **11**(2), 244–272 (2001)
6. Alesker, S.: Hard Lefschetz theorem for valuations, complex integral geometry, and unitarily invariant valuations. J. Diff. Geom. **63**(1), 63–95 (2003)
7. Alesker, S.: Hard Lefschetz theorem for valuations and related questions of integral geometry. In: Geometric aspects of functional analysis, vol. 1850 of Lecture Notes in Math., pp. 9–20. Springer, Berlin (2004)
8. Alesker, S.: SU(2)-invariant valuations. In: Milman, V.D. et al. (eds.) Geometric Aspects of Functional Analysis. Papers from the Israel seminar (GAFA) 2002–2003. Lecture Notes in Mathematics, vol. 1850, pp. 21–29. Springer, Berlin (2004)
9. Alesker, S.: The multiplicative structure on continuous polynomial valuations. Geom. Funct. Anal. **14**(1), 1–26 (2004)
10. Alesker, S.: Theory of valuations on manifolds. I: Linear spaces. Isr. J. Math. **156**, 311–339 (2006)
11. Alesker, S.: Theory of valuations on manifolds. II. Adv. Math. **207**(1), 420–454 (2006)
12. Alesker, S.: Theory of valuations on manifolds: A survey. Geom. Funct. Anal. **17**(4), 1321–1341 (2007)

13. Alesker, S.: Theory of valuations on manifolds. IV. New properties of the multiplicative structure. In: Geometric aspects of functional analysis, vol. 1910 of Lecture Notes in Math., pp. 1–44. Springer, Berlin (2007)
14. Alesker, S.: A Fourier type transform on translation invariant valuations on convex sets. Israel J. Math. **181**, 189–294 (2011)
15. Alesker, S., Bernig, A.: The product on smooth and generalized valuations. Am. J. Math. (to appear)
16. Alesker, S., Bernstein, J.: Range characterization of the cosine transform on higher Grassmannians. Adv. Math. **184**(2), 367–379 (2004)
17. Alesker, S., Fu, J.H.G.: Theory of valuations on manifolds. III. Multiplicative structure in the general case. Trans. Am. Math. Soc. **360**(4), 1951–1981 (2008)
18. Bernig, A.: Integral geometry under G_2 and $Spin(7)$. Israel J. Math. **184**, 301–316 (2011)
19. Bernig, A.: Invariant valuations on quaternionic vector spaces. J. Inst. Math. Jussieu (to appear)
20. Bernig, A.: The normal cycle of a compact definable set. Israel J. Math. **159**, 373–411 (2007)
21. Bernig, A.: A Hadwiger type theorem for the special unitary group. Geom. Funct. Anal. **19**, 356–372 (2009)
22. Bernig, A.: A product formula for valuations on manifolds with applications to the integral geometry of the quaternionic line. Comment. Math. Helv. **84**(1), 1–19 (2009)
23. Bernig, A., Bröcker, L.: Valuations on manifolds and Rumin cohomology. J. Differ. Geom. **75**(3), 433–457 (2007)
24. Bernig, A., Fu, J.H.G.: Convolution of convex valuations. Geom. Dedicata **123**, 153–169 (2006)
25. Bernig, A., Fu, J.H.G.: Hermitian integral geometry. Ann. Math. **173**, 907–945 (2011)
26. Besse, A.L.: Einstein manifolds. Classics in Mathematics. Springer, Berlin (2008) Reprint of the 1987 edition
27. Borel, A.: Some remarks about Lie groups transitive on spheres and tori. Bull. Am. Math. Soc. **55**, 580–587 (1949)
28. Fu, J.H.G.: Algebraic integral geometry. Preprint. available on http://arxiv.org/abs/1103.6256
29. Fu, J.H.G.: Kinematic formulas in integral geometry. Indiana Univ. Math. J. **39**(4), 1115–1154 (1990)
30. Fu, J.H.G.: Curvature measures of subanalytic sets. Am. J. Math. **116**(4), 819–880 (1994)
31. Fu, J.H.G.: Integral geometry and Alesker's theory of valuations. In: Integral geometry and convexity, pp. 17–27. World Sci. Publ., Hackensack, NJ (2006)
32. Fu, J.H.G.: Structure of the unitary valuation algebra. J. Diff. Geom. **72**(3), 509–533 (2006)
33. Fu, J.H.G.: The two faces of Blaschkean integral geometry. Lecture notes from the Summer school. New approaches to curvature, Les Diablerets September 2008, available on http://www.math.uga.edu/fu/research/research.html
34. Howard, R.: The kinematic formula in Riemannian homogeneous spaces. Mem. Am. Math. Soc. **106**(509), vi+69 (1993)
35. Hug, D., Schneider, R., Schuster, R.: The space of isometry covariant tensor valuations. Algebra i Analiz **19**(1), 194–224 (2007)
36. Hug, D., Schneider, R., Schuster, R.: Integral geometry of tensor valuations. Adv. Appl. Math. **41**(4) 482–509 (2008)
37. Huybrechts, D.: Complex geometry. An introduction. Universitext, Springer, Berlin (2005)
38. Kazarnovskiĭ, B.Ja.: On zeros of exponential sums. Dokl. Akad. Nauk SSSR **257**(4), 804–808 (1981)
39. Klain, D.A.: Even valuations on convex bodies. Trans. Am. Math. Soc. **352**(1), 71–93 (2000)
40. Klain, D.A., Rota, G.-C.: Introduction to geometric probability. Lezioni Lincee. [Lincei Lectures]. Cambridge University Press, Cambridge (1997)
41. Ludwig, M.: Intersection bodies and valuations. Am. J. Math. **128**(6), 1409–1428 (2006)
42. Ludwig, M.: Valuations in the affine geometry of convex bodies. In: Integral geometry and convexity, World Sci. Publ., Hackensack, NJ, pp. 49–65 (2006)
43. Ludwig, M., Reitzner M.: A characterization of affine surface area. Adv. Math. **147**(1), 138–172 (1999)

44. Lutwak, E.: Extended affine surface area. Adv. Math. **85**(1), 39–68 (1991)
45. McMullen, P.: Valuations and Euler-type relations on certain classes of convex polytopes. Proc. London Math. Soc. (3) **35**(1), 113–135 (1977)
46. McMullen, P.: Continuous translation-invariant valuations on the space of compact convex sets. Arch. Math. (Basel) **34**(4), 377–384 (1980)
47. McMullen, P.: Isometry covariant valuations on convex bodies. Rend. Circ. Mat. Palermo (2) Suppl. (50), 259–271 (1997); II International Conference in "Stochastic Geometry, Convex Bodies and Empirical Measures" (Agrigento, 1996)
48. Montgomery, D., Samelson, H.: Transformation groups of spheres. Ann. Math. (2) **44**, 454–470 (1943)
49. Nijenhuis, A.: On Chern's kinematic formula in integral geometry. J. Diff. Geom. **9**, 475–482 (1974)
50. Park, H.: Kinematic formulas for the real subspaces of complex space forms of dimension 2 and 3. PhD-thesis University of Georgia 2002
51. Rumin, M.: Differential forms on contact manifolds. (Formes différentielles sur les variétés de contact.). J. Differ. Geom. **39**(2), 281–330 (1994)
52. Santaló, L.A.: Integral geometry and geometric probability. In: With a foreword by Mark Kac, Encyclopedia of Mathematics and its Applications, vol. 1. Addison-Wesley, Reading (1976)
53. Schneider, R.: Convex bodies: the Brunn-Minkowski theory, vol. 44 of Encyclopedia of Mathematics and its Applications. Cambridge University Press, Cambridge (1993)
54. Schneider, R.: Simple valuations on convex bodies. Mathematika **43**(1), 32–39 (1996)
55. Schuster, F.E.: Valuations and Busemann-Petty type problems. Adv. Math. **219**(1), 344–368 (2008)
56. Tasaki, H.: Generalization of Kähler angle and integral geometry in complex projective spaces. In: Steps in differential geometry (Debrecen, 2000), pp. 349–361. Inst. Math. Inform., Debrecen (2001)
57. Tasaki, H.: Generalization of Kähler angle and integral geometry in complex projective spaces. II. Math. Nachr. **252**, 106–112 (2003)
58. Valuations and Integral Geometry. Abstracts from the mini-workshop held January 17–23, 2010. Organized by Semyon Alesker, Andreas Bernig and Franz Schuster, Oberwolfach Reports. Vol. 7, no. 1, pp. 141–178 (2010)
59. van den Dries, L.: Tame topology and o-minimal structures, vol. 248 of London Mathematical Society Lecture Note Series. Cambridge University Press, Cambridge (1998)

Metric Properties of Euclidean Buildings

Linus Kramer

Abstract This is a survey on nondiscrete euclidean buildings, with a focus on metric properties of these spaces.

Euclidean buildings are higher dimensional generalizations of trees. Indeed, the euclidean product X of two (leafless) metric trees T_1, T_2 is already a good "toy example" of a 2-dimensional euclidean building. The space X contains lots of copies of the euclidean plane $\mathbb{E}^2$ and has at the same time a complicated local branching.

Euclidean building were invented by Jacques Tits in the seventies. Similarly as in the case of spherical buildings, their definition was motivated by group theoretic questions. While spherical buildings are by now a standard tool in the structure theory of reductive algebraic groups over arbitrary fields, euclidean buildings are important for the advanced structure theory of reductive groups over fields with valuations. In particular, they are very much linked to number theory and arithmetic geometry.

In the last 25 years, however, euclidean buildings have also become important in geometry. This is due to the fact that euclidean buildings are spaces of nonpositive curvature. But more is true. Together with the Riemannian symmetric spaces of nonpositive curvature, euclidean buildings could be called the islands of high symmetry in the world of CAT(0) spaces. This claim will be made more precise below. Almost inevitably, questions about symmetry, rigidity, or higher rank for CAT(0) spaces lead to these geometries.

I thank Alexander Lytchak, Koen Struyve and Richard Weiss for helpful comments.

L. Kramer (✉)
Mathematisches Institut, Universität Münster, Einsteinstr. 62, 48149 Münster, Germany
e-mail: linus.kramer@uni-muenster.de

C. Bär et al. (eds.), *Global Differential Geometry*, Springer Proceedings in Mathematics 17, DOI 10.1007/978-3-642-22842-1_6,

1 The Definition of Euclidean Buildings

We first recall Tits' definition of a euclidean building. For more details, proofs and further results see Tits [36], Kleiner and Leeb [17], Kramer and Weiss [22] and in particular Parreau [26]. (The axioms used by Kleiner and Leeb [17] look somewhat different from Tits' definition. They were shown to be equivalent to Tits' by Parreau.)

1.1 Euclidean Buildings

Let W be a spherical Coxeter group acting in its natural orthogonal representation on euclidean space $\mathbb{E}^m$. We call the semidirect product $W\mathbb{R}^m$ of W and $(\mathbb{R}^m, +)$ the *affine Weyl group*. From the reflection hyperplanes of W we obtain a decomposition of $\mathbb{R}^m$ into *walls*, *half spaces*, *Weyl chambers* (a Weyl chamber is a fundamental domain for W – these are Tits' *chambres vectorielles*) and *Weyl simplices* (Tits' *facettes vectorielles*). The $W\mathbb{R}^m$-translates of these in $\mathbb{E}^m$ are also called walls, half spaces and Weyl chambers.

Let now X be a metric space. A *chart* is an isometric embedding $\varphi : \mathbb{E}^m \longrightarrow X$, and its image is called an (affine) *apartment*. We call two charts φ, ψ *W-compatible* if $Y = \varphi^{-1}(\psi(\mathbb{E}^m))$ is convex (in the Euclidean sense) and if there is an element $w \in W\mathbb{R}^m$ such that $\psi \circ w|_Y = \varphi|_Y$ (this condition is void if $Y = \varnothing$). We call a metric space X together with a collection $\mathcal{A}$ of charts a *Euclidean building* if it has the following five properties:

(A1) For all $\varphi \in \mathcal{A}$ and $w \in W\mathbb{R}^m$, the composition $\varphi \circ w$ is in $\mathcal{A}$.
(A2) The charts are W-compatible.
(A3) Any two points $p, q \in X$ are contained in some affine apartment.

The charts allow us to map Weyl chambers, walls and half spaces into X; their images are also called Weyl chambers, walls and half spaces. The first three axioms guarantee that these notions are coordinate independent.

(A4) If $C, D \subseteq X$ are Weyl chambers, then there is an affine apartment A such that the intersections $A \cap C$ and $A \cap D$ contain Weyl chambers.
(A5') For every apartment $A \subseteq X$ and every $p \in A$ there is a 1-Lipschitz retraction $h : X \longrightarrow A$ with $h^{-1}(p) = \{p\}$.

Condition (A5') may be replaced by the following condition:

(A5) If A_1, A_2, A_3 are affine apartments which intersect pairwise in half spaces, then $A_1 \cap A_2 \cap A_3 \neq \varnothing$.

See [26] for a thorough discussion of different sets of axioms.

Let p be a point in the apartment A. Axiom (A5') yields a 1-Lipschitz map $v_p : X \longrightarrow \mathbb{E}^m/W$ as follows. We identify (A, p) by means of a coordinate chart

φ with $(\mathbb{E}^m, 0)$, and then we quotient out the W-action. The resulting vector $v_p(q)$ is called the *vector distance* between p and q.

The definition of a euclidean building that we use here is the metric, "nondiscrete" version. It appeared implicitly in [6], and in detail in [36]. There is also the (older) notion of a simplicial affine building; see [1] and in particular [38]. The geometric realization of such a combinatorial affine building is always a euclidean building in our sense (but not vice versa); see Sect. 11.2 in [1].

An important invariant of a euclidean building is its *spherical building at infinity*, $\partial_{\mathcal{A}} X$. This is a combinatorial simplicial complex which is defined as follows. The simplices are equivalence classes of Weyl simplices. Two Weyl simplices are considered to be equivalent if they have finite Hausdorff distance. It turns out that $\partial_{\mathcal{A}} X$ is a (weak) *spherical building* of dimension $m - 1$ (resp., of rank m) [26, Sect. 1.5]. We refer to [34] and [1] for the definition of a simplicial spherical building.

Another important fact is that a euclidean building is always a CAT(0) space; see [26, Sect. 2.3]. This was first observed by Bruhat and Tits in [6], where they proved the CN-inequality.

Automorphisms of euclidean buildings are defined in the obvious way; they are bijections which preserve the charts in the given atlas. Clearly, every automorphism is an isometry of X.

2 Basic and Less Basic Properties

We assume that X is a euclidean building with m-dimensional apartments. First of all, we remark that the atlas $\mathcal{A}$ is by no means unique. However, Parreau proved that there is a unique maximal atlas $\mathcal{A}_{max}$ containing $\mathcal{A}$; see [26, Sect. 2.6]. The apartments in the maximal atlas have a simple characterization.

Theorem 2.1 ([26, Sect. 2.6] and [17, Sect. 4.6]). *Let X be a euclidean building with m-dimensional apartments. Suppose that $F \subseteq X$ is a subspace isometric to some $\mathbb{E}^\ell$. Then there exists an apartment in the maximal atlas containing F. In particular, $\ell \leq m$ and the apartments of the maximal atlas are precisely the maximal flats in X.*

The metric realization of the spherical building $\partial_{\mathcal{A}_{max}} X$ can be identified in a canonical way with the Tits boundary of the CAT(0) space X. We remark that the dimension m of the apartments coincides with the covering dimension of X as a topological space; see [23, Prop. 3.3] or [20, Thm. B]. Moreover, X is an AR (an absolute retract for the class of metric spaces).

Next, we note that the Weyl group may be "too big": there might be types of walls which never appear as branchings between apartments. A wall M in a euclidean building X is called *thick* if it can be written as the intersection of three apartments. We call a point $p \in X$ *thick* if every wall passing through p is thick. Now we can make the statement about the Weyl group being too big more precise: if X contains no thick points, then there is a (unique) euclidean building X_{th} (with a smaller Weyl group) containing a thick point, and X is a euclidean product $X \cong X_{th} \times \mathbb{E}^k$, for some $k \geq 0$; see [17, Sect. 4.9] and [22, Sect. 10]. For the thick part X_{th}, there is the following trichotomy.

Proposition 2.2 ([22, Sect. 10]). *Let X be an irreducible euclidean building of dimension $m \geq 2$ containing a thick point. Then there are the following three possibilities.*

(I) There is a unique thick point which is contained in every affine apartment of X. In this case X is a euclidean cone over a spherical building.
(II) The set of thick points is a closed, discrete and cobounded subset in X and in every apartment of X. Then X is the geometric realization of a simplicial affine building.
(III) The set of thick points is dense in X and in every apartment of X.

A simplicial affine building (type II) is called *thick* if every vertex of the simplicial structure is thick.

There are many 2-dimensional euclidean buildings. In fact, there are "free constructions" which show that it is impossible to classify these spaces. In higher dimensions, the picture is completely different. We call a euclidean building X a *Bruhat-Tits building* if the spherical building at infinity is a *Moufang building*; see [37]. Roughly speaking, the Moufang condition says that there are certain automorphisms, called *root automorphisms*, that fix a large subset pointwise, and yet act transitively on another subset. The following deep result is again due to Tits [36].

Theorem 2.3. *Let X be an irreducible euclidean building of dimension $m \geq 3$ containing a thick point. Then $\partial_{\mathcal{A}} X$ is Moufang, and all root automorphisms of $\partial_{\mathcal{A}} X$ extend to isometries of X. In particular, the isometry group of X is transitive on the apartments of X.*

Tits' article [36] contains in fact a complete classification of these buildings in terms of algebraic data. We remark that if a Bruhat-Tits building is not of type (I), the group generated by the root automorphisms acts with cobounded orbits on X.

It is by no means clear that every combinatorial automorphism of $\partial_{\mathcal{A}} X$ extends to an isometry of X. Surprisingly, the following is true.

Theorem 2.4 ([38, Sect. 27.6]). *Let X be a thick irreducible simplicial Bruhat-Tits building of dimension $m \geq 2$. Then every automorphism of $\partial_{\mathcal{A}_{max}} X$ extends to an isometry of X. Moreover, $\partial_{\mathcal{A}_{max}} X$ determines X up to isomorphism.*

The proof depends on the purely algebraic fact that a field admits at most one discrete complete valuation. It would be interesting to have a geometric proof for this. More generally, there is the following open problem.

Question 2.5. Is a thick irreducible simplicial affine building of dimension $m \geq 2$ uniquely determined by the spherical building $\partial_{\mathcal{A}_{max}} X$? Does every combinatorial automorphism of $\partial_{\mathcal{A}_{max}} X$ extend to X?

The answer is negative if X is not assumed to be simplicial. For locally finite simplicial thick irreducible affine buildings, the answer is positive [24].

3 Characterizations

The following very general characterization of locally finite (simplicial) euclidean buildings is due to Kleiner.

Theorem 3.1. *Let X be a locally compact CAT(0) space of dimension m. Suppose that any two points $x, y \in X$ are contained in some flat $A \cong \mathbb{E}^m$. Then X is a euclidean building.*

This result was not published by Kleiner; a proof was given by Balser and Lytchak in [3, Cor. 1.7]. The dimension may be taken to be the covering dimension; since X is locally compact, the covering dimension coincides with Kleiner's geometric dimension [16]. The following example shows that local compactness is crucial.

Example 3.2. Let Γ_n, for $n \geq 3$, be a family of thick generalized n-gons (1-dimensional spherical buildings whose Weyl group is dihedral of order $2n$). Such generalized n-gons exist by Tits' free construction [35], see also [33]. Let X_n be the euclidean cone over Γ_n, with cone point o_n. Then X_n is a 2-dimensional euclidean building with precisely one thick point. Now consider the asymptotic cone (or ultralimit) X over the family $\{(X_n, o_n) \mid n \geq 3\}$ (with respect to a constant scaling sequence and a nonprincipal ultrafilter μ on the index set $\mathbb{N}_{\geq 3}$). Then X is a complete CAT(0) space. Any two points in X are contained in some copy of $\mathbb{E}^2$. The "spherical Weyl group" W that describes the transition functions between these "apartments" is, however, the orthogonal group $W = \mathrm{O}(2)$. Using a similar argument as in [20, Sect. 7] (or by Kleiner's results in [16], see also Lytchak [25, Sect. 11.3]) one can show that X is 2-dimensional. But X is certainly not a euclidean building. Instead of the cones X_n, one could also use the euclidean buildings constructed recently by Berenstein and Kapovich [4] in order to get a more interesting asymptotic cone X.

In a somewhat more combinatorial setting, there is the following result of Charney and Lytchak. A CAT(0) space X has the *discrete extension property* if for every geodesic $\gamma = [a, b] \subseteq X$, the set of the directions of geodesics extending γ beyond b is nonempty and discrete.

Theorem 3.3. *Let X be a CAT(0) space of dimension $m \geq 2$ which is a piecewise euclidean cell complex. If X has the discrete extension property, then X is a euclidean building.*

We remark that a locally compact euclidean building always admits a euclidean cell structure. This is not true for general euclidean buildings. Finally, we should mention here the following result by Leeb [24].

Theorem 3.4. *Let X be a locally compact CAT(0) space with extendible geodesics. If the Tits boundary of X is an irreducible spherical building of rank at least 2, then X is either a Riemannian symmetric space of noncompact type or a simplicial euclidean building.*

4 Isometries and Automorphisms

If X is a euclidean building containing a thick point, then an isometry of X is almost the same as an automorphism.

Theorem 4.1 ([26, Sect. 4]). *Let g be an isometry of a euclidean building X. Assume that X contains a thick point. Then there exists an element $\gamma \in \mathrm{Nor}_{\mathrm{O}(m)}(W)$ such that $g \circ \psi \circ \gamma \in \mathcal{A}_{max}$ holds for all $\psi \in \mathcal{A}_{max}$.*

Such a map γ induces a diagram automorphism of the Coxeter group W; one also calls such a g a *non-type-preserving automorphism.*

Suppose that g is an isometry of a metric space (X, d). The *displacement function* of g is the nonnegative real function $d_g : x \longmapsto d(x, g(x))$. The infimum of $d_g(X)$ is the *translation length* l_g of g. We call an isometry g

elliptic if g has a fixed point.
hyperbolic if d_g attains a positive minimum.
parabolic if d_g does not attain its minimum.

If X is a Riemannian symmetric space of nonpositive curvature, then all three types of isometries appear in the isometry group. This is not true for euclidean buildings.

Theorem 4.2 ([26, Sect. 4]). *Let g be an isometry of a complete euclidean building X containing a thick point. Then g is either elliptic or hyperbolic.*

(Struyve informed me that he can prove this also for noncomplete euclidean buildings.) The next result was proved by Rapoport and Zink [28] for the Bruhat-Tits building of GL_n over a field with discrete valuation, and then extended using Landvogt's Embedding Theorem to other Bruhat-Tits buildings. However, there is a much simpler proof using CAT(0) geometry, which applies to all euclidean buildings, cp. [30] – the author found a somewhat simpler proof (unpublished). We put $X_r = \{q \in X \mid d_g(q) \leq r\}$. These sublevel sets form a filtration of X by convex sets.

Theorem 4.3. *Let g be an isometry of a complete euclidean building X containing a thick point. There exists a positive constant $c > 0$ (depending only on the Weyl group W) such that the following holds. If p is a point with $d(p, X_r) = t > 0$, then*

$$c \cdot t + r \leq d_g(p) \leq 2t + r.$$

The second inequality is trivial, the interesting fact is the lower estimate. We finally note the following (completely elementary) fact.

Lemma 4.4. *Let g be a nontrivial isometry of a euclidean building X containing a thick point. Then* $\sup d_g(X) = \infty$.

Proof. Suppose $r = \sup d_g(X) < \infty$. If A is an apartment in X, then $g(A)$ has Hausdorff distance at most r from A. Then A and $g(A)$ have the same boundary at infinity. By [26, p. 10], $A = g(A)$. Thus g fixes all apartments setwise, and therefore all thick walls and thick points. Since every apartment contains a thick point, g fixes every apartment pointwise. Thus $g = id_X$. □

We end this section with some remarks on noncomplete euclidean buildings. Struyve recently proved the following generalization of the Bruhat-Tits Fixed Point Theorem. If a finitely generated group acts acts isometrically and with bounded orbits on a euclidean building, then it has a fixed point [32]. Moreover, he showed that the main rigidity results in [22] also hold if the completeness assumptions on the euclidean buildings are dropped (unpublished). Finally, he and Martin, Schillewaert and Steinke extended results in [6] about noncomplete Bruhat-Tits buildings, by giving algebraic conditions on the underlying fields (unpublished).

5 Kostant Convexity

We first recall the statement of Kostant's Convexity Theorem [55] for Riemannian symmetric spaces. Let G be a simple noncompact Lie group with Iwasawa decomposition $G = KAU$. The group K is maximal compact, A is diagonizable, and U is unipotent. The group $W = \mathrm{Nor}_K(A)/\mathrm{Cen}_K(A)$ is the associated Weyl group.

The solvable group AU acts regularly on the Riemannian symmetric space $X = G/K$. Let $o \in X$ denote the point stabilized by K. The A-orbit $E = A(o) \subseteq X$ is a maximal flat in X. The projection $AU \longrightarrow AU/U \cong A$ induces a natural 1-Lipschitz map $\rho_U : X \longrightarrow E$ which we call the *Iwasawa projection*. Let $p \in E$. The Convexity Theorem says that

$$\rho_U(K(p)) = \mathrm{conv}(W(p)),$$

the image of the K-orbit of p in X under the Iwasawa projection is the convex hull of the W-orbit of p in E.

Geometrically, the Iwasawa decomposition can also be described as follows. The group U determines a chamber C of the spherical building at infinity of X. The maximal flats in X containing C in their boundary form a foliation of X. The Iwasawa projection identifies each leaf by means of the U-action with the leaf $A(o)$.

Suppose now that X is a euclidean building and that C is a chamber at infinity. We fix an apartment $E \subseteq X$ containing C in its boundary. If $E' \subseteq X$ is any other apartment containing C in its boundary, then $E \cap E'$ contains a Weyl chamber representing C. Thus, there is a canonical isometry $E' \longrightarrow E$ fixing $E \cap E'$ pointwise. These isometries fit together to a 1-Lipschitz retraction $\rho_C : X \longrightarrow E$.

Suppose now that $o, p \in E$ are special vertices. (A vertex $p \in E$ is called special if the reflections along the thick walls in E passing through p generate the spherical Weyl group W.) Let $S \subseteq X$ denote the set of all special vertices in X that have the same vector distance from o as p. This set S corresponds to the orbit $K(p)$ in the Riemannian symmetric case. If the euclidean building X happens to be a Bruhat-Tits building, then S is indeed the K-orbit of p, where K is the stabilizer of o. The following result was proved by Hitzelberger [13] in 2007.

Theorem 5.1. *Suppose that X is a thick simplicial euclidean building. With the same notation as above, suppose that o, p are special vertices (see [6] for the definition of a special vertex). Then*

$$\rho_C(S) = \{q \in \mathrm{conv}(W(p)) \mid q \text{ has the same type as } p\},$$

the image of S is the set of all vertices in E which are in the convex hull of the W-orbit of p in E and have the same type as p.

This result had been announced by Silberger [31] for the special case that X is the Bruhat-Tits building of a simple p-adic algebraic group (but the proof, which relied on a case-by-case analysis, was never published). The difficult part of the proof is to show that the map is onto. For the special case of Bruhat-Tits buildings, the theorem may be restated as a fact about intersections of certain double cosets in the group. The result was recently extended by Hitzelberger to general euclidean buildings [14].

6 Rigidity

We first recall some notions from coarse geometry [29]. A map $f : X \longrightarrow Y$ between metric spaces is called *controlled* if there is a monotonic real function $\rho : \mathbb{R}_{\geq 0} \longrightarrow \mathbb{R}_{\geq 0}$ such that

$$d_Y(f(x), f(y)) \leq \rho(d_X(x, y))$$

holds for all $x, y \in X$. If in addition the preimage of every bounded set is bounded, then f is called a *coarse map*. Neither f nor ρ is required to be continuous. Note that the image of a bounded set under a controlled map is bounded. Two maps $g, f : X \rightrightarrows Y$ between metric spaces have *finite distance* if the set $\{d_Y(f(x), g(x)) \mid x \in X\}$ is bounded. This is an equivalence relation which leads to the *coarse metric category* whose objects are metric spaces and whose morphisms are equivalence classes of coarse maps. A *coarse equivalence* is an isomorphism in this category. We remark that a coarse equivalence between geodesic metric spaces is the same as a quasi-isometric equivalence.

Prasad proved in 1978 the following analog of Mostow's Rigidity Theorem.

Theorem 6.1 ([27]). *Let X and Y be thick simplicial, irreducible and locally finite Bruhat-Tits buildings of rank at least 2. Suppose that a group Γ acts cocompactly and properly discontinuously on both spaces. Then there is a Γ-equivariant simplicial isomorphism between X and Y.*

The group Γ appearing in Prasad's Theorem is finitely presentable. From the Γ-action, one obtains a Γ-equivariant coarse equivalence $f : X \longrightarrow Y$ which plays a crucial role in the proof. About twenty years later, Kleiner and Leeb [17] proved the following generalization of Prasad's Theorem.

Theorem 6.2 ([17]). *Let X and Y be complete Bruhat-Tits buildings whose de Rham factors all have rank at least 2. Suppose that $f : X \longrightarrow Y$ is a coarse equivalence. Then there is an isometry $\bar{f} : X \longrightarrow Y$ (possibly after rescaling the metrics on the de Rham factors of Y) which has finite distance from f.*

The strategy of their proof is roughly as follows. Using *asymptotic cones*, Kleiner and Leeb show that the f-image of a maximal flat $E \subseteq X$ has finite Hausdorff distance from a (necessarily unique) maximal flat $E' \subseteq Y$. This fact is then used to set up a one-to-one correspondence between the maximal bounded subgroups of the isometry groups of the two Bruhat-Tits buildings. The maximal bounded subgroups, in turn, correspond to (certain) points in the buildings. In this way, they construct an equivariant isometry.

Weiss and the author proved in 2009 a more general result which is valid for all euclidean buildings.

Theorem 6.3 ([22, Thm. III]). *Let X and Y be complete euclidean buildings containing thick points, and without rank 1 de Rham factors. Suppose that f : $X \longrightarrow Y$ is a coarse equivalence. Then there is an isometry $\bar{f} : X \longrightarrow Y$. If no de Rham factor of X is a euclidean cone, then f has finite distance from $\bar{f}$.*

The proof relies, among other things, on the following result about trees.

Theorem 6.4 ([22, Thm. I]). *Let T, T' be two complete $\mathbb{R}$-trees without leaves. Suppose that a group G acts isometrically on both trees, and that this action is 2-transitive on the ends. Suppose that $f : T \longrightarrow T'$ is a coarse equivalence whose induced boundary map $\partial T \longrightarrow \partial T'$ is G-equivariant. Then T and T' are G-equivariantly isometric.*

The proof of Theorem 6.3 proceeds roughly as follows. The first step is a result due to Kleiner and Leeb which was already mentioned: the f-image of an apartment $E \subseteq X$ has finite Hausdorff distance from a (unique) apartment $E' \subseteq Y$. But then we follow a different line. We show directly that f induces a combinatorial isomorphism f_* between the Tits boundaries of X and Y. (For the case of simplicial Bruhat-Tits buildings, this implies by Theorem 2.4 already that X and Y are combinatorially isomorphic.) Next, we show that we obtain a coarse bijection between the so-called *wall trees* of X and Y. Since these trees have large holonomy groups, we may apply Theorem 6.4. In this way we get equivariant isomorphisms between the wall trees, and thus, by Tits [36], an isometry between the euclidean buildings. We remark that the main results in [36] also enter as important ingredients into the proof of [17].

7 Locally Compact Bruhat-Tits Buildings

In the mid-nineties, Grundhöfer, Knarr and the author completed the classification of all compact connected spherical buildings admitting a chamber transitive automorphism group. Such buildings arise for example as boundaries of Riemannian symmetric spaces. The proof and the method of the classification built on earlier work by Salzmann, Löwen, Burns and Spatzier. Briefly, it may be stated as follows.

Theorem 7.1 ([10, 11, 19]). *Let B be a compact spherical building (in the sense of [7]) without rank 1 factors. Suppose that B is (locally) connected and admits a chamber transitive group of continuous automorphisms. Then B is the Tits boundary of a Riemannian symmetric space of noncompact type.*

There should be an analog of this result, corresponding to the boundaries of locally compact euclidean buildings. The following conjecture is wide open (even for buildings of type A_2, i.e. compact projective planes).

Conjecture 7.2. Let B be a compact spherical building (in the sense of [7]) without rank 1 factors. Suppose that B is totally disconnected and admits a chamber transitive groups of continuous automorphisms. Then B is the Tits boundary of a locally finite simplicial Bruhat-Tits building.

The problem is that in comparison to Theorem 7.1, no homotopy theory is available. Presently, a proof of this conjecture seems to be out of reach. Assuming the Moufang property, we showed however the following.

Theorem 7.3 ([12]). *Let B be a compact spherical building (in the sense of [7]) without rank 1 factors. Suppose that B is totally disconnected and Moufang. Then B is the Tits boundary of a locally finite simplicial Bruhat-Tits building.*

We recall that the Moufang property is automatically satisfied if all irreducible factors of B have rank at least 3; see [34] and [37]. The proof of Theorem 7.3 relies very much on the classification of spherical Moufang buildings due to Tits and Weiss.

8 Lattices

Let X be a complete and locally compact CAT(0) space and let Γ be a group of isometries. We call Γ a *uniform lattice* if Γ acts properly discontinuously and cocompactly on X (such groups are also called CAT(0) groups). Borel's Density

Theorem says that Riemannian symmetric spaces of noncompact type admit (many) uniform lattices. Such a uniform lattice is always finitely presentable. However, very few presentations of lattices are known. Essert observed the following correspondence between uniform lattices acting regularly on the 1-simplices of a given type of a 2-dimensional locally finite simplicial euclidean building and Singer groups. A *Singer group* is a subgroup of the automorphism group a finite generalized polygon (a 1-dimensional spherical building) which acts regularly on the vertices of a given type. Singer groups are studied by finite geometers and group theorists, and quite a few constructions are known. Essert showed that from a collection of Singer groups, one can construct a 2-dimensional *complex of groups* which unfolds to a lattice Γ acting on such a 2-dimensional euclidean building. Specific examples are presentations such as

$$\langle a,b,c \mid a^7 = b^7 = c^7 = abc = a^3b^3c^3 = 1\rangle$$

or

$$\langle a,b,c \mid a^{13} = b^{13} = c^{13} = ab^3c^9 = a^3b^9c = a^9bc^3 = 1\rangle.$$

These explicit representations allow, for example, to compute the group homology of the lattices. It is clear that "most" of the buildings X that he constructed in this way are "exotic", i.e. they are not Bruhat-Tits buildings. There are presently many open questions about these lattices Γ, e.g., about commensurabilty, quasi-isometric type, or the covolume. We refer to [9] for details and more results.

9 Noncrystallographic Weyl Groups

The Weyl group of a Bruhat-Tits building arising from a reductive algebraic group over a field with valuation is always crystallographic. Also, the Weyl group of a simplicial euclidean building is necessarily a crystallographic group. But in the definition of a euclidean buildings, there is no reason to assume that W satisfies the crystallographic condition. It was remarked (without giving details) by Tits [36] that there are Bruhat-Tits buildings with non-crystallographic Weyl groups. An explicit construction of such euclidean buildings, defined over certain, very special fields, can be found in [15]. Their Weyl groups are dihedral groups of order 16, and their Tits boundaries are so-called Moufang generalized octagons.

In a completely different way, Berenstein and Kapovich constructed "wild" 2-dimensional euclidean buildings whose Weyl groups are dihedral groups of arbitrary order [4]. It would be interesting to see if the construction can be done in such a way that it yields highly transitive automorphism groups, as was the case for the 1-dimensional spherical buildings constructed by Tent in [33].

Acknowledgement The work by Essert, Hitzelberger, Kramer and Weiss that we report on was supported by the DFG through SPP 1154 *Differentialgeometrie*, GK *Analytische Topologie and Metageometrie*, SFB 478 and SFB 878.

References

1. Abramenko, P., Brown, K.S.: Buildings, Graduate Texts in Mathematics, vol. 248. Springer, New York (2008)
2. Ballmann, W., Gromov, M., Schroeder, V.: Manifolds of nonpositive curvature. Progress in Mathematics, vol. 61. Birkhäuser Boston, Boston, MA (1985)
3. Balser, A., Lytchak, A.: Building-like spaces. J. Math. Kyoto Univ. **46**(4), 789–804 (2006)
4. Berenstein, A., Kapovich, M.: Affine buildings for dihedral groups. To appear in: Geom. Dedicata.
5. Bridson, M.R., Haefliger, A.: Metric spaces of non-positive curvature. Grundlehren der Mathematischen Wissenschaften, vol. 319. Springer, Berlin (1999)
6. Bruhat, F., Tits, J.: Groupes réductifs sur un corps local. Inst. Hautes Études Sci. Publ. Math. **41**, 5–251 (1972)
7. Burns, K., Spatzier, R.: On topological Tits buildings and their classification. Inst. Hautes Études Sci. Publ. Math. **65**, 5–34 (1987)
8. Charney, R., Lytchak, A.: Metric characterizations of spherical and Euclidean buildings. Geom. Topol. **5**, 521–550 (2001)
9. Essert, J.: A geometric construction of panel-regular lattices in buildings of types $\tilde{A}_2$ and $\tilde{C}_2$. Preprint 2009, submitted. aXiv:0908.2713
10. Grundhöfer, T., Knarr, N., Kramer, L.: Flag-homogeneous compact connected polygons. Geom. Dedicata **55**(1), 95–114 (1995)
11. Grundhöfer, T., Knarr, N., Kramer, L.: Flag-homogeneous compact connected polygons. II. Geom. Dedicata **83**(1–3), 1–29 (2000)
12. Grundhöfer, T., Kramer, L., Van Maldeghem, H., Weiss, R.M.: Compact totally disconnected Moufang buildings. To appear in Tohoku Math. Journal
13. Hitzelberger, P.: Kostant convexity for affine buildings. Forum Math. **22**, 959–971 (2010)
14. Hitzelberger, P.: Non-discrete affine buildings and convexity. Advances in Mathematics **227**(1), 210–244 (2011)
15. Hitzelberger, P., Kramer, L., Weiss, R.M.: Non-discrete Euclidean buildings for the Ree and Suzuki groups. Am. J. Math. **132**(4), 1113–1152 (2010)
16. Kleiner, B.: The local structure of length spaces with curvature bounded above. Math. Z. **231**(3), 409–456 (1999)
17. Kleiner, B., Leeb, B.: Rigidity of quasi-isometries for symmetric spaces and Euclidean buildings. Inst. Hautes Études Sci. Publ. Math. **86**, 115–197 (1997/1998)
18. Kostant, B.: On convexity, the Weyl group and the Iwasawa decomposition. Ann. Sci. École Norm. Sup. (4) **6**, 413–455 (1973/1974)
19. Kramer, L.: Homogeneous spaces, Tits buildings, and isoparametric hypersurfaces. Mem. Am. Math. Soc. **158**(752), xvi+114 (2002)
20. Kramer, L.: On the local structure and the homology of CAT(κ) spaces and euclidean buildings. Adv. Geom. **11**, 347–369 (2011)
21. Kramer, L., Tent, K.: Asymptotic cones and ultrapowers of Lie groups. Bull. Symbolic Logic **10**(2), 175–185 (2004)
22. Kramer, L., Weiss, R.M.: Coarse rigidity of euclidean buildings. Preprint 2009, submitted. arXiv:0902.1332
23. Lang, U., Schlichenmaier, T.: Nagata dimension, quasisymmetric embeddings, and Lipschitz extensions. Int. Math. Res. Not. **2005**(58), 3625–3655.
24. Leeb, B.: A characterization of irreducible symmetric spaces and Euclidean buildings of higher rank by their asymptotic geometry. Bonner Mathematische Schriften, vol. 326. University of Bonn, Bonn (2000)
25. Lytchak, A.: Rigidity of spherical buildings and joins. Geom. Funct. Anal. **15**(3), 720–752 (2005)
26. Parreau, A.: Immeubles affines: construction par les normes et étude des isométries. In: Crystallographic Groups and Their Generalizations (Kortrijk), pp. 263–302. Contemp. Math., vol. 262. Amer. Math. Soc., Providence, RI (1999)

27. Prasad, G.: Lattices in semisimple groups over local fields. In: Studies in Algebra and Number Theory, pp. 285–356. Adv. in Math. Suppl. Stud., vol. 6. Academic, New York.
28. Rapoport, M., Zink, T.: A finiteness theorem in the Bruhat-Tits building: an application of Landvogt's embedding theorem. Indag. Math. (N.S.) **10**(3), 449–458 (1999)
29. Roe, J.: Lectures on coarse geometry. University Lecture Series, vol. 31, Amer. Math. Soc., Providence, RI (2003)
30. Rousseau, G.: Exercices métriques immobiliers. Indag. Math. (N.S.) **12**(3), 383–405 (2001)
31. Silberger, A.J.: Convexity for a simply connected p-adic group. Bull. Am. Math. Soc. **81**(5), 910–912 (1975)
32. Struyve, K.: (Non-)completeness of $\mathbb{R}$-buildings and fixed point theorems. Groups, Geometry and Dynamics **5**(1), 177–188 (2011)
33. Tent, K.: Very homogeneous generalized n-gons of finite Morley rank. J. London Math. Soc. (2) **62**(1), 1–15 (2000)
34. Tits, J.: Buildings of spherical type and finite BN-pairs. Lecture Notes in Mathematics, vol. 386. Springer, Berlin (1974)
35. Tits, J.: Endliche Spiegelungsgruppen, die als Weylgruppen auftreten. Invent. Math. **43**(3), 283–295 (1977)
36. Tits, J.: Immeubles de type affine. In: Buildings and the geometry of diagrams (Como), pp. 159–190. Lecture Notes in Math., vol. 1181. Springer, Berlin (1984)
37. Weiss, R.M.: The structure of spherical buildings. Princeton University Press, Princeton, NJ (2003)
38. Weiss, R.M.: The structure of affine buildings. Annals of Mathematics Studies, vol. 168. Princeton University Press, Princeton, NJ (2009)

27. Prasad, G.: Lattices in semisimple groups over local fields. In: Studies in Algebra and Number Theory, pp. 285–356. Adv. in Math. Suppl. Stud., vol. 6. Academic, New York
28. Rapoport, M., Zink, Th.: A finiteness theorem in the Bruhat–Tits building: an application of Landvogt's embedding theorem. Indag. Math. (N.S.) 10(3), 449–458 (1999)
29. Rosen, J.: Lectures on ... Lecture Notes Series ... Amer. Math. Soc., Providence, RI (...)

Part II
Geometric Analysis

Holonomy Groups of Lorentzian Manifolds: A Status Report

Helga Baum

Abstract In this survey we review the state of art in Lorentzian holonomy theory. We explain the recently completed classification of connected Lorentzian holonomy groups, we describe local and global metrics with special Lorentzian holonomy and some topological properties, and we discuss the holonomy groups of Lorentzian manifolds with parallel spinors as well as Lorentzian Einstein metrics with special holonomy.

1 Introduction

The holonomy group of a semi-Riemannian manifold (M, g) is the group of all parallel displacements along curves which are closed in a fixed point $x \in M$. This group is a Lie subgroup of the group of orthogonal transformations of (T_xM, g_x). The concept of holonomy group was probably first successfully applied in differential geometry by Cartan [31–33], who used it to classify symmetric spaces. Since then, it has proved to be a very important concept. In particular, it allows to describe parallel sections in geometric vector bundles associated to (M, g) as holonomy invariant objects and therefore by purely algebraic tools. Moreover, geometric properties like curvature properties can be read off if the holonomy group is special, i.e., a proper subgroup of $\mathrm{O}(T_xM, g_x)$. One of the important consequences of the holonomy notion is its application to the "classification" of special geometries that are compatible with Riemannian geometry. For each of these geometries an own branch of differential geometry has developed, for example Kähler geometry (holonomy $\mathrm{U}(n)$), geometry of Calabi–Yau manifolds ($\mathrm{SU}(n)$), hyper-Kähler geometry ($\mathrm{Sp}(n)$), quaternionic Kähler geometry ($\mathrm{Sp}(n) \cdot \mathrm{Sp}(1)$), geometry of G_2-manifolds or of

H. Baum (✉)
Humboldt Universität Berlin, Institut für Mathematik, Sitz: Rudower Chaussee 25, 12489 Berlin, Germany
e-mail: baum@mathematik.hu-berlin.de

C. Bär et al. (eds.), *Global Differential Geometry*, Springer Proceedings in Mathematics 17, DOI 10.1007/978-3-642-22842-1_7,

Spin(7)-manifolds. In physics there is much interest in semi-Riemannian manifolds with special holonomy, since they often allow to construct spaces with additional supersymmetries (Killing spinors). The development of holonomy theory has a long history. We refer for details to [19, 23, 24].

Whereas the holonomy groups of simply connected Riemannian manifolds are completely known since the 50th of the last century, the classification of holonomy groups for pseudo-Riemannian manifolds is widely open, only the irreducible holonomy representations of simply connected pseudo-Riemannian manifolds are known [16, 17]. The difficulty in case of indefinite metrics is the appearance of degenerate holonomy invariant subspaces. Such holonomy representations are hard to classify.

The holonomy groups of 4-dimensional Lorentzian manifolds were classified by physicists working in General Relativity [76, 77], the general dimension was long time ignored. Due to the development of supergravity and string theory in the last decades physicists as well as mathematicians became more interested in higher dimensional Lorentzian geometry. The search for special supersymmetries required the classification of holonomy groups in higher dimension. In the beginning of the 90th, Berard-Bergery and his students began a systematic study of Lorentzian holonomy groups. They discovered many special features of Lorentzian holonomy. Their groundbreaking paper [15] on the algebraic structure of subgroups $H \subset \mathrm{SO}(1, n-1)$ acting with a degenerate invariant subspace was the starting point for the classification. Leistner [65, 68] completed the classification of the (connected) Lorentzian holonomy groups by the full description of the structure of such $H \subset \mathrm{SO}(1, n-1)$ which can appear as holonomy groups. It remained to show that any of the groups in Leistner's holonomy list can be realized by a Lorentzian metric. Many realizations were known before but some cases were still open until Galaev [40] finally found a realization for all of the groups.

The aim of this review is to describe these classification results and the state of art in Lorentzian holonomy theory. In Sect. 2 we give a short basic introduction to holonomy theory and recall the classification of Riemannian holonomy groups. In Sect. 3 the algebraic classification of (connected) Lorentzian holonomy groups is explained. Section 4 is devoted to the realization of the Lorentzian holonomy groups by local metrics, which completes the classification of these groups. Section 5 deals with global aspects of Lorentzian holonomy theory. First we describe Lorentzian symmetric spaces and their holonomy groups. After that we discuss the holonomy group of Lorentzian cones and a construction of Lorentzian metrics with special holonomy on non-trivial torus bundles. Furthermore, we describe a class of examples of geodesically complete resp. globally hyperbolic Lorentzian manifolds with special holonomy. We close the section with some results on topological properties of Lorentzian manifolds with special holonomy. In Sect. 6 we consider the relation between holonomy groups and parallel spinors, derive the Lorentzian holonomy groups which allow parallel spinors and discuss a construction of globally hyperbolic Lorentzian manifolds with complete Cauchy surfaces and parallel spinors. The final part deals with Lorentzian Einstein manifolds with special holonomy. We describe their holonomy groups and the local structure of the metrics.

2 Holonomy Groups of Semi-Riemannian Manifolds

Let (M, g) be a connected[1] n-dimensional manifold with a metric g of signature[2] (p, q), and let ∇^g be the Levi-Civita connection of (M, g). If $\gamma : [a, b] \to M$ is a piecewise smooth curve connecting two points x and y of M, then for any $v \in T_x M$ there is a uniquely determined parallel vector field X_v along γ with initial value v:

$$\frac{\nabla^g X_v}{dt}(t) = 0 \quad \forall\, t \in [a, b], \quad X_v(a) = v.$$

Since the Levi-Civita connection is metric, the parallel displacement

$$\begin{aligned} \mathscr{P}^g_\gamma : T_x M &\longrightarrow T_y M \\ v &\longmapsto X_v(b) \end{aligned}$$

defined by X_v is a linear isometry between $(T_x M, g_x)$ and $(T_y M, g_y)$. In particular, if γ is closed, $\mathscr{P}^g_\gamma$ is an orthogonal map on $(T_x M, g_x)$. The *holonomy group of* (M, g) *with respect to* $x \in M$ is the Lie group

$$\mathrm{Hol}_x(M, g) := \{\mathscr{P}^g_\gamma : T_x M \to T_x M \mid \gamma \in \Omega(x)\} \subset \mathrm{O}(T_x M, g_x),$$

where $\Omega(x)$ denotes the set of piecewise smooth curves closed in x. If we restrict ourself to null homotopic curves, we obtain the *reduced holonomy group of* (M, g) *with respect to* $x \in M$:

$$\begin{aligned} \mathrm{Hol}^0_x(M, g) &:= \{\mathscr{P}^g_\gamma : T_x M \to T_x M \mid \gamma \in \Omega(x) \text{ null homotopic}\} \\ &\subset \mathrm{Hol}_x(M, g). \end{aligned}$$

$\mathrm{Hol}^0_x(M, g)$ is the connected component of the Identity in the Lie group $\mathrm{Hol}_x(M, g)$. Hence, the holonomy group of a simply connected manifold is connected. The holonomy groups of two different points are conjugated: If σ is a smooth curve connecting x with y, then

$$\mathrm{Hol}_y(M, g) = \mathscr{P}^g_\sigma \circ \mathrm{Hol}_x(M, g) \circ \mathscr{P}^g_{\sigma^-}.$$

Therefore, we often omit the reference point and consider the holonomy groups of (M, g) as class of conjugated subgroups of the (pseudo)orthogonal group $\mathrm{O}(p, q)$ (fixing an orthonormal basis in $(T_x M, g_x)$).

---

[1] In this paper all manifolds are supposed to be connected.

[2] In this paper p denotes the number of -1 and q the number of $+1$ in the normal form of the metric. We call (M, g) *Riemannian* if $p = 0$, *Lorentzian* if $p = 1 < n$ and *pseudo-Riemannian* if $1 \le p < n$. If we do not want to specify the signature we say *semi-Riemannian manifold.*

If $\pi : (\widetilde{M}, \tilde{g}) \to (M, g)$ is the universal semi-Riemannian covering, then

$$\mathrm{Hol}^0_{\tilde{x}}(\widetilde{M}, \tilde{g}) = \mathrm{Hol}_{\tilde{x}}(\widetilde{M}, \tilde{g}) \simeq \mathrm{Hol}^0_{\pi(\tilde{x})}(M, g).$$

For a semi-Riemannian product $(M, g) = (M_1, g_1) \times (M_2, g_2)$ and $(x_1, x_2) \in M_1 \times M_2$, the holonomy group is the product of its factors

$$\mathrm{Hol}_{(x_1,x_2)}(M, g) = \mathrm{Hol}_{x_1}(M_1, g_1) \times \mathrm{Hol}_{x_2}(M_2, g_2).$$

An important result, which relates the holonomy group to the curvature of (M, g), is the Holonomy Theorem of Ambrose and Singer. We denote by R^g the curvature tensor of (M, g). Due to the symmetry properties of the curvature tensor, for all $x \in M$ and $v, w \in T_xM$ the endomorphism $R^g_x(v, w) : T_xM \to T_xM$ is skew-symmetric with respect to g_x, hence an element of the Lie algebra $\mathfrak{so}(T_xM, g_x)$ of $\mathrm{O}(T_xM, g_x)$. Let γ be a piecewise smooth curve from x to y and $v, w \in T_xM$. We denote by $(\gamma^* R^g)_x(v, w)$ the endomorphism

$$(\gamma^* R^g)_x(v, w) := \mathcal{P}^g_{\gamma^-} \circ R^g_y\big(\mathcal{P}^g_\gamma(v), \mathcal{P}^g_\gamma(w)\big) \circ \mathcal{P}_\gamma \;\in\; \mathfrak{so}(T_xM, g_x).$$

The Lie algebra of the holonomy group is generated by the curvature operators $(\gamma^* R^g)_x$, more exactly, the *Holonomy Theorem of Ambrose and Singer* states:

Theorem 2.1 (Holonomy Theorem of Ambrose and Singer). *The Lie algebra of the holonomy group* $\mathrm{Hol}_x(M, g)$ *is given by*

$$\mathfrak{hol}_x(M, g) = \mathrm{span}\left\{(\gamma^* R^g)_x(v, w) \;\middle|\; \begin{array}{l} v, w \in T_xM\,, \\ \gamma \text{ curve with initial point } x \end{array}\right\}.$$

This Theorem provides a tool for the calculation of the holonomy algebra of a manifold which determines the connected component $\mathrm{Hol}^0_x(M, g)$ of the holonomy group. Therefore it is the starting point in the classification of holonomy groups. It is enough to describe holonomy groups of simply connected manifolds, whereas the description of the full holonomy group in the general case is a much more complicated problem.

Another important property of holonomy groups is stated in the following *holonomy principle*, which relates parallel tensor fields on M to fixed elements under the action of the holonomy group.

Theorem 2.2 (Holonomy Principle). *Let $\mathcal{T}$ be a tensor bundle on (M, g) and let ∇^g be the covariant derivative on $\mathcal{T}$ induced by the Levi-Civita connection. If $T \in \Gamma(\mathcal{T})$ is a tensor field with $\nabla^g T = 0$, then* $\mathrm{Hol}_x(M, g)\, T(x) = T(x)$, *where* $\mathrm{Hol}_x(M, g)$ *acts in the canonical way on the tensors $\mathcal{T}_x$. Contrary, if $T_x \in \mathcal{T}_x$ is a tensor with* $\mathrm{Hol}_x(M, g)\, T_x = T_x$, *then there is an uniquely determined tensor field $T \in \Gamma(\mathcal{T})$ with $\nabla^g T = 0$ and $T(x) = T_x$. T is given by parallel displacement of T_x, i.e., $T(y) := \mathcal{P}^{\nabla^g}_\gamma(T_x)$, where $y \in M$ and γ is a curve connecting x with y.*

The holonomy group $\mathrm{Hol}_x(M, g)$ acts as group of orthogonal mappings on the tangent space $(T_x M, g_x)$. This representation is called the *holonomy representation of* (M, g), we denote it in the following by ρ. The holonomy representation $\rho\colon \mathrm{Hol}_x(M, g) \to \mathrm{O}(T_x M, g_x)$ is called *irreducible* if there is no proper holonomy invariant subspace $E \subset T_x M$. ρ is called *weakly irreducible*, if there is no proper *non-degenerate* holonomy invariant subspace $E \subset T_x M$. To be short, we say that the holonomy group resp. its Lie algebra is *irreducible (weakly irreducible)*, if the holonomy representation has this property. If (M, g) is a Riemannian manifold, any weakly irreducible holonomy representation is irreducible. In the pseudo-Riemannian case there are weakly irreducible holonomy representations which admit degenerate holonomy invariant subspaces, i.e., which are not irreducible. This causes the problems in the classification of the holonomy groups of pseudo-Riemannian manifolds.
For a subspace $E \subset T_x M$ we denote by

$$E^{\perp} = \{v \in T_x M \mid g_x(v, E) = 0\} \subset T_x M$$

its orthogonal complement. If E is holonomy invariant, then $E^{\perp}$ is holonomy invariant as well. If E is in addition non-degenerate, then $E^{\perp}$ is non-degenerate as well and $T_x M$ is the direct sum of these holonomy invariant subspaces:

$$T_x M = E \oplus E^{\perp}.$$

For that reason we call weakly irreducible representations also *indecomposable* (meaning, that they do not decompose into the direct sum of non-degenerate subrepresentations). A semi-Riemannian manifold (M, g) is called *irreducible* if its holonomy representation is irreducible. (M, g) is called *weakly irreducible* or *indecomposable* if its holonomy representation is weakly irreducible.

If the holonomy representation of (M, g) has a proper non-degenerate holonomy invariant subspace, then the reduced holonomy group decomposes into a product of groups. Moreover, the manifold itself spits locally into a semi-Riemannian product. More exactly:

Theorem 2.3 (Local Decomposition Theorem). *Let* $E \subset T_x M$ *be a* k*-dimensional proper non-degenerate holonomy invariant subspace, then the groups*

$$H_1 := \{\mathscr{P}^g_\gamma \in \mathrm{Hol}^0_x(M, g) \mid (\mathscr{P}^g_\gamma)_{|E^{\perp}} = \mathrm{Id}_{E^{\perp}}\} \qquad \textit{and}$$
$$H_2 := \{\mathscr{P}^g_\gamma \in \mathrm{Hol}^0_x(M, g) \mid (\mathscr{P}^g_\gamma)_{|E} = \mathrm{Id}_{E}\}$$

are normal subgroups of $\mathrm{Hol}^0_x(M, g)$ *and*

$$\mathrm{Hol}^0_x(M, g) \simeq H_1 \times H_2.$$

Moreover, (M, g) is locally isometric to a semi-Riemannian product, i.e., for any point $p \in M$ there exists a neighborhood $U(p)$ and two semi-Riemannian manifolds (U_1, g_1) and (U_2, g_2) of dimension k and $(n-k)$, respectively, such that

$$(U(p), g) \overset{isometric}{\simeq} (U_1, g_1) \times (U_2, g_2).$$

The local decomposition of (M, g) follows from the Frobenius Theorem. If $E \subset T_xM$ is a non-degenerate, holonomy invariant subspace, then

$$\mathscr{E} : y \in M \longrightarrow \mathscr{E}_y := \mathscr{P}_\sigma^g(E) \subset T_yM,$$

where σ is a piecewise smooth curve from x to y, is an involutive distribution on M, the *holonomy distribution defined by E*. The maximal connected integral manifolds of $\mathscr{E}$ are totally geodesic submanifolds of M, which are geodesically complete if (M, g) is so. The manifolds (U_1, g_1) and (U_2, g_2) in Theorem 2.3 can be chosen as small open neighborhood of p in the integral manifold $M_1(p)$ of the holonomy distribution $\mathscr{E}$ defined by E and the integral manifold $M_2(p)$ of the holonomy distribution $\mathscr{E}^\perp$ defined by $E^\perp$, respectively, with the metric induced by g. If (M, g) is simply connected and geodesically complete, (M, g) is even globally isometric to the product of the two integral manifolds $(M_1(p), g_1)$ and $(M_2(p), g_2)$. Now, we decompose the tangent space T_xM into a direct sum of non-degenerate, orthogonal and holonomy invariant subspaces

$$T_xM = E_0 \oplus E_1 \oplus \ldots \oplus E_r,$$

where $\mathrm{Hol}_x(M, g)$ acts weakly irreducible on $E_1, \ldots, E_r$ and E_0 is a maximal subspace (possibly 0-dimensional), on which the holonomy group $\mathrm{Hol}_x(M, g)$ acts trivial. Applying the global version of Theorem 2.3 to this decomposition we obtain the *Decomposition Theorem of de Rham and Wu* [35, 81].

Theorem 2.4 (Decomposition Theorem of de Rham und Wu). *Let (M, g) be a simply connected, geodesically complete semi-Riemannian manifold. Then (M, g) is isometric to a product of simply connected, geodesically complete semi-Riemannian manifolds*

$$(M, g) \simeq (M_0, g_0) \times (M_1, g_1) \times \cdots \times (M_r, g_r),$$

where (M_0, g_0) is a (possibly null-dimensional) (pseudo-)Euclidian space and the factors $(M_1, g_1), \ldots, (M_r, g_r)$ are indecomposable and non-flat.

Theorem 2.4 reduces the classification of reduced holonomy groups of geodesically complete semi-Riemannian manifolds to the study of weakly irreducible holonomy representations. This classification is widely open, but the subcase of *irreducible* holonomy representations is completely solved. First of all, let us mention that the holonomy group of a symmetric space is given by its isotropy representation.

Theorem 2.5. *Let (M, g) be a symmetric space, and let $G(M) \subset \mathrm{Isom}(M, g)$ be its transvection group. Furthermore, let $\lambda : H(M) \longrightarrow \mathrm{GL}(T_{x_0}M)$ be the isotropy representation of the stabilizer $H(M) = G(M)_{x_0}$ of a point $x_0 \in M$. Then,*

$$\lambda(H(M)) = \mathrm{Hol}_{x_0}(M, g).$$

In particular, the holonomy group $\mathrm{Hol}_{x_0}(M, g)$ is isomorphic to the stabilizer $H(M)$ and, using this isomorphism, the holonomy representation ρ is given by the isotropy representation λ.

Therefore, the holonomy groups of symmetric spaces can be read off from the classification lists of symmetric spaces, which describe the pair $(G(M), H(M))$ and the isotropy representation λ. For *irreducible* symmetric spaces these lists can be found in [19, Chap. 10], in [54] and in [17]. In order to classify the *irreducible* holonomy representations, the classification of the non-symmetric case remains. This was done by Berger in 1955 (cf. [16]. He proved that there is only a short list of groups which can appear as holonomy groups of *irreducible* non-locally symmetric simply connected semi-Riemannian manifolds. This list is now called the *Berger list*. The Berger list of Riemannian manifolds is widely known. There appear only six special holonomy groups and due to the holonomy principle (cf. Theorem 2.2) each of these groups is related to a special, rich and interesting geometry, described by the corresponding parallel geometric object.

Theorem 2.6 (Riemannian Berger List). *Let (M^n, g) be an n-dimensional, simply connected, irreducible, non-locally symmetric Riemannian manifold. Then the holonomy group* $\mathrm{Hol}(M, g)$ *is up to conjugation in* $\mathrm{O}(n)$ *either* $\mathrm{SO}(n)$ *or one of the following groups with its standard representation:*

n	*Holonomy group*	*Special geometry*
$2m \geq 4$	$\mathrm{U}(m)$	*Kähler manifold*
$2m \geq 4$	$\mathrm{SU}(m)$	*Ricci-flat Kähler manifold*
$4m \geq 8$	$\mathrm{Sp}(m)$	*Hyperkähler manifold*
$4m \geq 8$	$\mathrm{Sp}(m) \cdot \mathrm{Sp}(1)$	*quaternionic Kähler manifold*
7	G_2	G_2*-manifold*
8	$\mathrm{Spin}(7)$	$\mathrm{Spin}(7)$*-manifold*

The Berger list for pseudo-Riemannian manifolds is given in the following Theorem.

Theorem 2.7 (Pseudo-Riemannian Berger List). *Let (M, g) be a simply connected, irreducible, non-locally symmetric semi-Riemannian manifold of signature (p, q). Then the holonomy group of (M, g) is up to conjugation in $\mathrm{O}(p, q)$ either $\mathrm{SO}^0(p, q)$ or one of the following groups with its standard representation:*

Dimension	*Signature*	*Holonomy group*
$2m \geq 4$	$(2r, 2s)$	$\mathrm{U}(r,s)$ *und* $\mathrm{SU}(r,s)$
$2m \geq 4$	(r, r)	$\mathrm{SO}(r,\mathbb{C})$
$4m \geq 8$	$(4r, 4s)$	$\mathrm{Sp}(r,s)$ *und* $\mathrm{Sp}(r,s)\cdot\mathrm{Sp}(1)$
$4m \geq 8$	$(2r, 2r)$	$\mathrm{Sp}(r,\mathbb{R})\cdot\mathrm{SL}(2,\mathbb{R})$
$4m \geq 16$	$(4r, 4r)$	$\mathrm{Sp}(r,\mathbb{C})\cdot\mathrm{SL}(2,\mathbb{C})$
7	$(4, 3)$	$\mathrm{G}^*_{2(2)}$
14	$(7, 7)$	$\mathrm{G}_2^{\mathbb{C}}$
8	$(4, 4)$	$\mathrm{Spin}(4,3)$
16	$(8, 8)$	$\mathrm{Spin}(7,\mathbb{C})$

As one easily sees, this list does not contain a group in Lorentzian signature. This reflects a special algebraic fact concerning irreducibly acting connected subgroups of the Lorentzian group $\mathrm{O}(1, n-1)$ (cf. [36]).

Theorem 2.8. *If $H \subset \mathrm{O}(1, n-1)$ is a connected Lie subgroup acting irreducibly on $\mathbb{R}^{1,n-1}$, then $H = \mathrm{SO}^0(1, n-1)$.*

The proofs of the basic Theorems stated in this section can be found in [11,56, 75]. We refer to [56,75] also for constructions of Riemannian manifolds with special holonomy.

3 Lorentzian Holonomy Groups: The Algebraic Classification

In this section we will describe the algebraic classification of the reduced holonomy groups of Lorentzian manifolds.

In dimension 4 there are 14 types of Lorentzian holonomy groups which were discovered by Schell [76] and Shaw [77]. We will not recall this list here, besides to the original papers we refer to [52,53], [19, Chap. 10], and [43]. In the following we will consider arbitrary dimension.

Due to Theorems 2.4 and 2.8 the Decomposition Theorem for Lorentzian manifolds can be formulated as follows:

Theorem 3.1. *Let (N, h) be a simply connected, geodesically complete Lorentzian manifold. Then (N, h) is isometric to the product*

$$(N,h) \simeq (M,g) \times (M_1,g_1) \times \cdots \times (M_r,g_r),$$

where (M_i, g_i) are either flat or irreducible Riemannian manifolds and (M, g) is either

1. $(\mathbb{R}, -dt^2)$,
2. *an irreducible Lorentzian manifold with holonomy group $\mathrm{SO}^0(1, n-1)$ or*

3. *a Lorentzian manifold with weakly irreducible holonomy representation which admits a degenerate invariant subspace.*

Since the holonomy groups of the Riemannian factors are known, it remains to classify the weakly irreducible Lorentzian holonomy representations which admit an invariant degenerate subspace.

Let (M, g) be a weakly irreducible, but non-irreducible Lorentzian manifold, and let $x \in M$. Then the holonomy representation $\rho : \mathrm{Hol}_x(M, g) \to \mathrm{O}(T_xM, g_x)$ admits a degenerate invariant subspace $W \subset T_xM$. The intersection $V := W \cap W^\perp \subset T_xM$ is a holonomy invariant light-like line. Hence the holonomy group $\mathrm{Hol}_x(M, g)$ lies in the stabilizer $\mathrm{O}(T_xM, g_x)_V$ of V in $\mathrm{O}(T_xM, g_x)$. Let us first describe this stabilizer more in detail.[3]

We fix a basis $(f_1, \ldots, f_n)$ in T_xM such that $f_1 \in V$ and

$$\big(g_x(f_i, f_j)\big) = \begin{pmatrix} 0 & 0 & 1 \\ 0 & I_{n-2} & 0 \\ 1 & 0 & 0 \end{pmatrix},$$

identify (T_xM, g_x) with the Minkowski space and write the elements of $\mathrm{O}(T_xM, g_x)$ as matrices with respect to this basis. The stabilizer of the isotropic line $\mathbb{R}f_1 \subset \mathbb{R}^{1,n-1}$ is a semidirect product and given by the matrices

$$\begin{aligned} \mathrm{O}(1, n-1)_{\mathbb{R}f_1} &= \big(\mathbb{R}^* \times \mathrm{O}(n-2)\big) \ltimes \mathbb{R}^{n-2} \\ &= \left\{ \begin{pmatrix} a^{-1} & x^t & -\frac{1}{2}a\|x\|^2 \\ 0 & A & -aAx \\ 0 & 0 & a \end{pmatrix} \,\middle|\, \begin{array}{l} a \in \mathbb{R}^* \\ x \in \mathbb{R}^{n-2} \\ A \in \mathrm{O}(n-2) \end{array} \right\}. \end{aligned}$$

The Lie algebra of $\mathrm{O}(1, n-1)_{\mathbb{R}f_1}$ is

$$\begin{aligned} \mathfrak{so}(1, n-1)_{\mathbb{R}f_1} &= \big(\mathbb{R} \oplus \mathfrak{so}(n-2)\big) \ltimes \mathbb{R}^{n-2} \\ &= \left\{ \begin{pmatrix} \alpha & y^t & 0 \\ 0 & X & -y \\ 0 & 0 & -\alpha \end{pmatrix} \,\middle|\, \begin{array}{l} \alpha \in \mathbb{R} \\ y \in \mathbb{R}^{n-2} \\ X \in \mathfrak{so}(n-2) \end{array} \right\}. \end{aligned}$$

Let us denote a matrix in the Lie algebra $\mathfrak{so}(1, n-1)_{\mathbb{R}f_1}$ by (α, X, y) (in the obvious way). The commutator is given by

[3] The connected component of the stabilizer $\mathrm{O}(1, n-1)_{\mathbb{R}f_1} \subset \mathrm{O}(1, n-1)$ of a light-like line $\mathbb{R}f_1$ is isomorphic to the group of similarity transformation of the Euclidian space $\mathbb{R}^{n-2}$, i.e., to the group generated by translations, dilatations and rotations of $\mathbb{R}^{n-2}$. Therefore, in some papers this group is denoted by $\mathrm{Sim}(n-2)$.

$$[(\alpha, X, y), (\beta, Y, z)] = \big(0, [X, Y], (X + \alpha \,\mathrm{Id})z - (Y + \beta \,\mathrm{Id})y\big),$$

which describes the semi-direct structure. In particular, $\mathbb{R}$, $\mathbb{R}^{n-2}$ and $\mathfrak{so}(n-2)$ are subalgebras of $\mathfrak{so}(1,n-1)_{\mathbb{R}f_1}$. Now, one can assign to any subalgebra $\mathfrak{h} \subset \mathfrak{so}(1,n-1)_{\mathbb{R}f_1}$ the projections $\mathrm{pr}_{\mathbb{R}}(\mathfrak{h})$, $\mathrm{pr}_{\mathbb{R}^{n-2}}(\mathfrak{h})$ and $\mathrm{pr}_{\mathfrak{so}(n-2)}(\mathfrak{h})$ onto these parts. The subalgebra

$$\mathfrak{g} := \mathrm{pr}_{\mathfrak{so}(n-2)}(\mathfrak{h}) \subset \mathfrak{so}(n-2)$$

is called the *orthogonal part of* $\mathfrak{h}$. $\mathfrak{g}$ is reductive, i.e. its Levi decomposition is given by $\mathfrak{g} = \mathfrak{z}(\mathfrak{g}) \oplus [\mathfrak{g},\mathfrak{g}]$, where $\mathfrak{z}(\mathfrak{g})$ is the center of $\mathfrak{g}$ and the commutator $[\mathfrak{g},\mathfrak{g}]$ is semi-simple.

The first step in the classification of weakly irreducible holonomy representations is a result due to Berard-Bergery and Ikemakhen (cf. [15]), who described the possible algebraic types of weakly irreducibly acting subalgebras $\mathfrak{h}$ of the stabilizer $\mathfrak{so}(1,n-1)_{\mathbb{R}f_1} = \big(\mathbb{R} \oplus \mathfrak{so}(n-2)\big) \ltimes \mathbb{R}^{n-2}$. A geometric proof of this result was later given by Galaev in [41].

Theorem 3.2. *Let $f_1 \in \mathbb{R}^{1,n-1}$ be a light-like vector in the Minkowski space and let*

$$\mathfrak{h} \subset \mathfrak{so}(1,n-1)_{\mathbb{R}f_1} = (\mathbb{R} \oplus \mathfrak{so}(n-2)) \ltimes \mathbb{R}^{n-2}$$

be a weakly irreducible subalgebra of the stabilizer of $\mathbb{R}f_1$ in $\mathfrak{so}(1,n-1)$. We denote by $\mathfrak{g} := \mathrm{pr}_{\mathfrak{so}(n-2)}(\mathfrak{h}) \subset \mathfrak{so}(n-2)$ the orthogonal part of $\mathfrak{h}$. Then $\mathfrak{h}$ is of one of the following four types:

1. $\mathfrak{h} = \mathfrak{h}^1(\mathfrak{g}) := (\mathbb{R} \oplus \mathfrak{g}) \ltimes \mathbb{R}^{n-2}$.
2. $\mathfrak{h} = \mathfrak{h}^2(\mathfrak{g}) := \mathfrak{g} \ltimes \mathbb{R}^{n-2}$.
3. $\mathfrak{h} = \mathfrak{h}^3(\mathfrak{g},\varphi) := \big\{\big(\varphi(X), X+Y, z\big) \mid X \in \mathfrak{z}(\mathfrak{g}),\ Y \in [\mathfrak{g},\mathfrak{g}],\ z \in \mathbb{R}^{n-2}\big\}$,
   where $\varphi : \mathfrak{z}(\mathfrak{g}) \to \mathbb{R}$ is a linear and surjective map.
4. $\mathfrak{h} = \mathfrak{h}^4(\mathfrak{g},\psi) := \big\{\big(0, X+Y, \psi(X)+z\big) \mid X \in \mathfrak{z}(\mathfrak{g}),\ Y \in [\mathfrak{g},\mathfrak{g}],\ z \in \mathbb{R}^{k}\big\}$,

   where $\mathbb{R}^{n-2} = \mathbb{R}^m \oplus \mathbb{R}^k$, $0 < m < n-2$,
   $\mathfrak{g} \subset \mathfrak{so}(\mathbb{R}^k)$,
   $\psi : \mathfrak{z}(\mathfrak{g}) \to \mathbb{R}^m$ *linear and surjective.*

In the following we will refer to theses cases as the Lie algebras $\mathfrak{h}$ of types 1–4. The types 1 and 2 are called *uncoupled types*, the types 3 and 4 *coupled types*, since the $\mathfrak{so}(n-2)$- part is coupled by φ and ψ with the $\mathbb{R}$- and the $\mathbb{R}^{n-2}$-part, respectively.

Theorem 3.2 reduces the classification of Lorentzian holonomy algebras to the description of the *orthogonal part* $\mathfrak{g} = \mathrm{pr}_{\mathfrak{so}(n-2)}\,\mathfrak{hol}(M,g)$. First, let us look at the geometric meaning of $\mathfrak{g}$. Let $\mathscr{V} \subset TM$ be the holonomy distribution, defined by $\mathbb{R}f_1 \subset T_xM$. The orthogonal complement $\mathscr{V}^{\perp} \subset TM$ is parallel as well and contains $\mathscr{V}$. Hence, $(\mathscr{V}^{\perp}/\mathscr{V}, \tilde{g}, \widetilde{\nabla}^g)$ is a vector bundle of rank $(n-2)$ on M equipped with a positive definite bundle metric $\tilde{g}$, induced by g, and a metric covariant derivative $\widetilde{\nabla}^g$, induced by the Levi-Civita-connection of g. It is not difficult to check, that the holonomy algebra of $(\mathcal{V}^\perp/\mathcal{V}, \widetilde{\nabla}^g)$ coincides with $\mathfrak{g}$:

Proposition 3.1 ([65]). *Let (M,g) be a Lorentzian manifold with a parallel light-like distribution $\mathcal{V} \subset TM$. Then,*

$$\mathfrak{hol}_x(\mathcal{V}^\perp/\mathcal{V}, \widetilde{\nabla}^g) = \mathrm{pr}_{\mathfrak{so}(n-2)}\, \mathfrak{hol}_x(M,g).$$

Moreover, the different types of holonomy algebras translate into special curvature properties of the light-like hypersurface of M, defined by the involutive distribution $\mathcal{V}^\perp$. For details we refer to [20].

Thomas Leistner studied the orthogonal part of $\mathfrak{hol}(M,g)$ and obtained the following deep result, cf. [65, 68].

Theorem 3.3. *Let (M^n, g) be a Lorentzian manifold with a weakly irreducible but non-irreducible holonomy group* $\mathrm{Hol}^0(M,g)$. *Then the orthogonal part* $\mathfrak{g} = \mathrm{pr}_{\mathfrak{so}(n-2)}(\mathfrak{hol}(M,g))$ *of the holonomy algebra is the holonomy algebra of a Riemannian manifold.*

Leistner's proof of this Theorem is based on the observation of a special algebraic property of the orthogonal part $\mathfrak{g}$ of a Lorentzian holonomy algebra. It is a so-called *weak Berger algebra* – a notion, which was introduced and studied by Leistner in [63] (see also [40, 65, 68]). We will explain this notion here shortly.
Let $\mathfrak{g} \subset \mathfrak{gl}(V)$ be a subalgebra of the linear maps of a finite dimensional real or complex vector space V with scalar product $\langle\cdot,\cdot\rangle$. Then we consider the following spaces:

$$\mathcal{K}(\mathfrak{g}) := \{R \in \Lambda^2(V^*) \otimes \mathfrak{g} \mid R(x,y)z + R(y,z)x + R(z,x)y = 0\},$$
$$\mathcal{B}(\mathfrak{g}) := \{B \in V^* \otimes \mathfrak{g} \mid \langle B(x)y, z\rangle + \langle B(y)z, x\rangle + \langle B(z)x, y\rangle = 0\}.$$

The space $\mathcal{K}(\mathfrak{g})$ is called *the space of curvature tensors*[4] *of* $\mathfrak{g}$. $\mathcal{B}(\mathfrak{g})$ is called *space of weak curvature tensors of* $\mathfrak{g}$. Now, let $\mathfrak{g}$ be an orthogonal Lie algebra, i.e., $\mathfrak{g} \subset \mathfrak{so}(V, \langle\cdot,\cdot\rangle)$. Then any curvature tensor R of $\mathfrak{g}$ satisfies in addition the symmetry properties

$$\langle R(x,y)u, v\rangle = -\langle R(x,y)v, u\rangle,$$
$$\langle R(x,y)u, v\rangle = +\langle R(u,v)x, y\rangle.$$

Hence, for each $R \in \mathcal{K}(\mathfrak{g})$ and $x \in V$ we have $R(x,\cdot) \in \mathcal{B}(\mathfrak{g})$.
A Lie algebra $\mathfrak{g} \subset \mathfrak{gl}(V)$ is called *Berger algebra* if there are enough curvature tensors to generate $\mathfrak{g}$, i.e., if

[4]This notation is motivated by the fact, that the condition which defines $\mathcal{K}(\mathfrak{g})$ is just the Bianchi identity for the curvature tensor R^∇_x of a torsion free covariant derivative ∇.

$$\mathfrak{g} = \text{span}\{R(x, y) \mid x, y \in V,\ R \in \mathcal{K}(\mathfrak{g})\}.$$

An orthogonal Lie algebra $\mathfrak{g} \subset \mathfrak{so}(V, \langle\cdot,\cdot\rangle)$ is called *weak Berger algebra* if there are enough weak curvature tensors to generate $\mathfrak{g}$, i.e., if

$$\mathfrak{g} = \text{span}\{B(x) \mid x \in V,\ B \in \mathcal{B}(\mathfrak{g})\}.$$

Obviously, every orthogonal Berger algebra is a weak Berger algebra. For an *Euclidian* space V, the Bianchi identity defining $\mathcal{B}(\mathfrak{g})$ is used to prove the following decomposition property of the space of weak curvature tensors $\mathcal{B}(\mathfrak{g})$:

Proposition 3.2. *Let V be an Euclidian space and let $\mathfrak{g} \subset \mathfrak{so}(V)$ be a weak Berger algebra. Then V decomposes into orthogonal $\mathfrak{g}$-invariant subspaces*

$$V = V_0 \oplus V_1 \oplus \ldots \oplus V_s,$$

where $\mathfrak{g}$ acts trivial on V_0 (possibly 0-dimensional) and irreducible on V_j, $j = 1, \ldots, s$. Moreover, $\mathfrak{g}$ is the direct sum of ideals

$$\mathfrak{g} = \mathfrak{g}_1 \oplus \ldots \oplus \mathfrak{g}_s, \tag{1}$$

where $\mathfrak{g}_j$ acts irreducible on V_j and trivial on V_i if $i \neq j$. $\mathfrak{g}_j \subset \mathfrak{so}(V_j)$ is a weak Berger algebra and $\mathcal{B}(\mathfrak{g}) = \mathcal{B}(\mathfrak{g}_1) \oplus \ldots \oplus \mathcal{B}(\mathfrak{g}_s)$.

Now, let (M, g) be a Lorentzian manifold with holonomy group $\text{Hol}_x(M, g)$. From the Ambrose–Singer Theorem 2.1 it follows that the holonomy algebra $\mathfrak{hol}_x(M, g) \subset \mathfrak{so}(T_xM, g_x)$ is a Berger algebra. Moreover, looking more carefully at the curvature endomorphisms one obtains:

Proposition 3.3. *Let (M^n, g) be a Lorentzian manifold with a weakly irreducible but non-irreducible holonomy group $\text{Hol}_x^0(M, g)$. Then the orthogonal part $\mathfrak{g} = \text{pr}_{\mathfrak{so}(n-2)}(\mathfrak{hol}_x(M, g))$ of the holonomy algebra is a weak Berger algebra on an Euclidean space. Hence it decomposes into a direct sum of irreducibly acting weak Berger algebras.*

Using representation and structure theory of semi-simple Lie algebras, Leistner proved the following central Theorem which implies Theorem 3.3.

Theorem 3.4. *Any irreducible weak Berger algebra on an Euclidian space is the holonomy algebra of an irreducible Riemannian manifold.*

All together we obtain the following classification Theorem for Lorentzian holonomy groups.

Theorem 3.5 (The Connected Holonomy Groups of Lorentzian Manifolds). *Let (M, g) be an n-dimensional, simply connected, indecomposable Lorentzian manifold. Then either (M, g) is irreducible and the holonomy group is the Lorentzian group $SO^0(1, n-1)$, or the holonomy group lies in the stabilizer*

$\mathrm{SO}^0(1,n-1)_V = (\mathbb{R}^+ \times \mathrm{SO}(n-2)) \ltimes \mathbb{R}^{n-2}$ *of a light-like line* V. *In the second case, let* $G' \subset G \subset \mathrm{SO}(n-2)$ *be the closed subgroups with Lie algebras* $\mathfrak{g}' := [\mathfrak{g},\mathfrak{g}] \subset \mathfrak{g} := \mathrm{pr}_{\mathfrak{so}(n-2)}\, \mathfrak{hol}_x(M,g) \subset \mathfrak{so}(n-2)$, *respectively. Then* $G \subset \mathrm{SO}(n-2)$ *is the holonomy group of a Riemannian manifold and* $\mathrm{Hol}_x(M,g)$ *is of one of the following types:*

1. $(\mathbb{R}^+ \times G) \ltimes \mathbb{R}^{n-2}$.
2. $G \ltimes \mathbb{R}^{n-2}$.
3. $L \cdot G' \ltimes \mathbb{R}^{n-2}$, *where* $L \subset \mathbb{R} \times \mathrm{SO}(n-2)$ *is the connected Lie group with Lie algebra* $\mathfrak{l} := \{(\varphi(X), X, 0) \mid X \in \mathfrak{z}(\mathfrak{g})\}$ *for a surjective linear map* $\varphi : \mathfrak{z}(\mathfrak{g}) \to \mathbb{R}$.
4. $\hat{L} \cdot G' \ltimes \mathbb{R}^{n-2-m}$, *where* $\hat{L} \subset \mathrm{SO}(n-2) \ltimes \mathbb{R}^m$ *is the connected Lie-group with Lie algebra* $\hat{\mathfrak{l}} := \{(0, X, \psi(X)) \mid X \in \mathfrak{z}(\mathfrak{g})\}$ *for a surjective linear map* $\psi : \mathfrak{z}(\mathfrak{g}) \longrightarrow \mathbb{R}^m$.

4 Local Realization of Lorentzian Holonomy Groups

In this section we will show, that *any* of the groups in the list of Theorem 3.5 can be realized as holonomy group of a Lorentzian manifold.

Lorentzian metrics with holonomy group of uncoupled types 1 and 2 are rather easy to construct:

Proposition 4.1. *Let* (F,h) *be a connected* $(n-2)$*-dimensional Riemannian manifold, and let* $H \in C^\infty(\mathbb{R} \times F \times \mathbb{R})$ *be a smooth function such that the Hessian of* $H(0,\cdot,0) \in C^\infty(F)$ *is non-degenerate in* $x \in F$. *Then the holonomy group of the Lorentzian manifold* (M,g)

$$M := \mathbb{R} \times F \times \mathbb{R}\ ,\quad g := 2dvdu + H\,du^2 + h, \tag{2}$$

where v,u *denote the coordinates of the* $\mathbb{R}$*-factors, is given by*

$$\mathrm{Hol}_{(0,x,0)}(M,g) = \begin{cases} \mathrm{Hol}_x(F,h) \ltimes \mathbb{R}^{n-2} & if \quad \frac{\partial H}{\partial v} = 0, \\ (\mathbb{R}^+ \times \mathrm{Hol}_x(F,h)) \ltimes \mathbb{R}^{n-2} & if \quad \frac{\partial^2 H}{\partial v^2} \neq 0. \end{cases}$$

This Theorem can be proved by direct calculation of the group of parallel displacements (cf. for example [11, Chap. 5]). It is more difficult to produce metrics with the holonomy groups of coupled types 3 and 4.

The basic observation for constructing local metrics is the existence of adapted coordinates for Lorentzian manifolds with special holonomy, called *Walker coordinates*. Let (M,g) be a Lorentzian manifold with holonomy group acting weakly irreducible, but non-irreducible. Then, as we know from the previous section, there exists a 1-dimensional parallel light-like distribution $\mathscr{V} \subset TM$. Locally, the distribution $\mathscr{V}$ is spanned by a recurrent light-like vector field ξ, where a vector field ξ on (M, g) is called *recurrent* if there is a 1-form ω such that

$$\nabla^g \xi = \omega \otimes \xi.$$

A.G. Walker ([78]) proved the existence of adapted coordinates in the presence of a parallel light-like line.

Proposition 4.2 (Walker Coordinates). *Let (M, g) be an n-dimensional Lorentzian manifold with a parallel light-like line $\mathscr{V} \subset TM$. Then around any point $p \in M$ there are coordinates $(U, (v, x_1, \dots, x_{n-2}, u))$ such that $g_{|U}$ has the form*

$$g_{|U} = 2dvdu + 2\sum_{i=1}^{n-2} A_i dx_i du + H\,du^2 + \sum_{j,k=1}^{n-2} h_{jk} dx_j dx_k, \tag{3}$$

where A_i, h_{jk} are smooth functions of $x_1, \dots, x_{n-2}, u$ and the function H depends smoothly on the coordinates $v, x_1, \dots, x_{n-2}, u$.

In these coordinates $\frac{\partial}{\partial v}$ generates the distribution $\mathscr{V}$ and $\frac{\partial}{\partial v}, \frac{\partial}{\partial x_1}, \dots, \frac{\partial}{\partial x_{n-2}}$ generate $\mathscr{V}^\perp$. The vector field $\frac{\partial}{\partial v}$ is parallel if H does not depend on v. We call a metric of the form (3) a *Walker metric*. The metrics (2) are special cases of Walker metrics, the holonomy group in this example is produced by the function H and the Riemannian metric h. In the following construction due to Galaev (cf. [42,43]), the functions H and A_i in the Walker metric (3) are used to produce the holonomy groups in Theorem 3.5.

We consider the situation as described in Sect. 3. Let $\mathfrak{g} \subset \mathfrak{so}(n-2)$ be the holonomy algebra of a Riemannian manifold. We will describe a Walker metric g of the form (3) on $\mathbb{R}^n$, such that $\mathfrak{hol}_0(\mathbb{R}^n, g)$ is of the form $\mathfrak{h}^1(\mathfrak{g})$, $\mathfrak{h}^2(\mathfrak{g})$, $\mathfrak{h}^3(\mathfrak{g}, \varphi)$ and $\mathfrak{h}^4(\mathfrak{g}, \psi)$, respectively, as described in Theorem 3.2. As we know, $\mathbb{R}^{n-2}$ has a decomposition into orthogonal subspaces

$$\mathbb{R}^{n-2} = \mathbb{R}^{n_0} \times \mathbb{R}^{n_1} \times \cdots \times \mathbb{R}^{n_s}, \tag{4}$$

where $\mathfrak{g}$ acts trivial on $\mathbb{R}^{n_0}$ and irreducible on $\mathbb{R}^{n_1}, \dots, \mathbb{R}^{n_s}$ and $\mathfrak{g}$ spits into a direct sum of ideals

$$\mathfrak{g} = \mathfrak{g}_1 \oplus \ldots \oplus \mathfrak{g}_s,$$

where $\mathfrak{g}_i \subset \mathfrak{so}(n_i)$ are holonomy algebras of irreducible Riemannian manifolds. Now, let $(e_1, \dots, e_{n-2})$ be an orthonormal basis of $\mathbb{R}^{n-2}$ adapted to the decomposition (4). We choose weak curvature endomorphisms $Q_I \in \mathscr{B}(\mathfrak{g})$, $I = 1, \dots, N$, which generate $\mathscr{B}(\mathfrak{g})$. Note that $Q_I(e_i) = 0$ for $i = 1, \dots, n_0$.

Let $\varphi : \mathfrak{z}(\mathfrak{g}) \to \mathbb{R}$ and $\psi : \mathfrak{z}(\mathfrak{g}) \to \mathbb{R}^m \subset \mathbb{R}^{n_0}$ be surjective linear maps. We extend φ and ψ to $\mathfrak{g}$ by setting $\varphi|_{[\mathfrak{g},\mathfrak{g}]} = 0$, $\psi|_{[\mathfrak{g},\mathfrak{g}]} = 0$ and define the numbers:

$$\varphi_{Ii} := \frac{1}{(I-1)!}\varphi(Q_I(e_i)), \tag{5}$$

$$\psi_{Iij} := \frac{1}{(I-1)!}\Big\langle \psi(Q_I(e_i)), e_j \Big\rangle_{\mathbb{R}^{n-2}}, \tag{6}$$

where $I = 1, \dots, N$, $i = n_0 + 1, \dots, n-2$, $j = 1, \dots, m$. Then, one can realize any connected Lorentzian holonomy group by a Walker metric with polynomials as coefficients in the metric (3).

Theorem 4.1 ([42]). *Let $\mathfrak{h} \subset \mathfrak{so}(1, n-1)$ be one of the Lie algebras $\mathfrak{h}^1(\mathfrak{g})$, $\mathfrak{h}^2(\mathfrak{g})$, $\mathfrak{h}^3(\mathfrak{g}, \varphi)$, $\mathfrak{h}^4(\mathfrak{g}, \psi)$ in the list of Theorem 3.2, where $\mathfrak{g} = \mathrm{pr}_{\mathfrak{so}(n-2)}\,\mathfrak{h} \subset \mathfrak{so}(n-2)$ is the holonomy algebra of a Riemannian manifold. We consider the following Walker metric g on $\mathbb{R}^n$:*

$$g = 2dvdu + 2\sum_{i=1}^{n-2} A_i dx_i du + H\, du^2 + \sum_{i=1}^{n-2} dx_i^2,$$

where the functions A_i are given by

$$A_i(x_1, \dots, x_{n-2}, u) := \sum_{I=1}^{N} \sum_{k,l=1}^{n-2} \times \frac{1}{3(I-1)!}\Big\langle Q_I(e_k)e_l + Q_I(e_l)e_k, e_i \Big\rangle_{\mathbb{R}^{n-2}} x_k x_l u^I,$$

and the function $H(v, x_1, \dots, x_{n-2}, u)$ is defined in the following list, corresponding to the type of $\mathfrak{h}$:

$\mathfrak{h}$	H
Type 1: $\mathfrak{h}^1(\mathfrak{g}) = (\mathbb{R} \oplus \mathfrak{g}) \ltimes \mathbb{R}^{n-2}$	$v^2 + \sum_{i=1}^{n_0} x_i^2$
Type 2: $\mathfrak{h}^2(\mathfrak{g}) = \mathfrak{g} \ltimes \mathbb{R}^{n-2}$	$\sum_{i=1}^{n_0} x_i^2$
Type 3: $\mathfrak{h}^3(\mathfrak{g}, \varphi)$	$2v \sum_{I=1}^{N} \sum_{i=n_0+1}^{n-2} \varphi_{Ii} x_i u^{I-1} + \sum_{k=1}^{n_0} x_k^2$
Type 4: $\mathfrak{h}^4(\mathfrak{g}, \psi)$	$2 \sum_{I=1}^{N} \sum_{i=n_0+1}^{n-2} \sum_{j=1}^{m} \psi_{Iib} x_i x_j u^{I-1} + \sum_{k=m+1}^{n_0} x_k^2$

Then, $\mathfrak{h}$ is the holonomy algebra of $(\mathbb{R}^n, g)$ with respect to the point $0 \in \mathbb{R}^n$.

The proof of this Theorem uses that g is analytic. In this case, the holonomy algebra $\mathfrak{hol}_0(\mathbb{R}^n, g)$ is generated by the curvature tensor and its derivatives in the point $0 \in \mathbb{R}^n$. One calculates for the derivatives of the curvature tensor R:

$$\begin{aligned}
\mathrm{pr}_{\mathfrak{so}(n-2)}\left[\left(\nabla^{I-1}_{\partial_u} R\right)_0\left(\partial_{x_i}, \partial_u\right)\right] &= Q_I(e_i),\\
\mathrm{pr}_{\mathbb{R}}\left[R_0\left(\partial_v, \partial_u\right)\right] &= \frac{1}{2}\frac{\partial^2 H}{(\partial v)^2}(0),\\
\mathrm{pr}_{\mathbb{R}}\left[\left(\nabla^{I-1}_{\partial_u} R\right)_0\left(\partial_{x_i}, \partial_u\right)\right] &= \frac{1}{2}\frac{\partial^{I+1} H}{\partial v \partial x_i (\partial u)^{I-1}}(0),\\
\mathrm{pr}_{\mathbb{R}^{n-2}}\left[\left(\nabla^{I-1}_{\partial_u} R\right)_0\left(\partial_{x_a}, \partial_u\right)\right] &= \frac{1}{2}\sum_{j=1}^{n_0}\frac{\partial^{I+1} H}{\partial x_a \partial x_j (\partial u)^{I-1}}(0)\cdot e_j.
\end{aligned}$$

Hereby $I = 1, \ldots, N$, $i = n_0 + 1, \ldots, n$ and $a = 1, \ldots, n_0$. The first formula describes the only non-vanishing orthogonal parts of the curvature tensor and its derivatives in $0 \in \mathbb{R}^n$. Since $\mathfrak{g}$ is a weak Berger algebra and $\{Q_I \mid I = 1, \ldots, N\}$ generate $\mathscr{B}(\mathfrak{g})$, the orthogonal part of $\mathfrak{hol}_0(\mathbb{R}^n, g)$ coincide with $\mathfrak{g}$. The inclusion $\mathbb{R}^{n_0} \subset \mathfrak{hol}_0(\mathbb{R}^n, g)$ follows from the last formula, the proof that $\mathbb{R}^{n_1}, \ldots, \mathbb{R}^{n_s} \subset \mathfrak{hol}_0(\mathbb{R}^n, g)$ uses the fact, that $\mathfrak{g}_i$ acts irreducible on $\mathbb{R}^{n_i}$. The $\mathbb{R}$-part in $\mathfrak{hol}_0(\mathbb{R}^n, g)$ is generated by H, as the second and third formula show.

Recently, Bazaikin [13] constructed coupled holonomy groups of types 3 and 4 using Walker metrics on $M := \mathbb{R} \times F \times \mathbb{R}$ of the form

$$g := 2dvdu + Hdu^2 + 2A \odot du + h, \tag{7}$$

where (F, h) is a Riemannian manifold, A is an 1-form on F and H is a function on M. Moreover, he discussed causality properties, in particular global hyperbolicity of such metrics (see also Sect. 5).

Finally, let us discuss Lorentzian manifolds (M, g) with the abelian and solvable reduced holonomy group

$$\mathrm{Hol}^0(M, g) = \begin{cases} \mathbb{R}^{n-2} \\ \mathbb{R}^+ \ltimes \mathbb{R}^{n-2}. \end{cases}$$

A Lorentzian manifold is called a *pp-wave*, if it admits a light-like parallel vector field ξ and if its curvature tensor R satisfies

$$R(X_1, X_2) = 0 \quad \text{for all } X_1, X_2 \in \xi^{\perp}. \tag{8}$$

A Lorentzian manifold is called a *pr-wave*, if it admits a light-like recurrent vector field ξ and if the curvature tensor satisfies (8). There are several equivalent conditions to (8), for which we refer to [66]. A pp-wave resp. a pr-wave is locally isometric to $(\mathbb{R}^n, g_H)$, where g_H is the Walker metric

$$g_H = 2dvdu + Hdu^2 + \sum_{i=1}^{n-2} dx_i^2$$

and, in case of a pp-wave, the function $H = H(v, x_1, \ldots, x_{n-2}, u)$ does not depend on v.

Proposition 4.3 ([66, 67]). *Let (M, g) be a Lorentzian manifold with a light-like parallel (resp., recurrent) vector field. Then (M, g) is a pp-wave (resp. pr-wave) if and only if its holonomy group $\mathrm{Hol}^0(M, g)$ is contained in the abelian subgroup $\mathbb{R}^{n-2} \subset \mathrm{SO}(1, n-1)$ (resp., the solvable subgroup $\mathbb{R}^+ \ltimes \mathbb{R}^{n-2} \subset \mathrm{SO}(1, n-1)$).*

5 Global Models with Special Lorentzian Holonomy

In the previous section we showed that any Lorentzian holonomy group can be realized by a local metric which is polynomial in the coordinates. In this section we will discuss some global constructions.

5.1 Lorentzian Symmetric Spaces

In the Riemannian case, the holonomy list is divided into the symmetric and the non-symmetric case. In Lorentzian signature, this distinction plays no essential role, since there are only few isometry classes of simply connected indecomposable Lorentzian symmetric spaces. We will shortly discuss these spaces and their holonomy groups.

Let (M^n, g) be a Lorentzian symmetric space. One has the following structure result:

Theorem 5.1 ([28]). *Let (M^n, g) be an indecomposable Lorentzian symmetric space of dimension $n \geq 2$. Then the transvection group of (M^n, g) is either semi-simple or solvable.*

We call a symmetric space *solvable* or *semi-simple* if its transvection group has this property. First we describe the *solvable* Lorentzian symmetric spaces.

Let $\underline{\lambda} = (\lambda_1, \ldots, \lambda_{n-2})$ be an $(n-2)$-tupel of real numbers $\lambda_j \in \mathbb{R}\backslash\{0\}$ and let us denote by $M^n_{\underline{\lambda}}$ the Lorentzian space $M^n_{\underline{\lambda}} := (\mathbb{R}^n, g_{\underline{\lambda}})$, where $g_{\underline{\lambda}}$ is the Walker metric

$$g_{\underline{\lambda}} := 2dv\,du + \sum_{i=1}^{n-2} \lambda_i x_i^2\, du^2 + \sum_{i=1}^{n-2} dx_i^2. \tag{9}$$

If $\underline{\lambda}_\pi = (\lambda_{\pi(1)}, \dots, \lambda_{\pi(n-2)})$ is a permutation of $\underline{\lambda}$ and $c > 0$, then $M^n_{\underline{\lambda}}$ is isometric to $M^n_{c\underline{\lambda}_\pi}$. A direct calculation shows, that the space $M_{\underline{\lambda}}$ is geodesically complete and its curvature tensor is parallel. Hence the pp-wave $M_{\underline{\lambda}}$ is a Lorentzian symmetric space. These symmetric spaces were first described by Cahen and Wallach and are called now *Cahen–Wallach-spaces.* From Proposition 4.1 follows that the holonomy group of $M_{\underline{\lambda}}$ is the abelian subgroup $\mathbb{R}^{n-2} \subset (\mathbb{R}^+ \times \mathrm{SO}(n-2)) \ltimes \mathbb{R}^{n-2}$. The transvection group of $M_{\underline{\lambda}}$ is solvable. For a description of this group we refer to [73].

Theorem 5.2 ([27, 28]). *Let (M^n, g) be an indecomposable solvable Lorentzian symmetric space of dimension $n \geq 3$. Then (M^n, g) is isometric to $M^n_{\underline{\lambda}}/\Gamma$, where $\underline{\lambda} \in (\mathbb{R}\backslash\{0\})^{n-2}$ and Γ is a discrete subgroup of the centralizer $Z_{\underline{\lambda}}$ of the transvection group of $M_{\underline{\lambda}}$ in its isometry group.*

For the centralizer $Z_{\underline{\lambda}}$ Cahen and Kerbrat proved:

Theorem 5.3 ([26]). *Let $\underline{\lambda} = (\lambda_1, \dots, \lambda_{n-2})$ be a tupel of non-zero real numbers.*

1. *If there is a positive λ_i or if there are two numbers λ_i, λ_j such that $\frac{\lambda_i}{\lambda_j} \notin \mathbb{Q}^2$, then $Z_{\underline{\lambda}} \simeq \mathbb{R}$ and $\varphi \in Z_{\underline{\lambda}}$ if and only if $\varphi(v, x, u) = (v + \alpha, x, u)$, $\alpha \in \mathbb{R}$.*
2. *Let $\lambda_i = -k_i^2 < 0$ and $\frac{k_i}{k_j} \in \mathbb{Q}$ for all $i, j \in \{1, \dots, n-2\}$. Then $\varphi \in Z_{\underline{\lambda}}$ if and only if*

$$\varphi(v, x, u) = (v + \alpha, (-1)^{m_1} x_1, \dots, (-1)^{m_{n-2}} x_{n-2}, u + \beta),$$

where $\alpha \in \mathbb{R}$, $m_1, \dots, m_{n-2} \in \mathbb{Z}$ and $\beta = \frac{m_i \cdot \pi}{k_i}$ for all $i = 1, \dots, n-2$.

Next, let us describe the *semi-simple* Lorentzian symmetric spaces. We denote by $S^n_1(r)$ the pseudo-sphere

$$S^n_1(r) := \{x \in \mathbb{R}^{1,n} \mid \langle x, x\rangle_{1,n} = -x_1^2 + x_2^2 + \ldots + x_{n+1}^2 = r^2\},$$

and by $H^n_1(r)$ the pseudo-hyperbolic space

$$H^n_1(r) := \{x \in \mathbb{R}^{2,n-1} \mid \langle x, x\rangle_{2,n-1} = -x_1^2 - x_2^2 + x_3^2 + \ldots + x_{n+1}^2 = -r^2\}$$

with the Lorentzian metrics induced by $\langle\cdot,\cdot\rangle_{1,n}$ and $\langle\cdot,\cdot\rangle_{2,n-1}$, respectively. $S^n_1(r)$ and $H^n_1(r)$ are semi-simple symmetric spaces of constant sectional curvature with full holonomy group $\mathrm{SO}^0(1, n-1)$. Moreover,

Theorem 5.4 ([29, 80]). *Let (M^n, g) be an indecomposable semi-simple Lorentzian symmetric space of dimension $n \geq 3$. Then (M^n, g) has constant sectional curvature $k \neq 0$. Therefore, it is isometric to $S^n_1(r)/_{\{\pm I\}}$ or $S^n_1(r)$ if $k = \frac{1}{r^2} > 0$, or to a Lorentzian covering of $H^n_1(r)/_{\{\pm I\}}$ if $k = -\frac{1}{r^2} < 0$.*

5.2 Holonomy of Lorentzian Cones

Cone constructions are often used to reduce a geometric problem on a manifold to a holonomy problem of the cone over that manifold. For example, Bär [2] used this method to describe all Riemannian geometries with real Killing spinors. Other applications can be found in [12, Chap. 2] and in [70]. It is a classical result of Gallot [48] that the holonomy group of the cone over a complete Riemannian manifold (N,h) is either irreducible or (N,h) has constant sectional curvature (which implies that the cone is flat). In the pseudo-Riemannian situation this is not longer true. The pseudo-Riemannian case was recently studied in [1]. We will describe the results of this paper for the Lorentzian cases here. There are two types of Lorentzian cones, the time-like cone $C_-(N,h)$ over a Riemannian manifold (N,h) and the space-like cone $C_+(N,h)$ over a Lorentzian manifold (N,h):

$$C_\varepsilon(N,h) := (\mathbb{R}^+ \times N,\ g_\varepsilon = \varepsilon\, dt^2 + t^2 h), \quad \varepsilon = \pm 1.$$

First, let us illustrate the difference to the Riemannian case with two examples.

1. Let (F,r) be a complete Riemannian manifold of dimension at least 2 which is not of constant sectional curvature. Then the Lorentzian manifold

$$N := \mathbb{R} \times F, \quad h := -ds^2 + \cosh^2(s) r$$

is complete and not of constant sectional curvature, and the holonomy representation of its space-like cone $C_+(N,h)$ decomposes into proper non-degenerate invariant subspaces.

2. Let $(F,r) = (F_1,r_1) \times (F_2,r_2)$ be a product of a flat and a non-flat complete Riemannian manifold. Then the Riemannian manifold

$$N := \mathbb{R} \times F, \quad h := ds^2 + e^{-2s} r$$

is complete and its time-like cone $C_-(N,h)$ is non-flat and has a light-like parallel vector field as well as a non-degenerate proper holonomy invariant subspace.

The structure of geodesically complete simply connected manifolds with non-irreducible Lorentzian cone is described in the following Theorem.

Theorem 5.5 ([1]). *Let (N,h) be a geodesically complete, simply connected Riemannian or Lorentzian manifold of dimension at least 2 with the corresponding Lorentzian cone $C_\varepsilon(N,h)$, $\varepsilon = \pm 1$.*

1. If the holonomy representation of the cone $C_\varepsilon(N,h)$ is decomposable, then

$$Hol_x(N,h) = \mathrm{SO}^0(T_x N, h_x).$$

If (N,h) is Riemannian (i.e. $\varepsilon = -1$), then (N,h) has either constant sectional curvature ε or is isometric to the product

$$(\mathbb{R}^+ \times N_1 \times N_2\,,\, -\varepsilon ds^2 + \cosh^2(s)h_1 + \sinh^2(s)h_2),$$

where (N_i, h_i) are Riemannian manifolds and (N_2, h_2) has constant curvature $-\varepsilon$ or dimension ≤ 1. If (N,h) is a Lorentzian manifold (i.e. $\varepsilon = 1$), then the same result is true on each connected component of a certain open dense set of N.

2. *If the holonomy representation of $C_\varepsilon(N,h)$ is indecomposable but non-irreducible, then the cone admits a parallel light-like vector field.*

If (N,h) is a Riemannian manifold, then (N,h) is isometric to

$$(\mathbb{R} \times F, -\varepsilon ds^2 + e^{-2s} r),$$

where (F, r) is a complete Riemannian manifold, and the holonomy group of the cone is given by

$$Hol(C_-(N,h)) = Hol(F,r) \ltimes \mathbb{R}^{\dim F}.$$

If (N,h) is a Lorentzian manifold, the same result is true for any connected component of a certain open sense set of N.

Note, that a compact pseudo-Riemannian manifold need not to be geodesically complete. A stronger result hold for *compact* manifolds (N,h).

Theorem 5.6 ([70]). *Let $C_\varepsilon(N,h)$ be the Lorentzian cone over a compact connected Riemannian or Lorentzian manifold (N,h). Then the holonomy representation of $C_\varepsilon(N,h)$ is indecomposable.*

In [1, 69] Theorem 5.6 is proved under the additional assumption that (N,g) is geodesically complete. It was first shown that decomposability implies that (N,h) has constant sectional curvature ε. Since there are no compact de Sitter spaces, (N,h) has to be Riemannian with flat, but non-simply connected cone. In [70, Proposition 4.1], Matveev and Mounoud gave a nice short argument using only the compactness of (N,h) to show that the metric of a decomposable cone is definite.

5.3 *Lorentzian Metrics with Special Holonomy on Non-trivial Torus Bundles*

In this section we describe a construction of Lorentzian metrics with special holonomy on non-trivial torus bundles which is due to Lärz ([60, 62]). The basic idea is to consider Lorentzian metrics on S^1-bundles which look like a Walker metric (see Sect. 4). For that, let (N, h_N) be a Riemannian manifold, $\omega \in H^2(N,\mathbb{Z})$ and $\pi : M \to N$ the S^1-bundle with $c_1(M) = \omega$. For any closed 2-form ψ on N representing ω in the de Rham cohomology, there is a connection form

$A : TM \to i\mathbb{R}$ on M with curvature $dA = -2\pi i \pi^* \psi$ (see, e.g. [11]). For a smooth function $f \in C^\infty(M)$ and a nowhere vanishing closed 1-form η on N we consider the following Lorentzian metric on M:

$$g := 2i\, A \odot \pi^*\eta + f \cdot (\pi^*\eta)^2 + \pi^* h_N. \tag{10}$$

The vertical fundamental vector field ξ of the S^1-action on M is light-like. Using that η is closed, one obtains for the covariant derivative of ξ

$$\nabla^g_Z \xi = -\xi(f) \cdot \eta\big(d\pi(Z)\big) \cdot \xi, \quad Z \in \mathfrak{X}(M).$$

This shows that the vertical tangent bundle $\mathscr{V} := \mathbb{R}\xi \subset TM$ is a parallel distribution, and that ξ is parallel iff f is constant on the fibers of π. Moreover, if $\xi(f) \neq 0$, the distribution $\mathscr{V}$ does not contain a parallel vector field. Hence, the holonomy representation of (M, g) has an invariant light-like vector resp. line.

First, let us mention that there are special cases of this construction where M is totally twisted, i.e., where M is not homeomorphic to $Y \times \mathbb{R}$ or $Y \times S^1$. Thereby, M can be compact as well as non-compact (cf. [60]).

Here we will consider a special case of this construction, where a 1-dimensional factor splits up: We take $N := B \times L$, with a 1-dimensional manifold L, $\eta := du$, where u is the coordinate of L, and $\omega \in H^2(B, \mathbb{Z})$. Then $M = \widetilde{M} \times L$, where $\widetilde{\pi} : \widetilde{M} \to B$ is the S^1-bundle on B with 1. Chern class ω. Now, let $\tilde{A}$ be a connection form on $\widetilde{M}$, h a Riemannian metric on B and f a smooth function on M. Then the metric (10) has the special form

$$g := g_{f,\tilde{A},h} := 2i \tilde{A} \odot du + f \cdot du^2 + \widetilde{\pi}^* h.$$

We call $(M, g_{f,\tilde{A},h})$ *a manifold of toric type over* (B, h).[5] One can use this construction to produce Lorentzian manifolds with non-trivial topology and holonomy group

$$\mathrm{Hol}(M, g) = \begin{cases} G \ltimes \mathbb{R}^{n-2} \\ (\mathbb{R}^+ \times G) \ltimes \mathbb{R}^{n-2} \end{cases},$$

where G is one of the groups $\mathrm{SO}(n-2)$, $\mathrm{U}(m)$, $\mathrm{SU}(m)$ or $\mathrm{Sp}(k)$. The horizontal lift $TB^* \subset TM$ of TB with respect to $\tilde{A}$ is isomorphic to the vector bundle $\mathscr{V}^\perp/\mathscr{V}$. Looking at the parallel displacement along the horizonal lifts of curves in B, one can check that $\mathrm{Hol}(B, h) \subset \mathrm{Hol}(\mathscr{V}^\perp/\mathscr{V}, \widetilde{\nabla}^g)$. The projection $G := \mathrm{pr}_{\mathrm{O}(n-2)} \mathrm{Hol}(M, g) \subset \mathrm{O}(n-2)$ coincides with $\mathrm{Hol}(\mathscr{V}^\perp/\mathscr{V}, \widetilde{\nabla}^g)$. Hence, to ensure that the holonomy group $\mathrm{Hol}(\mathscr{V}^\perp/\mathscr{V}, \widetilde{\nabla}^g)$ is not larger then $\mathrm{Hol}(B, h)$, the bundle $\mathscr{V}^\perp/\mathscr{V}$ has to admit an additional $\widetilde{\nabla}^g$-parallel structure corresponding to the group G is question. This is possible for appropriate classes $\omega \in H^2(B, \mathbb{Z})$ defining

[5] Note that if $L = S^1$, M is a torus bundle with one trivial direction.

the topological type of the S^1-bundle $\widetilde{M} \to B$ and appropriate closed 2-forms ψ representing ω and defining the connection form $\tilde{A}$. We quote some of the results of K. Lärz.

Theorem 5.7 ([60,62]). *With the notations above and a sufficient generic function f in every case, we have:*

1. *Let (B, h) be an $(n-2)$-dimensional Riemannian manifold such that* $\mathrm{Hol}(B, h) = \mathrm{SO}(n-2)$. *If $(M, g_{f,\tilde{A},h})$ is of toric type over (B, h), then*

$$\mathrm{Hol}(M, g_{f,\tilde{A},h}) = \begin{cases} \mathrm{SO}(n-2) \ltimes \mathbb{R}^{n-2} & f \text{ fiber-constant on } \widetilde{P} \\ (\mathbb{R}^+ \times \mathrm{SO}(n-2)) \ltimes \mathbb{R}^{n-2} & \text{otherwise.} \end{cases}$$

2. *Let (B^{2m}, h, J) be a compact, simply connected, irreducible Kähler manifold with $c_1(B, J) < 0$ and let h be its Kähler–Einstein metric. Then, for any Hodge class $\omega \in H^{1,1}(B, \mathbb{Z}) := \mathrm{Im}\left(H^2(B, \mathbb{Z}) \to H^2(B, \mathbb{C})\right) \cap H^{1,1}(B, J)$,*

$$\mathrm{Hol}(M, g_{f,\tilde{A},h}) = \begin{cases} \mathrm{U}(m) \ltimes \mathbb{R}^{2m} & f \text{ fiber-constant on } \widetilde{P} \\ (\mathbb{R}^+ \times \mathrm{U}(m) \ltimes \mathbb{R}^{2m} & \text{otherwise.} \end{cases}$$

3. *Let (B^{2m}, J, h) be a Calabi–Yau manifold, i.e., a compact Kähler manifold with holonomy group* $\mathrm{SU}(m)$. *Choose $\omega \in H^{1,1}(B, \mathbb{Z})$ and a harmonic representative $\psi \in \omega$ which in the case $L = S^1$ has integer values under the dual Lefschetz operator. Then*

$$\mathrm{Hol}(M, g_{f,\tilde{A},h}) = \begin{cases} \mathrm{SU}(m) \ltimes \mathbb{R}^{2m} & f \text{ fiber-constant on } \widetilde{P} \\ (\mathbb{R}^+ \times \mathrm{SU}(m) \ltimes \mathbb{R}^{2m} & \text{otherwise.} \end{cases}$$

4. *Let (B^{4k}, J) be a holomorphic symplectic manifold with $b_2 \geq 4$ and Picard number $\rho(B, J) = b_2 - 2$. Then there exists an irreducible hyperkähler structure (B, J, J_2, J_3, h) with Kähler class in $H^2(B, \mathbb{Q})$ and $0 \neq \omega \in H^{1,1}(B, J) \cap H^{1,1}(B, J_2) \cap H^2(M, \mathbb{Z})$. Let $\psi \in \omega$ be a harmonic representative. Then*

$$\mathrm{Hol}(M, g_{f,\tilde{A},h}) = \begin{cases} \mathrm{Sp}(k) \ltimes \mathbb{R}^{4k} & f \text{ fiber-constant on } \widetilde{P} \\ (\mathbb{R}^+ \times \mathrm{Sp}(k)) \ltimes \mathbb{R}^{4k} & \text{otherwise.} \end{cases}$$

For a proof we refer to [60,62]. There one can also find lots of concrete examples of the type described in the Theorem. In particular, this methods allows to construct spaces with disconnected holonomy groups.

Another bundle construction was considered by Krantz in [59]. He studied S^1-bundles $\pi : M \to N$ over Riemannian manifolds (N, h) with Lorentzian Kaluza–Klein metrics on the total space of the form

$$g := A \odot A + \pi^* h,$$

where A is a connection form on M. In this case the fiber is time-like. Hereby, a parallel light-like distribution on (M, g) can occur only if the S^1-bundle admits a flat connection.

5.4 *Geodesically Complete and Globally Hyperbolic Models*

The bundle construction in Sect. 5.3 produces compact as well as non-compact Lorentzian manifolds with special holonomy. Moreover, this bundle construction gives us complete compact examples: Let T^{n-1} be the flat torus with standard coordinates $(x_1, \ldots, x_{n-2}, u)$ and take $\eta := du$ and $\psi := dx_1 \wedge du$. Consider the S^1-bundle $\pi : M \to T$ over T defined by $c_1(M) = [\psi]$, a connection form A on M with curvature $dA = -2\pi i \pi^* \psi$ and a smooth function f on T. Then the Lorentzian metric

$$g := 2iA \odot du + (f \circ \pi + 1)du^2 + \sum_{i=1}^{n-2} dx_i^2$$

on M is a geodesically complete and, if f is sufficient generic, (M, g) is indecomposable with abelian holonomy algebra $\mathbb{R}^{n-2}$ (cf. [60, Cor. 5.3]).

Examples of non-compact geodesically complete Lorentzian manifolds with special holonomy of type 2 can be found in papers of Sanchez, Candela and Flores (see [30, 39]). These authors studied geodesics as well as causality properties for Lorentzian manifolds (M, g) of the form

$$M = \mathbb{R} \times F \times \mathbb{R}\,, \; g = 2dvdu + H(x,u)du^2 + h, \tag{11}$$

where (F, h) is a connected $(n-2)$-dimensional Riemannian manifold and H is a non-trivial smooth function. They call such manifolds *general plane-fronted waves (PFW)*. As we know from Proposition 4.1, if H is sufficient generic in a point, the holonomy group of the general plane-fronted wave is $\mathrm{Hol}(M, g) = \mathrm{Hol}(F, h) \ltimes \mathbb{R}^{n-2}$.

Proposition 5.1 ([30]). *A general plane-fronted wave (11) is geodesically complete if and only if (F, h) is a complete Riemannian manifold and the maximal solutions $s \to x(s) \in F$ of the equation*

$$\frac{\nabla^F \dot{x}(s)}{ds} = \frac{1}{2}\,(\mathrm{grad}^F\, H)(x(s), s) \tag{12}$$

are defined on $\mathbb{R}$.

Equation (12) is studied in several cases. For example, if $H = H(x)$ is at most quadratic, i.e., if there is a point $x_0 \in F$ and constants $r > 0$ and $C > 0$ such that

$$H(x) \leq C\, d(x, x_0)^2 \quad \text{for all } x \in F \text{ with } d(x, x_0) \geq r,$$

where $d(x, x_0)$ denotes the geodesic distance on (F, h), then the solutions of (12) are defined on $\mathbb{R}$. More on this subject can be found in [30].

Another property, which is of special interest in Lorentzian geometry and analysis, is global hyperbolicity. A Lorentzian manifold is called *globally hyperbolic* if it is connected and time-oriented and admits a Cauchy surface, i.e., a subset S which is met by each inextendible time-like piecewise C^1-curve exactly once. For an introduction to this kind of Lorentzian manifolds, its relevance and equivalent definitions we refer to [4, 14, 71, 74]. In [18], Bernal and Sánchez proved a characterization of globally hyperbolic manifolds which is very useful for geometric purposes.

Proposition 5.2 ([18]). *A Lorentzian manifold is globally hyperbolic if and only if it is isometric to*

$$(\mathbb{R} \times S,\ g = -\beta dt^2 + g_t), \tag{13}$$

where β is a smooth positive function, g_t is a family of Riemannian metrics on S smoothly depending on $t \in \mathbb{R}$, and each $\{t\} \times S$ is a smooth space-like Cauchy hypersurface in M.

Under special conditions a general plane-fronted wave is globally hyperbolic.

Proposition 5.3 ([39]). *A general plane-fronted wave (11) is globally hyperbolic if (F, h) is complete and if the function $-H(x, u)$ is subquadratic at spacial infinity, i.e., if there is a point $x_0 \in F$ and continuous functions $C_1(u) \geq 0$, $C_2(u) \geq 0$, $p(u) < 2$ such that*

$$-H(x, u) \leq C_1(u)\, d(x, x_0)^{p(u)} + C_2(u) \quad \textit{for all } (x, u) \in F \times \mathbb{R}.$$

In [13] Bazaikin constructed globally hyperbolic metrics of the more general form (7) and gave examples with holonomy of types 3 and 4. In Sect. 6.1 we will discuss globally hyperbolic metrics with complete Cauchy surface and parallel spinors.

5.5 Topological Properties

In [61, 62], Kordian Lärz studied topological properties of Lorentzian manifolds with special holonomy using Hodge theory of Riemannian foliations. We will briefly describe his results. Let (M, g) be a time-oriented Lorentzian manifolds with a 1-dimensional parallel light-like distribution $\mathcal{V} \subset TM$. Then there is a *global* recurrent vector field $\xi \in \Gamma(\mathcal{V})$. In [61], a Lorentzian manifold is called *decent*, if the vector field ξ can be chosen such that $\nabla_X \xi = 0$ for all $X \in \xi^\perp$. Now, fix a vector field Z on M satisfying

$$g(Z,Z) = 0 \text{ and } g(\xi, Z) = 1,$$

and denote by $S \subset TM$ the subbundle $S := \mathrm{span}(\xi, Z)^{\perp}$. Using the vector field Z we can define a Riemannian metric g^R on M by

$$g^R(\xi,\xi) := 1,\ g^R(Z,Z) := 1,\ g^R(\xi,Z) := 0,\ g^R_{|S\times S} := g_{|S\times S},\ \mathrm{span}(\xi,Z)\perp_{g^R} S.$$

Let $\mathscr{L}$ be the foliation of M in light-like curves given by the parallel line $\mathscr{V} \subset TM$ and let $\mathscr{L}^{\perp}$ be the foliation of M in light-like hypersurfaces given by the parallel subbundle $\mathscr{V}^{\perp} \subset TM$. If (M, g, ξ) is a decent spacetime, the Riemannian metric g^R is bundle-like with respect to the foliation $(M, \mathscr{L}^{\perp})$. Moreover, if $L^{\perp}$ is a leaf of $\mathscr{L}^{\perp}$, then $g^R_{|TL^{\perp}\times TL^{\perp}}$ is bundle like with respect to $(L^{\perp}, \mathscr{L}_{|L^{\perp}})$ as well. Then, an application of Hodge theory and Weitzenböck formula for the twisted basic Hodge–Laplacian of Riemannian foliations yields the following result:

Proposition 5.4 ([61,62]). *Let (M, g) be a decent spacetime and suppose that the foliation $\mathscr{L}^{\perp}$ of M contains a compact leaf $L^{\perp}$ with $Ric(X,X) \geq 0$ for all $X \in TL^{\perp}$. Let $b_1(M)$ be the first Betti number of M.*

1. *If M is compact, then $1 \leq b_1(M) \leq \dim M$.*
2. *If M is non-compact and all leaves of $\mathscr{L}^{\perp}$ are compact, then $0 \leq b_1(M) \leq \dim M - 1$.*

Moreover, if $Ric_q(X,X) > 0$ for some $q \in L^{\perp}$ and all $X \in S_q$, the bounds are $1 \leq b_1(M) \leq 2$ and $0 \leq b_1(M) \leq 1$, respectively.

Explicit examples show, that the bounds for the 1. Betti number in Proposition 5.4 are sharp. If the foliation $\mathscr{L}^{\perp}$ of M admits a compact leaf with finite fundamental group, the holonomy algebra of (M, g) can only be of type 1, 2 or 3 (cf. Theorem 3.2), where the orthogonal part $\mathfrak{g}$ has an additional property. In special situations estimates for higher Betti numbers are possible (cf. [61,62]).

6 Lorentzian Manifolds with Special Holonomy and Additional Structures

6.1 Parallel Spinors

Now, let us consider a semi-Riemannian spin manifold (M, g) of signature (p, q) with spinor bundle S und spinor derivative ∇^S. We suppose in this review, that spin manifolds are space-and time-oriented. For a detailed introduction to pseudo-Riemannian spin geometry and the formulas for the spin representation see [5, 8] or [7]. In spin geometry one is interested in the description of all manifolds which admit parallel spinors, i.e., with spinor fields $\varphi \in \Gamma(S)$ such that $\nabla^S \varphi = 0$. This question is closely related to the holonomy group of (M, g), since the existence of parallel spinors restricts the holonomy group of (M, g). Let us explain this shortly. The spinor bundle is given by $S = Q \times_{(\mathrm{Spin}(p,q),\kappa)} \Delta_{p,q}$, where (Q, f) is a spin structure of (M, g) and $\kappa : \mathrm{Spin}(p, q) \to \mathrm{GL}(\Delta_{p,q})$ denotes the spinor representation in signature (p, q). Furthermore, let $\lambda : \mathrm{Spin}(p, q) \to \mathrm{SO}(p, q)$ denote the double covering of the special orthogonal group by the spin group. We consider the holonomy group $\mathrm{Hol}_x(M, g)$ of (M, g) as a subgroup of $\mathrm{SO}(p, q)$ (by fixing a basis in $T_x M$). Using, that the spinor derivative is induced by the Levi-Civita connection, the holonomy principle gives:

Proposition 6.1. *1. If (M, g) admits a non-trivial parallel spinor, then there is an embedding $\iota : \mathrm{Hol}(M, g) \hookrightarrow \mathrm{Spin}(p, q)$ such that $\lambda \circ \iota = \mathrm{Id}_{\mathrm{Hol}(M,g)}$. Moreover, there exists a vector $v \in \Delta_{p,q}$ such that $\iota(\mathrm{Hol}(M, g)) \subset \mathrm{Spin}(p, q)_v$, where $\mathrm{Spin}(p, q)_v$ denotes the stabilizer of v under the action of the spin group. On the other hand, if there is an embedding $\iota : \mathrm{Hol}(M, g) \hookrightarrow \mathrm{Spin}(p, q)$ such that $\lambda \circ \iota = \mathrm{Id}_{\mathrm{Hol}(M,g)}$, then (M, g) admits a spin structure whose holonomy group is $\iota(\mathrm{Hol}(M, g))$. Moreover, if there is a spinor $v \in \Delta_{p,q}$ such that $\iota(\mathrm{Hol}(M, g)) \subset \mathrm{Spin}(p, q)_v$, than (M, g) admits a non-trivial parallel spinor field.*

2. If (M, g) is simply connected, then there is a bijective correspondence between the space of parallel spinors and the kernel of the action of the subalgebra $\lambda_*^{-1}(\mathfrak{hol}(M, g)) \subset \mathfrak{spin}(p, q)$ on $\Delta_{p,q}$:

$$\{\varphi \in \Gamma(S) \mid \nabla^S \varphi = 0\} \overset{1:1}{\Longleftrightarrow} \{v \in \Delta_{p,q} \mid \lambda_*^{-1}(\mathfrak{hol}(M, g))\, v = 0\}.$$

Using this Proposition, one can easily check which groups in the holonomy list allow the existence of parallel spinors. Let us first recall the results for Riemannian manifolds.

Theorem 6.1. *Let (M, g) be a Riemannian spin manifold of dimension $n \geq 2$ with non-trivial parallel spinor. Then (M, g) is Ricci-flat and non-locally symmetric. If (M, g) is irreducible and simply connected, the holonomy group is one of the groups $\mathrm{SU}(m)$ if $n = 2m \geq 4$, $\mathrm{Sp}(k)$ if $n = 4k \geq 8$, G_2 if $n = 7$, or $\mathrm{Spin}(7)$ if $n = 8$, with its standard representation.*

The list of holonomy groups in Theorem 6.1 was found by Wang [79]. A list of the holonomy groups of irreducible, non-simply connected Riemannian spin manifolds with parallel spinors can be found in [72].

The situation in the Lorentzian case is a bit different. First of all, note that there are non-Ricci-flat as well as symmetric Lorentzian manifolds which admit parallel spinors. For example, let us consider the symmetric Cahen–Wallach spaces $M_{\underline{\lambda}} := (\mathbb{R}^n, g_{\underline{\lambda}})$ (cf. Sect. 5.1, formula (9)). The Ricci-curvature of $M_{\underline{\lambda}}$ is given by

$$Ric(X) = -\sum_{j=1}^{n-2} \lambda_j \cdot g_{\underline{\lambda}}\Big(X, \frac{\partial}{\partial v}\Big) \cdot \frac{\partial}{\partial v}, \qquad X \in \mathfrak{X}(M_{\underline{\lambda}}).$$

If $\underline{\lambda} \neq (\lambda, \ldots, \lambda)$, i.e., if $M_{\underline{\lambda}}$ is not locally conformally flat, then the space of parallel spinors on $M_{\underline{\lambda}}$ is $2^{[n/2]-1}$-dimensional (cf. [9]).
Now, let us consider a Lorentzian spin manifold (M, g). Since (M, g) is time- and space-oriented, there is an indefinite hermitian bundle metric $\langle \cdot, \cdot \rangle$ on the spinor bundle S such that

$$\langle X \cdot \varphi, \psi \rangle = \langle \varphi, X \cdot \psi \rangle,$$
$$X(\langle \varphi, \psi \rangle) = \langle \nabla^S_X \varphi, \psi \rangle + \langle \varphi, \nabla^S_X \psi \rangle$$

for all vector fields X and spinor fields φ, ψ. If φ is a spinor field, the vector field V_φ defined by

$$g(X, V_\varphi) = -\langle X \cdot \varphi, \varphi \rangle$$

is future-directed and causal, i.e., $g(V_\varphi, V_\varphi) \leq 0$. Moreover, V_φ has the same zeros as φ.

Proposition 6.2. *Let (M, g) be a Lorentzian spin manifold with a non-trivial parallel spinor field φ. Then the vector field V_φ is parallel and either time-like or light-like. Moreover, the Ricci-tensor of (M, g) satisfies*

$$Ric(X) \cdot \varphi = 0\,, \qquad X \in \mathfrak{X}(M).$$

Therefore, the Ricci-tensor is totally isotropic[6] *and the scalar curvature of (M, g) vanishes.*

Proposition 6.2 shows that the holonomy representation of a Lorentzian spin manifold with a parallel spinor acts trivial on a time-like or a light-like 1-dimensional subspace. Since a product of spin manifolds admits a parallel spinor if and only if its factors admit one, we obtain from the Decomposition Theorem of de Rham and Wu (Theorem 2.4):

Proposition 6.3. *Let (M, g) be a simply connected, geodesically complete Lorentzian spin manifold with non-trivial parallel spinor φ. Then (M, g) is isometric to the product*

$$(\mathbb{R}, -dt^2) \times (M_1, g_1) \times \ldots \times (M_k, g_k) \text{ if } V_\varphi \text{ is time-like}$$
$$\text{or} \qquad (N, h) \times (M_1, g_1) \times \ldots \times (M_k, g_k) \text{ if } V_\varphi \text{ is light-like,}$$

where (M_1, g_1), …, (M_k, g_k) are flat or irreducible Riemannian spin manifolds with a parallel spinor and (N, h) is a weakly irreducible, but non-irreducible Lorentzian spin manifold with a parallel spinor.

---

[6]This means $Ric(TM) \subset TM$ is a totally isotropic subspace.

Let us now consider a weakly irreducible Lorentzian spin manifolds (M, g) with parallel spinor. For small dimension, by studying the orbit structure of the spinor modul, Bryant [25] and Figueroa-O'Farrill [38] proved

Proposition 6.4. *The maximal stabilizer groups of a spinor $v \in \Delta_{1,n-1}$ with a light-like associated vector under the spin representation are*

$$\begin{aligned}
n \leq 5 &: 1 \ltimes \mathbb{R}^{n-2}\\
n = 6 &: \mathrm{Sp}(1) \ltimes \mathbb{R}^4\\
n = 7 &: (\mathrm{Sp}(1) \times 1) \ltimes \mathbb{R}^5\\
n = 8 &: \mathrm{SU}(3) \ltimes \mathbb{R}^6\\
n = 9 &: \mathrm{G}_2 \ltimes \mathbb{R}^7\\
n = 10 &: \mathrm{Spin}(7) \ltimes \mathbb{R}^8 \textit{ and } \mathrm{SU}(4) \ltimes \mathbb{R}^8\\
n = 11 &: (\mathrm{Spin}(7) \times 1) \ltimes \mathbb{R}^9 \textit{ and } (\mathrm{SU}(4) \times 1) \ltimes \mathbb{R}^8.
\end{aligned}$$

Next, we explain the results of Leistner [64, 65], who was able to determine all possible holonomy algebras of a weakly irreducible Lorentzian spin manifold (M, g) admitting a non-trivial parallel spinor using his holonomy classification. Since there is a parallel light-like vector field on (M, g), the holonomy algebra $\mathfrak{hol}(M, g)$ is of type 2 or 4, in particular, $\mathfrak{hol}(M, g) \subset \mathfrak{so}(n-2) \ltimes \mathbb{R}^{n-2}$. In order to determine $\mathfrak{hol}(M, g)$, one has to calculate the subalgebra $\lambda_*^{-1}(\mathfrak{hol}(M, g)) \subset \mathfrak{spin}(1, n-1)$ and the space

$$\{v \in \Delta_{1,n-1} \mid \lambda_*^{-1}(\mathfrak{hol}(M, g))\, v = 0\},$$

see Proposition 6.1. Let $\mathbb{R}^{1,n-1} = \mathbb{R} f_1 \oplus \mathbb{R}^{n-2} \oplus \mathbb{R} f_n$ be the decomposition of the Minkowski space as in Sect. 3. We can identify the spinor moduls

$$\begin{aligned}
\Delta_{1,n-1} &= \Delta_{n-2} \otimes \Delta_{1,1},\\
v &= v_1 \otimes u_1 + v_2 \otimes u_2,
\end{aligned}$$

where (u_1, u_2) is a basis of $\Delta_{1,1} = \mathbb{C}^2$ and the Clifford multiplication with the isotropic vectors f_1 and f_n and with $x \in \mathbb{R}^{n-2}$ is given by

$$\begin{aligned}
f_1 \cdot (v_1 \otimes u_1 + v_2 \otimes u_2) &= \sqrt{2} v_1 \otimes u_1,\\
f_n \cdot (v_1 \otimes u_1 + v_2 \otimes u_2) &= -\sqrt{2} v_2 \otimes u_2,\\
x \cdot (v_1 \otimes u_1 + v_2 \otimes u_2) &= (-x \cdot v_1) \otimes u_1 + (x \cdot v_2) \otimes u_2.
\end{aligned}$$

For the covering map λ one calculates

$$\lambda_*^{-1}(\mathfrak{so}(n-2) \ltimes \mathbb{R}^{n-2}) = \mathfrak{spin}(n-2) + \{x \cdot f_1 \mid x \in \mathbb{R}^{n-2}\} \subset \mathfrak{spin}(1, n-1).$$

If $\mathfrak{h} \subset \mathfrak{so}(n-2) \ltimes \mathbb{R}^{n-2}$ acts weakly irreducible, there is a non-trivial vector $x \in \mathbb{R}^{n-2} \cap \mathfrak{h}$. Hence,

$$\{v \in \Delta_{1,n-1} \mid \lambda_*^{-1}(\mathfrak{h})v = 0\} = \{v_2 \otimes u_2 \mid v_2 \in \Delta_{n-2} \text{ with } \lambda_*^{-1}(\mathfrak{g})v_2 = 0\},$$

where $\mathfrak{g} = \mathrm{pr}_{\mathfrak{so}(n-2)}\,\mathfrak{h}$ is the orthogonal part of $\mathfrak{h}$. In view of Proposition 3.3, Theorem 3.4 and Proposition 6.1 this shows that the orthogonal part $\mathfrak{g}$ of the holonomy algebra $\mathfrak{hol}(M,g)$ of a Lorentzian manifold with parallel spinor (together with its representation) coincides with the holonomy representation of a Riemannian manifold with parallel spinors. By Theorem 6.1, this representation splits into a trivial part and irreducible factors, which can be the standard representations of $\mathfrak{su}(m)$, $\mathfrak{sp}(k)$, $\mathfrak{g}_2$ or $\mathfrak{spin}(7)$. These Lie algebras have trivial center. In particular, $\mathfrak{hol}(M,g)$ can not be of type 4. We obtain finally

Theorem 6.2 ([65]). *Let (M,g) be an indecomposable, simply connected Lorentzian manifold with non-trivial parallel spinor. Then the holonomy group is*

$$\mathrm{Hol}(M,g) = G \ltimes \mathbb{R}^{n-2},$$

where $G \subset \mathrm{SO}(n-2)$ is a product of Lie groups of the form $\{1\} \subset \mathrm{SO}(n_0)$, $\mathrm{SU}(m)$, $\mathrm{Sp}(k)$, G_2 or $\mathrm{Spin}(7)$ and the representation of G on $\mathbb{R}^{n-2}$ is the direct sum of the standard representations of these groups.

The calculation of the spinor derivative of a general plane-fronted wave (11) shows easily, that such waves admit parallel spinors if and only if the Riemannian manifold (F,h) admits such, and the number of independent parallel spinors on (M,g) is the same as on (F,h).

Bryant discussed local normal forms for pseudo-Riemannian metrics with parallel spinors in small dimension $n \leq 11$ (cf. [25]). For the special case of abelian holonomy group $\mathrm{Hol}^0(M,g) = \mathbb{R}^{n-2}$ we know already the local normal form of such a metric. g is locally isometric to $(\mathbb{R}^n, g_H)$ with

$$g_H = 2dv\,du + H\,du^2 + \sum_{i=1}^{n-2} dx_i^2,$$

where $H = H(x_1,\dots,x_{n-2},u)$ is an arbitrary smooth function. In view of Theorem 6.2 or Proposition 6.4 this is the only possible normal form for indecomposable Lorentzian manifolds with parallel spinors in dimension $n \leq 5$. For local metrics in dimension $6 \leq n \leq 11$ we refer to [21, 22, 25, 34, 37, 38, 55] and the references therein.

We will address here to a global problem, namely to the question, whether one can realize the holonomy groups $G \ltimes \mathbb{R}^{n-2}$ which allow a parallel spinor by a globally hyperbolic manifold *with complete Cauchy surface*. In [10] we proved:

Theorem 6.3. *Any Lorentzian holonomy group of the form*

$$G \ltimes \mathbb{R}^{n-2} \subset \mathrm{SO}(1, n-1),$$

where $G \subset \mathrm{SO}(n-2)$ *is a product of Lie groups of the form* $\{1\} \subset \mathrm{SO}(n_0)$, $\mathrm{SU}(m)$, $\mathrm{Sp}(k)$, G_2 *or* $\mathrm{Spin}(7)$ *with its standard representations, can be realized by a globally hyperbolic Lorentzian manifold* (M^n, g) *with a complete Cauchy surface and a non-trivial parallel spinor.*

For the proof we use the characterization (13) of globally hyperbolic manifolds by Bernal and Sanchez and ideas from the paper [3] of Bär, Gauduchon and Moroianu, who studied the spin geometry of generalized pseudo-Riemannian cylinders. First, we consider a special kind of spinor fields. Let (M_0, g_0) be a Riemannian spin manifold with a Codazzi tensor A, i.e., with a symmetric (1,1)-tensor field satisfying

$$(\nabla^{g_0}_X A)(Y) = (\nabla^{g_0}_Y A)(X) \qquad \text{for all vector fields } X, Y.$$

A spinor field φ on (M_0, g_0) is called *A-Codazzi spinor* if

$$\nabla^S_X \varphi = i\, A(X) \cdot \varphi \qquad \text{for all vector fields } X. \tag{14}$$

If A is uniformally bounded, we denote by $\mu_+(A)$ the supremum of the positive eigenvalues of A or zero if all eigenvalues are non-positive, and by $\mu_-(A)$ the infimum of the negative eigenvalues of A or zero if all eigenvalues are non-negative.

Proposition 6.5. *Let* (M_0, g_0) *be a complete Riemannian spin manifold with a uniformally bounded Codazzi tensor* A *and a non-trivial* A*-Codazzi spinor. Then the Lorentzian cylinder*

$$C := I \times M_0\ ,\ g_C := -dt^2 + (1 - 2tA)^* g_0,$$

with the interval $I = ((2\mu_-(A))^{-1}, (2\mu_+(A))^{-1})$ *is globally hyperbolic with complete Cauchy surface and with a parallel spinor.*

In order to obtain such cylinders, we have to ensure the existence of Codazzi spinors (14). Using our classification of Riemannian manifolds with imaginary Killing spinors ([6]), we obtain:

Proposition 6.6. *Let* (M_0, g_0) *be a complete Riemannian manifold with an* A*-Codazzi spinor and let all eigenvalues of the Codazzi tensor* A *be uniformally bounded away from zero. Then* (M_0, g_0) *is isometric to*

$$(\mathbb{R} \times F\ ,\ (A^{-1})^*(ds^2 + e^{-4s} g_F)),$$

where (F,h) is a complete Riemannian manifold with parallel spinors, and A^{-1} is a Codazzi-tensor on the warped product $(\mathbb{R}\times F, ds^2+e^{-4s}g_F)$.
Vice versa, let (F,h) be a complete Riemannian manifold with parallel spinors and a Codazzi tensor T which has eigenvalues uniformally bounded from below. Then there is a Codazzi tensor B on the warped product

$$M_0=\mathbb{R}\times F\ ,\ g_{wp}=ds^2+e^{-4s}\,g_F$$

with eigenvalues uniformally bounded away from zero. Moreover, B^{-1} is a Codazzi tensor on $(M_0,g_0:=(B^{-1})^*g_{wp})$, the Riemannian manifold (M_0,g_0) is complete and has B^{-1}–Codazzi spinors.

A Codazzi tensor B on a warped product

$$M_0=\mathbb{R}\times F\ ,\ g_{wp}=ds^2+f(s)^2\,g_F$$

with properties mentioned in Proposition 6.6 can be constructed from a Codazzi tensor T on (F,g_F) in the following way. We set

$$B:=\begin{pmatrix} b\cdot\mathrm{Id} & 0\\ 0 & E\end{pmatrix}$$

with respect to the decomposition $TM=\mathbb{R}\oplus TF$, where b is a function depending only on s and E is given by

$$E(s)=\frac{1}{f(s)}\Big(T+\int_0^s b(\sigma)\dot f(\sigma)d\sigma\cdot\mathrm{Id}_F\Big).$$

This yields a construction principle for globally hyperbolic manifolds with complete Cauchy surface and special holonomy.

Proposition 6.7 ([10]). *Let (F,g_F) be a complete Riemannian manifold with parallel spinors and a Codazzi tensor T with eigenvalues bounded from below. Then there are Codazzi tensors B on $(\mathbb{R}\times F, ds^2+e^{-4s}g_F)$ with eigenvalues uniformally bounded away from zero. Let*

$$C(F,B):=I\times\mathbb{R}\times F\ ,\ g_C:=-dt^2+(B-2t)^*(ds^2+e^{-4s}g_F).$$

Then

1. *(C,g_C) is globally hyperbolic with a complete Cauchy surface, it admits a parallel light-like vector field as well as a parallel spinor.*
2. *If (F,h) has a flat factor, then $C(F,B)$ is decomposable.*
3. *If (F,h) is (locally) a product of irreducible factors, then $C(F,B)$ is weakly irreducible and*

$$\mathrm{Hol}^0_{(0,0,x)}(C, g_C) = (B^{-1} \circ \mathrm{Hol}^0_x(F, g_F) \circ B) \ltimes \mathbb{R}^{\dim F}.$$

Our construction is based on the existence of Codazzi tensors on Riemannian manifolds with parallel spinors. Let us finally discuss some examples for that.

Example 1. On the flat space $\mathbb{R}^k$ the endomorphism $T_h^{\mathbb{R}^k}$,

$$T_h^{\mathbb{R}^k}(X) := \nabla_X^{\mathbb{R}^k}(\mathrm{grad}(h)) = X(\partial_1 h, \ldots, \partial_k h),$$

is a Codazzi tensor for any function h on $\mathbb{R}^k$, and every Codazzi tensor is of this form. In this case the cylinder $C(F, B)$ is flat for any Codazzi tensor B on the warped product that is constructed out of T as described above.

Example 2. Let (F_1, g_{F_1}) be a complete simply connected irreducible Riemannian spin manifold with parallel spinors and (F, g_F) its Riemannian product with a flat $\mathbb{R}^k$. Then (F, g_F) is complete and has parallel spinors. Let B be a Codazzi tensor on the warped product $\mathbb{R} \times_{e^{-2s}} F$ constructed out of the Codazzi tensor $\lambda\, \mathrm{Id}_{F_1} + T_h^{\mathbb{R}^k}$ of F, where $T_h^{\mathbb{R}^k}$ is taken from Example 1. Then the cylinder $C(F, B)$ is globally hyperbolic with complete Cauchy surface, it is decomposable and has the holonomy group

$$\mathrm{Hol}(F_1, g_{F_1}) \ltimes \mathbb{R}^{\dim F_1}.$$

Example 3. Let us consider the metric cone

$$(F^{n-2}, g_F) := (\mathbb{R}^+ \times N, dr^2 + r^2 g_N),$$

where (N, g_N) is simply connected and a Riemannian Einstein–Sasaki manifold, a nearly Kähler manifold, a 3-Sasakian manifold or a 7-dimensional manifold with vector product. Then (F, g_F) is irreducible and has parallel spinors (but fails to be complete). Furthermore, $T := \nabla^F \partial_r$ is a Codazzi tensor on (F, g_F). The cylinder $C(F, B)$, where the Codazzi tensor B is constructed out of T as described above, has the holonomy group

$$\mathrm{Hol}(C, g_C) \simeq G \ltimes \mathbb{R}^{n-2},$$

where

$$G = \begin{cases} \mathrm{SU}((n-2)/2) & \text{if } N \text{ is Einstein-Sasaki} \\ \mathrm{Sp}((n-2)/4) & \text{if } N \text{ is 3-Sasakian} \\ \mathrm{G}_2 & \text{if } N \text{ is nearly Kähler} \\ \mathrm{Spin}(7) & \text{if } N \text{ 7-dimensional with vector product.} \end{cases}$$

Example 4. Let $(F, g_F) = (F_1, g_{F_1}) \times \cdots \times (F_k, g_{F_k})$ be a Riemannian product of simply connected complete irreducible Riemannian manifolds with parallel spinors. Let T be the Codazzi tensor $T = \lambda_1\, \mathrm{Id}_{F_1} + \cdots + \lambda_k\, \mathrm{Id}_{F_k}$ and B constructed out of T as described above. Then $C(F, B)$ is globally hyperbolic with complete Cauchy surface, it is weakly irreducible and the holonomy group is isomorphic to

$$\big(\mathrm{Hol}(F_1, g_{F_1}) \times \cdots \times \mathrm{Hol}(F_k, g_{F_k})\big) \ltimes \mathbb{R}^{\dim F}.$$

Example 5. *Eguchi–Hansen space.* Eguchi–Hansen spaces are complete, irreducible Riemannian 4-manifolds with holonomy SU(2). They have two linearly independent parallel spinors. Any Codazzi tensor on a Eguchi–Hansen space has the form $T = \lambda \cdot \mathrm{Id}$ for a constant λ.

6.2 Einstein Metrics

In the final section we want to discuss recent results concerning Lorentzian Einstein spaces with special holonomy. As a first example let us look at the general plane-fronted wave (11). The Ricci tensor of such metric is given by

$$Ric = Ric^F - \frac{1}{2}\Delta^F H \cdot du^2.$$

Hence, for any Ricci-flat Riemannian manifold (F, h) and any family of harmonic functions $H(\cdot, u)$ on F, the general plane-fronted wave (11) is Ricci-flat with special holonomy.

Now, let (M, g) be a Lorentzian Einstein-space with Einstein constant Λ:

$$Ric = \Lambda \cdot g.$$

We suppose, that (M, g) admits a 1-dimensional parallel light-like distribution, i.e., that the holonomy group is contained in $(\mathbb{R}^* \times O(n-2)) \ltimes \mathbb{R}^{n-2}$. First we discuss the possible holonomy groups for such Einstein metrics. After that we review some results concerning the local structure of such metrics. We follow the papers of Gibbons and Pope [50, 51] as well as the results of Leistner and Galaev [43, 44, 47]. Recall, that an irreducible Riemannian manifold with holonomy algebra different from $\mathfrak{so}(n)$ and $\mathfrak{u}(n/2)$ is Einstein. The determination of the possible holonomy algebras for Lorentzian Einstein spaces is based on a detailed study of the space of curvature endomorphisms $\mathscr{K}(\mathfrak{h})$ of a weakly irreducible subalgebra $\mathfrak{h} \subset (\mathbb{R} \oplus \mathfrak{so}(n-2)) \ltimes \mathbb{R}^{n-2}$, which one can find in the papers of Galaev [40, 45, 47].

Theorem 6.4. *Let (M, g) be a weakly irreducible, but non-irreducible Lorentzian Einstein manifold. Then its holonomy algebra $\mathfrak{hol}(M, g)$ is of type 1 or type 2.*

1. *If (M, g) is Ricci-flat, then the holonomy algebra is either $(\mathbb{R} \oplus \mathfrak{g}) \ltimes \mathbb{R}^{n-2}$ and in the decomposition (1) of the orthogonal part $\mathfrak{g}$ at least one of the ideals $\mathfrak{g}_i \subset \mathfrak{so}(n_i)$ coincides with one of the Lie algebras $\mathfrak{so}(n_i)$, $\mathfrak{u}(n_i/2)$, $\mathfrak{sp}(n_i/4) \oplus \mathfrak{sp}(1)$ or with a symmetric Berger algebra, or the holonomy algebra is $\mathfrak{g} \ltimes \mathbb{R}^{n-2}$ and*

each ideal $\mathfrak{g}_i \subset \mathfrak{so}(n_i)$ *in the decomposition of* $\mathfrak{g}$ *coincides with one of the Lie-algebras* $\mathfrak{so}(n_i)$, $\mathfrak{su}(n_i/2)$, $\mathfrak{sp}(n_i/4)$, $\mathfrak{g}_2 \subset \mathfrak{so}(7)$, $\mathfrak{spin}(7) \subset \mathfrak{so}(8)$.

2. *If* (M, g) *is an Einstein space with non-zero Einstein constant* Λ, *then the holonomy algebra is* $(\mathbb{R}\oplus\mathfrak{g})\ltimes\mathbb{R}^{n-2}$, $\mathfrak{g}$ *has no trivial invariant subspace and each ideal* $\mathfrak{g}_i \subset \mathfrak{so}(n_i)$ *in the decomposition of* $\mathfrak{g}$ *is one of the Lie algebras* $\mathfrak{so}(n_i)$, $\mathfrak{u}(n_i/2)$, $\mathfrak{sp}(n_i/4) \oplus \mathfrak{sp}(1)$ *or a symmetric Berger algebra.*

We remark that contrary to the Riemannian situation, any of the holonomy algebras in Theorem 6.4 can be realized also by non-Einstein metrics.

Now, let us look at the local structure of a Lorentzian Einstein metric with a parallel light-like line and dimension at least 4. As we know from Sect. 4, locally such metric is a Walker metric. Around any point $p \in M$ there are coordinates $(U, (v, x_1, \dots, x_{n-2}, u)$ such that

$$g_{|U} = 2dv\,du + H\,du^2 + 2A(u) \odot du + h(u), \tag{15}$$

where $h(u) = h_{ij}(x_1, \dots, x_{n-2}, u)dx_i dx_j$ is an u-depending family of Riemannian metrics, H is a smooth function on U and $A(u) = A_i(x_1, \dots, x_{n-2}, u)dx_i$ is a u-depending family of 1-forms on U. Of course, the Einstein condition imposes conditions on the data A, H and h in the Walker metric. These conditions were derived by Gibbons and Pope in [50].

Theorem 6.5 ([50]). *Let* (M, g) *be a Lorentzian manifold with a parallel light-like line and assume that* (M, g) *is Einstein with Einstein constant* Λ. *Then the function* H *in the Walker metric (15) has the form*

$$H = \Lambda v^2 + vH_1 + H_0, \tag{16}$$

where H_1 *and* H_0 *are smooth functions on* U *which do not depend on* v, *and* H_0, H_1, $A(u)$ *and* $h(u)$ *satisfy the following system of differential equations:*

$$\begin{aligned} \Delta H_0 - \frac{1}{2}F^{ij}F_{ij} - 2A^i\partial_i H_1 - H_1\nabla^i A_i + 2\Lambda A^i A_i - 2\nabla^i \dot{A}_i & \\ + \frac{1}{2}\dot{h}^{ij}\dot{h}_{ij} + h^{ij}\ddot{h}_{ij} + \frac{1}{2}h^{ij}\dot{h}_{ij}H_1 &= 0, \\ \nabla^j F_{ij} + \partial_i H_1 - 2\Lambda A_i + \nabla^j \dot{h}_{ij} - \partial_i(h^{jk}\dot{h}_{jk}) &= 0, \\ \Delta H_1 - 2\Lambda\nabla^i A_i + \Lambda h^{ij}\dot{h}_{ij} &= 0, \\ Ric_{ij} &= \lambda h_{ij}. \end{aligned}$$

Hereby i, j, k *run from* 1 *to* $n-2$, *the dot denotes the derivative with respect to* u *and* ∂_i *the derivative with respect to* x_i, Δ *is the Laplace–Beltrami operator for the metrics* $h(u)$ *and* $F_{ij} = \partial_i A_j - \partial_j A_i$ *are the coefficients of the differential of the 1-form* $A(u)$. *Conversely, any Walker metric (15) satisfying these equations is an Einstein metric with Einstein constant* Λ.

In [44] Galaev and Leistner simplified this system of equations. They proved that one can always find Walker coordinates (15) with $A(u) = 0$. Moreover, using the special form of the curvature endomorphisms $\mathscr{K}((\mathbb{R} \oplus \mathfrak{so}(n-2)) \ltimes \mathbb{R}^{n-2})$ and the condition (16), they were able to show that for an Einstein manifold there exist Walker coordinates with $A(u) = 0$ and $H_0 = 0$, and furthermore, if $\Lambda \neq 0$ one can choose Walker coordinates with $A(u) = 0$ and $H_1 = 0$. If the Einstein manifold admits not only a parallel light-like line, but a parallel light-like vector field, then by Theorem 6.4 the Einstein constant Λ is zero.
In [50] one can find a lot of concrete Lorentzian Einstein metrics with a parallel light-like line, which are of physical relevance (time-dependent multi-center solutions). The case of 4-dimensional Einstein spaces was previously discussed for example in [49,57,58]. More concrete solutions in dimension 4 are obtained in [46].

References

1. Alekseevsky, D., Cortés, V., Galaev, A., Leistner, T.: Cones over pseudo-Riemannian manifolds and their holonomy. J. Reine Angew. Math. **635**, 23–69 (2009)
2. Bär, C.: Real Killing spinors and holonomy. Comm. Math. Phys. **154** (3), 509–521 (1993)
3. Bär, C., Gauduchon, P., Moroianu, A.: Generalized cylinders in semi-Riemannian and spin geometry. Math. Z. **249**, 545–580 (2005)
4. Bär, C., Ginoux, N, Pfäffle, F.: Wave Equations on Lorentzian Manifolds and Quantization. EMS Publishing House, Switzerland (2007)
5. Baum, H.: Spin-Strukturen und Dirac-Operatoren über pseudo-Riemannschen Mannigfaltigkeiten. Teubner, Leipzig (1981)
6. Baum, H.: Complete Riemannian manifolds with imaginary Killing spinors. Ann. Global Anal. Geom. **7**, 205–226 (1989)
7. Baum, H., Friedrich, T., Grunewald, R., Kath, I.: Twistors and Killing Spinors on Riemannian Manifolds. Teubner-Texte zur Mathematik, vol. 124. Teubner, Stuttgart/Leipzig (1991)
8. Baum, H., Kath, I.: Parallel spinors and holonomy groups on pseudo-Riemannian spin manifolds. Ann. Global Anal. Geom. **17**, 1–17 (1999)
9. Baum, H.: Twistor spinors on Lorentzian symmetric spaces. J. Geom. Phys. **34**, 270–286 (2000)
10. Baum, H., Müller, O.: Codazzi spinors and globally hyperbolic manifolds with special holonomy. Math. Z. **258** (1), 185–211 (2008)
11. Baum, H.: Eichfeldtheorie. Eine Einführung in die Differentialgeometrie auf Faserbündeln. Springer, Hiedelberg (2009)
12. Baum, H, Juhl, A.: Conformal Differential Geometry. Q-Curvature and Holonomy. Oberwolfach-Seminars, vol. 40, Birkhäuser, Basel (2010)
13. Bazaikin, Ya.V.: Globally hyperbolic Lorentzian manifolds with special holonomy. Siberian Math. J. **50** (4), 567–579 (2009)
14. Beem, J.K., Ehrlich, P.E., Easley, K.L.: Global Lorentzian Geometry. Monographs and Textbooks in Pure and Applied Mathematics, vol. 202. Dekker Inc., New York (1996)
15. Berard-Bergery, L., Ikemakhen, A.: On the holonomy of Lorentzian manifolds. In: Greene, R., Yau, S.T. (eds.) Differential Geometry: Geometry in Mathematical Physics and Related Topics, Proceedings of Symposia in Pure Mathematics, vol. 54, pp. 27–40. American Mathematical Society, Providence (1993)
16. Berger, M.: Sur les groupes d'holonomie homogène des variétés à connexion affine et des variétés riemanniennes. Bull. Soc. Math. France **83**, 279–330 (1955)

17. Berger, M.: Les espaces symétriques non compactes. Ann. École Norm. Sup. **74**, 85–177 (1957)
18. Bernal, A.N., Sánchez, M.: Smoothness of time function and the metric spitting of globally hyperbolic spacetimes. Comm. Math. Phys. **257**, 43–50 (2005)
19. Besse, A.L.: Einstein Manifolds. Springer, New York (1987)
20. Bezvitnaya, N.: Lightlike foliations on Lorentzian manifolds with weakly irreducible holonomy algebra. ArXiv:math/0506101v1 (2005)
21. Brannlund, J., Coley, A., Hervik, S.: Supersymmetry, holonomy and Kundt spacetimes. Classical Quant. Grav. **25**, 195007, 10 pp (2008)
22. Brannlund, J., Coley, A., Hervik, S.: Holonomy, decomposability, and relativity. Can. J. Phys. **87**, 241–243 (2009)
23. Bryant, R.: Classical, exceptional, and exotic holonomies: a status report. Séminaires & Congrès, Soc. Math. France **1**, 93–165 (1996)
24. Bryant, R.: Recent advances in the theorey of holonomy. Séminaire Bourbaki, Asterisque **51**, 525–576 (1999)
25. Bryant, R.: Pseudo-Riemannian metrics with parallel spinor fields and vanishing Ricci tensor. Global Analysis and Harmonic Analysis, Seminairs et Congress **4**, 53–93 (2000)
26. Cahen, M., Kerbrat, Y.: Champs de vecteurs conformes et transformations conformes des spaces Lorentzian symetriques. J. Math. Pure Appl. **57**, 99–132 (1978)
27. Cahen, M., Parker, M.: Pseudo-Riemannian symmetric spaces. Memoir. Am. Math. Soc. **24**, 108 (1980)
28. Cahen, M., Wallach, N.: Lorentzian symmetric spaces. Bull. Am. Math. Soc. **76**, 585–591 (1970)
29. Cahen, M., Leroy, J., Parker, M., Tricerri, F., Vanhecke, L.: Lorentzian manifolds modelled on a Lorentz symmetric space. J. Geom. Phys. **7**(4), 571–581 (1990)
30. Candela, A.M., Flores, J.L., Sánchez, M.: On general plane fronted waves. Geodesics. Gen. Relat. Gravit. **35**(4), 631–649 (2003)
31. Cartan, E.: La Geometrie des Espaces de Riemann. Memorial des Sciences Mathematiques, Fasc. IX, Ch. VII, Sec. II (1925)
32. Cartan, E.: Sur une classe remarquable d'espaces de Riemann. Bull. Soc. Math. France **54**, 214–264 (1926)
33. Cartan, E.: Les groups d'holonomie des espaces généralisés. Acta Math. **48**, 1–42 (1926)
34. Coley, A., Fuster, A., Hervik, S.: Supergravity solutions with constant scalar invariants. Int. J. Mod. Phys. **24**, 1119–1133 (2009)
35. De Rham, G.: Sur la réductibilité d'un espace de Riemann. Comm. Math. Helv. **26**, 328–344 (1952)
36. Di Scala, A., Olmos, C.: The geometry of homogeneous submanifolds of hyperbolic space. Math. Z. **237**, 199–209 (2001)
37. Figueroa-O'Farrill, J.: More Ricci-flat branes. Phys. Lett. B **471**, 128–132 (1999)
38. Figueroa-O'Farrill, J.: Breaking the M-wave. Classical Quant. Grav. **17**, 2925–2947 (2000)
39. Flores, J.L., Sánchez, M.: Causality and conjugated points in general plane waves. Classical Quant. Grav. **20**, 2275–2291 (2003)
40. Galaev, A.: The space of curvature tensors for holonomy algebras of Lorentzian manifolds. Differ. Geom. Appl. **22**(1), 1–18 (2005)
41. Galaev, A.: Isometry groups of Lobachewskian spaces, similarity transformation groups of Euclidian spaces and Lorentzian holonomy groups. Rend. Circ. Mat. Palermo (2)(Suppl. 79), 87–97 (2006).
42. Galaev, A.: Metrics that realize all Lorentzian holonomy algebras. Int. J. Geomet. Meth. Mod. Phys. **3**, 1025–1045 (2006)
43. Galaev, A., Leistner, T.: Holonomy groups of Lorentzian manifolds: classification, examples, and applications. In: Alekseevsky, D., Baum, H. (eds.) Recent Developments in Pseudo-Riemannian Geometry, pp. 53–96. EMS Publishing House, Switzerland (2008)
44. Galaev, A., Leistner, T.: On the local structure of Lorentzian Einstein manifolds with parallel null line. Classical Quant. Grav. **27**, 225003 (2010)

45. Galaev, A.: On one component of the curvature tensor of a Lorentzian manifold. J. Geom. Phys. **60**, 962–971 (2010)
46. Galaev, A.: Examples of Einstein spacetimes with recurrent null vector fields. Class. Quant. Grav. **28**, 175022 (2011)
47. Galaev, A.: Holonomy of Einstein Lorentzian manifolds. Class. Quant. Grav. **27**, 075008, 13 pp (2010)
48. Gallot, S.: Équations differentielles caractéristiques de la sphère. Ann. Sci. École Norm. Sup. (4) **12**(2), 235–267 (1979)
49. Ghanam, R., Thompson, G.: Two special metrics with R_{14}-type holonomy. Classical Quant. Grav. **18**(11), 2007–2014 (2001)
50. Gibbons, G.W., Pope, C.N.: Time-dependent multi-center solutions from new metrics with holonomy $Sim(n-2)$. Classical Quant. Grav. **25**, 125015, 21 pp (2008)
51. Gibbons, G.W.: Holonomy old and new. Progr. Theor. Phys. Suppl. (177), 33–41 (2009)
52. Hall, G.S.: Space-times and holonomy groups. In: Differential Geometry and Its Applications. Proceedings of the 5th International Conference, Opava, August 24–28, pp. 201–210. Silesian University (1993)
53. Hall, G.S., Lonie, D.P.: Holonomy groups and spacetimes. Classical Quant. Grav. **17**(6): 1369–1382 (2000)
54. Helgason, S.: Differential Geometry, Lie Groups, and Symmetric Spaces. Graduate Studies in Mathematics, vol. 34. American Mathematical Society, Providence (2001)
55. Hernandez, R., Sfetsos, K., Zoakos, D.: Supersymmetry and Lorentzian holonomy in varies dimensions. J. High Energy Phys. **9** 010, 17 pp (2004)
56. Joyce, D.: Compact Manifolds with Special Holonomy. Oxford University Press, Oxford (2000)
57. Kerr, R.P., Goldberg, J.N.: Some applications of the infinitesimal holonomy group to the Petrov classification of Einstein spaces. J. Math. Phys. **2**, 327–332 (1961)
58. Kerr, R.P., Goldberg, J.N.: Einstein spaces with four parameter holonomy group. J. Math. Phys. **2**, 332–336 (1961)
59. Krantz, T.: Kaluza–Klein-type metrics with special Lorentzian holonomy. J. Geom. Phys. **60**, 74–80 (2010)
60. Lärz, K.: A class of Lorentzian manifolds with indecomposable holonomy groups. arXiv:0803.4494v3 (2010)
61. Lärz, K.: Riemannian foliations and the topology of Lorentzian manifolds. arXiv:1010.2194v1 (2010)
62. Lärz, K.: Global aspects of holonomy in pseudo-Riemannian geometry. Ph.D. Thesis, Humboldt University of Berlin (2011)
63. Leistner, T.: Berger algebras, weak Berger algebras and Lorentzian holonomy. SFB 288-Preprint No. 567, 131–159 (2002)
64. Leistner, T.: Lorentzian manifolds with special holonomy and parallel spinors. Rend. Circ. Mat. Pal. II **69**, 131–159 (2002)
65. Leistner, T.: Holonomy and parallel spinors in Lorentzian geometry. Ph.D. Thesis, Humboldt University of Berlin (2003). Logos, Berlin (2004)
66. Leistner, T.: Conformal holonomy of C-spaces, Ricci-flat, and Lorentzian manifolds. Differ. Geom. Appl. **24**(5), 458–478 (2006)
67. Leistner, T.: Screen bundles of Lorentzian manifolds and some generalization of pp-waves. J. Geom. Phys. **56**(10), 2117–2134 (2006)
68. Leistner, T.: On the classification of Lorentzian holonomy groups. J. Differ. Geom. **76**, 423–484 (2007)
69. Matveev, V.: Gallot–Tanno theorem for pseudo-Riemannian metrics and a proof that decomposable cones over closed complete pseudo-Riemannian manifolds do not exist. J. Diff. Geom. Appl. **28**, 236–240 (2010)
70. Matveev, V, Mounoud, P.: Gallot–Tanno-theorem for closed incomplete pseudo-Riemannian manifolds and applications. Glob. Anal. Geom. **38**(3), 259–271 (2010)

71. Minguzzi, E., Sánchez, M.: The causal hierarchy of spacetimes. In: Alekseevsky, D., Baum, H. (eds.) Recent Developments in pseudo-Riemannian Geometry, pp. 299–358. EMS Publishing House, Switzerland (2008)
72. Moroianu, A., Semmelmann, U.: Parallel spinors and holonomy groups. J. Geom. Phys. **41**, 2395–2402 (2000)
73. Neukirchner, Th.: solvable pseudo-Riemannian symmetric spaces. arXiv:math.0301326 (2003)
74. Pfäffle, F.: Lorentzian manifolds. In: Bär, C., Fredenhagen, K. (eds.) Quantum Field Theory on Curved Spacetimes—Concepts and Mathematical Foundations, pp. 39–58, Lecture Notes in Physics, vol. 786, Springer, Heidelberg, 2009
75. Salamon, S.: Riemannian Geometry and Holonomy Groups. Longman Scientific & Technical, Harlow, Essex, England (1989)
76. Schell, J.F.: Classification of 4-dimensional Riemannian spaces. J. Math. Phys. **2**, 202–206 (1960)
77. Shaw, R.: The subgroup structure of the homogeneous Lorentz group. Q. J. Math. Oxford **21**, 101–124 (1970)
78. Walker, A.G.: Canonical form for a Riemannian metric with a parallel field of null planes. Q. J. Math. Oxford Ser. **1**(2), 69–79 (1950)
79. Wang, MKY.: Parallel spinors and parallel forms. Ann. Global Anal. Geom. **7**, 59–68 (1989)
80. Wolf, J.A.: Spaces of Constant Curvature. Publish or Perish, Inc., Wilmington (1984)
81. Wu, H.: On the de Rham decomposition theorem. Illinois J. Math. **8**, 291–311 (1964)

Some Curvature Problems in Semi-Riemannian Geometry

Felix Finster and Marc Nardmann

Abstract In this survey article we review several results on the curvature of semi-Riemannian metrics which are motivated by the positive mass theorem and have been obtained within the Priority Program "Globale Differentialgeometrie" of the Deutsche Forschungsgemeinschaft. The main themes are estimates of the Riemann tensor of an asymptotically flat manifold and the construction of Lorentzian metrics which satisfy the dominant energy condition.

In this survey article we review recent progress on several curvature problems in semi-Riemannian geometry, each of which has a certain relation to the positive mass theorem (PMT). The focus is on work in which we were involved within the Priority Program "Globale Differentialgeometrie."

The time-symmetric version of the PMT says in particular that an asymptotically flat Riemannian manifold with zero mass is flat. Section 1 investigates whether an asymptotically flat manifold whose mass is *almost zero* must be almost flat in a suitable sense. The general, not necessarily time-symmetric, situation is considered as well. The main tool in this work is the spinor which occurs in Witten's proof of the PMT.

The PMT implies that if the energy E and the momentum P of an asymptotically flat spacelike hypersurface M of a Lorentzian manifold in which the dominant energy condition holds satisfy $E = |P|$, then the Lorentzian metric is flat along M. Schoen–Yau proved that in this situation, M with its given second fundamental form can be isometrically embedded as a spacelike graph into Minkowski spacetime. The short Sect. 2 presents an alternative proof of this fact, based on the Lorentzian version of the fundamental theorem of hypersurface theory due to Bär–Gauduchon–Moroianu.

F. Finster (✉) · M. Nardmann
NWF I - Mathematik, Universität Regensburg, D-93040 Regensburg, Germany
e-mail: Felix.Finster@mathematik.uni-r.de; Marc.Nardmann@mathematik.uni-r.de

C. Bär et al. (eds.), *Global Differential Geometry*, Springer Proceedings in Mathematics 17, DOI 10.1007/978-3-642-22842-1_8,

Section 3 deals with the question which smooth manifolds admit a Lorentzian metric that satisfies the dominant energy condition. Since every closed or asymptotically flat spacelike hypersurface of a Lorentzian manifold can potentially yield a PMT-like obstruction to the dominant energy condition, one should avoid in the construction of dominant energy metrics that such spacelike hypersurfaces exist at all. This can indeed be accomplished in many situations.

1 Analysis of Asymptotically Flat Manifolds via Witten Spinors

Asymptotically flat Lorentzian manifolds describe isolated gravitating systems (like a star or a galaxy) in the framework of general relativity. As discovered by Arnowitt, Deser and Misner [1], to an asymptotically flat Lorentzian manifold one can associate the total energy and the total momentum, defined globally via the asymptotic behavior of the metric near infinity. Moreover, the energy-momentum tensor gives a local concept of energy and momentum. These global and local quantities are linked by Einstein's field equations, giving rise to an interesting interplay between local curvature and the global geometry of space-time. The first result which shed some light on the nature of this interplay is the positive energy theorem [29,30], which states that if the local energy density is positive (in the sense that the dominant energy condition holds), then the total energy is also positive. More recently, the proof of the Riemannian Penrose inequalities [5, 19] showed that in the time-symmetric situation, the total energy is not only positive, but it is even larger than the energy of the black holes, as measured by the surface area of their horizons. Despite this remarkable progress, many important problems remain open (see for example [22]).

The aim of our research project was to get a better understanding of how total energy and momentum control the geometry of space-time. In the special case that energy and momentum vanish, the positive energy theorem yields that the space-time manifold is flat [27, 30]. This suggests that if total energy and momentum are small, then the manifold should be almost flat, meaning that curvature should be small. But is this conjecture really correct? Suppose we consider a sequence of space-time metrics such that total energy and momentum tend to zero. In which sense do the metrics converge to the flat Minkowski metric?

Although our considerations could not give definitive answers to these questions, at least they led to a few inequalities giving some geometric insight, as we will outline in what follows. For simplicity, we begin in the Riemannian setting (in general dimension n), whereas the generalizations to include the second fundamental form will be explained in Sect. 1.7. All our methods use the Witten spinor as introduced in [37]. But in contrast to the spinor proof of the positive energy theorem [27], we consider second derivatives of the Witten spinor ψ. Our starting point is a basic inequality involving the L^2-norm of the second derivatives of ψ (Sect. 1.2).

Using Sobolev techniques, we deduce curvature estimates, which however involve the isoperimetric constant of the manifold (Sect. 1.3). An analysis of the level sets of $|\psi|$ allows us to get estimates which are independent of the isoperimetric constant but instead involve a volume bound (Sect. 1.4). In the case of an asymptotically Schwarzschild space-time, we then derive weighted L^2-estimates of ψ which involve the lowest eigenvalue λ of the Dirac operator on a conformal compactification (Sect. 1.5). These weighted L^2-estimates finally give rise to curvature estimates which involve the global geometry of the manifold only via λ (Sect. 1.6). We conclude by an outlook and a discussion of open problems (Sect. 1.8).

1.1 The Riemannian Setting, Asymptotically Flat Manifolds

We briefly recall how the Riemannian setting arises within the framework of general relativity. Suppose that space-time is described by the Lorentzian manifold $(N^4, \bar{g})$, which for simplicity we will assume to be orientable and time-orientable. To describe the splitting into space and time as experienced by an observer, one chooses a foliation of N^4 by spacelike hypersurfaces. Considering the situation at a fixed observer time, one restricts attention to one hypersurface M^3 of this foliation. Then $\bar{g}$ induces on M^3 a Riemannian metric g_{ij}. Furthermore, choosing on M a future-directed normal unit vector field ν, we obtain on M the second fundamental form $h_{ij} = (\bar{\nabla}_j \nu)_k$. The *time-symmetric* situation is obtained by assuming that h vanishes identically. This condition is in particular satisfied if the unit normal ν is a Killing field, meaning that the system is static. In this special case, the geometry at the fixed observer time is completely described by the Riemannian metric g on M^3.

The physical condition that the local energy density should be positive gives rise to the dominant energy condition (see [16, Sect. 4.3] and Sect. 3.2 in the present article) for the energy-momentum tensor. Using the Einstein equations, it can also be expressed in terms of the Ricci tensor on N^4. In the time-symmetric situation, the dominant energy condition reduces to the condition that (M^3, g) should have *non-negative scalar curvature*. Every orientable, three-dimensional Riemannian manifold is spin (see for example [20]). Therefore, it is a sensible mathematical generalization to consider in what follows a *spin* manifold (M^n, g) of dimension $n \geq 3$ of non-negative scalar curvature. Moreover, in order to exclude singularities, we shall assume that (M^n, g) is *complete*.

Having isolated gravitating systems in mind, we next want to impose that the Riemannian metric should approach the Euclidean metric in the "asymptotic ends" describing space near infinity. More precisely, considering for simplicity one asymptotic end, the manifold (M^n, g) is said to be *asymptotically flat* if there is a compact set $K \subset M$ and a diffeomorphism $\Phi : M \setminus K \to \mathbb{R}^n \setminus B_\rho(0)$, $\rho > 0$, such that

$$(\Phi_* g)_{ij} = \delta_{ij} + \mathscr{O}(r^{2-n}), \quad \partial_k (\Phi_* g)_{ij} = \mathscr{O}(r^{1-n}), \quad \partial_{kl} (\Phi_* g)_{ij} = \mathscr{O}(r^{-n}). \tag{1}$$

These decay conditions imply that scalar curvature is of the order $\mathcal{O}(r^{-n})$. We need to make the stronger assumption that *scalar curvature is integrable*. In the Riemannian setting, the total energy is also referred to as the *total mass m* of the manifold (whereas total momentum vanishes). It is defined by

$$m = \frac{1}{c(n)} \lim_{\rho\to\infty} \int_{S_\rho} (\partial_j (\Phi_* g)_{ij} - \partial_i (\Phi_* g)_{jj}) d\Omega^i, \tag{2}$$

where $c(n) > 0$ is a normalization constant and $d\Omega^i$ denotes the product of the volume form on $S_\rho \subset \mathbb{R}^n$ by the ith component of the normal vector on S_ρ (also we use the Einstein summation convention and sum over all indices which appear twice). The definition (2) was first given in [1]. In [3] it is proved that the definition is independent of the choice of Φ. The *positive mass theorem* [29] states that $m \geq 0$ in the case $n \leq 7$ (working even for non-spin manifolds). An alternative proof using spinors is given by [27, 37] and in general dimension in [3].

1.2 An L^2-Estimate for the Second Derivatives of the Witten Spinor

Before introducing our methods, we briefly recall the spinor proof of the positive mass theorem. The basic reason why spinors are very useful for the analysis of asymptotically flat spin manifolds is the *Lichnerowicz-Weitzenböck formula*

$$\mathcal{D}^2 = -\nabla^2 + \frac{s}{4}, \tag{3}$$

which actually goes back to Schrödinger [31]. Here $\mathcal{D}$ is the Dirac operator, ∇ is the spin connection, and s denotes scalar curvature. Witten [37] considered solutions of the Dirac equation with constant boundary values ψ_0 in the asymptotic end,

$$\mathcal{D}\psi = 0, \qquad \lim_{|x|\to\infty} \psi(x) = \psi_0 \quad \text{with} \quad |\psi_0| = 1, \tag{4}$$

where ψ is a smooth section of the spinor bundle SM. In [3, 27] it is proved that for any ψ_0, this boundary value problem has a unique solution. We refer to ψ as the *Witten spinor* with boundary values ψ_0. For a Witten spinor, the Lichnerowicz-Weitzenböck formula implies that

$$\nabla_i \langle \psi, \nabla^i \psi \rangle = |\nabla\psi|^2 + \frac{s}{4}|\psi|^2. \tag{5}$$

Integrating over M, applying Gauss' theorem and relating the boundary values at infinity to the total mass (where we choose $c(n)$ in (2) appropriately), one obtains the identity [3, 27, 37]

$$\int_M \left(|\nabla\psi|^2 + \frac{s}{4}|\psi|^2 \right) d\mu_M = m. \tag{6}$$

As the integrand is obviously non-negative, this identity immediately implies the positive mass theorem for spin manifolds.

We now outline the derivation of an L^2-estimate of $\nabla^2\psi$ (for details see [6] and [10]). We consider similar to (5) a divergence, but now of an expression involving higher derivatives,

$$\nabla_i \langle \nabla_j\psi, \nabla^i\nabla^j\psi\rangle = |\nabla^2\psi|^2 + \langle \nabla_j\psi, \nabla_i\nabla^i\nabla^j\psi\rangle.$$

In the third derivative term, we commute ∇^j to the left,

$$\nabla_i\nabla^i\nabla^j\psi = \left[\nabla_i\nabla^i, \nabla^j\right]\psi + \nabla^j\left(\nabla_i\nabla^i\psi\right).$$

Then in the last summand we can again apply the Lichnerowicz-Weitzenböck formula, whereas the commutator gives rise to curvature terms. We integrate the resulting equation over M. Using the faster decay of the higher derivatives of ψ, integrating by parts does not give boundary terms. Using the the Hölder inequality together with the inequality

$$\int_M |\nabla\psi|^2 d\mu_M \leq m \tag{7}$$

(which is obvious from (6)), we obtain the estimate

$$\boxed{\int_M |\nabla^2\psi|^2 \, d\mu_M \leq m C_1(n) \sup_M |R| + \sqrt{m} C_2(n) \|\nabla R\|_{L^2(M)} \sup_M |\psi|,} \tag{8}$$

where $|R| = \sqrt{R_{ijkl}R^{ijkl}}$ denotes the norm of the Riemann tensor. We remark that in [6, 10, 11] a more general inequality for $\int_M \eta|\nabla^2\psi|^2 \, d\mu_M$ with an arbitrary smooth function η is considered. By choosing η to be a test function, this makes it possible to "localize" the inequality to obtain curvature estimates on the support of η. For simplicity, in this survey article the function η will always be omitted.

1.3 Curvature Estimates Involving the Isoperimetric Constant

In short, curvature estimates are obtained from (8) by estimating the spinors by suitable a-priori bounds. We first outline how to treat the second derivative term $|\nabla^2\psi|^2$ (for details see [6] and [10]). The Schwarz inequality yields

$$\langle [\nabla_i, \nabla_j]\psi, [\nabla_i, \nabla_j]\psi\rangle \leq 4\,|\nabla^2\psi|^2.$$

Rewriting the commutators by curvature, we obtain an expression which is quadratic in the Riemann tensor. In dimension $n = 3$, one can use the properties of the Clifford multiplication to obtain

$$|R|^2\,|\psi|^2 \le c(n)|\nabla^2\psi|^2 .$$

In dimension $n > 3$, this inequality is in general wrong. But we get a similar inequality for a family $\psi_1, \ldots, \psi_N$ of Witten spinors,

$$\sum_{i=1}^{N} |R|^2\,|\psi_i|^2 \le c(n) \sum_{i=1}^{N} |\nabla^2\psi_i|^2, \tag{9}$$

where the boundary values $\lim_{|x|\to\infty} \psi_i(x)$ form an orthonormal basis of the spinors at infinity. The family of Witten spinors can be handled most conveniently by forming the so-called *spinor operator* (for details see [10]).

We next consider the term $\sup_M |\psi|$ in (8). A short calculation using the Lichnerowicz-Weitzenböck formula shows that $|\psi|$ is *subharmonic*, (see [9, Sect. 2]),

$$\Delta|\psi| \ge \frac{s}{4}|\psi| \ge 0. \tag{10}$$

Thus the maximum principle yields that $|\psi|$ has no interior maximum, and in view of the boundary conditions at infinity (4) we conclude that

$$\sup_M |\psi| = 1. \tag{11}$$

Using (9) and (11) in (8), we obtain the estimate

$$\int_M |R|^2 \left(\sum_{i=1}^N |\psi_i|^2\right) d\mu_M \le m C_1(n) \sup_M |R| + \sqrt{m} C_2(n) \|\nabla R\|_{L^2(M)}. \tag{12}$$

The remaining task is to estimate the norm of the spinors *from below*. Such estimates are difficult to obtain, partly because the norm of the spinor depends sensitively on the unknown geometry of M. We now begin with the simplest estimates, whereas more refined methods will be explained in Sects. 1.4 and 1.6.

The inequality (7) tells us that, for small m, the derivative of the spinor is small in the L^2-sense, suggesting that in this case the spinor should be almost constant, implying that $|\psi|$ should be bounded from below. In order to make this argument precise, we set $f = 1 - |\psi|$ and use the Kato inequality $|\nabla f| \le |\nabla\psi|$ to obtain $\|\nabla f\|_{L^2(M)} \le m$. The Sobolev inequality (see [10, Sect. 4]) for details) implies that

$$\|f\|_{L^q(M)} \le \frac{q}{k} m \qquad \text{where} \qquad q = \frac{2n}{n-2},$$

and k denotes the *isoperimetric constant.* Thus we only get an integral estimate of f. But this integral bound also implies that f is pointwise small, except on a set of small measure. We thus obtain the following result (see [10, Theorem 1.2]).

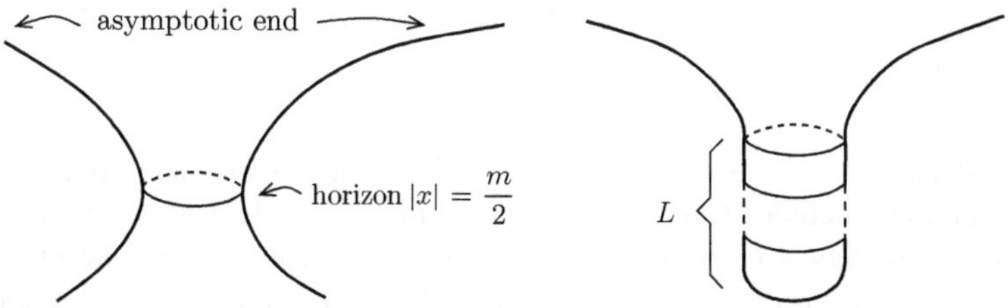

Fig. 1 The Schwarzschild metric (*left*) and the manifold after gluing (*right*)

Theorem 1.1. *Let (M^n, g), $n \geq 3$, be a complete asymptotically flat Riemannian spin manifold of non-negative scalar curvature. Then there is a set $\Omega \subset M$ with*

$$\mu(\Omega) \leq \left(\frac{c_3 m}{k^2}\right)^{\frac{n}{n-2}} \tag{13}$$

such that the following inequality holds,

$$\int_{M\setminus\Omega} |R|^2 d\mu_M \leq m c_1(n) \sup_M |R| + \sqrt{m} c_2(n) \|\nabla R\|_{L^2(M)}. \tag{14}$$

This theorem quantifies that the manifold indeed becomes flat in the limit $m \searrow 0$, provided that $\sup_M |R|$ and $\|\nabla R\|_{L^2(M)}$ are uniformly bounded and that the isoperimetric constant is bounded away from zero. The appearance of the isoperimetric constant and of the exceptional set Ω can be understood from the following simple example. We choose on $M^3 = \mathbb{R}^3$ the Schwarzschild metric $g_{ij}(x) = (1 - 2m/|x|)^4 \delta_{ij}$ (in order to clarify the connection to the construction in Sect. 1.1, we remark that this g_{ij} is isometric to the induced Riemannian metric on the $t = \text{const}$ slice of the standard Schwarzschild space-time). For the geometric understanding, it is helpful to isometrically embed M^3 into the Euclidean $\mathbb{R}^4$ (see the left of Fig. 1). This shows that M^3 has two asymptotic ends, one as $|x| \to \infty$ and the other as $|x| \to 0$. The minimal hypersurface $r = m/2$ has the interpretation as the *event horizon*.

In order to get a manifold with one asymptotic end, we cut M^3 at the event horizon and glue in a cylinder of length L as well as a spherical cap (see the right of Fig. 1). This manifold clearly has non-negative scalar curvature. In the limit $m \searrow 0$, the resulting manifold becomes flat outside the event horizon. The region inside the event horizon, however, does not become flat, because the radius of the cylinder shrinks to zero. This explains why we need an exceptional set. In the limit $L \to \infty$, the volume of this exceptional set necessarily tends to infinity. This is in agreement with (13) because in this limit, the isoperimetric constant tends to zero.

1.4 A Level Set Analysis, Curvature Estimates Involving a Volume Bound

The last example explains why working with the volume of an exceptional set might not be the best method. Namely, in the situation of Fig. 1, it seems preferable to consider the *surface area* of the exceptional set. Then cutting at the event horizon, the long cylinder has disappeared, and we no longer need to worry about the limit when L gets large. Working with the surface area also seems preferable for physical reasons. First, as the interior of a black hole is not accessible to measurements, our estimates should not depend on the geometry inside the event horizon. Therefore, choosing the exceptional set Ω such that it contains the interior of the event horizon, our estimates should not depend on the volume of Ω, only on its surface area. Second, the Riemannian Penrose inequalities yield that if the total mass is small, the area of the event horizon is also small. Thus we can hope that there should be an exceptional set of small surface area.

The basic question is how to choose the exceptional set Ω. In view of the estimate (12), it is tempting to choose the exceptional set as the set where the Witten spinor (or similarly the spinor operator) is small, i.e.

$$\Omega(\tau) = \{x \in M \text{ with } |\psi(x)| < \tau\} \tag{15}$$

for some $\tau \in (0, 1]$. This has the advantage that in the region $M \setminus \Omega$, the Witten spinor is by construction bounded from below by τ, so that (11) immediately gives rise to a curvature estimate. Clearly, the resulting estimates are of use only if the exceptional set is small, for example in the sense that it has small surface area. This consideration was our motivation for analyzing the *level sets* of the Witten spinor [9]. We here outline a few results of this analysis.

We set $\phi = |\psi|$ and introduce the functional

$$F(\tau) = \int_{\Omega(\tau)} |D\phi|^2 \, d\mu_M . \tag{16}$$

Using the Lichnerowicz-Weitzenböck formula, it is straightforward to verify that this functional is convex. Moreover, combining the co-area formula and the Schwarz inequality, one finds that for all t_0, t_1 with $0 < t_0 < t_1 < 1$, the area A and the volume V of the sets $\Omega(\tau)$ and $\Omega(\tau')$ are related by

$$\int_{t_0}^{t_1} A(\sigma)\, d\sigma \leq \sqrt{(V(t_1) - V(t_0))\,(F(t_1) - F(t_1))}.$$

Using the mean value theorem, there is $t \in [t_0, t_1]$ with

$$A(t) \leq \sqrt{F(t_1) - F(t_0)}\,\frac{\sqrt{V(t_1) - V(t_0)}}{t_1 - t_0}.$$

Furthermore, Sard's lemma can be used to arrange that $A(t)$ is a hypersurface. Choosing the exceptional set $\Omega = \Omega(t)$, the inequality (12) gives rise to the following curvature estimate.

Theorem 1.2. *Let (M^n, g), $n \geq 4$, be a complete, asymptotically flat manifold whose scalar curvature is non-negative and integrable. Suppose that for an interval $[t_0, t_1] \subset (0, 1]$ there is a constant C such that every Witten spinor (4) satisfies the volume bound*

$$V(t_1) - V(t_0) \leq C. \tag{17}$$

Then there is an open set $\Omega \subset M$ with the following properties. The $(n-1)$-dimensional Hausdorff measure μ_{n-1} of the boundary of Ω is bounded by

$$\mu_{n-1}(\partial\Omega) \leq \sqrt{m}\, c_0(n, t_0) \frac{\sqrt{C}}{t_1 - t_0}.$$

On the set $M \setminus \Omega$, the Riemann tensor satisfies the inequality

$$\int_{M \setminus \Omega} |R|^2 \leq m c_1(n, t_0) \sup_M |R| + \sqrt{m}\, c_2(n, t_0) \|\nabla R\|_{L^2(M)}.$$

For clarity, we point out that (17) only involves the volume of the region

$$\Omega(t_1) \setminus \Omega(t_0) = \{x \text{ with } t_0 \leq |\psi| < t_1\}. \tag{18}$$

Thus in order to apply our theorem to the example of Fig. 1, we can choose t_0 such that $\Omega(t_0)$ just includes the region inside the event horizon. Then the statement of the theorem no longer depends on the parameter L. This consideration also explains how it is possible that Theorem 1.2 no longer involves the isoperimetric constant.

1.5 Weighted L^2-Estimates of the Witten Spinor

The curvature estimate in Theorem 1.2 has the disadvantage that it involves the a-priori bound (17) on the volume of the region $\Omega(t_1) \setminus \Omega(t_0)$. Since in this region, the spinors are bounded from above and below, the volume bound could be obtained from an L^p-estimate of the Witten spinor for any $p < \infty$. Our search for such estimates led to the weighted L^2-estimates in [12], which we now outline. A point of general interest is that these estimates involve the smallest eigenvalue of the Dirac operator on a conformal compactification of M, thus giving a connection to spectral geometry.

For technical simplicity, the weighted L^2-estimates were derived under the additional assumption that space-time is *asymptotically Schwarzschild*. Thus we assume that there is a compact set $K \subset M$ and a diffeomorphism $\Phi : M \setminus K \to \mathbb{R}^n \setminus B_\rho(0)$, $\rho > 0$, such that

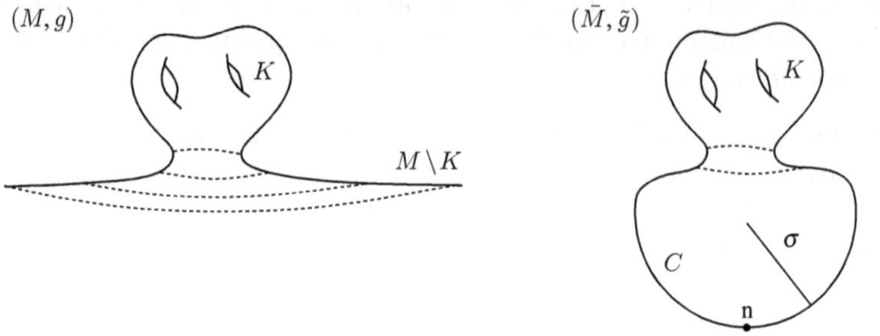

Fig. 2 The asymptotically Schwarzschild manifold (M,g) and its conformal compactification $(\bar{M},\tilde{g})$

$$(\Phi_* g)_{ij} = \left(1 + \frac{1}{|x|^{n-2}}\right)^{\frac{4}{n-2}} \delta_{ij}\,.$$

Then outside the compact set, the metric is conformally flat, and thus by a conformal transformation

$$\tilde{g} = \lambda^2 g \tag{19}$$

with a smooth function λ with $\lambda|_K \equiv 1$ we can arrange that $\tilde{g}|_{M\setminus K}$ is isometric to a spherical cap of radius σ with the north pole removed. By adding the north pole, we obtain the complete manifold $(\bar{M},\tilde{g})$, being a conformal one-point compactification of (M,g) (see Fig. 2).

The manifold $(\bar{M},\tilde{g})$ is again spin. We denote its Dirac operator by $\tilde{\mathcal{D}}$.

In order to improve the decay properties of the spinor at infinity, in the asymptotic end we subtract from ψ a constant spinor multiplied by a function coming from the conformal weight of the sphere,

$$\delta\psi(x) = \psi(x) - \left(1 + \frac{1}{|x|^{n-2}}\right)^{-\frac{n-1}{n-2}} \psi_0 \qquad \text{on } M \setminus K.$$

Under these assumptions, in [12] we prove the following theorem.

Theorem 1.3. *Every Witten spinor satisfies the inequality*

$$\int_K \|\psi(x)\|^2 dx + \int_{M\setminus K} \|\delta\psi(x)\|^2 \lambda(x) dx \le c(n) \frac{(\rho+1)^n}{\sigma^2 \inf \operatorname{spec}(\tilde{\mathcal{D}}^2)}.$$

We now sketch the main steps of the proof, also explaining how the infimum of the spectrum of the operator $\tilde{\mathcal{D}}^2$ enters. Our first step is to get a connection between the conformally transformed spinor operator and a quadratic expression in the Dirac Green's function on $\bar{M}$. After subtracting suitable counter terms, we can integrate this expression over $\bar{M}$ to obtain the Green's function G of the square of the Dirac operator minus suitable counter terms. Then our task becomes to analyze the behavior of G near the pole $\mathfrak{n}$ of the spherical cap. This is accomplished by taking the difference of G and the Green's function on the sphere and using Sobolev techniques inside the spherical cap. In this analysis, we need to estimate the sup-norm of G in the Hilbert space $L^2(\bar{M}, S\bar{M})$ by

$$\|G\| = \sup \operatorname{spec}(G) = \frac{1}{\inf \operatorname{spec}(\tilde{\mathcal{D}}^2)}.$$

The theorem then follows by using a positivity argument for the Witten operator and a similar operator built up of the corresponding wave functions $\delta\psi_i$.

1.6 Curvature Estimates Involving the Lowest Eigenvalue on a Conformal Compactification

We now outline how Theorem 1.3 can be used to satisfy the volume bound (17) in Theorem 1.2. For simplicity, we choose $t_0 = 1/4$ and $t_1 = 1/2$. The main step is to prove that choosing the radius

$$r_1 := c(n)\,\sigma\,\left(\sigma \inf \operatorname{spec}|\tilde{D}|\right)^{-\frac{1}{n-1}},$$

the Witten spinor is bounded from below by

$$|\psi(x)| \geq \frac{1}{2} \qquad \text{for all } x \in M \setminus K \text{ with } |x| > r_1. \tag{20}$$

This is achieved by combining elliptic estimates in the spherical cap with spectral estimates for $\mathcal{D}^2$. Then the inequality (20) allows us to estimate the volume difference in (17) by

$$\begin{aligned} V\Big(\frac{1}{2}\Big) - V\Big(\frac{1}{4}\Big) &\leq 16 \int_{\Omega(1/2)} |\psi|^2 d\mu_M \leq 16 \int_{B_{r_1}(0)} |\psi|^2 d\mu_M \\ &\leq 16 \int_K |\psi|^2 d\mu_M + \int_{\{x \in M \setminus K \text{with} |x| > r_1\}} |\psi|^2 d\mu_M. \end{aligned}$$

Using the upper bound (11), we obtain

$$V\Big(\frac{1}{2}\Big) - V\Big(\frac{1}{4}\Big) \leq \int_K |\psi|^2 d\mu_M + \mu\Big(\{x \in M \setminus K \text{with} |x| \leq r_1\}\Big).$$

The first summand can be estimated by Theorem 1.3, whereas the second summand can be bounded by the volume of a Euclidean ball of radius r_1. This method gives the following results (see [9, Theorems 1.4 and 4.5]).

Theorem 1.4. *Let (M^n, g), $n \geq 3$, be a complete manifold of non-negative scalar curvature such that $M \setminus K$ is isometric to the Schwarzschild geometry. Then there is an open set $\Omega \subset M$ with the following properties. The $(n-1)$-dimensional Hausdorff measure μ_{n-1} of the boundary of Ω is bounded by*

$$\mu_{n-1}(\partial\Omega) \leq c_0(n)\,\sqrt{m}\,\frac{\left(\rho + m^{\frac{1}{n-2}}\right)^{\frac{n}{2}}}{\sigma\ \inf \operatorname{spec}|\tilde{\mathcal{D}}|}. \tag{21}$$

On the set $M \setminus \Omega$, the Riemann tensor satisfies the inequality

$$\int_{M\setminus\Omega} |R|^2 \leq m c_1(n) \sup_M |R| + \sqrt{m} c_2(n) \|\nabla R\|_{L^2(M)}.$$

Note that this theorem involves the surface area of the exceptional set (21). The geometry of K enters the estimate only via the smallest eigenvalue of the Dirac operator on $\bar{M}$. This is a weaker and apparently more practicable condition than working with the isoperimetric constant, in particular because eigenvalue estimates can be obtained with spectral methods for the Dirac operator on a compact manifold (see for example [14]).

1.7 Results in the Setting with Second Fundamental Form

We now return to the setting of general relativity. Thus we again let $(N^4, \bar{g})$ be a Lorentzian manifold and (M^3, g, h) a spacelike hypersurface with induced Riemannian metric g and second fundamental form h. Now asymptotic flatness involves in addition to (1) decay assumptions for the second fundamental form,

$$(\Phi_* h)_{ij} = \mathcal{O}(r^{-2}), \qquad \partial_k (\Phi_* h)_{ij} = \mathcal{O}(r^{-3}).$$

Total energy and momentum are defined by

$$E = \frac{1}{16\pi} \lim_{R\to\infty} \sum_{i,j=1}^{3} \int_{S_R} (\partial_j (\Phi_* g)_{ij} - \partial_i (\Phi_* g)_{jj}) d\Omega^i$$

$$P_k = \frac{1}{8\pi} \lim_{R\to\infty} \sum_{i=1}^{3} \int_{S_R} ((\Phi_* h)_{ki} - \sum_{j=1}^{3} \delta_{ki} (\Phi_* h)_{jj}) d\Omega^i.$$

The spinor proof of the positive mass theorem as outlined in (3–6) works similarly in the case with second fundamental form, if $\mathcal{D}$ is replaced by the so-called *hypersurface Dirac operator*, which uses the spin connection $\bar{\nabla}$ of the ambient space-time N^4, but acts only in directions tangential to the hypersurface M^3 (see [27, 37]). The Lichnerowicz-Weitzenböck formula becomes

$$\mathcal{D}^2 = \bar{\nabla}_i^* \bar{\nabla}^i + \mathfrak{R},$$

where now the dominant energy condition ensures that $\mathfrak{R}$ is a positive semi-definite multiplication operator on the spinors. The existence of a solution of the hypersurface Dirac equation with constant boundary values in the asymptotic end is proved in [27]. The integration-by-parts argument (5) gives in analogy to (7) the inequality

$$\int_M |\psi|^2 d\mu_M \le 4\pi \left(E + \langle \psi_0, P \cdots \psi_0 \rangle\right)$$

Choosing ψ_0 appropriately, one gets the *positive energy theorem* $E - |P| \ge 0$.

We now outline the method for deriving curvature estimates (for details see [11]). As the space-time dimension is larger than three, we again need to work with the spinor operator. Then one can derive an identity similar to (12), but additional terms involving h arise. Moreover, $|R|^2$ is to be replaced by the norm of all components of the Riemann tensor which are determined by the Gauss-Codazzi equations,

$$|\bar{R}_M|^2 = \sum_{i,j=1}^{3} \sum_{\alpha,\beta=0}^{3} (\bar{R}_{ij\alpha\beta})^2$$

(where the sums run over orthonormal or pseudo-orthonormal frames). The presence of the second fundamental form leads to the difficulty that the function $|\psi|^2$ is no longer subharmonic, making it impossible to estimate the norm of the spinor with the maximum principle. In order get around this difficulty, we first construct a barrier function F, which is a solution of a suitable Poisson equation. We then derive Sobolev estimates for F, and these finally give us control of $\| |\psi|^2 - 1 \|_{L^6(M)}$. This leads to the following result (see [11, Theorem 1.3]).

Theorem 1.5. *We choose $L \ge 3$ such that*

$$(L^\alpha - 1)^2 \ge C \frac{4\pi E + \|h\|_2}{k^2 (k + 24\|h\|_3)^2} \| |h|^2 + |\nabla h| \|_3$$

where

$$\alpha = \left(1 + 24 \frac{\|h\|_3}{k}\right)^{-1}.$$

Then there are numerical constants $c_1, \ldots, c_4$ and a set $\Omega \subset M$ with measure bounded by

$$\mu(\Omega) \le c_1 \frac{L^6}{k^2} (4\pi E + \|h\|_2^2)$$

such that on $M \setminus \Omega$ the following inequality holds,

$$\begin{aligned}
\int_{M \setminus \Omega} |\overline{R}_M|^2 d\mu_M &\le c_2 \sup_M \left(|h| + (|R| + |h|^2 + |\overline{\nabla} h|\right) E \\
&+ c_3 L \sup_M \left(|\overline{\nabla} \overline{R}_M| + |h||\overline{R}_M|\right) \sqrt{E} \\
&+ c_4 \frac{\sqrt{L+1}}{k} \sqrt{\| |h|^2 + |\nabla h| \|_{6/5}} \left\| |\overline{\nabla} \overline{R}_M| + |h||\overline{R}_M| \right\|_{5/12} \sqrt{E}.
\end{aligned}$$

This theorem is the analog of Theorem 1.1 for a spacelike hypersurface (M^3, g, h) of a Lorentzian manifold N^4. Unfortunately, the second fundamental form enters the theorem in a rather complicated way. It is conceivable that the theorem could be simplified by improving our method of proof.

The results described in Sects. 1.4–1.6 in the Riemannian setting so far have not been worked out in the setting with second fundamental form. Many results could be extended. However, with the present methods, the proofs and the statements of the results would be rather involved.

1.8 Outlook

We now give a brief outlook on open problems and outline possible directions for future research. The following problems seem interesting and promising; they have not yet been studied by us only due to other obligations.

- Explore the *convexity* of F: In [9, Sect. 1], it was shown and briefly discussed that the functional F, (16), is convex. However, the geometric meaning of this convexity has not yet been analyzed. It also seems promising to search for potential applications.
- *Extend the weighted L^2-estimates* to more general asymptotically flat manifolds: The weighted L^2-estimates of the Witten spinor [12] were worked out under the assumption that the manifold is asymptotically Schwarzschild. As a consequence, we could arrange that the point compactification of the asymptotic end was isometric to a spherical cap (see Fig. 2), simplifying the elliptic estimates considerably (see [12, Sect. 5]). However, our methods also seem to apply to more general asymptotically flat manifolds, possibly with more general compactifications.

Moreover, it seems a challenging problem to extend the results outlined in Sects. 1.4–1.6 to the *setting with second fundamental form*. As mentioned at the end of Sect. 1.7, the main difficulty is to improve our methods so as to obtain simple and clean results.

Our long-term goal is to study the limiting behavior of the manifold as total energy and momentum tend to zero. Thus, stating the problem for simplicity in the Riemannian setting, we consider a sequence (M_ℓ, g_ℓ) of asymptotically flat manifolds with $m_\ell \searrow 0$. In order to get better control of the global geometry, one could make the further assumptions that the manifolds are all asymptotically Schwarzschild and that the Dirac operators on the conformal compactifications satisfy the uniform spectral bound

$$\inf \operatorname{spec}|\mathcal{D}_\ell| \geq \varepsilon \qquad \text{for all } \ell.$$

Then one could hope that after cutting out exceptional sets Ω_ℓ of small surface area (21), the manifolds $M_\ell \setminus \Omega_\ell$ converge to flat $\mathbb{R}^n$ in a suitable sense, for example in a Gromov-Hausdorff sense. In our attempts to prove results in this direction, we faced the difficulty that convergence can be established only in suitable charts. Thus on $M_\ell \setminus \Omega_\ell$ one would like to choose suitable canonical charts, in which the metrics g^ℓ_{ij} converge to the flat metric δ_{ij}. Unfortunately, the chart (1) is defined only in the asymptotic end, and thus it would be necessary to extend this chart to $M_\ell \setminus \Omega_\ell$. As an alternative, one could hope that the vector fields associated to the Witten spinors ψ_i form a suitable frame of the tangent bundle. However, it seems difficult to get global control of this frame. As another alternative, we tried to construct orthonormal frames (e_i) by minimizing a corresponding Dirichlet energy,

$$\int_M \sum_{i=1}^n |\nabla e_i|^2 d\mu_M \to \min.$$

Unfortunately, it seems difficult to rule out that the corresponding minimizer has singularities. These difficulties were our main obstacle for making substantial progress towards a proof of Gromov-Hausdorff convergence. But once the problem of choosing a canonical chart is settled, the limiting behavior of sequences of asymptotically flat manifolds could be attacked.

2 Minkowski Embeddability of Hypersurfaces in Flat Space-Times

The positive energy theorem makes two statements on the energy E and the momentum P (see Sect. 1.7 above) of an asymptotically flat spacelike hypersurface M of a Lorentzian manifold $(\bar{M}, \bar{g})$ which satisfies the dominant energy condition (see Sect. 3.2 below) at every point of M. The first statement is that the inequality $E \geq |P|$ holds. The second statement is that if $E = |P|$ holds, then the Riemann tensor of $\bar{g}$ vanishes at every point of M. This latter "rigidity statement" has been proved by Parker–Taubes [27] in the case when M admits a spin structure – and under the assumption that M is 3-dimensional, but the argument generalizes to higher dimensions. (The original proof of Witten [37] deduced the rigidity statement from the stronger assumption that $(\bar{M}, \bar{g})$ satisfies the dominant energy condition on a neighborhood of M.)

Another proof of the rigidity statement was given by Schoen–Yau [30], without the spin assumption, but only in the case $\dim M \leq 7$. However, Schoen–Yau proved more than Parker–Taubes: they showed that if $E = |P|$ holds, then the Riemannian n-manifold M with its given second fundamental form can be embedded isometrically into Minkowski space-time $\mathbb{R}^{n,1} = \mathbb{R}^n \times \mathbb{R}$ as the graph of a function $\mathbb{R}^n \to \mathbb{R}$; in particular, M is diffeomorphic to $\mathbb{R}^n$.

It is natural to ask whether one can decouple the proof of embeddability into Minkowski space-time from the proof of the rigidity statement. That is, when we already know (e.g. from the Parker–Taubes proof) that $\bar{g}$ is flat along M, can we deduce in a simple way that M with its second fundamental form admits an embedding of the desired form and is in particular diffeomorphic to $\mathbb{R}^n$?

This is indeed possible. The proof works in all dimensions and without topological (e.g. spin) conditions. Moreover, it generalises directly to the embeddability of asymptotically hyperbolic hypersurfaces into anti-de Sitter space-time in the rigidity case. This situation is considered in the work of Maerten [21], to which we refer for the definition of the rigidity case in that context. Like Parker–Taubes in the asymptotically flat case, Maerten makes a spin assumption. His proof allows him to obtain an embedding into anti-de Sitter space-time via an explicit construction. Our argument below works differently, without any topological condition.

For $c \le 0$, let $\mathcal{M}^{n,1}_c$ denote Minkowski space-time if $c = 0$, and anti-de Sitter space-time of curvature c if $c < 0$. In each case, the underlying smooth manifold of $\mathcal{M}^{n,1}_c$ is $\mathbb{R}^n \times \mathbb{R}$. The metric on $\mathcal{M}^{n,1}_c$ is $\sum_{i=1}^n dx_i^2 - dx_{n+1}^2$ if $c = 0$; for $c < 0$, it is induced by the embedding $\mathbb{R}^n \times \mathbb{R} \to \mathbb{R}^{n+2} = \mathbb{R}^n \times \mathbb{R} \times \mathbb{R}$ given by $(x,t) \mapsto \left(x, \cos t\sqrt{|x|^2 - 1/c}, \sin t\sqrt{|x|^2 - 1/c}\right)$, where $\mathbb{R}^{n+2}$ is equipped with the metric $\sum_{i=1}^n dx_i^2 - \sum_{i=n+1}^{n+2} dx_i^2$. Let $\mathrm{pr}\colon \mathcal{M}^{n,1}_c \to \mathbb{R}^n$ denote the projection $(x,t) \mapsto x$.

Stated with minimal assumptions, our result is the following [25]:

Theorem 2.1. *Let $n \ge 0$ and $c \in \mathbb{R}_{\le 0}$, let M be a connected n-manifold which contains a simply connected noncompact n-dimensional submanifold-with-boundary that is closed in M and has compact boundary. Let (M, g, K) be a complete Riemannian manifold with second fundamental form which satisfies the Gauss and Codazzi equations for constant curvature c. Then:*

1. *(M, g, K) admits an isometric embedding f into $\mathcal{M}^{n,1}_c$ such that* $\mathrm{pr} \circ f\colon M \to \mathbb{R}^n$ *is a diffeomorphism.*
2. *When $\tilde{f}$ is an isometric immersion of (M, g, K) into $\mathcal{M}^{n,1}_c$, then there exists an isometry $A\colon \mathcal{M}^{n,1}_c \to \mathcal{M}^{n,1}_c$ with $\tilde{f} = A \circ f$; in particular, $\tilde{f}$ is an embedding.*

In this theorem, the second fundamental form K is allowed to be a field of symmetric bilinear forms on M with values in an arbitrary (not necessarily trivial) normal bundle of rank 1. That is, we do not *assume* the normal bundle to be trivial, we get its triviality as a *conclusion* of the theorem (because every line bundle over $\mathbb{R}^n$ is trivial). To understand this, consider the manifold $M = S^1 \times \mathbb{R}^{n-1}$ and the flat Riemannian metric g on M. It admits an isometric embedding into the flat Lorentzian manifold $\mathfrak{M} \times \mathbb{R}^{n-1}$, where $\mathfrak{M}$ is the Möbius strip, regarded a line bundle over S^1 with timelike fibers. The second fundamental form K of this embedding vanishes identically, but the normal bundle is not trivial. (M, g, K) is not a counterexample to Theorem 2.1 because the simply-connectedness assumption is not satisfied. Replacing $\mathfrak{M}$ by the trivial line bundle over S^1 shows that the simply-connectedness assumption in Theorem 2.1 is also needed when the normal bundle is trivial.

In the rigidity case of the positive energy theorem, the assumptions of Theorem 2.1 are satisfied because $n \ge 3$ is assumed in the positive energy theorem and (M, g) is noncompact and complete and contains a compact n-submanifold-with-boundary C such that every connected component of $M \setminus (C \setminus \partial C)$ is diffeomorphic to $S^{n-1} \times \mathbb{R}_{\geq 0}$ (and is thus a simply connected noncompact n-dimensional submanifold-with-boundary which is closed in M and has compact boundary).

Let us sketch the proof of Theorem 2.1. Much of the necessary work is already contained in the semi-Riemannian version of the fundamental theorem of hypersurface theory due to Bär–Gauduchon–Moroianu [2, Sect. 7]:

Theorem 2.2 (Bär–Gauduchon–Moroianu)**.** *Let $c \in \mathbb{R}$, let (M, g, K) be a Riemannian manifold with second fundamental form which satisfies the Gauss and Codazzi equations for constant curvature c. Assume that M is simply connected. Then (M, g, K) admits an isometric immersion into $\mathcal{M}_c^{n,1}$. When f_0, f_1 are isometric immersions of (M, g, K) into $\mathcal{M}_c^{n,1}$, then there exists an isometry $A: \mathcal{M}_c^{n,1} \to \mathcal{M}_c^{n,1}$ with $f_1 = A \circ f_0$.*

Recall that a map $f: M \to N$ to a Lorentzian manifold (N, h) is *spacelike* iff for every $x \in M$ the image of $T_x f: T_x M \to T_{f(x)} N$ is spacelike. A spacelike map $f: (M, g) \to (N, h)$ from a Riemannian manifold to a Lorentzian manifold is *long* iff for every interval $I \subseteq \mathbb{R}$ and every smooth path $w: I \to M$, the g-length of w is finite if the h-length of $f \circ w$ is finite. For example, every spacelike isometric immersion is long. The second ingredient for the proof of Theorem 2.1 is the following fact:

Proposition 2.3. *Let (M, g) be a nonempty connected complete Riemannian n-manifold, let $f: (M, g) \to \mathcal{M}_c^{n,1}$ be a spacelike long immersion into Minkowski space. Then $f: M \to \mathcal{M}_c^{n,1}$ is a smooth embedding, and* $\mathrm{pr} \circ f: M \to \mathbb{R}^n$ *is a diffeomorphism.*

The idea of the proof of 2.3 is as follows. Let us call a map $\phi: M \to \mathbb{R}^n$ a *quasicovering* iff it is a local embedding and for all paths $\gamma: [0, 1] \to \mathbb{R}^n$ and $\tilde{\gamma}: [0, 1[\to M$ with $\phi \circ \tilde{\gamma} = \gamma | [0, 1[$, there exists an extension of $\tilde{\gamma}$ to a path $[0, 1] \to M$. One can show that every quasicovering $\phi: M \to \mathbb{R}^n$ is a diffeomorphism. This is done in the same way in which one proves the well-known fact that every covering map $M \to \mathbb{R}^n$ is a diffeomorphism (because $\mathbb{R}^n$ is simply connected and M is nonempty and connected).

Now one verifies that $\mathrm{pr} \circ f$ is a quasicovering: It is an immersion (thus a local embedding) because f is a spacelike immersion. For the extension property of a quasicovering, one notes that $f \circ \tilde{\gamma}$ has finite length because $\mathrm{pr} \circ f \circ \tilde{\gamma} = \gamma | [0, 1[$ has finite length. Since f is long, $\tilde{\gamma}$ has finite length. Completeness implies that $\tilde{\gamma}$ can be extended to $[0, 1]$. Thus $\mathrm{pr} \circ f$ is a quasicovering. (In contrast, it is difficult to show directly that $\mathrm{pr} \circ f$ is a covering map.)

Hence $\mathrm{pr} \circ f$ is a diffeomorphism. Since every proper injective immersion is an embedding, so is f. This completes the proof of Proposition 2.3. (Cf. [25] for details.)

Theorem 2.1 is now easy to prove: We pull back g and K by the universal covering map $\pi: \tilde{M} \to M$ and apply Theorem 2.2. To the resulting spacelike isometric immersion $\tilde{M} \to \mathcal{M}_c^{n,1}$ we apply Proposition 2.3. This shows in particular that $\tilde{M}$ is diffeomorphic to $\mathbb{R}^n$. For an n-submanifold-with-boundary Z of M with the properties assumed in Theorem 2.1, a simple topological argument yields then that the covering $\pi|\pi^{-1}(Z): \pi^{-1}(Z) \to Z$ has only one sheet. Thus π is a diffeomorphism. Now all statements of Theorem 2.1 follow immediately.

3 Spacelike Foliations and the Dominant Energy Condition

3.1 Pseudo-Riemannian Manifolds Without Spacelike Foliations

When Lorentzian manifolds are considered in general relativity, it is often assumed that they have nice causality properties like stable causality or even global hyperbolicity. Such manifolds admit a smooth real-valued function with timelike gradient [4] and thus a spacelike foliation of codimension 1, by the level sets of the function. Let us call spacelike foliations of codimension 1 on a Lorentzian manifold *space foliations* for simplicity. A few years ago, Christian Bär asked us whether *every* Lorentzian manifold admits a space foliation.

The answer is not obvious, for the following reasons. First, clearly every point in a Lorentzian manifold has an open neighborhood which admits a space foliation.

Second, the tangent bundle of every semi-Riemannian manifold has an orthogonal decomposition $V \oplus H$ into a timelike sub vector bundle V and a spacelike sub vector bundle H. (At every point of an n-dimensional manifold M which is equipped with a semi-Riemannian metric of index q, the choice of a time/space splitting corresponds to a point in the contractible space $\mathrm{O}(n)/(\mathrm{O}(q) \times \mathrm{O}(n-q))$. Thus a global time/space splitting of the tangent bundle TM exists if a certain fiber bundle over M with contractible fibers admits a smooth section. Obstruction theory tells us that such a section exists for every manifold and metric.)

The question is therefore whether the spacelike bundle H can always be chosen *integrable*, i.e. tangent to a foliation. Since *every* sub vector bundle of rank 1 of a tangent bundle is integrable, it is clear that every 2-dimensional Lorentzian manifold admits a space foliation. (There are many quite complicated examples of Lorentzian 2-manifolds, because every noncompact connected smooth 2-manifold admits a Lorentzian metric.)

Third, a theorem of W. Thurston says that every connected component of the space of $(n-1)$-plane distributions on an n-manifold M contains an integrable distribution [33]. Here we use the word *distribution* in the differential-topological sense: a *k-plane distribution* on a manifold M is a sub vector bundle of rank k of TM. Distributions can be viewed as sections in the bundle $\mathrm{Gr}_k(TM) \to M$ whose fiber over x is the Grassmann manifold $\mathrm{Gr}_k(T_xM)$ of k-dimensional sub vector spaces of T_xM. Connected components of the set of k-plane distributions on M are considered with respect to the compact-open topology on the space of sections in $\mathrm{Gr}_k(TM) \to M$. In contrast to the situation for Riemannian metrics, the space of Lorentzian metrics on a given manifold can be empty or have several connected components.

Thurston's theorem implies that every connected component of the space of Lorentzian metrics on a manifold contains metrics which admit space foliations: The set of connected components of the space of $(n-q)$-plane distributions on an n-manifold M is in canonical bijective correspondence to the set of connected components of the space of semi-Riemannian metrics of index q on M. The correspondence maps the connected component of each distribution H to the connected component of a metric which makes H spacelike.

These facts show that there are no *topological* obstructions to the existence of space foliations on Lorentzian manifolds (in contrast to the situation on semi-Riemannian manifolds of higher index: the analog of Thurston's theorem is in general false for distributions of codimension ≥ 2). Nevertheless, the answer to Bär's question is negative. Counterexamples exist even on topologically trivial manifolds like $\mathbb{R}^n$ (see [24, Theorem 0.1]):

Theorem 3.1. *Let (M, g) be an n-dimensional pseudo-Riemannian manifold of index $q \in \{1, \ldots, n-2\}$ (e.g. a Lorentzian manifold of dimension $n \geq 3$). Let $A \neq M$ be a closed subset of M. Then there exists a metric g' of index q on M such that:*

1. *$g = g'$ on A;*
2. *Every g-timelike vector in TM is g'-timelike;*
3. *$M \setminus A$ does not admit any codimension-q foliation none of whose tangent vectors is g'-timelike; in particular, (M, g') does not admit any space foliation.*

Here and in the following, our conventions are such that $v \in TM$ is g-spacelike resp. g-timelike resp. g-causal iff $g(v,v) > 0$ resp. $g(v,v) < 0$ resp. $g(v,v) \leq 0$; such that the index of a metric is the maximal dimension of timelike sub vector spaces of tangent spaces; and such that Lorentzian metrics have index 1.

The idea of the proof of Theorem 3.1 is simple: We choose a g-spacelike $(n-q)$-plane distribution H on M and modify it on $M \setminus A$ in such a way that the new distribution H' is not integrable on $M \setminus A$ but still g-spacelike; this is possible because $2 \leq n-q \leq n-1$. We construct a sequence $(g_k)_{k\in\mathbb{N}}$ of semi-Riemannian metrics of index q on M with $g_0 = g$ such that each g_k is equal to g on A; and such that on some compact ball B in $M \setminus A$, the g_k-lightcones become wider and wider as k tends to ∞, and the g_k-spacelike regions "converge" to H' as they become smaller with increasing k (cf. Fig. 3).

We claim that for sufficiently large k, the restriction of g_k to B does not admit a space foliation. Otherwise we would obtain a sequence $(H_k)_{k\in\mathbb{N}}$ of integrable distributions on B such that every H_k is g_k-spacelike. By our construction of the metrics g_k, this sequence would converge in the C^0-topology to the nonintegrable distribution H'. But C^0-limits of integrable distributions are always integrable; cf. [24] for details (or [36] for a slightly different proof sketch in the case $n-q = n-1$). This contradiction proves our claim. Now the proof of Theorem 3.1 is complete: we can take $g' = g_k$ for any sufficiently large k.

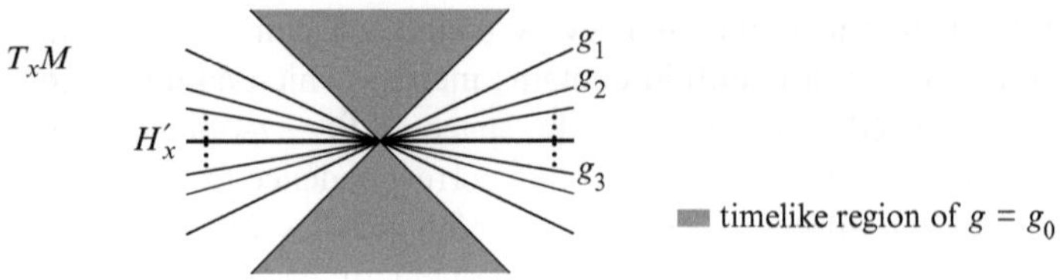

Fig. 3 The lightcones at a point $x \in B$ of the metrics g_k in the proof of Theorem 3.1

3.2 *Existence of Dominant Energy Metrics*

One might be tempted to regard Lorentzian metrics without space foliations as objects of little physical relevance, things which could only serve as examples of the strange phenomena that occur when the standard causality assumptions in general relativity are dropped. But there is another side of the medal: certain physically *desirable* properties – which from the geometric viewpoint are conditions on the Ricci curvature – can in general be satisfied *only* by Lorentzian metrics without space foliations.

This holds in particular for the dominant energy condition, which plays an important role in the positive energy theorem (cf. Sects. 1 and 2 above). In the following discussion we will use the version with arbitrary cosmological constant:

Definition 3.2. Let (M, g) be a Lorentzian manifold, let $\Lambda \in \mathbb{R}$. The *energy-momentum tensor of* (M, g) *with respect to (the cosmological constant)* Λ is the symmetric $(0, 2)$-tensor $T = \mathrm{Ric} - \frac{1}{2}sg + \Lambda g$; here s and Ric are the scalar and Ricci curvatures of g, respectively. (This means that we interpret Einstein's field equation as the definition of the energy-momentum tensor when g and Λ are given.)

(M, g) *satisfies the dominant energy condition with respect to* Λ iff for every $x \in M$ and every g-timelike vector $v \in T_xM$, the vector $-T^a{}_b v^b$ lies in the closure of the connected component of $\{u \in T_xM \mid g(u, u) < 0\}$ which contains v. (The abstract index notation $-T^a{}_b v^b$ describes the vector which is the g-dual of the linear form $T(., v)$ on T_xM.)

In other words, (M, g) satisfies the dominant energy condition with respect to Λ iff every g-timelike vector $v \in TM$ satisfies $T(v, v) \geq 0$ and $g(w, w) \leq 0$, where $w^a = -T^a{}_b v^b$.

In general relativity, every physically reasonable space-time metric g should satisfy the dominant energy condition: Timelike vectors v are tangents to observer wordlines. Every observer should see a nonnegative energy density at each space-time point she passes through; that is expressed by the condition $T(v, v) \geq 0$. And she should see that matter does not move faster than light; that is what $g(w, w) \leq 0$ means (w is the momentum density observed by v).

In view of the physical importance of the dominant energy condition, a natural geometric question arises: *For given cosmological constant* Λ, *which manifolds* M *admit a Lorentzian metric that satisfies the dominant energy condition with respect to* Λ*?* A trivially necessary condition is that M admits a Lorentzian metric at all, but are there other conditions? (Note that every noncompact connected manifold admits a Lorentzian metric.)

Riemannian geometry offers many nonexistence results for metrics of nonnegative scalar curvature, most notably the positive energy theorem and obstructions to Riemannian metrics of nonnegative scalar curvature on closed manifolds. On every spacelike hypersurface S in a Lorentzian manifold which satisfies the dominant energy condition for some Λ, the Gauss equation yields an inequality $s \geq \ldots$, where s is the scalar curvature of the induced Riemannian metric on S and "…" depends on Λ and the second fundamental form of S.

One might therefore suspect that there exist obstructions to the existence of dominant energy metrics. For instance, when M has the form $S^1 \times N$ for some closed manifold N which does not admit a Riemannian metric of nonnegative scalar curvature, then it is difficult to satisfy a $\Lambda \geq 0$ dominant energy condition with a Lorentzian metric which makes every submanifold $N_t = \{t\} \times N$ spacelike: for every $t \in S^1$, properties of the second fundamental form of N_t would have to compensate the negative scalar curvature of N_t that occurs unavoidably on some subset of N_t.

However, one can construct dominant energy metrics in a way that circumvents such problems. In our $S^1 \times N$ example, we can even arrange that the vector field ∂_t along the S^1 factor becomes timelike (see [24, Theorem 0.5]):

Theorem 3.3. *Let* (M, g) *be a connected Lorentzian manifold of dimension* $n \geq 4$*, let* K *be a compact subset of* M*, let* $\Lambda \in \mathbb{R}$*. If* $n = 4$*, assume that* (M, g) *is time- and space-orientable, and that either* M *is noncompact, or compact with intersection form signature divisible by* 4*. Then there exists a Lorentzian metric* g' *on* M *such that:*

1. *Every* g*-causal vector in* TM *is* g'*-timelike;*
2. g' *satisfies on* K *the dominant energy condition with cosmological constant* Λ*;*
3. (M, g') *does not admit a space foliation.*

Recall the definition of the intersection form signature of a closed oriented 4-manifold M: On the de Rham cohomology $\mathbb{R}$-vector space $V = H^2_{\mathrm{dR}}(M)$, we have a symmetric nondegenerate bilinear form $\omega\colon V \times V \to \mathbb{R}$ given by $\omega([\alpha], [\beta]) = \int_M \alpha \wedge \beta$. Diagonalization of ω yields a diagonal matrix with p positive and q negative entries. The intersection form signature of M is $p - q$. When we reverse the orientation of M, then ω and the signature change their signs. Thus the condition that the signature be divisible by 4 is well-defined for a closed orientable 4-manifold. If a closed 4-manifold admits a Lorentzian metric (this happens iff the Euler characteristic vanishes) and is orientable, then its signature is automatically even.

Theorem 3.3 generalizes to dimension 3 when one assumes that M is orientable [24, Theorem 8.6]. However, property (1) must then be replaced by the weaker property that g' lies in the same connected component of the space of Lorentzian metrics as g.

Let us consider again the case where M has the form $S^1 \times N$ for some closed manifold N which does not admit a Riemannian metric of nonnegative scalar curvature – the connected sum of two 3-tori, say, for which $S^1 \times N$ is orientable and has intersection form signature 0. Then we can choose any Riemannian metric g_N on N and consider the product Lorentzian metric $g = -dt^2 \oplus g_N$ on M. This g does certainly not satisfy the dominant energy condition for any $\Lambda \geq 0$. It is time- and space-orientable and makes the vector field ∂_t timelike. For every Λ, Theorem 3.3 applied to $K = M$ yields a metric g' on M which satisfies the dominant energy condition with respect to Λ and still makes ∂_t timelike.

We will sketch the proof of Theorem 3.3 in a moment, but already at this point property (3) provides a hint how the difficulties arising from Riemannian nonnegative scalar curvature obstructions can be avoided: the foliation given by the leaves N_t will not be spacelike for the metric g'. (Probably no compact hypersurface of M will be g'-spacelike, but that is not obvious from the proof.)

For the proof of Theorem 3.3, we need a pointwise measure of the nonintegrability of a distribution [34, p. 176]:

Definition 3.4. Let H be a k-plane distribution on a manifold M. The *twistedness* Tw_H *of* H is a TM/H-valued 2-form on M (i.e. a section in $\bigwedge^2(H^*) \otimes (TM/H)$) which is defined as follows. Let $[.,.]_x$ denote the Lie bracket of vector fields on M evaluated in a point $x \in M$, and let $\pi\colon TM \to TM/H$ denote the projection. For $x \in M$ and $v, w \in H_x$, we define

$$Tw_H(v,w) = \pi([v,w]_x),$$

where on the right-hand side we have extended the vectors v, w to sections in H. The definition does not depend on the extension, because it is antisymmetric and $C^\infty(M,\mathbb{R})$-linear in v: $\pi(v) = 0$ implies that $\pi([fv,w]) = \pi(f[v,w] - df(w)v) = f\pi([v,w])$ holds for every $f \in C^\infty(M,\mathbb{R})$.

By Frobenius' theorem, a distribution H is integrable (i.e. tangent to a foliation) if and only if Tw_H vanishes identically. In contrast to the proof of Theorem 3.1 above, we need for the proof of Theorem 3.3 distributions H which are not just not integrable, but even pointwise nonintegrable in the sense that Tw_H does not vanish in any point $x \in M$; i.e., we need that for every $x \in M$, there exist sections v, w in H such that the Lie bracket value $[v,w]_x$ is not contained in H_x. Let us call a distribution with this property *twisted*.

The first step in the proof is to choose for the given Lorentzian metric g a spacelike twisted $(n-1)$-plane distribution H. In order to find such a distribution, we start with an arbitrary spacelike distribution and prove that it can be approximated in the fine C^0-topology by twisted distributions. (For the definition of the *fine*, also known as *Whitney*, C^0*-topology*, see [17, p. 35]. Because Theorem 3.3 arranges the dominant energy condition only on a compact set K, it would suffice here to prove that H can be approximated *on* K in the fine C^0-topology, which is the same as the compact-open topology because K is compact. But we want to emphasize that this first step of the proof works also on noncompact manifolds.)

The proof of this approximation employs M. Gromov's h-principle for ample open partial differential relations, also known as the convex integration method; cf. [15] or [32]. In dimensions $n \geq 5$, one can even use R. Thom's jet transversality theorem [8, Theorem 2.3.2], which shows that twisted distributions lie not only C^0-dense but even C^∞-dense in the space of distributions. The situation in dimension 4 is more subtle and requires the additional assumptions in Theorem 3.3. For instance, if the bundles H and TM/H are orientable and the intersection form signature of M is congruent to 2 modulo 4, then the connected component of H does not contain any twisted distribution, not even far away from H. Theorems of Hirzebruch–Hopf [18] and Donaldson [7] are applied to solve the problem when the manifold is noncompact or the signature is divisible by 4. For details of these differential-topological considerations see Chap. 5 of the second author's PhD thesis [23].

Since every distribution which is sufficiently C^0-close to our spacelike start distribution is spacelike as well, we obtain a spacelike twisted distribution H, as desired. Let V be the (timelike) g-orthogonal complement of H. For every $f \in C^\infty(M, \mathbb{R}_{>0})$, we can now consider the Lorentzian metric g' on M which is given by

$$g'(v_0 + h_0, v_1 + h_1) = \frac{1}{f^2} g(v_0, v_1) + g(h_0, h_1) \tag{22}$$

for all $x \in M$ and $h_0, h_1 \in H_x$ and $v_0, v_1 \in V_x$. Using similar arguments as in the proof of Theorem 3.1, we see that there exists a constant $\varepsilon_0 > 0$ such that whenever $f \leq \varepsilon_0$, then the metric g' has the properties (1) and (3) of Theorem 3.3.

Now we have to compute the Ricci tensor of g' and check whether g' satisfies the dominant energy condition. Let $x \in M$. If $(v, w) \in H_x \times H_x$, then

$$\begin{aligned}
\mathrm{Ric}_{g'}(v,w) &= \mathrm{Ric}_g(v,w) + \frac{1}{f}\mathrm{Hess}_g f\,(v,w) - \frac{2}{f^2} df(v) df(w) \\
&\quad + \frac{1}{f}\mathrm{div}_g^V(w) df(v) + \frac{1}{f}\mathrm{div}_g^V(v) df(w) + \frac{1+f^2}{2f} \mathrm{Sw}_g^H(v,w,a) df(a) \\
&\quad - (1-f^2)\Phi_g^H(v,w) - \frac{1-f^2}{2f^2}\mathrm{Tw}_g^H(v,a,b)\mathrm{Tw}_g^H(w,a,b).
\end{aligned} \tag{23}$$

If $(v, w) \in V_x \times V_x$, then

$$\begin{aligned}
f^2 \mathrm{Ric}_{g'}(v,w) &= f^2 \mathrm{Ric}_g(v,w) - f\,\mathrm{Hess}_g f(v,w) + \Big(\frac{1}{f}\Delta_g^H f + f \Delta_g^V f\Big) g(v,w) \\
&\quad - \frac{2}{f^2}|df|^2_{g,H}\, g(v,w) - f\,\mathrm{div}_g^H(w) df(v) - f\,\mathrm{div}_g^H(v) df(w) \\
&\quad + \Big(\frac{1}{f}\mathrm{div}_g^V(a) df(a) + f\,\mathrm{div}_g^H(a) df(a)\Big) g(v,w)
\end{aligned}$$

$$- \frac{1+f^2}{2f} Sw_g^V(v,w,a)df(a)$$

$$+ (1-f^2)\Phi_g^V(v,w) + \frac{(1-f^2)^2}{4f^2} Tw_g^H(a,b,v)Tw_g^H(a,b,w).$$

If $(v,w) \in V_x \times H_x$, then

$$f\mathrm{Ric}_{g'}(v,w) = f\mathrm{Ric}_g(v,w) + \frac{3}{2f^2} Tw_g^H(w,a,v)df(a) - \mathrm{div}_g^H(v)df(w)$$

$$+ \mathrm{div}_g^V(w)df(v) + \sigma_g^H(v,w,a)df(a)$$

$$+ \left(\frac{1-f^2}{2f}\tilde{\Theta}_g^H - \frac{f(1-f^2)}{2}\tilde{\Theta}_g^V + \frac{(1-f^2)^2}{4f}\tilde{\Xi}_g^H\right)(v,w).$$

In these formulas, $\mathrm{div}_g^?$ are certain $(0,1)$-tensors, $\Phi_g^?, \tilde{\Theta}_g^?, \tilde{\Xi}_g^H$ are $(0,2)$-tensors, and $Sw_g^?, \sigma_g^H$ are $(0,3)$-tensors, induced by the metric g and the distributions V, H. The precise definitions are not relevant for our discussion. $\mathrm{Hess}_g f$ is the g-Hessian of f, $\Delta_g^U f$ is the g-contraction of the restriction of Hess_g to the subbundle U, and $|df|_{g,H}^2$ is the g-contraction of the restriction of $df \otimes df$ to the subbundle H. Arguments a, b occur always pairwise in the formulas; they have to be interpreted in the sense of a summation convention, i.e., a g-contraction is performed in these tensor indices.

Finally, Tw_g^H is the $(0,3)$-tensor on M which is defined as follows: We identify TM/H with V. Using the projection $\pi_H : TM = V \oplus H \to H$, we let

$$Tw_g^H(v,w,z) = g\big(Tw_H(\pi_H v, \pi_H w), z\big).$$

Although the formula for $\mathrm{Ric}_{g'}$ is quite complicated, it is relatively easy to see what happens on a compact set $K \subseteq M$ when the function f is a very *small* positive *constant*: All summands containing derivatives of f are zero, and at every point $x \in K$ where Tw_g^H does not vanish, the summands containing $Tw_g^H \otimes Tw_g^H$ dominate the Ricci tensor because their coefficients have f^2 in the denominator. In this situation one can determine from the tensor field Tw_g^H alone whether g' satisfies the dominant energy condition for a given Λ.

The result is that if H is twisted, then for every Λ and every compact subset K of M there exists a constant $c_{K,\Lambda} > 0$ such that for every constant f with $0 < f < c_{K,\Lambda}$, the metric g' satisfies on K the dominant energy condition with respect to Λ. This completes the proof of Theorem 3.3.

One can apply an analogous "stretching" by a function f as in equation (22) to an arbitrary semi-Riemannian metric g of index q and an arbitrary g-spacelike

k-plane distribution H. The effect is similar: For every comparison of $\mathrm{Ric}_{g'}$ with g', the most important contribution to $\mathrm{Ric}_{g'}$ comes from the nonintegrability properties of H when f is a small constant. This principle can also be employed to prove new results in *Riemannian* geometry [26].

3.3 Lorentz Cobordisms and Topology Change

A striking example of the difference between metrics without space foliations and space-foliated metrics occurs in the classical problem of "topology change" in general relativity. This problem deals with the question whether the spatial topology of our universe could change as time goes by. It was discussed by several authors in the 1960s and 1970s; cf. e.g. [13, 28, 35, 38, 39]. In order to describe it in detail, let us adopt the following terminology (see also Fig. 4):

Definition 3.5. Let S_0, S_1 be $(n-1)$-dimensional compact manifolds. A *weak Lorentz cobordism between* S_0 *and* S_1 is a compact n-dimensional Lorentzian manifold-with-boundary (M, g) whose boundary is the disjoint union $S_0 \sqcup S_1$, such that M admits a g-timelike vector field which is inward-directed on S_0 and outward-directed on S_1. A *Lorentz cobordism between* S_0 *and* S_1 is a weak Lorentz cobordism (M, g) between S_0 and S_1 such that ∂M is g-spacelike. S_0 is [*weakly*] *Lorentz cobordant* to S_1 iff there exists a [weak] Lorentz cobordism between S_0 and S_1.

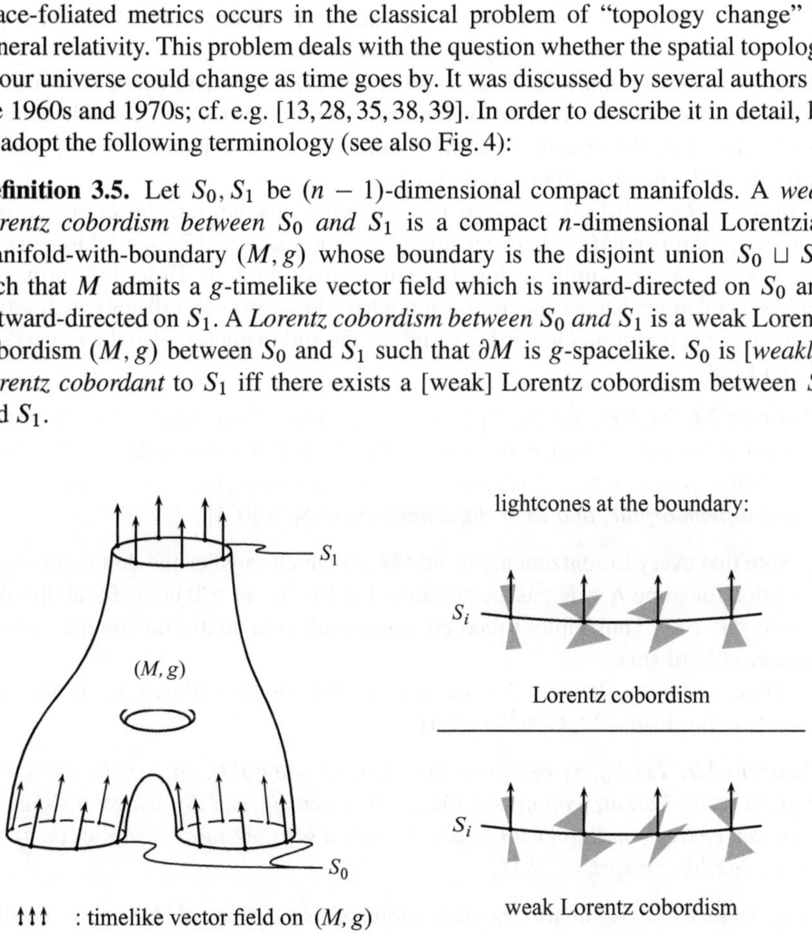

Fig. 4 A [weak] Lorentz cobordism. The picture on the left is not optimal because the vector field cannot be extended nonvanishingly to the whole cobordism. The viewer is supposed to imagine that it can. (It is impossible to draw a good picture, because nontrivial 2-dimensional [weak] Lorentz cobordisms do not exist.)

One can generalize this definition to noncompact manifolds, but for reasons of simplicity and mathematical elegance we will discuss only the compact case here. Moreover, we will concentrate on the dimension which is relevant in general relativity, $n = 4$.

[Weak] Lorentz cobordance is an equivalence relation. Clearly two manifolds are Lorentz cobordant if and only if they are weakly Lorentz cobordant. But when we require the cobordism metric g to satisfy the dominant energy condition – recall that every physically realistic space-time metric should have this property –, then the two cobordance relations become very different, as we will see.

In 1963, Reinhart [28] proved (using the well-known computations of the unoriented and oriented cobordism rings due to R. Thom and C.T.C. Wall) that every two closed 3-manifolds S_0, S_1 are Lorentz cobordant; and that for every two closed *oriented* 3-manifolds, there exists a Lorentz cobordism (M, g) between S_0 and S_1 and an orientation of M which turns M into an oriented cobordism in the usual sense (i.e., the orientation of M induces the given orientation on S_1 and the opposite of the given orientation on S_0).

In 1967, Geroch [13] observed that topology change, i.e. the existence of a Lorentz cobordism (M, g) between nondiffeomorphic closed 3-manifolds, can only occur when (M, g) admits a closed timelike curve. Finally, Tipler [35] proved in 1977 several nonexistence theorems for nontrivial Lorentz cobordisms which satisfy some energy condition. One of his results is the following (cf. also the remarks on p. 29 of [24]):

Theorem 3.6 (Tipler). *Let S_0, S_1 be closed connected 3-manifolds, let (M, g) be a connected Lorentz cobordism between S_0 and S_1 which satisfies $\mathrm{Ric}_g(v, v) > 0$ for all lightlike vectors $v \in TM$ (i.e. nonzero vectors v with $g(v, v) = 0$). Then S_0 and S_1 are diffeomorphic, and M is diffeomorphic to $S_0 \times [0, 1]$.*

Note that every Lorentzian manifold (M, g) which satisfies the dominant energy condition for some $\Lambda \in \mathbb{R}$ has the property that $\mathrm{Ric}_g(v, v) \geq 0$ holds for all lightlike vectors $v \in TM$. Thus Tipler's theorem *almost* rules out nontrivial dominant energy Lorentz cobordisms.

When we apply Theorem 3.3, we obtain a completely different result for *weak* Lorentz cobordisms [24, Corollary 9.4]:

Theorem 3.7. *Let S_0, S_1 be closed orientable 3-manifolds, let $\Lambda \in \mathbb{R}$. Then there exists a weak Lorentz cobordism (M, g) between S_0 and S_1 which satisfies the dominant energy condition with respect to Λ and satisfies, moreover, $\mathrm{Ric}_g(v, v) > 0$ for all lightlike vectors $v \in TM$.*

Proof (sketch). By Reinhart's theorem, there exists an orientable Lorentz cobordism (M, g) between S_0 and S_1. Since every (weak) Lorentz cobordism is time-orientable by definition, we can apply Theorem 3.3 (with $K = M$) and get a Lorentz metric g' on M which satisfies the dominant energy condition with respect to Λ and makes every g-timelike vector timelike. The latter property implies that (M, g') is a weak Lorentz cobordism between S_0 and S_1. The former property yields already $\mathrm{Ric}_{g'}(v, v) \geq 0$ for all lightlike vectors v. A closer look at the curvature estimates in the proof of Theorem 3.3 reveals that one can even arrange that $\mathrm{Ric}_{g'}(v, v) > 0$ holds for all lightlike v. □

If S_0, S_1 in the previous theorem are not diffeomorphic, then Tipler's theorem shows that the weak Lorentz cobordism produced by our theorem *must* make some tangent vectors to the boundary lightlike or timelike; see also Fig. 4. (Nevertheless, there exists a timelike vector field which is transverse to the boundary.) The flexibility required for topology change via a dominant energy metric can only be obtained from the absence of compact spacelike hypersurfaces.

3.4 Outlook

We conclude the discussion with some remarks and open problems.

Two obvious questions are whether the topological conditions on the 4-manifold in Theorem 3.3 can be removed, and whether one can always arrange that the dominant energy condition holds not only on a compact subset but globally on a noncompact manifold. We expect that the correct answer to both questions is *yes*. The method of proof will be essentially the same as in Theorem 3.3, but the start metric g and the function f in the proof have to be chosen more carefully; for instance, one cannot expect that a *constant* f will suffice. Unfortunately a non-constant f makes the necessary estimates much more difficult, as the complicated formulas 23 indicate.

Several points related to the nonexistence of space foliations remain to be clarified. First, it would be nice to have a quantitative criterion saying that if the spacelike region of a Lorentzian metric g is on an open set U in a suitable sense sufficiently close – here we want an explicit estimate – to a nonintegrable distribution, then U does not admit a g-space foliation. Whereas in the previous sections, our constructions of metrics without space foliations always used a sequence $(g_k)_{k\in\mathbb{N}}$ of metrics and stated that there exists a k_0 such that for each $k \geq k_0$, the metric g_k does not admit a space foliation; but we did not know explicitly how large k_0 had to be chosen.

Second, we would like to prove more than absence of space foliations, namely even absence of single spacelike hypersurfaces with certain properties. For example, consider in Minkowski space-time $\mathbb{R}^{3,1}$ an infinitely long cylinder Z with timelike axis and, say, Euclidean radius 1. When we perform our standard space foliation-removing procedure inside of Z, can we construct a Lorentzian metric which is equal to the Minkowski metric outside of Z, makes every Minkowski-timelike vector timelike, but does not admit any asymptotically flat spacelike hypersurface diffeomorphic to $\mathbb{R}^3$?

As to the physical relevance of the metrics we considered, astronomical observations suggest that the cosmological constant Λ is positive in the universe we live in. Since $\Lambda > 0$ is the hardest case for the construction of dominant energy metrics, the physically most interesting case coincides with the geometrically most interesting case.

Lorentzian metrics without space foliations violate the usual causality assumptions of general relativity, but these causality assumptions seem to be true for the space-time metric of our universe, within the range of currently available experimental data. Thus metrics without space foliations can be physically realistic only if the causality violations occur on length scales which are too small (or too large) to be observed. For instance, if in Theorem 3.1 the set A is the complement of a tiny ball in Minkowski space-time, then the metric g' violates e.g. global hyperbolicity, but since g' is equal to the Minkowski metric outside of the tiny ball, this violation could hardly be detected.

In order to find dominant energy metrics g' that violate causality only on sets which are small in some sense, one would like to prove a "relative" version of the existence theorem 3.3 for dominant energy metrics, in which the metric is not changed on a given set (similar to Theorem 3.1): If A is a suitable closed subset of a (say causally well-behaved) Lorentzian manifold (M, g) such that g satisfies the dominant energy condition on A, one wants to find a metric g' on M which is equal to g on A and satisfies the dominant energy condition everywhere.

A closely related problem is whether topology change can occur when the space-time metric violates causality only on a very small set. One would like to construct a weak Lorentz cobordism (M, g) between nondiffeomorphic 3-manifolds which satisfies the dominant energy condition and is "almost a Lorentz cobordism" in the sense that only a tiny set of ∂M is not g-spacelike.

Finally, an important question is whether there exists a Lorentzian Λ-*vacuum manifold* (M, g) – i.e., a Lorentzian manifold with $\mathrm{Ric} - \frac{1}{2}sg + \Lambda g = 0$; in other words, a Lorentzian Einstein manifold – which does not admit a space foliation.

Acknowledgements We would like to thank Margarita Kraus for helpful comments on the manuscript. Supported in part by the Deutsche Forschungsgemeinschaft within the Priority Program "Globale Differentialgeometrie."

References

1. Arnowitt, R., Deser, S., Misner, C.W.: Energy and the criteria for radiation in general relativity. Phys. Rev. **118**(2), 1100–1104 (1960)
2. Bär, C., Gauduchon, P., Moroianu, A.: Generalized cylinders in semi-riemannian and spin geometry. Math. Z. **249**, 545–580 (2005)
3. Bartnik, R.: The mass of an asymptotically flat manifold. Comm. Pure Appl. Math. **39**(5), 661–693 (1986)
4. Bernal, A.N., Sánchez, M.: Smoothness of time functions and the metric splitting of globally hyperbolic spacetimes. Comm. Math. Phys. **257**, 43–50 (2005)
5. Bray, H.L.: Proof of the riemannian penrose inequality using the positive mass theorem. J. Differential Geom. **59**(2), 177–267 (2001)
6. Bray, H.L., Finster, F.: Curvature estimates and the positive mass theorem, arXiv:math/9906047 [math.DG]. Comm. Anal. Geom. **10**(2), 291–306 (2002)

7. Donaldson, S.K.: The orientation of yang-mills moduli spaces and 4-manifold topology. J. Differential Geom. **26**, 397–428 (1987)
8. Eliashberg, Y., Mishachev, N.: Introduction to the h-principle. vol. 48 of Graduate Studies in Mathematics, American Mathematical Society, Providence, RI (2002)
9. Finster, F.: A level set analysis of the Witten spinor with applications to curvature estimates. arXiv:math/0701864 [math.DG], Math. Res. Lett. **16**(1), 41–55 (2009)
10. Finster, F., Kath, I.: Curvature estimates in asymptotically flat manifolds of positive scalar curvature. arXiv:math/0101084 [math.DG], Comm. Anal. Geom. **10**(5), 1017–1031 (2002)
11. Finster, F., Kraus, M.: Curvature estimates in asymptotically flat Lorentzian manifolds. arXiv:math/0306159 [math.DG], Canad. J. Math. **57**(4), 708–723 (2005)
12. Finster, F., Kraus, M.: A weighted L^2-estimate of the Witten spinor in asymptotically Schwarzschild manifolds, arXiv:math/0501195 [math.DG], Canad. J. Math. **59**(5), 943–965 (2007)
13. Geroch, R.P.: Topology in general relativity, J. Math. Phys. **8**, 782–786 (1967)
14. Ginoux, N.: The Dirac spectrum. Lecture Notes in Mathematics, vol. 1976, Springer, Berlin (2009)
15. Gromov, M.: Partial differential relations, vol. 9 of Ergebnisse der Mathematik und ihrer Grenzgebiete (3), Springer, Berlin (1986)
16. Hawking, S.W., Ellis, G.F.R.: The large scale structure of space-time, Cambridge Monographs on Mathematical Physics, No. 1. Cambridge University Press, London (1973)
17. Hirsch, M.W.: Differential topology, vol. 33 of Graduate Texts in Mathematics. Springer, New York (1994); Corrected reprint of the 1976 original.
18. Hirzebruch, F., Hopf, H.: Felder von Flächenelementen in 4-dimensionalen Mannigfaltigkeiten. Math. Ann. **136**, 156–172 (1958)
19. Huisken, G., Ilmanen, T.: The inverse mean curvature flow and the Riemannian Penrose inequality. J. Differential Geom. **59**(3), 353–437 (2001)
20. Lawson, H.B. Jr., Michelsohn, M.-L.: Spin geometry. Princeton Mathematical Series, vol. 38, Princeton University Press, Princeton, NJ (1989)
21. Maerten, D.: Positive energy-momentum theorem for AdS-asymptotically hyperbolic manifolds, Ann. Henri Poincaré **7**, 975–1011 (2006)
22. Mars, M.: Present status of the penrose inequality. arXiv:0906.5566 [gr-qc] (2009)
23. Nardmann, M.: Pseudo-Riemannian metrics with prescribed scalar curvature, PhD thesis at the University of Leipzig. arXiv:math/0409435 [math.DG]
24. Nardmann, M.: Nonexistence of spacelike foliations and the dominant energy condition in Lorentzian geometry. arXiv:math/0702311 [math.DG]
25. Nardmann, M.: A remark on the rigidity case of the positive mass theorem. arXiv:1004.5430 [math.DG]
26. Nardmann, M.: A $C^{0,\alpha}$-dense h-principle for negative scalar curvature of semi-Riemannian metrics, in preparation
27. Parker, T., Taubes, C.H.: On Witten's proof of the positive energy theorem, Comm. Math. Phys. **84**(2), 223–238 (1982)
28. Reinhart, B.L.: Cobordism and the euler number. Topology **2**, 173–177 (1963)
29. Schoen, R., Yau, S.-T.: On the proof of the positive mass conjecture in general relativity, Comm. Math. Phys. **65**(1), 45–76 (1979)
30. Schoen, R., Yau, S.-T.: Proof of the positive mass theorem. II. Comm. Math. Phys. **79**(2), 231–260 (1981)
31. Schrödinger, E.: Diracsches Elektron im Schwerefeld I. Sitzungsbericht der Preussischen Akademie der Wissenschaften Phys.-Math. Klasse 1932, Verlag der Akademie der Wissenschaften, 436–460 (1932)
32. Spring, D.: Convex integration theory. Solutions to the h-principle in geometry and topology. vol. 92 of Monographs in Mathematics, Birkhäuser, Basel (1998)
33. Thurston, W.P.: Existence of codimension-one foliations. Ann. of Math. **104**(2), 249–268 (1976)

34. Thurston, W.P.: Three-dimensional geometry and topology. Vol. 1, vol. 35 of Princeton Mathematical Series, Princeton University Press, Princeton, NJ (1997)
35. Tipler, F.J.: Singularities and causality violation, Ann. Phys. **108**, 1–36 (1977)
36. Varela, F.: Formes de Pfaff, classe et perturbations. Ann. Inst. Fourier (Grenoble) **26**, 239–271 (1976)
37. Witten, E.: A new proof of the positive energy theorem. Comm. Math. Phys. **80**(3), 381–402 (1981)
38. Yodzis, P.: Lorentz cobordism. Comm. Math. Phys. **26**, 39–52 (1972)
39. Yodzis, P.: Lorentz cobordism. II, General relativity and gravitation. **4**, 299–307 (1973)

Mean Curvature Flow in Higher Codimension: Introduction and Survey

Knut Smoczyk

Abstract In this text we outline the major techniques, concepts and results in mean curvature flow with a focus on higher codimension. In addition we include a few novel results and some material that cannot be found elsewhere.

1 Mean Curvature Flow

Mean curvature flow is perhaps the most important geometric evolution equation of submanifolds in Riemannian manifolds. Intuitively, a family of smooth submanifolds evolves under mean curvature flow, if the velocity at each point of the submanifold is given by the mean curvature vector at that point. For example, round spheres in euclidean space evolve under mean curvature flow while concentrically shrinking inward until they collapse in finite time to a single point, the common center of the spheres.

Mullins [63] proposed mean curvature flow to model the formation of grain boundaries in annealing metals. Later the evolution of submanifolds by their mean curvature has been studied by Brakke [10] from the viewpoint of geometric measure theory. Among the first authors who studied the corresponding nonparametric problem were Temam [82] in the late 1970s and Gerhardt [36] and Ecker [26] in the early 1980s. Pioneering work was done by Gage [35], Gage & Hamilton [34] and Grayson [37] who proved that the curve shortening flow (more precisely, the "mean" curvature flow of curves in $\mathbb{R}^2$) shrinks embedded closed curves to "round" points. In his seminal paper Huisken [48] proved that closed convex hypersurfaces in euclidean space $\mathbb{R}^{m+1}, m > 1$ contract to single round points in finite time

K. Smoczyk (✉)
Leibniz Universität Hannover, Institut für Differentialgeometrie, Welfengarten 1, 30167 Hannover, Germany
e-mail: smoczyk@math.uni-hannover.de

C. Bär et al. (eds.), *Global Differential Geometry*, Springer Proceedings in Mathematics 17, DOI 10.1007/978-3-642-22842-1_9,

(later he extended his result to hypersurfaces in Riemannian manifolds that satisfy a suitable stronger convexity, see [49]). Then, until the mid 1990s, most authors who studied mean curvature flow mainly considered hypersurfaces, both in euclidean and Riemannian manifolds, whereas mean curvature flow in higher codimension did not play a great role. There are various reasons for this, one of them is certainly the much different geometric situation of submanifolds in higher codimension since the normal bundle and the second fundamental tensor are more complicated. But also the analysis becomes more involved and the algebra of the second fundamental tensor is much more subtle since for hypersurfaces there usually exist more scalar quantities related to the second fundamental form than in case of submanifolds in higher codimension. Some of the results previously obtained for mean curvature flow of hypersurfaces carry over without change to submanifolds of higher codimension but many do not and in addition even new phenomena occur.

Among the first results in this direction are the results on mean curvature flow of space curves by Altschuler and Grayson [2, 3], measure-theoretic approaches to higher codimension mean curvature flows by Ambrosio and Soner [4], existence and convergence results for the Lagrangian mean curvature flow [72, 74, 75, 83], mean curvature flow of symplectic surfaces in codimension two [17, 87] and long-time existence and convergence results of graphic mean curvature flows in higher codimension [18, 76, 78, 87, 95]. Recently there has been done quite some work on the formation and classification of singularities in mean curvature flow [5, 11, 12, 23, 25, 39, 47, 54, 56, 58, 59, 61, 71], partially motivated by Hamilton's and Perelman's [45, 66–68] work on the Ricci flow that in many ways behaves akin to the mean curvature flow and vice versa.

The results in mean curvature flow can be roughly grouped into two categories: The first category contains results that hold (more or less) in general, i.e. that are independent of dimension, codimension or the ambient space. In the second class we find results that are adapted to more specific geometric situations, like results for hypersurfaces, Lagrangian or symplectic submanifolds, graphs, etc..

Our aim in this article is twofold. We first want to summarize the most important properties of mean curvature flow that hold in any dimension, codimension and ambient space (first category). In the second part of this exposition we will give a – certainly incomplete and not exhaustive –, overview on more specific results in higher codimension, like an overview on the Lagrangian mean curvature flow or the mean curvature flow of graphs (part of the second category). Graphs and Lagrangian submanifolds certainly form the best understood subclasses of mean curvature flow in higher codimension.

In addition this article is intended as an introduction to mean curvature flow for the beginner and we will derive the most relevant geometric structure and evolution equations in a very general but consistent form that is rather hard to find in the literature. However, there are several nice monographs on mean curvature flow, a well written introduction to the regularity of mean curvature flow of hypersurfaces is [28]. For the curve shortening flow see [22]. For mean curvature flow in higher codimension there exist some lecture notes by Wang [92].

Let us now turn our attention to the mathematical definition of mean curvature flow. Suppose M is a differentiable manifold of dimension m, $T > 0$ a real number and $F : M \times [0, T) \to (N, g)$ a smooth time dependent family of immersions of M into a Riemannian manifold (N, g) of dimension n, i.e. F is smooth and each

$$F_t : M \to N\,, \quad F_t(p) := F(p,t)\,, \quad t \in [0,T)$$

is an immersion. If F satisfies the evolution equation

$$\frac{dF}{dt}(p,t) = \overrightarrow{H}(p,t)\,, \quad \forall p \in M, t \in [0,T), \tag{MCF}$$

where $\overrightarrow{H}(p,t) \in T_{F(p,t)}N$ is the mean curvature vector of the immersion F_t at p (or likewise of the submanifold $U_t := F_t(U)$ at $F_t(p)$, if for some $U \subset M$, $F_{t|U}$ is an embedding), then we say that M evolves by mean curvature flow in N with initial data $F_0 : M \to N$. As explained in Sect. 2.1, the mean curvature vector field can be defined for any immersion into a Riemannian manifold (or more generally for any space-like immersion into a pseudo-Riemannian manifold; in this survey we will restrict to the Riemannian mean curvature flow) and it is the negative L^2-gradient of the volume functional vol : $\mathscr{I} \to \mathbb{R}$ on the space $\mathscr{I}$ of immersions of M into (N, g). Hence mean curvature flow is the steepest descent or negative L^2-gradient flow of the volume functional and formally (MCF) makes sense for any immersed submanifold in a Riemannian manifold. Therefore, following Hadamard, given an initial immersion $F_0 : M \to N$ one is interested in the well-posedness of (MCF) in the sense of

1. Does a solution exist?
2. Is it unique?
3. Does it behave continuously in some suitable topology?

In addition, once short-time existence is established on some maximal time interval $[0, T)$, $T \in (0, \infty]$, one wants to study the behavior of the flow and in particular of the evolving immersed submanifolds $M_t := F_t(M)$ as $t \to T$. Either singularities of some kind will form and one might then study the formation of singularities in more details – with possible significant geometric implications – or the flow has a long-time solution. In such a case convergence to some nice limit (e.g. stationary, i.e. a limit with vanishing mean curvature) would be rather expected but in general will not hold a priori.

In the most simplest case, i.e. if the dimension of M is one, mean curvature flow is called curve shortening flow. In many contributions to the theory of mean curvature flow one assumes that M is a smooth *closed* manifold. The reason is, that one key technique in mean curvature flow (or more generally in the theory of parabolic geometric evolution equations) is the application of the maximum principle and in absence of compactness the principle of "first time violation" of a stated inequality simply does not hold. But even for complete non-compact submanifolds there are powerful techniques, similar to the maximum principle, that can be applied in some situations. In the complete case one of the most important tools is the monotonicity formula found by Huisken [50], Ecker and Huisken [29] and Hamilton [44] and that equally well applies to mean curvature flow in higher codimension. Ecker [27] proved a beautiful local version of the monotonicity formula for hypersurfaces and another local monotonicity for evolving Riemannian manifolds has been found recently by Ecker et al. [31].

There are some very important contributions to the regularity theory of mean curvature flow by White [93,94] that apply in all codimensions. For example in [93] he proves uniform curvature bounds of the euclidean mean curvature flow in regions of space-time where the Gaussian density ratios are close to 1. With this result one can often exclude finite time singularities and prove long-time existence of the flow (see for example [62, 87]).

For simplicity and since some techniques and results do not hold for complete non-compact manifolds we will always assume in this article, unless otherwise agreed, that M is an oriented *closed* smooth manifold.

The organization of the survey is as follows: In Sect. 2 we will review the geometric structure equations for immersions in Riemannian manifolds and we will introduce most of our terminology and notations that will be used throughout the paper. In particular we will mention the explicit formulas in the case of Lagrangian submanifolds in Kähler–Einstein manifolds. For most computations we will use the Ricci calculus and apply the Einstein convention to sum over repeated indices. In Sect. 3 we will summarize those results that hold in general (first category). The section is subdivided into four subsections. In the first Sect. 3.1 we will show that the mean curvature flow is a quasilinear (degenerate) parabolic system and we will treat the existence and uniqueness problem. In Sect. 3.2 we derive the evolution equations of the most important geometric quantities in the general situation, i.e. for immersions in arbitrary Riemannian manifolds. In this general form these formulas are hard to find in the literature and one can later easily derive all related evolution equations from them that occur in special situations like evolution equations for tensors that usually appear in mean curvature flow of hypersurfaces, Lagrangian submanifolds or graphs. In Sect. 3.3 we recall general results concerning long-time existence of solutions. In the final Sect. 3.4 of this section we explain the two types of singularities that appear in mean curvature flow and discuss some rescaling techniques. Moreover we will recall some of the results that have been obtained in the classification of solitons. Section 4 is on more specific results in higher codimension, the first subsection treats the Lagrangian mean curvature flow and in the last and final subsection of this article we give an overview of the results in mean curvature flow of graphs.

2 The Geometry of Immersions

2.1 *Second Fundamental Form and Mean Curvature Vector*

In this subsection we recall the definition of the second fundamental form and mean curvature vector of an immersion and we will introduce most of our notation.

Let $F : M \to (N, g)$ be an immersion of an m-dimensional differentiable manifold M into a Riemannian manifold (N, g) of dimension n, i.e. F is smooth and the pull-back F^*g defines a Riemannian metric on M. The number $k := n - m \geq 0$ is called the codimension of the immersion.

For $p \in M$ let

$$T_p^{\perp}M := \{\nu \in T_{F(p)}N : g(\nu, DF_{|p}(W)) = 0,\ \forall W \in T_pM\}$$

denote the normal space of M at p and $T^{\perp}M$ the associated normal bundle. By definition, the normal bundle of M is a sub-bundle of rank k of the pull-back bundle $F^*TN = \bigcup_{p\in M} T_{F(p)}N$ over M. Using the differential of F we thus have a splitting

$$T_{F(p)}N = DF_{|p}(T_pM) \oplus T_p^{\perp}M\,.$$

The differential DF can be considered as a 1-form on M with values in F^*TN, i.e.

$$DF \in \Gamma(F^*TN \otimes T^*M) =: \Omega^1(M, F^*TN),$$

$$T_pM \ni V \mapsto DF_{|p}(V) \in T_{F(p)}N.$$

The Riemannian metric F^*g is also called the first fundamental form on M. In an obvious way the metrics g and F^*g induce Riemannian metrics on all bundles formed from products of $TM, T^*M, T^{\perp}M, F^*TN, TN$, and T^*N and in the sequel we will often denote all such metrics simply by the usual brackets $\langle\cdot,\cdot\rangle$ for an inner product.

Similarly the Levi-Civita connection ∇ on (N, g) induces connections on the bundles $TM, T^*M, T^{\perp}M, F^*TN$ and products hereof. Since the precise definition of these connections will be crucial in the understanding of the second fundamental form, the mean curvature vector and later also of the evolution equations, we will briefly recall them. The connection ∇^{TM} on TM can be obtained in two equivalent ways: either as the Levi-Civita connection of the induced metric F^*g on TM or else by projection of the ambient connection to the tangent bundle, more precisely via the formula

$$DF(\nabla_X^{TM}Y) := \nabla_{DF(X)}^{\top}\overline{DF(Y)}, \quad X, Y \in TM\,,$$

where $^{\top}$ denotes the projection onto $DF(TM)$ and $\overline{DF(Y)}$ is an arbitrary (local) smooth extension of $DF(Y)$. The connection ∇^{T^*M} on T^*M is then simply given by the dual connection of ∇^{TM}. Similarly one obtains the connection ∇^{F^*TN} on F^*TN via the formula

$$\nabla_X^{F^*TN} V := \nabla_{DF(X)}\overline{V},$$

for any smooth section $V \in \Gamma(F^*TN)$ and finally the connection $\nabla^\perp$ on the normal bundle is given by projection

$$\nabla_X^\perp \nu := \left(\nabla_X^{F^*TN}\nu\right)^\perp$$

for $\nu \in \Gamma(T^\perp M) \subset \Gamma(F^*TN)$. Since the connections ∇^{TM}, ∇^{T^*M}, ∇^{F^*TN} and their associated product connections on product bundles over M formed from the factors TM, T^*M, F^*TN are induced by ∇, it is common (and sometimes confusing) to denote all of them by the same symbol ∇. Since $T^\perp M$ is a subbundle of F^*TN, one can consider a section $\nu \in \Gamma(T^\perp M)$ also as an element of $\Gamma(F^*TN)$ and hence one can apply both connections $\nabla^\perp$ and $\nabla = \nabla^{F^*TN}$ to them, i.e. we will write $\nabla_X \nu$ $(= \nabla_X^{F^*TN}\nu)$, if we consider ν as a section in F^*TN and $\nabla_X^\perp \nu$, if ν is considered as a section in the normal bundle $T^\perp M$. The same holds, if we consider sections in product bundles that contain $T^\perp M$ as a factor.

If we apply the resulting connection ∇ on $F^*TN \otimes T^*M$ to DF, we obtain – by definition – the second fundamental tensor

$$A := \nabla DF \in \Gamma(F^*TN \otimes T^*M \otimes T^*M).$$

It is then well known that the second fundamental tensor is symmetric

$$A(X,Y) = (\nabla_X DF)(Y) = (\nabla_Y DF)(X) = A(Y,X) \tag{1}$$

and normal in the sense that

$$\langle A(X,Y), DF(Z)\rangle = 0, \quad \forall X,Y,Z \in TM. \tag{2}$$

Therefore in particular $A \in \Gamma(T^\perp M \otimes T^*M \otimes T^*M)$.

Taking the trace of A gives the mean curvature vector field

$$\vec{H} := \operatorname{trace} A = \sum_{i=1}^m A(e_i, e_i), \tag{3}$$

where $(e_i)_{i=1,\dots,m}$ is an arbitrary orthonormal frame of TM. Hence, since A is normal, we obtain a canonical section $\vec{H} \in \Gamma(T^\perp M)$ in the normal bundle of the immersion $F : M \to N$.

2.2 Structure Equations

The second fundamental tensor is a curvature quantity that determines how curved the immersed submanifold $F(M)$ given by an immersion $F : M \to N$ lies within the ambient manifold (N, g). According to this we have a number of geometric equations that relate the second fundamental tensor to the intrinsic curvatures of (M, F^*g) and (N, g).

Let ∇ be a connection on a vector bundle E over a smooth manifold M. Our convention for the curvature tensor $R^{E,\nabla} \in \Omega^2(M, E)$ w.r.t. ∇ is

$$R^{E,\nabla}(X, Y)\sigma := (\nabla_X \nabla_Y - \nabla_Y \nabla_X - \nabla_{[X,Y]})\sigma\,, \quad \forall X, Y \in TM, \sigma \in \Gamma(E)\,.$$

Moreover, if E is a bundle with bundle metric $\langle \cdot, \cdot \rangle$, then we set

$$R^{E,\nabla}(\mu, \sigma, X, Y) := \langle \mu, R^{E,\nabla}(X, Y)\sigma \rangle\,, \quad \forall X, Y \in TM,\ \sigma, \mu \in E\,.$$

We denote the curvature tensors $R^{TM,\nabla}$ and $R^{TN,\nabla}$ by R^M resp. R^N. Letting

$$(\nabla_X A)(Y, V) := \nabla_X(A(Y, V)) - A(\nabla_X Y, V) - A(Y, \nabla_X V),$$

the Codazzi equation is

$$\begin{aligned}&(\nabla_X A)(Y, V) - (\nabla_Y A)(X, V)\\&= R^N(DF(X), DF(Y))DF(V) - DF(R^M(X, Y)V)\,.\end{aligned} \tag{4}$$

Note that ∇ denotes the full connection, i.e. here we consider A as a section in $F^*TN \otimes T^*M \otimes T^*M$ and not in $T^\perp M \otimes T^*M \otimes T^*M$. Later we will sometimes consider A as a section in $T^\perp M \otimes T^*M \otimes T^*M$ and then we will also use the connection on the normal bundle instead, so that in this case we write $(\nabla^\perp_X A)(Y, V) = ((\nabla_X A)(Y, V))^\perp$. In terms of $\nabla^\perp$ the Codazzi equation becomes

$$(\nabla^\perp_X A)(Y, V) - (\nabla^\perp_Y A)(X, V) = \Big(R^N(DF(X), DF(Y))DF(V)\Big)^\perp\,. \tag{5}$$

From

$$\langle A(Y, V), DF(W) \rangle = 0\,, \quad \forall Y, V, W \in TM$$

we get

$$\langle (\nabla_X A)(Y, V), DF(W) \rangle = -\langle A(Y, V), A(X, W) \rangle\,. \tag{6}$$

From these equations we obtain Gauß equation (Theorema Egregium):

$$\begin{aligned}R^M(X, Y, V, W) &= R^N(DF(X), DF(Y), DF(V), DF(W))\\&\quad + \langle A(X, V), A(Y, W) \rangle - \langle A(X, W), A(Y, V) \rangle\,.\end{aligned} \tag{7}$$

Finally, we have Ricci's equation. If $\nu \in T^\perp M$ and $X, Y \in TM$ then the following holds:

$$\begin{aligned} R^\perp(X,Y)\nu &= (R^N(DF(X), DF(Y))\nu)^\perp \\ &\quad - \sum_{i=1}^m \big(\langle \nu, A(X,e_i)\rangle A(Y,e_i) - \langle \nu, A(Y,e_i)\rangle A(X,e_i)\big)\,, \end{aligned} \tag{8}$$

where $(e_i)_{i=1,\dots,m}$ is an arbitrary orthonormal frame of TM and $R^\perp = R^{T^\perp M, \nabla^\perp}$ denotes the curvature tensor of the normal bundle of M. Note that the Codazzi equation is useless in dimension one (i.e. for curves) and that Ricci's equation is useless for hypersurfaces, i.e. in codimension one.

2.3 *Tensors in Local Coordinates*

For computations one often needs local expressions of tensors. Whenever we use local expressions and $F : M \to N$ is an immersion we make the following general assumptions and notations:

(1) (U, x, Ω) and (V, y, Λ) are local coordinate charts around $p \in U \subset M$ and $F(p) \in V \subset N$ such that $F_{|U} : U \to F(U)$ is an embedding and such that $F(U) \subset V$.
(2) From the coordinate functions

$$(x^i)_{i=1,\dots,m} : U \to \Omega \subset \mathbb{R}^m\,, \quad (y^\alpha)_{\alpha=1,\dots,n} : V \to \Lambda \subset \mathbb{R}^n,$$

we obtain a local expression for F,

$$y \circ F \circ x^{-1} : \Omega \to \Lambda\,, \quad F^\alpha := y^\alpha \circ F \circ x^{-1}, \quad \alpha = 1, \dots, n.$$

(3) The Christoffel symbols of the Levi-Civita connections on M resp. N will be denoted

$$\Gamma^i_{jk}, \quad i,j,k = 1,\dots,m\,, \quad \text{resp.} \quad \Gamma^\alpha_{\beta\gamma}, \quad \alpha,\beta,\gamma = 1,\dots,n\,.$$

(4) All indices referring to M will be denoted by Latin minuscules and those related to N by Greek minuscules. Moreover, we will always use the Einstein convention to sum over repeated indices from 1 to the respective dimension.

Then the local expressions for g, DF, F^*g and A are

$$g = g_{\alpha\beta} dy^\alpha \otimes dy^\beta,$$

$$DF = F^{\alpha}_{\ i}\frac{\partial}{\partial y^{\alpha}} \otimes dx^i, \quad F^{\alpha}_{\ i} := \frac{\partial F^{\alpha}}{\partial x^i},$$

$$F^*g = g_{ij}dx^i \otimes dx^j, \quad g_{ij} := g_{\alpha\beta}F^{\alpha}_{\ i}F^{\beta}_{\ j},$$

and

$$A = A_{ij}dx^i \otimes dx^j = A^{\alpha}_{\ ij}\frac{\partial}{\partial y^{\alpha}} \otimes dx^i \otimes dx^j,$$

where the coefficients $A^{\alpha}_{\ ij}$ are given by Gauß' formula

$$A^{\alpha}_{\ ij} = \frac{\partial^2 F^{\alpha}}{\partial x^i \partial x^j} - \Gamma^k_{ij}\frac{\partial F^{\alpha}}{\partial x^k} + \Gamma^{\alpha}_{\beta\gamma}\frac{\partial F^{\beta}}{\partial x^i}\frac{\partial F^{\gamma}}{\partial x^j}. \tag{9}$$

Let (g^{ij}) denote the inverse matrix of (g_{ij}) so that $g^{ik}g_{kj} = \delta^i_{\ j}$ gives the Kronecker symbol. (g^{ij}) defines the metric on T^*M dual to F^*g. For the mean curvature vector we get

$$\vec{H} = H^{\alpha}\frac{\partial}{\partial y^{\alpha}}, \quad H^{\alpha} := g^{ij}A^{\alpha}_{\ ij}. \tag{10}$$

Gauß' equation (7) now becomes

$$R_{ijkl} = R_{\alpha\beta\gamma\delta}F^{\alpha}_{\ i}F^{\beta}_{\ j}F^{\gamma}_{\ k}F^{\delta}_{\ l} + g_{\alpha\beta}(A^{\alpha}_{\ ik}A^{\beta}_{\ jl} - A^{\alpha}_{\ il}A^{\beta}_{\ jk}), \tag{11}$$

where the notation should be obvious, e.g.

$$R_{ijkl} = R^M\left(\frac{\partial}{\partial x^i}, \frac{\partial}{\partial x^j}, \frac{\partial}{\partial x^k}, \frac{\partial}{\partial x^l}\right)$$

and

$$R_{\alpha\beta\gamma\delta} = R^N\left(\frac{\partial}{\partial y^{\alpha}}, \frac{\partial}{\partial y^{\beta}}, \frac{\partial}{\partial y^{\gamma}}, \frac{\partial}{\partial y^{\delta}}\right).$$

Note that the choice of the indices already indicates which curvature tensor is used. In addition we write

$$\nabla A = \nabla_i A^{\alpha}_{\ jk}\frac{\partial}{\partial y^{\alpha}} \otimes dx^i \otimes dx^j \otimes dx^k,$$

so that

$$(\nabla_{\frac{\partial}{\partial x^i}} A)\left(\frac{\partial}{\partial x^j}, \frac{\partial}{\partial x^k}\right) = \nabla_i A^{\alpha}_{\ jk}\frac{\partial}{\partial y^{\alpha}}.$$

Similar notations will be used for other covariant derivatives, e.g. $\nabla_i\nabla_j T^k_{\ l}$ will denote the coefficients of the tensor $\nabla^2 T$ with $T \in \Gamma(TM \otimes T^*M) = \mathrm{End}(TM)$. The Codazzi equation in local coordinates is

$$\nabla_i A^\alpha{}_{jk} - \nabla_j A^\alpha{}_{ik} = R^\alpha{}_{\beta\gamma\delta} F^\beta{}_k F^\gamma{}_i F^\delta{}_j - R^l{}_{kij} F^\alpha{}_l\,, \tag{12}$$

where here and in the following all indices will be raised and lowered using the metric tensors, e.g.

$$R^\alpha{}_{\beta\gamma\delta} = g^{\alpha\epsilon} R_{\epsilon\beta\gamma\delta}\,, \quad R_k{}^i{}_l{}^j = g^{ip} g^{jq} R_{kplq}\,.$$

Finally, if $(\nu_A)_{A=1,\dots,k:=n-m}$, $\nu_A = \nu^\alpha_A \frac{\partial}{\partial y^\alpha}$, is a local trivialization of $T^\perp M$, then

$$R^\perp\left(\frac{\partial}{\partial x^i}, \frac{\partial}{\partial x^j}\right)\nu_A =: (R^\perp)^B{}_{Aij}\nu_B$$

and Ricci's equation becomes

$$\begin{aligned}(R^\perp)^B{}_{Aij}\nu^\alpha_B &= R^\alpha{}_{\beta\gamma\delta}\nu^\beta_A F^\gamma{}_i F^\delta{}_j - g^{kl} R^\epsilon{}_{\beta\gamma\delta} g_{\epsilon\sigma}\nu^\beta_A F^\gamma{}_i F^\delta{}_j F^\sigma{}_k \\ &\quad - g_{\beta\gamma} g^{kl}(\nu^\beta_A A^\gamma_{ik} A^\alpha_{jl} - \nu^\beta_A A^\gamma_{jk} A^\alpha_{il})\,. \end{aligned}\tag{13}$$

Using the rule for interchanging covariant derivatives and the structure equations one obtains Simons' identity

$$\begin{aligned}\nabla_k\nabla_l H^\alpha &= \Delta A^\alpha{}_{kl} + \left(\nabla_\epsilon R^\alpha{}_{\beta\gamma\delta} + \nabla_\gamma R^\alpha{}_{\delta\beta\epsilon}\right) F^\epsilon{}_i F^\beta{}_l F^\gamma{}_k F^{\delta i} \\ &\quad + R^\alpha{}_{\beta\gamma\delta}\Big(2A^\beta{}_{ik} F^\gamma{}_l F^{\delta i} + 2A^\beta{}_{il} F^\gamma{}_k F^{\delta i} \\ &\quad + H^\delta F^\beta{}_l F^\gamma{}_k + A^\gamma{}_{lk} F^\beta{}_i F^{\delta i}\Big) \\ &\quad - \left(\nabla_k R^p{}_l + \nabla_l R^p{}_k - \nabla^p R_{kl}\right) F^\alpha{}_p \\ &\quad + 2R_k{}^i{}_l{}^j A^\alpha{}_{ij} - R^p{}_k A^\alpha{}_{pl} - R^p{}_l A^\alpha{}_{pk}\,, \end{aligned}\tag{14}$$

where $R_{ij} = g^{kl} R_{ikjl}$ denotes the Ricci curvature on M. If one multiplies Simons' identity (14) with $2A_\alpha{}^{kl} = 2g_{\alpha\epsilon} g^{km} g^{ln} A^\epsilon{}_{mn}$, one gets

$$\begin{aligned}2\langle A, \nabla^2\vec{H}\rangle &= \Delta|A|^2 - 2|\nabla A|^2 \\ &\quad + 2\left(\nabla_\epsilon R_{\alpha\beta\gamma\delta} + \nabla_\gamma R_{\alpha\delta\beta\epsilon}\right) F^\epsilon{}_i F^{\beta l} F^{\gamma k} F^{\delta i} A^\alpha{}_{kl} \\ &\quad + 2R_{\alpha\beta\gamma\delta} A^{\alpha kl}\left(4A^\beta{}_{ik} F^\gamma{}_l F^{\delta i} + H^\delta F^\beta{}_l F^\gamma{}_k + A^\gamma{}_{lk} F^\beta{}_i F^{\delta i}\right) \\ &\quad + 4R^{kilj}\langle A_{ij}, A_{kl}\rangle - 4R^{ij}\langle A_{ik}, A_j{}^k\rangle \end{aligned}$$

and then since

$$\begin{aligned}\nabla_i A_{kl} &= \nabla_i^\perp A_{kl} + g^{pq}\langle \nabla_i A_{kl}, F_p\rangle F_q\\ &= \nabla_i^\perp A_{kl} - g^{pq}\langle A_{kl}, \nabla_i F_p\rangle F_q\\ &= \nabla_i^\perp A_{kl} - g^{pq}\langle A_{kl}, A_{ip}\rangle F_q\end{aligned}$$

implies

$$|\nabla A|^2 = |\nabla^\perp A|^2 + \langle A^{ij}, A^{kl}\rangle\langle A_{ij}, A_{kl}\rangle$$

we obtain with Gauß' equation the second Simons' identity

$$\begin{aligned}2\langle A, \nabla^2\vec{H}\rangle =\ & \Delta|A|^2 - 2|\nabla^\perp A|^2\\ & +2\langle A^{ij}, A^{kl}\rangle\langle A_{ij}, A_{kl}\rangle - 4\langle A^{kj}, A^{il}\rangle\langle A_{ij}, A_{kl}\rangle\\ & -4\langle \vec{H}, A^{ij}\rangle\langle A_{ik}, A_j{}^k\rangle + 4\langle A^{il}, A_l{}^j\rangle\langle A_{ik}, A_j{}^k\rangle\\ & +4R_{\alpha\beta\gamma\delta}F^\alpha{}_k F^\beta{}_i F^\gamma{}_l F^\delta{}_j\left(\langle A^{ij}, A^{kl}\rangle - g^{kl}\langle A^{ip}, A_p{}^j\rangle\right)\\ & +2R_{\alpha\beta\gamma\delta}A^{\alpha kl}\left(4A^\beta{}_{ik}F^\gamma{}_l F^{\delta i} + F^\beta{}_l F^\gamma{}_k H^\delta + F^\beta{}_i A^\gamma{}_{lk}F^{\delta i}\right)\\ & +2\left(\nabla_\epsilon R_{\alpha\beta\gamma\delta} + \nabla_\gamma R_{\alpha\delta\beta\epsilon}\right)F^\epsilon_i F^\beta{}_l F^\gamma{}_k F^{\delta i}A^{\alpha kl}.\end{aligned}$$

The second and third line can be further simplified , so that we get

$$\begin{aligned}2\langle A, \nabla^2\vec{H}\rangle =\ & \Delta|A|^2 - 2|\nabla^\perp A|^2\\ & +\left|\langle A_{ij}, A_{kl}\rangle - \langle A_{il}, A_{jk}\rangle\right|^2 + \left|A^\alpha{}_{ik}A^\beta{}_j{}^k - A^\beta{}_{ik}A^\alpha{}_j{}^k\right|^2\\ & +2\left|\langle \vec{H}, A_{ij}\rangle - \langle A_{ik}, A_j{}^k\rangle\right|^2 - 2\left|\langle \vec{H}, A_{ij}\rangle\right|^2\\ & +4R_{\alpha\beta\gamma\delta}F^\alpha{}_k F^\beta{}_i F^\gamma{}_l F^\delta{}_j\left(\langle A^{ij}, A^{kl}\rangle - g^{kl}\langle A^{ip}, A_p{}^j\rangle\right)\\ & +2R_{\alpha\beta\gamma\delta}A^{\alpha kl}\left(4A^\beta{}_{ik}F^\gamma{}_l F^{\delta i} + F^\beta{}_l F^\gamma{}_k H^\delta + F^\beta{}_i A^\gamma{}_{lk}F^{\delta i}\right)\\ & +2\left(\nabla_\epsilon R_{\alpha\beta\gamma\delta} + \nabla_\gamma R_{\alpha\delta\beta\epsilon}\right)F^\epsilon_i F^\beta{}_l F^\gamma{}_k F^{\delta i}A^{\alpha kl}.\end{aligned} \tag{15}$$

This last equation is useful to substitute terms in the evolution equation of $|A|^2$ (see Sect. 3.2 below).

2.4 Special Situations

2.4.1 Hypersurfaces

If $F : M \to N$ is an immersion of a hypersurface, then $n = m + 1$ and one can define a number of scalar curvature quantities related to the second fundamental tensor of M. For simplicity assume that both M and N are orientable (otherwise the following computations are only local). Then there exists a unique normal vector field $\nu \in \Gamma(T^{\perp}M)$ – called the principle normal – such that for all $p \in M$:

(1) $|\nu_{|p}| = 1$, $\nu_{|p} \in T^{\perp}_p M$,
(2) If $e_1, \ldots, e_m$ is a positively oriented basis of T_pM, then

$$DF(e_1), \ldots, DF(e_m), \nu_{|p}$$

forms a positively oriented basis of $T_{F(p)}N$.

Using the principle normal ν, one defines the (scalar) second fundamental form $h \in \Gamma(T^*M \otimes T^*M)$ by

$$h(X, Y) := \langle A(X, Y), \nu \rangle$$

and the scalar mean curvature H by

$$H := \operatorname{trace} h,$$

so that

$$A = \nu \otimes h, \quad \vec{H} = H\nu.$$

The map

$$\flat : TM \to T^*M, \quad V \mapsto V_\flat := \langle V, \cdot \rangle$$

is a bundle isomorphism with inverse denoted by

$$\sharp : T^*M \to TM.$$

This musical isomorphism can be used to define the Weingarten map

$$\mathscr{W} \in \operatorname{End}(TM), \quad \mathscr{W}(X) := (h(X, \cdot))^\sharp.$$

Since h is symmetric, the Weingarten map is self-adjoint and the real eigenvalues of $\mathscr{W}$ are called principle curvatures, often denoted by $\lambda_1, \ldots, \lambda_m$, so that e.g. $H = \lambda_1 + \cdots + \lambda_m$. Note, that in the theory of mean curvature flow H is not the arithmetic means $\frac{1}{m}\sum_{i=1}^m \lambda_i$ (which would justify its name) as is often the case in classical books on differential geometry. In local coordinates we have

$$A^\alpha{}_{ij} = \nu^\alpha h_{ij}$$

and then the equations of Gauß and Codazzi can be rewritten in terms of h_{ij}. For example, since $|\nu|^2 = 1$ we have $\langle \nabla_i \nu, \nu \rangle = 0$ and then

$$\nabla_i \nu = \langle \nabla_i \nu, F^m \rangle F_m = -\langle \nu, \nabla_i F^m \rangle F_m = -h_i{}^m F_m .$$

This implies

$$\begin{aligned}\nabla_i A^\alpha{}_{jk} &= \nabla_i(\nu^\alpha h_{jk})\\ &= -h_i^{\ m} h_{jk} F^\alpha{}_m + \nabla_i h_{jk}\nu^\alpha .\end{aligned}$$

Multiplying with ν_α yields

$$\langle \nabla_i A_{jk}, \nu\rangle = \nabla_i h_{jk} .$$

Interchanging i, j and subtracting gives

$$\begin{aligned}\nabla_i h_{jk} - \nabla_j h_{ik} &= \langle \nabla_i A_{jk} - \nabla_j A_{ik}, \nu\rangle\\ &\overset{(12)}{=} R_{\alpha\beta\gamma\delta}\nu^\alpha F^\beta{}_k F^\gamma{}_i F^\delta{}_j = R^N(\nu, F_k, F_i, F_j) .\end{aligned}$$

Similarly we get Gauß equation in the form

$$R_{ijkl} = R^N(F_i, F_j, F_k, F_l) + h_{ik}h_{jl} - h_{il}h_{jk}$$

and since the codimension is one, we do not have a Ricci equation in this case.

2.4.2 Lagrangian Submanifolds

Let $(N, g = \langle\cdot,\cdot\rangle, J)$ be a Kähler manifold, i.e. $J \in \mathrm{End}(TN)$ is a parallel complex structure compatible with g. Then N becomes a symplectic manifold with the symplectic form ω given by the Kähler form $\omega(V, W) = \langle JV, W\rangle$. An immersion $F : M \to N$ is called Lagrangian, if $F^*\omega = 0$ and $n = \dim N = 2m = 2\dim M$. For a Lagrangian immersion we define a section

$$\nu \in \Gamma(T^\perp M \otimes T^*M), \quad \nu := JDF ,$$

where J is applied to the F^*TN-part of DF. ν is a 1-form with values in $T^\perp M$ since by the Lagrangian condition J induces a bundle isomorphism (actually even a bundle isometry) between $DF(TM)$ and $T^\perp M$. In local coordinates ν can be written as

$$\nu = \nu_i dx^i = \nu^\alpha{}_i \frac{\partial}{\partial y^\alpha} \otimes dx^i$$

with

$$\nu_i = JF_i = J^\alpha{}_\beta F^\beta{}_i \frac{\partial}{\partial y^\alpha}, \quad \nu^\alpha{}_i = J^\alpha{}_\beta F^\beta{}_i .$$

Since J is parallel, we have

$$\nabla\nu = J\nabla DF = JA .$$

In contrast to hypersurfaces, we may now define a second fundamental form as a tri-linear form

$$h(X, Y, Z) := \langle \nu(X), A(Y, Z) \rangle .$$

It turns out that h is fully symmetric. Moreover, taking a trace, we obtain a 1-form $H \in \Omega^1(M)$, called the mean curvature form,

$$H(X) := \operatorname{trace} h(X, \cdot, \cdot) .$$

In local coordinates

$$h = h_{ijk} dx^i \otimes dx^j \otimes dx^k , \quad H = H_i dx^i , \quad H_i = g^{kl} h_{ikl} .$$

The second fundamental tensor A and the mean curvature vector $\vec{H}$ can be written in the form

$$A^\alpha{}_{ij} = h_{ij}{}^k \nu^\alpha{}_k , \quad \vec{H} = H^k \nu_k .$$

Since J gives an isometry between the normal and tangent bundle of M, the equations of Gauß and Ricci coincide, so that we get the single equation

$$R_{ijkl} = R^N(F_i, F_j, F_k, F_l) + h_{ikm} h_{jl}{}^m - h_{ilm} h_{jk}{}^m .$$

Since $\nabla J = 0$ and $J^2 = -\operatorname{Id}$ we also get

$$\nabla_i \nu^\alpha{}_j = \nabla_i (J^\alpha{}_\beta F^\beta{}_{,j}) = J^\alpha{}_\beta \nabla_i F^\beta{}_{,j} = J^\alpha{}_\beta A^\beta{}_{ij} = J^\alpha{}_\beta \nu^\beta{}_k h_{ij}{}^k = -h_{ij}{}^k F^\alpha{}_k .$$

Similarly as above we conclude

$$\begin{aligned} \nabla_i h_{jkl} - \nabla_j h_{ikl} &= \nabla_i \langle A_{jk}, \nu_l \rangle - \nabla_j \langle A_{ik}, \nu_l \rangle \\ &\overset{(\nabla \nu_l \in DF(TM))}{=} \langle \nabla_i A_{jk} - \nabla_j A_{ik}, \nu_l \rangle \\ &\overset{(12)}{=} R^N(\nu_l, F_k, F_i, F_j) . \end{aligned}$$

Taking a trace over k and l, we deduce

$$\nabla_i H_j - \nabla_j H_i = R^N(\nu_k, F^k, F_i, F_j)$$

and if we take into account that N is Kähler and M Lagrangian, then the RHS is a Ricci curvature, so that the exterior derivative dH of the mean curvature form H is given by

$$(dH)_{ij} = \nabla_i H_j - \nabla_j H_i = -\operatorname{Ric}^N(\nu_i, F_j).$$

If (N, g, J) is Kähler–Einstein, then H is closed (since $\operatorname{Ric}^N(\nu_i, F_j) = c \cdot \omega(F_i, F_j) = 0$) and defines a cohomology class on M. In this case any (in general only locally defined) function α with $d\alpha = H$ is called a Lagrangian angle. In some sense the Lagrangian condition is an integrability condition. If we represent a Lagrangian submanifold locally as the graph over its tangent space, then the m "height" functions are not completely independent but are related to a common potential. An easy way to see this, is to consider a locally defined 1-form λ on M (in a neighborhood of some point of $F(M)$) with $d\lambda = \omega$. Then by the Lagrangian condition

$$0 = F^*\omega = F^*d\lambda = dF^*\lambda\,.$$

So $F^*\lambda$ is closed and by Poincaré's Lemma locally integrable. By the implicit function theorem this potential for λ is related to the height functions of M (cf. [74]). Note also that by a result of Weinstein for any Lagrangian embedding $M \subset N$ there exists a tubular neighborhood of M which is symplectomorphic to T^*M with its canonical symplectic structure $\omega = d\lambda$ induced by the Liouville form λ.

2.4.3 Graphs

Let (M, g^M), (K, g^K) be two Riemannian manifolds and $f : M \to K$ a smooth map. f induces a graph

$$F : M \to N := M \times K\,, \quad F(p) := (p, f(p))\,.$$

Since N is also a Riemannian manifold equipped with the product metric $g = g^M \times g^K$ one may consider the geometry of such graphs. It is clear that the geometry of F must be completely determined by f, g^M and g^K. Local coordinates $(x^i)_{i=1,\ldots,m}$, $(z^A)_{A=1,\ldots,k}$ for M resp. K induce local coordinates $(y^\alpha)_{\alpha=1,\ldots,n=m+k}$ on N by $y = (x, z)$. Then locally

$$F_i(x) = \frac{\partial}{\partial x^i} + f^A_{\;i}(x)\frac{\partial}{\partial z^A}\,,$$

where similarly as before $f^A = z^A \circ f \circ x^{-1}$ and $f^A_{\;i} = \frac{\partial f^A}{\partial x^i}$. For the induced metric $F^*g = g_{ij}dx^i \otimes dx^j$ we get

$$g_{ij} = g^M_{ij} + g^K_{AB} f^A_{\;i} f^B_{\;j}\,.$$

Since this is obviously positive definite and F is injective, graphs $F : M \to M \times K$ of smooth mappings $f : M \to K$ are always embeddings. From the formula for $DF = F_i dx^i$ and the Gauß formula one may then compute the second fundamental tensor $A = \nabla DF$. Since the precise formula for A is not important in this article, we leave the details as an exercise to the reader.

3 General Results in Higher Codimension

In this section we focus on results in mean curvature flow that are valid in any dimension and codimension and that do not depend on specific geometric situations.

3.1 Short-Time Existence and Uniqueness

Consider the mean curvature vector field $\overrightarrow{H} = \overrightarrow{H}[F]$ as an operator on the class of smooth immersions

$$\mathscr{I} := \{F : M \to N : F \text{ is a smooth immersion}\}.$$

We want to compute the linearized operator belonging to $\overrightarrow{H}$. To this end we need to look at the symbol and therefore we consider the locally defined expression

$$L_\beta^{\alpha;ij} := \frac{\partial H^\alpha}{\partial F_{ij}^\beta},$$

where F_{ij}^β is shorthand for $\frac{\partial^2 F^\beta}{\partial x^i \partial x^j}$ and locally $\overrightarrow{H} = H^\alpha \frac{\partial}{\partial y^\alpha}$.

Let $g_{ki,j} := \partial g_{ki}/\partial x^j$. We start with

$$\begin{aligned}\frac{\partial g_{kt,m}}{\partial F_{ij}^\beta} &= \frac{\partial}{\partial F_{ij}^\beta}\left(g_{\delta\epsilon,\rho} F^\delta{}_k F^\epsilon{}_t F^\rho{}_m + g_{\delta\epsilon}(F_{km}^\delta F^\epsilon{}_t + F^\delta{}_k F^\epsilon_{tm})\right) \\ &= g_{\beta\epsilon}\delta^j{}_m (F^\epsilon{}_t \delta^i{}_k + F^\epsilon{}_k \delta^i{}_t).\end{aligned}$$

From this we then obtain

$$\begin{aligned}\frac{\partial \Gamma_{km}^s}{\partial F_{ij}^\beta} &= \frac{1}{2} g^{st} g_{\beta\epsilon}\Big((\delta^i{}_k \delta^j{}_m + \delta^i{}_m \delta^j{}_k) F^\epsilon{}_t \\ &\quad + (\delta^i{}_t \delta^j{}_m - \delta^i{}_m \delta^j{}_t) F^\epsilon{}_k + (\delta^i{}_t \delta^j{}_k - \delta^i{}_k \delta^j{}_t) F^\epsilon{}_m\Big).\end{aligned}$$

Since by Gauß' formula

$$H^\alpha = g^{km} A^\alpha{}_{km} = g^{km}(F_{km}^\alpha - \Gamma_{km}^s F^\alpha{}_s + \Gamma_{\beta\gamma}^\alpha F^\beta{}_k F^\gamma{}_m)$$

we obtain

$$\begin{aligned}L^{\alpha;ij}_{\beta} &= g^{km}\Big(\delta^{\alpha}{}_{\beta}\delta^{i}{}_{k}\delta^{j}{}_{m} - \frac{1}{2}g^{st}g_{\beta\epsilon}\big((\delta^{i}{}_{k}\delta^{j}{}_{m} + \delta^{i}{}_{m}\delta^{j}{}_{k})F^{\epsilon}_{t}\\&\quad +(\delta^{i}{}_{t}\delta^{j}{}_{m} - \delta^{i}{}_{m}\delta^{j}{}_{t})F^{\epsilon}_{k} + (\delta^{i}{}_{t}\delta^{j}{}_{k} - \delta^{i}{}_{k}\delta^{j}{}_{t})F^{\epsilon}_{m}\big)F^{\alpha}_{s}\Big)\\&= \delta^{\alpha}{}_{\beta}g^{ij} - g^{st}g_{\beta\epsilon}g^{ij}F^{\epsilon}_{t}F^{\alpha}_{s} - (g^{kj}g^{si} - g^{ki}g^{sj})g_{\beta\epsilon}F^{\epsilon}_{k}F^{\alpha}_{s}\,.\end{aligned}$$

For an arbitrary nonzero 1-form $\xi = \xi_i dx^i$ we define the endomorphism $L = (L^{\alpha}{}_{\beta})_{\alpha,\beta=1,\dots,n}$ by

$$L^{\alpha}{}_{\beta} := L^{\alpha;ij}_{\beta}\xi_i\xi_j\,.$$

We compute

$$L^{\alpha}{}_{\beta} = (\delta^{\alpha}{}_{\beta} - g_{\beta\epsilon}g^{st}F^{\epsilon}_{t}F^{\alpha}_{s})|\xi|^2\,.$$

Applying this to a tangent vector $F_l = F^{\beta}{}_{l}\frac{\partial}{\partial y^{\beta}}$ we get

$$L^{\alpha}{}_{\beta}F^{\beta}{}_{l} = 0\,.$$

If $\nu = \nu^{\beta}\frac{\partial}{\partial y^{\beta}}$ is normal, then

$$g_{\beta\epsilon}\nu^{\beta}F^{\epsilon}_{t} = 0$$

and hence

$$L^{\alpha}{}_{\beta}\nu^{\beta} = |\xi|^2\nu^{\alpha}\,.$$

Consequently L is degenerate along tangent directions of F and elliptic along normal directions, more precisely for $\xi \in T^*_pM$ we have

$$L_{|p} = |\xi|^2\pi_{|p}\,,$$

where $\pi_{|p} : T_{F(p)}N \to T^{\perp}_{p}M$ is the projection of $T_{F(p)}N$ onto $T^{\perp}_{p}M$. The reason for the m degeneracies is the following: Writing a solution $F : M \to N$ of $\vec{H} = 0$ locally as the graph over its tangent plane at $F(p)$, we see that we need as many height functions as there are codimensions, i.e. we need $k = n-m$ functions. On the other hand the system $H^{\alpha} = 0, \alpha = 1,\dots,n$ consists of n coupled equations and is therefore overdetermined with a redundancy of m equations. These m redundant equations correspond to the diffeomorphism group of the underlying m-dimensional manifold M. This means the following:

Proposition 3.1 (Invariance under the diffeomorphism group). *If $F : M \times [0,T) \to N$ is a solution of the mean curvature flow, and $\phi \in \mathrm{Diff}(M)$ a fixed diffeomorphism of M, then $\tilde{F} : M \times [0,T) \to N$, $\tilde{F}(p,t) := F(\phi(p),t)$ is* *another solution. In particular, the (immersed) submanifolds* $\tilde{M}_t := \tilde{F}(M,t)$ *and* $M_t := F(M,t)$ *coincide for all* t.

Thus the mean curvature flow is a (degenerate) quasilinear parabolic evolution equation. The following theorem is well known and in particular forms a special case of a theorem by Richard Hamilton [42], based on the Nash–Moser implicit function theorem treated in another paper by Hamilton [41].

Proposition 3.2 (Short-time existence and uniqueness). *Let* M *be a smooth closed manifold and* $F_0 : M \to N$ *a smooth immersion into a smooth Riemannian manifold* (N, g). *Then the mean curvature flow admits a unique smooth solution on a maximal time interval* $[0, T)$, $0 < T \leq \infty$.

Besides the invariance of the equation under the diffeomorphism group of M the flow is isotropic, i.e. invariant under isometries of the ambient space. This property follows from the invariance of the first and second fundamental forms under isometries.

Proposition 3.3 (Invariance under isometries). *Suppose* $F : M \times [0, T) \to N$ *is a smooth solution of the mean curvature flow and assume that* ϕ *is an isometry of the ambient space* (N, g). *Then the family* $\tilde{F} := \phi \circ F$ *is another smooth solution of the mean curvature flow. In particular, if the initial immersion is invariant under* ϕ, *then it will stay invariant for all* $t \in [0, T)$.

We note that the short-time existence and uniqueness result stated above is not in the most general form. For example, it is not necessary to assume smoothness initially, it suffices to assume Lipschitz continuity. We note also that in general the short-time existence and uniqueness result for non-compact complete manifolds M is open but there exist important contributions in special cases. Based on interior estimates, Ecker and Huisken [30] proved – requiring only a local Lipschitz condition for the initial hypersurface –, a short-time existence result for the mean curvature flow of complete hypersurfaces. In that paper the authors also show that the mean curvature flow smoothes out Lipschitz hypersurfaces (i.e. the solution becomes smooth for $t > 0$). This short-time existence result has been improved in a paper by Colding and Minicozzi [24] where one only needs to assume a local bound for the initial height function. The smoothing out result by Ecker and Huisken has been extended by Wang to any dimension and codimension in [89] provided the submanifolds have a small local Lipschitz norm (which cannot be improved by an example of Lawson and Osserman) and the ambient space has bounded geometry. Recently Chen and Yin [14] proved that uniqueness for complete manifolds M still holds within the class of smooth solutions with bounded second fundamental tensor, if the ambient Riemannian manifold (N, g) has bounded geometry in a certain sense. Chen and Pang [19] considered uniqueness of unbounded solutions of the Lagrangian mean curvature flow equation for graphs.

3.2 Evolution Equations

Suppose $F : M \times [0,T) \to N$ is a smooth solution of the mean curvature flow

$$\frac{d}{dt}F = \vec{H}\,.$$

In this subsection we want to state and prove evolution equations of the most important geometric quantities on M, like the first and second fundamental forms.

To this end we will compute evolution equations for various sections σ in vector bundles E over M. We will use the index notation introduced in Sect. 2.3. In particular, we will consider those cases, where σ is a section in a vector bundle E_t which itself depends on time t. If for example ν_t is the principal normal vector field of a hypersurface $F : M \to N$, then ν_t is a section in $E_t := F_t^*TN$. In this case the mere computation of the total derivative of ν_t w.r.t. t will be insufficient since this would only make sense in local coordinates (local in space and time). To overcome this difficulty we just need to define a connection ∇ on F^*TN, where F is now considered as a smooth map (in general no immersion) from the space-time manifold $M \times [0,T)$ to N. A time derivative then becomes a covariant derivative in direction of $\frac{d}{dt}$, for example for a time dependent section $\nu \in F^*TN$ we have in local coordinates

$$\nu(x,t) = \nu^\alpha(x,t)\frac{\partial}{\partial y^\alpha}$$

$$\nabla_{\frac{d}{dt}}\nu = \left(\frac{d\nu^\alpha}{dt} + \Gamma^\alpha_{\beta\delta}\frac{dF^\beta}{dt}\nu^\delta\right)\frac{\partial}{\partial y^\alpha} = \left(\frac{d\nu^\alpha}{dt} + \Gamma^\alpha_{\beta\delta}H^\beta\nu^\delta\right)\frac{\partial}{\partial y^\alpha}\,,$$

where $\Gamma^\alpha_{\beta\delta}$ are the Christoffel symbols of the Levi-Civita connection on N and (y^α) are local coordinates on N. On the other hand, if σ is a section in a bundle E and E does not depend on t, then the covariant derivative $\nabla_{\frac{d}{dt}}\sigma$ coincides with $\frac{d}{dt}\sigma$. For example for the induced metric $F_t^*g \in \Gamma(T^*M \otimes T^*M)$ we have

$$F_t^*g = g_{ij}(x,t)dx^i \otimes dx^j$$

and

$$\nabla_{\frac{d}{dt}}F_t^*g = \frac{d}{dt}g_{ij}(x,t)dx^i \otimes dx^j$$

since T^*M does not depend on t. Likewise, for the second fundamental tensor A (considered as a section in $F^*TN \otimes T^*M \otimes T^*M$, which makes sense since for $\tilde{M} = M \times [0,T)$ we have $T^*\tilde{M} = T^*M \oplus T^*\mathbb{R}$) we get

$$\nabla_{\frac{d}{dt}}A^\alpha{}_{ij} = \frac{d}{dt}A^\alpha{}_{ij} + \Gamma^\alpha_{\beta\gamma}\frac{dF^\beta}{dt}A^\gamma{}_{ij} = \frac{d}{dt}A^\alpha{}_{ij} + \Gamma^\alpha_{\beta\gamma}H^\beta A^\gamma{}_{ij}\,. \tag{16}$$

Lemma 3.4. *If $F : M \times [0,T) \to (N,g)$ evolves under the mean curvature flow, then the induced Riemannian metrics $F_t^* g = g_{ij}(x,t)dx^i \otimes dx^j \in \Gamma(T^*M \otimes T^*M)$ evolve according to*

$$\nabla_{\frac{d}{dt}} g_{ij} = \frac{d}{dt} g_{ij} = -2\langle \overrightarrow{H}, A_{ij} \rangle . \tag{17}$$

Proof. We have

$$g_{ij} = g_{\alpha\beta} F^\alpha_{\ i} F^\beta_{\ j}$$

and thus

$$\begin{aligned}
\nabla_{\frac{d}{dt}} g_{ij} &= \underbrace{\nabla_\gamma g_{\alpha\beta}}_{=0} \frac{dF^\gamma}{dt} F^\alpha_{\ i} F^\beta_{\ j} + g_{\alpha\beta}\left(\nabla_{\frac{d}{dt}} F^\alpha_{\ i} F^\beta_{\ j} + F^\alpha_{\ i} \nabla_{\frac{d}{dt}} F^\beta_{\ j}\right) \\
&= g_{\alpha\beta}\left(\nabla_i \frac{dF^\alpha}{dt} F^\beta_{\ j} + F^\alpha_{\ i} \nabla_j \frac{dF^\beta}{dt}\right) \\
&= g_{\alpha\beta}\left(\nabla_i H^\alpha F^\beta_{\ j} + F^\alpha_{\ i} \nabla_j H^\beta\right), \qquad (18)
\end{aligned}$$

where we have used that $\nabla_\gamma g_{\alpha\beta} = 0$ (since ∇ is metric) and $\nabla_{\frac{d}{dt}} F^\alpha_{\ i} = \nabla_i \frac{dF^\alpha}{dt}$. This last identity holds since the second fundamental tensor $\tilde{A} \in \Gamma(F^*TN \otimes T^*\tilde{M} \otimes T^*\tilde{M})$ of the map $F : \tilde{M} \to N$ is symmetric, so that

$$\tilde{A}\left(\frac{\partial}{\partial x^i}, \frac{d}{dt}\right) = \nabla_i \frac{dF^\alpha}{dt} \frac{\partial}{\partial y^\alpha} = \nabla_{\frac{d}{dt}} F^\alpha_{\ i} \frac{\partial}{\partial y^\alpha} = \tilde{A}\left(\frac{d}{dt}, \frac{\partial}{\partial x^i}\right).$$

Now since $g_{\alpha\beta} H^\alpha F^\beta_{\ j} = 0$, we get

$$\begin{aligned}
0 &= \nabla_i (g_{\alpha\beta} H^\alpha F^\beta_{\ j}) \\
&= \nabla_\gamma g_{\alpha\beta} F^\gamma_{\ i} H^\alpha F^\beta_{\ j} + g_{\alpha\beta}(\nabla_i H^\alpha F^\beta_{\ j} + H^\alpha \nabla_i F^\beta_{\ j}) \\
&= g_{\alpha\beta}(\nabla_i H^\alpha F^\beta_{\ j} + H^\alpha A^\beta_{\ ij})
\end{aligned}$$

since $\nabla_i F^\beta_{\ j} = A^\beta_{\ ij}$. If we insert this into (18), then we obtain the result. □

Corollary 3.5. *The induced volume form $d\mu_t$ on M evolves according to*

$$\nabla_{\frac{d}{dt}} d\mu_t = \frac{d}{dt} d\mu_t = -|\overrightarrow{H}|^2 d\mu_t . \tag{19}$$

Proof. In local coordinates we have

$$d\mu_t = \sqrt{\det g_{kl}}\,dx^1 \wedge \cdots \wedge dx^m .$$

Since

$$\frac{d}{dt}\,(\det g_{kl}) = \left(g^{ij}\,\frac{d}{dt}\,g_{ij}\right)\det g_{kl}$$

the claim follows easily. □

Corollary 3.6. *The Christoffel symbols* Γ_{ij}^k *of the Levi-Civita connection on* M *evolve according to*

$$\frac{d}{dt}\,\Gamma_{ij}^k = -g^{kl}\Big(\nabla_i\langle\vec{H},A_{jl}\rangle + \nabla_j\langle\vec{H},A_{il}\rangle - \nabla_l\langle\vec{H},A_{ij}\rangle\Big)\,. \tag{20}$$

Proof. This follows directly from

$$\Gamma_{ij}^k = \frac{1}{2}g^{kl}\left(g_{il,j} + g_{jl,i} - g_{ij,l}\right),$$

the evolution equation of the metric and the fact that $\frac{d}{dt}\,\Gamma_{ij}^k$ is a tensor (though Γ_{ij}^k is not). □

Next we compute the evolution equation for the second fundamental tensor $A = A^\alpha{}_{ij}\frac{\partial}{\partial y^\alpha}\otimes dx^i\otimes dx^j$

Lemma 3.7. *The second fundamental tensor* A *evolves under the mean curvature flow by*

$$\nabla_{\frac{d}{dt}} A^\alpha{}_{ij} = \nabla_i\nabla_j H^\alpha - C_{ij}^k F^\alpha{}_k + R^\alpha{}_{\delta\gamma\epsilon}F^\delta{}_j H^\gamma F^\epsilon{}_i\,, \tag{21}$$

where $C_{ij}^k = \frac{d}{dt}\,\Gamma_{ij}^k$.

Proof. Since

$$A^\alpha{}_{ij} = \frac{\partial^2 F^\alpha}{\partial x^i\partial x^j} - \Gamma_{ij}^k F^\alpha{}_k + \Gamma_{\beta\gamma}^\alpha F^\beta{}_i F^\gamma{}_j$$

we get

$$\begin{aligned}\frac{d}{dt}\,A_{ij}^\alpha = {} & \frac{\partial^2 H^\alpha}{\partial x^i\partial x^j} - \Gamma_{ij}^k\frac{\partial H^\alpha}{\partial x^k} + \Gamma_{\beta\gamma}^\alpha\left(\frac{\partial H^\beta}{\partial x^i}F^\gamma{}_j + F^\beta{}_i\frac{\partial H^\gamma}{\partial x^j}\right)\\ & -\frac{d}{dt}\,\Gamma_{ij}^k F^\alpha{}_k + \Gamma_{\beta\gamma,\delta}^\alpha H^\delta F^\beta{}_i F^\gamma{}_j\,.\end{aligned} \tag{22}$$

To continue we need some covariant expressions. For a section $V = V^\alpha\frac{\partial}{\partial y^\alpha} \in \Gamma(F^*TN)$ we have

$$\nabla_j V^\alpha = \frac{\partial V^\alpha}{\partial x^j} + \Gamma_{\beta\gamma}^\alpha F^\beta{}_j V^\gamma$$

and then

$$
\begin{aligned}
\nabla_i\nabla_j V^\alpha &= \frac{\partial}{\partial x^i}\left(\frac{\partial V^\alpha}{\partial x^j} + \Gamma^\alpha_{\beta\gamma}F^\beta{}_j V^\gamma\right) - \Gamma^k_{ij}\left(\frac{\partial V^\alpha}{\partial x^k} + \Gamma^\alpha_{\beta\gamma}F^\beta{}_k V^\gamma\right)\\
&\quad + \Gamma^\alpha_{\beta\gamma}F^\beta{}_i\left(\frac{\partial V^\gamma}{\partial x^j} + \Gamma^\gamma_{\delta\epsilon}F^\delta{}_j V^\epsilon\right)\\
&= \frac{\partial^2 V^\alpha}{\partial x^i\partial x^j} + \Gamma^\alpha_{\beta\gamma,\delta}F^\delta{}_i F^\beta{}_j V^\gamma + \Gamma^\alpha_{\beta\gamma}\frac{\partial^2 F^\beta}{\partial x^i\partial x^j}V^\gamma + \Gamma^\alpha_{\beta\gamma}F^\beta{}_j\frac{\partial V^\gamma}{\partial x^i}\\
&\quad - \Gamma^k_{ij}\left(\frac{\partial V^\alpha}{\partial x^k} + \Gamma^\alpha_{\beta\gamma}F^\beta{}_k V^\gamma\right) + \Gamma^\alpha_{\beta\gamma}F^\beta{}_i\left(\frac{\partial V^\gamma}{\partial x^j} + \Gamma^\gamma_{\delta\epsilon}F^\delta{}_j V^\epsilon\right)\\
&= \frac{\partial^2 V^\alpha}{\partial x^i\partial x^j} - \Gamma^k_{ij}\frac{\partial V^\alpha}{\partial x^k} + \Gamma^\alpha_{\beta\gamma}\left(\frac{\partial V^\beta}{\partial x^i}F^\gamma{}_j + F^\beta{}_i\frac{\partial V^\gamma}{\partial x^j}\right)\\
&\quad + \Gamma^\alpha_{\beta\gamma}\frac{\partial^2 F^\beta}{\partial x^i\partial x^j}V^\gamma - \Gamma^k_{ij}\Gamma^\alpha_{\beta\gamma}F^\beta{}_k V^\gamma + \Gamma^\alpha_{\beta\epsilon}\Gamma^\beta_{\delta\gamma}F^\epsilon{}_i F^\delta{}_j V^\gamma\\
&\quad + \Gamma^\alpha_{\beta\gamma,\delta}F^\delta{}_i F^\beta{}_j V^\gamma\\
&= \frac{\partial^2 V^\alpha}{\partial x^i\partial x^j} - \Gamma^k_{ij}\frac{\partial V^\alpha}{\partial x^k} + \Gamma^\alpha_{\beta\gamma}\left(\frac{\partial V^\beta}{\partial x^i}F^\gamma{}_j + F^\beta{}_i\frac{\partial V^\gamma}{\partial x^j}\right)\\
&\quad + \Gamma^\alpha_{\beta\gamma}V^\gamma A^\beta{}_{ij} + \left(\Gamma^\alpha_{\beta\epsilon}\Gamma^\beta_{\delta\gamma} - \Gamma^\alpha_{\beta\gamma}\Gamma^\beta_{\delta\epsilon}\right)F^\epsilon{}_i F^\delta{}_j V^\gamma + \Gamma^\alpha_{\beta\gamma,\delta}F^\delta{}_i F^\beta{}_j V^\gamma\,,
\end{aligned}
$$

where we have used $\Gamma^\alpha_{\beta\gamma} = \Gamma^\alpha_{\gamma\beta}$ several times.

Applying this to $V^\alpha = H^\alpha$ we conclude

$$
\begin{aligned}
\frac{d}{dt}A^\alpha{}_{ij} &= \nabla_i\nabla_j H^\alpha - \frac{d}{dt}\Gamma^k_{ij}F^\alpha{}_k + \Gamma^\alpha_{\beta\gamma,\delta}H^\delta F^\beta{}_i F^\gamma{}_j\\
&\quad - \Gamma^\alpha_{\beta\gamma}H^\gamma A^\beta{}_{ij} - \left(\Gamma^\alpha_{\beta\epsilon}\Gamma^\beta_{\delta\gamma} - \Gamma^\alpha_{\beta\gamma}\Gamma^\beta_{\delta\epsilon}\right)F^\epsilon{}_i F^\delta{}_j H^\gamma - \Gamma^\alpha_{\beta\gamma,\delta}F^\delta{}_i F^\beta{}_j H^\gamma\\
&= \nabla_i\nabla_j H^\alpha - \frac{d}{dt}\Gamma^k_{ij}F^\alpha{}_k - \Gamma^\alpha_{\beta\gamma}H^\gamma A^\beta{}_{ij}\\
&\quad + \left(\Gamma^\alpha_{\epsilon\delta,\gamma} - \Gamma^\alpha_{\gamma\delta,\epsilon} - \Gamma^\alpha_{\beta\epsilon}\Gamma^\beta_{\delta\gamma} + \Gamma^\alpha_{\beta\gamma}\Gamma^\beta_{\delta\epsilon}\right)F^\epsilon{}_i F^\delta{}_j H^\gamma\\
&= \nabla_i\nabla_j H^\alpha - \frac{d}{dt}\Gamma^k_{ij}F^\alpha{}_k - \Gamma^\alpha_{\beta\gamma}H^\gamma A^\beta{}_{ij} + R^\alpha{}_{\delta\gamma\epsilon}F^\epsilon{}_i F^\delta{}_j H^\gamma\,.
\end{aligned}
$$

The result then follows from (16). □

Corollary 3.8. *Under the mean curvature flow the mean curvature satisfies the following evolution equations:*

$$
\nabla_{\frac{d}{dt}} H^\alpha = \Delta H^\alpha - g^{ij}C^k_{ij}F^\alpha{}_k + R^\alpha{}_{\delta\gamma\epsilon}F^\epsilon{}_i F^{\delta i}H^\gamma + 2\langle A_{kl}, \overrightarrow{H}\rangle A^{\alpha kl} \tag{23}
$$

$$
\begin{aligned}
\nabla_{\frac{d}{dt}} |\vec{H}|^2 &= \Delta|\vec{H}|^2 - 2|\nabla\vec{H}|^2 + 4\langle A^{ij}, \vec{H}\rangle\langle A_{ij}, \vec{H}\rangle \\
&\quad + 2R_{\alpha\beta\gamma\delta} H^\alpha F^\beta{}_i H^\gamma F^{\delta i} \qquad (24) \\
&= \Delta|\vec{H}|^2 - 2|\nabla^\perp\vec{H}|^2 + 2\langle A^{ij}, \vec{H}\rangle\langle A_{ij}, \vec{H}\rangle \\
&\quad + 2R_{\alpha\beta\gamma\delta} H^\alpha F^\beta{}_i H^\gamma F^{\delta i} \qquad (25)
\end{aligned}
$$

Proof. The first equation follows from $H^\alpha = g^{ij} A^\alpha{}_{ij}$, (17), (21) and

$$\nabla_{\frac{d}{dt}} g^{ij} = -g^{ik} g^{jl} \nabla_{\frac{d}{dt}} g_{kl} \,.$$

The second equation then follows from $|\vec{H}|^2 = g_{\alpha\beta} H^\alpha H^\beta$, $g_{\alpha\beta} F^\alpha{}_i H^\beta = 0$ and

$$\nabla_{\frac{d}{dt}} g_{\alpha\beta} = \nabla_\gamma g_{\alpha\beta} H^\gamma = 0 \,.$$

Finally, (25) follows from

$$
\begin{aligned}
\nabla_k \vec{H} &= \nabla_k^\perp \vec{H} + g^{ij} \langle \nabla_k \vec{H}, F_i\rangle F_j \\
&= \nabla_k^\perp \vec{H} - g^{ij} \langle \vec{H}, \nabla_k F_i\rangle F_j \\
&= \nabla_k^\perp \vec{H} - g^{ij} \langle \vec{H}, A_{ki}\rangle F_j
\end{aligned}
$$

and $\langle \nabla_k^\perp \vec{H}, F_j\rangle = 0$. □

From the evolution equation of $A^\alpha{}_{ij}$ we obtain in the same way

$$
\begin{aligned}
\nabla_{\frac{d}{dt}} |A|^2 &= 2\langle \nabla^2 \vec{H}, A\rangle + 4\langle \vec{H}, A^{ij}\rangle\langle A_{ik}, A_j{}^k\rangle \\
&\quad + 2R_{\alpha\beta\gamma\delta} A^{\alpha kl} F^\beta{}_k H^\gamma F^\delta{}_l . \qquad (26)
\end{aligned}
$$

Applying Simons' identity (15) we get

$$
\begin{aligned}
\nabla_{\frac{d}{dt}} |A|^2 = \Delta|A|^2 - 2|\nabla^\perp A|^2 \\
&+ \big|\langle A_{ij}, A_{kl}\rangle - \langle A_{il}, A_{jk}\rangle\big|^2 + \big|A^\alpha{}_{ik} A^\beta{}_j{}^k - A^\beta{}_{ik} A^\alpha{}_j{}^k\big|^2 \\
&+ 2\big|\langle \vec{H}, A_{ij}\rangle - \langle A_{ik}, A_j{}^k\rangle\big|^2 - 2\big|\langle \vec{H}, A_{ij}\rangle\big|^2 \\
&+ 4R_{\alpha\beta\gamma\delta} F^\alpha{}_k F^\beta{}_i F^\gamma{}_l F^\delta{}_j \left(\langle A^{ij}, A^{kl}\rangle - g^{kl}\langle A^{ip}, A_p{}^j\rangle\right) \\
&+ 2R_{\alpha\beta\gamma\delta} A^{\alpha kl} \left(4A^\beta{}_{ik} F^\gamma{}_l F^{\delta i} + F^\beta{}_l F^\gamma{}_k H^\delta + F^\beta{}_i A^\gamma{}_{lk} F^{\delta i}\right)
\end{aligned}
$$

$$
\begin{aligned}
&+2\left(\nabla_\epsilon R_{\alpha\beta\gamma\delta}+\nabla_\gamma R_{\alpha\delta\beta\epsilon}\right)F^\epsilon_{\ i}F^\beta_{\ l}F^\gamma_{\ k}F^{\delta i}A^{\alpha kl}\\
&+4\langle\vec{H},A^{ij}\rangle\langle A_{ik},A_j^{\ k}\rangle\\
&+2R_{\alpha\beta\gamma\delta}A^{\alpha kl}F^\beta_{\ k}H^\gamma F^\delta_{\ l}\\
=\ &\Delta|A|^2-2|\nabla^\perp A|^2\\
&+2\left|\langle A_{ij},A_{kl}\rangle\right|^2+\left|A^\alpha_{\ ik}A^\beta_{\ j}{}^k-A^\beta_{\ ik}A^\alpha_{\ j}{}^k\right|^2\\
&+4R_{\alpha\beta\gamma\delta}F^\alpha_{\ k}F^\beta_{\ i}F^\gamma_{\ l}F^\delta_{\ j}\left(\langle A^{ij},A^{kl}\rangle-g^{kl}\langle A^{ip},A_p^{\ j}\rangle\right)\\
&+2R_{\alpha\beta\gamma\delta}A^{\alpha kl}\left(4A^\beta_{\ ik}F^\gamma_{\ l}F^{\delta i}+F^\beta_{\ i}A^\gamma_{\ lk}F^{\delta i}\right)\\
&+2\left(\nabla_\epsilon R_{\alpha\beta\gamma\delta}+\nabla_\gamma R_{\alpha\delta\beta\epsilon}\right)F^\epsilon_{\ i}F^\beta_{\ l}F^\gamma_{\ k}F^{\delta i}A^{\alpha kl}.
\end{aligned}
$$

Thus we have shown

Corollary 3.9. *Under the mean curvature flow the quantity $|A|^2$ satisfies the following evolution equation:*

$$
\begin{aligned}
\nabla_{\frac{d}{dt}}|A|^2=\ &\Delta|A|^2-2|\nabla^\perp A|^2\\
&+2\left|\langle A_{ij},A_{kl}\rangle\right|^2+\left|A^\alpha_{\ ik}A^\beta_{\ j}{}^k-A^\beta_{\ ik}A^\alpha_{\ j}{}^k\right|^2\\
&+4R_{\alpha\beta\gamma\delta}F^\alpha_{\ k}F^\beta_{\ i}F^\gamma_{\ l}F^\delta_{\ j}\left(\langle A^{ij},A^{kl}\rangle-g^{kl}\langle A^{ip},A_p^{\ j}\rangle\right)\\
&+2R_{\alpha\beta\gamma\delta}A^{\alpha kl}\left(4A^\beta_{\ ik}F^\gamma_{\ l}F^{\delta i}+F^\beta_{\ i}A^\gamma_{\ lk}F^{\delta i}\right)\\
&+2\left(\nabla_\epsilon R_{\alpha\beta\gamma\delta}+\nabla_\gamma R_{\alpha\delta\beta\epsilon}\right)F^\epsilon_{\ i}F^\beta_{\ l}F^\gamma_{\ k}F^{\delta i}A^{\alpha kl}.
\end{aligned}
\tag{27}
$$

These general evolution equations simplify in more special geometric situations. E.g., if the codimension is one, then $A^\alpha_{\ ij}=\nu^\alpha h_{ij}$ (cf. Sect. 2.4.1) implies $|\nabla^\perp A|^2=|\nabla h|^2$, $|A|^2=|h|^2$ and

$$
\begin{aligned}
\nabla_{\frac{d}{dt}}|h|^2=\ &\Delta|h|^2-2|\nabla h|^2+2|h|^2(|h|^2+\overline{\mathrm{Ric}}(\nu,\nu))\\
&-4(h^{ij}h_j^{\ m}\bar{R}_{mli}{}^l-h^{ij}h^{lm}\bar{R}_{milj})\\
&+2h^{ij}(\bar{\nabla}_j\bar{R}_{0li}{}^l+\bar{\nabla}_l\bar{R}_{0ij}{}^l),
\end{aligned}
\tag{28}
$$

where

$$
\bar{R}_{milj}:=R_{\alpha\beta\gamma\delta}F^\alpha_{\ m}F^\beta_{\ i}F^\gamma_{\ l}F^\delta_{\ j},\quad \overline{\mathrm{Ric}}(\nu,\nu):=R_{\alpha\beta\gamma\delta}\nu^\alpha F^\beta_{\ i}\nu^\gamma F^{\delta i}
$$

and

$$\bar{\nabla}_l \bar{R}_{0ij}{}^{l} := \nabla_\alpha R_{\beta\gamma\delta\epsilon} F^\alpha{}_l \nu^\beta F^\gamma{}_i F^\delta{}_j F^{\epsilon l}\,.$$

Equation (28) is Corollary 3.5 (ii) in [49]. Note that there is a plus sign in the last line of (28) since our unit normal is inward pointing and the unit normal in [49] is outward directed.

3.3 Long-Time Existence

In general long-time existence of solutions cannot be expected as the following well-known theorem shows:

Proposition 3.10. *Suppose $F_0 : M \to \mathbb{R}^n$ is a smooth immersion of a closed m-dimensional manifold M. Then the maximal time T of existence of a smooth solution $F : M \times [0,T) \to \mathbb{R}^n$ of the mean curvature flow with initial immersion F_0 is finite.*

Proof. The proof easily follows by applying the parabolic maximum principle to the function $f := |F|^2 + 2mt$ which satisfies the evolution equation

$$\frac{d}{dt} f = \Delta f\,.$$

Hence $T \leq \frac{1}{2m} \max |F_0|^2$ and the inequality is sharp since equality is attained for round spheres centered at the origin. □

This result is no longer true for complete submanifolds since for example for entire m-dimensional graphs in $\mathbb{R}^{m+1}$ one has long-time existence (see [29]). In addition, the result can fail, if the ambient space is a Riemannian manifold since in some cases one gets long-time existence and convergence (for example in [38, 75, 76, 78, 84, 87]).

The next well known theorem holds in any case:

Proposition 3.11. *Let M be a closed manifold and $F : M \times [0,T) \to (N, g)$ a smooth solution of the mean curvature flow in a complete (compact or non-compact) Riemannian manifold (N, g). Suppose the maximal time of existence T is finite. Then*

$$\limsup_{t \to T} \max_{M_t} |A|^2 = \infty\,.$$

Here, $M_t := F(M,t)$.

Remark 3.12. The same result also holds in some other situations. For example one can easily see that under suitable assumptions on the solution one can allow N to have boundary.

Proof. The theorem is one of the "folklore" results in mean curvature flow for which a rigorous proof in all dimensions and codimensions has not been written up in detail but can be carried out in the same way as the corresponding proof for hypersurfaces.

This has been done by Huisken in [48, 50] and is again based on the maximum principle. The key observation is, that all higher derivatives $\nabla^k A$ of the second fundamental tensor are uniformly bounded, once A is uniformly bounded. This can be shown by induction and has originally been carried out for hypersurfaces using L^p-estimates in [48]. For compact hypersurfaces there exists a more direct argument involving the maximum principle applied to the evolution equations of $|A|^2$ in (27) and $|\nabla^k A|^2$. The method can be found in the proof of Proposition 2.3 in [50] and works in the same way in any codimension and in any ambient Riemannian manifold with bounded geometry. □

A corollary is

Corollary 3.13. *Let M be a closed manifold and $F : M \times [0, T) \to N$ a smooth solution of the mean curvature flow on a maximal time interval in a complete Riemannian manifold (N, g). If $\sup_{t\in[0,T)} \max_{M_t} |A| < \infty$, then $T = \infty$.*

Note that long-time existence does not automatically imply convergence. For example, consider the surface of revolution $N \subset \mathbb{R}^3$ generated by the function $f(x) = 1 + e^{-x}$. A circle γ of revolution moving by curve shortening flow on N will then exist for all $t \in [0, \infty)$ with uniformly bounded curvature but it will not converge since it tends off to infinity. Some results on the regularity of curve shortening flow in high codimension have been derived in [15].

However, in some geometries once long-time existence is established one can use the Arzela–Ascoli theorem to extract convergent subsequences.

3.4 Singularities

If a solution $F : M \times [0, T) \to N$ of the mean curvature flow exists only for finite time, then Proposition 3.11 implies the formation of a singularity. The question then arises how to understand the geometric and analytic nature of these singularities. From Proposition 3.11 we know that

$$\limsup_{t\to T} \max_{M_t} |A|^2 = \infty .$$

One possible approach to classify singularities is to distinguish them by the blow-up rate of $\max_{M_t} |A|^2$. The next definition originally appeared in [50] in the context of hypersurfaces in $\mathbb{R}^{m+1}$ but can be stated in the same way for arbitrary mean curvature flows.

Definition 3.14. Suppose $F : M \times [0, T) \to N$ is a smooth solution of the mean curvature flow with $T < \infty$ and

$$\limsup_{t\to T} \max_{M_t} |A|^2 = \infty .$$

(a) A point $q \in N$ is called a blow-up point, if there exists a point $p \in M$ such that

$$\lim_{t\to T} F(p,t) = q\,, \quad \lim_{t\to T} |A(p,t)| = \infty\,.$$

(b) One says that M develops a singularity of Type I, if there exists a constant $c > 0$ such that

$$\max_{M_t} |A|^2 \leq \frac{c}{T-t}\,, \forall t \in [0,T)\,.$$

Otherwise one calls the singularity of Type II.

So if q is a blow-up point then for $t \to T$ a singularity of Type I or Type II will form at $q \in N$ (and perhaps at other points as well).

In this context it is worth noting that the flow need not have a blow-up point in the sense of Definition 3.14, even if the second fundamental form blows up, e.g. the ambient space might have boundary or the singularity might form at spatial infinity. For this and other reasons it is appropriate to come up with more definitions. In [81], Stone introduced special and general singular points.

Definition 3.15. (a) A point $p \in M$ is called a special singular point of the mean curvature flow, as $t \to T$, if there exists a sequence of times $t_k \to T$, such that

$$\limsup_{k\to\infty} |A|(p,t_k) = \infty.$$

(b) A point $p \in M$ is called a general singular point of the mean curvature flow, as $t \to T$, if there exists a sequence of times $t_k \to T$ and a sequence of points $p_k \to p$, such that

$$\limsup_{k\to\infty} |A|(p_k,t_k) = \infty.$$

The reason to introduce the blow-up rate in Definition 3.14 is that for closed submanifolds in euclidean space one always has an analog inequality in the other direction, i.e.

$$\max_{M_t} |A|^2 \geq \frac{\tilde{c}}{T-t} \tag{29}$$

for some positive number $\tilde{c}$ (note that this does not necessarily hold, if the ambient space N differs from $\mathbb{R}^n$). So in some sense singularities of Type I have the best controlled blow-up rate of $|A|^2$. Because of (29) one may actually refine the definition of special and general singular points for the mean curvature flow in $\mathbb{R}^n$, as was originally done by Stone in [81]. Instead of requiring $\limsup_{k\to\infty} |A|(p_k,t_k) = \infty$ one can define a general singular point $p \in M$ such that there exists some $\delta > 0$ and a sequence $(p_k,t_k) \to (p,T)$ with

$$|A|^2(p_k,t_k) \geq \frac{\delta}{T-t_k}.$$

A sequence (p_k, t_k) with this property is called an essential blow-up sequence. Although (29) gives a minimum blow-up rate for $\max_{p\in M}|A|^2(p,t)$ in the euclidean space, as t approaches T, this does not rule out the possibility that, while $|A|^2(p,t) \geq \frac{\delta}{T-t}$ in some part of M, the blow-up of $|A|^2$ might simultaneously occur at some slower rate (say like $(T-t)^{-\alpha}, \alpha \in (0,1)$) somewhere else. Such "slowly forming singularities" would not be detected by a Type I blow-up procedure (see below) since the rescaling would be too fast. It is therefore interesting to understand, if this phenomenon occurs at all. As was recently shown by Le and Sesum [58] this does not happen in the case of Type I singularities of hypersurfaces in $\mathbb{R}^{m+1}$ and all notions of singular sets defined in [81] coincide. In particular they prove that the blow-up rate of the mean curvature must coincide with the blow-up rate of the second fundamental form, if a singularity of Type I is forming. We also mention that there exist many similarities between the formation of singularities in mean curvature flow and Ricci flow (see [32] for a nice overview on Type I singularities in Ricci flow).

Type I: Let us now assume that $q \in \mathbb{R}^n$ is a blow-up point of Type I of $F : M \times [0,T) \to \mathbb{R}^n$ and that $\dim M = m$. Huisken introduced the following rescaling technique in [50] for hypersurfaces, but obviously it can be done in the same way for any codimension in $\mathbb{R}^n$: Define an immersion $\tilde{F} : M \times [-1/2 \log T, \infty) \to \mathbb{R}^n$ by

$$\tilde{F}(\cdot, s) := (2(T-t))^{-1/2}(F(\cdot,t) - q)\,, \quad s(t) = -\frac{1}{2}\log(T-t)\,.$$

One can then compute that $\tilde{F}$ satisfies the rescaled flow equation

$$\frac{d}{ds}\tilde{F} = \vec{\tilde{H}} + \tilde{F}\,.$$

Since by assumption $|A|^2 \leq c/(T-t)$ the second fundamental tensor $\tilde{A}$ of the rescaling is uniformly bounded in space and time. To study the geometric and analytic behavior of the rescaled immersions $\tilde{M}_s = \tilde{F}(M,s)$, Huisken proved a monotonicity formula for hypersurfaces in $\mathbb{R}^n$ moving by mean curvature. The corresponding result in arbitrary dimension and codimension is as follows: For $t_0 \in \mathbb{R}$ let

$$\rho : \mathbb{R}^n \times \mathbb{R} \setminus \{t_0\} := \frac{1}{(4\pi(t_0-t))^{\frac{m}{2}}} e^{-\frac{|y|^2}{4(t_0-t)}}\,.$$

Then $\rho_{|\mathbb{R}^m\times\mathbb{R}\setminus\{t_0\}}$ is the backward heat kernel of $\mathbb{R}^m$ at $(0,t_0)$ and the following monotonicity formula holds

Proposition 3.16 (Monotonicity formula (cf. Huisken [50])). *Let $F : M \times [0,T) \to \mathbb{R}^n$ be a smooth solution of the mean curvature flow and let M be closed and m-dimensional. Then*

$$\frac{d}{dt} \int_M \rho(F(p,t),t)d\mu(p,t)$$

$$= -\int_M \left| \vec{H}(p,t) + \frac{F^\perp(p,t)}{2(t_0-t)} \right|^2 \rho(F(p,t),t)d\mu(p,t),$$

where $d\mu(\cdot,t)$ denotes the volume element on M induced by the immersion $F(\cdot,t)$ and $F^\perp$ denotes the normal part of the position vector F.

The proof is a simple consequence of

$$\frac{d}{dt}\rho = \left(\frac{m}{2(t_0-t)} - \frac{|F|^2}{4(t_0-t)^2} - \frac{\langle F, \vec{H} \rangle}{2(t_0-t)} \right) \rho$$

and

$$\Delta \rho = \left(-\frac{m}{2(t_0-t)} + \frac{|F^\top|^2}{4(t_0-t)^2} - \frac{\langle F, \vec{H} \rangle}{2(t_0-t)} \right) \rho$$

so that by the divergence theorem and from $\frac{d}{dt} d\mu = -|\vec{H}|^2 d\mu$ we get

$$\frac{d}{dt} \int_M \rho d\mu = \int_M (\frac{d}{dt}\rho + \Delta\rho - |\vec{H}|^2\rho) d\mu = -\int_M \left| \vec{H} + \frac{F^\perp}{2(t_0-t)} \right|^2 \rho d\mu .$$

Though the proof is easy, it is not obvious to look at the backward heat kernel when studying the mean curvature flow. This nice formula was used by Huisken to study the asymptotic behavior of the Type I blow-up and he proved the following beautiful theorem for hypersurfaces which again holds in arbitrary codimension

Proposition 3.17 (Type I blow-up (cf. Huisken [50])). *Suppose $F : M \times [0,T) \to \mathbb{R}^n$ is a smooth solution of the mean curvature flow of a closed m-dimensional smooth manifold M. Further assume that $T < \infty$ is finite and that $0 \in \mathbb{R}^n$ is a Type I blow-up point as $t \to T$. Then for any sequence s_j there is a subsequence s_{j_k} such that the rescaled immersed submanifolds $\tilde{M}_{s_{j_k}}$ converge smoothly to an immersed nonempty limiting submanifold $\tilde{M}_\infty$. Any such limit satisfies the equation*

$$\vec{\tilde{H}} + \tilde{F}^\perp = 0 . \tag{30}$$

Note that by Proposition 3.3 it is no restriction to assume that the blow-up point coincides with the origin. In general the limiting submanifold $\tilde{M}_\infty$ need not have the same topology as M, for example compactness might no longer hold. In addition it is unclear, if all solutions of (30) occur as blow-up limits of Type I singularities of compact submanifolds.

A solution of (30) is called a self-similar shrinking solution (or self-shrinker for short) of the mean curvature flow. Namely, one easily proves that a solution of (30) shrinks homothetically under the mean curvature flow and that there is a smooth positive function c explicitly computable from the initial data and depending on the rescaled time s such that

$$\vec{\tilde{H}}_s + c(s)\tilde{F}_s^{\perp} = 0\,.$$

There exists another interesting class of self-similar solutions of the mean curvature flow. These are characterized by the elliptic equation

$$\vec{H} - F^{\perp} = 0 \tag{31}$$

and are called self-expanders. In [29] Ecker and Huisken proved that entire graphs in $\mathbb{R}^{m+1}$ (in codimension 1) approach asymptotically expanding self-similar solutions if they satisfy a certain growth condition at infinity. Later Stavrou [80] proved the same result under the weaker assumption that the graph has bounded gradient and a unique cone at infinity. Furthermore, he gave a characterization of expanding self-similar solutions to mean curvature flow with bounded gradient.

A classification of self-shrinking or self-expanding solutions is far from being complete. However there are some special situations for which one can say something. Self-shrinking curves have been completely classified by Abresch and Langer in [1]. Though their proof has been carried out for the curve shortening flow in $\mathbb{R}^2$ the result also applies to arbitrary codimension since (30) becomes an ODE for $m = 1$ and the solutions are uniquely determined by their position and velocity vectors so that all 1-dimensional solutions of (30) must be planar. For hypersurfaces there exists a beautiful theorem by Huisken in [51] that describes all self-shrinking hypersurfaces with nonnegative (scalar) mean curvature. Later this result could be generalized by the author in the following sense

Proposition 3.18 ([77]). *For a closed immersion $M^m \subset \mathbb{R}^n$, $m \geq 2$ are equivalent:*

(a) M is a self-shrinker of the mean curvature flow with nowhere vanishing mean curvature vector $\vec{H}$ and the principal normal vector $\nu := \vec{H}/|\vec{H}|$ is parallel in the normal bundle.
(b) M is a minimal immersion in a round sphere.

In the same paper one finds a similar description for the non-compact case.

Type I singularities usually occur when there exists some kind of pinching of the second fundamental form and such situations occur quite often (cf. Sect. 4). It is therefore surprising that there are situations, where one can exclude Type I singularities at all. In [74, Theorem 2.3.5] it was shown that there do not exist any compact Lagrangian solutions of (30) with trivial Maslov class $m_1 = [H/\pi] = 0$. Wang [86] and Chen and Li [17] observed that finite time Type I singularities of the Lagrangian mean curvature flow of closed Lagrangian submanifolds can be excluded, if the initial Lagrangian is almost calibrated in the sense that $*\,\mathrm{Re}(dz_{|M}) > 0$. The condition to be almost calibrated is equivalent to the assumption that the Maslov class is trivial and that the Lagrangian angle α satisfies $\cos\alpha > 0$. The difference of the results of Wang, Chen and Li in [17, 86] w.r.t. the result in [74] is, that the blow-up need not be compact any more. Later Neves [64] extended this result to the case of zero Maslov class, i.e. to the case where a globally defined Lagrangian angle α exists on M, thus removing the almost calibrated condition. In [39, Theorem 1.9] we proved a classification result for Lagrangian self-shrinkers and expanders in case they are entire graphs with a growth condition at infinity. In these cases Lagrangian self-similar solutions must be minimal Lagrangian cones.

Therefore when we study the Lagrangian mean curvature flow of closed Lagrangian submanifolds with trivial Maslov class we need to consider singularities of Type II only.

Type II: To study the shape of the submanifold near a singularity of Type II one can define a different family of rescaled flows. Following an idea of Hamilton [45] one can choose a sequence (p_k, t_k) as follows: For any integer $k \geq 1$ let $t_k \in [0, T - 1/k]$, $p_k \in M$ be such that

$$|A(p_k, t_k)|^2 (T - \frac{1}{k} - t_k) = \max_{\substack{t \leq T - 1/k \\ p \in M}} |A(p,t)|^2 (T - \frac{1}{k} - t)\,.$$

Furthermore one chooses

$$L_k = |A(p_k, t_k)|\,, \quad \alpha_k = -L_k^2 t_k\,, \quad \omega_k = L_k^2 (T - t_k - 1/k)\,.$$

If the singularity is of Type II then one has

$$t_k \to T\,, \quad L_k \to \infty\,, \quad \alpha_k \to -\infty\,, \quad \omega_k \to \infty\,.$$

Instead of $|A|$ one may use other quantities in the definition of these sequences, if it's known that they blow-up with a certain rate as $t \to T$. For example, in [53] the mean curvature H was used in the case of mean convex hypersurfaces in $\mathbb{R}^{m+1}$.

Then one can consider the following rescaling: For any $k \geq 1$, let $M_{k,\tau}$ be the family of submanifolds defined by the immersions

$$F_k(\cdot, \tau) := L_k(F(\cdot, L_k^{-2}\tau + t_k) - F(p_k, t_k))\,, \quad \tau \in [\alpha_k, \omega_k]\,.$$

The proper choice of the blow-up quantity ($|A|$, H or similar) in the definition of the rescaling will be essential to describe its behavior. Besides this rescaling technique there exist other methods to rescale singularities and the proper choice of the rescaling procedure depends on the particular situation in which the flow is considered. A nice reference for some of the scaling techniques is [28].

If M is compact and develops a Type II singularity then a subsequence of the flows $M_{k,\tau}$ converges smoothly to an eternal mean curvature flow $\tilde{M}_\tau$ defined for all $\tau \in \mathbb{R}$. Then a classification of Type II singularities depends on the classification of eternal solutions of the mean curvature flow.

In $\mathbb{R}^2$ the only convex eternal solution (up to scaling) of the mean curvature flow is given by the "grim reaper"

$$y = -\log\cos x/\pi\,.$$

The grim reaper is a translating soliton of the mean curvature flow, i.e. it satisfies the geometric PDE

$$\vec{H} = V^{\perp}$$

for some fixed vector $V \in \mathbb{R}^n$. A translating soliton moves with constant speed in direction of V.

In [8] the authors constructed some particular solutions of the mean curvature flow that develop Type II singularities. In $\mathbb{R}^2$ examples of curves that develop a Type II singularity are given by some cardioids [7]. Using a Harnack inequality, Hamilton [46] proved that any eternal convex solution of the mean curvature flow of hypersurfaces in $\mathbb{R}^{m+1}$ must be a translating soliton, if it assumes its maximal curvature at some point in space-time. In [20] the authors study whether such convex translating solutions are rotationally symmetric, and if every 2-dimensional rotationally symmetric translating soliton is strictly convex.

Various different notions of weak solution have been developed to extend the flow beyond the singular time T, including the geometric measure theoretic solutions of Brakke [10] and the level set solutions of Chen et al. [21] and Evans and Spruck [33], which were subsequently studied further by Ilmanen [55]. In [54] Huisken and Sinestrari define such a notion based on a surgery procedure.

4 Special Results in Higher Codimension

In this chapter we mention the most important results in mean curvature flow that depend on more specific geometric situations and we will focus on results in higher codimension, especially on graphs and results in Lagrangian mean curvature flow.

4.1 Preserved Classes of Immersions

Definition 4.1. Let $\mathscr{I}$ be the class of smooth m-dimensional immersions into a Riemannian manifold (N, g) and suppose $\mathscr{F} \subset \mathscr{I}$ is a subclass. We say that $\mathscr{F}$ is a preserved class under the mean curvature flow, if for any solution $F_t : M \to N$, $t \in [0, T)$ of the mean curvature flow with $(F_0 : M \to N) \in \mathscr{F}$ we also have $(F_t : M \to N) \in \mathscr{F}$ for all $t \in [0, T)$.

Preserved classes of the mean curvature flow are very important since one can often prove special results within these classes. Many classes can be expressed in terms of algebraic properties of the second fundamental form and in general it is a hard problem to detect those classes. We give a number of examples

Example 4.2. (a) $\mathscr{F}_1 :=$ {Convex hypersurfaces in $\mathbb{R}^{m+1}$}
(b) $\mathscr{F}_2 :=$ {Mean convex hypersurfaces in $\mathbb{R}^{m+1}$, i.e. $H > 0$}
(c) $\mathscr{F}_3 :=$ {Embedded hypersurfaces in Riemannian manifolds}
(d) $\mathscr{F}_4 :=$ {Hypersurfaces in $\mathbb{R}^{m+1}$ as entire graphs over a flat plane}
(e) $\mathscr{F}_5 :=$ {Lagrangian immersions in Kähler–Einstein manifolds}

To prove that classes are preserved one often uses the parabolic maximum principle (at least in the compact case). Besides the classical maximum principle for scalar quantities there exists an important maximum principle for bilinear forms due to Richard Hamilton that was originally proven in [42] and improved in [43].

Another very important property is the pinching property of certain classes of immersions in $\mathbb{R}^n$.

Definition 4.3. Let $F : M \to \mathbb{R}^n$ be a (smooth) immersion. We say that the second fundamental form A of F is δ-pinched, if the inequality

$$|A|^2 \leq \delta|\vec{H}|^2$$

holds everywhere on M.

From

$$0 \leq \left| A - \frac{1}{m}\vec{H} \otimes F^*g \right|^2 = |A|^2 - \frac{1}{m}|\vec{H}|^2$$

with $m = \dim M$ we immediately obtain that δ is bounded from below by $1/m$.

For hypersurfaces in $\mathbb{R}^{m+1}$ it is known:

Proposition 4.4. *Let $\delta \geq 1/m$. The class of closed δ-pinched hypersurfaces in $\mathbb{R}^{m+1}$ is a preserved class under the mean curvature flow.*

Proof. This easily follows from the maximum principle and the evolution equation for $f := |A|^2/H^2$. □

It can be shown that an m-dimensional submanifold in $\mathbb{R}^n$ is $1/m$-pinched, if and only if it is either a part of a round sphere or a flat subspace. Therefore closed pinched submanifolds are in some sense close to spheres. In some cases this pinching can improve under the mean curvature flow. To explain this in more detail, we make the following definition: Let $\mathscr{F}$ be a nonempty class of smooth m-dimensional immersions $F : M \to \mathbb{R}^n$, where M is not necessarily fixed, and set

$$\delta_{\mathscr{F}} := \sup\{\delta \in \mathbb{R} : |A_F(p)|^2 \geq \delta|\vec{H}_F(p)|^2, \forall p \in M, \forall (F : M \to \mathbb{R}^n) \in \mathscr{F}\},$$

where A_F and $\overrightarrow{H}_F$ denote the second fundamental form and mean curvature vector of the immersion $F : M \to \mathbb{R}^n$. Then $\delta_{\mathscr{F}} \geq \frac{1}{m}$ and $\delta_{\mathscr{F}}$ is finite, if and only if $\mathscr{F}$ contains an immersion $F : M \to \mathbb{R}^n$ for which $\overrightarrow{H}_F$ does not vanish completely.

Definition 4.5. Let $\mathscr{F}$ be a preserved class of smooth m-dimensional immersions with $\delta_{\mathscr{F}} < \infty$ and suppose δ is some real number with $\delta > \delta_{\mathscr{F}}$. We say that $\mathscr{F}$ is δ-pinchable, if for any ϵ with $0 \leq \epsilon < \delta - \delta_{\mathscr{F}}$ the class

$$\mathscr{F}_\epsilon := \{(F : M \to \mathbb{R}^n) \in \mathscr{F} : |A_F(p)|^2 \leq (\delta_{\mathscr{F}} + \epsilon)|\overrightarrow{H}_F(p)|^2 \,, \forall p \in M\}$$

is a preserved class under the mean curvature flow.

Example 4.6. (a) It follows from Theorem 4.4 that the class $\mathscr{F}(m, m+1)$ of smooth m-dimensional closed immersions into $\mathbb{R}^{m+1}$ is δ-pinchable for any $\delta \geq 1/m = \delta_{\mathscr{F}(m,m+1)}$ and that the pinching constant $\delta_{\mathscr{F}(m,m+1)}$ is attained if and only if the immersion $F : M \to \mathbb{R}^n$ is a round sphere or a flat plane (or part of).
(b) A beautiful result recently obtained by Andrews and Baker [6] shows that the class $\mathscr{F}(m, m+k)$ of smooth m-dimensional closed immersions into $\mathbb{R}^{m+k}$ is δ-pinchable with $\delta = 1/(m-1)$, if $m \geq 4$ and with $\delta = 4/3m$ for $2 \leq m \leq 4$. Here $\delta_{\mathscr{F}(m,m+k)} = 1/m$. They prove that δ-pinched immersions contract to round points. Thus for such immersions one has $M = S^m$ and they are smoothly homotopic to hyperspheres.

We will now show that the class $\mathscr{L}(m)$ of smooth closed Lagrangian immersions into $\mathbb{C}^m$ is not δ-pinchable for any δ.

Theorem 4.7. *Let $\mathscr{L}(m)$ be the class of smooth closed Lagrangian immersions into $\mathbb{C}^m$, $m > 1$. Then $\delta_{\mathscr{L}(m)} = 3/(m+2)$ and $\mathscr{L}(m)$ is not δ-pinchable for any δ.*

Proof. Given a Lagrangian immersion $F : M \to \mathbb{C}^m$ we have

$$0 \leq \left| h_{ijk} - \frac{1}{m+2}(H_i g_{jk} + H_j g_{ki} + H_k g_{ij}) \right|^2 = |A|^2 - \frac{3}{m+2}|\overrightarrow{H}|^2 \,,$$

where $H_i dx^i$ is the mean curvature form. This implies $\delta_{\mathscr{L}(m)} \geq \frac{3}{m+2}$. On the other hand equality is attained for flat Lagrangian planes and for the Whitney spheres. These are given by restricting the immersions

$$\tilde{F}_r : \mathbb{R}^{m+1} \to \mathbb{C}^m \,, \quad \tilde{F}_r(x^1, \ldots, x^{m+1}) := \frac{r(1 + i x^{m+1})}{1 + (x^{m+1})^2}(x^1, \ldots, x^m), \quad r > 0$$

to $S^m \subset \mathbb{R}^{m+1}$, i.e. $F_r := \tilde{F}_{r|S^m} : S^m \to \mathbb{C}^m$ is a Lagrangian immersion of the sphere with $|A|^2 = \frac{3}{m+2}|\overrightarrow{H}|^2$. The number r is called the radius of the Whitney sphere. This shows $\delta_{\mathscr{L}(M)} = \frac{3}{m+2}$. It has been shown by Ros and Urbano in [69] that Whitney spheres and flat Lagrangian planes are the only Lagrangian submanifolds in $\mathbb{C}^m$, $m > 1$, for which $|A|^2 = \frac{3}{m+2}|\vec{H}|^2$. Now if $\mathscr{L}(M)$ would be δ-pinchable for some δ, then in particular the Lagrangian mean curvature flow would preserve the identity $|A|^2 = \frac{3}{m+2}|\vec{H}|^2$. This is certainly true for the flat planes but for the Whitney sphere this cannot be true. Because the result of Ros and Urbano implies that under the assumption of δ-pinchability a Whitney sphere would then stay a Whitney sphere under the Lagrangian mean curvature flow and the radius of the spheres would decrease. In other words, the Whitney sphere would have to be a self-similar shrinking solution of the Lagrangian mean curvature flow. This is a contradiction to the well-known result (first shown in [74, Corollary 2.3.6]), that there are no self-shrinking Lagrangian spheres in $\mathbb{C}^m$, if $m > 1$. □

4.2 Lagrangian Mean Curvature Flow

In this subsection we will assume that $F : M \to N$ is a closed smooth Lagrangian immersion into a Kähler manifold (N, g, J). It has been shown in [72] that the Lagrangian condition is preserved, if the ambient Kähler manifold is Einstein. This includes the important case of Calabi–Yau manifolds, i.e. of Ricci flat Kähler manifolds. Recently a generalized Lagrangian mean curvature flow in almost Kähler manifolds with Einstein connections has been defined by Wang and the author in [79]. This generalizes an earlier result by Behrndt [9]. The Einstein condition is relevant in view of the Codazzi equation which implies that the mean curvature form is closed, a necessary condition to guarantee that the deformation is Lagrangian. To explain this in more detail, observe that the symplectic form ω induces an isomorphism between the space of smooth normal vector fields along M, and the space of smooth 1-forms on M. Namely, given $\theta \in \Omega^1(M)$ there exists a unique normal vector field $V \in \Gamma(T^\perp M)$ with $\theta = \omega(\cdot, V)$. If $F : M \times [0, T) \to N$ is a smooth family of Lagrangian immersions evolving in normal direction driven by some smooth time depending 1-forms $\theta \in \Omega^1(M)$ we have

$$0 = \frac{d}{dt} F^*\omega = d(\omega(\frac{d}{dt} F, \cdot)) = -d\theta$$

and consequently θ must be closed. Since the mean curvature form is given by

$$H = \omega(\cdot, \vec{H})$$

we obtain that the closeness of H is necessary to guarantee that the mean curvature flow preserves the Lagrangian condition, and it is indeed sufficient ([72, 74]). In the non-compact case this is open in general, but in some cases (like graphs over complete Lagrangian submanifolds with bounded geometry) this can be reduced to the existence problem of solutions to a parabolic equation of Monge–Ampère type. The Lagrangian condition can be interpreted as an integrability condition. For example, if M is a graph in $\mathbb{C}^m = \mathbb{R}^m \oplus i\mathbb{R}^m$ over the real part, i.e. if M is the image of some embedding

$$F : \mathbb{R}^m \to \mathbb{C}^m , \quad F(x) = x + iy(x) ,$$

where $y = y_i dx^i$ is a smooth 1-form on $\mathbb{R}^m$, then M is Lagrangian if and only if y is closed. Consequently there exists a smooth function u (called a generating function) such that $y = du$. Assuming that M evolves under the mean curvature flow and that all subsequent graphs M_t are still Lagrangian one can integrate the evolution equation of $y = du$ and obtains a parabolic evolution equation of Monge–Ampère type for u. Conversely, given a solution u of this parabolic Monge–Ampère type equation on $\mathbb{R}^m$ one can generate Lagrangian graphs $F = (x, du)$ and it can be shown that these graphs move under the mean curvature flow (cf. [74]). The same principle works in a much more general context, namely if the initial Lagrangian submanifold lies in some Kähler–Einstein manifold and the Lagrangian has bounded geometry. The boundedness of the geometry is essential for the proof since this allows to exploit the implicit function theorem to obtain the existence of a Monge–Ampère type equation similar as above.

This integrability property has one important consequence. In general, given a second order parabolic equation, one would need uniform $C^{2,\alpha}$-bounds of the solution in space and uniform $C^{1,\alpha}$-estimates in time to ensure long-time existence, as follows from Schauder theory. For the mean curvature flow these estimates are already induced by a uniform estimate of the second fundamental form A (see Corollary 3.13), so essentially by C^2-estimates. In the Lagrangian mean curvature flow $F : M \times [0, T) \to N$ one may instead use the parabolic equation of Monge–Ampère type for the generating function u and consequently one just needs $C^{1,\alpha}$-estimates in space and $C^{0,\alpha}$ estimates in time for F which itself is of first order in u. In some situations this principle has been used successfully, for example in [76, 78]. There it was shown that Lagrangian tori $M = T^m$ in flat tori $N = T^{2m}$ converge to flat Lagrangian tori, if the universal cover possesses a convex generating function u. We also mention a recent generalization to the complete case by Chau et al. [13].

The evolution equations for the Lagrangian mean curvature flow have been derived in [74] (see also [72]) and can also be obtained directly from our general evolution equations stated in Sect. 3.2. Besides the evolution equation for the induced metric the equation for the mean curvature form $H = H_i dx^i$ is perhaps the most important and is given by

$$\nabla_{\frac{d}{dt}} H = dd^\dagger H + \frac{S}{2m} H , \tag{32}$$

where S denotes the scalar curvature of the ambient Kähler–Einstein manifold, m is the dimension of the Lagrangian immersion and $d^\dagger H = \nabla^i H_i$. In particular it follows that the cohomology class $[He^{-\frac{S}{2m}t}]$ is invariant under the Lagrangian mean curvature flow and in a Calabi–Yau manifold the Lagrangian immersions with trivial first Maslov class m_1 (we have $m_1 = \frac{1}{\pi}[H]$) form a preserved class. This also shows that if the scalar curvature S is nonnegative, then a necessary condition to have long-time existence and smooth convergence of the Lagrangian mean curvature flow to a minimal Lagrangian immersion is that the initial mean curvature form is exact. Exactness of the mean curvature form will then be preserved and a globally defined Lagrangian angle α with $d\alpha = H$ exists for all t. This last result also holds for general scalar curvature S and after choosing a proper gauge for α one can prove [73, Lemma 2.4] that α satisfies the evolution equation

$$\frac{d}{dt}\alpha = \Delta\alpha + \frac{S}{2m}\alpha\,. \tag{33}$$

It is then a simple consequence of the maximum principle that on compact Lagrangian submanifolds M with trivial Maslov class in a Calabi–Yau manifold there exist uniform upper and lower bounds for the Lagrangian angle given by its initial maximum resp. minimum. In particular, the condition to be almost calibrated, i.e. $*\operatorname{Re}(dz_{|M}) = \cos\alpha > 0$ is preserved. Here dz denotes the complex volume form on the Calabi–Yau manifold and it is well known that the Lagrangian angle α satisfies

$$dz_{|M} = e^{i\alpha}d\mu\,,$$

where $d\mu$ is the volume form on M. Almost calibrated Lagrangian submanifolds in Calabi–Yau manifolds have some nice properties under the mean curvature flow. As was mentioned earlier, from the results in [17, 64, 74, 86] we know that singularities of the Lagrangian mean curvature flow of compact Lagrangian immersions with trivial Maslov class in Calabi–Yau manifolds cannot be of Type I and therefore a big class of singularities is excluded. So far one cannot say much about singularities of Type II and in particular, one does not know if they occur at all in the case of compact almost calibrated Lagrangians (though some authors have some rather heuristic arguments for the existence of such singularities). It is worth noting that there do not exist any compact almost calibrated Lagrangian immersions in $\mathbb{R}^{2m}$ (but in $\mathbb{T}^{2m}$ they exist). In [75, Theorem 1.3] it was shown that there exists a uniform (in time) lower bound for the volume of a compact almost calibrated Lagrangian evolving by its mean curvature in a Calabi–Yau (and more generally in a Kähler–Einstein manifold of non-positive scalar curvature).

An interesting class of Lagrangian immersions is given by monotone Lagrangians. A Lagrangian immersion $F: M \to \mathbb{R}^{2m}$ is called monotone, if

$$[H] = \epsilon[F^*\lambda]\,, \tag{34}$$

for some positive constant ϵ (called monotonicity constant). Here λ is the Liouville form on $\mathbb{R}^{2m} = T\mathbb{R}^m$. In [39] we proved several theorems concerning monotone Lagrangian immersions. ¿From the evolution equations of H and $F^*\lambda$ one derives that monotonicity is preserved with a time dependent monotonicity constant $\epsilon(t)$.

Gromov [40] proved that given an embedded Lagrangian submanifold M in $\mathbb{R}^{2m}$ there exists a holomorphic disk with boundary on M. On the other hand, from the evolution equations of H and $F^*\lambda$ we get that the area of holomorphic disks with boundary representing some fixed homology class in M is shrinking linearly in time. If the Lagrangian is monotone, then the shrinking rate for the area of holomorphic disks is the same for all homology classes.

Unfortunately it is unknown, if embeddedness of Lagrangian submanifolds is preserved under mean curvature flow (in general, embeddedness in higher codimension is not preserved but self-intersection numbers might be). Suppose $F : M \times [0, T) \to \mathbb{R}^{2m}$ is a Lagrangian mean curvature flow of a compact monotone Lagrangian with initial monotonicity constant $\epsilon > 0$ and suppose $0 < T_e \le T$ is the embedding time, i.e. the maximal time such that $F_t : M \to \mathbb{R}^{2m}$ is an embedding for all $0 \le t < T_e$. Then we proved [39, Theorem 1.6 and Theorem 1.11] that $T_e \le \frac{1}{\epsilon}$. Moreover

$$T = \frac{1}{\epsilon},$$

in case $T_e = T$ and if M develops a Type I singularity as $t \to T$. We note that this result is rather unique in mean curvature flow. Usually it is not possible to explicitly determine the span of life of a solution and to determine it in terms of its initial data. In the same paper we also proved the existence of compact embedded monotone Lagrangian submanifolds (even with some additional symmetry) that develop Type II singularities and consequently it is not true that monotone embedded Lagrangian submanifolds must develop Type I singularities, as was conjectured earlier by some people.

Lagrangian submanifolds appear naturally in another context. If

$$f : M \to K$$

is a symplectomorphism between two symplectic manifolds (M, ω^M), (K, ω^K) then the graph

$$F : M \to M \times K, \quad F(p) = (p, f(p))$$

is a Lagrangian embedding in $(M \times K, (\omega^M, -\omega^K))$.

If (M, ω^M, J^M, g^M) and (K, ω^K, J^K, g^K) are both Kähler–Einstein, then the product manifold is Kähler–Einstein as well and one can use the Lagrangian mean curvature flow to deform a symplectomorphism. In [75] symplectomorphisms between Riemann surfaces of the same constant curvature S have been studied and it was shown (Lemmas 10 and 14) that Lagrangian graphs that come from symplectomorphisms stay graphs for all time. The same result was obtained independently by Wang in [85] (the quantities r in [75, Lemma 10] and η in [85, Proposition 2.1] are the same up to some positive constant). In [75] the graphical condition was then used in the case of non-positive curvature S and under the angle condition $\cos\alpha > 0$ (almost calibrated) to derive explicit bounds for the second fundamental form and to establish long-time existence and smooth convergence to a minimal Lagrangian surface. Wang used the graphical condition in [85] to obtain long-time existence without a sign condition on S by methods related to White's regularity theorem and then proved convergence of subsequences to minimal Lagrangian surfaces. Later he refined his result and proved smooth convergence in [91]. In a recent paper by Medos and Wang [62] it is shown that symplectomorphisms of $\mathbb{CP}^m$ for which the singular values satisfy some pinching condition can be smoothly deformed into a biholomorphic isometry of $\mathbb{CP}^m$.

In a joint paper [78] (see also [76]) Wang and the author studied Lagrangian graphs in the cotangent bundle of a flat torus and proved that Lagrangian tori with a convex generating function converge smoothly to a flat Lagrangian torus. In this case the convexity of the generating function u implies that the Monge–Ampère type operator that appears in the evolution equation of u becomes concave and then results of Krylov [57] imply uniform $C^{2,\alpha}$-estimates in space and $C^{1,\alpha}$-estimates in time and long-time existence and convergence follows. A similar result holds for non-compact graphs [13].

4.3 Mean Curvature Flow of Graphs

As the results mentioned at the end of the last subsection show, mean curvature flow of graphs behaves much "nicer" than in the general case. There are many results for graphs moving under mean curvature flow. The first result in this direction was the paper by Ecker and Huisken [29] where long-time existence of entire graphs in $\mathbb{R}^{m+1}$ (hypersurfaces) was shown. Convergence to flat subspaces follows, if the growth rate at infinity is linear. Under a different growth rate they prove that the hypersurfaces converge asymptotically to entire self-expanding solutions of the mean curvature flow. The crucial observation in their paper was that the angle function $v := \langle \nu, Z\rangle$ (scalar product of the unit normal and the height vector Z) satisfies a very useful evolution equation that can be exploited to bound the second fundamental form appropriately.

Many results in mean curvature flow of graphs have been obtained by Wang. For example in [87] he studied the graph induced by a map $f : M \to K$ between to Riemannian manifolds of constant sectional curvatures. Under suitable assumptions on the differential of f and the curvatures of M resp. K he obtained long-time existence and convergence to constant maps. In [84] the authors consider a graph in the product $M \times K$ of two Riemannian manifolds of constant sectional curvatures. A map $f : M \to K$ for which the singular values λ_i of f satisfy the condition $\lambda_i\lambda_j < 1$ for all $i \neq j$ is called an area decreasing map. The main theorem in their paper states long-time existence of the mean curvature flow and convergence to a constant map under the following assumptions:

1. the initial graph of f is area-decreasing;
2. $\sigma^M \geq |\sigma^K|$, $\sigma^M + \sigma^K > 0$ and $\dim M \geq 2$,

where σ^M, σ^K denote the sectional curvatures of M resp. K. In particular area decreasing maps from S^m to S^k are homotopically trivial for $m \geq 2$.

In [60] graphs in Riemannian products of two space forms have been studied and under certain assumptions on the initial graph long-time existence was established. In [90] two long-time existence and convergence results for the mean curvature flow of graphs induced by maps $f : M \to K$ between two compact Riemannian manifolds of dimension $m = \dim M \geq 2$ and $\dim K = 2$ are given. In the first theorem M and K are assumed to be flat, and in the second theorem, $M = S^m$ is an m-sphere of constant curvature $k_1 > 0$ and K a compact surface of constant curvature k_2 with $|k_2| \leq k_1$. The key assumption on the graph is expressed in terms of the Gauß map, i.e. the map which assigns to a point p its tangent space. The latter is an element of the bundle of m-dimensional subspaces of TN, $N = M \times K$ and it is shown that there exists a sub-bundle $\mathfrak{G}$ of TN which is preserved along the mean curvature flow. The same author proved a beautiful general theorem for the Gauß map under the mean curvature flow (see [88]).

Acknowledgements This survey is a contribution within the framework of the priority program "Globale Differentialgeometrie", DFG-SPP 1154, supported by the German Science Foundation (DFG).

References

1. Abresch, U., Langer, J.: The normalized curve shortening flow and homothetic solutions. J. Differ. Geom. **23**(2), 175–196 (1986)
2. Altschuler, S.J.: Singularities of the curve shrinking flow for space curves. J. Differ. Geom. **34**(2), 491–514 (1991)
3. Altschuler, S.J., Grayson, M.A.: Shortening space curves and flow through singularities. J. Differ. Geom. **35**(2), 283–298 (1992)
4. Ambrosio, L., Soner, H.M.: A measure-theoretic approach to higher codimension mean curvature flows. Ann. Scuola Norm. Sup. Pisa Cl. Sci. (4) **25**(1-2), 27–49 (1997). Dedicated to Ennio De Giorgi
5. Anciaux, H.: Construction of lagrangian self-similar solutions to the mean curvature flow in c^n. Geom. Dedicata **120**, 37–48 (2006)
6. Andrews, B., Baker, C.: Mean curvature flow of pinched submanifolds to spheres. J. Differ. Geom. **85**(3), 357–395 (2010)
7. Angenent, S.: On the formation of singularities in the curve shortening flow. J. Differ. Geom. **33**(3), 601–633 (1991)
8. Angenent, S.B., Velázquez, J.J.L.: Degenerate neckpinches in mean curvature flow. J. Reine Angew. Math. **482**, 15–66 (1997)
9. Behrndt, T.: Generalized lagrangian mean curvature flow in kähler manifolds that are almost einstein. arXiv:0812.4256, to appear in Proceedings of CDG 2009, Leibniz Universität Hannover (2008)
10. Brakke, K.A.: The motion of a surface by its mean curvature, Mathematical Notes, vol. 20. Princeton University Press, Princeton (1978)
11. Castro, I., Lerma, A.M.: Hamiltonian stationary self-similar solutions for lagrangian mean curvature flow in the complex euclidean plane. Proc. Amer. Math. Soc. **138**(5), 1821–1832 (2010)
12. Chau, A., Chen, J.: Entire self-similar solutions to lagrangian mean curvature flow. arXiv:0905.3869 (2009)

13. Chau, A., Chen, J., He, W.: Lagrangian mean curvature flow for entire lipschitz graphs. arXiv:0902.3300 (2009)
14. Chen, B.L., Yin, L.: Uniqueness and pseudolocality theorems of the mean curvature flow. Comm. Anal. Geom. **15**(3), 435–490 (2007)
15. Chen, D., Ma, L.: Curve shortening in a riemannian manifold. Ann. Mat. Pura Appl. (4) **186**(4), 663–684 (2007)
16. Chen, J., Li, J.: Mean curvature flow of surface in 4-manifolds. Adv. Math. **163**(2), 287–309 (2001)
17. Chen, J., Li, J.: Singularity of mean curvature flow of lagrangian submanifolds. Invent. Math. **156**(1), 25–51 (2004)
18. Chen, J., Li, J., Tian, G.: Two-dimensional graphs moving by mean curvature flow. Acta Math. Sin. (Engl. Ser.) **18**(2), 209–224 (2002)
19. Chen, J., Pang, C.: Uniqueness of unbounded solutions of the lagrangian mean curvature flow equation for graphs. C. R. Math. Acad. Sci. Paris **347**(17-18), 1031–1034 (2009)
20. Chen, X., Jian, H., Liu, Q.: Convexity and symmetry of translating solitons in mean curvature flows. Chinese Ann. Math. Ser. B **26**(3), 413–422 (2005)
21. Chen, Y., Giga, Y., Goto, S.: Uniqueness and existence of viscosity solutions of generalized mean curvature flow equations. J. Differ. Geom. **33**(3), 749–786 (1991)
22. Chou, K.S., Zhu, X.P.: The curve shortening problem. Chapman & Hall/CRC, Boca Raton, FL (2001)
23. Clutterbuck, J., Schnürer, O., Schulze, F.: Stability of translating solutions to mean curvature flow. Calc. Var. Part. Differ. Equat. **29**(3), 281–293 (2007)
24. Colding, T.H., Minicozzi W. P., I.: Sharp estimates for mean curvature flow of graphs. J. Reine Angew. Math. **574**, 187–195 (2004)
25. Colding, T.H., Minicozzi W. P., I.: Generic mean curvature flow i; generic singularities. arXiv:0908.3788 (2009)
26. Ecker, K.: Estimates for evolutionary surfaces of prescribed mean curvature. Math. Z. **180**(2), 179–192 (1982)
27. Ecker, K.: A local monotonicity formula for mean curvature flow. Ann. Math. (2) **154**(2), 503–525 (2001)
28. Ecker, K.: Regularity theory for mean curvature flow. Progress in Nonlinear Differential Equations and their Applications, 57. Birkhäuser Boston Inc. (2004)
29. Ecker, K., Huisken, G.: Mean curvature evolution of entire graphs. Ann. of Math. (2) **130**(3), 453–471 (1989)
30. Ecker, K., Huisken, G.: Interior estimates for hypersurfaces moving by mean curvature. Invent. Math. **105**(3), 547–569 (1991)
31. Ecker, K., Knopf, D., Ni, L., Topping, P.: Local monotonicity and mean value formulas for evolving riemannian manifolds. J. Reine Angew. Math. **616**, 89–130 (2008)
32. Enders, J., Müller, R., Topping, P.: On type i singularities in ricci flow. arXiv:1005.1624 (2010)
33. Evans, L.C., Spruck, J.: Motion of level sets by mean curvature. i. J. Differ. Geom. **33**(3), 635–681 (1991)
34. Gage, M., Hamilton, R.S.: The heat equation shrinking convex plane curves. J. Differ. Geom. **23**(1), 69–96 (1986)
35. Gage, M.E.: Curve shortening makes convex curves circular. Invent. Math. **76**(2), 357–364 (1984)
36. Gerhardt, C.: Evolutionary surfaces of prescribed mean curvature. J. Differential Equations **36**(1), 139–172 (1980)
37. Grayson, M.A.: The heat equation shrinks embedded plane curves to round points. J. Differ. Geom. **26**(2), 285–314 (1987)
38. Grayson, M.A.: Shortening embedded curves. Ann. of Math. (2) **129**(1), 71–111 (1989)
39. Groh, K., Schwarz, M., Smoczyk, K., Zehmisch, K.: Mean curvature flow of monotone lagrangian submanifolds. Math. Z. **257**(2), 295–327 (2007)

40. Gromov, M.: Pseudoholomorphic curves in symplectic manifolds. Invent. Math. **82**(2), 307–347 (1985)
41. Hamilton, R.S.: The inverse function theorem of nash and moser. Bull. Amer. Math. Soc. (N.S.) **7**(1), 65–222 (1982)
42. Hamilton, R.S.: Three-manifolds with positive ricci curvature. J. Differ. Geom. **17**(2), 255–306 (1982)
43. Hamilton, R.S.: Four-manifolds with positive curvature operator. J. Differ. Geom. **24**(2), 153–179 (1986)
44. Hamilton, R.S.: Monotonicity formulas for parabolic flows on manifolds. Comm. Anal. Geom. **1**(1), 127–137 (1993)
45. Hamilton, R.S.: The formation of singularities in the ricci flow. In: Surveys in differential geometry Vol. II, 1993, pp. 7–136. International Press, Cambridge, MA (1995)
46. Hamilton, R.S.: Harnack estimate for the mean curvature flow. J. Differ. Geom. **41**(1), 215–226 (1995)
47. Han, X., Li, J.: Translating solitons to symplectic and lagrangian mean curvature flows. Internat. J. Math. **20**(4), 443–458 (2009)
48. Huisken, G.: Flow by mean curvature of convex surfaces into spheres. J. Differ. Geom. **20**(1), 237–266 (1984)
49. Huisken, G.: Contracting convex hypersurfaces in riemannian manifolds by their mean curvature. Invent. Math. **84**(3), 463–480 (1986)
50. Huisken, G.: Asymptotic behavior for singularities of the mean curvature flow. J. Differ. Geom. **31**(1), 285–299 (1990)
51. Huisken, G.: Local and global behaviour of hypersurfaces moving by mean curvature. In: Differential geometry: partial differential equations on manifolds (Los Angeles, CA, 1990), Proc. Sympos. Pure Math., vol. 54, pp. 175–191. American Mathematical Society, Providence, RI (1993)
52. Huisken, G., Sinestrari, C.: Convexity estimates for mean curvature flow and singularities of mean convex surfaces. Acta Math. **183**(1), 45–70 (1999)
53. Huisken, G., Sinestrari, C.: Mean curvature flow singularities for mean convex surfaces. Calc. Var. Part. Differ. Equat. **8**(1), 1–14 (1999)
54. Huisken, G., Sinestrari, C.: Mean curvature flow with surgeries of two-convex hypersurfaces. Invent. Math. **175**(1), 137–221 (2009)
55. Ilmanen, T.: Generalized flow of sets by mean curvature on a manifold. Indiana Univ. Math. J. **41**(3), 671–705 (1992)
56. Joyce, D., Lee, Y.I., Tsui, M.P.: Self-similar solutions and translating solitons for lagrangian mean curvature flow. J. Differ. Geom. **84**(1), 127–161 (2010)
57. Krylov, N.V.: Nonlinear elliptic and parabolic equations of the second order, Mathematics and its Applications (Soviet Series), vol. 7, pp.xiv+462, D. Reidel Publishing Co., Dordrecht (1987)
58. Le, N.Q., Sesum, N.: Blow-up rate of the mean curvature during the mean curvature flow and a gap theorem for self-shrinkers. arXiv:1011.5245v1 (2010)
59. Le, N.Q., Sesum, N.: The mean curvature at the first singular time of the mean curvature flow. Ann. Inst. H. Poincaré Anal. Non Linéaire **27**(6), 1441–1459 (2010)
60. Li, J., Li, Y.: Mean curvature flow of graphs in $\sigma_1 \times \sigma_2$. J. Part. Differ. Equat. **16**(3), 255–265 (2003)
61. Liu, K., Xu, H., Ye, F., Zhao, E.: The extension and convergence of mean curvature flow in higher codimension. arXiv:1104.0971v1 (2011)
62. Medos, I., Wang, M.T.: Deforming symplectomorphisms of complex projective spaces by the mean curvature flow. Preprint (2009)
63. Mullins, W.W.: Two-dimensional motion of idealized grain boundaries. J. Appl. Phys. **27**, 900–904 (1956)
64. Neves, A.: Singularities of lagrangian mean curvature flow: zero-maslov class case. Invent. Math. **168**(3), 449–484 (2007)
65. Neves, A.: Singularities of lagrangian mean curvature flow: monotone case. Math. Res. Lett. **17**(1), 109–126 (2010)

66. Perelman, G.: The entropy formula for the ricci flow and its geometric applications. arXiv:math/0211159 (2002)
67. Perelman, G.: Finite extinction time for the solutions to the ricci flow on certain three-manifolds. arXiv:math/0307245 (2003)
68. Perelman, G.: Ricci flow with surgery on three-manifolds. arXiv:math/0303109 (2003)
69. Ros, A., Urbano, F.: Lagrangian submanifolds of $\mathbf{c}^n$ with conformal maslov form and the whitney sphere. J. Math. Soc. Japan **50**(1), 203–226 (1998)
70. Schoen, R., Wolfson, J.: Minimizing area among lagrangian surfaces: the mapping problem. J. Differ. Geom. **58**(1), 1–86 (2001)
71. Schoen, R., Wolfson, J.: Mean curvature flow and lagrangian embeddings. Preprint (2003)
72. Smoczyk, K.: A canonical way to deform a lagrangian submanifold. arXiv:dg-ga/9605005 (1996)
73. Smoczyk, K.: Harnack inequality for the lagrangian mean curvature flow. Calc. Var. Part. Differ. Equat. **8**(3), 247–258 (1999)
74. Smoczyk, K.: The lagrangian mean curvature flow (der lagrangesche mittlere Krümmungsfluß. Leipzig: Univ. Leipzig (Habil.), 102 S. (2000)
75. Smoczyk, K.: Angle theorems for the lagrangian mean curvature flow. Math. Z. **240**(4), 849–883 (2002)
76. Smoczyk, K.: Longtime existence of the lagrangian mean curvature flow. Calc. Var. Part. Differ. Equat. **20**(1), 25–46 (2004)
77. Smoczyk, K.: Self-shrinkers of the mean curvature flow in arbitrary codimension. Int. Math. Res. Not. (48), 2983–3004 (2005)
78. Smoczyk, K., Wang, M.T.: Mean curvature flows of lagrangians submanifolds with convex potentials. J. Differ. Geom. **62**(2), 243–257 (2002)
79. Smoczyk, K., Wang, M.T.: Generalized lagrangian mean curvature flows in symplectic manifolds. Asian J. Math. **15**(1), 129–140 (2011)
80. Stavrou, N.: Selfsimilar solutions to the mean curvature flow. J. Reine Angew. Math. **499**, 189–198 (1998)
81. Stone, A.: A density function and the structure of singularities of the mean curvature flow. Calc. Var. Part. Differ. Equat. **2**(4), 443–480 (1994)
82. Temam, R.: Applications de l'analyse convexe au calcul des variations. In: Nonlinear operators and the calculus of variations (Summer School, Univ. Libre Bruxelles, Brussels, 1975). Lecture Notes in Math., Vol. 543, pp. 208–237, Springer, Berlin (1976)
83. Thomas, R.P., Yau, S.T.: Special lagrangians, stable bundles and mean curvature flow. Comm. Anal. Geom. **10**(5), 1075–1113 (2002)
84. Tsui, M.P., Wang, M.T.: Mean curvature flows and isotopy of maps between spheres. Comm. Pure Appl. Math. **57**(8), 1110–1126 (2004)
85. Wang, M.T.: Deforming area preserving diffeomorphism of surfaces by mean curvature flow. Math. Res. Lett. **8**(5-6), 651–661 (2001)
86. Wang, M.T.: Mean curvature flow of surfaces in einstein four-manifolds. J. Differ. Geom. **57**(2), 301–338 (2001)
87. Wang, M.T.: Long time existence and convergence of graphic mean curvature flow in arbitrary codimension. Invent. Math. **148**(3), 525–543 (2002)
88. Wang, M.T.: Gauss maps of the mean curvature flow. Math. Res. Lett. **10**(2-3), 287–299 (2003)
89. Wang, M.T.: The mean curvature flow smoothes lipschitz submanifolds. Comm. Anal. Geom. **12**(3), 581–599 (2004)
90. Wang, M.T.: Subsets of grassmannians preserved by mean curvature flows. Comm. Anal. Geom. **13**(5), 981–998 (2005)
91. Wang, M.T.: A convergence result of the lagrangian mean curvature flow. In: Third International Congress of Chinese Mathematicians. Part 1, 2, AMS/IP Stud. Adv. Math., 42, pt. 1, vol. 2, pp. 291–295. American Mathematical Society, Providence, RI (2008)
92. Wang, M.T.: Lectures on mean curvature flows in higher codimensions. In: Handbook of geometric analysis. No. 1. Adv. Lect. Math. (ALM), vol. 7, pp. 525–543. International Press, Somerville, MA (2008)

93. White, B.: A local regularity theorem for mean curvature flow. Ann. of Math. (2) **161**(3), 1487–1519 (2005)
94. White, B.: Currents and flat chains associated to varifolds, with an application to mean curvature flow. Duke Math. J. **148**(1), 41–62 (2009)
95. Xin, Y.: Mean curvature flow with convex gauss image. Chin. Ann. Math. Ser. B **29**(2), 121–134 (2008)

Positive Scalar Curvature, K-area and Essentialness

Bernhard Hanke

Abstract The Lichnerowicz formula yields an index theoretic obstruction to positive scalar curvature metrics on closed spin manifolds. The most general form of this obstruction is due to Rosenberg and takes values in the K-theory of the group C^*-algebra of the fundamental group of the underlying manifold.

We give an overview of recent results clarifying the relation of the Rosenberg index to notions from large scale geometry like enlargeability and essentialness. One central topic is the concept of K-homology classes of infinite K-area. This notion, which in its original form is due to Gromov, is put in a general context and systematically used as a link between geometrically defined large scale properties and index theoretic considerations. In particular, we prove essentialness and the non-vanishing of the Rosenberg index for manifolds of infinite K-area.

1 Introduction and Summary

One of the fundamental problems in Riemannian geometry is to investigate the types of Riemannian metrics that exist on a given closed smooth manifold. It turns out that the signs of the associated curvature invariants distinguish classes of Riemannian manifolds with considerably different geometric and topological properties. Usually the class of manifolds admitting metrics with negative curvature is "big" and the one with positive curvature is "small". The general existence theorems for negative Ricci curvature metrics [29] and negative scalar curvature metrics [45], the classical theorem of Bonnet-Myers on the finiteness of the fundamental groups of closed Riemannian manifolds with positive Ricci curvature, Gromov's Betti number theorem for closed manifolds of non-negative sectional curvature [17], the recent

---

B. Hanke (✉)
Institut für Mathematik, Universität Augsburg, 86135 Augsburg, Germany
e-mail: hanke@math.uni-augsburg.de

C. Bär et al. (eds.), *Global Differential Geometry*, Springer Proceedings in Mathematics 17, DOI 10.1007/978-3-642-22842-1_10,

classification of manifolds with positive curvature operators [4] and the proof of the differentiable sphere theorem [5,6] are prominent illustrations of this empirical fact.

In this context one may formulate two goals. The first is to develop methods to construct Riemannian metrics with distinguished properties on general classes of smooth manifolds. Important examples are the powerful tools in the theory of geometric partial differential equations, the surgery method due to Gromov-Lawson [15] and Schoen-Yau [40] for the construction of positive scalar curvature metrics, and methods based on geometric flow equations. The second deals with the formulation of (computable) obstructions to the existence of Riemannian metrics with specific properties. Often this happens in connection with topological invariants associated to the given manifold like homology and homotopy groups and related data. These two goals are usually not completely seperate from each other in that they can result in overlapping questions, concepts and methods. For example the Ricci flow is used to produce metrics with special properties, which a posteriori determine the topological type of the underlying manifold.

Here we shall concentrate on the most elementary curvature invariant associated to a Riemannian manifold (M, g), the scalar curvature $\mathrm{scal}_g : M \to \mathbb{R}$. This is usually defined by a twofold contraction of the Riemannian curvature tensor of (M, g), but also has a geometric interpretation in terms of the deviation of the volume growth of geodesic balls in M compared to geodesic balls in Euclidean space:

$$\frac{\mathrm{vol}_{(M^n,g)}(B_p(\epsilon))}{\mathrm{vol}_{(\mathbb{R}^n, g_{eucl})}(B_0(\epsilon))} = 1 - \frac{\mathrm{scal}_g(p)}{6(n+2)} \cdot \epsilon^2 + O(\epsilon^4)\,.$$

Given a closed smooth manifold M we shall study whether M admits a Riemannian metric g of positive scalar curvature, i.e. satisfying $\mathrm{scal}_g(p) > 0$ for all $p \in M$. In view of the preceding description and the previous remarks it is on the one hand plausible that the resulting "inside bending of M" at every point might put topological restrictions on M. On the other hand the scalar curvature involves an averaging process over sectional curvatures of M so that a certain variability of the precise geometric shape and the topological properties of M can be expected.

In connection with the positive scalar curvature question both aspects, the obstructive and constructive side, play important roles and have led to a complex body of mathematical insight with connections to index theory, geometric analysis, non-commutative geometry, surgery theory, bordism theory and stable homotopy theory. The paper [37] gives a comprehensive survey of the subject. As such it represents not only an interesting geometric field of its own, but serves as a unifying link between several well established areas in geometry, topology and analysis.

For metrics of positive scalar curvature there are two important obstructions, whose relation to each other is still not completely understood. One is based on the method of minimal hypersurfaces [40] and the other on the analysis of the Dirac operator and index theory [27].

In some sense the former obstruction is more elementary than the latter as it can be shown by a direct calculation [40] that a nonsingular minimal hypersurface in a positive scalar curvature manifold admits itself a metric of positive scalar curvature. In connection with results from geometric measure theory that provide nonsingular minimal hypersurfaces representing codimension one homology classes in manifolds of dimension at most eight [41], this can inductively be used to exclude the existence of positive scalar curvature metrics on tori up to dimension eight, for instance. In higher dimensions the discussion of singularities on minimal hypersurfaces representing codimension one homology classes is a subtle topic and the subject of recent work of Lohkamp [8, 30, 31]. This theme, which has important connections to the positive mass theorem in general relativity, will not be pursued further in our paper.

The second, index theoretic, obstruction is both more restrictive as it requires a spin structure on the underlying manifold (or at least its universal cover), and less elementary as it is based on the Atiyah-Singer index theorem. In its most basic form it says that closed spin manifolds with non-vanishing $\hat{A}$-genus do not admit metrics of positive scalar curvature, the $\hat{A}$-genus being an integer (in the spin case) which depends on the rational Pontrjagin classes of the underlying manifold and its orientation class and hence only on its oriented homeomorphism type.

This obstruction was refined by Hitchin [25] and Rosenberg [35] and in its most general form takes values in $KO_*(C^*_{\mathbb{R},max}\pi_1(M))$, the K-theory of the real maximal group C^*-algebra of the fundamental group of the underlying manifold. It therefore touches important questions in noncommutative geometry linked to the Baum-Connes and Novikov conjectures. The Gromov-Lawson-Rosenberg conjecture predicts that for closed spin manifolds of dimension at least five the vanishing of this index obstruction is not only necessary, but also sufficient for the existence of a positive scalar curvature metric. Despite the fact that this conjecture is wrong in general [38], the index obstruction being surpassed by the minimal hypersurface obstruction in some cases, it is remarkable that it holds for simply connected manifolds [42] and – in a stable sense – for all spin manifolds for which the assembly map with values in the K-theory of the real group C^*-algebra of the fundamental group is injective [43], see Theorem 2.4 below. It is up to date unknown whether this conjecture in its original, unstable, form is true for spin manifolds with finite fundamental groups, although in this case the injectivity of the assembly map is known. The index theoretic obstruction to positive scalar curvature will be recalled in Sect. 2 of our paper.

Gromov and Lawson used the index of the usual Dirac operator on closed spin manifolds twisted with bundles of small curvature to prove that some manifolds with vanishing $\hat{A}$-genus do still not admit positive scalar curvature metrics. For this aim they introduced several kinds of largeness properties for Riemannian manifolds, the most important ones being the notion of enlargeability [16, 18] and infinite K-area [14]. These properties have an asymptotic character in that they require, for each $\epsilon > 0$, the existence of a certain geometric structure attached to the underlying manifold which is ϵ-small in an appropriate sense. Precise definitions will be given in Sect. 2 below.

In light of the common index theoretic origin of these obstructions it is reasonable to expect that they are related to the Rosenberg index. In the papers [19–21] it is proved that the Rosenberg obstruction indeed subsumes the enlargeability obstruction in the sense that the former is non-zero for enlargeable spin manifolds. Moreover, it was shown in the cited papers that enlargeable manifolds are *essential*, i.e. the classifying maps of their universal covers map the homological fundamental classes to non-zero classes in the homology of the fundamental groups. This notion was introduced by Gromov in [13] in connection with the systolic inequality giving an upper bound of the length of the shortest noncontractible loop in a Riemannian manifold M in terms of the volume of M. In particular it follows from these results that enlargeable manifolds obey Gromov's systolic inequality.

The methods introduced in [20, 21] were applied in [22] to prove some cases of the strong Novikov conjecture. This is implied by the Baum-Connes conjecture and predicts that for discrete groups G the rational assembly map

$$K_*(BG) \otimes \mathbb{Q} \to K_*(C^*_{max}G) \otimes \mathbb{Q}$$

is injective. In loc. cit. it is shown that this map is indeed non-zero on all classes in $K_*(BG) \otimes \mathbb{Q}$ which are detected by classes in the subring generated by $H^{\leq 2}(BG;\mathbb{Q})$. As a corollary higher signatures associated to elements in this subring of $H^*(BG;\mathbb{Q})$ are oriented homotopy invariants, a fact which had been proven first by Mathai [64].

It turns out that the methods of [20, 22] fit very nicely the concept of K-area introduced by Gromov in [14]. It is one purpose of the paper at hand to elaborate on this connection. Our main result, Theorem 3.9, states that K-homology classes of *infinite K-area* in closed manifolds M map nontrivially to $K_*(C^*_{max}\pi_1(M)) \otimes \mathbb{Q}$ under the assembly map. Generalizing the original concept of Gromov we call a K-homology class of *infinite K-area*, if it can be detected by bundles of finitely generated Hilbert A-modules equipped with holonomy representations which are arbitrarily close to the identity, where A is some C^*-algebra with unit. Precise definitions are given in Sect. 3 below, see in particular Definition 3.5.

From Theorem 3.9 the main results of the papers [19–22] follow quite directly. Apart from this we will demonstrate that closed spin manifolds whose K-theoretic fundamental classes are of infinite K-area have non-vanishing Rosenberg index (Corollary 3.10) and oriented manifolds with fundamental classes of infinite K-area are essential (Theorem 4.9). The first result solves a problem stated in the introduction of [28].

In [7] essentialness is discussed from a purely homological point of view. Among other things it is proved that the property of being enlargeable depends only on the image of the homological fundamental class of the underlying manifold in the rational homology of its fundamental group. This flexible formulation allows the construction of manifolds which are essential, but not enlargeable. We will briefly review these results in Sect. 5. We do not know whether a proof of Theorem 4.9 is feasible which avoids the "infinite product construction" laid out in Sect. 3. Also, we do not know an essential manifold whose fundamental class is not of infinite K-area, see Conjecture 5.6.

This paper is intended on the one hand as a report on recent results pertaining to the positive scalar curvature question in the light of methods from index theory,

K-theory and asymptotic geometry as obtained by the author and his coauthors. On the other hand it is meant to establish the point of view that the notion of infinite K-area may serve as a unifying principle for these results, which sometimes allows short and conceptual proofs.

I am grateful to the DFG Schwerpunkt "Globale Differentialgeometrie" for financial support during the last years. Special thanks go to Thomas Schick for a very fruitful and pleasant collaboration. Most of the material in these notes is based on ideas developed during this collaboration.

Daniel Pape carefully read the first version of this manuscript and helped to improve the presentation with many useful comments.

2 Index Obstruction to Positive Scalar Curvature

The Gauß-Bonnet formula implies that closed surfaces with nonpositive Euler characteristic do not admit positive scalar curvature metrics. These comprise all closed surfaces apart from the two sphere and the real projective plane. The mechanism behind this obstruction is the fact that a topological invariant, the Euler characteristic, may be expressed as an integral over a curvature quantity, the Gauß curvature.

In higher dimensions obstructions to positive scalar curvature metrics can be obtained in a more indirect way by use of the Atiyah-Singer index theorem. Let M be a closed smooth oriented manifold of dimension divisible by four. The $\hat{A}$-genus $\hat{A}(M) \in \mathbb{Q}$ of M is obtained by evaluating the $\hat{A}$-polynomial

$$\hat{A}(M) = 1 - \frac{p_1(M)}{24} + \frac{-4p_2(M) + 7p_1^2(M)}{2^7 \cdot 3^2 \cdot 5} + \cdots$$

in the Pontrjagin classes of M on the fundamental class of M. This is an invariant of the oriented homeomorphism type of M by the topological invariance of rational Pontrjagin classes. It is an integer, if M is equipped with a spin structure. This is implied by the fact that in this case the Atiyah-Singer index theorem provides an equation

$$\hat{A}(M) = \operatorname{ind}(D_g^+) = \dim_{\mathbb{C}}(\ker D_g^+) - \dim_{\mathbb{C}}(\operatorname{coker} D_g^+)$$

where

$$D_g^{\pm} : \Gamma(S^{\pm}) \to \Gamma(S^{\mp})$$

is the Dirac operator on the complex spinor bundle $S = S^+ \oplus S^- \to M$ of (M, g). Here g is an arbitrary Riemannian metric on M. Due to the appearance of g in the definition of D_g^+, the Atiyah-Singer index theorem relates topological to geometric properties of M. Detailed information on the definition of D_g^+ and spin geometry in general can be found in [26].

The Bochner-Lichnerowicz-Weitzenböck formula [27]

$$D_g^- \circ D_g^+ = \nabla^*\nabla + \frac{\mathrm{scal}_g}{4}$$

implies that if $\mathrm{scal}_g(M) > 0$, then the Dirac operator D_g^+ is invertible and hence $\mathrm{ind}(D_g^+) = 0$. From this we obtain the following fundamental result, see [27, Theorème 2].

Theorem 2.1. *Let M be a closed spin manifold with $\hat{A}(M) \neq 0$. Then M does not admit a metric of positive scalar curvature.*

However, the vanishing of this obstruction is not sufficient for the existence of positive scalar curvature metrics. For example, the $\hat{A}$-genus of the $4k$-dimensional torus T^{4k} vanishes for all $k > 0$, because these manifolds are parallelizable.

The index theoretic approach explained above can be refined by considering the twisted Dirac operator

$$D_{g,E}^+ : \Gamma(S^+ \otimes E) \to \Gamma(S^- \otimes E)$$

where $E \to M$ is some finite dimensional Hermitian vector bundle equipped with a Hermitian connection, cf. [26, Prop. II.5.10]. The Atiyah-Singer index theorem computes the index of this operator as

$$\mathrm{ind}(D_{g,E}^+) = \langle \hat{A}(M) \cup \mathrm{ch}(E), [M] \rangle \in \mathbb{Z}.$$

Due to the appearance of the Chern character $\mathrm{ch}(E) \in H^{ev}(M;\mathbb{Q})$ this number can be non-zero even though $\hat{A}(M)$ vanishes. Unfortunately, the nonvanishing of $\mathrm{ind}(D_{g,E}^+)$ does not obstruct positive scalar curvature metrics on M as the following example shows.

Example 2.2. Let $M^n = S^{4k+2}$. Because the Chern character defines an isomorphism

$$\mathrm{ch} : K^0(M) \otimes \mathbb{Q} \cong H^{ev}(M;\mathbb{Q}),$$

there is a finite dimensional Hermitian bundle $E \to M$ with $\mathrm{ch}_{2k+1}(E) \neq 0 \in H^n(M;\mathbb{Q})$. Hence, for any connection on E and any choice of Riemannian metric g on M, we get $\mathrm{ind}(D_{g,E}^+) \neq 0$ although M admits a metric of positive scalar curvature.

This is due to the fact that now the Bochner-Lichnerowicz-Weitzenböck formula

$$D_{g,E}^- \circ D_{g,E}^+ = \nabla^*\nabla + \frac{\mathrm{scal}_g}{4} + R^E$$

contains an additional operator $R^E : \Gamma(S^\pm \otimes E) \to \Gamma(S^\pm \otimes E)$ of order zero which depends on the curvature of the bundle E, cf. [26, Theorem 8.17], so that even in the case when $\mathrm{scal}_g > 0$, the operator $D_{g,E}^+$ may not be invertible.

Gromov and Lawson observed in [16] that this method does still lead to an effective obstruction to positive scalar curvature metrics on M in case that for all ϵ there is a twisting bundle $E \to M$ which satisfies $\|R^E\| < \epsilon$ and whose Chern character contributes nontrivially to $\operatorname{ind}(D^+_{g,E})$. If in this case M carried a metric g satisfying $\operatorname{scal}_g > 0$ we would find a twisting bundle E with

$$\|R^E\| < \frac{\min_{p\in M} \operatorname{scal}_g(p)}{4}$$

and the Bochner-Lichnerowicz-Weitzenböck formula would then imply that $\operatorname{ind} D^+_{g,E} = 0$, a contradiction.

For example this reasoning can be used to show that the tori T^n do not admit metrics of positive scalar curvature [16].

A general class of manifolds where twisting bundles with the described property can be found are *enlargeable* manifolds, which were introduced in loc. cit., and manifolds of infinite K-area in the sense of [14]. We will discuss these notions and put them in a general context in Sect. 3.

The index theoretic point of view was refined by Rosenberg [35, 36]. For any discrete group G the group C^*-algebra C^*G is constructed by completing the group algebra $\mathbb{C}[G]$ with respect to some pre-C^*-norm coming from unitary representations of G on a Hilbert space and taking the induced embedding of $\mathbb{C}[G]$ into the bounded operators on this Hilbert space. More specifically, if one starts with the regular representation of G on the space of square summable functions $l^2(G)$ this leads to the *reduced group C^*-algebra* $C^*_{red}G$ and taking all unitary representations of G into account one arrives at the *maximal group C^*-algebra* $C^*_{max}G$. For more details we refer to [3, 24, 44]. These C^*-algebras and their K-theories are in general different [24, Exercise 12.7.7], but the following construction works for both variants, and this is why we drop the subscript from our notation. Note that the left translation action of G on $\mathbb{C}[G]$ induces a left G-action on C^*G.

Let M be a closed spin manifold of even dimension. The Mishchenko-Fomenko bundle $E \to M$ is defined as

$$E = \widetilde{M} \times_{\pi_1(M)} C^*\pi_1(M).$$

It is a locally trivial bundle of free right Hilbert $C^*\pi_1(M)$-modules of rank one in the sense of [39, 44]. The fibrewise inner product is induced by the canonical inner product

$$\begin{aligned} C^*\pi_1(M) \times C^*\pi_1(M) &\to C^*\pi_1(M) \\ (x, y) &\mapsto x^* \cdot y. \end{aligned}$$

By construction the bundle $E \to \mathrm{M}$ can be equipped with a flat connection. Depending on the choice of a metric g on M we obtain a twisted Dirac operator

$$D^+_{g,E} : \Gamma(S^+ \otimes E) \to \Gamma(S^- \otimes E)$$

with an index

$$\alpha(M) := \mathrm{ind}(D^+_{g,E}) = \ker(D^+_{g,E}) - \mathrm{coker}\,(D^+_{g,E}) \in K_0(C^*\pi_1(M)).$$

The group $K_0(C^*\pi_1(M))$ consists of formal differences of finitely generated projective $C^*\pi_1(M)$-modules, cf. [3]. For the infinite dimensional twisting bundle E the modules $\ker(D^+_{g,E})$ and $\mathrm{coker}\,(D^+_{g,E})$ are not in this class in general, but this holds after a $C^*\pi_1(M)$-compact perturbation of $D^+_{g,E}$ which makes this operator a $C^*\pi_1(M)$-Fredholm operator. For precise formulations and more details on the involved theory we refer the reader to [33], in particular to Theorem 3.4. in loc. cit.

It follows again from the Bochner-Lichnerowicz-Weitzenböck formula (which does not contain a curvature term R^E as E is flat) that the index $\alpha(M) \in K_0(C^*\pi_1(M))$ vanishes, if $\mathrm{scal}_g > 0$. Moreover, the Mishchenko-Fomenko index theorem [33] implies that – similar to the invariant $\hat{A}(M)$ – the obstruction $\alpha(M)$ does not depend on the choice of a Riemannian metric on M, but only on the oriented homeomorphism type of M.

There is an alternative construction of $\alpha(M)$ based on analytic K-homology [3, 24]. As before let M be a closed spin manifold. We do no longer assume that $n := \dim M$ is even (this only simplified the above considerations).

In this setting $\alpha(M)$ is defined as the image of the K-theoretic fundamental class $[M]_K \in K_n(M)$ which is induced by the given spin structure under the composition

$$K_n(M) = K_n^{\pi_1(M)}(\widetilde{M}) \to K_n^{\pi_1(M)}(\underline{E}\pi_1(M)) \stackrel{\mu}{\to} K_n(C^*\pi_1(M)).$$

Here the first map is induced by the $\pi_1(M)$-equivariant classifying map $\widetilde{M} \to \underline{E}\pi_1(M)$ from the universal cover of M to the universal contractible $\pi_1(M)$-space with finite isotropy groups and the second map is the Baum-Connes assembly map, cf. [3].

There is a real analogue $\alpha_{\mathbb{R}}(M)$ of the index obstruction $\alpha(M)$ which, for simply connected manifolds, was introduced in the paper [25] and is defined as the image of the KO-theoretic fundamental class $[M]_{KO} \in KO_n(M)$ under the composition

$$KO_n(M) = KO_n^{\pi_1(M)}(\widetilde{M}) \to KO_n^{\pi_1(M)}(\underline{E}\pi_1(M)) \stackrel{\mu}{\to} KO_n(C^*\pi_1(M)).$$

The invariant $\alpha_{\mathbb{R}}(M)$ is more sensitive to differential topological properties of M than $\alpha(M)$. For example it is different from zero on some exotic spheres [25]. A refined variant of the Bochner-Lichnerowicz-Weitzenböck argument shows that $\alpha_{\mathbb{R}}(M) = 0$, if M admits a metric of positive scalar curvature.

In case we are dealing with the reduced group C^*-algebra $C^*_{red}\pi_1(M)$, the vanishing of the α-obstruction is closely linked to properties of the Baum-Connes assembly map

$$\mu_{\mathbb{R}} : KO^G_*(\underline{E}G) \to KO_*(C^*_{red}G)$$

and its complex analogue

$$\mu_{\mathbb{C}} : K_*^G(\underline{E}G) \to K_*(C^*_{red}G).$$

According to the Baum-Connes conjecture [3], a central open problem in noncommutative geometry, these two maps are isomorphisms for all discrete groups G.

The following conjecture has played a prominent role in the subject. It expresses the expectation that the Rosenberg obstruction is in some sense optimal.

Conjecture 2.3 (Gromov-Lawson-Rosenberg conjecture). Let M be a closed spin manifold of dimension at least five and with $\alpha_{\mathbb{R}}(M) = 0$. Then M admits a metric of positive scalar curvature.

This is true, if M is simply connected [42], but wrong in general [38]. In dimensions two and three analogues of the Gromov-Lawson-Rosenberg conjecture are true [34], but in dimension four there are additional obstructions coming from Seiberg-Witten theory. However, the following stable version of the conjecture conditionally holds in the following sense.

Theorem 2.4 ([43]). *Assume that the real Baum-Connes assembly map $\mu_{\mathbb{R}}$ is injective for $\pi_1(M)$ and that $\alpha_{\mathbb{R}}(M) = 0$. Then some manifold of the form $M \times B^8 \times \ldots \times B^8$ admits a metric of positive scalar curvature, where B^8 is an arbitrary eight dimensional closed spin manifold with $\hat{A}(M) = 1$.*

This result is remarkable, because it is not understood how it can happen that a manifold N does not admit a positive scalar curvature metric, but $N \times B^8$ does. Notice that the vanishing or non-vanishing of $\alpha_{\mathbb{R}}(M)$ is not affected, when M is multiplied with copies of B^8. In this respect Theorem 2.4 establishes $\alpha_{\mathbb{R}}(M)$ as the universal stable index theoretic obstruction to positive scalar curvature metrics.

If the assembly map for the maximal complex group C^*-algebra is injective, then also the rational assembly map

$$K_*^G(\underline{E}G) \otimes \mathbb{Q} = K_*(BG) \otimes \mathbb{Q} \to K_*(C^*_{max}G) \otimes \mathbb{Q}$$

is injective. The strong Novikov conjecture [3] states that here injectivity holds for all discrete groups G.

Therefore it makes sense to single out those manifolds M whose fundamental classes map nontrivially to $K_*(B\pi_1(M)) \otimes \mathbb{Q}$. This motivates the next definition.

Definition 2.5. A closed spin^c manifold M^n is called *(rationally) K-theoretic essential*, if the classifying map $\phi : M \to B\pi_1(M)$ for the universal cover of M satisfies

$$\phi_*([M]_K) \neq 0 \in K_n(B\pi_1(M)) \otimes \mathbb{Q},$$

where $[M]_K \in K_n(M)$ is the K-theoretic fundamental class of M.

Conjecture 2.6. A K-theoretic essential spin manifold does not admit a metric of positive scalar curvature.

It follows from the previous remarks that this conjecture holds, if the rational assembly map for the associated fundamental group is injective. An important consequence of Conjecture 2.6 is the following

Conjecture 2.7 ([16]). Let M be a closed aspherical spin manifold. Then M does not admit a metric of positive scalar curvature.

The following is a variation of Definition 2.5 for singular homology.

Definition 2.8 ([13]). A closed oriented manifold M^n is called *(rationally) essential*, if the classifying map $\phi : M \to B\pi_1(M)$ satisfies

$$\phi_*([M]_H) \neq 0 \in H_n(B\pi_1(M); \mathbb{Q}),$$

where $[M]_H$ is the fundamental class of M in singular homology.

Recall that the homological Chern character defines an isomorphism

$$\mathrm{ch} : K_{(*)}(M) \otimes \mathbb{Q} \cong H_{(*)}(M; \mathbb{Q}),$$

where the brackets in the subscripts indicate that we regard both theories as $\mathbb{Z}/2$-graded. Keeping in mind that for a closed spin^c manifold M^n we have

$$\mathrm{ch}([M]_K) = [M]_H + c$$

where $c \in H_{<n}(M; \mathbb{Q})$, we see that essential spin^c manifolds are also K-theoretic essential. Hence it makes sense to formulate the following conjecture.

Conjecture 2.9. An essential manifold does not admit a metric of positive scalar curvature.

This seems especially intriguing, if the universal cover of this manifold is not spin (so that index theoretic obstructions are not available). Evidence for the conjecture in this case is provided by the fact that sometimes essential manifolds satisfy a weak form of enlargeability [11, 12].

3 K-area for Hilbert Module Bundles

All manifolds in this section are closed, smooth and connected. We recall the following definition from [18].

Definition 3.1. Let (M^n, g) be an orientable Riemannian manifold.

- We call M *enlargeable*, if for every $\epsilon > 0$ there is a Riemannian cover $(\overline{M}, \overline{g})$ of (M, g) together with an ϵ-Lipschitz map $f_\epsilon : \overline{M} \to S^n$ which is constant outside of a compact subset of $\overline{M}$ and of non-zero degree.
- We call (M, g) *area-enlargeable*, if for every $\epsilon > 0$ there is a Riemannian cover $(\overline{M}, \overline{g})$ of (M, g) together with a smooth map $f_\epsilon : \overline{M} \to S^n$ which is

ϵ-contracting on 2-forms, constant outside of a compact subset of $\overline{M}$ and of nonzero degree.

Because M is compact, all Riemannian metrics on M are in bi-Lipschitz correspondence and hence both of the above properties are independent of the specific choice of the metric g on M. Enlargeability is therefore a purely topological property of M. Indeed, whether M is enlargeable depends only on the image of the fundamental class of M in the rational group homology of $\pi_1(M)$ under the classifying map, see [7, Corollary 3.5] restated as Theorem 5.3 below. We do not know whether a similiar result holds for area-enlargeability.

Examples for enlargeable manifolds are manifolds which admit Riemannian metrics of nonpositive sectional curvature. This follows from the Cartan-Hadamard theorem.

Area-enlargeable spin manifolds allow the construction of finite dimensional Hermitian twisting bundles for the Dirac operator as described after Example 2.2. We remark that the index theoretic setting explained there needs to be slightly generalized (relative index theory on open manifolds, see [18]), if infinite covers of M are involved (this case is not excluded in Definition 3.1). These considerations lead to the following theorem.

Theorem 3.2 ([16, 18]). *Let M be an area-enlargeable spin manifold. Then M does not admit a metric of positive scalar curvature.*

At this point one might ask whether the enlargeability obstruction is reflected by the Rosenberg obstruction.

The twisting bundles $E \to M$ of arbitrarily small curvature going into the obstruction expressed in Theorem 3.2 motivate the notion of *K-area*, see [14].

In this section we will introduce a related property for K-homology classes of M. Examples of such K-homology classes are K-theoretic fundamental classes of area-enlargeable spin manifolds, see Proposition 3.8. The main result in this section, Theorem 3.9, shows that classes in $K_0(M) \otimes \mathbb{Q}$ of infinite K-area are mapped to non-zero classes in $K_0(C^*_{max}\pi_1(M))$ under the assembly map. Together with Proposition 3.8 this implies that the Rosenberg obstruction subsumes the enlargeability obstruction of Gromov and Lawson:

Theorem 3.3 ([20, 21]). *Let M^n be an area-enlargeable spin manifold. Then the Rosenberg index $\alpha(M) \in K_n(C^*_{max}\pi_1(M))$ is different from zero.*

A convenient setting for our discussion is provided by Kasparov's KK-theory, cf. [3], which associates to any pair of separable C^*-algebras A and B an abelian group $KK(A, B)$. We work over the field of complex numbers and will restrict attention to the special cases $A = C(M)$, $B = \mathbb{C}$ and $A = \mathbb{C}$, $B = C(M) \otimes S$ for a seperable unital C^*-algebra S. Here we will work only with ungraded KK-groups.

According to the analytic description of K-homology [24] we have a canonical identification

$$KK(C(M), \mathbb{C}) \cong K_0(M)$$

the 0-th K-homology of M which, for example, can be defined homotopy theoretically as the homology theory dual to topological K-theory [1].

Elements in $KK(A, B)$ are represented by *Fredholm triples* (E, ϕ, F) where E is a countably generated graded Hilbert B-module, $\phi : A \to \mathcal{B}(E)$ is a graded $*$-homomorphism (here $\mathcal{B}(E)$ is the graded C^*-algebra of graded adjointable bounded B-module homomorphisms $E \to E$) and $F \in \mathcal{B}(E)$ is an operator of degree one such that the commutator $[F, \phi(a)]$ and the operators $(F^2 - \mathrm{id}_E)\phi(a)$ and $(F - F^*)\phi(a)$ are B-compact for all $a \in A$. In our context we will be dealing with Fredholm triples of very special forms which will be specified in a moment. The reader who is interested in more information on the notion of Hilbert modules and the construction of Kasparov KK-theory can consult the sources [3, 44].

A typical situation arises when M is a spin manifold of even dimension equipped with a Riemannian metric g. The Dirac operator from Sect. 2

$$D_g : \Gamma(S^{\pm}) \to \Gamma(S^{\mp})$$

is a symmetric graded first-order elliptic differential operator. It therefore gives rise to an element $[D_g] \in KK(C(M), \mathbb{C})$ represented by the Fredholm triple $(L^2(S), \phi, F)$ where $L^2(S)$ is the space of L^2-sections of the bundle $S^+ \oplus S^-$, the map $\phi : C(M) \to \mathcal{B}(L^2(S))$ is the standard representation as multiplication operators and $F \in \mathcal{B}(L^2(S))$ is a bounded operator which is obtained from D_g by functional calculus.

The construction works more generally for symmetric graded elliptic differential operators on graded smooth Hermitian vector bundles over M, cf. [24, Theorem 10.6.5]. In this way we may think of elements in $KK(C(M), \mathbb{C}) = K_0(M)$ as a kind of generalized symmetric elliptic differential operators over M. In this picture the index of a graded elliptic differential operator corresponds to the image of the KK-class represented by this operator under the map

$$K_0(M) \to K_0(*) = \mathbb{Z}$$

which is induced by the unique map $M \to *$.

If $E \to M$ is a (finite dimensional) Hermitian bundle with a Hermitian connection we obtain the twisted Dirac operator

$$D_{g,E} : \Gamma(S^{\pm} \otimes E) \to \Gamma(S^{\mp} \otimes E)$$

which is again a symmetric graded elliptic differential operator and has an index $\mathrm{ind}(D_{g,E}) \in \mathbb{Z}$.

The index of the twisted operator $D_{g,E}$ has the following description in KK-theory, cf. [3]. The bundle $E \to M$ represents a class $[E]$ in topological K-theory $K^0(M)$, which can be canonically identified with $KK(\mathbb{C}, C(M))$. The element $[E] \in KK(\mathbb{C}, C(M))$ is represented by the Fredholm triple $(\Gamma(E), \phi, 0)$ where $\Gamma(E)$ is the $C(M)$-module of continuous sections $M \to E$ equipped with the

$C(M)$-valued inner product given by fibrewise application of the Hermitian inner product on E and $\phi : \mathbb{C} \hookrightarrow \mathcal{B}(\Gamma(E))$ is the standard embedding.

Under the Kasparov product map [3]

$$KK(\mathbb{C}, C(M)) \times KK(C(M), \mathbb{C}) \to KK(\mathbb{C}, \mathbb{C}) = \mathbb{Z}$$

which in this case corresponds to the usual Kronecker product pairing of K-homology and topological K-theory (i.e. K-cohomology)

$$\begin{aligned} K^0(M) \times K_0(M) &\to \mathbb{Z} \\ (c, h) &\mapsto \langle c, h \rangle \end{aligned}$$

the pair $([E], [D_g])$ is sent to $\mathrm{ind}(D_{g,E}) \in \mathbb{Z}$.

This point of view may be generalized by allowing twisting bundles $E \to M$ which are locally trivial bundles of finitely generated right Hilbert A-modules where A is a unital C^*-algebra.

We recall [39, 44] that each finitely generated Hilbert A-module bundle $E \to M$ is isomorphic to an orthogonal direct summand of a trivial A-module bundle $M \times A^n \to M$ where A^n carries the canonical A-valued inner product

$$\langle (a_1, \ldots, a_n), (b_1, \ldots, b_n) \rangle \mapsto a_1^* b_1 + \ldots + a_n^* b_n.$$

We can take this description as definition of finitely generated Hilbert A-module bundles.

Let $E \to M$ be a finitely generated Hilbert A-module bundle. We associate to $E \to M$ a KK-class $[E] \in KK(\mathbb{C}, C(M) \otimes A)$ as follows. First note that the space $\Gamma(E)$ of continuous sections in E is a finitely generated Hilbert $(C(M) \otimes A)$-module and the identity $\Gamma(E) \to \Gamma(E)$ is a $(C(M) \otimes A)$-compact (indeed finite rank) operator by a partition of unity argument. Therefore the triple $(\Gamma(E), \phi, 0)$, where $\phi : \mathbb{C} \to \mathcal{B}(\Gamma(E))$ is the standard embedding, defines an element in $KK(\mathbb{C}, C(M) \otimes A)$.

Using the Kasparov product (which we again interprete as a Kronecker product pairing)

$$\begin{aligned} KK(\mathbb{C}, C(M) \otimes A) \times KK(C(M), \mathbb{C}) &\to KK(\mathbb{C}, A) = K_0(A) \\ (c, h) &\mapsto \langle c, h \rangle \end{aligned}$$

we have a pairing of generalized elliptic differential operators on M and finitely generated Hilbert A-module bundles.

If M is a Riemannian spin manifold of even dimension, then the element in $\langle [E], [D_g] \rangle \in K_0(A)$ can be interpreted as the index of the Dirac operator D_g twisted with the bundle E, cf. [3]. Hence, for the special case when $E \to M$ is the Mishchenko-Fomenko bundle, the class $\langle [E], [D_g] \rangle$ coincides with the Rosenberg index $\alpha(M)$ defined in Sect. 2.

We will now single out those K-homology classes $h \in K_0(M)$ which can be detected by finitely generated Hilbert A-module bundles of arbitrarily small curvature. In the following let M be a closed smooth Riemannian manifold. In order to avoid the discussion of smooth bundles and curvature notions for infinite dimensional bundles we proceed as follows.

Recall that the *path groupoid* $\mathcal{P}_1(M)$ of M has as objects the points in M and as morphisms $\mathcal{P}_1(M)(x, y)$ the set of piecewise smooth paths $[0, 1] \to M$ connecting x and y. This is a topological category, in particular both the sets of objects and morphisms are topological spaces.

Let A be a unital C^*-algebra and let $E \to M$ be a finitely generated Hilbert A-module bundle. The *transport category* $\mathcal{T}(E)$ has as objects the points in M and as set of morphisms

$$\mathcal{T}(E)(x, y) := \mathrm{Iso}_A(E_x, E_y).$$

This is again a topological category where the set of morphisms is topologized by choosing local trivializations in order to identify nearby fibres of $E \to M$ and the set of Hilbert A-module isomorphisms $\mathrm{Iso}_A(E_x, E_y)$ is topologized as a subset of the Banach space $\mathrm{Hom}_A(E_x, E_y)$.

A *holonomy representation* on $E \to M$ is a continuous functor

$$\mathcal{H} : \mathcal{P}_1(M) \to \mathcal{T}(E).$$

It is called *ϵ-close to the identity at scale ℓ*, if for each $x \in M$ and each closed loop $\gamma \in \mathrm{Mor}(\mathcal{P}_1(M))$ based at $x \in M$ and of length $\ell(\gamma) \leq \ell$ we have

$$\|\mathcal{H}(\gamma) - \mathrm{id}_{E_x}\| < \epsilon \cdot \ell(\gamma).$$

Here we use the operator norm on the left hand side.

The following proposition establishes a link to the notion to parallel transport in differential geometry.

Proposition 3.4. *Depending on M^n there are a real constants $C, \ell > 0$ so that the following holds. Let $E \to M$ be a finite dimensional smooth Hermitian bundle of rank d equipped with a smooth Hermitian connection ∇ whose curvature $\eta \in \Omega^2(M; \mathfrak{u}(d))$ is norm bounded by ϵ. Then the parallel transport with respect to ∇ is $(C \cdot \epsilon)$-close to the identity at scale ℓ.*

Proof. By a Lebesgue number argument there is a small $\ell > 0$ and a cover of M^n by finitely many closed subsets $D_1, \ldots, D_k \subset M$ so that the following holds: Each D_i is diffeomorphic to the n-dimensional unit cube $[0, 1]^n \subset \mathbb{R}^n$ and each closed loop in M of length at most ℓ is contained in a subset D_i. It is hence enough to prove the assertion for a closed loop $\gamma \in \mathrm{Mor}(\mathcal{P}_1(M))$ contained in one of these subsets $D_i \subset M$ and based at a point $x \in D_i$. In the following we write D instead of D_i and identify D and $[0, 1]^n$ by a fixed diffeomorphism.

Let $E \to M$ be a Hermitian bundle of rank d as described in the proposition. We construct a trivialization of $E|_D \to M$ by choosing an isomorphism $E|_{(0,\ldots,0)} \cong \mathbb{C}^d$

and extending the trivialization inductively into each of the n coordinate directions by parallel transport. We denote the induced connection one form with respect to this trivialization by $\omega \in \Omega^1(D; \mathfrak{u}(d))$.

Now an argument similiar to [20, Lemma 2.3], but using the Riemannian metric on $[0,1]^n$ induced by M, shows that there is a number $C > 0$, which depends on D, but not on the bundle $E \to M$, so that

$$\|\omega|_D\| \leq C \cdot \|\eta|_D\|,$$

where we use the operator norm on $\mathfrak{u}(d)$ and the maximum norms on the unit sphere bundles of T^*D and $\Lambda^2 T^* D$.

Let $\phi : [0,1] \to E$ be a parallel vector field along a piecewise smooth (not necessarily closed) path $\zeta : [0,1] \to D \subset M$. By virtue of the given trivialization consider ϕ as a smooth map $[0,1] \to \mathbb{C}^d$. As such it satisfies the differential equation

$$\phi'(t) + (\omega_{\gamma(t)}(\gamma'(t))) \cdot \phi(t) = 0$$

and it follows that

$$\|\phi(1) - \phi(0)\| \leq \exp\left(\ell(\zeta) \cdot \|\omega|_D\|\right) \cdot \|\phi(0)\|.$$

Because we started with a Hermitian connection on E we get $\|\phi(1)\| = \|\phi(0)\|$ which implies that we can assume (by subdividing ζ into small pieces and appealing to the triangle inequality) that $\ell(\zeta)$ is arbitrarily small. Because $\exp : \mathbb{C}^d \to \mathbb{C}^d$ is uniformly Lipschitz continuous on each bounded neighbourhood of 0 with Lipschitz constant arbitrarily close to 1 we hence obtain

$$\|\phi(1) - \phi(0)\| \leq 1.5 \cdot \ell(\zeta) \cdot \|\omega|_D\| \cdot \|\phi(0)\|$$

from which the claim of the proposition follows. □

Definition 3.5. Let M be a closed smooth manifold and let $h \in K_0(M) \otimes \mathbb{Q}$. We say that h has *infinite K-area*, if there is a Riemannian metric on M and a number $\ell > 0$ so that the following holds: For each $\epsilon > 0$ there is a unital C^*-algebra A and a finitely generated Hilbert A-module bundle $E \to M$ which carries a holonomy representation which is ϵ-close to the identity at scale ℓ and satisfies

$$\langle [E], h \rangle \neq 0 \in K_0(A) \otimes \mathbb{Q}$$

where $[E] \in KK(\mathbb{C}, C(M) \otimes A)$ is the element represented by $E \to M$. If h is not of infinite K-area, we say that it is *of finite K-area*.

A class $h \in H_{ev}(M; \mathbb{Q})$ is defined to be of infinite K-area, if the class $\mathrm{ch}^{-1}(h) \in K_0(M) \otimes \mathbb{Q}$ is of infinite K-area.

By adapting the involved scale appropriately it is clear that for testing whether h is of infinite K-area or not any Riemannian metric on M can be used.

The notion of finitely generated Hilbert A-module bundles can be generalized to C^*-algebras without unit. However, in the context of Definition 3.5, this does not result in a wider class of K-homology classes of infinite K-area, since any finitely generated Hilbert A-module bundle is in a trivial way also a finitely generated Hilbert A^+-module bundle over the unitalization A^+ of A. This procedure does not change the property of $\langle [E], h \rangle$ being zero or not (in the rationalization of the K-homology of A and A^+ respectively).

Our Definition 3.5 is inspired by the preprint [28] where the property of finite K-area is investigated from a homological perspective. In contrast to the approach in loc. cit. and in the original source [14] we do not further quantify classes of finite K-area, since we will be concentrating on the property of infinite K-area as one instance of a largeness property besides enlargeability and essentialness. The discussion in [28] and other previous papers is restricted to finite dimensional smooth Hermitian vector bundles as twisting bundles $E \to M$ occuring in our Definition 3.5. Our more general setting is needed in connection with enlargeability questions and applications to the strong Novikov conjecture, see Sect. 4.

By a suspension procedure we can also define classes in $h \in K_1(M) \otimes \mathbb{Q}$ of infinite K-area by requiring that the class $h \times [S^1]_K \in K_0(M \times S^1) \otimes \mathbb{Q}$ be of infinite K-area, with an arbitrary choice of a K-theoretic fundamental class $[S^1]_K \in K_1(S^1)$. Note that with this definition the class $[S^1]_K \in K_1(S^1) \otimes \mathbb{Q}$ is of infinite K-area. The following discussion can be extended to K-homology classes of odd degree, but we restrict our exposition to classes in $K_0(M) \otimes \mathbb{Q}$ for simplicity.

The following two facts are similiar to Propositions 2 and 3 in [28], cf. also Proposition 3.4. and Theorem 3.6 in [7].

Proposition 3.6. *The elements of finite K-area in $K_0(M) \otimes \mathbb{Q}$ form a rational vector subspace.*

Proof. Obviously $0 \in K_0(M) \otimes \mathbb{Q}$ is of finite K-area. If $h \in K_0(M) \otimes \mathbb{Q}$ is of infinite K-area, then the same is true for any nonzero rational multiple of h. This implies that the set of elements of finite K-area is closed under scalar multiplication. Now assume that $h + h'$ is of infinite K-area. It follows from Definition 3.5 that either h or h' are of infinite K-area (choose $\epsilon := \frac{1}{k}$ with $k = 1, 2, \ldots$). This shows that the set of elements of finite K-area is closed under addition. □

Proposition 3.7. *If $f : M \to M'$ is a continuous map, then $f_* : K_0(M) \otimes \mathbb{Q} \to K_0(M') \otimes \mathbb{Q}$ restricts to a map between vector subspaces consisting of elements of finite K-area. In particular, the vector subspace of elements of finite K-area in $K_0(M) \otimes \mathbb{Q}$ is an invariant of the homotopy type of M.*

We will return to homological aspects of largeness properties in Sect. 5. The notion of infinite K-area is illustrated by the following examples.

Assume that M is an oriented manifold of even dimension $2n$ which has infinite K-area in the sense of Gromov [14]. By definition this means that for each $\epsilon > 0$ there is a finite dimensional smooth Hermitian vector bundle $V \to M$ with a Hermitian connection whose curvature form in $\Omega^2(M; \mathfrak{u}(d))$ (where $d = \operatorname{rk} V$) has norm smaller than ϵ and with at least one nonvanishing Chern number.

Using linear combinations of tensor products and exterior products of V one can show that there is a Hermitian bundle $E \to M$ with Hermitian connection whose curvature has norm smaller than $C \cdot \epsilon$ (where C is a bound which depends only on $\dim M$) and which satisfies

$$\langle \mathrm{ch}(E), \mathrm{PD}(\hat{A}(M)) \rangle \neq 0 \in H_0(M;\mathbb{Q}),$$

where $\mathrm{PD}(\hat{A}(M))$ is the Poincaré dual in $H_{ev}(M;\mathbb{Q})$ of the $\hat{A}$-polynomial of M.

The precise argument is carried out in [10] where the following fact is shown. There is a number N depending only on $\dim M$ with the following property: Assume that $V \to M$ is a complex vector bundle and assume that all bundles $V' \to M$ which may be constructed out of V by at most N operations of the form direct sum, tensor product and exterior product satisfy

$$\langle \mathrm{ch}(V'), \mathrm{PD}(\hat{A}(M)) \rangle = 0 \in H_0(M;\mathbb{Q}).$$

Then all Chern numbers of $V \to M$ are zero.

Considering Hermitian vector bundles as finitely generated Hilbert $\mathbb{C}$-module bundles this means in the language of Definition 3.5 that the class $\mathrm{PD}(\hat{A}(M)) \in H_{ev}(M;\mathbb{Q})$ has infinite K-area (here we use that the Chern character is compatible with the Kronecker pairing). If M is equipped with a spin structure, this element is equal to $\mathrm{ch}([M]_K)$, the Chern character applied to the K-theoretic fundamental class of M, and hence we have shown that under the stated assumptions the class $[M]_K$ has infinite K-area in our sense.

By a similar argument one shows that if M has infinite K-area in the sense of Gromov, then

$$[M]_H \in H_{2n}(M;\mathbb{Q})$$

has infinite K-area, where $[M]_H \in H_{2n}(M;\mathbb{Q})$ is the homological fundamental class of M.

As a second example, cf. [20, Sect. 4], assume that M is area-enlargeable and that the covers $\overline{M} \to M$ in Definition 3.1 can always be assumed to be finite. By pulling back a suitable Hermitian bundle $V \to S^{2n}$ with connection to these covers along the maps $f_\epsilon : \overline{M} \to S^{2n}$ and wrapping these bundles up to get finite dimensional Hermitian bundles $E \to M$ with small curvature, one can show that the classes $[M]_H \in H_{2n}(M;\mathbb{Q})$ and $[M]_K \in K_0(M) \otimes \mathbb{Q}$ (if M is spin) have infinite K-area.

More generally assume that M^{2n} is area-enlargeable with no restriction on the covers $\overline{M} \to M$. Then [21, Proposition 1.5] implies that the classes $[M]_H$ and $[M]_K$, respectively, have infinite K-area. In this case we need infinite dimensional bundles $E \to M$ which shows the usefulness of Definition 3.5 in the general context of Hilbert A-module bundles where A is a C^*-algebra different from $\mathbb{C}$.

For later reference we state the last observation seperately.

Proposition 3.8. *Let M be area-enlargeable and of even dimension. Then the K-area of $[M]_H$ is infinite. If M is equipped with a spin structure, then also the K-area of $[M]_K$ is infinite.*

We denote by

$$\alpha : K_0(M) \to K_0(B\pi_1(M)) \xrightarrow{\mu} K_0(C^*_{max}\pi_1(M))$$

the composition of the map induced by the classifying map $M \to B\pi_1(M)$ and the assembly map. If M is a spin manifold of even dimension, note the equations

$$\alpha(M) = \alpha([M]_K)$$

(the left hand side coincides with the Rosenberg index) and – more generally –

$$\alpha(h) = \langle [E], h \rangle \in K_0(C^*_{max}\pi_1(M)) \otimes \mathbb{Q}$$

for all $h \in K_0(M) \otimes \mathbb{Q}$ where $E \to M$ is the Mishchenko-Fomenko bundle for $C^*_{max}\pi_1(M)$.

The following is the main result of our paper.

Theorem 3.9. *Let M be a closed connected smooth manifold and let $h \in K_0(M)\otimes \mathbb{Q}$ be of infinite K-area. Then*

$$\alpha(h) \neq 0 \in K_0(C^*_{max}\pi_1(X)) \otimes \mathbb{Q}.$$

We note the following implication for the Rosenberg index.

Corollary 3.10. *Let M be a closed spin manifold of even dimension whose K-theoretic fundamental class has infinite K-area. Then*

$$\alpha(M) \neq 0 \in K_0(C^*_{max}\pi_1(M)).$$

In particular, closed even-dimensional spin manifolds of infinite K-area in the sense of Gromov [14] have nonvanishing Rosenberg index. (A similar result holds, if M is odd dimensional.)

The proof of Theorem 3.9 is based on the construction of "infinite product bundles" from [20]. We shall explain how this construction fits the setting of the paper at hand.

Let $(E_k)_{k\in\mathbb{N}}$ be a sequence of finitely generated Hilbert A_k-module bundles over M, where (A_k) is a sequence of unital C^*-algebras. We assume that the fibre of E_k is isomorphic (as a Hilbert A_k-module) to $q_k A_k$ where $q_k \in A_k$ is a (self-adjoint) projection. This assumption is important for our construction. In general the fibre of E_k is of the form $q \cdot (A_k)^n$ for some n with a projection $q \in \mathrm{Mat}(A_k, n)$. In this case we use the same transition functions as for E_k to construct a Hilbert $\mathrm{Mat}(A_k, n)$-module bundle of the required form. By Morita equivalence of A_k and $\mathrm{Mat}(A_k, n)$ this does not affect the K-theoretic considerations relevant for our discussion.

We consider the unital C^*-algebra A consisting of norm bounded sequences

$$(a_k)_{k\in\mathbb{N}} \in \prod_{k=1}^{\infty} A_k$$

and wish to construct a Hilbert A-module bundle $E \to M$ with fibre qA, where $q = (q_k)$ is the product of the projections q_k, by taking the "infinite product" of the bundles E_k. However, taking the infinite product of the transition functions for the bundles E_k may not result in continuous transition functions for the infinite product bundle. The following example indeed shows that an infinite product construction of this kind may be obstructed by topological properties of the bundles E_k.

Example 3.11. Let $E_k \to S^2$ be the complex line bundle with Chern number k. Assume we have a Hilbert A-module bundle $E \to S^2$ over the C^*-algebra $A = \prod_k \mathbb{C}$ (which is equal to the standard seperable Hilbert space) with typical fibre $V = \prod_k \mathbb{C}$ and Lipschitz continous transition functions in diagonal form so that the kth component of this bundle is isomorphic to E_k as a complex line bundle.

Restricting the transition functions of E to the single factors leads to trivializations for the bundles $E_k \to S^2$ whose transition functions have uniformly (in k) bounded Lipschitz constants. This implies that the Euler numbers of the bundles E_k are bounded, contrary to our assumption.

This example indicates that we need to choose Lipschitz trivializations of the bundles E_k so that the resulting transition functions have uniformly bounded Lipschitz constants. This can be achieved as follows.

Proposition 3.12. *Assume that each bundle $E_k \to M$ is equipped with a holonomy representation $\mathcal{H}_k$ so that $\mathcal{H}_k$ is ϵ-close to the identity at scale ℓ where the constants ϵ and ℓ are independent of k, and M is equipped with a fixed Riemannian metric. Then there is a finitely generated Hilbert A-module bundle $V \to M$ with transition functions in diagonal form and so that the kth component of this bundle is isomorphic to E_k as an A_k-Hilbert module bundle.*

Proof. We start with a cover of M^n by finitely many closed subsets $(D_i)_{i\in I}$ each of which is diffeomorphic to the n-dimensional unit cube $[0,1]^n \subset \mathbb{R}^n$ and so that the interiors of these subsets still cover M. The size of each D_i can be assumed to be small compared to ℓ.

For each k, using the holonomy representation $\mathcal{H}_k$, we trivialize the bundle E_k over each subset D_i inductively into each of the n coordinate directions (compare the proof of Proposition 3.4).

This leads to local trivializations of $E_k|_{D_i}$ whose transition maps (for fixed k, but varying i) have uniformly bounded (in i and k) Lipschitz constants. Hence the product of these transition maps can be used to define the Hilbert A-module bundle $V \to M$ as required. □

We remark that the product bundle $V \to M$ is a bundle of finitely generated Hilbert A-modules isomorphic to qA by our assumption that E_k has typical fibre $q_k A_k$.

For the proof of Theorem 3.9 we assume that $h \in K_0(M) \otimes \mathbb{Q}$ and that (E_k) is a sequence of Hilbert A_k-module bundles with fibres $q_k A_k$ so that $\langle [E_k], h\rangle \neq 0 \in K_0(A_k) \otimes \mathbb{Q}$ for all k. Furthermore, we assume that E_k is equipped with a holonomy representation $\mathcal{H}_k$ which is $1/k$-close to the identity at some scale ℓ which is independent of k.

We consider the Hilbert A-module bundle $V \to M$ constructed in Proposition 3.12.

Starting from V we can construct various other Hilbert module bundles over M as follows. Let

$$\psi_k : A \to A_k$$

denote the projection onto the kth component. Moreover, we denote by

$$A' := \bigoplus_{k=1}^{\infty} A_k \subset A$$

the closed two sided ideal consisting of sequences in A tending to zero and by

$$Q := A/A'$$

the quotient C^*-algebra. Finally, let

$$\psi : A \to Q$$

be the quotient map.

We obtain Hilbert A_k-bundle isomorphisms

$$E_k \cong V \otimes A_k$$

and a Hilbert Q-module bundle

$$W := V \otimes Q$$

with typical fibre qQ, where we identify $q \in A$ and its image in Q.

The following fact is crucial

Proposition 3.13. *The bundle W has local trivializations with locally constant transition maps. More precisely, it can be written as an associated bundle*

$$W = \widetilde{M} \times_{\pi_1(M)} qQ$$

for some unitary representation $\pi_1(M) \to \mathrm{Hom}_Q(qQ, qQ)$.

Proof. The family of holonomy representations $(\mathcal{H}_k)$ induces a holonomy representation on W which is equal to the identity on each closed loop of length at most ℓ in M (and hence on contractible loops of arbitrary length), because the holonomy representation $\mathcal{H}_k$ is $1/k$-close to the identity at scale ℓ. Using this holonomy representation on W we construct the desired local trivializations of W. □

These facts in combination with naturality properties of Kasparov KK-theory allow us to show that $\alpha(h) \neq 0 \in K_0(C^*_{max}\pi_1(M)) \otimes \mathbb{Q}$. The holonomy representation for the bundle W induces an involutive map

$$\pi_1(M) \to \mathrm{Hom}_Q(qQ, qQ) = qQq$$

with values in the unitaries of the C^*-algebra qQq. Hence, by the universal property of $C^*_{max}\pi_1(M)$ we get an induced map of C^*-algebras

$$\phi : C^*_{max}\pi_1(M) \to qQq \hookrightarrow Q.$$

Note that this step is not possible in general, if we use the reduced C^*-algebra $C^*_{red}\pi_1(M)$ instead. Let $E = \widetilde{M} \times_{\pi_1(M)} C^*_{max}\pi_1(M) \to M$ be the Mishchenko-Fomenko bundle.

We study the commutative diagram

$$\begin{array}{ccccc} K_0(M) & \xrightarrow{\langle [E], - \rangle} & K_0(C^*_{max}\pi_1(M)) & \xrightarrow{\phi_*} & K_0(Q) \\ \downarrow = & & & & \downarrow = \\ K_0(M) & \xrightarrow{\langle [V], - \rangle} & K_0(A) & \xrightarrow{\psi_*} & K_0(Q) \end{array}$$

The composition

$$K_0(M) \stackrel{\langle [V], - \rangle}{\longrightarrow} K_0(A) \stackrel{(\psi_k)_*}{\longrightarrow} K_0(A_k)$$

sends the element h to $\langle [E_k], h \rangle \in K_0(A_k)$ which is different from zero by assumption. This implies that under the map

$$\chi : K_0(A) \to \prod_k K_0(A_k)$$

$$z \mapsto ((\psi_k)_*(z))_{k=1,2,\ldots}$$

the element $z := \langle [V], h \rangle$ is sent to a sequence all of whose components are different from zero. We will conclude from this that also $\psi_*(z) \neq 0$ finishing the proof of Theorem 3.9.

Consider the long exact sequence in K-theory induced by the short exact sequence

$$0 \to A' \to A \to Q \to 0.$$

Using the fact that K-theory commutes with direct limits we have a canonical isomorphism

$$K_0(A') \cong \bigoplus_k K_0(A_k).$$

Assume that $\psi_*(z) = 0$. This implies that χ maps z to a sequence $(z_k) \in \prod_k K_0(A_k)$ with only finitely many nonzero entries. But this contradicts the calculation that we carried out before. Hence $\psi_*(z) \neq 0$.

4 The Strong Novikov Conjecture

The method presented in the previous paragraph can be used to prove a special case of the strong Novikov conjecture. Let G be a discrete group and let $\Lambda^*(G) \subset H^*(BG;\mathbb{Q})$ be the subring generated by $H^{\leq 2}(BG;\mathbb{Q})$

Theorem 4.1 ([22]). *Let $h \in K_0(BG) \otimes \mathbb{Q}$ be a K-homology class with the following property: There is a class $c \in \Lambda^*(G)$ so that $\langle c, \mathrm{ch}(h)\rangle \neq 0 \in H_0(BG;\mathbb{Q}) = \mathbb{Q}$. Then under the assembly map*

$$K_0(BG) \otimes \mathbb{Q} \to K_0(C^*_{max} G) \otimes \mathbb{Q}$$

the element h is sent to a a non-zero class.

As a corollary one obtains the following special case of the classical Novikov conjecture.

Corollary 4.2 ([9,64]). *Let M be a connected closed oriented manifold, let G be a discrete group and let $f : M \to BG$ be a continuous map. Then for all $c \in \Lambda^*(G)$ the higher signature $\langle \mathcal{L}(M) \cup f^*(c), [M]\rangle$ is an oriented homotopy invariant, where $\mathcal{L}(M)$ denotes the Hirzebruch L-polynomial.*

We will establish Theorem 4.1 as a fairly straightforward consequence of Theorem 3.9. It illustrates again the flexibility of the notion of infinite K-area in Definition 3.5 based on Hilbert module bundles. For simplicity we restrict to the case when there is a class $c \in H^2(BG;\mathbb{Q})$ with $\langle c, \mathrm{ch}(h)\rangle \neq 0$. Furthermore, without loss of generality, we can assume that G is finitely presented. The general case follows by applying a direct limit argument.

Using the description of K-homology due to Baum and Douglas [2] there is a closed connected spin manifold M of even dimension (which can be chosen arbitrarily large) together with a finite dimensional complex vector bundle $V \to M$ and a continuous map $f : M \to BG$ so that

$$f_*([V] \cap [M]_K) = h.$$

Here we regard again $V \to M$ as an element in $K^0(M)$ and use the cap product pairing

$$\cap : K^0(M) \times K_0(M) \to K_0(M).$$

As G is finitely presented we can assume that f induces an isomorphism of fundamental groups. In view of Theorem 3.9 we need to show that the class $[V] \cap [M]_K \in K_0(M)$ is of infinite K-area.

Let $L \to M$ be the complex line bundle classified by $f^*(c)$. We pick a Hermitian connection on L and denote by $\eta \in \Omega^2(M; i\mathbb{R})$ the associated curvature form. Because the universal cover of BG is contractible, the pull back $\pi^*(L) \to \widetilde{M}$ of L to the universal cover $\pi : \widetilde{M} \to M$ is trivial. We fix a trivialization and denote the 1-form associated to the pull back connection by $\omega \in \Omega^1(\widetilde{M}; i\mathbb{R})$. The curvature form $\pi^*(\eta)$ is equal to $d\omega$, since U(1) is abelian. However, the connection 1-form ω is in general not invariant under the action of the deck transformation group on $\widetilde{M}$, because in this case the curvature form η would be exact and hence $L \to M$ would be the trivial line bundle.

We will now "flatten" the bundle $L \to M$ by scaling its curvature by a constant $0 < t < 1$. Unfortunately, this cannot be done directly, because the first Chern class of L would no longer be integral.

The following construction originating from [22] gives a solution to this problem by considering infinite dimensional bundles. At first we consider the Hilbert space bundle

$$E = \widetilde{M} \times_G l^2(G) \to M$$

where $l^2(G)$ is the set of square summable complex valued functions on G and G acts on the left of $l^2(G)$ by the formula

$$(\gamma\psi)(x) = \psi(x\gamma)$$

and on the right of $\widetilde{M}$ by $(x, g) \mapsto g^{-1}x$. Let $0 < t < 1$. We consider the G-invariant connection 1-form on $\widetilde{M} \times l^2(G)$ which on the subbundle

$$\widetilde{M} \times \mathbb{C} \cdot 1_g \subset \widetilde{M} \times l^2(G)$$

concides with $(g^{-1})^*(t\omega)$. Here $1_g \in l^2(G)$ is the characteristic function of $g \in G$. Because this one form is G-invariant, we obtain an induced connection ∇^t on the Hilbert space bundle E whose curvature form is norm bounded by $t \cdot \|\eta\|$. In other words, the Hilbert space bundle E can be equipped with holonomy representations which are arbitrarily close to the identity (at some fixed scale). It hence remains to show that E detects the K-homology class $[V] \cap [M]_K$.

However, by Kuiper's theorem, any Hilbert space bundle is trivial. Therefore we will first reduce the structure group of E in a canonical way. This will result in finitely generated Hilbert A_t-module bundles $E_t \to M$ with appropriate unital C^*-algebras A_t, where $t \in (0, 1]$. The algebras A_t will depend on t.

We fix a base point $p \in M$ and choose a point $q \in \widetilde{M}$ above p. The fibre over p is then identified with the Hilbert space $l^2(G)$. Now we define

$$A_t \subset B(l^2(G))$$

as the norm-linear closure of all maps $l^2(G) \to l^2(G)$ arising from parallel transport with respect to ∇^t along piecewise smooth loops in M based at p. We furthermore define a bundle $E_t \to M$ whose fibre over $x \in M$ is given by the norm-linear closure in $\mathrm{Hom}(E|_p, E|_x)$ of all Hilbert space isomorphisms $E|_p \to E|_x$ aring from parallel transport with respect to ∇^t along piecewise smooth curves connecting p with x. In this way we obtain, for each $t \in (0, 1]$, a free Hilbert A_t-module bundle of rank 1 where the A_t-module structure on each fibre is induced by precomposition with parallel transport along piecewise smooth loops based at p.

Now, on the one hand, parallel transport with respect to ∇^t induces a holonomy representation on $E_t \to M$ which, for small enough t, is arbitrarily closed to the identity (at a fixed scale which is independent of t).

On the other hand, each of the algebras A_t carries a canonical trace

$$\tau_t : A_t \to \mathbb{C}, \; \tau_t(\psi) = \langle \psi(1_e), 1_e \rangle$$

where $1_e \in l^2(G)$ is the characteristic function of the neutral element $e \in G$ and $\langle -, - \rangle$ is the inner product on $l^2(G)$. For details we refer to [22, Lemma 2.2]. Using the Chern-Weil calculus from [39] we obtain

$$\tau_t(\langle [E_t], [V] \cap [M]_K \rangle) = \langle \exp(tc), \mathrm{ch}(h) \rangle \in \mathbb{R}[t].$$

See also [22]. The last polynomial is nonzero by our assumption $\langle c, \mathrm{ch}(h) \rangle \neq 0$. In particular, for infinitely many $k \in \mathbb{N}$ we have

$$\langle [E_{1/k}], [V] \cap [M]_K \rangle \neq 0 \in K_0(A_{1/k}) \otimes \mathbb{Q}.$$

This implies that $[V] \cap [M]_K$ is a class of infinite K-area and together with Theorem 3.9 finishes the proof of Theorem 4.1.

5 Homological Invariance of Essentialness

Recall from Definition 2.8 that a closed oriented manifold M^n is called *essential*, if the classifying map $\phi : M \to B\pi_1(M)$ satisfies

$$\phi_*([M]_H) \neq 0 \in H_n(B\pi_1(M); \mathbb{Q}).$$

Essential manifolds obey Gromov's systolic inequality:

Theorem 5.1 ([13]). *Let M be an essential Riemannian manifold of dimension n. Then there is a noncontractible loop $\gamma : [0,1] \to M$ satisfying*

$$\ell(\gamma) \leq C(n) \cdot \mathrm{vol}(M)^{1/n}$$

where the constant $C(n)$ depends only on n.

We show the following implication.

Theorem 5.2. *Let M be an oriented manifold of even dimension $2n$. If the class $[M]_H \in H_{2n}(M;\mathbb{Q})$ has infinite K-area, then M is essential.*

Proof. Let $E \to M$ be the Mishchenko-Fomenko bundle. The proof of Theorem 4.9 is based on the commutative diagram

$$\begin{array}{ccccc}
K_0(M)\otimes\mathbb{Q} & & \xrightarrow{\langle [E],-\rangle} & & K_0(C^*_{max}\pi_1(M))\otimes\mathbb{Q} \\
\downarrow = & & & & \downarrow = \\
K_0(M)\otimes\mathbb{Q} & \xrightarrow{\phi_*} & K_0(B\pi_1(M))\otimes\mathbb{Q} & \xrightarrow{\mu} & K_0(C^*_{max}\pi_1(M))\otimes\mathbb{Q} \\
\cong \downarrow \mathrm{ch} & & \cong \downarrow \mathrm{ch} & & \\
H_{ev}(M;\mathbb{Q}) & \xrightarrow{\phi_*} & H_{ev}(B\pi_1(M),\mathbb{Q}) & &
\end{array}$$

Indeed, by Theorem 3.9 the image of $\mathrm{ch}^{-1}([M]_H)$ under the map in the first line is non-zero. □

This theorem implies:

- Closed manifolds of infinite K-area in the sense of Gromov are essential.
- ([20, 21]) Area-enlargeable manifolds are essential (use Proposition 3.8).

The second implication can be obtained without referring to K-theoretic considerations. This is carried out in [7], where several largeness properties of Riemannian manifolds are investigated from a purely homological point of view. The best results can be obtained for enlargeable manifolds, for which we have the following homological invariance result.

Theorem 5.3 ([7]). *Let G be a finitely presented group. Then there is a rational vector subspace*

$$H_*^{sm}(BG;\mathbb{Q}) \subset H_*(BG;\mathbb{Q})$$

with the following property: Let M be a closed oriented manifold of dimension n. Then M is enlargeable, if and only if under the classifying map $\phi : M \to B\pi_1(M)$ we have

$$\phi_*([M]) \notin H_n^{sm}(B\pi_1(M);\mathbb{Q})$$

This result indeed implies that enlargeable manifolds are essential, because $0 \in H_n(B\pi_1(M);\mathbb{Q})$ is contained in every vector subspace of $H_n(B\pi_1(M);\mathbb{Q})$.

Theorem 5.3 can be seen as a form of homological invariance of enlargeability. The proof is based on the following definition of enlargeable homology classes in simplicial complexes.

Definition 5.4 ([7]). Let C be a connected simplicial complex with finitely generated fundamental group. A homology class $h \in H_n(C;\mathbb{Q})$ is called enlargeable, if the following holds: Let $S \subset C$ be a finite subcomplex carrying h and inducing a surjection on π_1. Then, for every $\epsilon > 0$, there is a cover $\overline{C} \to C$ and an ϵ-Lipschitz map $\overline{S} \to S^n$ which is constant outside a compact subset of $\overline{S}$ and sends the transfer $\mathrm{tr}(h) \in H_n^{lf}(\overline{S};\mathbb{Q})$ in the locally finite homology of $\overline{S}$ to a nonzero class in the reduced homology $\widetilde{H}_n(S^n;\mathbb{Q})$. Here $\overline{S}$ is the preimage of S under the covering map $\overline{C} \to C$.

It is shown in [7] that the condition for c described in this definition is independent of the finite subcomplex $S \subset C$ carrying c and inducing a surjection on π_1. Using this property it is not difficult to prove the following fact, see [7, Prop. 3.4.].

Proposition 5.5. *Let $f : C \to D$ be a continuous map inducing an isomorphism of (finitely generated) fundamental groups. Then a class $h \in H_*(C;\mathbb{Q})$ is enlargeable, if and only if the class $f_*(h) \in H_*(D;\mathbb{Q})$ is enlargeable.*

From this Theorem 5.3 follows, if we define $H_n^{sm}(BG;\mathbb{Q})$ as the subset consisting of all homology classes which are not enlargeable.

Theorem 5.3 transforms the problem of determining enlargeable manifolds to a problem in group homology: Given a finitely generated group G, determine $H_*^{sm}(BG;\mathbb{Q})$, the "small" group homology of G. In light of Theorem 5.3 and the fact that the fundamental classes of enlargeable manifolds are of infinite K-area (see Proposition 3.8) it is desirable to decide whether $H_*^{sm}(BG;\mathbb{Q})$ can be non-zero. This is answered in the positive in [7, Theorem 4.8] by use of the Higman 4-group [23]. Together with Theorem 5.3 this implies that there are essential manifolds which are not enlargeable, see [7, Theorem 1.5].

In contrast to these positive results we do not know, whether there are essential manifolds which are not area-enlargeable. The following conjecture is implied by the strong Novikov conjecture.

Conjecture 5.6. Let M be an essential manifold. Then its fundamental class in singular homology $[M]_H$ is of infinite K-area.

6 Rosenberg Index and the Reduced Group C^*-Algebra

Let M^n be a closed spin manifold. The method of Sect. 2 can be used equally well to construct an index obstruction to positive scalar curvature

$$\alpha(M) \in K_n(C^*_{red}\pi_1(M)).$$

The reduced group C^*-algebra does not share the universal property of the maximal group C^*-algebra which we used in the proof of Theorem 3.9.

Exploiting the connection of $C^*_{red}\pi_1(M)$ to coarse geometry [24] we can still prove

Theorem 6.1 ([19]). *Let M^n be an enlargeable spin manifold. Then*

$$\alpha(M) \neq 0 \in K_n(C^*_{red}\pi_1(M)).$$

We do not know whether the same conclusion holds for area-enlargeable spin manifolds. This would be implied by an affirmative answer to the following question.

Question 6.2. Does Theorem 3.9 remain true for the reduced group C^*-algebra?

References

1. Atiyah, M.: K-theory, Advanced Books Classics, Westview Press (1994)
2. Baum, P., Douglas, R.: K-homology and index theory. Operator algebras and applications, Part I (Kingston, Ont., 1980), pp. 117-173, Proc. Sympos. Pure Math. **38**, Amer. Math. Soc., Providence, RI (1982)
3. Blackadar, B.: K-Theory for Operator Algebras, MSRI Publications **5**, Springer
4. Böhm, Ch., Wilking, B.: Manifolds with positive curvature operators are space forms. Ann. of Math. (2) **167**(3), 1079–1097 (2008)
5. Brendle, S., Schoen, R.: Classification of manifolds with weakly $1/4$-pinched curvatures. Acta Math. **200**(1), 1–13 (2008)
6. Brendle, S., Schoen, R.: Manifolds with $1/4$-pinched curvature are space forms. J. Amer. Math. Soc. **22**(1), 287–307 (2009)
7. Brunnbauer, M., Hanke, B.: Large and small group homology. J. Topology **3**, 463–486 (2010)
8. Christ, U., Lohkamp, J.: Singular Minimal Hypersurfaces and Scalar Curvature. arXiv math.DG/0609338
9. Connes, A., Gromov, M., Moscovici, H.: Group cohomology with Lipschitz control and higher signatures. Geom. Funct. Anal. **3**(1), 1–78 (1993)
10. Davaux, H.: La K-aire selon M. Gromov. Séminaire de Théorie Spectrale et Géométrie. Vol. 21. Année 2002-2003, 9–35, Sémin. Théor. Spectr. Géom. **21**, Univ. Grenoble I, Saint-Martin-d'Hères, 2003
11. Dranishnikov, A.N.: On hypersphericity of manifolds with finite asymptotic dimension, Trans. Amer. Math. Soc. **355**, 155–167 (2003)
12. Dranishnikov, A.N.: On hypereuclidean manifolds, Geom. Dedicata **117**, 215–231 (2006)
13. Gromov, M.: Filling Riemannian manifolds, J. Differential Geom. **18**, 1–147 (1983)
14. Gromov, M.: Positive Curvature, Macroscopic Dimension, Spectral gaps and Higher Signatures. In: Functional Analysis on the Eve of the 21st Century, vol. II (eds.: S. Gindikin, J. Lepowsky, R. Wilson), Birkhäuser (1995)
15. Gromov, M., Lawson, H.B.: The classification of simply connected manifolds of positive scalar curvature, Ann. of Math. (2) **111**(3), 423–434 (1980)
16. Gromov, M., Lawson, H.B.: Spin and scalar curvature in the presence of a fundamental group. I. Ann. of Math. (2) **111**, 209–230 (1980)

17. Gromov, M.: Curvature, diameter and Betti numbers. Comment. Math. Helv. **56**(2), 179–195 (1981)
18. Gromov, M., Lawson, H.B.: Positive scalar curvature and the Dirac operator on complete Riemannian manifolds. Inst. Hautes Études Sci. Publ. Math. **58**, 83–196 (1984)
19. Hanke, B., Kotschick, D., Roe, J., Schick, T.: Coarse topology, enlargeability, and essentialness. Ann. Sci. École Norm. Sup. **41**, 471–493 (2008)
20. Hanke, B., Schick, T.: Enlargeability and index theory, J. Differential Geom. **74**, 293–320 (2006)
21. Hanke, B., Schick, T.: Enlargeability and index theory: infinite covers. K-Theory **38**, 23–33 (2007)
22. Hanke, B., Schick, T.: The strong Novikov conjecture for low degree cohomology. Geom. Dedicata **135**, 119–127 (2008)
23. Higman, G.: A finitely generated infinite simple group. J. London Math. Soc. **26**, 61–64 (1951)
24. Higson, N., Roe, J.: Analytic K-homology. Oxford Mathematical Monographs, Oxford University Press, UK (2004)
25. Hitchin, N.: Harmonic spinors, Advances in Math. **14**, 1–55 (1974)
26. Lawson, H.B., Michelsohn, M.-L.: Spin geometry, Princeton University Press, Princeton, NJ (1989)
27. Lichnerowicz, A.: Spineurs harmoniques. C. R. Acad. Sci. Paris **257**, 7–9 (1963)
28. Listing, M.: Homology of finite K-area. arXiv:1007.3166
29. Lohkamp, J.: Metrics of negative Ricci curvature. Ann. of Math. (2) **140**(3), 655–683 (1994)
30. Lohkamp, J.: The Higher Dimensional Positive Mass Theorem I. arXiv math.DG/0608795
31. Lohkamp, J.: Inductive Analysis on Singular Minimal Hypersurfaces. arXiv:0808.2035
32. Mathai, V.: The Novikov conjecture for low degree cohomology classes. Geom. Dedicata **99**, 1–15 (2003)
33. Mishchenko, A., Fomenko, A.: The index of elliptic operators over C^*-algebras, Izv. Akad.Nauk SSSR, Ser. Mat. **43**, 831-0859 (1979); English Translation: Mathematics of the USSR Izvestija **15**, 87–112 (1980)
34. Matthey, M., Oyono-Oyono, H., Pitsch, W.: Homotopy invariance .of higher signatures and 3-manifold groups. Bull. Soc. Math. France **136**(1), 1–25 (2008)
35. Rosenberg, J.: C^*-algebras, positive scalar curvature, and the Novikov conjecture, Pub. Math. IHES **58**, 197–212 (1983)
36. Rosenberg, J.: C^*-algebras, positive scalar curvature, and the Novikov conjecture. III.. Topology **25**(3), 319–336 (1986)
37. Rosenberg, J.: Manifolds of positive scalar curvature: a progress report. In: Surveys on Differerential Geometry, vol. XI: Metric and Comparison Geometry, International Press, London, UK (2007)
38. Schick, T.: A counterexample to the (unstable) Gromov-Lawson-Rosenberg conjecture. Topology **37**(6), 1165–1168 (1998)
39. Schick, T.: L^2-index theorems, KK-theory, and connections. New York J. Math. **11**, 387–443 (2005)
40. Schoen, R., Yau, S.T.: On the structure of manifolds with positive scalar curvature. Manuscripta Math. **28**, 159–183 (1979)
41. Smale, N.: Generic regularity of homologically area minimizing hypersurfaces in eight-dimensional manifolds. Comm. Anal. Geom. **1**(2), 217–228 (1993)
42. Stolz, S.: Simply connected manifolds of positive scalar curvature. Ann. of Math. (2) **136**(3), 511–540 (1992)
43. Stolz, S.: Manifolds of positive scalar curvature, In: Farrell, T., et al. (eds.) Topology of high dimensional manifolds (Trieste 2001), ICTP Lecture Notes, vol. **9**, 661–709 (2001)
44. Wegge-Olsen, N.E.: K-theory and C^*-algebras: A Friendly Approach. Oxford Science Publications, Oxford University Press, UK (2004)
45. Yamabe, H.: On a deformation of riemannian structures on compact manifolds. Osaka Math. J. **12**, 21–37 (1960)

Differential K-Theory: A Survey

Ulrich Bunke and Thomas Schick

Abstract Generalized differential cohomology theories, in particular differential K-theory (often called "smooth K-theory"), are becoming an important tool in differential geometry and in mathematical physics.

In this survey, we describe the developments of the recent decades in this area. In particular, we discuss axiomatic characterizations of differential K-theory (and that these uniquely characterize differential K-theory). We describe several explicit constructions, based on vector bundles, on families of differential operators, or using homotopy theory and classifying spaces. We explain the most important properties, in particular about the multiplicative structure and push-forward maps and will state versions of the Riemann–Roch theorem and of Atiyah–Singer family index theorem for differential K-theory.

1 Introduction

The most classical differential cohomology theory is ordinary differential cohomology with integer coefficients. It has various realizations, e.g. as smooth Deligne cohomology (compare [18]) or as Cheeger–Simons differential characters [32]. In the last decade, differential extensions of generalized cohomology theories, in particular of K-theory, have been studied intensively. In part, this is motivated by its application in mathematical physics, for the description of fields with quantization anomalies in abelian gauge theories, suggested by Freed in [36], compare also [44].

U. Bunke (✉)
Johannes-Kepler-Universität Regensburg, 93040 Regensburg, Germany
e-mail: ulrich.bunke@mathematik.uni-regensburg.de

T. Schick
Georg-August-Universität Göttingen, 37073 Göttingen, Germany
e-mail: schick@uni-math.gwdg.de

C. Bär et al. (eds.), *Global Differential Geometry*, Springer Proceedings in Mathematics 17, DOI 10.1007/978-3-642-22842-1_11,

The basic idea is that a differential cohomology theory should combine cohomological information with differential form information. More precisely, given a generalized cohomology theory E together with a natural transformation $\mathrm{ch}\colon E(X) \to H(X;N)$, to cohomology with coefficients in a graded real vector space N, and using an appropriate setup one can define the differential refinement $\hat{E}$ of E as a *homotopy* pullback

$$\begin{array}{ccc} \hat{E}(X) & \xrightarrow{I} & E(X) \\ \downarrow{\scriptstyle R} & & \downarrow{\scriptstyle \mathrm{ch}} \\ \Omega_{d=0}(X;N) & \xrightarrow{\mathrm{Rham}} & H(X;N). \end{array}$$

The natural transformations I (the underlying cohomology class) and R (the characteristic closed differential form) are essential parts of the picture. With slight abuse of notation, we call R the *curvature* homomorphism. This is a bit of a misnomer, as in a geometric situation R will be determined by the honest curvature, but not vice versa. $\hat{E}$ is *not* a generalized cohomology theory and not meant to be one: it contains differential form information and as a consequence is not homotopy invariant.

If E is ordinary integral cohomology, ch is just induced by the inclusion of coefficients $\mathbb{Z} \to \mathbb{R}$. For K-theory, the situation we are mainly discussing in this article, ch is the ordinary Chern character.

The *flat part* $\hat{E}_{\text{flat}}(X)$ of $\hat{E}(X)$ is defined as the kernel of the curvature morphism:

$$\hat{E}_{\text{flat}}(X) := \ker\Big(R\colon \hat{E}(X) \to \Omega(X;N)\Big).$$

It turns out that $\hat{E}_{\text{flat}}(X)$ is a cohomology theory, usually just $E\mathbb{R}/\mathbb{Z}[-1]$, the generalized cohomology with $\mathbb{R}/\mathbb{Z}$-coefficients with a degree shift: $\hat{E}^k_{\text{flat}}(X) = E\mathbb{R}/\mathbb{Z}^{k-1}(X)$. An original interest in differential K-theory (before it even was introduced as such) was its role as a geometric model for $K\mathbb{R}/\mathbb{Z}$. Karoubi in [46, Sect. 7.5] defined $K^{-1}\mathbb{C}/\mathbb{Z}$ using essentially the flat part of a cycle model for $\hat{K}^0$, compare also Lott [54, Definition 5, Definition 7] where also $K\mathbb{R}/\mathbb{Z}^{-1}$ is introduced. Homotopy theory provides a universal construction of $E\mathbb{R}/\mathbb{Z}$ for a generalized cohomology theory E. However, this is in general hard to combine with geometry.

K-theory is the home for index theory. The differential K-theory (in particular its different cycle models) and also its flat part naturally are the home for index problems, taking more of the geometry into account. Indeed, in suitable models it is built into the definitions that geometric families of Dirac operators, parameterized by X, give rise to classes in $\hat{K}^*(X)$, where $*$ is the parity of the dimension of the fiber. A submersion $p\colon X \to Z$ with closed fibers with fiberwise geometric spin^c-structure (the precise meaning of geometry will be discussed below) is oriented for differential K-theory and one has an associated push-forward

$$\hat{p}_!\colon \hat{K}^*(X) \to \hat{K}^{*-d}(Y)$$

(with $d = (\dim(X) - \dim(Z))$). The same data also gives rise to a push-forward in Deligne cohomology

$$\hat{p}_!\colon \hat{H}^*(X) \to \hat{H}^{*-d}(Y).$$

There is a unique lift of the Chern character to a natural transformation

$$\hat{\mathrm{ch}}\colon \hat{K}^*(X) \to \hat{H}_{\mathbb{Q}}^{*+2\mathbb{Z}}(X).$$

Here, the right hand side is the differential extension of $H^*(X;\mathbb{Q})$ and one of the main results of [26] is a refinement of the classical Riemann–Roch theorem to a differential Riemann–Roch theorem which identifies the correction for the compatibility of $\hat{\mathrm{ch}}$ with the push-forwards. In [39], Bismut superconnection techniques are used to define the analytic index of a geometric family: it can be understood as a particular representative of the differential K-theory class of a geometric family as above, determined by the analytic solution of the index problem. Moreover, they develop a geometric refinement of the topological index construction of Atiyah–Singer [5] based on geometrically controlled embeddings into Euclidean space (which does not require deep spectral analysis) and prove that topological and analytic index in differential K-theory coincide.

Finally, we observe that, in suitable special situations, we can easily construct classes in differential K-theory which turn out not to depend on the special geometry, but only on the underlying differential-topological data. Typically, these live in the flat part of differential K-theory and are certain (generalizations of) secondary index invariants. Examples are rho-invariants of the Dirac operator twisted with two flat vector bundles (and family versions hereof), or the $\mathbb{Z}/k\mathbb{Z}$-index of Lott [54] for a manifold W whose boundary is identified with the disjoint union of k copies of a given manifold M.

Similar to smooth Deligne cohomology, there is a counterpart of differential K-theory in the holomorphic setting [41] and there is an arithmetic Riemann–Roch for these groups. This, however, will not be discussed in this survey.

1.1 Differential Cohomology and Physics

A motivation for the introduction of differential K-theory comes from quantum physics. The fields of abelian gauge theories are described by objects which carry the local field strength information of a closed differential form (assuming that there are no sources). Dirac quantization, however, requires that their de Rham classes lie in an integral lattice in de Rham cohomology. For Maxwell theory the field strength simply is a 2-form which is the curvature of a complex line bundle and therefore lies in the image of ordinary integral cohomology. For Ramond–Ramond fields in type II string theories it is a differential form of higher degree which lies in the image under the Chern character of K-theory, as suggested by [37, 57]. Indeed, Witten suggests that D-brane charges in the low energy limit of type IIA/B superstring theory are classified by K-theory. In this case, even if the field strength differential form is zero, the fields or D-brane charges can contain some *global* information, corresponding to torsion in K-theory.

It is suggested by Freed [36] that these Ramond–Ramond fields are described by classes in *differential* K-theory (or other generalized differential cohomology theories, depending on the particular physical model). Given a space-time background X and a field represented by a class $F \in \hat{E}^*(X)$, this field contains the differential form information $R(F)$ (as expected for an abelian gauge field). The field equations (generalizing Maxwell's equations) require that $dR(F) = 0$ if there are no sources (which we assume here). However, there is a quantization condition: the de Rham class represented by $R(F)$ is not arbitrary, but lies in an integral lattice, namely in $\mathrm{im}(ch)$. Indeed, F also contains the integral (and possibly torsion) information of the class $I(F) \in E^*(X)$. Finally, even $I(F)$ and $R(F)$ together don't determine F entirely, there is extra information, corresponding to a physically significant potential or holonomy. More precisely, $\hat{E}^*(X)$ is the configuration space with a gauge group action. Details of such a gauge field theory are studied, e.g. in [50], where it is shown that the free part of $E^*(X)$ is an obstruction to a global gauge fixing. Nonetheless, [50] proposes a partition function and among others computes the vacuum expectation value.

All discussed so far describes the situation without any background field or flux. However, such background fields are an important ingredient of the theory. Depending on the chosen model and the precise situation, a background field can be defined in many different ways. In the classical situation where fields are just given by differential forms, a background field is a closed 3-form Ω. It creates an extra term in the field equations. Correspondingly, the relevant charges are even or odd forms (depending on the type of the theory) which are closed for the differential d^Ω with $d^\Omega\omega := d\omega + \omega \wedge \Omega$. And they are classified up to equivalence by the *Ω-twisted de Rham cohomology*

$$H^{*+\Omega}_{dR}(X) := \ker(d^\Omega)/\operatorname{im}(d^\Omega).$$

When looking at charges in the presence of a background B-field (producing an H-flux) which are classified topologically by K-theory, we need to work with *twisted K-theory*, compare in particular [17, 67]. We will give a short introduction to twisted K-theory in Sect. 7.1. The role of twisted K-theory is discussed a lot in the case of T-duality. T-duality predicts an isomorphism of string theories on different background manifolds which are T-dual to each other, and in particular an isomorphism of the K-theory groups which classify the D-brane charges.

It turns out, however, that the topology of one of the partners in duality dictates a background B-field on the other, and the required isomorphism can only hold in twisted K-theory, compare e.g. [16, 23]. In those papers, mainly the *topological* classification of D-brane charges is considered. A new picture now arises when one wants to move to T-duality for Ramond–Ramond fields described by *differential K-theory* as explained above. One has to construct and study *twisted* differential K-theory. A first step toward this is carried out in [31]. Now, physicists try to understand T-duality at the level of Ramond–Ramond fields, compare e.g. [9] where the ideas are discussed explicitly without mathematical rigor. With mathematical rigor, the T-duality isomorphism in (twisted) differential K-theory has been worked out by Kahle and Valentino in [45]. We will describe these results in more detail in Sect. 7.4.

2 Axioms for Differential Cohomology

A fruitful approach to generalized cohomology theories is based on the Eilenberg–Steenrod axioms. It turns out that many of the basic properties of smooth Deligne cohomology and differential K-theory also are captured by a rather small set of axioms, proposed in [26, Sect. 1.2.2] (and motivated by [36]). Therefore, we want to base our treatment of differential K-theory on those axioms, as well.

The starting point is a generalized cohomology theory E, together with a natural transformation $\mathrm{ch}\colon E(X) \to H(X;N)$, where N is a graded coefficient $\mathbb{R}$-vector space. The two basic examples are

- $E(X) = H(X;\mathbb{Z})$, ordinary cohomology with integer coefficients, where $N = \mathbb{R}$ and ch is induced by the inclusion of coefficients $\mathbb{Z} \to \mathbb{R}$. More generally, $\mathbb{Z}$ can be replaced by any subring of $\mathbb{R}$, e.g. by $\mathbb{Q}$.
- $E(X) = K(X)$, K-theory, where ch is the usual Chern character, and $N = \mathbb{R}[u, u^{-1}]$ with u of degree 2. Multiplication with u corresponds to Bott periodicity.

Definition 2.1. A differential extension of the pair (E, ch) is a functor $X \to \hat{E}(X)$ from the category of compact smooth manifolds (possibly with boundary) to $\mathbb{Z}$-graded groups together with natural transformations

1. $R\colon \hat{E}(X) \to \Omega_{d=0}(E;N)$ (curvature)
2. $I\colon \hat{E}(X) \to E(X)$ (underlying cohomology class)
3. $a\colon \Omega(X;N)/\operatorname{im}(d) \to \hat{E}(X)$ (action of forms).

Here $\Omega(E;N) := \Omega(E) \otimes_{\mathbb{R}} N$[1] denote the smooth differential forms with values in N, $d\colon \Omega(E;N) \to \Omega(E;N)[1]$ the usual de Rham differential and $\Omega_{d=0}(E;N)$ the space of closed differential forms ([1] stands for degree-shift by 1).

The transformations I, a, R are required to satisfy the following axioms:

1. The following diagram commutes

$$\begin{array}{ccc} \hat{E}(X) & \xrightarrow{I} & E(X) \\ \downarrow R & & \downarrow \mathrm{ch} \\ \Omega_{d=0}(X,N) & \xrightarrow{\text{Rham}} & H(X;N). \end{array} \tag{1}$$

[1]This definition has to be modified in a generalization to non-compact manifolds!

2.
$$R \circ a = d. \tag{2}$$

3. a is of degree 1.
4. The following sequence is exact:

$$E^{*-1}(X) \xrightarrow{ch} \Omega^{*-1}(X,N)/\operatorname{im}(d) \xrightarrow{a} \hat{E}^*(X) \xrightarrow{I} E^*(X) \to 0. \tag{3}$$

Alternatively, when dealing with K-theory one can and often (e.g. in [26]) does consider the whole theory as $\mathbb{Z}/2\mathbb{Z}$-graded with the obvious adjustments. Note that with $N = K^*(pt) \otimes \mathbb{R} = \mathbb{R}[u,u^{-1}]$ we have natural and canonical isomorphism $\Omega^*(X;N) = \oplus_{k\in\mathbb{Z}}\Omega^{*+2k}(X)$ and $H^*(X;N) = \oplus_{k\in\mathbb{Z}}H^{*+2k}(X;\mathbb{R})$. The associated $\mathbb{Z}/2\mathbb{Z}$-graded ordinary cohomology is therefore given by the direct sum of even or odd degree forms.

Corollary 2.2. *If $\hat{E}$ is a differential extension of (E,ch), then we have a second exact sequence*

$$E^{*-1}(X) \xrightarrow{\mathrm{ch}} H^{*-1}(X;N) \xrightarrow{a} \hat{E}^*(X) \xrightarrow{R} \Omega^*_{d=0}(X;N) \times_{\mathrm{ch}} E^*(X) \to 0, \tag{4}$$

where $\Omega_{d=0}(X;N) \times_{\mathrm{ch}} E(X) = \{(\omega,x) \mid \mathrm{Rham}(\omega) = \mathrm{ch}(x)\}$ is the pullback of abelian groups.

Proof. This is a direct consequence of (3), (1) and (2). □

Definition 2.3. Given a differential extension $\hat{E}$ of a cohomology theory (E,ch), we define the associated *flat functor*

$$\hat{E}_{flat}(X) := \ker(R\colon \hat{E}(X) \to \Omega_{d=0}(X;N)).$$

Remark 2.4. The naturality of R indeed implies that $X \mapsto \hat{E}_{flat}(X)$ is a contravariant functor on the category of smooth manifolds. Actually, this functor by Corollary 2.7 is always homotopy invariant and extends to a cohomology theory in many examples, as we will discuss in Sect. 2.3. Typically, there is a natural isomorphism $\hat{E}^*_{flat}(X) \cong E\mathbb{R}/\mathbb{Z}^{*-1}(X)$, but we still don't know whether this is necessarily always the case (compare the discussion in [27, Sects. 5 and 7]).

The most interesting cases are not just group valued cohomology functors, but multiplicative cohomology theories, for example K-theory and ordinary cohomology. We therefore want typically a differential extension which carries a compatible product structure.

Definition 2.5. Assume that E is a multiplicative cohomology theory, that N is a $\mathbb{Z}$-graded algebra over $\mathbb{R}$, and that ch is compatible with the ring structures. A differential extension $\hat{E}$ of (E,ch) is called multiplicative if $\hat{E}$ together with the transformations R, I, a is a differential extension of (E,ch), and in addition

1. $\hat{E}$ is a functor to $\mathbb{Z}$-graded rings,
2. R and I are multiplicative,
3. $a(\omega) \cup x = a(\omega \wedge R(x))$ for all $x \in \hat{E}(B)$ and $\omega \in \Omega(B; N)/\operatorname{im}(d)$.

Deligne cohomology is multiplicative [18, Chap. 1], [35, Sect. 6], [20, Sect. 4]. In this paper, we will consider multiplicative extensions $\hat{K}$ of K-theory.

2.1 Variations of the Axiomatic Approach

Our list of axioms for differential cohomology theories seems particularly natural: it allows for efficient constructions and to derive the conclusions we are interested in. However, for the differential refinements of integral cohomology, a slightly different system of axioms has been proposed in [62]. The main point there is that the requirement of a given natural isomorphism $\hat{E}^*_{flat}(X) \to E\mathbb{R}/\mathbb{Z}^{*-1}(X)$ between the flat part of Definition 2.3 and E with coefficients in $\mathbb{R}/\mathbb{Z}$. It turns out that, for differential extensions of ordinary cohomology, both sets of axioms imply that there is a unique natural isomorphism to Deligne cohomology (compare [62] and [27, Sect. 7]). In particular, for ordinary cohomology they are equivalent. The corresponding result holds in general under extra assumptions, which are satisfied for K-theory, compare Sect. 2.3.

2.2 Homotopy Formula

A simple, but important consequence of the axioms is the homotopy formula. If one differential cohomology class can be deformed to another, this formula allows to compute the difference of the two classes entirely in terms of differential form information. In a typical application, one will deform an unknown class to one which is better understood and that way get one's hands on the complicated class one started with.

Theorem 2.6 (Homotopy formula). *Let $\hat{E}$ be a differential extension of (E, ch). If $x \in \hat{E}([0,1] \times B)$ and $i_k \colon B \to [0,1] \times B; b \mapsto (k, b)$ are the inclusions then*

$$i_1^* x - i_0^* x = a\left(\int_{[0,1]\times B/B} R(x)\right),$$

where $\int_{[0,1]\times B/B}$ denotes integration of differential forms over the fiber of the projection $p\colon [0,1] \times B/B$ with the canonical orientation of the fiber $[0,1]$.

Proof. Note that, if $x = p^* y$ for some $y \in \hat{E}(B)$ then by naturality the left hand side of the equation is zero. Moreover, in this case $R(x) = p^* R(x)$ so that by the properties of integration over the fiber the right hand side vanishes, as well.

In general, observe that p is a homotopy equivalence, so that we always find $\bar{y} \in E(B)$ with $I(x) = p^*\bar{y}$. Using surjectivity of I, we find $y \in \hat{E}(B)$ with $I(y) = \bar{y}$, and then $I(p^*y - x) = p^*\bar{y} - \bar{y} = 0$. Since the sequence (3) is exact, there is $\omega \in \Omega(B;N)$ with $a(\omega) = x - p^*y$. Stokes' theorem applied to ω yields

$$i_1^*\omega - i_0^*\omega = \int_{[0,1]\times B/B} d\omega.$$

On the other hand, because of (2), $d\omega = R(a(\omega))$. Substituting, we get

$$\int_{[0,1]\times B/B} d\omega = \int_{[0,1]\times B/B} R(a(\omega)) = \int_{[0,1]\times B/B} R(x - p^*y) = \int_{[0,1]\times B/B} R(x)$$

and (using again vanishing of our expressions for p^*y)

$$i_1^*x - i_0^* = i_1^*a(\omega) = i_0^*a(\omega) = a(i_1^*\omega - i_0^*\omega) = a\left(\int_{[0,1]\times B/B} R(x)\right).$$

□

Corollary 2.7. *Given a differential extension $\hat{E}$ of (E, ch), the associated flat functor $\hat{E}_{flat}$ of Definition 2.3 is homotopy invariant.*

Proof. Let $H\colon [0,1] \times X \to Y$ be a homotopy between $f = H_0$ and $g = H_1$. We have to show that $f^* = g^*\colon \hat{E}_{flat}(Y) \to \hat{E}_{flat}(X)$. By functoriality, it suffices to show that $i_0^* = i_1^*\colon \hat{E}_{flat}([0,1]\times X) \to \hat{E}_{flat}(X)$, as $f^* = i_0^* \circ H^*$ and $g^* = i_1^* \circ H^*$. This, however, follows immediately from Theorem 2.6 once $R(x) = 0$. □

We have seen above that E^*_{flat} is a homotopy invariant functor. Ideally, it should extend to a generalized cohomology theory (compare the discussion of Sect. 2.1). For this, we need a bit of extra structure which corresponds to the suspension isomorphism, and which is typically easily available. We formulate this in terms of integration over the fiber for $X \times S^1$, originally defined in [27, Definition 1.3]. Note that the projection $X \times S^1 \to X$ is canonically oriented for an arbitrary cohomology theory because the tangent bundle of S^1 is canonically trivialized. For a push-down in differential cohomology, in general we expect that one has to choose geometric data of the fibers, which again we can assume to be canonically given for the fiber S^1. Orientations and push-forward homomorphisms for differential cohomology theories, in particular for differential K-theory are discussed in Sect. 5.

Definition 2.8. We say that a differential extension $\hat{E}$ of a cohomology theory (E, ch) has S^1-*integration* if there is a natural transformation $\int_{X\times S^1/X}\colon \hat{E}^*(X \times S^1) \to \hat{E}^{*-1}(X)$ which is compatible with the transformations R, I and the "integration over the fiber" $\Omega^*(X \times S^1; N) \to \Omega^{*-1}(X;N)$ as well as $E^*(X \times S^1) \to E^{*-1}(X)$. In addition, we require that $\int_{X\times S^1/X} p^*x = 0$ for each $x \in \hat{E}^*(X)$ and $\int_{X\times S^1/X} (\mathrm{id} \times t)^*x = -x$ for each $x \in \hat{E}^*(X \times S^1)$, where $t\colon S^1 \to S^1$ is complex conjugation.

In [27, Corollary 4.3] we prove that in many situations, e.g. for ordinary cohomology or for K-theory, there is a canonical choice of integration transformation.

Theorem 2.9. *If $\hat{E}$ is a multiplicative differential extension of (E, ch) and if $E^{-1}(pt)$ is a torsion group, then $\hat{E}$ has a canonical S^1-integration as in Definition 2.8.*

2.3 $\hat{E}^*_{flat}$ as Generalized Cohomology Theory $E\mathbb{R}/\mathbb{Z}$

Let E be a generalized cohomology theory. In the present section we consider a universal differential extension $\hat{E}$, i.e. we take $N := E(*) \otimes_{\mathbb{Z}} \mathbb{R}$ and let $\mathrm{ch} : E(X) \to H(X; N)$ be the canonical transformation.

To E there is an associated generalized cohomology theory $E\mathbb{R}/\mathbb{Z}$ (E with coefficients in $\mathbb{R}/\mathbb{Z}$). It is constructed with the help of stable homotopy theory: the cohomology theory E is given by a spectrum (in the sense of stable homotopy theory) $\mathbf{E}$, and $E\mathbb{R}/\mathbb{Z}$ is given by the spectrum $\mathbf{E}M(\mathbb{R}/\mathbb{Z})$, where $M(\mathbb{R}/\mathbb{Z})$ is the Moore spectrum of the abelian group $\mathbb{R}/\mathbb{Z}$. $E\mathbb{R}/\mathbb{Z}$ is constructed in such a way that one has natural long exact sequences

$$\to E^*(X) \to E\mathbb{R}^*(X) \to E\mathbb{R}/\mathbb{Z}^*(X) \to E^{*+1}(X) \to \dots.$$

Note that for a finite CW-complex X one can alternatively write $E^*(X) \otimes \mathbb{R} = E\mathbb{R}^*(X)$ with $E\mathbb{R}$ defined by smash product with $M(\mathbb{R})$.

In the fundamental paper [44], Hopkins and Singer construct a specific differential extension for any generalized cohomology theory E. For this particular construction, one has by [44, (4.57)] a natural isomorphism

$$E\mathbb{R}/\mathbb{Z}^{*-1}(X) \to E^*_{flat}(X).$$

However, this is a consequence of the particular model used in [44]. It is therefore an interesting question to which extend the axioms alone imply that $\hat{E}_{flat}$ is a generalized cohomology theory. Here, some extra structure about the suspension isomorphism seems to be necessary, implemented by the transformation "integration over S^1" of Definition 2.8. With a surprisingly complicated proof one gets [27, Theorem 7.11]:

Theorem 2.10. *If $(\hat{E}, R, I, a, \int)$ is a differential extension of E with integration over S^1, then $\hat{E}^*_{flat}$ has natural long exact Mayer–Vietoris sequences. It is equivalent to the restriction to compact manifolds of a generalized cohomology theory represented by a spectrum. Moreover, $a\colon E\mathbb{R}^* \to \hat{E}^{*+1}_{flat}$ and $I\colon \hat{E}^*_{flat} \to E^*$ are natural transformations of cohomology theories and one obtains a natural long exact sequence for each finite CW-complex X*

$$\to E\mathbb{R}^{*-1}(X) \xrightarrow{a} \hat{E}^*_{flat}(X) \xrightarrow{I} E^*(X) \xrightarrow{ch} E\mathbb{R}^*(X) \to . \qquad (5)$$

Proof. The long exact sequence (5) is not stated like that in [27] and we will need it below, therefore we explain how this is achieved. Because of (2), the restriction of a to $E\mathbb{R}^{*-1}(X)$ (realized as de Rham cohomology, i.e. as $\ker(d\colon \Omega^{*-1}(X;N)/\operatorname{im}(d) \to \Omega^*(X;N)))$ hits exactly $\hat{E}_{flat}(X)$. The exactness at $\hat{E}_{flat}(X)$ therefore is a direct consequence of the exactness of (3). (3) also implies immediately that $\ker(a) = \operatorname{im}(ch)$ in (5). Because of (1), $ch \circ I = 0$ in (5). Finally, as $I\colon \hat{E}^*(X) \to E(X)$ is surjective, any $x \in \ker(ch)$ can be written as $I(y)$ with $y \in \hat{E}^*(X)$ and such that $R(y) \in \operatorname{im}(d)$, i.e. $R(y) = d(\omega)$ for some $\omega \in \Omega^{*-1}(X)/\operatorname{im}(d)$. But then $x = I(y - d\omega)$ and $y - d\omega \in E^*_{flat}(X)$, which implies that (5) is also exact at $E^*(X)$. □

Corollary 2.11. *Let $(\hat{E}, R, I, a, \int)$ be a differential extension of E with integration over S^1, or more generally assume that $\hat{E}_{flat}$ has natural long exact Mayer–Vietoris sequences. Assume that $X_1, X_2, X_0 := X_1 \cap X_2 \subseteq X$ are closed submanifolds of codimension 0 with boundary (and corners) such the interiors of X_1 and X_2 cover X. Then one has a long exact Mayer–Vietoris sequence*

$$\hat{E}^{n+1}_{flat}(X_0) \xrightarrow{i\circ\delta} \hat{E}^n(X) \to \hat{E}^n(X_1) \oplus \hat{E}^n(X_2) \to \hat{E}^n(X_0) \xrightarrow{\delta\circ I} E^{n-1}(X) \cdots,$$

which continues to the right with the Mayer–Vietoris sequence for E and to the left with the Mayer–Vietoris sequence for $\hat{E}$.

Proof. The proof is an standard diagram chase, using the Mayer–Vietoris sequences for E and E_{flat}, the short exact sequence

$$0 \to \Omega^*(X) \to \Omega^*(X_1) \oplus \Omega^*(X_2) \to \Omega^*(X_0) \to 0,$$

the homotopy formula, and the exact sequences (3). □

Theorem 2.12. *If, in addition to the assumption of Theorem 2.10, E^k is finitely generated for each $k \in \mathbb{Z}$ and the torsion subgroup $E^*_{tors}(pt) = 0$ then there is an isomorphism of cohomology theories $\hat{E}^*_{flat} \xrightarrow{\cong} E\mathbb{R}/\mathbb{Z}^{*+1}$.*

Proof. We claim this statement as [27, Theorem 7.12]. However, the proof given there is not correct, and the assertion of [27, Theorem 7.12] unfortunately is slightly stronger than the one we can actually prove, namely Theorem 2.12.

As by Theorem 2.10 $\hat{E}_{flat}$ is a generalized cohomology theory, it is represented by a spectrum U, and the natural transformation a by a map of spectra. We extend this to a fiber sequence of spectra $F \to E\mathbb{R} \xrightarrow{a} U$, inducing for each compact CW-complex X an associated long exact sequence

$$\to F^*(X) \to E\mathbb{R}^*(X) \xrightarrow{a} \hat{E}^*_{flat}(X) \to . \qquad (6)$$

Comparison with (5) implies that the image of $F(X)$ in $E\mathbb{R}(X)$ coincides with the image of $E(X)$. This means by definition that the composed map of spectra $F \to E\mathbb{R} \to E\mathbb{R}/\mathbb{Z}$ is a phantom map. However, under the assumption that the E^k is finitely generated for each k, it is shown in [27, Sect. 8] that such a phantom map is automatically trivial. Using the triangulated structure of the homotopy category of spectra, we can choose ϕ, ϕ_F to obtain a map of fiber sequences (distinguished triangles)

$$\begin{array}{ccccc} F & \longrightarrow & E\mathbb{R} & \xrightarrow{a} & U \\ \downarrow{\scriptstyle \phi_F} & & \downarrow{\scriptstyle =} & & \downarrow{\scriptstyle \phi} \\ E & \longrightarrow & E\mathbb{R} & \longrightarrow & E\mathbb{R}/\mathbb{Z} \end{array}$$

with associated diagram of exact sequences which because of the knowledge about image and kernel of a specializes to

$$\begin{array}{ccccccc} E^*(X) & \longrightarrow & E\mathbb{R}^*(X) & \xrightarrow{a} & \hat{E}^{*+1}_{flat}(X) & \longrightarrow & E^{*+1}_{tors}(X) \to 0 \\ \downarrow{\scriptstyle =} & & \downarrow{\scriptstyle =} & & \downarrow{\scriptstyle \phi} & & \\ E^*(X) & \longrightarrow & E\mathbb{R}^*(X) & \longrightarrow & E\mathbb{R}/\mathbb{Z}^*(X) & \longrightarrow & E^{*+1}_{tors}(X) \to 0. \end{array} \tag{7}$$

Note that we do not claim (and don't know) whether the diagram can be completed to a commutative diagram by id: $E^{*+1}_{tors}(X) \to E^{*+1}_{tors}(X)$.

If $E^*_{tors}(pt) = 0$, the 5-lemma implies that $\phi\colon U \to E\mathbb{R}/\mathbb{Z}$ induces an isomorphism on the point and therefore for all finite CW-complexes. □

3 Uniqueness of Differential Extensions

Given the many different models of differential extensions of K-theory, many of which we are going to described in Sect. 4, it is reassuring that the resulting theory is uniquely determined. The corresponding statement for differential extensions of ordinary cohomology has been established in [62] by Simons and Sullivan. For K-theory and many other generalized cohomology theories we will establish this in the current section.

As in Sect. 2.3 we consider universal differential extensions of E. Given two extensions $\hat{E}$ and $\hat{E}'$ of E with corresponding natural transformations a, I as in 2.1, we are looking transformations for a natural isomorphism $\Phi\colon \hat{E} \to \hat{E}'$ compatible with the natural transformation. Provided such a natural transformation exists, we ask whether it is unique. We have seen that it often is natural to have additionally a transformation "integration over S^1" as in Definition 2.8, and we require that Φ is compatible with this transformation, as well.

To construct the transformation Φ in degree k there is the following basic strategy:

- Find a classifying space B for E^k with universal element $u \in E^k(B)$. This means that for any space X and $x \in E^k(X)$, there is a map $f: X \to B$ (unique up to homotopy) such that $x = f^* u$.
- Lift this universal element to $\hat{u} \in \hat{E}^k(B)$ and show that this class is universal for $\hat{E}^k$, or at least that for a class $\hat{x} \in \hat{E}^k(X)$ one can find $f: X \to B$ such that the difference $\hat{x} - f^* \hat{u}$ is under good control.
- Obtain similarly $\hat{u}' \in (\hat{E}')^k(B)$. Define the transformation by $\Phi(\hat{u}) = \hat{u}'$ and extend by naturality.
- Check that Φ has all desired properties.
- Uniqueness of Φ does follow if the lifts $\hat{u}, \hat{u}'$ are uniquely determined by u once their curvature is also fixed.

There are a couple of obvious difficulties implementing this strategy. The first is the fact, that the classifying space B almost never has the homotopy type of a finite dimensional manifold. Therefore, $\hat{E}(B)$ is not defined. This is solved by replacing B by a sequence of manifolds approximating B with a compatible sequence of classes in $\hat{E}^k$ which replace $\hat{u}$. Then, the construction of Φ as indicated indeed is possible. However, a priori this has a big flaw. Φ is not necessarily a transformation of abelian groups. Because of the compatibilities with a, R and I the deviation from being additive is rather restricted and in the end is a class in $E\mathbb{R}^{k-1}(B \times B)/\operatorname{im}(ch)$. The different possible transformation are by naturality and the compatibility conditions determined by the different lift $\hat{u}'$ with fixed curvature. This indeterminacy is given by an element in $E\mathbb{R}^{k-1}(B)/\operatorname{im}(ch)$. If k is even and E is rationally even, then so is B as classifying space of E^k. It then follows that $E\mathbb{R}^{k-1}(B \times B)/\operatorname{im}(ch) = \{0\}$ and $E\mathbb{R}^{k-1}(B)/\operatorname{im}(ch) = \{0\}$, i.e. Φ automatically is additive and unique.

For k odd, the transformation can then be defined (and is uniquely determined) by the requirement that it is compatible with $\int_{X \times S^1/X}$. This construction has been carried out in detail in [27] and we arrive at the following theorem.

Theorem 3.1. *Assume that $(\hat{E}, R, I, a, \int)$ and $(\hat{E}', R', I', a', \int')$ are two differential extensions with S^1-integration of a generalized cohomology theory E which is rationally even, i.e. $E^{2k+1}(pt) \otimes \mathbb{Q} = 0$ for all $k \in \mathbb{Z}$. Assume furthermore that $E^k(pt)$ is a finitely generated abelian group for each $k \in \mathbb{Z}$. Then there is a unique natural isomorphism between these differential extensions compatible with the S^1-integrations.*

If no S^1-orientation is given, the natural isomorphism can still be constructed on the even degree part.

If $\hat{E}$ and $\hat{E}'$ are multiplicative, the transformation is automatically multiplicative. Note that the assumptions imply by Theorem 2.9 that then there is a canonical integration.

If $\hat{E}$ and $\hat{E}'$ are defined on all manifold, not only on compact manifolds (possibly with boundary), then it suffices to require that $E^k(pt)$ is countably generated for each $k \in \mathbb{Z}$, and the same assertions hold.

Proof. The proof is given in [27] and we don't plan to repeat it here. However, there we made the slightly stronger assumption that $E^{2k+1}(pt) = 0$ if $\hat{E}$ is only defined on the category of compact manifolds. Let us therefore indicate why the stronger result also holds. As described in the strategy, the first task is to approximate the classifying spaces B for E^{2k} by spaces on which we can evaluate $\hat{E}^*$. These approximations are constructed inductively by attaching handles to obtain the correct homotopy groups (introducing new homotopy, but also killing superfluous homotopy). To construct a compact manifold, we are only allowed to attach finitely many handles. Therefore, we have to know a priori that we have to kill only a finitely generated homotopy groups. In [27] we assume that $\pi_1(B) = E^{2k-1}(pt)$ is zero, to be allowed to start with a simply connected approximation. Then we use that all homotopy groups of a simply connected finite CW-space are finitely generated. However, exactly the same holds for finite CW-spaces with finite fundamental group, because the higher homotopy groups are the homotopy groups of the universal covering, which in this case is a finite simply connected CW-complex. Note that a finitely generated abelian group A with $A \otimes \mathbb{Q} = 0$ is automatically finite. □

Remark 3.2. The general strategy leading toward Theorem 3.1 has been developed by Moritz Wiethaup 2006/07. However, his work has not been published yet. This was then taken up and developed further in [27].

3.1 Uniqueness of Differential K-Theory

We observe that all the assumptions of Theorem 3.1 are satisfied by K-theory, and also by real K-theory. Therefore, we have the following theorem:

Theorem 3.3. *Given two differential extensions $\hat{K}$ and $\hat{K}'$ of complex K-theory, there is a unique natural isomorphism $\hat{K}^{ev} \to \hat{K'}^{ev}$ compatible with all the structure. If the extensions are multiplicative, this transformation is compatible with the products.*

If both extensions come with S^1-integration as in Definition 2.8 there is a unique natural isomorphism $\hat{K} \to \hat{K}'$ compatible with all the structure, including the integration.

In other words, all the different models for differential K-theory of Sect. 4 define the same groups – up to a canonical isomorphism.

Remark 3.4. In Theorem 3.3 we really have to require the existence of S^1-integration. In [27, Theorem 6.2] an infinite family of "exotic" differential extensions of K-theory are constructed. Essentially, the abelian group structure is modified in a subtle way in these examples to produce non-isomorphic functors which all satisfy the axioms of Sect. 2.

4 Models for Differential K-Theory

4.1 Vector Bundles with Connection

The most obvious attempt to construct differential cohomology (at least for $\hat{K}^0$) is to use vector bundles *with connection*. It is technically convenient to also add an odd differential form to the cycles. This, indeed is the classical picture already used by Karoubi in [46, Sect. 7] for his definition of "multiplicative K-theory", which we would call the flat part of differential K-theory.

Definition 4.1. A cycle for vector bundle K-theory $\hat{K}^0(M)$ is a triple (E, ∇, ω), where E is a smooth complex Hermitean vector bundle over M, ∇ a Hermitean connection on E and $\omega \in \Omega^{odd}(M)/\mathrm{im}(d)$ a class of a differential form of odd degree. The *curvature* of a cycle essentially is defined as the Chern–Weil representative $\mathrm{ch}(\nabla) := \mathrm{tr}(e^{-\frac{\nabla^2}{2\pi i}})$ of the Chern character of E, computed using the connection ∇:

$$R(E, \nabla, \omega) := \mathrm{ch}(\nabla) - d\omega.$$

Remark 4.2. We require the use of Hermitean connections to obtain real valued curvature forms. Alternatively, one would have to use as target of ch cohomology with complex instead of cohomology with real coefficients. A slightly more extensive discussion of this matter can be found in [54, Sect. 2].

We define in the obvious way the sum $(E, \nabla_E, \omega) + (F, \nabla_F, \eta) := (E \oplus F, \nabla_E \oplus \nabla_F, \omega + \eta)$. Two cycles (E, ∇, ω) and (E', ∇', ω') are equivalent if there is a third bundle with connection (F, ∇_F) and an isomorphism $\Phi\colon E \oplus F \to E' \oplus F$ such that

$$\widetilde{\mathrm{ch}}(\nabla \oplus \nabla_F, \Phi^{-1}(\nabla' \oplus \nabla_F)\Phi) = \omega - \omega',$$

where $\widetilde{\mathrm{ch}}(\nabla, \nabla')$ denotes the transgression Chern form between the two connections such that $\mathrm{ch}(\nabla) - \mathrm{ch}(\nabla') = d\widetilde{\mathrm{ch}}(\nabla, \nabla')$.

Remark 4.3. In [39, Sect. 9], a model for differential $\hat{K}^1$ is given where the cycles are Hermitean vector bundles with connection and a unitary automorphism, and an additional form (modulo the image of d) of even degree. The relations include in particular a rule for the composition of the unitary automorphisms: if U_1 and U_2 are two unitary automorphisms of E then $(E, \nabla, U_1, \omega_1) + (E, \nabla, U_2, \omega_2) = (E, \nabla, U_2 \circ U_1, CS(\nabla, U_1, U_2) + \omega_1 + \omega_2)$, where $CS(\nabla, U_1, U_2)$ is the Chern–Simons form relating ∇, $U_2 \nabla U_2^{-1}$ and $U_2 U_1 \nabla U_1^{-1} U_2^{-1}$. We do not discuss this model in detail here.

4.2 Classifying Maps

Hopkins and Singer, in their ground breaking paper [44] give a cocycle model of a differential extension $\hat{E}$ of any cohomology theory (with transformation) (E, ch), based on classifying maps. Here, for the construction of $\hat{E}^n(X)$ one has to choose two fundamental ingredients:

1. A classifying space X_n for E^n; note that no smooth structure for X_n is required (and could be expected).
2. A cocycle c representing ch_n, so that, whenever $f: X \to X_n$ represents a class $x \in E^n(X)$, then f^*c represents $\mathrm{ch}(x) \in H^n(X;N)$. We can think of this cocycle as an N-valued singular cocycle, although variations are possible.

A Hopkins–Singer cycle for $\hat{E}(X)$ then is a so called *differential function*, which by definition is a triple (f, h, ω) consisting of a continuous map $f: X \to X_n$, a closed differential n-form ω with values in N and an $(n-1)$-cochain h satisfying

$$\delta h = \omega - f^*c.$$

In other words, f is an explicit representative for a class $x \in E^n(X)$, and we are of course setting $I(f, h, \omega) = [f] \in E^n(X)$. ω is a de Rham representative of $\mathrm{ch}(x)$, and we are setting $R(f, h, \omega) := \omega$. This data actually gives two explicit representatives for $\mathrm{ch}(x)$, namely ω and f^*c (here, we have to map both ω and f^*c to a common cocycle model for $H^n(X;N)$ like smooth singular cochains. ω defines such a cocycle by the de Rham homomorphism "integrate ω over the chain", f^*c by restriction).

By definition, (f_1, h_1, ω_1) and (f_0, h_0, ω_0) are equivalent if $\omega_0 = \omega_1$ and there is $(f, h, \mathrm{pr}^*\,\omega_1)$ on $X \times [0,1]$ which restricts to the two cycles on $X \times \{0\}$ and $X \times \{1\}$.

The advantage of this approach is its complete generality. A disadvantage is that the cycles don't have a nice geometric interpretation. Moreover, operations (like the addition and multiplication) rely on the choice of corresponding maps between the classifying spaces realizing those. These maps have then to be used to define the same operations on the differential cohomology groups. This is typically not very explicit. Moreover, properties like associativity, commutativity, etc. will not hold on the nose for these classifying maps but are implemented by homotopies which have to be taken into account when establishing the same properties for the generalized differential cohomology. This can quickly get quite cumbersome and we refrain from carrying this out in any detail.

4.3 Geometric Families of Elliptic Operators

In [26], a cycle model for differential K-theory (there called "smooth K-theory") is developed which is based on local index theory. In spirit, it is similar to the passage of the classical model of K-theory via vector bundles to the Kasparov KK-model, where all families of generalized index problems are cycles.

Similarly, the cycles of [26] are geometric families of Dirac operators. It is clear that a lot of differential structure has to be present to obtain *differential* K-theory, so the definition has to be more restrictive than in Kasparov's model. There are many advantages of an approach with very general cycles:

- First and most obvious, it is very easy to construct elements of differential K-theory if one has a broad class of cycles.
- The approach allows for a unified treatment of even and odd degrees.
- The flexibilities of the cycles allows for explicit constructions in many contexts. In particular, it is easy to explicitly define the product and also the push-forward along a fiber-bundle.

It might seems as a disadvantage that one necessarily has a broad equivalence relation and that it is therefore hard to construct homomorphisms out of differential K-theory. To do this, one has to use the full force of local index theory. However, this is very well developed and one can make use of many properties as black box and then efficiently carry out the relevant constructions.

Definition 4.4. Let X be a compact manifold, possibly with boundary. A cycle for a $\hat{K}(X)$ is a pair $(\mathcal{E}, \rho)$, where $\mathcal{E}$ is a geometric family, and $\rho \in \Omega(X)/\operatorname{im}(d)$ is a class of differential forms.

A geometric family over X (introduced in [19]) consists of the following data:

1. A proper submersion with closed fibers $\pi\colon E \to X$
2. A vertical Riemannian metric $g^{T^v\pi}$, i.e. a metric on the vertical bundle $T^v\pi \subset TE$, defined as $T^v\pi := \ker(d\pi\colon TE \to \pi^*TX)$
3. A horizontal distribution $T^h\pi$, i.e. a bundle $T^h\pi \subseteq TE$ such that $T^h\pi \oplus T^v\pi = TE$
4. A family of Dirac bundles $V \to E$
5. An orientation of $T^v\pi$

Here, a family of Dirac bundles consists of

1. A Hermitean vector bundle with connection (V, ∇^V, h^V) on E,
2. A Clifford multiplication $c\colon T^v\pi \otimes V \to V$, and
3. On the components where $\dim(T^v\pi)$ has even dimension a $\mathbb{Z}/2\mathbb{Z}$-grading z.

We require that the restrictions of the family Dirac bundles to the fibers $E_b := \pi^{-1}(b)$, $b \in X$, give Dirac bundles in the usual sense (see [19, Def. 3.1]):

1. The vertical metric induces the Riemannian structure on E_b.
2. The Clifford multiplication turns $V_{|E_b}$ into a Clifford module (see [11, Def. 3.32]) which is graded if $\dim(E_b)$ is even.
3. The restriction of the connection ∇^V to E_b is a Clifford connection (see [11, Def. 3.39]).

A geometric family is called even or odd, if $\dim(T^v\pi)$ is even-dimensional or odd-dimensional, respectively and the form ρ has the corresponding opposite parity.

Example 4.5. The cycles for $\hat{K}^0(X)$ of Definition 4.1 are special cases of the cycles of Definition 4.4: $p\colon E \to X$ is just $\mathrm{id}\colon X \to X$, i.e. the fibers consist of the 0-dimensional manifold $\{pt\}$.

There are obvious notions of isomorphism (preserving all the structure) and of direct sum of cycles. We now introduce the structure maps I and R and the equivalence relation on the semigroup of isomorphism classes.

Definition 4.6. The opposite $\mathcal{E}^{op}$ of a geometric family $\mathcal{E}$ is obtained by reversing the signs of the Clifford multiplication and the grading (in the even case) of the underlying family of Clifford bundles, and of the orientation of the vertical bundle.

Definition 4.7. The usual construction of Dirac type operators of a Clifford bundle (compare [11, 53]), applied fiberwise, assigns to a geometric family $\mathcal{E}$ over X a family of Dirac type operators parameterized by X, and this is indeed the main idea behind the geometric families. Then, the classical construction of Atiyah–Singer assigns to this family its (analytic) index $\operatorname{ind}(\mathcal{E}) \in K^*(B)$, where $*$ is equal to the parity of the dimension of the fibers. In the special case of Example 4.5 – a vector bundle with connection over X – this is exactly the K-theory class of the underlying $\mathbb{Z}/2$-graded vector bundle.

We define $I(\mathcal{E}, \omega) := \operatorname{ind}(\mathcal{E}) \in K^*(X)$.

Remark 4.8. We define $\mathcal{E}^{op}$ in such a way that $\operatorname{ind}(\mathcal{E}^{op}) = -\operatorname{ind}(\mathcal{E})$. Moreover, ind is additive under sums of geometric families. The equivalence relation we are going to define will be compatible with ind.

We now proceed toward the definition of $R(\mathcal{E})$. It is based on the notion of local index form, an explicit de Rham representative of $\operatorname{ch}(\operatorname{ind}(\mathcal{E})) \in H^*_{dR}(X)$. It is one of the important points of the data collected in a geometric family that such a representative can be constructed canonically. For a detailed definition we refer to [19, Def. 4.8], but we briefly formulate its construction as follows. The vertical metric $T^v\pi$ and the horizontal distribution $T^h\pi$ together induce a connection $\nabla^{T^v\pi}$ on $T^v\pi$. Locally on E we can assume that $T^v\pi$ has a spin structure. We let $S(T^v\pi)$ be the associated spinor bundle. Then we can write the family of Dirac bundles V as $V = S \otimes W$ for a twisting bundle (W, h^W, ∇^W, z^W) with metric, metric connection, and $\mathbb{Z}/2\mathbb{Z}$-grading which is determined uniquely up to isomorphism. The form $\hat{A}(\nabla^{T^v\pi}) \wedge \operatorname{ch}(\nabla^W) \in \Omega(E)$ is globally defined, and we get the local index form by applying the integration over the fiber $\int_{E/B}: \Omega(E) \to \Omega(B)$:

$$\Omega(\mathcal{E}) := \int_{E/B} \hat{A}(\nabla^{T^v\pi}) \wedge \operatorname{ch}(\nabla^W). \tag{8}$$

The characteristic class version of the index theorem for families is

Theorem 4.9 ([6]). $\operatorname{ch}_{dR}(\operatorname{ind}(\mathcal{E})) = [\Omega(\mathcal{E})] \in H^*_{dR}(X)$.

A proof using methods of local index theory has been given in [12], compare [11].

The equivalence relation we impose is based on the following idea: a geometric family $\mathcal{E}$ with index 0 should potentially be equivalent to the cycle 0, but in general only up to some differential form (with degree of shifted parity). Moreover, in this case the local index form $R(\mathcal{E})$ will be exact, but it is important to find an explicit primitive η with $d\eta = R(\mathcal{E})$. Therefore, we identify geometric *reasons* why the index is zero and which provide such a primitive.

Definition 4.10. A pre-taming of $\mathcal{E}$ is a family $(Q_b)_{b\in B}$ of self-adjoint operators $Q_b \in B(H_b)$ given by a smooth fiberwise integral kernel $Q \in C^\infty(E \times_B E, V \boxtimes V^*)$. In the even case we assume in addition that Q_b is odd with respect to the grading. The pre-taming is called a taming if $D(\mathcal{E}_b)+Q_b$ is invertible for all $b \in B$. In this case, by definition $\operatorname{ind}(D(\mathcal{E})+Q)$ is zero. However, as the index is unchanged by the smoothing perturbation Q, also $\operatorname{ind}(\mathcal{E}) = 0$ if $\mathcal{E}$ admits a taming.

Let $\mathcal{E}_t$ be the notation for geometric family $\mathcal{E}$ with a chosen taming. In [19, Def. 4.16], the η-form $\eta(\mathcal{E}_t) \in \Omega(X)$ is defined. By [19, Theorem 4.13] it satisfies

$$d\eta(\mathcal{E}_t) = \Omega(\mathcal{E}). \tag{9}$$

We skip the considerable analytic difficulties in the construction of the eta-form and use it and its properties as a black box.

Definition 4.11. We call two cycles $(\mathcal{E}, \rho)$ and $(\mathcal{E}', \rho')$ paired if there exists a taming $(\mathcal{E} \sqcup_B \mathcal{E}'^{op})_t$ such that

$$\rho - \rho' = \eta((\mathcal{E} \sqcup_B \mathcal{E}'^{op})_t).$$

We let $\sim$ denote the equivalence relation generated by the relation "paired".

We define the differential K-theory $\hat{K}^*(B)$ of B to be the group completion of the abelian semigroup of equivalence classes of cycles as in Definition 4.4 with fiber dimension congruent $*$ modulo 2.

Theorem 4.12. *Definition 4.11 of $\hat{K}^*$ indeed defines (with the obvious notion of pullback) a contravariant functor on the category of smooth manifolds which is a differential extension of K-theory.*

The necessary natural transformations are induced by

1. $I\colon \hat{K}^*(X) \to K^*(X)$; $\quad (\mathcal{E}, \omega) \mapsto \operatorname{ind}(\mathcal{E})$ *of Definition 4.7*
2. $a\colon \Omega^{*-1}(X)/\operatorname{im}(d) \to \hat{K}^*(X)$; $\quad \omega \mapsto (0, -\omega)$
3. $R\colon \hat{K}^*(X) \to \Omega^*_{d=0}(X)$; $\quad (\mathcal{E}, \omega) \mapsto \Omega(\mathcal{E}) + d\omega$ *of* (8).

Proof. This is carried out in [26, Sect. 2.4]. □

Definition 4.13. Given two geometric families $\mathcal{E}, \mathcal{F}$ over a base X, there is an obvious geometric way to define their fiber product $\mathcal{E} \times_B \mathcal{F}$, with underlying fiber bundle the fiber product of bundles of manifolds, etc. Details are spelled out in [26, Sect. 4.1]. We then define the product of two cycles $(\mathcal{E}, \rho)$ and $(\mathcal{F}, \theta)$ (homogeneous of degree e and f, respectively) as

$$(\mathcal{E}, \rho) \cup (\mathcal{F}, \theta) := [\mathcal{E} \times_B \mathcal{F}, (-1)^e \Omega(\mathcal{E}) \wedge \theta + \rho \wedge \Omega(\mathcal{F}) - (-1)^e d\rho \wedge \theta].$$

Theorem 4.14. *The product of Definition 4.13 turns $\hat{K}^*$ into a multiplicative differential extension of K-theory.*

Proof. This is carried out in [26, Sect. 4]. □

Remark 4.15. Note that it is much more cumbersome to define a product structure with the other models described: for the vector bundle model the inclusion of the odd part is problematic, for the homotopy theoretic model one would have to realize the product structure using maps between (products of) classifying spaces. This is certainly a worthwhile task, but does not seem to be carried out in detail so far.

In the geometric families model, it is also rather easy to construct "integration over the fiber" and in particular S^1-integration as defined in 2.8. This will be explained in Sect. 5.

4.4 Differential Characters

One of the first models for a differential extension of integral cohomology are the differential characters of Cheeger–Simons [32]. A differential character is a pair (ϕ, ω), consisting of a homomorphism $\phi\colon Z_*(X) \to \mathbb{R}/\mathbb{Z}$ defined on the group of (smooth) singular cycles on X such that for a boundary $b = dc$, $c \in C_{*+1}(X)$ $\phi(b) = [\int_c \omega]_{\mathbb{R}/\mathbb{Z}}$. The set of differential characters on $C_*(X)$ is a model for $\hat{H}^{*+1}(X)$. Along similar lines, in [10] the cycle model for K-homology due to Baum and Douglas [7] – recently worked out in detail in [8] – is used to define $\hat{K}^*(X)$ as the group of $\mathbb{R}/\mathbb{Z}$-valued homomorphism from the set of Baum–Douglas cycles for K-homology which, on boundaries, are given as pairing with a differential form. [10] also define this pairing of a cycle and a differential form. To do this, in contrast to [7] the cycles have to carry additional geometric structure.

4.5 Multiplicative K-Theory

The first work toward differential K-theory probably has been carried out by Karoubi. We already mentioned his approach to $\mathbb{R}/\mathbb{Z}$-K-theory as the flat part of differential K-theory of [46] with a model based on vector bundles with connection.

This research has been deepened by Karoubi in subsequent publications [47,48]. The "multiplicative" K-theory introduced and studied there (and related to work of Connes and Karoubi on the multiplicative character of a Fredholm module [33], which explains the terminology) is associated to the usual K-theory and a filtration of the de Rham complex.

This theory looks very much like differential K-theory. With some normalizations, it seems that one should be able to obtain differential K-theory from the multiplicative one by considering the filtration of the de Rham complex given by the subcomplex of closed differential forms which are Chern characters of vector bundles (with a connection) on the given space. Karoubi uses multiplicative K-theory to construct characteristic classes for holomorphic, foliated or flat bundles.

We thank Max Karoubi for explanations of his work on multiplicative and differential K-theory.

4.6 Currential K-Theory

Freed and Lott in [39, 2.28] introduce a variant of differential K-theory where they replace the differential forms throughout by currents. This *changes* the theory in us much as the curvature homomorphism then also takes values in currents, i.e. forms are replaced by currents throughout. Because the multiplication of currents is not well defined, in this new theory one loses the ring structure. The advantage of this variant, on the other hand, is that push-forward maps can be described very directly also for embeddings: for the differential form part, the image under push-forward will be a current supported on the submanifold. The currential theory has the advantage that push-forwards can be defined easily. Moreover, they can be defined for arbitrary maps. However, pullbacks are not always defined (and not discussed at all for this theory in [39]). Consequently, the theory should (up to a degree shift given by Poincaré duality) better be considered as differential K-homology, twisted by some kind of spin^c-orientation twist.

To describe differential extensions of bordism theories as in [20, 28], currents are also used – but there only as an intermediate tool. The axiomatic setup for those theories is the one described in 2.1. In particular, the curvature homomorphism takes values in smooth differential forms.

4.7 Differential K-Theory via Bordism

In [28, Sect. 4], a geometric model of the differential extension $\hat{MU}^*$ of MU^* is constructed, where MU^* is cohomology theory dual to complex bordism. Cycles are given by pairs $(\tilde{c}, \alpha)$. Here $\tilde{c}$ is a geometric cycle for $MU^n(X)$, i.e. a proper smooth map $W \to X$ with $n = \dim(X) - \dim(W)$ and an explicit geometric model of the stable normal bundle with $U(N)$-structure. Moreover, α is a differential $(n-1)$-form with distributional coefficients whose differential differs from a certain characteristic current of $\tilde{c}$ by a smooth differential form. We impose on these cycles the equivalence relation generated by an obvious notion of bordism together with stabilization of the normal bundle data.

The functor $\hat{M}U^*$ is a multiplicative extension of complex cobordism. In particular, it takes values in algebras over $\hat{MU}^{ev}(*)$, as $\hat{M}U^{ev}(*)$ is a subring of $\hat{M}U^*(*)$. However, as $MU^{odd}(*) = 0$, the long exact sequence (3) give the canonical isomorphism

$$\hat{M}U^{ev}(*) \xrightarrow{\cong} MU^{ev}(*) = MU^*(*).$$

Assume now that E^* is a generalized cohomology theory which is *complex oriented*, i.e. which comes with a natural transformation $MU^* \to E^*$. An example is complex K-theory, with the usual orientation $MU \to MSpin^c \to K$. If the complex oriented theory E^* is Landweber exact (i.e. the condition of [52, Theorem 2.6$_{MU}$] is satisfied), then Landweber [52, Corollary 2.7] proves that $E^*(X) \cong MU^*(X) \otimes_{MU} E^*$ is obtained from $MU^*(X)$ by just taking the tensor product with the coefficients E^* (in the graded sense). Complex K-theory is Landweber exact [52, Example 3.3] so that this principle gives a simple bordism definition of K-theory.

In [28, Theorem 2.5] it is shown that Landweber's results extends directly to differential extensions. If E^* is a Landweber exact complex oriented generalized cohomology theory, then

$$\hat{E}^*(X) := \hat{MU}^*(X) \otimes_{MU} E$$

is a multiplicative differential extension of $E^*(X)$. Moreover, as explained above, we can apply this to complex K-theory and obtain a bordism description of differential K-theory:

$$\hat{K}^*(X) = \hat{M}U^*(X) \otimes_{MU} K.$$

By [28, Sect. 2.3], this is indeed a multiplicative differential extension of complex K-theory. Furthermore, it has a natural S^1-integration as in Definition 2.8 and even general "integration over the fiber" as discussed in Sect. 5.

4.8 Geometric Cycle Models for $\mathbb{R}/\mathbb{Z}$-K-Theory via the Flat Part of Differential K-Theory

As K-theory satisfies the conditions of Theorem 2.9, any multiplicative differential extension canonically comes with a transformation "integration over S^1". Therefore, by Theorem 2.10 its flat subfunctor is a generalized cohomology theory. Moreover, K-theory satisfies the conditions of Theorem 2.12 and we therefore get a natural isomorphism

$$\hat{K}_{flat} \to K\mathbb{R}/\mathbb{Z}.$$

In other words, any of the models for differential K-theory described above provides a model for $\mathbb{R}/\mathbb{Z}$-K-theory. In particular, we recover the original result [46, Sect. 7.21] that the flat part of the vector bundle model of Sect. 4.1 describes $K\mathbb{R}/\mathbb{Z}^{-1}$.

4.9 Real K-Theory

It seems to be not difficult to modify the above models to obtain differential extensions $\hat{KO}$ of real K-theory. This is particularly clear with the vector bundle model of Sect. 4.1. Complex Hermitean vector bundles with Hermitean connection have to be replaced by real Euclidean vector bundles with metric connection. This describes $\hat{KO}^0$. Of course, $\hat{KO}^*$ now is eight-periodic. The full model in terms of geometric families, following Sect. 4.3 replaces the families of Dirac bundles by their real analogs. For the analysis of η-forms it seems to be most suitable to implement the real structures via additional conjugate linear operations on the complex Dirac bundle (as explained, e.g. in [24, Sec. 2.2]). Alternatively one could possibly work with $\dim(T^v\pi)$-multigradings in the sense of, e.g. [43, A.3] (in other words, a compatible Cl_k-module structure, compare also [53, Sect. 16]). The details of a model for $\hat{KO}$ based on geometric families have still to be worked out. It is actually worthwhile to work out a Real differential K-theory in the usual sense (i.e. for spaces with additional involution). However, we will not carry this out here.

4.10 Differential K-Homology and Bivariant Differential K-Theory

An important aspect of any cohomology theory is the dual homology theory and the pairing between the cohomology theory and the homology theory. This applies in particular to K-theory, where the dual homology theory can be described with three different flavors:

- homotopy theoretically, using the K-theory spectrum
- with a geometric cycle model introduced by Baum and Douglas [7]
- an analytic model proposed by Atiyah [4] and made precise by Kasparov [49], compare also [43].

Definition 4.16. The cycles for $K_*(X)$ are triples (M, E, f) where M is a closed manifold with a given spin^c-structure with dimension congruent to $*$ modulo 2, $f: M \to X$ is a continuous map and $E \to X$ is a complex vector bundle. On the set of isomorphism classes of triples we put the equivalence relation generated by bordism, *vector bundle modification* and the relation that $(M, E_1, f) \amalg (M, E_2, f) \sim (M, E_1 \oplus E_2, f)$.

Here, vector bundle modification means that, given a complex vector bundle $V \to M$, $(M, E, f) \sim (S(V \oplus \mathbb{R}), T(E), f \circ p)$ where $S(V \oplus \mathbb{R})$ is the sphere bundle of the real bundle $V \oplus \mathbb{R}$, $p: V \oplus \mathbb{R} \to M$ is the bundle projection and $T(E)$ is a vector bundle which (up to addition of a bundle which extends over the disk bundle) represents the push-forward of E along the "north pole inclusion" $i: M \to S(E \oplus \mathbb{R}); x \mapsto (x, 0, 1)$. There is an explicit clutching construction of this bundle.

It took over 20 years before in [8] the equivalence of the geometric cycle model with the other models has been worked out in full detail.

In the geometric model, the pairing between K-homology and K-theory has a very transparent description, related to index theory. Given a K-homology cycle (M, E, f) and a K-theory cycle represented by the vector bundle $V \to X$, the pairing is the index of the spin^c Dirac operator $D_{E\otimes f^*V}$ on M, twisted with

$E \otimes f^*V$. In the analytic model, this Dirac operator by itself essentially already is a K-homology cycle.

The analytic theory has two very important extensions: first of all, it extends from a (co)homology theory for spaces to one for arbitrary C^*-algebras. "Extension" here in the sense that the category of compact Hausdorff spaces is equivalent to the opposite category of commutative unital C^*-algebras via the functor $X \mapsto C(X)$. Secondly, Kasparov's theory really is bivariant, with $KK(X,*) = K_*(X)$ being K-homology, and $KK(*,X) = K^*(X)$ K-theory. Most important is that KK-theory comes with an associative composition product $KK(X,Y) \otimes KK(Y,Z) \to KK(X,Z)$. This is an extremely powerful tool with many applications in index theory and beyond.

In this context, it is very desirable to have differentiable extensions also of K-homology and ideally of bivariant KK-theory (of course only restricted to manifolds). This should, e.g. provide a powerful home for refined index theory.

Definition 4.17. Indeed, guided by the model of differential K-theory, we see that differential K-homology or more general $\hat{KK}$, differential bivariant KK-theory, should be a functor from smooth manifolds to graded abelian groups with the following additional structure:

1. A transformation $I: \hat{KK}(X,Y) \to KK(X,Y)$.
2. A transformation R to the appropriate version of differential forms. For K-homology, the best bet for this are differential currents (essentially the dual space of differential forms) and for the bivariant theory the closed (i.e. commuting with the differentials) continuous homomorphisms from differential forms of the first space to differential forms of the second: $R: \hat{KK}(X,Y) \to \mathrm{Hom}(\Omega(X), \Omega(Y))$.
3. Action of "forms": a degree-1 transformation $a: \mathrm{Hom}(\Omega(X), \Omega(Y)) \to \hat{KK}(X,Y)$.

These have to satisfy relations and exact sequences which are direct generalizations of those of Definition 2.1:

$$\hat{KK}(X,Y) \xrightarrow{R} [-1]\,\mathrm{Hom}(\Omega^*(X), \Omega^*(Y))[-1]/\operatorname{im}(d) \xrightarrow{a} \hat{KK}(X,Y) \xrightarrow{I} KK(X,Y) \to 0,$$

$$\alpha \circ R = d.$$

Here, we adopt the tradition that KK-theory is $\mathbb{Z}/2\mathbb{Z}$-graded and use the even/odd grading of forms throughout.

Additionally, there should be a natural associative "composition product" $\hat{KK}(X,Y) \otimes \hat{KK}(Y,Z) \to \hat{KK}(X,Z)$ such that I maps this product to the Kasparov product and R to the composition.

Note that (up to Poincaré duality degree shifts), the currential K-theory of Sect. 4.6 actually is a model for differential K-homology, which is the specialization of the bivariant theory to $\hat{KK}(X,*)$.

An early preprint version of [26] contains a section which develops a bivariant theory as in Definition 4.17 and its applications. The groups are defined in a cycle model where a cycle ϕ for $KK(X,Y)$ consists of

1. A $\hat{K}$-oriented bundle $p\colon W \to X$, using $\hat{K}$-orientations as in Sect. 5.1
2. A class $x \in \hat{K}(W)$
3. A smooth map $f\colon W \to Y$
4. A continuous homomorphism $\Phi \in \mathrm{Hom}(\Omega^*(X), \Omega^*(Y))[-1]$

On the set of isomorphism classes of cycles one puts the equivalence relation generated by

1. Compatibility of the sum operations (one given by disjoint union, the other by addition in $\hat{K}(W)$ and $\mathrm{Hom}(\Omega^*(X), \Omega^*(Y))$)
2. Bordism
3. A suitable definition of vector bundle modification
4. Change of Φ by the image of d

Details of such a bivariant construction, or a construction of differential K-homology seem not to exist in the published literature. Note that this approach is not tied to K-theory. It provides a construction of a bivariant differential theory from a differential extension of a cohomology theory together with a theory of differential orientation and integration satisfying a natural set of axioms.

5 Orientation and Integration

5.1 *Differential K-Orientations*

Let $p\colon W \to B$ a proper submersion with fiber-dimension n. An important aspect of cohomology is a map "integration along the fiber" or "push-forward" $p_!\colon E^*(W) \to E^{*-n}(B)$. There is a general theory for this, and it requires the extra structure of an E-*orientation* of p (one could also say an E-orientation of the fibers of p). If E is ordinary cohomology, this comes down to an ordinary orientation and in de Rham cohomology $p_!$ literally is "integration along the fiber".

For K-theory, it is a spin^c-structure of π which gives rise to a K-theory orientation, which in turn defines a topological push-forward map $p_!\colon K^*(W) \to K^{*-n}(B)$.

To extend this to *differential* K-theory, one has to add additional geometric data to the spin^c-structure.

This is a pattern which applies to a general differential cohomology theory $\hat{E}$: a differential orientation will consist of an E-orientation in the usual sense together with extra differential and geometric data. There is one notable exception, though. In ordinary integral differential cohomology (Cheeger–Simons differential characters), a topological orientation (which is an ordinary orientation) lifts uniquely to a differential orientation. Therefore, in the literature treating ordinary differential cohomology and push-forward in that context, differential orientations are not discussed [35, 42].

We now describe a model for differential K-orientations for a proper submersion $p: W \to B$ which particularly suits the analytic model for $\hat{K}$ of Sect. 4.3. We will describe the $\mathbb{Z}$-graded version of orientations in order to make clear in which precise degrees the form constituents live. This is important if one wants to set up a similar theory in the case of other, non-two-periodic theories like complex bordism.

Fix an underlying topological K-orientation of p by choosing a spin^c-reduction of the structure group of the vertical tangent bundle $T^v p$ of p. In order to make this precise we choose along the way an orientation and a metric on $T^v p$.

We now consider the set $\mathcal{O}$ of tuples $(g^{T^v p}, T^h p, \tilde{\nabla}, \sigma)$ where

1. $g^{T^v p}$ is the Riemannian metric on the vertical bundle $\ker Tp$ (and the K-orientation is given by a spin^c-reduction of the resulting $O(n)$-principal bundle P, i.e. a spin^c-principal bundle Q lifting P).
2. $T^h p$ is a horizontal distribution.
3. From the horizontal distribution, we get a connection $\nabla^{T^v p}$ which restricts to the Levi-Civita connection along the fibers as follows. First one chooses a metric g^{TB} on B. It induces a horizontal metric $g^{T^h p}$ via the isomorphism $dp_{|T^h p}: T^h p \xrightarrow{\sim} p^*TB$. We get a metric $g^{T^v p} \oplus g^{T^h p}$ on $TW \cong T^v p \oplus T^h p$ which gives rise to a Levi-Civita connection. Its projection to $T^v p$ is $\nabla^{T^v p}$. $\tilde{\nabla}$ is a spin^c-reduction of $\nabla^{T^v p}$, i.e. a connection on the spin^c-principal bundle Q which reduces to $\nabla^{T^v p}$. If we think of the connections $\nabla^{T^v p}$ and $\tilde{\nabla}$ in terms of horizontal distributions $T^h SO(T^v p)$ and $T^h Q$, then we say that $\tilde{\nabla}$ reduces to $\nabla^{T^v p}$ if $d\pi(T^h Q) = \pi^*(T^h SO(T^v p))$.
4. $\sigma \in \Omega^{-1}(W; \mathbb{R}[u, u^{-1}])/\operatorname{im}(d)$.

We introduce the globally defined complex line bundle

$$L^2 := Q \times_\lambda \mathbb{C} \to W \tag{10}$$

associated to the spin^c-bundle Q via the representation $\lambda: Spin^c(n) \to U(1)$. The connection $\tilde{\nabla}$ thus induces a connection on ∇^{L^2}.

We introduce the form

$$c_1(\tilde{\nabla}) := \frac{1}{2}(2\pi i u)^{-1} R^{L^2} \in \Omega^0(W; \mathbb{R}[u, u^{-1}]). \tag{11}$$

Let $R^{\nabla^{T^v p}} \in \Omega^2(W, \operatorname{End}(T^v p))$ denote the curvature of $\nabla^{T^v p}$. The closed form

$$\hat{\mathbf{A}}(\nabla^{T^v p}) := \det{}^{1/2}\left(\frac{\frac{u^{-1} R^{\nabla^{T^v p}}}{4\pi}}{\sinh\left(\frac{u^{-1} R^{\nabla^{T^v p}}}{4\pi}\right)}\right) \in \Omega^0(W; \mathbb{R}[u, u^{-1}])$$

represents the $\hat{\mathbf{A}}$-class of $T^v p$.

Definition 5.1. The relevant differential form for local index theory in the spin^c-case is

$$\hat{\mathbf{A}}^c(\tilde{\nabla}) := \hat{\mathbf{A}}(\nabla^{T^v p}) \wedge e^{c_1(\tilde{\nabla})}. \tag{12}$$

If we consider $p: W \to B$ with the geometry $(g^{T^v p}, T^h p, \tilde{\nabla})$ and the Dirac bundle $S^c(T^v p)$ as a geometric family $\mathcal{W}$ over B, then

$$\int_{W/B} \hat{\mathbf{A}}^c(\tilde{\nabla}) = \Omega(\mathcal{W}) \in \Omega^{-n}(B; \mathbb{R}[u, u^{-1}])$$

is the local index form of $\mathcal{W}$.

We introduce a relation $\sim$ on $\mathcal{O}$: Two tuples $(g_i^{T^v p}, T_i^h p, \tilde{\nabla}_i, \sigma_i)$, $i = 0, 1$ are related if and only if $\sigma_1 - \sigma_0 = \tilde{\hat{\mathbf{A}}}^c(\tilde{\nabla}_1, \tilde{\nabla}_0)$, the transgression form of $\hat{\mathbf{A}}^c(\tilde{\nabla})$. In [26, Sect. 3.1], it is shown that $\sim$ is an equivalence relation.

Definition 5.2. The set of differential K-orientations which refine a fixed underlying topological K-orientation of $p: W \to B$ is the set of equivalence classes $\mathcal{O}/\sim$. The vector space $\Omega^{-1}(W; \mathbb{R}[u, u^{-1}])/\operatorname{im}(d)$ acts on the set of differential K-orientations by translation of the differential form entry.

In [26, Corollary 3.6] we show:

Proposition 5.3. *The set of differential K-orientations refining a fixed underlying topological K-orientation is a torsor over $\Omega^{-1}(W; \mathbb{R}[u, u^{-1}])/\operatorname{im}(d)$.*

Remark 5.4. Proposition 5.3 should generalize to differential extensions of a general cohomology theory E: the differential orientations refining a fixed E-orientation should form a torsor over $\Omega^{-1}(W; E\mathbb{R})$. If we apply this to the special case of ordinary cohomology, this group is trivial as the coefficients are concentrated in degree 0. This explains why there is a unique lift of an ordinary orientation to a differential orientation in case of ordinary cohomology.

5.2 Integration

We have set up our model for differential K-orientations in such a way that we have a natural description of the integration homomorphism in differential K-theory using the analytic model of Sect. 4.3. From now on we reduce to the two-periodic case (by setting $u = 1$).

Given $p: W \to B$ with a differential K-orientation as in Sect. 5.1 and a class $x = [\mathcal{E}, \rho] \in \hat{K}(W)$, the basic idea for the construction of $p_!(x) \in \hat{K}(B)$ is to form the representative for $p_!(x)$ as the geometric family $p_!\mathcal{E}$ obtained by simply composing the family over W with p to obtain a family over B. The geometry on p given by the differential orientation allows to put the required geometry on the composition in a canonical way. For example, the fiberwise metric is obtained as direct sum of the fiberwise metric on $\mathcal{E}$ and on the fibers of p. We omit the details which are given in [26, Sect. 3.2].

The main remaining question is to define the correct differential form; this requires the study of adiabatic limits. Indeed, in the construction of the geometry on $p_!\mathcal{E}$ we can introduce an additional parameter $\lambda \in (0,\infty)$ by scaling the metric on the fibers of $\mathcal{E}$ by λ^2 (and adjusting the remaining geometry) to obtain $p_!^\lambda\mathcal{E}$. In total, this gives an adiabatic deformation family $\mathcal{F}$ over $(0,\infty)\times B$ which restricts to $p_!^\lambda\mathcal{E}$ on $\{\lambda\}\times B$.

Although the vertical metrics of $\mathcal{F}$ and $p_!^\lambda\mathcal{E}$ collapse as $\lambda \to 0$ the induced connections and the curvature tensors on the vertical bundle T^vq converge and simplify in this limit. This fact is heavily used in local index theory, and we refer to [11, Sect. 10.2] for details. In particular, the integral

$$\tilde{\Omega}(\lambda,\mathcal{E}) := \int_{(0,\lambda)\times B/B} \Omega(\mathcal{F}) \tag{13}$$

converges.

We now define

$$\hat{p}_!(\mathcal{E},\rho) := [p_!\mathcal{E}, \int_{W/B} \hat{\mathbf{A}}^c(\tilde{\nabla})\wedge\rho + \tilde{\Omega}(1,\mathcal{E}) + \int_{W/B} \sigma\wedge R([\mathcal{E},\rho])] \in \hat{K}(B). \tag{14}$$

This push-forward has all the expected properties, listed below. The proofs use the explicit model and a heavy dose of local index theory in order to verify that this cycle-level construction is compatible with the equivalence relation involving tamings and η-forms. They can be found in [26, Sects. 3.2, 3.3, 4.2].

Theorem 5.5. *The push-forward/integration for differential K-theory defined in* (14) *has the following properties.*

1. *Given a proper submersion $p\colon W \to B$ with differential K-orientation and with fiber dimension n, the differential push-forward is $\hat{p}_!\colon \hat{K}^*(W) \to \hat{K}^{*-n}(B)$ a well defined homomorphism which only depends on the differential orientation, not its particular representative.*
2. *If $q\colon Z \to W$ is another proper submersion with differential K-orientation, there is a canonical way to put a composed differential K-orientation on the composition $p\circ q$, and push-forward is functorial: $(p\hat{\circ} q)_! = \hat{p}_!\circ\hat{q}_!$.*
3. *Fix a Cartesian diagram*

$$\begin{array}{ccc} W' & \xrightarrow{F} & W \\ \downarrow{\scriptstyle f^*p} & & \downarrow{\scriptstyle p} \\ B' & \xrightarrow{f} & B \end{array}$$

where as before $p\colon W \to B$ is a proper submersion with differential K-orientation. There is a canonical way to pull back the differential K-orientation of p to $p' := f^*p$. One then has compatibility of pull-back and push-forward

$$f^*\hat{p}_! = \hat{p}'_! F^*.$$

4. *The projection formula relating push-forward and product holds in differential K-theory*

$$\hat{p}_!(p^* y \cup x) = y \cup \hat{p}_!(x); \quad x \in \hat{K}(W),\ y \in \hat{K}(B).$$

5. *The differential push-forward is via the structure maps compatible with the push-forward in ordinary K-theory and on differential forms. However, for the differential forms we have to define the modified integration, depending on the differential K-orientation o, as*

$$p_!^o : \Omega(W) \to \Omega(B), \quad \omega \mapsto \int_{W/B} (\hat{\mathbf{A}}^c(\tilde{\nabla}) - d\sigma) \wedge \omega.$$

Indeed, this does not depend on the representative of the differential K-orientation. Moreover, it induces $p_!^o : \Omega(W)/\operatorname{im}(d) \to \Omega(B)/\operatorname{im}(d)$. *With this map, we get commutative diagrams*

$$\begin{array}{ccccccc} K(W) & \xrightarrow{\mathrm{ch}} & \Omega(W)/\operatorname{im}(d) & \xrightarrow{a} & \hat{K}(W) & \xrightarrow{I} & K(W) \\ \downarrow{\scriptstyle p_!} & & \downarrow{\scriptstyle p_!^o} & & \downarrow{\scriptstyle \hat{p}_!} & & \downarrow{\scriptstyle p_!} \\ K(B) & \xrightarrow{\mathrm{ch}} & \Omega(B)/\operatorname{im}(d) & \xrightarrow{a} & \hat{K}(B) & \xrightarrow{I} & K(B) \end{array} \tag{15}$$

$$\begin{array}{ccc} \hat{K}(W) & \xrightarrow{R} & \Omega_{d=0}(W) \\ \downarrow{\scriptstyle \hat{p}_!} & & \downarrow{\scriptstyle p_!^o} \\ \hat{K}(B) & \xrightarrow{R} & \Omega_{d=0}(B) \end{array} \tag{16}$$

5.3 S^1-Integration

We consider the projection $\mathrm{pr}_1 : B \times S^1 \to B$.

The projection pr_1 fits into the Cartesian diagram

$$\begin{array}{ccc} B \times S^1 & \xrightarrow{\mathrm{pr}_2} & S^1 \\ \downarrow{\scriptstyle \mathrm{pr}_1} & & \downarrow{\scriptstyle p} \\ B & \xrightarrow{r} & \{*\}. \end{array}$$

We choose the metric g^{TS^1} of unit volume and the bounding spin structure on TS^1. This spin structure induces a spin^c structure on TS^1 together with the connection $\tilde{\nabla}$. In this way we get a representative o of a differential K-orientation of p. By pull-back we get the representative r^*o of a differential K-orientation of pr_1 which is used to define $(\hat{\mathrm{pr}}_1)_!$.

Definition 5.6. We define S^1-integration for differential K-theory as in Definition 2.8 simply by setting

$$\int_{B\times S^1/B} := (p_1)_!\colon \hat{K}^*(B\times S^1) \to \hat{K}^{*-1}(B)$$

where we use the differential K-orientation of $p_1\colon B\times S^1 \to B$ just described. Note that by Theorem 5.5 it has the properties required of S^1-integration.

By [26, Corollary 4.6] we get

$$(\hat{\mathrm{pr}}_1)_! \circ \mathrm{pr}_1^* = 0.$$

6 Index Theory and Natural Transformations

It is well established that index theory of elliptic operators is closely related to K-theory and K-homology. Indeed, this is the reason why the analytic model of Sect. 4.3 can work at all. For a fiber bundle $p\colon W \to B$ with K-orientation, the push-forward in K-theory can be interpreted as the family index of the fiberwise spin^c Dirac operator, twisted with the bundle representing the K-theory class. Of course, one might define the push-forward in a different way – then this is a somewhat abstract statement.

Continuing on this formal level, the Chern character provides a way to compute K-theory in terms of cohomology. Now there is also the push-forward in cohomology. It is well known that Chern character and push-forward are not compatible. The Riemann–Roch formula provides the appropriate correction. In some sense, a Riemann–Roch theorem therefore is a cohomological index theorem.

On this level, we will now find a lift of index theory to differential K-theory. In Sect. 5 we have discussed the push-forward in differential K-theory. We will now describe how to lift the Chern character to a natural transformation from differential K-theory to differential cohomology and will then discuss a differential Riemann–Roch theorem correcting the defect that this Chern character is not compatible with integration.

A further refinement of this theorem is obtained by a direct analytic definition of a (family) index with values in differential K-theory, given in a natural way for geometric families of elliptic index problems. The goal of an index theorem is then to find a topological formula for this index, i.e. a formula which does not involve the explicit solution of differential equations. This has indeed been achieved by Lott and Freed in [39] and we discuss the details in Sect. 6.3.

6.1 Differential Chern Character

The classical Chern character has two fundamental properties. First, it is a certain characteristic class of vector bundles. As such, it is a certain explicit (rational) polynomial $p_{\mathrm{ch}}(c_1, c_2, \dots)$ in the Chern classes of the vector bundle. Secondly, it turns out that this characteristic class is compatible with direct sum and stabilization and with Bott periodicity in the appropriate way to define a natural transformation of cohomology theories

$$\mathrm{ch}\colon K^*(\cdot) \to H^*(\cdot; \mathbb{Q}[u, u^{-1}]), \quad (u \text{ of degree } 2).$$

Finally, after tensor product with $\mathbb{Q}$, $\mathrm{ch} \otimes \mathrm{id}_{\mathbb{Q}}$ becomes an isomorphism for finite CW-complexes.

We demand that the Chern character for differential K-theory should display the same properties: be implemented as characteristic class of vector bundles with connection, and then pass to a natural transformation between differential cohomology theories.

6.1.1 Characteristic Classes in Differential Cohomology

Early on, differential cohomology is closely related to characteristic classes of vector bundles with connection. Indeed, $\hat{H}^2(X; \mathbb{Z})$ even is isomorphic to the set of isomorphism classes of complex line bundles with Hermitean connection. This isomorphism is implemented by the differential version of the first Chern class.

Elaborating on this, Cheeger and Simons [32, Sect. 4] construct differential Chern classes of vector bundles with Hermitean connection with values in integral differential cohomology

$$\hat{c}_k(E, \nabla) \in \hat{H}^{2k}(X; \mathbb{Z})$$

with the following properties:

1. Naturality under pullback with smooth maps;
2. Compatibility with the classical Chern classes:

$$I(\hat{c}_k(E, \nabla)) = c_k(E) \in H^{2k}(X; \mathbb{Z});$$

3. Compatibility with Chern–Weil theory

$$R(\hat{c}_k(E, \nabla)) = C_k(\nabla^2) \in \Omega^{2k}(X),$$

   where $C_k(\nabla)$ is the Chern–Weil form of the connection ∇ associated to the invariant polynomial C_k which is (up to a scalar) the elementary symmetric polynomial in the eigenvalues.

Moreover, Cheeger and Simons show that the differential Chern classes are uniquely determined by these requirements. The proof is reminiscent of (and indeed inspired) the proof of uniqueness in Sect. 3. They use the universal example of $\mathbb{C}^n$-vector bundles with connection – known to exist by [58]. Once the differential Chern classes are chosen for the universal example, they are determined by naturality for every (E, ∇). The integral cohomology of the base space $BU(n)$ of the universal example is concentrated in even degrees. The long exact sequence (4) then implies that $I \oplus R\colon \hat{H}^{2k}(BU(n); \mathbb{Z}) \oplus \Omega^{2k}_{d=0}(BU(n))$ is injective, so that 2 and 3 determine the universal $\hat{c}_k$, and they exist by the defining properties of $C_k(\nabla^2)$. One only has to check that the construction is independent of the choice of the universal model, which essentially follows from uniqueness. Note that $BU(n)$ is not itself a finite dimensional manifold so that one has (as usual) to work with finite dimensional approximations.

Differential ordinary cohomology is defined with coefficients in any subring of $\mathbb{R}$, and the inclusion of coefficient rings induces natural maps with all the expected compatibility relations. In particular, we have differential cohomology with coefficients in $\mathbb{Q}$, $\hat{H}^*(\cdot; \mathbb{Q})$, taking values in the category of $\mathbb{Q}$-vector spaces, and commutative diagrams

$$\begin{array}{ccc} \hat{H}(\cdot;\mathbb{Z}) & \longrightarrow & \hat{H}(\cdot;\mathbb{Q}) \\ \downarrow R & & \downarrow R \\ \Omega(\cdot) & \xrightarrow{=} & \Omega(\cdot) \end{array} ; \qquad \begin{array}{ccc} \hat{H}(\cdot;\mathbb{Z}) & \longrightarrow & \hat{H}(\cdot;\mathbb{Q}) \\ \downarrow I & & \downarrow I \\ H^*(\cdot;\mathbb{Z}) & \longrightarrow & H^*(\cdot;\mathbb{Q}). \end{array}$$

Expressing the classical Chern character (uniquely) as a rational polynomial in the Chern classes

$$\mathrm{ch}(E) = P_{\mathrm{ch}}(c_1(E), c_2(E), \dots) \quad \text{with } P_{\mathrm{ch}} \in \mathbb{Q}[[x_1, x_2, \dots]],$$

we now define the differential Chern character

$$\hat{\mathrm{ch}}(E, \nabla) := P_{\mathrm{ch}}(\hat{c}_1(E, \nabla), \dots) \in \hat{H}^{2*}(X; \mathbb{Q}). \tag{17}$$

By construction, this is a natural characteristic class satisfying

$$I(\hat{\mathrm{ch}}(E, \nabla)) = \mathrm{ch}(E) \in H^{2*}(X; \mathbb{Q}),$$
$$R(\hat{\mathrm{ch}}(E, \nabla)) = P_{\mathrm{ch}}(C_1(\nabla^2), \dots) = \mathrm{tr}(-\exp(\nabla^2/2\pi i)) \in \Omega^{2*}(X).$$

6.1.2 Differential Chern Character Transformation

The second point of view of the differential Chern character is as a natural transformation between differential K-theory and differential cohomology. Indeed, we prove in [26, Sect. 6] the following theorem.

Theorem 6.1. *There is a unique natural transformation of differential cohomology theories*

$$\hat{\mathrm{ch}}\colon \hat{K}^*(\cdot) \to \hat{H}^*(\cdot;\mathbb{Q}[u,u^{-1}])$$

with the following properties:

1. *Compatibility with the Chern character in ordinary cohomology and with the action of forms, i.e. the following diagram commutes:*

$$\begin{array}{ccccc} \Omega(X)/\operatorname{im}(d) & \xrightarrow{a} & \hat{K}(X) & \xrightarrow{I} & K(X) \\ \downarrow{=} & & \downarrow{\hat{\mathrm{ch}}} & & \downarrow{\mathrm{ch}} \\ \Omega(X)/\operatorname{im}(d) & \xrightarrow{a} & \hat{H}(X;\mathbb{Q}[u,u^{-1}]) & \xrightarrow{I} & H(X:\mathbb{Q}[u,u^{-1}]). \end{array}$$

2. *Compatibility with the curvature homomorphism, i.e. the following diagram commutes*

$$\begin{array}{ccc} \hat{K}(X) & \xrightarrow{R} & \Omega_{d=0}(X;\mathbb{R}[u,u^{-1}]) \\ \downarrow{\hat{\mathrm{ch}}} & & \downarrow{=} \\ \hat{H}(X;\mathbb{Q}[u,u^{-1}]) & \xrightarrow{R} & \Omega_{d=0}(X;\mathbb{R}[u,u^{-1}]). \end{array}$$

3. *Compatibility with suspension, i.e. the following diagram commutes*

$$\begin{array}{ccc} \hat{K}(S^1\times X) & \xrightarrow{\hat{\mathrm{ch}}} & \hat{H}(S^1\times X;\mathbb{Q}[u,u^{-1}]) \\ \downarrow{\mathrm{pr}_{2!}} & & \downarrow{\mathrm{pr}_{2!}} \\ \hat{K}(X) & \xrightarrow{\hat{\mathrm{ch}}} & \hat{H}(X;\mathbb{Q}[u,u^{-1}]). \end{array}$$

Here, we use the differential K-orientation of pr_2 *as in Sect. 5.3.*

Moreover, if $x \in \hat{K}^0(X)$ *is represented (as in 4.1 or 4.3) by the zero-dimensional family* $((E,\nabla),\rho)$ *(with* ρ *the additional form of odd degree), then*

$$\hat{\mathrm{ch}}(x) = \hat{\mathrm{ch}}(E,\nabla) + a(\rho) \in \hat{H}^{2*}(X;\mathbb{Q}[u,u^{-1}]).$$

In addition, $\hat{\mathrm{ch}}$ *is a multiplicative natural transformation and becomes an isomorphism after tensor product with* $\mathbb{Q}$.

6.2 Differential Riemann–Roch Theorem

We are now in the situation to formulate the Riemann–Roch theorem, which describes the relations between push-forward in differential K-theory and differential cohomology and the differential Chern character. Let $p\colon W \to B$ be a proper submersion with a differential K-orientation o represented by $(g^{T^v p}, T^h p, \tilde{\nabla}, \sigma)$ as in Sect. 5.1.

The Riemann Roch theorem asserts the commutativity of a diagram

$$\begin{array}{ccc} \hat{K}(W) & \xrightarrow{\hat{\mathrm{ch}}} & \hat{H}(W,\mathbb{Q}) \\ \downarrow{\scriptstyle \hat{p}_!} & & \downarrow{\scriptstyle \hat{p}^A_!} \\ \hat{K}(B) & \xrightarrow{\hat{\mathrm{ch}}} & \hat{H}(B,\mathbb{Q}). \end{array}$$

Here $\hat{p}^A_!$ is the composition of the cup product with a differential rational cohomology class $\hat{\hat{\mathbf{A}}}^c(o)$ and the push-forward in differential rational cohomology (uniquely determined by an ordinary orientation of p, in particular by the K-orientation which is already fixed).

We have to define the refinement $\hat{\hat{\mathbf{A}}}(o) \in \hat{H}^{ev}(W,\mathbb{Q})$ of the form $\hat{\mathbf{A}}^c(\tilde{\nabla}) \in \Omega^{ev}(W; R[u,u^{-1}])$. The geometric data of o determines a connection $\nabla^{T^v p}$ and hence a geometric bundle $\mathbf{T^v p} := (T^v p, g^{T^v p}, \nabla^{T^v p})$. According to [32] we can define Pontryagin classes

$$\hat{p}_i(\mathbf{T^v p}) \in \hat{H}^{4i}(W,\mathbb{Z}), \quad i \geq 1 .$$

The spin^c-structure gives rise to a Hermitean line bundle $L^2 \to W$ with connection ∇^{L^2} (see (10)). We set $\mathbf{L}^2 := (L^2, h^{L^2}, \nabla^{L^2})$. Again using [32], we get a class

$$\hat{c}_1(\mathbf{L}^2) \in \hat{H}^2(W,\mathbb{Z}).$$

Inserting the classes $u^{-2i}\hat{p}_i(\mathbf{T^v p})$ into that $\hat{\mathbf{A}}$-series $\hat{\mathbf{A}}(p_1, p_2, \ldots) \in \mathbb{Q}[[p_1, p_2 \ldots]]$ we define

$$\hat{\hat{\mathbf{A}}}(\mathbf{T^v p}) := \hat{\mathbf{A}}(\hat{p}_1(\mathbf{T^v p}), \hat{p}_2(\mathbf{T^v p}), \ldots) \in \hat{H}^0(W, \mathbb{Q}[u,u^{-1}]). \tag{18}$$

Definition 6.2. We define

$$\hat{\hat{\mathbf{A}}}^c(o) := \hat{\hat{\mathbf{A}}}(\mathbf{T^v p}) \wedge e^{\frac{1}{2u}\hat{c}_1(\mathbf{L}^2)} - a(\sigma) \in \hat{H}^0(W, \mathbb{Q}[u,u^{-1}]).$$

Note that $R(\hat{\hat{\mathbf{A}}}^c(o)) = \hat{\mathbf{A}}^c(\tilde{\nabla})$, with $\hat{\mathbf{A}}^c(\tilde{\nabla})$ of (12).

By [26, Lemma 6.17], $\hat{\hat{\mathbf{A}}}^c(o)$ indeed does not depend on the particular representative of o. This follows from the homotopy formula.

We now define

$$\hat{p}^A_!\colon \hat{H}^*(W; \mathbb{Q}[u,u^{-1}]) \to \hat{H}^{*-n}(B; \mathbb{Q}[u,u^{-1}]);\ x \mapsto \hat{p}_!(\hat{\hat{\mathbf{A}}}^c(o) \cup x).$$

The differential Riemann–Roch now reads

Theorem 6.3. *The following square commutes*

$$\begin{array}{ccc} \hat{K}(W) & \xrightarrow{\hat{\mathrm{ch}}} & \hat{H}(W, \mathbb{Q}[u, u^{-1}]) \\ \downarrow{\hat{p}_!} & & \downarrow{\hat{p}_!^A} \\ \hat{K}(B) & \xrightarrow{\hat{\mathrm{ch}}} & \hat{H}(B, \mathbb{Q}[u, u^{-1}]). \end{array}$$

This diagram is compatible via the transformations I with the maps of the classical Riemann–Roch theorem.

6.3 Differential Atiyah–Singer Index Theorem

We want to understand the Atiyah–Singer index theorem as the equality of the analytic and the topological (family) index. To formulate a differential version of this, the first step is to define the differential analytic and topological (family) index.

We start with the proper submersion $p \colon W \to B$ with n-dimensional fibers, and we think of $W \to B$ as the underlying family of manifolds on which the family of elliptic operators shall be given. To be able to define a *differential* index, we have to choose geometry for $p \colon W \to B$, which amounts exactly to the choice of data representing a differential orientation of p as in Sect. 5.1, namely a tuple $(g^{T^v p}, T^h p, \tilde{\nabla}, \sigma)$ consisting of vertical metric, horizontal distribution, fiberwise spin^c-structure and compatible spin^c-connection. The form σ could of course be chosen equal to 0.

We follow the definition of the analytic index as given by Freed and Lott in [39]. We start with a vector bundle (E, ∇_E) with connection over W (representing a class in $\hat{K}^0(W)$ using either the model of Sect. 4.1 or 4.3). We then want to define the analytic index of the family of spin^c-Dirac operators twisted by (E, ∇_E). This is based on Bismut–Quillen superconnections. Indeed, it uses the Bismut–Cheeger eta-form which mediates between the Chern character of the finite dimensional index bundle (giving the naive analytic index) and the Chern character of the Bismut superconnection. For details see below.

The topological index of [39] is modelled closely after the classical definition of the topological index by Atiyah and Singer. One factors $p \colon W \to B$ as a fiberwise embedding $W \to S^N \times B$ and $\mathrm{pr}_2 \colon S^N \times B \to B$. One then uses very explicit formulas for the differential K-theory push-forward of the embedding $W \to S^N \times B$, based on a model of differential K-theory using currents. Finally, a Künneth decomposition of $\hat{K}(S^N \times B)$ gives an explicit push-forward for $\mathrm{pr}_2 \colon S^N \times B \to B$. The topological index is defined as a modification of the composition of these two push-forwards. It does not involve spectral analysis nor does it require the solution of differential equations. It does use differential forms, so the term "differential topological index" is indeed quite appropriate. For details again see below.

The main result of [39] is then

Theorem 6.4. *Differential analytic index and differential topological index both define homomorphism*

$$\mathrm{ind}^{an},\ \mathrm{ind}^{top}\colon \hat{K}^0(W) \to \hat{K}^{-n}(B).$$

Moreover, analytic and topological index coincide:

$$\mathrm{ind}^{an} = \mathrm{ind}^{top}$$

Finally, $\mathrm{ind}^{an} = \mathrm{ind}^{top}$ *fits into a commutative diagram*

$$\begin{array}{ccccccc} 0 & \longrightarrow & K\mathbb{R}/\mathbb{Z}^{-1}(W) & \longrightarrow & \hat{K}^0(W) & \xrightarrow{R} & \Omega^0_{d=0}(W;\mathbb{R}[u,u^{-1}]) \\ & & \Big\downarrow{\scriptstyle \mathrm{ind}^{an}=\mathrm{ind}^{top}} & & \Big\downarrow{\scriptstyle \mathrm{ind}^{an}=\mathrm{ind}^{top}} & & \Big\downarrow{\scriptstyle \omega\mapsto\int_{W/B}\hat{\mathbf{A}}^c(\tilde{\nabla})\wedge\omega} \\ 0 & \longrightarrow & K\mathbb{R}/\mathbb{Z}^{-1-n}(B) & \longrightarrow & \hat{K}^{-n}(B) & \xrightarrow{R} & \Omega^{-n}_{d=0}(B;\mathbb{R}[u,u^{-1}]). \end{array}$$

In the following, we explain the ingredients of this formula.

1. On the differential form level, $\hat{\mathbf{A}}^c(\tilde{\nabla})$ of (12) coincides with the form $Todd(\hat{\nabla}^W)$ of [39, (2.14)].
2. For K-theory with coefficients in $\mathbb{R}/\mathbb{Z}$, $\mathrm{ind}^{an}\colon K\mathbb{R}/\mathbb{Z}^{-1}(W) \to K\mathbb{R}/\mathbb{Z}^{-1-n}(B)$ has been constructed in and is the main theme of [54]. In particular, the equality of analytic and topological index in this context is proved there.

Proposition 6.5. *In the end, of course,* $\mathrm{ind}^{an} = \mathrm{ind}^{top}\colon \hat{K}^0(W) \to \hat{K}^{-n}(B)$ *coincide with* $\hat{p}_!\colon \hat{K}^0(W) \to \hat{K}^{-n}(B)$ *of Sect. 5.2.*

The main point is that ind^{an} is defined as an honest analytic index, whereas $\hat{p}_!$ only formally does so.

In the following, we will assume that the fiber dimension n is even. The case of odd n is easily reduced to this via suspension–desuspension constructions using products with S^1.

6.3.1 Analytic Index in Differential K-Theory

We now define the analytic index of a cycle $(\mathcal{E},\phi) = (E, h_E, \nabla_E, \phi)$ for $\hat{K}^0(W)$ given in terms of a vector bundle with connection and an auxiliary form ϕ as in Sect. 4.1. To get a cleaner picture, we assume that the family D_E of twisted spin^c-Dirac operators over B, constructed as the geometric Dirac operators provided by the differential K-orientation data twisted with (E,∇_E) has a kernel bundle $\ker(D_E)$ which comes with an induced Hermitean metric and compressed connection $\nabla_{\ker(D_E)}$. The Bismut–Cheeger eta-form in this situation is defined as

$$\tilde{\eta} := u^{-n/2} R_u \int_0^\infty STr\left(u^{-1}\frac{dA_s}{ds}e^{u^{-1}A_s^2}\right) ds \in \Omega^{-n-1}(B;\mathbb{R}[u,u^{-1}])/\operatorname{im}(d).$$

R_u is introduced to simplify notation, it is induced from the ring homomorphism $\mathbb{R}[u,u^{-1}] \to \mathbb{R}[u,u^{-1}]; u \mapsto (2\pi i)u$. A_s is the Bismut superconnection on the (typically infinite dimensional) bundle $\mathcal{H}$ over B whose fiber over $b \in B$ is the space of sections of the twisted spinor bundle over $W_b = p^{-1}(B)$, the bundle on which D_E acts. More precisely, for $s > 0$

$$A_s = su^{1/2}D_E + \nabla^{\mathcal{H}} - s^{-1}u^{-1/2}c(R)/4.$$

Here, $\nabla^{\mathcal{H}}$ is a canonical connection on $\mathcal{H}$ constructed out of the given connections, and $c(R)$ is Clifford multiplication by the curvature 2-form of $\mathcal{H}$. For details about this construction, see [11, Sect. 10]. Note that, following Freed-Lott, powers of s instead of powers of $s^{1/2}$ are used in the definition of the superconnection.

The eta-form provides an interpolation between the Chern character form of the kernel bundle and of E as follows (compare [39, (3.11)] and [26, (0.6)]):

$$d\tilde{\eta} = \int_{W/B} \hat{A}^c(\tilde{\nabla}) \wedge \mathrm{ch}(\nabla_E) - \mathrm{ch}(\nabla^{\ker(D_E)}). \tag{19}$$

Following [39, Definition 3.12], one now defines for $(\mathcal{E},\phi) = (E, h_E, \nabla_E, \phi)$

$$\mathrm{ind}^{an}(\mathcal{E},\phi) := \left((\ker(D_E), h_{\ker(D_E)}, \nabla^{ker(D_E)}), \int_{W/B} \hat{A}^c(\tilde{\nabla}) \wedge \phi + \tilde{\eta}\right), \tag{20}$$

given by the zero dimensional family $\ker(D_E)$ with its Hermitean metric and connection, and the differential form part which uses the eta form.

In [39, Theorem 6.2] it is shown that this formula indeed factors through a map ind^{an} on differential K-theory as stated in Theorem 6.4. Alternatively one could use [26, Corollary 5.5] which immediately implies that $\mathrm{ind}^{an}(\mathcal{E},\phi)$ is equal to the push-forward $\hat{p}^o_!(\mathcal{E},\phi)$ as defined in (14).

The construction of the analytic differential index in the general case, i.e. if the kernels do not form a bundle, is carried out by a perturbation to reduce to the special case treated so far.

6.3.2 Topological Index in Differential K-Theory

The topological index is defined in two steps. One chooses a *fiberwise isometric* fiberwise embedding $i: W \to S^N \times B$ of even codimension, where the target is equipped with the product structure. We assume that we have a differential spin^c-structure on the normal bundle ν of the embedding which is compatible with structures on W and on $S^N \times B$ (in the sense of [39, Sect. 5]).

Given a Hermitean bundle with connection (E, ∇_E) and a differential form ϕ on B, one now has to construct the differential push-forward i_* of $(\mathcal{E},\phi) = (E, h_E, \nabla_E, \phi)$. This is based on the Thom homomorphism in differential K-theory. More precisely, one has to choose a ($\mathbb{Z}/2\mathbb{Z}$-graded) Hermitean bundle F on $S^N \times B$

with connection together with an endomorphism V such that V is invertible outside $i(W)$ but the kernel of V is isomorphic (as geometric bundle) to the tensor product of E with the spinor bundle, and a suitably defined compression of the covariant derivative of V under this isomorphism becomes Clifford multiplication. In [13, Definition 1.3] in this situation a transgression form γ is defined which interpolated between the Chern form of F and the image of $\mathrm{ch}(\nabla_E)$ under the differential form spin^c-Thom homomorphism.

We then define

$$\hat{i}_!(\mathcal{E},\phi) := \Big((F, h_F, \nabla_F), \phi \wedge \hat{A}^o(\nabla_\nu)^{-1} \wedge \delta_X - \gamma - C'\Big) \in \hat{K}^{N-n}(S^N \times B), \tag{21}$$

where $\hat{A}^o(\nabla_\nu)^{-1} \wedge \delta_X$ is the current representing the spin^c-Thom form and C' is a further correction term which vanishes if the horizontal distribution of $W \to B$ is the restriction of the product horizontal distribution of $S^n \times B$ under the embedding i.

We now define

$$\mathrm{ind}^{top}(\mathcal{E},\phi) := \hat{p}_!(\hat{i}_!(\mathcal{E},\phi)) \in \hat{K}^{-n}(B), \tag{22}$$

where $p\colon S^N \times B \to B$ is the projection and we use the product structure to define the spin^c-orientation.

Finally, we observe that for the product $S^N \times B$ we can use the Künneth type formula to obtain an explicit formula for $\hat{p}_!(x)$. More precisely, use the Künneth formula in ordinary K-theory to write the underlying K-theory class $I(X) = p^*a + \mathrm{pr}_1^*(t) \cdot p^*b$ where t is a second additive generator (besides 1) of $K^*(S^N)$. We lift all the classes a, b, t to classes A, B, T in differential K-theory and then write $X = p^*A + pr_1^*(T) \cdot p^*B + a(\alpha)$ with a suitable differential form α. One then obtains by [39, (5.31)]

$$\hat{p}_!(X) = B + a\Big(\int_{S^N \times B/B} \hat{A}^o \wedge \alpha\Big), \tag{23}$$

where $\hat{A}^o \in \Omega^0(S^N \times B; \mathbb{R}[u,u^{-1}])$ is given by the product structure on $p\colon S^N \times B \to B$. By [39, Corollary 7.36] this construction indeed factors through a homomorphism ind^{top} on differential K-theory.

6.3.3 Proof of the Differential Index Theorem

It remains to prove that for any $(\mathcal{E},\phi) = (E, h_E, \nabla_E, \phi)$ as above, $\mathrm{ind}^{top}(\mathcal{E}) - \mathrm{ind}^{an}(\mathcal{E}) = 0$. The constructions are carried out in such a way that the desired identity holds for the images under R. Therefore $\mathrm{ind}^{top}(\mathcal{E}) - \mathrm{ind}^{an}(\mathcal{E}) =: T \in K^{-n-1}(B; \mathbb{R}/\mathbb{Z})$ is a flat differential K-theory class. Now Freed and Lott in [39] follow the method of proof of the $\mathbb{R}/\mathbb{Z}$-index theorem of [54]. One has to show that the pairing of T with $K_{-n-1}(B)$ (the ordinary K-homology of B) vanishes. These pairings are given by reduced η-invariants (provided we have a kernel bundle). In the end, one has to establish certain identities for η-invariants on W, on B and on $S^N \times B$. These follow from adiabatic limit considerations and – to deal with the embedding $i: W \to S^N \times B$ also the main result [13, Theorem 2.2].

For the general case, one has to follow the effect of the perturbations which reduce to the case of a kernel bundle.

6.3.4 Related Work

The thesis [51] discusses a model of even differential K-theory using vector bundles with connection and push-forward maps in this model. The work culminates in the special case of the index theorem for differential K-theory if B is the point.

7 Twisted Differential K-Theory

Usual cohomology theories often have severe limitations when dealing with situations in which orientations are required, but not present. This happens in particular when one wants to describe the cohomological properties of a fiber bundle which is not oriented for the cohomology theory one wants to study. Closely related is the non-existence of integration maps for non-orientable bundles or more generally non-orientable maps.

This problem is solved by using cohomology with twisted coefficients. For ordinary cohomology, this is just described by a local coefficient system, which one can easily implement, e.g. in a Čech description of cohomology, compare e.g. [14, Chap. 10]. Twists have been introduced for generalized cohomology theories and successfully used. We refer to [56] for the approach to twisted cohomology via parameterized spectra and to [1] for a construction using infinity categories.

In particular, twisted K-theory has been studied extensively, motivated by the classification of D-brane charges in the presence of a background B-field as discussed in Sect. 1.1.

7.1 Twists for Ordinary K-Theory

The most general twists for a multiplicative generalized cohomology theory represented by an E_∞-ring spectrum E are (up to equivalence) described by degree 0 cohomology classes with coefficients in $bgl_1(E)$, where $gl_1(E)$ is the spectrum of units of E and $bgl_1(E) := gl_1(E)[1]$ is its one-fold deloop. In the case of K-theory, the spectrum $bgl_1(K)$ contains the summand $H\mathbb{Z}[3]$. In other words, there is a subgroup of the isomorphism classes of twists for K-theory on a space B given by $H^3(B;\mathbb{Z})$. Most authors concentrate on these twists.

However, it is inappropriate to think only in terms of isomorphism classes of twists. The twists always form a pointed groupoid (with a trivial object). Technically, one can take the path groupoid of the mapping space

$$\mathrm{map}_{Spectra}(\Sigma^\infty X_+, bgl_1(E)). \tag{24}$$

The twisted cohomology theory is more appropriately understood as a functor from this groupoid to graded abelian groups. This path groupoid is the truncation of the ∞-groupoid given by the mapping space itself which should really be considered as right object. In the present paper we prefer the truncation since it is used in most applications and the generalization to the differential case.

In particular, a given twist usually will have non-trivial automorphisms, and these automorphisms act non-trivially on the twisted cohomology. In our case, the automorphisms of the trivial K-theory twist on B are given by $H^2(B;\mathbb{Z})$. Because of the non-trivial automorphisms, for an isomorphism class of twists, e.g. $c \in H^3(B;\mathbb{Z})$ it does in general not make sense to talk about "'the"' twisted K-theory group $K^c(B)$. Only the isomorphism class of this group is well defined, but this is not sufficient, e.g. if one wants to discuss functorial properties.

As indicated above one usually does not work with the most general kind of twists determined homotopy theoretically by $gl_1(E)$ but with a more explicit class closely tied to the relevant geometric situation. We therefore collect, following [38, Sects. 2 and 3.1], the standard properties of a twisted extension of the cohomology theory E in an axiomatic manner.

Definition 7.1. Let E be a generalized cohomology theory. An extension of E to cohomology with twists consists of the following data:

- For every space X of a (pointed) groupoid $\mathfrak{Twist}_X$.
- For every continuous map $f: Y \to X$ a functor $f^*: \mathfrak{Twist}_X \to \mathfrak{Twist}_Y$ which is (weakly) functorial in f. Even more: the association $X \to \mathfrak{Twist}_X$ should become a weak presheaf of groupoids. In most examples this presheaf of groupoids satisfies descend for open coverings and therefore forms a stack in topological spaces.
- Define then $\mathfrak{Twist}$ as the Grothendieck construction of the presheaf above, i.e. the category with objects (X, τ) where X is a space and $\tau \in \mathfrak{Twist}_X$ and morphisms from (X, τ_X) to (Y, τ_Y) consisting of a map $f: X \to Y$ together with an isomorphism $\tau_X \to f^*\tau_Y$. Define the category $\mathfrak{Twist}^2$ of *pairs in twists* with objects (X, A, τ) as before, but where $A \subset X$ and τ is a twist on X.
- The twisted version of E is then a contravariant functor from $\mathfrak{Twist}^2$ to graded abelian groups,
$$(X, A, \tau) \mapsto E^{\tau+n}(X, A)$$
together with natural transformation
$$\delta: E^{\tau+n+1}(X, A) \to E^{\tau+n}(A, \emptyset).$$

These have to satisfy the following properties:

- Homotopy invariance: here, a homotopy between $f, g\colon (X, \tau_X) \to (Y, \tau_Y)$ is a morphism $h\colon (X \times [0,1], \mathrm{pr}^* \tau_X) \to (Y, \tau_Y)$ such that $i_0 \circ h = f$ and $i_1 \circ h = g$. The morphism of pairs $i_k\colon (X, \tau_X) \to (X \times [0,1], \mathrm{pr}^* \tau_X)$ uses the identity morphism o twists $i_k^* \mathrm{pr}^* \tau_X \to \tau_X$.
- Long exact sequence of the pair:

$$\to E^{\tau+n}(X, A) \to E^{\tau+n}(X) \to E^{\tau+n}(A) \xrightarrow{\delta} E^{\tau+n-1}(X, A) \to .$$

- Excision isomorphism.
- Wedge axiom: if $(X, A, \tau) = \coprod_{i \in I}(X_i, A_i, \tau_i)$ then the natural map

$$E^{\tau+n}(X, A) \to \prod_{i \in I} E^{\tau+n}(X_i, A_i)$$

is an isomorphism.
- For the base point $0 \in \mathfrak{Twist}_X$, we require that $E^{0+n}(X, A) = E^n(X, A)$ with the given definition of E^n.

Often, one will require additional structure, in particular a monoidal structure on $\mathfrak{Twist}_X$ which one typically writes additively. Then one requires a natural bilinear product

$$E^{\tau_1+n}(X, A) \otimes E^{\tau_2+m}(X, A) \to E^{\tau_1+\tau_2+n+m}(X, A)$$

which should be associative and graded commutative up to the natural isomorphism of twistings coming from the monoidal structure.

In this situation, one would also require a functorial and compatible push-forward for a proper map between smooth manifolds $f\colon X \to Y$

$$f_!\colon E^{f^*\tau+o(f)+*}(X) \to E^{\tau+*-(\dim X - \dim Y)}(Y),$$

where $o(f)$ is an orientation twist associated to the map f. An E-orientation of the map f will give rise to a trivialization $o(f) \to 0$, so that for an oriented map one has a push-forward

$$f_!\colon E^{f^*\tau+*}(X) \to E^{\tau+*-(\dim X - \dim Y)}(Y).$$

As usual with cohomology theories, there are variants, depending on which category of spaces and pairs of spaces one considers, and for which situations precisely one requires excision.

In the approach of [1] these axioms can easily be realized.

Example 7.2. As mentioned in Sect. 1.1, one can twist de Rham cohomology, defined on the category of smooth manifolds (possibly with boundary), as follows. Let N be a graded commutative algebra, e.g. $N = \mathbb{R}[u, u^{-1}]$.

- $\mathfrak{Twist}_X := \Omega^1_{d=0}(X;N)$, the closed N-valued forms of total degree 1, with pullback the usual pullback. The base point is the form $\Omega = 0$.
- The morphisms from Ω_1 to Ω_2 are given by all forms $\eta \in \Omega^0(X;N)$ with $\Omega_2 = \Omega_1 + d\eta$. Composition is defined as the sum of differential forms.
- $H^{\Omega+n+ev}_{dR}(X,A) := \ker(d^\Omega|_{\bigoplus_{k\in\mathbb{Z}}\Omega^{n+2k}})/\operatorname{im}(d^\Omega)$, with differential $d^\Omega(\omega) := d\omega + \Omega$. Note that the fact that Ω is closed implies that this is indeed a differential.
- Given a morphism $\eta \in \Omega^0(X;N)$ from Ω_1 to $\Omega_1 + d\eta$, define the induced isomorphism of twisted de Rham groups

$$\eta^*\colon H^{\Omega_1+n}_{dR}(X;N) \to H^{\Omega_1+d\eta+n}_{dR}(X;N); [\omega] \mapsto [\omega \cup \exp(-\eta)].$$

- The sum of forms defines a strictly symmetric monoidal structure on $\mathfrak{Twist}$ and the cup product of differential forms induces an associative and graded commutative product structure on twisted de Rham cohomology.

Remark 7.3. A variant of Example 7.2 uses as twists only the differential forms $\mathfrak{Twist}'_X := \Omega^3_{d=0}(X) \cdot u \subset \Omega^1_{d=0}(X;N)$. This is particularly relevant for the comparison with K-theory.

Note that the for the isomorphism classes of twists one obtains $\pi_0(\mathfrak{Twist}'_X) = H^3_{dR}(X)$.

In the case of K-theory, there are many different models for the groupoid of twists we consider. A particularly simple version is the following: Let U be the unitary group of an infinite dimensional separable Hilbert space H (with norm topology) and $PU := U/S^1$ where S^1 is the center, the multiples of the identity. Because of Kuiper's theorem, U is contractible and PU has the homotopy type of the Eilenberg–MacLane space $K(\mathbb{Z},2)$. Let K be the C^*-algebra of compact operators on H. Conjugation defines an action of PU on K by C^*-algebra automorphism.

Example 7.4. Assume that B is compact. $\mathfrak{Twist}_B$, the groupoid of twists for K-theory on B is now defined as the category of principal PU-bundles over B, with morphisms the homotopy classes of bundle isomorphisms. If τ is such a bundle, we can form the associated bundle of C^*-algebras $\tau \times_{PU} K$. The sections of this bundle form themselves a C^*-algebra. Now set $K^{\tau+*}(B) := K^*(\Gamma(\tau \times_{PU} K))$. A isomorphism $\beta\colon \tau' \to \tau$ of PU-principal bundles induces an isomorphism of associated K-bundles and therefore also of the C^*-algebras of sections, which finally induces a (functorial) isomorphism $\beta^*\colon K^{*+\tau'}(B) \to K^{*+\tau}(B)$.

Algebras of sections of K-bundles like $\tau \times_{BU} K$ are called "continuous trace algebras" and are an important object of study in operator algebras, compare e.g. [61]. This point of view of twisted K-theory – using continuous trace algebras – is exploited and developed, e.g. in [55, 60, 64, 65].

Homotopy invariance of K-theory of C^*-algebras implies that β^* depends only on the homotopy class of β. The trivial twist is the trivial bundle $PU \times B$. Its automorphisms are given by maps from B to PU. As $PU = K(\mathbb{Z},2)$, the set of homotopy classes of such maps is $H^2(B;\mathbb{Z})$. For the special case of torsion classes in $H^3(B;\mathbb{Z})$, this model has first been considered in [34]. More precisely, this paper uses bundles of finite dimensional matrix algebras over B instead of K-bundles which is exactly the reason why only torsion twists occur. The general case is studied in [64, 65]. Another, closely related model for the twists is given by $U(1)$-bundle gerbes. This point of view is studied, e.g. in [15].

Note that PU also acts by conjugation on the space Fred of Fredholm operators on H. The latter is a model for the zeroth space of the K-theory spectrum. Given a twist τ, we can form the associated bundle $\tau \times_{PU}$ Fred. We can then define $K^{0+\tau}(B)$ alternatively as the homotopy classes of sections of $\tau \times_{PU}$ Fred. This model is used, e.g. in [2, 3]. One can define $K^{1+\tau}(B)$ by using an appropriate classifying space for K^1 instead of Fred which is a classifying space for K^0.

Obviously, a more refined version of this construction uses bundle of spectra (also called parameterized spectra) instead of bundles of spaces. A very precise version of such a model, with a satisfactory description of a product structure, of orientation and of the natural transformation from twisted spin^c-cobordism to twisted K-theory corresponding to the Atiyah orientation has been worked out in [66]. When dealing with bundles, it is necessary to deal with objects and maps on the nose, and not only up to homotopy.

Our description suggests a further "categorification" of the concept of (twisted) generalized cohomology theory. In the same way as twists have to be considered as a groupoid, one should also think of (twisted) generalized cohomology as a groupoid. The objects of this groupoid are the cocycles, and a cochain c of shifted degree (modulo boundaries) is a morphism from x to $x' = x + dc$. This would require a two-groupoid of twists, e.g. a two-truncation of the mapping space (24).

More explicitly, in our example we might think of $H^3(X;\mathbb{Z})$ as the groupoid whose objects are isomorphism classes of principal PU-bundles over X, morphisms are PU-bundle maps and 2-morphism are homotopies of PU-bundle maps. Similarly, we might think of $K^{0+\tau}(B)$ as the groupoid whose objects are sections of $\tau \times_{PU}$ Fred and with morphisms homotopies of sections. If one likes ∞-categories then one could consider twisted cohomology as an ∞-functor which associates to a twist $\tau \in \mathrm{map}_{Spectra}(\Sigma^\infty X_+, gl_1(E))$ the spectrum of sections of the associated bundle E_τ of spectra. This can be made precise using [1]. In this picture it is easy to implement additional structures like multiplication or push-forward.

In the truncated groupoid picture most of this has been carried out in the even more elaborate equivariant situation in [38, Sect. 3]. The model there is based on the construction of twisted K-theory spectra.

7.2 *Twisted Differential K-Theory*

To define the concept of a twisted differential generalized cohomology theory, one has to combine the concept of twist with the concept of differential extension (which is *not* a cohomology theory, but there the deviation is well under control). One does need groupoids of *differential twists* which contain differential form information. Along the way, one will need an appropriate Chern character to twisted de Rham cohomology.

The following definition, what in general a twisted differential cohomology theory should be, follows essentially [45, Appendix A.3].

Definition 7.5. A differential extension of a twisted cohomology theory as in Definition 7.1 consists of the following data:

- For each smooth manifold X a groupoid $\mathfrak{Twist}_{\hat{E},X}$, together with (weakly) functorial pullback along smooth maps. They form a weak presheaf of groupoids. As above, we can "combine" all these groupoids to the category $\mathfrak{Twist}_{\hat{E}}$.
- Natural functors

$$F\colon \mathfrak{Twist}_{\hat{E},X} \to \mathfrak{Twist}_X,$$
$$\mathrm{Curv}\colon \mathfrak{Twist}_{\hat{E},X} \to \Omega^1_{d=0}(X;N).$$

  Here and in the following, we lift the action of $\Omega^0(X;N)$ on twisted de Rham cohomology with the same formula to the twisted de Rham complexes.
- To each $\tau \in \mathfrak{Twist}_{\hat{E},X}$ we assign a Chern character

$$\mathrm{ch}^\tau\colon E^{F(\tau)+*}(X) \to H_{dR}^{\mathrm{Curv}(\tau)+*}(X;N),$$

  natural with respect to pullback.
- The differential twisted extension of E is then a functor from the category of differential twists $\mathfrak{Twist}_{\hat{E}}$ to graded abelian groups:

$$(X,\tau) \mapsto \hat{E}^{\tau+*}(X).$$

  Note that this includes functoriality with respect to pullback along smooth maps and along isomorphism of twists.
- There are natural (for pullback along smooth maps and along isomorphism of twists) transformations

$$I\colon \hat{E}^{\tau+*}(X) \to E^{F(\tau)+*}(X),$$
$$R\colon \hat{E}^{\tau+*}(X) \to \Omega^*_{d^{\mathrm{Curv}(\tau)}=0}(X;N),$$
$$a\colon \Omega^{*-1}(X;N)/\operatorname{im}(d^{\mathrm{Curv}(\tau)}) \to \hat{E}^{\tau+*}(X).$$

- They satisfy

$$R \circ a = d^{\mathrm{Curv}(\tau)} \qquad \text{and} \qquad \mathrm{ch}^\tau \circ I = \mathrm{pr} \circ R,$$

  where $\mathrm{pr}\colon \Omega^*_{d^{\mathrm{Curv}(\tau)}=0}(X;N) \to H_{dR}^{\mathrm{Curv}(\tau)}(X;N)$ is the canonical projection.

- Using these, we get exact sequences

$$E^{F(\tau)+*-1}(X) \to \Omega^{*-1}(X;N)/\operatorname{im}(d^{\mathrm{Curv}(\tau)}) \xrightarrow{a} \hat{E}^{\tau+*}(X) \xrightarrow{I} E^{F(\tau)+*}(X) \to 0.$$

Additionally, one would typically like to have a compatible product structure, as in Definition 7.1, with an adopted rule for the compatibility of the transformation a.

Finally, if one has a product structure, one would like to have a push-forward along smooth maps (or at least proper submersions) $f\colon X \to Y$ of the form

$$f_!\colon \hat{E}^{f^*\tau+o(f)+*}(X) \to \hat{E}^{\tau+*+(\dim X - \dim Y)}(Y),$$

where $o(f)$ is a differential orientation twist associated to f. A differential E-orientation should induce a trivialization $o(f) \to 0$ so that in this case one gets a push-forward

$$f_!\colon \hat{E}^{f^*\tau+*}(X) \to \hat{E}^{\tau+*+(\dim X - \dim Y)}(Y).$$

A first attempt toward a definition and description of twisted differential K-theory is given in [31], although not exactly in the setting of Definition 7.1. The main problems are of course:

1. construction of the groupoid of differential twists,
2. construction of the differential cohomology groups,
3. construction of the push-forward.

[31] works with $U(1)$-banded bundle gerbes with connection and curving as objects of the groupoid of twists, and the curvature 3-form of this connection and curving is the transformation Curv (on objects). Given such a twist, they construct a principal $PU(H)$-bundle. Their twisted differential K-theory is then based on sections of associated bundles of Fredholm operators and explicitly constructed locally defined vector bundles with connection. For the rather elaborate precise definition, we refer to [31, Sect. 3].

Definition 7.6. The twists for X used in [45, Definition A.1] are *geometric central extensions*. Such a geometric central extension is

1. a groupoid (P_0, P_1) with a local equivalence to the trivial groupoid (X, X),
2. a central $U(1)$-extension of groupoids $L \to P_1$,
3. in particular, $L \to P_1$ is a $U(1)$-principal bundle, and another part of the data is a connection ∇ on this principal bundle,
4. moreover, $L \to P_1$ being a central extension means one has over $P_2 = P_1 \times_{P_0} P_1$ an isomorphism of line bundles $\lambda\colon \mathrm{pr}_1^* L \otimes \mathrm{pr}_2^* L \to \circ^* L$ using the two projections and the composition of arrows $\mathrm{pr}_1, \mathrm{pr}_2, \circ\colon P_2 \to P_1$. λ should satisfy the cocycle condition, i.e. the two different ways to map $L_h \otimes L_g \otimes L_f$ to $L_{h\circ g\circ f}$ on $P_3 = P_1 \times_{P_0} P_1 \times_{P_0} P_1$ coincide.
5. a 2-form $\omega \in \Omega^2(P_0)$.

These ingredients have to satisfy certain compatibility conditions explained in [45, Definition A.1]. In particular, $p_1^*\omega - p_0^*\omega = \frac{\sqrt{-1}}{2\pi}\Omega^\nabla$, and λ is an isomorphism of line bundles with connection.

In [45] it is observed that no construction of twisted differential K-theory with their twists (the geometric central extensions) is available yet, but one certainly expects that such a construction is possible.

7.3 T-Duality

Motivated from string theory, T-duality is expected to be an equivalence of low energy limits of type IIA/B superstring theories on T-dual pairs. In particular, as D-brane charges are classified by twisted K-theory, T-duality predicts a canonical isomorphism between appropriate twisted K-theory groups of the underlying topological spaces of the T-dual pairs. This prediction has been made mathematically rigorous under the term "topological T-duality". It is investigated, e.g. in [16, 22, 23, 29].

We briefly introduce into the mathematical setup as proposed in [22, Sect. 2], compare also [29, Sect. 4].

Definition 7.7. We let $T^n = U(1)^n$ be the n-dimensional real torus, considered as Lie group. Let B be a topological space (often with some restrictions, e.g. to be a compact CW-complex).

A T-duality triple consists of two T^n-principal bundles E, E' over the common base space B, and twists $\tau \in \mathfrak{Twist}_E$, $\tau' \in \mathfrak{Twist}_{E'}$ for K-theory. The third ingredient of a T-duality triple is an isomorphism of twists $u\colon p^*\tau \to (p')^*\tau'$ over the *correspondence space* $E \times_B E'$ with the two canonical projections $p\colon E \times_B E' \to E$ and $p'\colon E \times_B E' \to E'$.

The twists and the isomorphism u of twists have to satisfy certain conditions. These are most transparent if $n = 1$. In this case, they simply say that

$$\int_{E/B} [\tau] = c_1(E'), \quad \int_{E'/B} [\tau'] = c_1(E),$$

where $[\tau] \in H^3(E;\mathbb{Z})$ is the characteristic class determined and determining the isomorphism class of the twist τ. Moreover, restricted to a point each $x \in B$, $\tau|_{E_x}$ and $\tau'_{E'_x}$ are canonically trivialized (because $E_x \cong U(1) \cong E'_x$ have vanishing H^2 and H^3). Consequently, using the induced trivializations, the restriction of u to the fiber over x becomes an automorphism of the trivial twist and therefore is classified by an element in $H^2(E_x \times_x E'_x) \cong H^2(U(1) \times U(1);\mathbb{Z}) \cong \mathbb{Z}$. We require that this element is the canonical generator.

For the details for general n, we refer to [30, Sect. 2], where this is again treated using cohomology, or [29, Defintion 4.1.3] where the language of stacks is used.

Of course, in this setting we first have to choose appropriate data for a twisted extension of K-theory, e.g. the model where twists are PU-principal bundles or $U(1)$-banded gerbes.

Definition 7.8. Given a T-duality triple $((E,\tau),(E',\tau'),u)$ as in Definition 7.7, we define the T-duality transformation of twisted K-theory

$$T := p'_! u^* p^* \colon K^{*+\tau}(E) \to K^{*-n+\tau'}(E'). \tag{25}$$

It is defined as the composition of pull-back to the correspondence space, using u to map τ-twisted K-theory to τ'-twisted K-theory and finally integration along p', where we use the fact that T^n-principal bundles are canonically oriented for any cohomology theory, in particular for K-theory.

The main results of [22] and [29] concern:

1. The classification of T-duality triples: there is, e.g. a universal T-duality triple over a classifying space $\mathbf{R}_n$ of such triples whose homotopy type is computed: it is the homotopy fiber in the sequence

$$\mathbf{R}_N \to K(\mathbb{Z}^n,2) \times K(\mathbb{Z}^n,2) \xrightarrow{\cup} K(\mathbb{Z},4), \tag{26}$$

   where the map $\cup$ is the composition of the map associated to the usual cup-product and the standard scalar product of the coefficient group $\mathbb{Z}^n \otimes \mathbb{Z}^n \to \mathbb{Z}$.
   Note, therefore, that up to equivalence all the information of a T-duality triple is given by two T^n-principal bundles P, P' with Chern classes $c_1,\dots,c_n$, $c'_1,\dots,c'_n$ with $\sum c_i c'_i = 0$, together with an explicit trivialization of the cycle representing this product (e.g. a lift of its classifying map to the homotopy fiber $\mathbf{R}_n$).
   In [22], we also discuss, which pairs (E,τ) can be part of a T-duality triple and in how many ways. For $n=1$, this is always the case, even in a unique way (up to equivalence). For $n>1$, both these assertions are wrong in general.
2. The T-duality transform T of (25) is always an isomorphism (compare [22, Theorem 6.2]).

7.4 T-Duality and Differential K-Theory

Alex Kahle and Alessandro Valentino [45] study the effect of T-duality in differential K-theory. They follow the approach of Sect. 7.3, i.e. they first make a precise definition of a differential T-duality triple [45, Definition 2.1] (there called "pair").

However, they make very efficient use of the higher structures of differential cohomology alluded to above.

Definition 7.9. Fix a base space B. A differential T-duality triple on B according to Kahle and Valentino [45, Definition 2.1] first consists of two objects

$\mathcal{P} = (P, \nabla), \mathcal{P}' = (P', \nabla')$ in the groupoid of cycles for differential cohomology with coefficients in $\mathbb{Z}^n$, given by two T^n-bundles with connection over B, $\pi: P \to B, \pi': P' \to B$.

Secondly, using a product on the level of the groupoid of cycles, form $\mathcal{P} \cdot \mathcal{P}'$, a cycle for $\hat{H}^4(B;\mathbb{Z})$ (on the level of the coefficients $\mathbb{Z}^n$, for this multiplication we use the standard inner product). The last ingredient for a T-duality triple is an isomorphism in the groupoid of cycles for differential H^4:

$$\sigma: 0 \to \mathcal{P} \cdot \mathcal{P}'.$$

Note that the existence of such a trivialization is a strong condition on $\mathcal{P}, \mathcal{P}'$.

Observe that this description is very much in line with the description of the homotopy type of the classifying space for topological T-duality triples (26) and the resulting description of T-duality triples.

To obtain the twists for differential K-theory one expects for a differential T-duality triple, Kahle and Valentino argue as follows:

The pullback of $\mathcal{P}$ to the total space P of the underlying bundle has a canonical trivialization, and similarly for $\mathcal{P}'$. This trivialization can be multiplied with $\pi^*\mathcal{P}'$ to give a trivialization of $\pi^*\mathcal{P} \cdot \pi^*\mathcal{P}'$. The composition of $\pi^*\sigma$ with the inverse of this is an automorphism of the trivial object 0 and therefore defines a cycle $\hat{\tau}$ for the third differential cohomology of P. Similarly, we obtain a cycle $\hat{\tau}'$ for the third differential cohomology of P'. Finally, in [45, Lemma 2.2], it is shown how the canonical trivializations of $\pi^*\mathcal{P}$ and $(\pi')^*\mathcal{P}'$ give rise to a morphism $\hat{u}$ in the groupoid of cycles for $\hat{H}^3(P \times_B P')$ from $p^*\hat{\tau}$ to $(p')^*\hat{\tau}'$, where $p: P \times_B P' \to P$, $p': P \times_B P' \to P'$ are the canonical projections.

The crucial points assumed by Kahle–Valentino is

1. to have a groupoid cycle model for differential cohomology where cycles for $\hat{H}^2$ are principal $U(1)$-bundles with connection and where one has a multiplication with good properties on the level of cycles;
2. to have a model for a twisted extension of differential K-theory, where the groupoid of cycles for $\hat{H}^3$ is exactly the groupoid of twists.

Let us repeat that, at the moment, no complete construction of twisted differential K-theory satisfying these requirements seems to be available.

With these assumptions, it is now immediate how to define a T-duality transformation in twisted differential K-theory (assuming that "integration along T^n-principal bundles with connection" is also defined for the twisted differential K-theory at hand):

$$\hat{T} := \hat{p}'_! \circ \hat{u}^* \circ p^*: \hat{K}^{*+\hat{\tau}}(P) \to \hat{K}^{*-n+\hat{\tau}'}(P'). \tag{27}$$

Here $\hat{u}^*$ is the isomorphism induced by the isomorphism of differential twists $\hat{u}$ of [45, Lemma 2.2] as above.

However, there is one observation to be made: upon application of the curvature transformation, simple calculations show that $\hat{T}$ can never be surjective, as the forms in the image of its differential form analog have a very specific invariance property under the action of T^n.

Definition 7.10. For a T^n-principal bundle like P, let $\hat{\tau}$ be a cycle for differential cohomology and twist for differential K-theory as above and assume that the differential form $R(\hat{\tau})$ is T^n-invariant. Define the *geometrically invariant subgroup* of twisted differential K-theory as

$$\hat{K}^{*+\hat{\tau}}(P)^{T^n} := \{x \in \hat{K}^{*+\hat{\tau}}(P) \mid g^* R(x) = R(x) \ \forall g \in T^n\}.$$

With this notion, Kahle and Valentino prove their main result [45, Theorem 2.4].

Theorem 7.11. *The differential T-duality transform $\hat{T}$ of* (27) *preserves the geometrically invariant subgroups and defines an isomorphism*

$$\hat{T}\colon \hat{K}^{*+\hat{\tau}}(P)^{T^n} \to \hat{K}^{*-n+\hat{\tau}'}(P').$$

The main point of the proof of this theorem is the construction of the transformation in such a way that it is compatible with all the transformations given for differential K-theory. One then has to check/use that the transformation is an isomorphism for geometrically invariant forms (the image under the curvature homomorphism of geometrically invariant differential K-theory) and for topological twisted K-theory. The proof then concludes using the five lemma.

8 Applications of Differential K-Theory

Differential K-theory is a natural home for many well known, and hopefully some new, typically secondary invariants. In this section, we want to present some examples of this kind. To be able to do this, we start with a couple of elementary calculations.

Lemma 8.1.

$$\hat{K}^1(*) = \mathbb{R}/\mathbb{Z}; \qquad \hat{K}^0(*) = \mathbb{Z}, \qquad \hat{K}^1_{flat}(*) = \mathbb{R}/\mathbb{Z}; \qquad \hat{K}^0_{flat}(*) = \{0\}$$

as follows directly from the short exact sequence (3).

8.1 Holonomy

Let (V, ∇) be a Hermitian vector bundle of rank n over S^1 with unitary connection and with holonomy ϕ (well defined modulo conjugation in $U(n)$). (V, ∇) defines a geometric family $\mathcal{V}$ and therefore an element in differential K-theory $[\mathcal{V}, 0]$. By [26, Lemma 5.3]

Lemma 8.2.

$$[\mathcal{V}, 0] = a(\frac{1}{2\pi i}\det(\phi)).$$

8.2 $\mathbb{Z}/k\mathbb{Z}$-Invariants

Recall that a $\mathbb{Z}/k\mathbb{Z}$-manifold is a manifold W with boundary together with a manifold X together with a diffeomorphism $f : \partial W \to \underbrace{X \amalg \ldots \amalg X}_{n \text{ copies}}$.

We now associate to a family of $\mathbb{Z}/k\mathbb{Z}$-manifolds over B a class in $\hat{K}_{flat}(B)$.

Definition 8.3. A *geometric family of $\mathbb{Z}/k\mathbb{Z}$-manifolds* is a triple $(\mathcal{W}, \mathcal{E}, \phi)$, where $\mathcal{W}$ is a geometric family with boundary, $\mathcal{E}$ is a geometric family without boundary, and $\phi : \partial\mathcal{W} \xrightarrow{\sim} k\mathcal{E}$ is an isomorphism of the boundary of $\mathcal{W}$ with k copies of $\mathcal{E}$.

We define $u(\mathcal{W}, \mathcal{E}, \phi) := [\mathcal{E}, -\frac{1}{k}\Omega(\mathcal{W})] \in \hat{K}(B)$.

Lemma 8.4. *We have $u(\mathcal{W}, \mathcal{E}, \phi) \in \hat{K}_{flat}(B)$. This class is a k-torsion class. It only depends on the underlying differential-topological data.*

Theorem 8.5. *Let $B = *$ and $\dim(\mathcal{W})$ be even. Then $u(\mathcal{W}, \mathcal{E}, \phi) \in \hat{K}^1_{flat}(*) \cong \mathbb{R}/\mathbb{Z}$. Let $i_k : \mathbb{Z}/k\mathbb{Z} \to \mathbb{R}/\mathbb{Z}$ the embedding which sends $1 + k\mathbb{Z}$ to $\frac{1}{k}$. Then*

$$i_k(\mathrm{ind}_a(\bar{W})) = u(\mathcal{W}, \mathcal{E}, \phi),$$

where $i_k(\mathrm{ind}_a(\bar{W})) \in \mathbb{Z}/k\mathbb{Z}$ is the index of the $\mathbb{Z}/k\mathbb{Z}$-manifold $\bar{W}$, and where we use the notation of [40].

8.3 Reduced Eta-Invariants

Let π be a finite group. We construct a transformation

$$\phi : \Omega_d^{\mathrm{spin^c}}(BU(n) \times B\pi) \to \hat{K}^{-d}_{flat}(*).$$

Let $f : M \to BU(n) \times B\pi$ represent $[M, f] \in \Omega^{Spin^c}(BU(n) \times B\pi)$. This map determines a π-covering $p : \tilde{M} \to M$ and an n-dimensional complex vector bundle $V \to M$. We choose a Riemannian metric g^{TM} and a spin^c-connection $\tilde{\nabla}$. These structures determine a differential K-orientation of $t : M \to *$. We further fix a metric h^V and a connection ∇^V in order to define a geometric bundle $\mathbf{V} := (V, h^V, \nabla^V)$ and the associated geometric family $\mathcal{V}$. The pull back of g^{TM} and $\tilde{\nabla}$ via $\tilde{M} \to M$ fixes a differential K-orientation of $\tilde{t} : \tilde{M} \to *$.

Then we set

$$\phi([M,f]) := \left[\tilde{t}_!(p^*\mathcal{V}) \sqcup_* |\pi| t_! \mathcal{V}^{op}, 0\right] \in \hat{K}_{flat}(*).$$

In [26, Sect. 5.10] it is shown that this class only depends on the bordism class of $[M,f]$.

More generally, without even the assumption that π is finite, choose two finite dimensional representations ρ_1, ρ_2 of π of the same dimension, with associated flat bundles F_1, F_2. Replace in the above $|\pi|\, t_!\mathcal{V}$ by $t_!(\mathcal{V} \otimes F_2)$ and $\tilde{t}_!(p^*\mathcal{V})$ by $t_!(\mathcal{V} \otimes F_1)$.

Note that this boils down to the previous case if ρ_1 is the regular representation of the finite group π, and ρ_2 is $\mathbb{C}^{|\pi|}$, where $\mathbb{C}$ stands for the trivial representation.

Proposition 8.6. *This construction defines a homomorphism*

$$\phi_{\rho_1,\rho_2}\colon \Omega_d^{\mathrm{spin}^c}(BU(n) \times B\pi) \to \hat{K}_{flat}^{-d}(*).$$

If d is even, the target group is trivial. If d is odd, $\hat{K}_{flat}^{-d}(*) \cong \mathbb{R}/\mathbb{Z}$. In this case, ϕ_{ρ_1,ρ_2} coincides with the reduced rho-invariant of Atiyah–Patodi–Singer.

The construction immediately generalizes to a parameterized version: to a smooth family of d-dimensional spin^c-manifolds parameterized by B, with a family of $\mathbb{C}^n$-vector bundles and also of π-coverings one associates in the same way a class in $\hat{K}_{flat}^{-d}(B)$.

For details of all of this, compare [26, Sect. 5.10].

8.4 e-Invariant

A framed manifold is a manifold M together with a trivialization of its tangent bundle.

[26, Proposition 5.22] states that a bundle of framed n-manifolds $\pi\colon E \to B$ has a canonical differential K-orientation, given by the fiberwise spin^c-structure which comes from the trivialization, and the spin^c-connection which again comes from the trivial connection (form part 0). We then define

$$e([\pi\colon E \to B]) := \hat{\pi}_!(1) \in \hat{K}_{flat}^{-n}(B).$$

The push-down is with respect to the canonical $\hat{K}$-orientation of π, and the flatness of the connection of this differential K-orientation in the end implies that $R(\hat{\pi}_1(1)) = 0$.

Proposition 8.7. *If $B = *$ and n is odd, $e([B]) \in K_{flat}^{-1}(*) = \mathbb{R}/\mathbb{Z}$.*

This class coincides with the Adam's classical e-invariant for the stable homotopy groups, identified with the framed bordism groups.

For details, compare [26, Sect. 5.11].

8.5 Secondary Invariants for String Bordism

In [21], using spectral invariants of Dirac operators, Bunke and Naumann construct a secondary Witten genus, a bordism invariant of string manifolds. They use differential cohomology to facilitate some of their calculations, compare e.g. [21, Lemma 2.2].

9 Equivariant Differential K-Theory and Orbifold Differential K-Theory

As explained in Sect. 1.1, one of the motivations for the study of differential K-theory comes from physics, where fields in abelian gauge theories are suggested to be modelled by cocycles for differential K-theory and where some of the main features are captured by the properties of differential cohomology theories. In Sect. 7.4 we have seen how this is successfully applied to T-duality, another important subject motivated by string theory.

In particular for the latter, however, mathematical physics also requires the study of singular spaces. Such singular spaces often arise as quotients of smooth manifolds by the action of a group, which is one reason why equivariant situations are important. We would therefore like to study differential K-theory for singular spaces and equivariant differential K-theory. Unfortunately, these theories are not well understood yet.

In [25], we construct differential K-theory of representable smooth orbifolds, i.e. global quotients of a manifold by a compact group. The construction is based on equivariant local index theory in the spirit of Sect. 4.3. The relevant Chern character takes values in delocalized de Rham cohomology of the orbifold. In case of a global quotient by a finite group, this is defined in terms of the de Rham complexes of the fixed point sets. We obtain a ring valued functor with the usual properties of a differential extension of a cohomology theory. For proper submersions (with smooth fibers) we construct a push-forward map in differential orbifold K-theory. Finally, we construct a non-degenerate intersection pairing with values in $\mathbb{C}/\mathbb{Z}$ for the subclass of smooth orbifolds which can be written as global quotients by a finite group action. We construct a real subfunctor of our theory, where the pairing restricts to a non-degenerate $\mathbb{R}/\mathbb{Z}$-valued pairing. Indeed, we use in that paper the language of (differentiable étale) stacks which turns out to be particularly convenient.

In [63, Sect. 5.4], a model for equivariant differential K-theory in the spirit of Sect. 4.2 is constructed. It uses the fact that there are very nice models for the classifying space for equivariant K-theory. As target for the Chern character on equivariant K-theory [63] uses Bredon cohomology with coefficients in the representation ring tensored with $\mathbb{R}$. Via a de Rham isomorphism, this is canonically isomorphic to delocalized de Rham cohomology. As in the non-equivariant case, this model is not so well suited to the construction of a product structure and of push-forwards, which are therefore not discussed in [63]. However, in [63, Sect. 6] is described how equivariant differential K-theory can be used to described Ramond–Ramond fields and their flux quantization condition in orbifolds of type II superstring theory.

The preprint [59] gives yet another construction of differential equivariant K-theory for finite group actions along the lines of [44], i.e. of Sect. 4.2. Moreover, it constructs a product and push-forward to a point. The constructions are mainly homotopy theoretical. Ortiz in [59] raises the interesting question [59, Conjecture 6.1] of identifying his push-forward in analytic terms. In the model of [25], in view of the geometric construction of the push-forward and the analytic nature of the relations, the conjectured relation is essentially a tautology. See [26, Corollary 5.5] for a more general statement in the non-equivariant case. [59, Conjecture 6.1] would be an immediate consequence of a theorem stating that any two models of equivariant differential K-theory for finite group actions are canonically isomorphic in a way compatible with integration. It seems to be plausible that the method of [27] extends to the equivariant case. In [59], the conjectured equivariant index formula is proved in a number of special cases, e.g. if Γ is trivial, or in case the G-manifold is a G- boundary.

Acknowledgements Thomas Schick was partially funded by the Courant Research Center "Higher order structures in Mathematics" within the German initiative of excellence.

References

1. Ando, M., Blumberg, A.J., Gepner, D.J., Hopkins, M.J., Rezk, C.: Units of ring spectra and Thom spectra. arXiv:0810.4535v3 (2008)
2. Atiyah, M., Segal, G.: Ukr. Mat. Visn. **1**(3), 287–330 (2004)
3. Atiyah, M., Segal, G.: Twisted K-theory and cohomology. In: Inspired by S.S. Chern, Nankai Tracts in Mathematics, vol. 11, pp. 5–43. World Scientific Publishing, Hackensack, NJ (2006)
4. Atiyah, M.F.: Global theory of elliptic operators. In: Proceedings of International Conference on Functional Analysis and Related Topics, Tokyo, 1969, pp. 21–30. University of Tokyo Press, Tokyo (1970)
5. Atiyah, M.F., Singer, I.M.: The index of elliptic operators. I. Ann. Math. (2) **87**, 484–530 (1968)
6. Atiyah, M.F., Singer, I.M.: The index of elliptic operators. IV. Ann. Math. (2) **93**, 119–138 (1971)
7. Baum, P., Douglas, R.G.: K-homology and index theory. In: Richard V. Kadison (ed.), Operator Algebras and Applications, Part I, Kingston, Ontario, 1980, Proceedings of Symposia in Pure Mathematics, vol. 38, pp. 117–173. American Mathematical Society, Providence, R.I. (1982)
8. Baum, P., Higson, N., Schick, T.: On the equivalence of geometric and analytic K-homology. Pure Appl. Math. Q. **3**(1, part 3), 1–24 (2007)
9. Becker, K., Bergman, A.: Geometric aspects of D-branes and T-duality. J. High Energ Phys. (11), 067, 22 pp (2009). doi: 10.1088/1126-6708/2009/11/067. URL http://dx.doi.org/10.1088/1126-6708/2009/11/067
10. Benameur, M.T., Maghfoul, M.: Differential characters in K-theory. Differ. Geom. Appl. **24**(4), 417–432 (2006). doi: 10.1016/j.difgeo.2005.12.008. URL http://dx.doi.org/10.1016/j.difgeo.2005.12.008

11. Berline, N., Getzler, E., Vergne, M.: Heat kernels and Dirac operators, Grundlehren der Mathematischen Wissenschaften [Fundamental Principles of Mathematical Sciences], vol. 298. Springer, Berlin (1992)
12. Bismut, J.M.: The Atiyah–Singer index theorem for families of Dirac operators: two heat equation proofs. Invent. Math. **83**(1), 91–151 (1985). doi: 10.1007/BF01388755. URL http://dx.doi.org/10.1007/BF01388755
13. Bismut, J.M., Zhang, W.: Real embeddings and eta invariants. Math. Ann. **295**, 661–684 (1993)
14. Bott, R., Tu, L.W.: Differential forms in algebraic topology, Graduate Texts in Mathematics, vol. 82. Springer, New York (1982)
15. Bouwknegt, P., Carey, A.L., Mathai, V., Murray, M.K., Stevenson, D.: Twisted K-theory and K-theory of bundle gerbes. Comm. Math. Phys. **218**, 17–45 (2002)
16. Bouwknegt, P., Evslin, J., Mathai, V.: T-duality: topology change from H-flux. Comm. Math. Phys. **249**(2), 383–415 (2004). doi: 10.1007/s00220-004-1115-6. URL http://dx.doi.org/10.1007/s00220-004-1115-6
17. Bouwknegt, P., Mathai, V.: D-branes, B-fields and twisted K-theory. J. High Energ. Phys. **3**, 7, 11 pp (2000). doi: 10.1088/1126-6708/2000/03/007. URL http://dx.doi.org/10.1088/1126-6708/2000/03/007
18. Brylinski, J.L.: Loop Spaces, Characteristic Classes and Geometric Quantization, Progress in Mathematics, vol. 107. Birkhäuser Boston Inc., Boston, MA (1993)
19. Bunke, U.: Index theory, eta forms, and Deligne cohomology. Memoir. Am. Math. Soc. **198**(928), vi+120 (2009)
20. Bunke, U., Kreck, M., Schick, T.: A geometric description of smooth cohomology. Ann. Math. Blaise Pascal **17**(1), 1–16 (2010)
21. Bunke, U., Naumann, N.: Secondary Invariants for String Bordism and tmf. arXiv:0912.4875 (2009)
22. Bunke, U., Rumpf, P., Schick, T.: The topology of T-duality for T^n-bundles. Rev. Math. Phys. **18**(10), 1103–1154 (2006). doi: 10.1142/S0129055X06002875. URL http://dx.doi.org/10.1142/S0129055X06002875
23. Bunke, U., Schick, T.: On the topology of T-duality. Rev. Math. Phys. **17**(1), 77–112 (2005)
24. Bunke, U., Schick, T.: Real secondary index theory. Algebr. Geomet. Topology **8**(2), 1093–1139 (2008). doi: 10.2140/agt.2008.8.1093. URL http://dx.doi.org/10.2140/agt.2008.8.1093
25. Bunke, U., Schick, T.: Differential orbifold K-theory (2009). ArXiv:0905.4181
26. Bunke, U., Schick, T.: Smooth K-theory. Astérisque (328), 45–135 (2007)
27. Bunke, U., Schick, T.: Uniqueness of smooth extensions of generalized cohomology theories. J. Topology **3**(1), 110–156 (2010). doi: 10.1112/jtopol/jtq002. URL http://dx.doi.org/10.1112/jtopol/jtq002
28. Bunke, U., Schick, T., Schröder, I., Wiethaup, M.: Landweber exact formal group laws and smooth cohomology theories. Algebr. Geomet. Topology **9**(3), 1751–1790 (2009). doi: 10.2140/agt.2009.9.1751. URL http://dx.doi.org/10.2140/agt.2009.9.1751
29. Bunke, U., Schick, T., Spitzweck, M., Thom, A.: Duality for topological abelian group stacks and T-duality. In: Guillermo Cortiñas, Joachim Cuntz, Max Karoubi, Ryszard Nest and Charles A. Weibel (eds.) K-Theory and Noncommutative Geometry, EMS Series of Congress Reports, pp. 227–347. European Mathematical Society, Zürich (2008). doi: 10.4171/060-1/10. URL http://dx.doi.org/10.4171/060-1/10
30. Bunke, U., Spitzweck, M., Schick, T.: Periodic twisted cohomology and T-duality. Astérisque (337), vi+134 (2011)
31. Carey, A.L., Mickelsson, J., Wang, B.L.: Differential twisted K-theory and applications. J. Geomet. Phys. **59**(5), 632–653 (2009). doi: 10.1016/j.geomphys.2009.02.002. URL http://dx.doi.org/10.1016/j.geomphys.2009.02.002
32. Cheeger, J., Simons, J.: Differential characters and geometric invariants. In: J. Alexander and J. Harer (eds.) Geometry and Topology (College Park, Md., 1983/84), Lecture Notes in Mathematics, vol. 1167, pp. 50–80. Springer, Berlin (1985). doi: 10.1007/BFb0075216. URL http://dx.doi.org/10.1007/BFb0075216

33. Connes, A., Karoubi, M.: Caractère multiplicatif d'un module de Fredholm. K-Theory **2**(3), 431–463 (1988). doi: 10.1007/BF00533391. URL http://dx.doi.org/10.1007/BF00533391
34. Donovan, P., Karoubi, M.: Graded Brauer groups and K-theory with local coefficients. Publ. Math. IHES **38**, 5–25 (1970)
35. Dupont, J.L., Ljungmann, R.: Integration of simplicial forms and Deligne cohomology. Math. Scand. **97**(1), 11–39 (2005)
36. Freed, D.S.: Dirac charge quantization and generalized differential cohomology. In: Surveys in Differential Geometry, vol. VII, pp. 129–194. International Press, Somerville, MA (2000)
37. Freed, D.S., Hopkins, M.: On Ramond–Ramond fields and K-theory. J. High Energ. Phys. **5**, 44 (2000)
38. Freed, D.S., Hopkins, M.J., Teleman, C.: Loop groups and twisted K-theory I (2007). ArXiv.org:0711.1906
39. Freed, D.S., Lott, J.: An index theorem in differential K-theory. Geomet. Topology **14**, 903–966 (2010)
40. Freed, D.S., Melrose, R.B.: A mod k index theorem. Invent. Math. **107**(2), 283–299 (1992). doi: 10.1007/BF01231891. URL http://dx.doi.org/10.1007/BF01231891
41. Gillet, H., Soulé, C.: Characteristic classes for algebraic vector bundles with Hermitian metric. II. Ann. Math. (2) **131**(2), 205–238 (1990). doi: 10.2307/1971493. URL http://dx.doi.org/10.2307/1971493
42. Gomi, K., Terashima, Y.: A fiber integration formula for the smooth Deligne cohomology. Int. Math. Res. Not. **13**, 699–708 (2000). doi: 10.1155/S1073792800000386. URL http://dx.doi.org/10.1155/S1073792800000386
43. Higson, N., Roe, J.: Analytic K-Homology. Oxford Mathematical Monographs. Oxford University Press, Oxford (2000)
44. Hopkins, M.J., Singer, I.M.: Quadratic functions in geometry, topology, and M-theory. J. Differ. Geom. **70**(3), 329–452 (2005). URL http://projecteuclid.org/getRecord?id=euclid.jdg/1143642908
45. Kahle, A., Valentino, A.: T-duality and differential K-theory (2009). URL http://www.citebase.org/abstract?id=oai:arXiv.org:0912.2516. ArXiv:0912.2516
46. Karoubi, M.: Homologie cyclique et K-théorie. Astérisque **149**, 147 (1987)
47. Karoubi, M.: Théorie générale des classes caractéristiques secondaires. K-Theory **4**(1), 55–87 (1990). doi: 10.1007/BF00534193. URL http://dx.doi.org/10.1007/BF00534193
48. Karoubi, M.: Classes caractéristiques de fibrés feuilletés, holomorphes ou algébriques. In: Proceedings of Conference on Algebraic Geometry and Ring Theory in honor of Michael Artin, Part II, Antwerp, 1992, vol. 8, pp. 153–211 (1994). doi: 10.1007/BF00961455. URL http://dx.doi.org/10.1007/BF00961455
49. Kasparov, G.G.: Topological invariants of elliptic operators. I. K-homology. Izv. Akad. Nauk SSSR Ser. Mat. **39**(4), 796–838 (1975)
50. Kelnhofer, G.: Functional integration and gauge ambiguities in generalized abelian gauge theories. J. Geomet. Phys. **59**(7), 1017–1035 (2009). doi: 10.1016/j.geomphys.2009.04.007. URL http://dx.doi.org/10.1016/j.geomphys.2009.04.007
51. Klonoff, K.R.: An index theorem in differential k-theory. Ph.D. Thesis, University of Texas at Austin (2008)
52. Landweber, P.S.: Homological properties of comodules over $mu_*(mu)$ and $bp_*(bp)$. Am. J. Math. **98**, 591–610 (1976)
53. Lawson Jr., H.B., Michelsohn, M.L.: Spin Geometry, Princeton Mathematical Series, vol. 38. Princeton University Press, Princeton, NJ (1989)
54. Lott, J.: $\mathbf{R}/\mathbf{Z}$ index theory. Comm. Anal. Geom. **2**(2), 279–311 (1994)
55. Mathai, V., Rosenberg, J.: T-duality for torus bundles with H-fluxes via noncommutative topology. Comm. Math. Phys. **253**(3), 705–721 (2005). doi: 10.1007/s00220-004-1159-7. URL http://dx.doi.org/10.1007/s00220-004-1159-7
56. May, J.P., Sigurdsson, J.: Parametrized Homotopy Theory, Mathematical Surveys and Monographs, vol. 132. American Mathematical Society, Providence, RI (2006)

57. Moore, G., Witten, E.: Self-duality, Ramond–Ramond fields, and K-theory. J. High Energ. Phys. **0005**, 032 (2000)
58. Narasimhan, M.S., Ramanan, S.: Existence of universal connections. Am. J. Math. **83**, 563–572 (1961)
59. Ortiz, M.L.: Differential equivariant K-theory (2009). ArXiv:0905.0476
60. Raeburn, I., Rosenberg, J.: Crossed products of continuous-trace C^*-algebras by smooth actions. Trans. Am. Math. Soc. **305**(1), 1–45 (1988). doi: 10.2307/2001039. URL http://dx.doi.org/10.2307/2001039
61. Raeburn, I., Williams, D.P.: Morita Equivalence and Continuous-Trace C^*-Algebras, Mathematical Surveys and Monographs, vol. 60. American Mathematical Society, Providence, RI (1998)
62. Simons, J., Sullivan, D.: Axiomatic characterization of ordinary differential cohomology. J. Topology **1**(1), 45–56 (2008). doi: 10.1112/jtopol/jtm006. URL http://dx.doi.org/10.1112/jtopol/jtm006
63. Szabo, R.J., Valentino, A.: Ramond–Ramond fields, fractional branes and orbifold differential K-theory. Comm. Math. Phys. **294**(3), 647–702 (2010). doi: 10.1007/s00220-009-0975-1. URL http://dx.doi.org/10.1007/s00220-009-0975-1
64. Tu, J.L., Xu, P.: The ring structure for equivariant twisted K-theory. J. Reine Angew. Math. **635**, 97–148 (2009). doi: 10.1515/CRELLE.2009.077. URL http://dx.doi.org/10.1515/CRELLE.2009.077
65. Tu, J.L., Xu, P., Laurent-Gengoux, C.: Twisted K-theory of differentiable stacks. Ann. Sci. École Norm. Sup. (4) **37**(6), 841–910 (2004). doi: 10.1016/j.ansens.2004.10.002. URL http://dx.doi.org/10.1016/j.ansens.2004.10.002
66. Waldmüller, R.: Products and push-forwards in twisted cohomology theories. Ph.D. Thesis, Georg-August-Universität Göttingen (2006)
67. Witten, E.: D-branes and K-theory. J. High Energ. Phys. (12), Paper 19, 41 pp. (electronic) (1998). doi: 10.1088/1126-6708/1998/12/019. URL http://dx.doi.org/10.1088/1126-6708/1998/12/019

Classical and Quantum Fields on Lorentzian Manifolds

Christian Bär and Nicolas Ginoux

Abstract We construct bosonic and fermionic locally covariant quantum field theories on curved backgrounds for large classes of fields. We investigate the quantum field and n-point functions induced by suitable states.

1 Introduction

Classical fields on spacetime are mathematically modeled by sections of a vector bundle over a Lorentzian manifold. The field equations are usually partial differential equations. We introduce a class of differential operators, called Green-hyperbolic operators, which have good analytical solubility properties. This class includes wave operators as well as Dirac type operators.

In order to quantize such a classical field theory on a curved background, we need local algebras of observables. They come in two flavors, bosonic algebras encoding the canonical commutation relations and fermionic algebras encoding the canonical anti-commutation relations. We show how such algebras can be associated to manifolds equipped with suitable Green-hyperbolic operators. We prove that we obtain locally covariant quantum field theories in the sense of [11]. There is a large literature where such constructions are carried out for particular examples of fields, see e.g. [14, 17, 18, 20, 26, 38]. In all these papers the well-posedness of the Cauchy problem plays an important role. We avoid using the Cauchy problem altogether

C. Bär (✉)
Universität Potsdam, Institut für Mathematik, Am Neuen Palais 10, Haus 8, 14469 Potsdam, Germany
e-mail: baer@math.uni-potsdam.de

N. Ginoux
Fakultät für Mathematik, Universität Regensburg, 93040 Regensburg, Germany
e-mail: nicolas.ginoux@mathematik.uni-regensburg.de

C. Bär et al. (eds.), *Global Differential Geometry*, Springer Proceedings in Mathematics 17, DOI 10.1007/978-3-642-22842-1_12,

and only make use of Green's operators. In this respect, our approach is similar to the one in [39]. This allows us to deal with larger classes of fields, see Sect. 2.7, and to treat them systematically. Much of the earlier work on constructing observable algebras for particular examples can be subsumed under this general approach.

It turns out that bosonic algebras can be obtained in much more general situations than fermionic algebras. For instance, for the classical Dirac field both constructions are possible. Hence, on the level of observable algebras, there is no spin-statistics theorem. In order to obtain results like Theorem 5.1 in [41] one needs more structure, namely representations of the observable algebras with good properties.

In order to produce numbers out of our quantum field theory that can be compared to experiments, we need states, in addition to observables. We show how states with suitable regularity properties give rise to quantum fields and n-point functions. We check that they have the properties expected from traditional quantum field theories on a Minkowski background.

2 Field Equations on Lorentzian Manifolds

2.1 Globally Hyperbolic Manifolds

We begin by fixing notation and recalling general facts about Lorentzian manifolds, see e.g. [30] or [4] for more details. Unless mentioned otherwise, the pair (M, g) will stand for a smooth m-dimensional manifold M equipped with a smooth Lorentzian metric g, where our convention for Lorentzian signature is $(- + \cdots +)$. The associated volume element will be denoted by dV. We shall also assume our Lorentzian manifold (M, g) to be time-orientable, i.e., that there exists a smooth timelike vector field on M. Time-oriented Lorentzian manifolds will be also referred to as *spacetimes*. Note that in contrast to conventions found elsewhere, we do not assume that a spacetime is connected nor do we assume that its dimension be $m = 4$.

For every subset A of a spacetime M we denote the causal future and past of A in M by $J_+(A)$ and $J_-(A)$, respectively. If we want to emphasize the ambient space M in which the causal future or past of A is considered, we write $J_\pm^M(A)$ instead of $J_\pm(A)$. Causal curves will always be implicitly assumed (future or past) oriented.

Definition 2.1. A *Cauchy hypersurface* in a spacetime (M, g) is a subset of M which is met exactly once by every inextensible timelike curve.

Cauchy hypersurfaces are always topological hypersurfaces but need not be smooth. All Cauchy hypersurfaces of a spacetime are homeomorphic.

Definition 2.2. A spacetime (M, g) is called *globally hyperbolic* if and only if it contains a Cauchy hypersurface.

A classical result of Geroch [21] says that a globally hyperbolic spacetime can be foliated by Cauchy hypersurfaces. It is a rather recent and very important result that this also holds in the smooth category:

Theorem 2.3 (Bernal and Sánchez [6, Thm. 1.1]). *Let (M, g) be a globally hyperbolic spacetime.*

Then there exists a smooth manifold Σ, a smooth one-parameter-family of Riemannian metrics $(g_t)_t$ on Σ and a smooth positive function β on $\mathbb{R} \times \Sigma$ such that (M, g) is isometric to $(\mathbb{R} \times \Sigma, -\beta dt^2 \oplus g_t)$. Each $\{t\} \times \Sigma$ corresponds to a smooth spacelike Cauchy hypersurface in (M, g).

For our purposes, we shall need a slightly stronger version of Theorem 2.3 where one of the Cauchy hypersurfaces $\{t\} \times \Sigma$ can be prescribed:

Theorem 2.4 (Bernal and Sánchez [7, Thm. 1.2]). *Let (M, g) be a globally hyperbolic spacetime and $\tilde{\Sigma}$ a smooth spacelike Cauchy hypersurface in (M, g).*

Then there exists a smooth splitting $(M, g) \cong (\mathbb{R} \times \Sigma, -\beta dt^2 \oplus g_t)$ as in Theorem 2.3 such that $\tilde{\Sigma}$ corresponds to $\{0\} \times \Sigma$.

We shall also need the following result which tells us that one can extend any compact acausal spacelike submanifold to a smooth spacelike Cauchy hypersurface. Here a subset of a spacetime is called *acausal* if no causal curve meets it more than once.

Theorem 2.5 (Bernal and Sánchez [7, Thm. 1.1]). *Let (M, g) be a globally hyperbolic spacetime and let $K \subset M$ be a compact acausal smooth spacelike submanifold with boundary.*

Then there exists a smooth spacelike Cauchy hypersurface Σ in (M, g) with $K \subset \Sigma$.

Definition 2.6. A closed subset $A \subset M$ is called spacelike compact if there exists a compact subset $K \subset M$ such that $A \subset J^M(K) := J^M_-(K) \cup J^M_+(K)$.

Note that a spacelike compact subset is in general not compact, but its intersection with any Cauchy hypersurface is compact, see e.g. [4, Cor. A.5.4].

Definition 2.7. A subset Ω of a spacetime M is called *causally compatible* if and only if $J^\Omega_\pm(x) = J^M_\pm(x) \cap \Omega$ for every $x \in \Omega$.

This means that every causal curve joining two points in Ω must be contained entirely in Ω.

2.2 *Differential Operators and Green's Functions*

A differential operator of order (at most) k on a vector bundle $S \to M$ over $\mathbb{K} = \mathbb{R}$ or $\mathbb{K} = \mathbb{C}$ is a linear map $P : C^\infty(M, S) \to C^\infty(M, S)$ which in local coordinates $x = (x^1, \ldots, x^m)$ of M and with respect to a local trivialization looks like

$$P = \sum_{|\alpha| \le k} A_\alpha(x) \frac{\partial^\alpha}{\partial x^\alpha}.$$

Here $C^\infty(M,S)$ denotes the space of smooth sections of $S \to M$, $\alpha = (\alpha_1,\ldots,\alpha_m) \in \mathbb{N}_0 \times \cdots \times \mathbb{N}_0$ runs over multi-indices, $|\alpha| = \alpha_1 + \ldots + \alpha_m$ and $\frac{\partial^\alpha}{\partial x^\alpha} = \frac{\partial^{|\alpha|}}{\partial(x^1)^{\alpha_1}\cdots\partial(x^m)^{\alpha_m}}$. The principal symbol σ_P of P associates to each covector $\xi \in T_x^*M$ a linear map $\sigma_P(\xi) : S_x \to S_x$. Locally, it is given by

$$\sigma_P(\xi) = \sum_{|\alpha|=k} A_\alpha(x)\xi^\alpha,$$

where $\xi^\alpha = \xi_1^{\alpha_1}\cdots\xi_m^{\alpha_m}$ and $\xi = \sum_j \xi_j dx^j$. If P and Q are two differential operators of order k and ℓ, respectively, then $Q \circ P$ is a differential operator of order $k+\ell$ and

$$\sigma_{Q\circ P}(\xi) = \sigma_Q(\xi) \circ \sigma_P(\xi).$$

For any linear differential operator $P : C^\infty(M,S) \to C^\infty(M,S)$ there is a unique formally dual operator $P^* : C^\infty(M,S^*) \to C^\infty(M,S^*)$ of the same order characterized by

$$\int_M \langle \varphi, P\psi\rangle \,\mathrm{dV} = \int_M \langle P^*\varphi, \psi\rangle \,\mathrm{dV}$$

for all $\psi \in C^\infty(M,S)$ and $\varphi \in C^\infty(M,S^*)$ with $\mathrm{supp}(\varphi)\cap\mathrm{supp}(\psi)$ compact. Here $\langle\cdot,\cdot\rangle : S^* \otimes S \to \mathbb{K}$ denotes the canonical pairing, i.e., the evaluation of a linear form in S_x^* on an element of S_x, where $x \in M$. We have $\sigma_{P^*}(\xi) = (-1)^k\sigma_P(\xi)^*$ where k is the order of P.

Definition 2.8. Let a vector bundle $S \to M$ be endowed with a non-degenerate inner product $\langle\cdot,\cdot\rangle$. A linear differential operator P on S is called *formally self-adjoint* if and only if

$$\int_M \langle P\varphi, \psi\rangle \,\mathrm{dV} = \int_M \langle \varphi, P\psi\rangle \,\mathrm{dV}$$

holds for all $\varphi, \psi \in C^\infty(M,S)$ with $\mathrm{supp}(\varphi) \cap \mathrm{supp}(\psi)$ compact.

Similarly, we call P *formally skew-adjoint* if instead

$$\int_M \langle P\varphi, \psi\rangle \,\mathrm{dV} = -\int_M \langle \varphi, P\psi\rangle \,\mathrm{dV}.$$

We recall the definition of advanced and retarded Green's operators for a linear differential operator.

Definition 2.9. Let P be a linear differential operator acting on the sections of a vector bundle S over a Lorentzian manifold M. An *advanced Green's operator* for P on M is a linear map

$$G_+ : C_c^\infty(M,S) \to C^\infty(M,S)$$

satisfying:

(G_1) $P \circ G_+ = \mathrm{id}_{C_c^\infty(M,S)}$;
(G_2) $G_+ \circ P|_{C_c^\infty(M,S)} = \mathrm{id}_{C_c^\infty(M,S)}$;
(G_3^+) $\mathrm{supp}(G_+\varphi) \subset J_+^M(\mathrm{supp}(\varphi))$ for any $\varphi \in C_c^\infty(M,S)$.

A *retarded Green's operator* for P on M is a linear map $G_- : C_c^\infty(M,S) \to C^\infty(M,S)$ satisfying (G_1), (G_2), and

(G_3^-) $\mathrm{supp}(G_-\varphi) \subset J_-^M(\mathrm{supp}(\varphi))$ for any $\varphi \in C_c^\infty(M,S)$.

Here we denote by $C_c^\infty(M,S)$ the space of compactly supported smooth sections of S.

Definition 2.10. Let $P : C^\infty(M,S) \to C^\infty(M,S)$ be a linear differential operator. We call P *Green-hyperbolic* if the restriction of P to any globally hyperbolic subregion of M has advanced and retarded Green's operators.

Remark 2.11. If the Green's operators of the restriction of P to a globally hyperbolic subregion exist, then they are necessarily unique, see Remark 3.7.

2.3 Wave Operators

The most prominent class of Green-hyperbolic operators are wave operators, sometimes also called normally hyperbolic operators.

Definition 2.12. A linear differential operator of second order $P : C^\infty(M,S) \to C^\infty(M,S)$ is called a *wave operator* if its principal symbol is given by the Lorentzian metric, i.e., for all $\xi \in T^*M$ we have

$$\sigma_P(\xi) = -\langle \xi, \xi \rangle \cdot \mathrm{id}.$$

In other words, if we choose local coordinates $x^1, \ldots, x^m$ on M and a local trivialization of S, then

$$P = -\sum_{i,j=1}^{m} g^{ij}(x)\frac{\partial^2}{\partial x^i \partial x^j} + \sum_{j=1}^{m} A_j(x)\frac{\partial}{\partial x^j} + B(x),$$

where A_j and B are matrix-valued coefficients depending smoothly on x and (g^{ij}) is the inverse matrix of (g_{ij}) with $g_{ij} = \langle \frac{\partial}{\partial x^i}, \frac{\partial}{\partial x^j} \rangle$. If P is a wave operator, then so is its dual operator P^*. In [4, Cor. 3.4.3] it has been shown that wave operators are Green-hyperbolic.

Example 2.13 (d'Alembert operator). Let S be the trivial line bundle so that sections of S are just functions. The d'Alembert operator $P = \Box = -\mathrm{div} \circ \mathrm{grad}$ is a formally self-adjoint wave operator, see e.g. [4, p. 26].

Example 2.14 (connection-d'Alembert operator). More generally, let S be a vector bundle and let ∇ be a connection on S. This connection and the Levi-Civita connection on T^*M induce a connection on $T^*M \otimes S$, again denoted ∇. We define the connection-d'Alembert operator $\Box^\nabla$ to be the composition of the following three maps:

$$C^\infty(M,S) \xrightarrow{\nabla} C^\infty(M, T^*M \otimes S) \xrightarrow{\nabla} C^\infty(M, T^*M \otimes T^*M \otimes S) \xrightarrow{-\mathrm{tr}\otimes \mathrm{id}_S} C^\infty(M,S),$$

where $\mathrm{tr} : T^*M \otimes T^*M \to \mathbb{R}$ denotes the metric trace, $\mathrm{tr}(\xi \otimes \eta) = \langle \xi, \eta \rangle$. We compute the principal symbol,

$$\sigma_{\Box^\nabla}(\xi)\varphi = -(\mathrm{tr} \otimes \mathrm{id}_S) \circ \sigma_\nabla(\xi) \circ \sigma_\nabla(\xi)(\varphi) = -(\mathrm{tr} \otimes \mathrm{id}_S)(\xi \otimes \xi \otimes \varphi) = -\langle \xi, \xi \rangle\, \varphi.$$

Hence $\Box^\nabla$ is a wave operator.

Example 2.15 (Hodge–d'Alembert operator). Let $S = \Lambda^k T^*M$ be the bundle of k-forms. Exterior differentiation $d : C^\infty(M, \Lambda^k T^*M) \to C^\infty(M, \Lambda^{k+1} T^*M)$ increases the degree by one while the codifferential $\delta = d^* : C^\infty(M, \Lambda^k T^*M) \to C^\infty(M, \Lambda^{k-1} T^*M)$ decreases the degree by one. While d is independent of the metric, the codifferential δ does depend on the Lorentzian metric. The operator $P = -d\delta - \delta d$ is a formally self-adjoint wave operator.

2.4 The Proca Equation

The Proca operator is an example of a Green-hyperbolic operator of second order which is not a wave operator. First we need the following observation:

Lemma 2.16. *Let M be globally hyperbolic, let $S \to M$ be a vector bundle and let P and Q be differential operators acting on sections of S. Suppose P has advanced and retarded Green's operators G_+ and G_-.*

If Q commutes with P, then it also commutes with G_+ and with G_-.

Proof. Assume $[P, Q] = 0$. We consider

$$\tilde{G}_\pm := G_\pm + [G_\pm, Q] : \; C^\infty_\mathrm{c}(M, s) \to C^\infty_\mathrm{sc}(M, S).$$

We compute on $C^\infty_\mathrm{c}(M, S)$:

$$\tilde{G}_\pm P = G_\pm P + G_\pm Q P - Q G_\pm P = \mathrm{id} + G_\pm P Q - Q = \mathrm{id} + Q - Q = \mathrm{id},$$

and similarly $P\tilde{G}_\pm = \mathrm{id}$. Hence $\tilde{G}_\pm$ are also advanced and retarded Green's operators, respectively. By Remark 2.11, Green's operators are unique, hence $\tilde{G}_\pm = G_\pm$ and therefore $[G_\pm, Q] = 0$. $\Box$

Example 2.17 (Proca operator). The discussion of this example follows [39, p. 116f], see also [20] where is the discussion is based on the Cauchy problem. The Proca equation describes massive vector bosons. We take $S = T^*M$ and let $m_0 > 0$. The Proca equation is

$$P\varphi := \delta d\varphi + m_0^2\varphi = 0, \tag{1}$$

where $\varphi \in C^\infty(M, S)$. Applying δ to (1) we obtain, using $\delta^2 = 0$ and $m_0 \neq 0$,

$$\delta\varphi = 0, \tag{2}$$

and hence

$$(d\delta + \delta d)\varphi + m_0^2\varphi = 0. \tag{3}$$

Conversely, (2) and (3) clearly imply (1).

Since $\tilde{P} := d\delta + \delta d + m_0^2$ is minus a wave operator, it has Green's operators $\tilde{G}_\pm$. We define

$$G_\pm : C_c^\infty(M, S) \to C_{sc}^\infty(M, S), \quad G_\pm := (m_0^{-2}d\delta + \mathrm{id}) \circ \tilde{G}_\pm = \tilde{G}_\pm \circ (m_0^{-2}d\delta + \mathrm{id}).$$

The last equality holds because d and δ commute with $\tilde{P}$. For $\varphi \in C_c^\infty(M, S)$ we compute

$$G_\pm P\varphi = \tilde{G}_\pm(m_0^{-2}d\delta + \mathrm{id})(\delta d + m_0^2)\varphi = \tilde{G}_\pm\tilde{P}\varphi = \varphi$$

and similarly $PG_\pm\varphi = \varphi$. Since the differential operator $m_0^{-2}d\delta + \mathrm{id}$ does not increase supports, the third axiom in the definition of advanced and retarded Green's operators holds as well.

This shows that G_+ and G_- are advanced and retarded Green's operators for P, respectively. Thus P is not a wave operator but Green-hyperbolic.

2.5 Dirac Type Operators

The most important Green-hyperbolic operators of first order are the so-called Dirac type operators.

Definition 2.18. A linear differential operator $D : C^\infty(M, S) \to C^\infty(M, S)$ of first order is called of Dirac type, if $-D^2$ is a wave operator.

Remark 2.19. If D is of Dirac type, then i times its principal symbol satisfies the Clifford relations

$$(i\sigma_D(\xi))^2 = -\sigma_{D^2}(\xi) = -\langle \xi, \xi \rangle \cdot \mathrm{id},$$

hence by polarization

$$(i\sigma_D(\xi))(i\sigma_D(\eta)) + (i\sigma_D(\eta))(i\sigma_D(\xi)) = -2\langle \xi, \eta \rangle \cdot \mathrm{id}.$$

The bundle S thus becomes a module over the bundle of Clifford algebras $\mathrm{Cl}(TM)$ associated with $(TM, \langle\cdot\,,\cdot\rangle)$. See [5, Sect. 1.1] or [60, Chap. I] for the definition and properties of the Clifford algebra $\mathrm{Cl}(V)$ associated with a vector space V with inner product.

Remark 2.20. If D is of Dirac type, then so is its dual operator D^*. On a globally hyperbolic region let G_+ be the advanced Green's operator for D^2 which exists since $-D^2$ is a wave operator. Then it is not hard to check that $D \circ G_+$ is an advanced Green's operator for D, see e.g. the proof of Theorem 2.3 in [14] or [29, Thm. 3.2]. The same discussion applies to the retarded Green's operator. Hence any Dirac type operator is Green-hyperbolic.

Example 2.21 (Classical Dirac operator). If the spacetime M carries a spin structure, then one can define the spinor bundle $S = \Sigma M$ and the classical Dirac operator

$$D : C^\infty(M, \Sigma M) \to C^\infty(M, \Sigma M), \quad D\varphi := i\sum_{j=1}^{m} \varepsilon_j e_j \cdot \nabla_{e_j}\varphi.$$

Here $(e_j)_{1\le j\le m}$ is a local orthonormal basis of the tangent bundle, $\varepsilon_j = \langle e_j, e_j\rangle = \pm 1$ and "$\cdot$" denotes the Clifford multiplication, see e.g. [5] or [3, Sect. 2]. The principal symbol of D is given by

$$\sigma_D(\xi)\psi = i\xi^\# \cdot \psi.$$

Here $\xi^\#$ denotes the tangent vector dual to the 1-form ξ via the Lorentzian metric, i.e., $\langle \xi^\#, Y\rangle = \xi(Y)$ for all tangent vectors Y over the same point of the manifold. Hence

$$\sigma_{D^2}(\xi)\psi = \sigma_D(\xi)\sigma_D(\xi)\psi = -\xi^\# \cdot \xi^\# \cdot \psi = \langle \xi, \xi\rangle\, \psi.$$

Thus $P = -D^2$ is a wave operator. Moreover, D is formally self-adjoint, see e.g. [3, p. 552].

Example 2.22 (Twisted Dirac operators). More generally, let $E \to M$ be a complex vector bundle equipped with a non-degenerate Hermitian inner product and a metric connection ∇^E over a spin spacetime M. In the notation of Example 2.21, one may define the Dirac operator of M twisted with E by

$$D^E := i\sum_{j=1}^{m} \varepsilon_j e_j \cdot \nabla^{\Sigma M\otimes E}_{e_j} : C^\infty(M, \Sigma M \otimes E) \to C^\infty(M, \Sigma M \otimes E),$$

where $\nabla^{\Sigma M\otimes E}$ is the tensor product connection on $\Sigma M \otimes E$. Again, D^E is a formally self-adjoint Dirac type operator.

Example 2.23 (Euler operator). In Example 2.15, replacing $\Lambda^k T^*M$ by $S := \Lambda T^*M \otimes \mathbb{C} = \oplus_{k=0}^n \Lambda^k T^*M \otimes \mathbb{C}$, the Euler operator $D = i(d - \delta)$ defines a formally self-adjoint Dirac type operator. In case M is spin, the Euler operator coincides with the Dirac operator of M twisted with ΣM if m is even and with $\Sigma M \oplus \Sigma M$ if m is odd.

Example 2.24 (Buchdahl operators). On a 4-dimensional spin spacetime M, consider the standard orthogonal and parallel splitting $\Sigma M = \Sigma_+ M \oplus \Sigma_- M$ of the complex spinor bundle of M into spinors of positive and negative chirality. The finite dimensional irreducible representations of the simply-connected Lie group $\mathrm{Spin}^0(3,1)$ are given by $\Sigma_+^{(k/2)} \otimes \Sigma_-^{(\ell/2)}$ where $k, \ell \in \mathbb{N}$. Here $\Sigma_+^{(k/2)} = \Sigma_+^{\odot k}$ is the k-th symmetric tensor product of the positive half-spinor representation Σ_+ and similarly for $\Sigma_-^{(\ell/2)}$. Let the associated vector bundles $\Sigma_\pm^{(k/2)} M$ carry the induced inner product and connection.

For $s \in \mathbb{N}$, $s \geq 1$, consider the twisted Dirac operator $D^{(s)}$ acting on sections of $\Sigma M \otimes \Sigma_+^{((s-1)/2)} M$. In the induced splitting

$$\Sigma M \otimes \Sigma_+^{((s-1)/2)} M = \Sigma_+ M \otimes \Sigma_+^{(s-1/2)} M \oplus \Sigma_- M \otimes \Sigma_+^{((s-1)/2)} M$$

the operator $D^{(s)}$ is of the form

$$\begin{pmatrix} 0 & D_-^{(s)} \\ D_+^{(s)} & 0 \end{pmatrix}$$

because Clifford multiplication by vectors exchanges the chiralities. The Clebsch–Gordan formulas [10, Prop. II.5.5] tell us that the representation $\Sigma_+ \otimes \Sigma_+^{(\frac{s-1}{2})}$ splits as

$$\Sigma_+ \otimes \Sigma_+^{(\frac{s-1}{2})} = \Sigma_+^{(\frac{s}{2})} \oplus \Sigma_+^{(\frac{s}{2}-1)}.$$

Hence we have the corresponding parallel orthogonal projections

$$\pi_s : \Sigma_+ M \otimes \Sigma_+^{(\frac{s-1}{2})} M \to \Sigma_+^{(\frac{s}{2})} M \quad \text{and} \quad \pi_s' : \Sigma_+ M \otimes \Sigma_+^{(\frac{s-1}{2})} M \to \Sigma_+^{(\frac{s}{2}-1)} M.$$

On the other hand, the representation $\Sigma_- \otimes \Sigma_+^{(\frac{s-1}{2})}$ is irreducible. Now *Buchdahl operators* are the operators of the form

$$B_{\mu_1,\mu_2,\mu_3}^{(s)} := \begin{pmatrix} \mu_1 \cdot \pi_s + \mu_2 \cdot \pi_s' & D_-^{(s)} \\ D_+^{(s)} & \mu_3 \cdot \mathrm{id} \end{pmatrix}$$

where $\mu_1, \mu_2, \mu_3 \in \mathbb{C}$ are constants. By definition, $B_{\mu_1,\mu_2,\mu_3}^{(s)}$ is of the form $D^{(s)} + b$, where b is of order zero. In particular, $B_{\mu_1,\mu_2,\mu_3}^{(s)}$ is a Dirac-type operator, hence it is Green-hyperbolic.

If M were Riemannian, then $D^{(s)}$ would be formally self-adjoint. Hence the operator $B_{\mu_1,\mu_2,\mu_3}^{(s)}$ would be formally self-adjoint if and only if the constants μ_1, μ_2, μ_3 are real. In Lorentzian signature, $\Sigma_+ M$ and $\Sigma_- M$ are isotropic for the natural inner product on ΣM, so that the bundles on which the Buchdahl operators act, carry no natural non-degenerate inner product.

For a definition of Buchdahl operators using indices we refer to [12, 13, 44] and to [28, Def. 8.1.4, p. 104].

2.6 The Rarita–Schwinger operator

For the Rarita–Schwinger operator on Riemannian manifolds, we refer to [43, Sect. 2], see also [8, Sect. 2]. In this section let the spacetime M be spin and consider the Clifford-multiplication $\gamma : T^*M \otimes \Sigma M \to \Sigma M, \theta \otimes \psi \mapsto \theta^\sharp \cdot \psi$, where ΣM is the complex spinor bundle of M. Then there is the representation theoretic splitting of $T^*M \otimes \Sigma M$ into the orthogonal and parallel sum

$$T^*M \otimes \Sigma M = \iota(\Sigma M) \oplus \Sigma^{3/2} M,$$

where $\Sigma^{3/2} M := \ker(\gamma)$ and $\iota(\psi) := -\frac{1}{m}\sum_{j=1}^m e_j^* \otimes e_j \cdot \psi$. Here again $(e_j)_{1 \le j \le m}$ is a local orthonormal basis of the tangent bundle. Let $\mathcal{D}$ be the twisted Dirac operator on $T^*M \otimes \Sigma M$, that is, $\mathcal{D} := i \cdot (\mathrm{id} \otimes \gamma) \circ \nabla$, where ∇ denotes the induced covariant derivative on $T^*M \otimes \Sigma M$.

Definition 2.25. The *Rarita–Schwinger operator* on the spin spacetime M is defined by $\mathcal{Q} := (\mathrm{id} - \iota \circ \gamma) \circ \mathcal{D} : C^\infty(M, \Sigma^{3/2} M) \to C^\infty(M, \Sigma^{3/2} M)$.

By definition, the Rarita–Schwinger operator is pointwise obtained as the orthogonal projection onto $\Sigma^{3/2} M$ of the twisted Dirac operator $\mathcal{D}$ restricted to a section of $\Sigma^{3/2} M$. Using the above formula for ι, the Rarita–Schwinger operator can be written down explicitly:

$$\mathcal{Q}\psi = i \cdot \sum_{\beta=1}^m e_\beta^* \otimes \sum_{\alpha=1}^m \varepsilon_\alpha \left(e_\alpha \cdot \nabla_{e_\alpha} \varphi_\beta - \frac{2}{m} e_\beta \cdot \nabla_{e_\alpha} \varphi_\alpha \right)$$

for all $\psi = \sum_{\beta=1}^m e_\beta^* \otimes \psi_\beta \in C^\infty(M, \Sigma^{3/2} M)$, where here ∇ is the standard connection on ΣM. It can be checked that $\mathcal{Q}$ is a formally self-adjoint linear differential operator of first order, with principal symbol

$$\sigma_{\mathcal{Q}}(\xi) : \psi \mapsto i \left\{ (\mathrm{id} \otimes \xi^\sharp \cdot)\psi - \frac{2}{m} \sum_{\beta=1}^m e_\beta^* \otimes e_\beta \cdot (\xi^\sharp \lrcorner \psi) \right\},$$

for all $\psi = \sum_{\beta=1}^m e_\beta^* \otimes \psi_\beta \in \Sigma^{3/2} M$. Here $X \lrcorner \psi$ denotes the insertion of the tangent vector X in the first factor, that is, $X \lrcorner \psi := \sum_{\beta=1}^m e_\beta^*(X) \psi_\beta$.

Lemma 2.26. *Let M be a spin spacetime of dimension $m \geq 3$. Then the characteristic variety of the Rarita–Schwinger operator of M coincides with the set of lightlike covectors.*

Proof. By definition, the characteristic variety of $\mathcal{Q}$ is the set of nonzero covectors ξ for which $\sigma_{\mathcal{Q}}(\xi)$ is not invertible. Fix an arbitrary point $x \in M$. Let $\xi \in T_x^*M \setminus \{0\}$ be non-lightlike. Without loss of generality we may assume that ξ is normalized and that the Lorentz orthonormal basis is chosen so that $\xi^\sharp = e_1$. Hence $\varepsilon_1 = 1$ if ξ is spacelike and $\varepsilon_1 = -1$ if ξ is timelike. Take $\psi = \sum_{\beta=1}^m e_\beta^* \otimes \psi_\beta \in \ker(\sigma_{\mathcal{Q}}(\xi))$. Then

$$\begin{aligned} 0 &= \sum_{\beta=1}^m e_\beta^* \otimes e_1 \cdot \psi_\beta - \frac{2}{m} \sum_{\beta=1}^m e_\beta^* \otimes e_\beta \cdot \psi_1 \\ &= \sum_{\beta=1}^m e_\beta^* \otimes (e_1 \cdot \psi_\beta - \frac{2}{m} e_\beta \cdot \psi_1), \end{aligned}$$

which implies $e_1 \cdot \psi_\beta = \frac{2}{m} e_\beta \cdot \psi_1$ for all $\beta \in \{1, \dots, m\}$. Choosing $\beta = 1$, we obtain $e_1 \cdot \psi_1 = 0$ because $m \geq 3$. Hence $\psi_1 = 0$, from which $\psi_\beta = 0$ follows for all $\beta \in \{1, \dots, m\}$. Hence $\psi = 0$ and $\sigma_{\mathcal{Q}}(\xi)$ is invertible.

If $\xi \in T_x^*M \setminus \{0\}$ is lightlike, then we may assume that $\xi^\sharp = e_1 + e_2$, where $\varepsilon_1 = -1$ and $\varepsilon_2 = 1$. Choose $\psi_1 \in \Sigma_x M \setminus \{0\}$ with $(e_1 + e_2) \cdot \psi_1 = 0$. Such a ψ_1 exists because Clifford multiplication by a lightlike vector is nilpotent. Set $\psi_2 := -\psi_1$ and $\psi := e_1^* \otimes \psi_1 + e_2^* \otimes \psi_2$. Then $\psi \in \Sigma_x^{3/2} M \setminus \{0\}$ and

$$-i\sigma_{\mathcal{Q}}(\xi)(\psi) = \sum_{j=1}^2 e_j^* \otimes \underbrace{(e_1 + e_2) \cdot \psi_j}_{=0} - \frac{2}{m} e_j^* \otimes e_j \cdot \underbrace{(\psi_1 + \psi_2)}_{=0} = 0.$$

This shows $\psi \in \ker(\sigma_{\mathcal{Q}}(\xi))$ and hence $\sigma_{\mathcal{Q}}(\xi)$ is not invertible. □

The same proof shows that in the Riemannian case the Rarita–Schwinger operator is elliptic.

Remark 2.27. Since the characteristic variety of the Rarita–Schwinger operator is exactly that of the Dirac operator, Lemma 2.26 together with [24, Thms. 23.2.4 and 23.2.7] imply that the Cauchy problem for $\mathcal{Q}$ is well-posed in case M is globally hyperbolic. This implies they $\mathcal{Q}$ has advanced and retarded Green's operators. Hence $\mathcal{Q}$ is not of Dirac type but it is Green-hyperbolic.

Remark 2.28. The equations originally considered by Rarita and Schwinger in [33] correspond to the twisted Dirac operator $\mathcal{D}$ restricted to $\Sigma^{3/2}M$ but not projected back to $\Sigma^{3/2}M$. In other words, they considered the operator

$$\mathcal{D}|_{C^\infty(M,\Sigma^{3/2}M)} : C^\infty(M, \Sigma^{3/2}M) \to C^\infty(M, T^*M \otimes \Sigma M).$$

These equations are over-determined. Therefore it is not a surprise that non-trivial solutions restrict the geometry of the underlying manifold as observed by Gibbons [22] and that this operator has no Green's operators.

2.7 Combining Given Operators into a New One

Given two Green-hyperbolic operators we can form the direct sum and obtain a new operator in a trivial fashion. It turns out that this operator is again Green-hyperbolic. Note that the two operators need not have the same order.

Lemma 2.29. *Let $S_1, S_2 \to M$ be two vector bundles over the globally hyperbolic manifold M. Let P_1 and P_2 be two Green-hyperbolic operators acting on sections of S_1 and S_2 respectively. Then*

$$P_1 \oplus P_2 := \begin{pmatrix} P_1 & 0 \\ 0 & P_2 \end{pmatrix} : C^\infty(M, S_1 \oplus S_2) \to C^\infty(M, S_1 \oplus S_2)$$

is Green-hyperbolic.

Proof. If G_1 and G_2 are advanced Green's operators for P_1 and P_2 respectively, then clearly $\begin{pmatrix} G_1 & 0 \\ 0 & G_2 \end{pmatrix}$ is an advanced Green's operator for $P_1 \oplus P_2$. The retarded case is analogous. □

It is interesting to note that P_1 and P_2 need not have the same order. Hence Green-hyperbolic operators need not be hyperbolic in the usual sense. Moreover, it is not obvious that Green-hyperbolic operators have a well-posed Cauchy problem. For instance, if P_1 is a wave operator and P_2 a Dirac-type operator, then along a Cauchy hypersurface one would have to prescribe the normal derivative for the S_1-component but not for the S_2-component.

3 Algebras of Observables

Our next aim is to quantize the classical fields governed by Green-hyperbolic differential operators. We construct local algebras of observables and we prove that we obtain locally covariant quantum field theories in the sense of [11].

3.1 Bosonic Quantization

In this section we show how a quantization process based on canonical commutation relations (CCR) can be carried out for formally self-adjoint Green-hyperbolic operators. This is a functorial procedure. We define the first category involved in the quantization process.

Definition 3.1. The category GlobHypGreen consists of the following objects and morphisms:

- An object in GlobHypGreen is a triple (M, S, P), where
  - ► M is a globally hyperbolic spacetime,
  - ► S is a real vector bundle over M endowed with a non-degenerate inner product $\langle \cdot\,,\cdot \rangle$ and
  - ► P is a formally self-adjoint Green-hyperbolic operator acting on sections of S.
- A morphism between two objects (M_1, S_1, P_1) and (M_2, S_2, P_2) of GlobHypGreen is a pair (f, F), where
  - ► f is a time-orientation preserving isometric embedding $M_1 \to M_2$ with $f(M_1)$ causally compatible and open in M_2,
  - ► F is a fiberwise isometric vector bundle isomorphism over f such that the following diagram commutes:

$$\begin{array}{ccc} C^\infty(M_2, S_2) & \xrightarrow{P_2} & C^\infty(M_2, S_2) \\ \Big\downarrow \text{res} & & \Big\downarrow \text{res} \\ C^\infty(M_1, S_1) & \xrightarrow{P_1} & C^\infty(M_1, S_1), \end{array} \tag{4}$$

    where $\text{res}(\varphi) := F^{-1} \circ \varphi \circ f$ for every $\varphi \in C^\infty(M_2, S_2)$.

Note that morphisms exist only if the manifolds have equal dimension and the vector bundles have the same rank. Note furthermore, that the inner product $\langle \cdot\,,\cdot \rangle$ on S is not required to be positive or negative definite.

The causal compatibility condition, which is not automatically satisfied (see e.g. [4, Fig. 33]), ensures the commutation of the extension and restriction maps with the Green's operators:

Lemma 3.2. *Let (f, F) be a morphism between two objects (M_1, S_1, P_1) and (M_2, S_2, P_2) in the category* GlobHypGreen *and let $(G_1)_\pm$ and $(G_2)_\pm$ be the respective Green's operators for P_1 and P_2. Denote by $\text{ext}(\varphi) \in C_c^\infty(M_2, S_2)$ the extension by 0 of $F \circ \varphi \circ f^{-1} : f(M_1) \to S_2$ to M_2, for every $\varphi \in C_c^\infty(M_1, S_1)$. Then*

$$\text{res} \circ (G_2)_\pm \circ \text{ext} = (G_1)_\pm .$$

Proof. Set $(\widetilde{G_1})_\pm := \mathrm{res} \circ (G_2)_\pm \circ \mathrm{ext}$ and fix $\varphi \in C_c^\infty(M_1, S_1)$. First observe that the causal compatibility condition on f implies that

$$\begin{aligned}
\mathrm{supp}((\widetilde{G_1})_\pm(\varphi)) &= f^{-1}(\mathrm{supp}((G_2)_\pm \circ \mathrm{ext}(\varphi))) \\
&\subset f^{-1}\big(J_\pm^{M_2}(\mathrm{supp}(\mathrm{ext}(\varphi)))\big) \\
&= f^{-1}\big(J_\pm^{M_2}(f(\mathrm{supp}(\varphi)))\big) \\
&= J_\pm^{M_1}(\mathrm{supp}(\varphi)).
\end{aligned}$$

In particular, $(\widetilde{G_1})_\pm(\varphi)$ has spacelike compact support in M_1 and $(\widetilde{G_1})_\pm$ satisfies Axiom (G_3). Moreover, it follows from (4) that $P_2 \circ \mathrm{ext} = \mathrm{ext} \circ P_1$ on $C_c^\infty(M_1, S_1)$, which directly implies that $(\widetilde{G_1})_\pm$ satisfies Axioms (G_1) and (G_2) as well. The uniqueness of the advanced and retarded Green's operators on M_1 yields $(\widetilde{G_1})_\pm = (G_1)_\pm$. □

Next we show how the Green's operators for a formally self-adjoint Green-hyperbolic operator provide a symplectic vector space in a canonical way. First we see how the Green's operators of an operator and of its formally dual operator are related.

Lemma 3.3. *Let M be a globally hyperbolic spacetime and G_+, G_- the advanced and retarded Green's operators for a Green-hyperbolic operator P acting on sections of $S \to M$. Then the advanced and retarded Green's operators G_+^* and G_-^* for P^* satisfy*

$$\int_M \langle G_\pm^* \varphi, \psi \rangle \, \mathrm{dV} = \int_M \langle \varphi, G_\mp \psi \rangle \, \mathrm{dV}$$

for all $\varphi \in C_c^\infty(M, S^*)$ and $\psi \in C_c^\infty(M, S)$.

Proof. Axiom (G_1) for the Green's operators implies that

$$\begin{aligned}
\int_M \langle G_\pm^* \varphi, \psi \rangle \, \mathrm{dV} &= \int_M \langle G_\pm^* \varphi, P(G_\mp \psi) \rangle \, \mathrm{dV} \\
&= \int_M \langle P^*(G_\pm^* \varphi), G_\mp \psi \rangle \, \mathrm{dV} \\
&= \int_M \langle \varphi, G_\mp \psi \rangle \, \mathrm{dV},
\end{aligned}$$

where the integration by parts is justified since $\mathrm{supp}(G_\pm^* \varphi) \cap \mathrm{supp}(G_\mp \psi) \subset J_\pm^M(\mathrm{supp}(\varphi)) \cap J_\mp^M(\mathrm{supp}(\psi))$ is compact. □

Proposition 3.4. *Let (M, S, P) be an object in the category* GlobHypGreen. *Set $G := G_+ - G_-$, where G_+, G_- are the advanced and retarded Green's operator for P, respectively.*

Then the pair $(\mathrm{SYMPL}(M,S,P),\omega)$ *is a symplectic vector space, where*

$$\mathrm{SYMPL}(M,S,P) := C_c^\infty(M,S)/\ker(G) \quad \textit{and} \quad \omega([\varphi],[\psi]) := \int_M \langle G\varphi, \psi\rangle \,\mathrm{dV}.$$

Here the square brackets $[\cdot]$ *denote residue classes modulo* $\ker(G)$.

Proof. The bilinear form $(\varphi,\psi) \mapsto \int_M \langle G\varphi,\psi\rangle \,\mathrm{dV}$ on $C_c^\infty(M,S)$ is skew-symmetric as a consequence of Lemma 3.3 because P is formally self-adjoint. Its null-space is exactly $\ker(G)$. Therefore the induced bilinear form ω on the quotient space $\mathrm{SYMPL}(M,S,P)$ is non-degenerate and hence a symplectic form. □

Put $C_{sc}^\infty(M,S) := \{\varphi \in C^\infty(M,S) \mid \mathrm{supp}(\varphi) \text{ is spacelike compact}\}$. The next result will in particular show that we can consider $\mathrm{SYMPL}(M,S,P)$ as the space of smooth solutions of the equation $P\varphi = 0$ which have spacelike compact support.

Theorem 3.5. *Let M be a Lorentzian manifold, let $S \to M$ be a vector bundle, and let P be a Green-hyperbolic operator acting on sections of S. Let $G_\pm$ be advanced and retarded Green's operators for P, respectively. Put*

$$G := G_+ - G_- : C_c^\infty(M,S) \to C_{sc}^\infty(M,S).$$

Then the following linear maps form a complex:

$$\{0\} \to C_c^\infty(M,S) \xrightarrow{P} C_c^\infty(M,S) \xrightarrow{G} C_{sc}^\infty(M,S) \xrightarrow{P} C_{sc}^\infty(M,S). \tag{5}$$

This complex is always exact at the first $C_c^\infty(M,S)$. If M is globally hyperbolic, then the complex is exact everywhere.

Proof. The proof follows the lines of [4, Thm. 3.4.7] where the result was shown for wave operators. First note that, by ($G_3^\mp$) in the definition of Green's operators, we have that $G_\pm : C_c^\infty(M,S) \to C_{sc}^\infty(M,S)$. It is clear from ($G_1$) and ($G_2$) that $PG = GP = 0$ on $C_c^\infty(M,S)$, hence (5) is a complex.

If $\varphi \in C_c^\infty(M,S)$ satisfies $P\varphi = 0$, then by (G_2) we have $\varphi = G_+P\varphi = 0$ which shows that $P|_{C_c^\infty(M,S)}$ is injective. Thus the complex is exact at the first $C_c^\infty(M,S)$.

From now on let M be globally hyperbolic. Let $\varphi \in C_c^\infty(M,S)$ with $G\varphi = 0$, i.e., $G_+\varphi = G_-\varphi$. We put $\psi := G_+\varphi = G_-\varphi \in C^\infty(M,S)$ and we see that $\mathrm{supp}(\psi) = \mathrm{supp}(G_+\varphi) \cap \mathrm{supp}(G_-\varphi) \subset J_+(\mathrm{supp}(\varphi)) \cap J_-(\mathrm{supp}(\varphi))$. Since (M,g) is globally hyperbolic $J_+(\mathrm{supp}(\varphi)) \cap J_-(\mathrm{supp}(\varphi))$ is compact, hence $\psi \in C_c^\infty(M,S)$. From $P\psi = PG_+\varphi = \varphi$ we see that $\varphi \in P(C_c^\infty(M,S))$. This shows exactness at the second $C_c^\infty(M,S)$.

It remains to show that any $\varphi \in C_{sc}^\infty(M,S)$ with $P\varphi = 0$ is of the form $\varphi = G\psi$ with $\psi \in C_c^\infty(M,S)$. Using a cut-off function decompose φ as $\varphi = \varphi_+ - \varphi_-$ where $\mathrm{supp}(\varphi_\pm) \subset J_\pm(K)$ where K is a suitable compact subset of M. Then $\psi := P\varphi_+ = P\varphi_-$ satisfies $\mathrm{supp}(\psi) \subset J_+(K) \cap J_-(K)$. Thus $\psi \in C_c^\infty(M,S)$. We check that $G_+\psi = \varphi_+$. Namely, for all $\chi \in C_c^\infty(M,S^*)$ we have by Lemma 3.3

$$\int_M \langle \chi, G_+ P\varphi_+ \rangle \, dV = \int_M \langle G_-^* \chi, P\varphi_+ \rangle \, dV = \int_M \langle P^* G_-^* \chi, \varphi_+ \rangle \, dV$$
$$= \int_M \langle \chi, \varphi_+ \rangle \, dV.$$

The integration by parts in the second equality is justified because $\mathrm{supp}(\varphi_+) \cap \mathrm{supp}(G_-^* \chi) \subset J_+(K) \cap J_-(\mathrm{supp}(\chi))$ is compact. Similarly, one shows $G_- \psi = \varphi_-$. Now $G\psi = G_+\psi - G_-\psi = \varphi_+ - \varphi_- = \varphi$ which concludes the proof. □

In particular, given an object (M, S, P) in GlobHypGreen, the map G induces an isomorphism from

$$\mathrm{SYMPL}(M, S, P) = C_c^\infty(M, S)/\ker(G) \xrightarrow{\cong} \ker(P) \cap C_{sc}^\infty(M, S).$$

Remark 3.6. Exactness at the first $C_c^\infty(M, S)$ in sequence (5) says that there are no non-trivial smooth solutions of $P\varphi = 0$ with compact support. Indeed, if M is globally hyperbolic, more is true.

If $\varphi \in C^\infty(M, S)$ solves $P\varphi = 0$ and $\mathrm{supp}(\varphi)$ *is future or past-compact, then* $\varphi = 0$.

Here a subset $A \subset M$ is called future-compact if $A \cap J_+(x)$ is compact for any $x \in M$. Past-compactness is defined similarly.

Proof. Let $\varphi \in C^\infty(M, S)$ solve $P\varphi = 0$ such that $\mathrm{supp}(\varphi)$ is future-compact. For any $\chi \in C_c^\infty(M, S^*)$ we have

$$\int_M \langle \chi, \varphi \rangle \, dV = \int_M \langle P^* G_+^* \chi, \varphi \rangle \, dV = \int_M \langle G_+^* \chi, P\varphi \rangle \, dV = 0.$$

This shows $\varphi = 0$. The integration by parts is justified because $\mathrm{supp}(G_+^* \chi) \cap \mathrm{supp}(\varphi) \subset J_+(\mathrm{supp}(\chi)) \cap \mathrm{supp}(\varphi)$ is compact, see [4, Lemma A.5.3]. □

Remark 3.7. Let M be a globally hyperbolic spacetime and (M, S, P) an object in GlobHypGreen. Then the Green's operators G_+ and G_- are unique. Namely, if G_+ and $\tilde{G}_+$ are advanced Green's operators for P, then for any $\varphi \in C_c^\infty(M, S)$ the section $\psi := G_+\varphi - \tilde{G}_+\varphi$ has past-compact support and satisfies $P\psi = 0$. By the previous remark, we have $\psi = 0$ which shows $G_+ = \tilde{G}_+$.

Now, let (f, F) be a morphism between two objects (M_1, S_1, P_1) and (M_2, S_2, P_2) in the category GlobHypGreen. For $\varphi \in C_c^\infty(M_1, S_1)$ consider the extension by zero $\mathrm{ext}(\varphi) \in C_c^\infty(M_2, S_2)$ as in Lemma 3.2.

Lemma 3.8. *Given a morphism (f, F) between two objects (M_1, S_1, P_1) and (M_2, S_2, P_2) in the category* GlobHypGreen, *extension by zero induces a symplectic linear map* $\mathrm{SYMPL}(f, F) : \mathrm{SYMPL}(M_1, S_1, P_1) \to \mathrm{SYMPL}(M_2, S_2, P_2)$.

Moreover,

$$\mathrm{SYMPL}(\mathrm{id}_M, \mathrm{id}_S) = \mathrm{id}_{\mathrm{SYMPL}(M,S,P)}, \tag{6}$$

and for any further morphism $(f', F') : (M_2, S_2, P_2) \to (M_3, S_3, P_3)$ *one has*

$$\mathrm{SYMPL}((f', F') \circ (f, F)) = \mathrm{SYMPL}(f', F') \circ \mathrm{SYMPL}(f, F). \tag{7}$$

Proof. If $\varphi = P_1\psi \in \ker(G_1) = P_1(C_c^\infty(M_1, S_1))$, then $\mathrm{ext}(\varphi) = P_2(\mathrm{ext}(\psi)) \in P_2(C_c^\infty(M_2, S_2)) = \ker(G_2)$. Hence ext induces a linear map

$$\mathrm{SYMPL}(f, F) : C_c^\infty(M_1, S_1)/\ker(G_1) \to C_c^\infty(M_2, S_2)/\ker(G_2).$$

Furthermore, applying Lemma 3.2, we have, for any $\varphi, \psi \in C_c^\infty(M_1, S_1)$

$$\int_{M_2} \langle G_2(\mathrm{ext}(\varphi)), \mathrm{ext}(\psi)\rangle\, \mathrm{dV} = \int_{M_1} \langle \mathrm{res} \circ G_2 \circ \mathrm{ext}(\varphi), \psi\rangle\, \mathrm{dV} = \int_{M_1} \langle G_1\varphi, \psi\rangle\, \mathrm{dV},$$

hence $\mathrm{SYMPL}(f, F)$ is symplectic. Equation (6) is trivial and extending once or twice by 0 amounts to the same, so (7) holds as well. □

Remark 3.9. Under the isomorphism $\mathrm{SYMPL}(M, S, P) \to \ker(P) \cap C_{sc}^\infty(M, S)$ induced by G, the extension by zero corresponds to an extension as a smooth solution of $P\varphi = 0$ with spacelike compact support. This follows directly from Lemma 3.2. In other words, for any morphism (f, F) from (M_1, S_1, P_1) to (M_2, S_2, P_2) in GlobHypGreen we have the following commutative diagram:

$$\begin{array}{ccc}
\mathrm{SYMPL}(M_1, S_1, P_1) & \xrightarrow{\ \mathrm{SYMPL}(f,F)\ } & \mathrm{SYMPL}(M_2, S_2, P_2) \\
\cong \downarrow & & \downarrow \cong \\
\ker(P_1) \cap C_{sc}^\infty(M_1, S_1) & \xrightarrow[\text{a solution}]{\text{extension as}} & \ker(P_2) \cap C_{sc}^\infty(M_2, S_2).
\end{array}$$

Let Sympl denote the category of real symplectic vector spaces with symplectic linear maps as morphisms. Lemma 3.8 says that we have constructed a covariant functor

$$\mathrm{SYMPL} : \mathsf{GlobHypGreen} \longrightarrow \mathsf{Sympl}.$$

In order to obtain an algebra-valued functor, we compose SYMPL with the functor CCR which associates to any symplectic vector space its Weyl algebra. Here "CCR" stands for "canonical commutation relations". This is a general algebraic construction which is independent of the context of Green-hyperbolic operators and which is carried out in Sect. A.2. As a result, we obtain the functor

$$\mathfrak{A}_{\mathrm{bos}} := \mathrm{CCR} \circ \mathrm{SYMPL} : \mathsf{GlobHypGreen} \longrightarrow \mathsf{C^*Alg},$$

where $\mathsf{C^*Alg}$ is the category whose objects are the unital C*-algebras and whose morphisms are the injective unit-preserving C*-morphisms.

In the remainder of this section we show that the functor CCR $\circ$ SYMPL is a bosonic locally covariant quantum field theory. We call two subregions M_1 and M_2 of a spacetime M *causally disjoint* if and only if $J^M(M_1) \cap M_2 = \emptyset$. In other words, there are no causal curves joining M_1 and M_2.

Theorem 3.10. *The functor* $\mathfrak{A}_{\mathrm{bos}} : \mathsf{GlobHypGreen} \longrightarrow \mathsf{C^*Alg}$ *is a bosonic locally covariant quantum field theory, i.e., the following axioms hold:*

(i) ***(Quantum causality)*** *Let* (M_j, S_j, P_j) *be objects in* $\mathsf{GlobHypGreen}$, $j = 1, 2, 3$, *and* (f_j, F_j) *morphisms from* (M_j, S_j, P_j) *to* (M_3, S_3, P_3), $j = 1, 2$, *such that* $f_1(M_1)$ *and* $f_2(M_2)$ *are causally disjoint regions in* M_3.
Then the subalgebras $\mathfrak{A}_{\mathrm{bos}}(f_1, F_1)(\mathfrak{A}_{\mathrm{bos}}(M_1, S_1, P_1))$ *and* $\mathfrak{A}_{\mathrm{bos}}(f_2, F_2)(\mathfrak{A}_{\mathrm{bos}}(M_2, S_2, P_2))$ *of* $\mathfrak{A}_{\mathrm{bos}}(M_3, S_3, P_3)$ *commute.*

(ii) ***(Time slice axiom)*** *Let* (M_j, S_j, P_j) *be objects in* $\mathsf{GlobHypGreen}$, $j = 1, 2$, *and* (f, F) *a morphism from* (M_1, S_1, P_1) *to* (M_2, S_2, P_2) *such that there is a Cauchy hypersurface* $\Sigma \subset M_1$ *for which* $f(\Sigma)$ *is a Cauchy hypersurface of* M_2. *Then*

$$\mathfrak{A}_{\mathrm{bos}}(f, F) : \mathfrak{A}_{\mathrm{bos}}(M_1, S_1, P_1) \to \mathfrak{A}_{\mathrm{bos}}(M_2, S_2, P_2)$$

is an isomorphism.

Proof. We first show (i). For notational simplicity we assume without loss of generality that f_j and F_j are inclusions, $j = 1, 2$. Let $\varphi_j \in C_c^\infty(M_j, S_j)$. Since M_1 and M_2 are causally disjoint, the sections $G\varphi_1$ and φ_2 have disjoint support, thus

$$\omega([\varphi_1], [\varphi_2]) = \int_M \langle G\varphi_1, \varphi_2 \rangle \, \mathrm{dV} = 0.$$

Now relation (iv) in Definition A.11 tells us

$$w([\varphi_1]) \cdot w([\varphi_2]) = w([\varphi_1] + [\varphi_2]) = w([\varphi_2]) \cdot w([\varphi_1]).$$

Since $\mathfrak{A}_{\mathrm{bos}}(f_1, F_1)(\mathfrak{A}_{\mathrm{bos}}(M_1, S_1, P_1))$ is generated by elements of the form $w([\varphi_1])$ and $\mathfrak{A}_{\mathrm{bos}}(f_2, F_2)(\mathfrak{A}_{\mathrm{bos}}(M_2, S_2, P_2))$ by elements of the form $w([\varphi_2])$, the assertion follows.

In order to prove (ii) we show that SYMPL(f, F) is an isomorphism of symplectic vector spaces provided f maps a Cauchy hypersurface of M_1 onto a Cauchy hypersurface of M_2. Since symplectic linear maps are always injective, we only need to show surjectivity of SYMPL(f, F). This is most easily seen by replacing SYMPL(M_j, S_j, P_j) by $\ker(P_j) \cap C_{\mathrm{sc}}^\infty(M_j, S_j)$ as in Remark 3.9. Again we assume without loss of generality that f and F are inclusions.

Let $\psi \in C_{\mathrm{sc}}^\infty(M_2, S_2)$ be a solution of $P_2\psi = 0$. Let φ be the restriction of ψ to M_1. Then φ solves $P_1\varphi = 0$ and has spacelike compact support in M_1 by Lemma 3.11 below. We will show that there is only one solution in M_2 with spacelike compact support extending φ. It will then follow that ψ is the image of φ under the extension map corresponding to SYMPL(f, F) and surjectivity will be shown. □

To prove uniqueness of the extension, we may, by linearity, assume that $\varphi = 0$. Then ψ_+ defined by

$$\psi_+(x) := \begin{cases} \psi(x), & \text{if } x \in J_+^{M_2}(\Sigma), \\ 0, & \text{otherwise}, \end{cases}$$

is smooth since ψ vanishes in an open neighborhood of Σ. Now ψ_+ solves $P_2\psi_+ = 0$ and has past-compact support. By Remark 3.6, $\psi_+ \equiv 0$, i.e., ψ vanishes on $J_+^{M_2}(\Sigma)$. One shows similarly that ψ vanishes on $J_-^{M_2}(\Sigma)$, hence $\psi = 0$. □

Lemma 3.11. *Let M be a globally hyperbolic spacetime and let $M' \subset M$ be a causally compatible open subset which contains a Cauchy hypersurface of M. Let $A \subset M$ be spacelike compact in M.*

Then $A \cap M'$ is spacelike compact in M'.

Proof. Fix a common Cauchy hypersurface Σ of M' and M. By assumption, there exists a compact subset $K \subset M$ with $A \subset J^M(K)$. Then $K' := J^M(K) \cap \Sigma$ is compact [4, Cor. A.5.4] and contained in M'.

Moreover $A \subset J^M(K')$: let $p \in A$ and let γ be a causal curve (in M) from p to some $k \in K$. Then γ can be extended to an inextensible causal curve in M, which hence meets Σ at some point q. Because of $q \in \Sigma \cap J^M(k) \subset K'$ one has $p \in J^M(K')$.

Therefore $A \cap M' \subset J^M(K') \cap M' = J^{M'}(K')$ because of the causal compatibility of M' in M. The lemma is proved. □

The quantization process described in this subsection applies in particular to formally self-adjoint wave and Dirac-type operators.

3.2 Fermionic Quantization

Next we construct a fermionic quantization. For this we need a functorial construction of Hilbert spaces rather than symplectic vector spaces. As we shall see this seems to be possible only under much more restrictive assumptions. The underlying Lorentzian manifold M is assumed to be a globally hyperbolic spacetime as before. The vector bundle S is assumed to be complex with Hermitian inner product $\langle \cdot , \cdot \rangle$ which may be indefinite. The formally self-adjoint Green-hyperbolic operator P is assumed to be of first order.

Definition 3.12. A formally self-adjoint Green-hyperbolic operator P of first order acting on sections of a complex vector bundle S over a spacetime M is of *definite type* if and only if for any $x \in M$ and any future-directed timelike tangent vector

$\mathfrak{n} \in T_x M$, the bilinear map

$$S_x \times S_x \to \mathbb{C}, \qquad (\varphi, \psi) \mapsto \langle i\sigma_P(\mathfrak{n}^\flat) \cdot \varphi, \psi \rangle,$$

yields a positive definite Hermitian scalar product on S_x.

Example 3.13. The classical Dirac operator P from Example 2.21 is, when defined with the correct sign, of definite type, see e.g. [5, Sect. 1.1.5] or [3, Sect. 2].

Example 3.14. If $E \to M$ is a semi-Riemannian or -Hermitian vector bundle endowed with a metric connection over a spin spacetime M, then the twisted Dirac operator from Example 2.22 is of definite type if and only if the metric on E is positive definite. This can be seen by evaluating the tensorized inner product on elements of the form $\sigma \otimes v$, where $v \in E_x$ is null.

Example 3.15. The operator $P = i(d - \delta)$ on $S = \Lambda T^*M \otimes \mathbb{C}$ is of Dirac type but not of definite type. This follows from Example 3.14 applied to Example 2.23, since the natural inner product on ΣM is not positive definite. An alternative elementary proof is the following: for any timelike tangent vector $\mathfrak{n}$ on M and the corresponding covector $\mathfrak{n}^\flat$, one has

$$\langle i\sigma_P(\mathfrak{n}^\flat)\mathfrak{n}^\flat, \mathfrak{n}^\flat \rangle = -\langle \mathfrak{n}^\flat \wedge \mathfrak{n}^\flat - \mathfrak{n} \lrcorner \mathfrak{n}^\flat, \mathfrak{n}^\flat \rangle = \langle \mathfrak{n}, \mathfrak{n} \rangle \langle 1, \mathfrak{n}^\flat \rangle = 0.$$

Example 3.16. The Rarita–Schwinger operator defined in Sect. 2.6 is not of definite type if the dimension of the manifolds is $m \geq 3$. This can be seen as follows. Fix a point $x \in M$ and a pointwise orthonormal basis $(e_j)_{1 \leq j \leq m}$ of $T_x M$ with e_1 timelike. The Lorentzian metric induces inner products on ΣM and on $\Sigma^{3/2} M$ which we denote by $\langle \cdot, \cdot \rangle$. Choose $\xi := e_1^\flat \in T_x^* M$ and $\psi \in \Sigma_x^{3/2} M$. Since $\sigma_{\mathcal{Q}}(\xi)$ is pointwise obtained as the orthogonal projection of $\sigma_{\mathcal{D}}(\xi)$ onto $\Sigma_x^{3/2} M$, one has

$$\begin{aligned} \langle -i\sigma_{\mathcal{Q}}(\xi)\psi, \psi \rangle &= \langle (\mathrm{id} \otimes \xi^\# \cdot)\psi, \psi \rangle - \frac{2}{m} \underbrace{\sum_{\beta=1}^{m} \langle e_\beta^* \otimes e_\beta \cdot \psi_1, \psi \rangle}_{=0} \\ &= \sum_{\beta=1}^{m} \varepsilon_\beta \langle e_1 \cdot \psi_\beta, \psi_\beta \rangle. \end{aligned}$$

Choose, as in the proof of Lemma 2.26, a $\psi \in \Sigma_x^{3/2} M$ with $\psi_k = 0$ for all $3 \leq k \leq m$. For such a ψ the condition $\psi \in \Sigma_x^{3/2} M$ becomes $e_1 \cdot \psi_1 = e_2 \cdot \psi_2$. As in the proof of Lemma 2.26 we obtain

$$\langle -i\sigma_{\mathcal{Q}}(\xi)\psi, \psi \rangle = -\langle e_1 \cdot \psi_2, \psi_2 \rangle + \langle e_1 \cdot \psi_2, \psi_2 \rangle = 0,$$

which shows that the Rarita–Schwinger operator cannot be of definite type.

We define the category GlobHypDef, whose objects are the triples (M, S, P), where M is a globally hyperbolic spacetime, S is a complex vector bundle equipped with a complex inner product $\langle \cdot\,, \cdot \rangle$, and P is a formally self-adjoint Green-hyperbolic operator of definite type acting on sections of S. The morphisms are the same as in the category GlobHypGreen.

We construct a covariant functor from GlobHypDef to HILB, where HILB denotes the category whose objects are complex pre-Hilbert spaces and whose morphisms are isometric linear embeddings. As in Sect. 3.1, the underlying vector space is the space of classical solutions to the equation $P\varphi = 0$ with spacelike compact support. We put

$$\mathrm{SOL}(M, S, P) := \ker(P) \cap C^\infty_{\mathrm{sc}}(M, S).$$

Here "SOL" stands for classical solutions of the equation $P\varphi = 0$ with spacelike compact support.

Lemma 3.17. *Let (M, S, P) be an object in* GlobHypDef. *Let $\Sigma \subset M$ be a smooth spacelike Cauchy hypersurface with its future-oriented unit normal vector field $\mathfrak{n}$ and its induced volume element* dA. *Then*

$$(\varphi, \psi) := \int_\Sigma \langle i\sigma_P(\mathfrak{n}^\flat) \cdot \varphi_{|\Sigma}, \psi_{|\Sigma} \rangle \, \mathrm{dA}, \tag{8}$$

yields a positive definite Hermitian scalar product on $\mathrm{SOL}(M, S, P)$ *which does not depend on the choice of* Σ.

Proof. First note that $\mathrm{supp}(\varphi) \cap \Sigma$ is compact since $\mathrm{supp}(\varphi)$ is spacelike compact, so that the integral is well-defined. We have to show that it does not depend on the choice of Cauchy hypersurface. Let Σ' be any other smooth spacelike Cauchy hypersurface. Assume first that Σ and Σ' are disjoint and let Ω be the domain enclosed by Σ and Σ' in M. Its boundary is $\partial\Omega = \Sigma \cup \Sigma'$. Without loss of generality, one may assume that $\Sigma' \subset J^M_+(\Sigma)$. By the Green's formula [40, p. 160, Prop. 9.1] we have for all $\varphi, \psi \in C^\infty_{\mathrm{sc}}(M, S)$,

$$\int_\Omega (\langle P\varphi, \psi \rangle - \langle \varphi, P\psi \rangle) \, \mathrm{dV} = \int_{\Sigma'} \langle \sigma_P(\mathfrak{n}^\flat)\varphi, \psi \rangle \, \mathrm{dA} - \int_\Sigma \langle \sigma_P(\mathfrak{n}^\flat)\varphi, \psi \rangle \, \mathrm{dA}. \tag{9}$$

For $\varphi, \psi \in \mathrm{SOL}(M, S, P)$ we have $P\varphi = P\psi = 0$ and thus

$$0 = \int_\Sigma \langle \sigma_P(\mathfrak{n}^\flat)\varphi, \psi \rangle \, \mathrm{dA} - \int_{\Sigma'} \langle \sigma_P(\mathfrak{n}^\flat)\varphi, \psi \rangle \, \mathrm{dA}.$$

This shows the result in the case $\Sigma \cap \Sigma' = \emptyset$.

If $\Sigma \cap \Sigma' \neq \emptyset$ consider the subset $I^M_-(\Sigma) \cap I^M_-(\Sigma')$ of M where, as usual, $I^M_+(\Sigma)$ and $I^M_-(\Sigma)$ denote the chronological future and past of the subset Σ in

M, respectively. This subset is nonempty, open, and globally hyperbolic. This follows e.g. from [4, Lemma A.5.8]. Hence it admits a smooth spacelike Cauchy hypersurface Σ'' by Theorem 2.3. By construction, Σ'' meets neither Σ nor Σ' and it can be easily checked that Σ'' is also a Cauchy hypersurface of M. The result follows from the argument above being applied first to the pair (Σ, Σ'') and then to the pair (Σ'', Σ'). □

Remark 3.18. If one drops the assumption that P be of definite type, then the above sesquilinear form $(\cdot\,,\cdot)$ on $\ker(P)\cap C^\infty_{\mathrm{sc}}(M,S)$ still does not depend on the choice of Σ, however it need no longer be positive definite and can even be degenerate. Pick for instance the spin Dirac operator D_g associated to the underlying Lorentzian metric g on a spin spacetime M (see Example 2.21) and, keeping the spinor bundle $\Sigma_g M$ associated to g, change the metric on M so that the new metric g' has larger future and past cones at each point. Note that this implies that any globally hyperbolic subregion of (M, g') is also globally hyperbolic in (M, g). Then, denoting by D_g^* the formal adjoint of D_g with respect to the metric g', the operator $\begin{pmatrix} 0 & D_g \\ D_g^* & 0 \end{pmatrix}$ on $\Sigma_g M \oplus \Sigma_g M$ remains Green-hyperbolic but it fails to be of definite type, since there exist timelike vectors for g' which are lightlike for g. Hence the principal symbol of the operator becomes non-invertible and the bilinear form in (8) becomes degenerate for these g'-timelike covectors.

For any object (M, S, P) in GlobHypDef we will from now on equip $\mathrm{SOL}(M, S, P)$ with the Hermitian scalar product in (8) and thus turn $\mathrm{SOL}(M, S, P)$ into a pre-Hilbert space.

Given a morphism (f, F) from (M_1, S_1, P_1) to (M_2, S_2, P_2) in GlobHypDef, then this is also a morphism in GlobHypGreen and hence induces a homomorphism $\mathrm{SYMPL}(f, F) : \mathrm{SYMPL}(M_1, S_1, P_1) \to \mathrm{SYMPL}(M_2, S_2, P_2)$. As explained in Remark 3.9, there is a corresponding extension homomorphism $\mathrm{SOL}(f, F) : \mathrm{SOL}(M_1, S_1, P_1) \to \mathrm{SOL}(M_2, S_2, P_2)$. In other words, $\mathrm{SOL}(f, F)$ is defined such that the diagram

$$\begin{array}{ccc} \mathrm{SYMPL}(M_1, S_1, P_1) & \xrightarrow{\mathrm{SYMPL}(f,F)} & \mathrm{SYMPL}(M_2, S_2, P_2) \\ \cong\downarrow & & \downarrow\cong \\ \mathrm{SOL}(M_1, S_1, P_1) & \xrightarrow{\mathrm{SOL}(f,F)} & \mathrm{SOL}(M_2, S_2, P_2) \end{array} \tag{10}$$

commutes. The vertical arrows are the vector space isomorphisms induced be the Green's propagators G_1 and G_2, respectively.

Lemma 3.19. *The vector space homomorphism* $\mathrm{SOL}(f, F) : \mathrm{SOL}(M_1, S_1, P_1) \to \mathrm{SOL}(M_2, S_2, P_2)$ *preserves the scalar products, i.e., it is an isometric linear embedding of pre-Hilbert spaces.*

Proof. Without loss of generality we assume that f and F are inclusions. Let Σ_1 be a spacelike Cauchy hypersurface of M_1. Let $\varphi_1, \psi_1 \in C^\infty_{sc}(M_1, S_1)$. Denote the extension of φ_1 by $\varphi_2 := \mathrm{SOL}(f, F)(\varphi_1)$ and similarly for ψ_1.

Let $K_1 \subset M_1$ be a compact subset such that $\mathrm{supp}(\varphi_2) \subset J^{M_2}(K_1)$ and $\mathrm{supp}(\psi_2) \subset J^{M_2}(K_1)$. We choose a compact submanifold $K \subset \Sigma_1$ with boundary such that $J^{M_1}(K_1) \cap \Sigma_1 \subset K$. Since Σ_1 is a Cauchy hypersurface in M_1, $J^{M_1}(K_1) \subset J^{M_1}(J^{M_1}(K_1) \cap \Sigma_1) \subset J^{M_1}(K)$.

By Theorem 2.5 there is a spacelike Cauchy hypersurface $\Sigma_2 \subset M_2$ containing K. Since Σ_i is a Cauchy hypersurface of M_i (where $i = 1, 2$), it is met by every inextensible causal curve [30, Lemma 14.29]. Moreover, by definition of a Cauchy hypersurface, Σ_i is achronal in M_i. Since it is also spacelike, Σ_i is even acausal [30, Lemma 14.42]. In particular, it is met *exactly once* by every inextensible causal curve in M_i.

This implies $J^{M_2}(K_1) \subset J^{M_2}(K)$ (see Fig. 1): namely, pick $p \in J^{M_2}(K_1)$ and a causal curve γ in M_2 from p to some $k_1 \in K_1$. Extend γ to an inextensible causal curve $\overline{\gamma}$ in M_2. Then $\overline{\gamma}$ meets Σ_2 at some point q_2, because Σ_2 is a Cauchy hypersurface in M_2. But $\overline{\gamma} \cap M_1$ is also an inextensible causal curve in M_1, hence it intersects Σ_1 at a point q_1, which must lie in K by definition of K. Because of $K \subset \Sigma_2$ and the uniqueness of the intersection point, one has $q_1 = q_2$. In particular, $p \in J^{M_2}(K)$.

We conclude $\mathrm{supp}(\varphi_2) \subset J^{M_2}(K)$. Since $K \subset \Sigma_2$, we have $\mathrm{supp}(\varphi_2) \cap \Sigma_2 \subset J^{M_2}(K) \cap \Sigma_2$ and $J^{M_2}(K) \cap \Sigma_2 = K$ using the acausality of Σ_2. This shows $\mathrm{supp}(\varphi_2) \cap \Sigma_2 = \mathrm{supp}(\varphi_1) \cap \Sigma_1$ and similarly for ψ_2. Now we get

$$(\varphi_2, \psi_2) = \int_{\Sigma_2} \langle i\sigma_{P_2}(\mathfrak{n}^\flat) \cdot \varphi_2, \psi_2 \rangle \, \mathrm{dA} = \int_{\Sigma_1} \langle i\sigma_{P_1}(\mathfrak{n}^\flat) \cdot \varphi_1, \psi_1 \rangle \, \mathrm{dA} = (\varphi_1, \psi_1)$$

and the lemma is proved. □

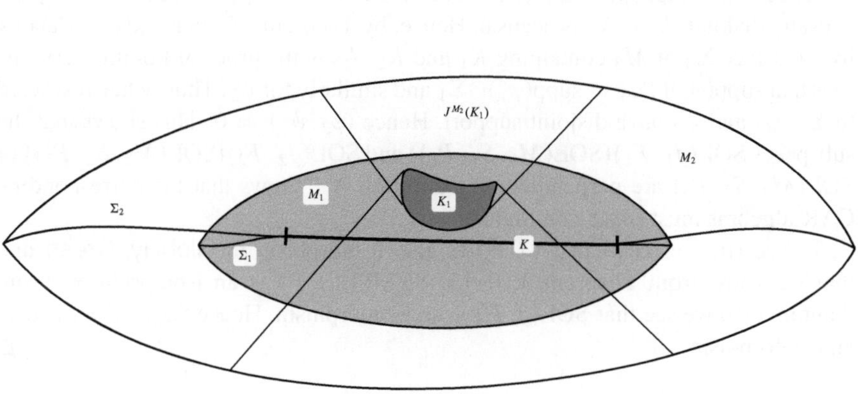

Fig. 1 $J^{M_2}(K_1) \subset J^{M_2}(K)$

The functoriality of SYMPL and diagram (10) show that SOL is a functor from GlobHypDef to HILB, the category of complex pre-Hilbert spaces with isometric linear embeddings. Composing with the functor CAR (see Sect. A.1), we obtain the covariant functor

$$\mathfrak{A}_{\mathrm{ferm}} := \mathrm{CAR} \circ \mathrm{SOL} : \mathsf{GlobHypDef} \longrightarrow \mathsf{C^*Alg}.$$

The fermionic algebras $\mathfrak{A}_{\mathrm{ferm}}(M, S, P)$ are actually $\mathbb{Z}_2$-graded algebras, see Proposition A.5 (A.5).

Theorem 3.20. *The functor* $\mathfrak{A}_{\mathrm{ferm}} : \mathsf{GlobHypDef} \longrightarrow \mathsf{C^*Alg}$ *is a fermionic locally covariant quantum field theory, i.e., the following axioms hold:*

(i) ***(Quantum causality)*** *Let* (M_j, S_j, P_j) *be objects in* GlobHypDef, $j = 1, 2, 3$, *and* (f_j, F_j) *morphisms from* (M_j, S_j, P_j) *to* (M_3, S_3, P_3), $j = 1, 2$, *such that* $f_1(M_1)$ *and* $f_2(M_2)$ *are causally disjoint regions in* M_3.
Then the subalgebras $\mathfrak{A}_{\mathrm{ferm}}(f_1, F_1)(\mathfrak{A}_{\mathrm{ferm}}(M_1, S_1, P_1))$ *and* $\mathfrak{A}_{\mathrm{ferm}}(f_2, F_2)(\mathfrak{A}_{\mathrm{ferm}}(M_2, S_2, P_2))$ *of* $\mathfrak{A}_{\mathrm{ferm}}(M_3, S_3, P_3)$ *super-commute*[1].

(ii) ***(Time slice axiom)*** *Let* (M_j, S_j, P_j) *be objects in* GlobHypDef, $j = 1, 2$, *and* (f, F) *a morphism from* (M_1, S_1, P_1) *to* (M_2, S_2, P_2) *such that there is a Cauchy hypersurface* $\Sigma \subset M_1$ *for which* $f(\Sigma)$ *is a Cauchy hypersurface of* M_2. *Then*

$$\mathfrak{A}_{\mathrm{ferm}}(f, F) : \mathfrak{A}_{\mathrm{ferm}}(M_1, S_1, P_1) \to \mathfrak{A}_{\mathrm{ferm}}(M_2, S_2, P_2)$$

is an isomorphism.

Proof. To show (i), we assume without loss of generality that f_j and F_j are inclusions. Let $\varphi_1 \in \mathrm{SOL}(M_1, S_1, P_1)$ and $\psi_1 \in \mathrm{SOL}(M_2, S_2, P_2)$. Denote the extensions to M_3 by $\varphi_2 := \mathrm{SOL}(f_1, F_1)(\varphi_1)$ and $\psi_2 := \mathrm{SOL}(f_2, F_2)(\psi_1)$. Choose a compact submanifold K_1 (with boundary) in a spacelike Cauchy hypersurface Σ_1 of M_1 such that $\mathrm{supp}(\varphi_1) \cap \Sigma_1 \subset K_1$ and similarly K_2 for ψ_1. Since M_1 and M_2 are causally disjoint, $K_1 \cup K_2$ is acausal. Hence, by Theorem 2.5, there exists a Cauchy hypersurface Σ_3 of M_3 containing K_1 and K_2. As in the proof of Lemma 3.19 one sees that $\mathrm{supp}(\varphi_2) \cap \Sigma_3 = \mathrm{supp}(\varphi_1) \cap \Sigma_1$ and similarly for ψ_2. Thus, when restricted to Σ_3, φ_2 and ψ_2 have disjoint support. Hence $(\varphi_2, \psi_2) = 0$. This shows that the subspaces $\mathrm{SOL}(f_1, F_1)(\mathrm{SOL}(M_1, S_1, P_1))$ and $\mathrm{SOL}(f_2, F_2)(\mathrm{SOL}(M_2, S_2, P_2))$ of $\mathrm{SOL}(M_3, S_3, P_3)$ are perpendicular. Definition A.1 shows that the corresponding CAR-algebras must super-commute.

To see (ii) we recall that (f, F) is also a morphism in GlobHypGreen and that we know from Theorem 3.10 that $\mathrm{SYMPL}(f, F)$ is an isomorphism. From diagram (10) we see that $\mathrm{SOL}(f, F)$ is an isomorphism. Hence $\mathfrak{A}_{\mathrm{ferm}}(f, F)$ is also an isomorphism. □

[1]This means that the odd parts of the algebras anti-commute while the even parts commute with everything.

Remark 3.21. Since causally disjoint regions should lead to commuting observables also in the fermionic case, one usually considers only the even part $\mathfrak{A}^{\mathrm{even}}_{\mathrm{ferm}}(M,S,P)$ (or a subalgebra thereof) as the observable algebra while the full algebra $\mathfrak{A}_{\mathrm{ferm}}(M,S,P)$ is called the field algebra.

There is a slightly different description of the functor $\mathfrak{A}_{\mathrm{ferm}}$. Let $\mathsf{HILB}_{\mathbb{R}}$ denote the category whose objects are the real pre-Hilbert spaces and whose morphisms are the isometric linear embeddings. We have the functor REAL : $\mathsf{HILB} \to \mathsf{HILB}_{\mathbb{R}}$ which associates to each complex pre-Hilbert space $(V,(\cdot\,,\cdot))$ its underlying real pre-Hilbert space $(V,\mathfrak{Re}(\cdot\,,\cdot))$. By Remark A.10,

$$\mathfrak{A}_{\mathrm{ferm}} = \mathrm{CAR}_{\mathrm{sd}} \circ \mathrm{REAL} \circ \mathrm{SOL}.$$

Since the self-dual CAR-algebra of a real pre-Hilbert space is the Clifford algebra of its complexification and since for any complex pre-Hilbert space V we have

$$\mathrm{REAL}(V) \otimes_{\mathbb{R}} \mathbb{C} = V \oplus V^*,$$

$\mathfrak{A}_{\mathrm{ferm}}(M,S,P)$ is also the Clifford algebra of $\mathrm{SOL}(M,S,P) \oplus \mathrm{SOL}(M,S,P)^* = \mathrm{SOL}(M,S\oplus S^*,P\oplus P^*)$. This is the way this functor is often described in the physics literature, see e.g. [39, p. 115f].

Self-dual CAR-representations are more natural for real fields. Let M be globally hyperbolic and let $S \to M$ be a real vector bundle equipped with a real inner product $\langle\cdot\,,\cdot\rangle$. A formally skew-adjoint[2] differential operator P acting on sections of S is called of definite type if and only if for any $x \in M$ and any future-directed timelike tangent vector $\mathfrak{n} \in T_xM$, the bilinear map

$$S_x \times S_x \to \mathbb{R}, \qquad (\varphi,\psi) \mapsto \langle \sigma_P(\mathfrak{n}^\flat)\cdot\varphi,\psi\rangle,$$

yields a positive definite Euclidean scalar product on S_x. An example is given by the real Dirac operator

$$D := \sum_{j=1}^{m} \varepsilon_j e_j \cdot \nabla_{e_j}$$

acting on sections of the real spinor bundle $\Sigma^{\mathbb{R}}M$.

Given a smooth spacelike Cauchy hypersurface $\Sigma \subset M$ with future-directed timelike unit normal field $\mathfrak{n}$, we define a scalar product on $\mathrm{SOL}(M,S,P) = \ker(P) \cap C^\infty_{\mathrm{sc}}(M,S,P)$ by

$$(\varphi,\psi) := \int_\Sigma \langle \sigma_P(\mathfrak{n}^\flat)\cdot\varphi_{|\Sigma},\psi_{|\Sigma}\rangle \,\mathrm{dA}.$$

[2] instead of self-adjoint!

With essentially the same proofs as before, one sees that this scalar product does not depend on the choice of Cauchy hypersurface Σ and that a morphism $(f, F) : (M_1, S_1, P_1) \to (M_2, S_2, P_2)$ gives rise to an extension operator $\mathrm{SOL}(f, F) : \mathrm{SOL}(M_1, S_1, P_1) \to \mathrm{SOL}(M_2, S_2, P_2)$ preserving the scalar product. We have constructed a functor

$$\mathrm{SOL} : \mathsf{GlobHypSkewDef} \longrightarrow \mathsf{HILB}_{\mathbb{R}},$$

where $\mathsf{GlobHypSkewDef}$ denotes the category whose objects are triples (M, S, P) with M globally hyperbolic, $S \to M$ a real vector bundle with real inner product and P a formally skew-adjoint, Green-hyperbolic differential operator of definite type acting on sections of S. The morphisms are the same as before.

Now the functor

$$\mathfrak{A}^{\mathrm{sd}}_{\mathrm{ferm}} := \mathrm{CAR}_{\mathrm{sd}} \circ \mathrm{SOL} : \mathsf{GlobHypSkewDef} \longrightarrow \mathsf{C^*Alg}$$

is a locally covariant quantum field theory in the sense that Theorem 3.20 holds with $\mathfrak{A}_{\mathrm{ferm}}$ replaced by $\mathfrak{A}^{\mathrm{sd}}_{\mathrm{ferm}}$.

4 States and Quantum Fields

In order to produce numbers out of our quantum field theory that can be compared to experiments, we need states, in addition to observables. We briefly recall the relation between states and representations via the GNS-construction. Then we show how the choice of a state gives rise to quantum fields and n-point functions.

4.1 States and Representations

Recall that a *state* on a unital C*-algebra A is a linear functional $\tau : A \to \mathbb{C}$ such that

(i) τ is positive, i.e., $\tau(a^*a) \geq 0$ for all $a \in A$;
(ii) τ is normed, i.e., $\tau(1) = 1$.

One checks that for any state the sesquilinear form $A \times A \to \mathbb{C}$, $(a, b) \mapsto \tau(b^*a)$, is a positive semi-definite Hermitian product and $|\tau(a)| \leq \|a\|$ for all $a \in A$. In particular, τ is continuous.

Any state induces a representation of A. Namely, the sesquilinear form $\tau(b^*a)$ induces a scalar product $\langle\cdot,\cdot\rangle_\tau$ on $A/\{a \in A \mid \tau(a^*a) = 0\}$. The Hilbert space completion of $A/\{a \in A \mid \tau(a^*a) = 0\}$ is denoted by $\mathscr{H}_\tau$. The action of A on $\mathscr{H}_\tau$ is induced by the multiplication in A,

$$\pi_\tau(a)[b]_\tau := [ab]_\tau,$$

where $[a]_\tau$ denotes the residue class of $a \in A$ in $A/\{a \in A \mid \tau(a^*a) = 0\}$. This representation is known as the *GNS-representation* induced by τ. The residue class $\Omega_\tau := [1]_\tau \in \mathscr{H}_\tau$ is called the *vacuum vector*. By construction, it is a cyclic vector, i.e., the orbit $\pi_\tau(A) \cdot \Omega_\tau = A/\{a \in A \mid \tau(a^*a) = 0\}$ is dense in $\mathscr{H}_\tau$.

The GNS-representation together with the vacuum vector allows to reconstruct the state since

$$\tau(a) = \tau(1^*a1) = \langle \pi_\tau(a)\Omega_\tau, \Omega_\tau \rangle_\tau. \tag{11}$$

If we look at the vector state $\tilde{\tau} : \mathscr{L}(\mathscr{H}_\tau) \to \mathbb{C}$, $\tilde{\tau}(\tilde{a}) = \langle \tilde{a}\,\Omega_\tau, \Omega_\tau \rangle_\tau$, on the C*-algebra $\mathscr{L}(\mathscr{H}_\tau)$ of bounded linear operators on $\mathscr{H}_\tau$, then (11) says that the diagram

$$\begin{array}{ccc} A & \xrightarrow{\ \pi_\tau\ } & \mathscr{L}(\mathscr{H}_\tau) \\ & {\scriptstyle\tau}\searrow \quad \swarrow{\scriptstyle\tilde{\tau}} & \\ & \mathbb{C} & \end{array}$$

commutes. One checks that $\|\pi_\tau\| \leq 1$, see [2, p. 20]. In particular, $\pi_\tau : A \to \mathscr{L}(\mathscr{H}_\tau)$ is continuous.

See e.g. [2, Sect. 1.4] or [9, Sect. 2.3] for details on states and representations of C*-algebras.

4.2 *Bosonic Quantum Field*

Now let (M, S, P) be an object in GlobHypGreen and τ a state on the corresponding bosonic algebra $\mathfrak{A}_{\mathrm{bos}}(M, S, P)$. Intuitively, the quantum field should be an operator-valued distribution Φ on M such that

$$e^{i\Phi(f)} = w([f])$$

for all test sections $f \in C_c^\infty(M, S)$. Here $[f]$ denotes the residue class in $\mathrm{SYMPL}(M, S, P) = C_c^\infty(M, S)/\ker G$ and $w : \mathrm{SYMPL}(M, S, P) \to \mathfrak{A}_{\mathrm{bos}}(M, S, P)$ is as in Definition A.11. This suggests the definition

$$\Phi(f) := -i \left.\frac{d}{dt}\right|_{t=0} w(t[f]).$$

The problem is that w is highly discontinuous so that this derivative does not make sense. This is where states and representations come into the play. We call a state τ on $\mathfrak{A}_{\mathrm{bos}}(M, S, P)$ regular if for each $f \in C_c^\infty(M, S)$ and each $h \in \mathscr{H}_\tau$ the map

$t \mapsto \pi_\tau(w(t[f]))h$ is continuous. Then $t \mapsto \pi_\tau(w(t[f]))$ is a strongly continuous one-parameter unitary group for any $f \in C_c^\infty(M, S)$ because

$$\begin{aligned}\pi_\tau(w((t+s)[f])) &= \pi_\tau(e^{i\omega(t[f],s[f])/2} w(t[f])w(s[f])) \\ &= \pi_\tau(w(t[f]))\pi_\tau(w(s[f])).\end{aligned}$$

Here we used Definition A.11 (iv) and the fact that ω is skew-symmetric so that $\omega(t[f], s[f]) = 0$. By Stone's theorem [34, Thm. VIII.8] this one-parameter group has a unique infinitesimal generator, i.e., a self-adjoint, generally unbounded operator $\Phi_\tau(f)$ on $\mathscr{H}_\tau$ such that

$$e^{it\Phi_\tau(f)} = \pi_\tau(w(t[f])).$$

For all h in the domain of $\Phi_\tau(f)$ we have

$$\Phi_\tau(f)h = -i \left.\frac{d}{dt}\right|_{t=0} \pi_\tau(w(t[f]))h.$$

We call the operator-valued map $f \mapsto \Phi_\tau(f)$ the quantum field corresponding to τ.

Definition 4.1. A regular state τ on $\mathfrak{A}_{\mathrm{bos}}(M, S, P)$ is called strongly regular if

(i) there is a dense subspace $\mathscr{D}_\tau \subset \mathscr{H}_\tau$ contained in the domain of $\Phi_\tau(f)$ for any $f \in C_c^\infty(M, S)$;
(ii) $\Phi_\tau(f)(\mathscr{D}_\tau) \subset \mathscr{D}_\tau$ for any $f \in C_c^\infty(M, S)$;
(iii) the map $C_c^\infty(M, S) \to \mathscr{H}_\tau$, $f \mapsto \Phi_\tau(f)h$, is continuous for every fixed $h \in \mathscr{D}_\tau$.

For a strongly regular state τ we have for all $f, g \in C_c^\infty(M, S)$, $\alpha, \beta \in \mathbb{R}$ and $h \in \mathscr{D}_\tau$:

$$\begin{aligned}\Phi_\tau(\alpha f + \beta g)h &= -i \left.\frac{d}{dt}\right|_{t=0} \pi_\tau(w(t[\alpha f + \beta g]))h \\ &= -i \left.\frac{d}{dt}\right|_{t=0} \left\{e^{i\alpha\beta t^2\omega([f],[g])/2} \pi_\tau(w(\alpha t[f]))\pi_\tau(w(\beta t[g]))h\right\} \\ &= -i \left.\frac{d}{dt}\right|_{t=0} \pi_\tau(w(\alpha t[f]))h - i \left.\frac{d}{dt}\right|_{t=0} \pi_\tau(w(\beta t[g]))h \\ &= \alpha\Phi_\tau(f)h + \beta\Phi_\tau(g)h.\end{aligned}$$

Hence $\Phi_\tau(f)$ depends linearly on f. The quantum field Φ_τ is therefore a distribution on M with values in self-adjoint operators on $\mathscr{H}_\tau$.

The n-point functions are defined by

$$\begin{aligned}\tau_n(f_1,\dots,f_n) &:= \langle \Phi_\tau(f_1)\cdots\Phi_\tau(f_n)\Omega_\tau,\Omega_\tau\rangle_\tau\\ &= \tilde{\tau}\,(\Phi_\tau(f_1)\cdots\Phi_\tau(f_n))\end{aligned}$$

$$\begin{aligned}&= \tilde{\tau}\left(\left(-i\,\frac{d}{dt_1}\Big|_{t_1=0}\pi_\tau(w(t_1[f_1]))\right)\cdots\left(-i\,\frac{d}{dt_n}\Big|_{t_n=0}\pi_\tau(w(t_n[f_n]))\right)\right)\\ &= (-i)^n\,\frac{\partial^n}{\partial t_1\cdots\partial t_n}\Big|_{t_1=\cdots=t_n=0}\tilde{\tau}\,(\pi_\tau(w(t_1[f_1]))\cdots\pi_\tau(w(t_n[f_n])))\\ &= (-i)^n\,\frac{\partial^n}{\partial t_1\cdots\partial t_n}\Big|_{t_1=\cdots=t_n=0}\tilde{\tau}\,(\pi_\tau(w(t_1[f_1])\cdots w(t_n[f_n])))\\ &= (-i)^n\,\frac{\partial^n}{\partial t_1\cdots\partial t_n}\Big|_{t_1=\cdots=t_n=0}\tau\,(w(t_1[f_1])\cdots w(t_n[f_n]))\,.\end{aligned}$$

For a strongly regular state τ the n-point functions are continuous separately in each factor. By the Schwartz kernel theorem [23, Thm. 5.2.1] the n-point function τ_n extends uniquely to a distribution on $M\times\cdots\times M$ (n times) in the following sense: Let $S^*\boxtimes\cdots\boxtimes S^*$ be the bundle over $M\times\cdots\times M$ whose fiber over $(x_1,\dots,x_n)$ is given by $S^*_{x_1}\otimes\cdots\otimes S^*_{x_n}$. Then there is a unique distribution on $M\times\cdots\times M$ in the bundle $S^*\boxtimes\cdots\boxtimes S^*$, again denoted τ_n, such that for all $f_j\in C^\infty_c(M,S)$,

$$\tau_n(f_1,\dots,f_n) = \tau_n(f_1\otimes\cdots\otimes f_n),$$

where $(f_1\otimes\cdots\otimes f_n)(x_1,\dots,x_n) := f_1(x_1)\otimes\cdots\otimes f_n(x_n)$.

Theorem 4.2. *Let (M,S,P) be an object in* GlobHypGreen *and τ a strongly regular state on the corresponding bosonic algebra $\mathfrak{A}_{\mathrm{bos}}(M,S,P)$. Then*

(i) $P\Phi_\tau = 0$ and $P\tau_n(f_1,\dots,f_{j-1},\cdot,f_{j+1},\dots,f_n) = 0$ hold in the distributional sense where $f_k\in C^\infty_c(M,S)$, $k\neq j$, are fixed;

(ii) the quantum field satisfies the canonical commutation relations, i.e.,

$$[\Phi_\tau(f),\Phi_\tau(g)]h = i\int_M \langle Gf,g\rangle\,\mathrm{dV}\cdot h$$

for all $f,g\in C^\infty_c(M,S)$ and $h\in\mathscr{D}_\tau$;

(iii) the n-point functions satisfy the canonical commutation relations, i.e.,

$$\begin{aligned}&\tau_{n+2}(f_1,\dots,f_{j-1},f_j,f_{j+1},\dots,f_{n+2})\\ &\quad-\tau_{n+2}(f_1,\dots,f_{j-1},f_{j+1},f_j,f_{j+2},\dots,f_{n+2})\\ &= i\int_M\langle Gf_j,f_{j+1}\rangle\,\mathrm{dV}\cdot\tau_n(f_1,\dots,f_{j-1},f_{j+2},\dots,f_{n+2})\end{aligned}$$

for all $f_1, \dots, f_{n+2} \in C_c^\infty(M, S)$.

Proof. Since P is formally self-adjoint and $GPf = 0$ for any $f \in C_c^\infty(M, S)$, we have for any $h \in \mathscr{D}_\tau$:

$$(P\Phi_\tau)(f)h = \Phi_\tau(Pf)h = -i\left.\frac{d}{dt}\right|_{t=0} \pi_\tau(w(t\underbrace{[Pf]}_{=0}))h = -i\left.\frac{d}{dt}\right|_{t=0} h = 0.$$

This shows $P\Phi_\tau = 0$. The result for the n-point functions follows and (i) is proved.

To show (ii) we observe that by Definition A.11 (iv) we have on the one hand

$$w([f+g]) = e^{i\omega([f],[g])/2}w([f])w([g]),$$

and on the other hand

$$w([f+g]) = e^{i\omega([g],[f])/2}w([g])w([f]),$$

hence

$$w([f])w([g]) = e^{-i\omega([f],[g])}w([g])w([f]).$$

Thus

$$\begin{aligned}
\Phi_\tau(f)\Phi_\tau(g)h &= -\left.\frac{\partial^2}{\partial t\,\partial s}\right|_{t=s=0} \pi_\tau(w(t[f])w(s[g]))h \\
&= -\left.\frac{\partial^2}{\partial t\,\partial s}\right|_{t=s=0} \pi_\tau(e^{-i\omega(t[f],s[g])}w(s[g])w(t[f]))h \\
&= -\left.\frac{\partial^2}{\partial t\,\partial s}\right|_{t=s=0} \left\{e^{-i\omega(t[f],s[g])}\cdot \pi_\tau(w(s[g])w(t[f]))h\right\} \\
&= i\omega([f],[g])h + \Phi_\tau(g)\Phi_\tau(f)h \\
&= i\int_M \langle Gf, g\rangle \,\mathrm{dV}\cdot h + \Phi_\tau(g)\Phi_\tau(f)h.
\end{aligned}$$

This shows (ii). Assertion (iii) follows from (ii). □

Remark 4.3. As a consequence of the canonical commutation relations we get

$$[\Phi_\tau(f), \Phi_\tau(g)] = 0$$

if the supports of f and g are causally disjoint, i.e., if there is no causal curve from $\mathrm{supp}(f)$ to $\mathrm{supp}(g)$. The reason is that in this case the supports of Gf and g are disjoint. A similar remark holds for the n-point functions.

Remark 4.4. In the physics literature one also finds the statement $\Phi(\overline{f}) = \Phi(f)^*$. This simply expresses the fact that we are dealing with a theory over the reals. We have encoded this by considering real vector bundles S, see Definition 3.1, and the fact that $\Phi_\tau(f)$ is always self-adjoint.

4.3 Fermionic Quantum Fields

Let (M, S, P) be an object in GlobHypDef and let τ be a state on the fermionic algebra $\mathfrak{A}_{\mathrm{ferm}}(M, S, P)$. For $f \in C_c^\infty(M, S)$ we put

$$\Phi_\tau(f) := -\pi_\tau(\mathbf{a}(Gf)^*),$$
$$\Phi_\tau^+(f) := \pi_\tau(\mathbf{a}(Gf)),$$

where $\mathbf{a}$ is as in Definition A.1 (compare [18, Sect. III.B, p. 141]). Since π_τ, $\mathbf{a}$, and G are sequentially continuous (for G see [4, Prop. 3.4.8]), so are Φ_τ and Φ_τ^+. In contrast to the bosonic case, no regularity assumption on τ is needed. Hence Φ_τ and Φ_τ^+ are distributions on M with values in the space of bounded operators on $\mathscr{H}_\tau$. Note that Φ_τ is linear while Φ_τ^+ is anti-linear.

Theorem 4.5. *Let (M, S, P) be an object in* GlobHypDef *and τ a state on the corresponding fermionic algebra $\mathfrak{A}_{\mathrm{ferm}}(M, S, P)$. Then*

(i) $P\Phi_\tau = P\Phi_\tau^+ = 0$ holds in the distributional sense;
(ii) the quantum fields satisfy the canonical anti-commutation relations, i.e.,

$$\{\Phi_\tau(f), \Phi_\tau(g)\} = \{\Phi_\tau^+(f), \Phi_\tau^+(g)\} = 0,$$
$$\{\Phi_\tau(f), \Phi_\tau^+(g)\} = i\Big(\int_M \langle Gf, g\rangle \, \mathrm{dV}\Big) \cdot \mathrm{id}_{\mathscr{H}_\tau}$$

for all $f, g \in C_c^\infty(M, S)$.

Proof. Since $GP = 0$ on $C_c^\infty(M, S)$, we have $P\Phi_\tau(f) = \Phi_\tau(Pf) = -\pi_\tau(\mathbf{a}(GPf)^*) = 0$ and similarly for Φ_τ^+. This proves assertion (i).

Using Definition A.1 (ii) we compute

$$\begin{aligned}\{\Phi_\tau(f), \Phi_\tau(g)\} &= \{\pi_\tau(\mathbf{a}(Gf)^*), \pi_\tau(\mathbf{a}(Gg)^*)\}\\ &= \pi_\tau(\{\mathbf{a}(Gf)^*, \mathbf{a}(Gg)^*\})\\ &= \pi_\tau(\{\mathbf{a}(Gg), \mathbf{a}(Gf)\}^*)\\ &= 0.\end{aligned}$$

Similarly one sees $\{\Phi_\tau^+(f), \Phi_\tau^+(g)\} = 0$. Definition A.1 (iii) also yields

$$\{\Phi_\tau(f), \Phi_\tau^+(g)\} = -\pi_\tau(\{\mathbf{a}(Gf)^*, \mathbf{a}(Gg)\}) = -(Gf, Gg) \cdot \mathrm{id}_{\mathscr{H}_\tau}.$$

To prove assertion (ii) we have to verify

$$(Gf, Gg) = -i \int_M \langle Gf, g \rangle \, \mathrm{dV} \tag{12}$$

Let $\Sigma \subset M$ be a smooth spacelike Cauchy hypersurface. Since $\mathrm{supp}(G_+ g)$ is past-compact, we can find a Cauchy hypersurface $\Sigma' \subset M$ in the past of Σ which does not intersect $\mathrm{supp}(G_+ g) \subset J_+^M(\mathrm{supp}(g))$. Denote the region between Σ and Σ' by Ω'. The Green's formula (9) yields

$$\begin{aligned}
(Gf, G_+ g) &= \int_\Sigma \langle i\sigma_P(\mathrm{n}^\flat) \cdot Gf, G_+ g \rangle \, \mathrm{dA} \\
&= \int_{\Sigma'} \langle i\sigma_P(\mathrm{n}^\flat) \cdot Gf, G_+ g \rangle \, \mathrm{dA} + i \int_{\Omega'} (\langle PGf, G_+ g \rangle - \langle Gf, PG_+ g \rangle) \, \mathrm{dV} \\
&= -i \int_{\Omega'} \langle Gf, g \rangle \, \mathrm{dV}
\end{aligned}$$

because $PG_+ g = g$ and $PGf = 0$. Since Σ' can be chosen arbitrarily to the past, this shows

$$(Gf, G_+ g) = -i \int_{J_-(\Sigma)} \langle Gf, g \rangle \, \mathrm{dV}. \tag{13}$$

A similar computation yields

$$(Gf, G_- g) = i \int_{J_+(\Sigma)} \langle Gf, g \rangle \, \mathrm{dV}. \tag{14}$$

Subtracting (14) from (13) yields (12) and concludes the proof of assertion (ii). □

Remark 4.6. Similarly to the bosonic case, we find

$$\{\Phi_\tau(f), \Phi_\tau^+(g)\} = 0,$$

if the supports of f and g are causally disjoint.

Remark 4.7. Using the anti-commutation relations in Theorem 4.5 (ii), the computation of n-point functions can be reduced to those of the form

$$\tau_{n,n'}(f_1, \dots, f_n, g_1, \dots, g_{n'}) = \langle \Omega_\tau, \Phi_\tau(f_1) \cdots \Phi_\tau(f_n) \Phi_\tau^+(g_1) \cdots \Phi_\tau^+(g_{n'}) \Omega_\tau \rangle_\tau.$$

As in the bosonic case, the n-point functions satisfy the field equation in the distributional sense in each argument and extend to distributions on $M \times \cdots \times M$.

If one uses the self-dual fermionic algebra $\mathfrak{A}^{\mathrm{sd}}_{\mathrm{ferm}}(M,S,P)$ instead of $\mathfrak{A}_{\mathrm{ferm}}(M,S,P)$, then one gets the quantum field

$$\Psi_\tau(f) := \pi_\tau(\mathbf{b}(Gf))$$

where $\mathbf{b}$ is as in Definition A.6. Then the analog to Theorem 4.5 is

Theorem 4.8. *Let (M,S,P) be an object in* GlobHypSkewDef *and τ a state on the corresponding self-dual fermionic algebra $\mathfrak{A}^{\mathrm{sd}}_{\mathrm{ferm}}(M,S,P)$. Then*

(i) $P\Psi_\tau = 0$ holds in the distributional sense;
(ii) the quantum field takes values in self-adjoint operators, $\Psi_\tau(f) = \Psi_\tau(f)^*$ for all $f \in C^\infty_c(M,S)$;
(iii) the quantum fields satisfy the canonical anti-commutation relations, i.e.,

$$\{\Psi_\tau(f), \Psi_\tau(g)\} = \int_M \langle Gf, g\rangle \,\mathrm{dV} \cdot \mathrm{id}_{\mathscr{H}_\tau}$$

for all $f, g \in C^\infty_c(M,S)$.

Remark 4.9. It is interesting to compare the concept of locally covariant quantum field theories as proposed in [11] to the axiomatic approach to quantum field theory on Minkowski space based on the Gårding–Wightman axioms as exposed in [35, Sect. IX.8]. Property 1 (relativistic invariance of states) and Property 6 (Poincaré invariance of the field) in [35] are replaced by functoriality (covariance). Property 4 (invariant domain for fields) and Property 5 (regularity of the field) have been encoded in strong regularity of the state used to define the quantum field in the bosonic case and are automatic in the fermionic case. Property 7 (local commutativity or microscopic causality) is contained in Theorems 4.2 and 4.5. Property 3 (existence and uniqueness of the vacuum) has no analog and is replaced by the *choice* of a state. Property 8 (cyclicity of the vacuum) is then automatic by the general properties of the GNS-construction.

There remains one axiom, Property 2 (spectral condition), which we have not discussed at all. It gets replaced by the Hadamard condition on the state chosen. It was observed by Radzikowski [32] that earlier formulations of this condition are equivalent to a condition on the wave front set of the 2-point function. Much work has been put into constructing and investigating Hadamard states for various examples of fields, see e.g. [15, 16, 19, 25, 36–38, 42] and the references therein.

Acknowledgements It is a pleasure to thank Alexander Strohmaier and Rainer Verch for very valuable discussion. The authors would also like to thank SPP 1154 "Globale Differentialgeometrie" and SFB 647 "Raum-Zeit-Materie", both funded by Deutsche Forschungsgemeinschaft, for financial support.

Appendix A. Algebras of Canonical (Anti-) Commutation Relations

We collect the necessary algebraic facts about CAR and CCR-algebras.

A.1 CAR Algebras

The symbol "CAR" stands for "canonical anti-commutation relations". These algebras are related to pre-Hilbert spaces. We always assume the Hermitian inner product $(\cdot\,,\cdot)$ to be linear in the first argument and anti-linear in the second.

Definition A.1. A CAR-*representation* of a complex pre-Hilbert space $(V, (\cdot\,,\cdot))$ is a pair $(\mathbf{a}, A)$, where A is a unital C*-algebra and $\mathbf{a} : V \to A$ is an anti-linear map satisfying:

(i) $A = C^*(\mathbf{a}(V))$,
(ii) $\{\mathbf{a}(v_1), \mathbf{a}(v_2)\} = 0$ and
(iii) $\{\mathbf{a}(v_1)^*, \mathbf{a}(v_2)\} = (v_1, v_2) \cdot 1$,

for all $v_1, v_2 \in V$.

We want to discuss CAR-representations in terms of C*-Clifford algebras, whose definition we recall. Given a complex pre-Hilbert vector space $(V, (\cdot\,,\cdot))$, we denote by $V_{\mathbb{C}} := V \otimes_{\mathbb{R}} \mathbb{C}$ the complexification of V considered as a real vector space and by $q_{\mathbb{C}}$ the complex-bilinear extension of $\mathfrak{Re}(\cdot\,,\cdot)$ to $V_{\mathbb{C}}$. Let $\mathrm{Cl}_{\mathrm{alg}}(V_{\mathbb{C}}, q_{\mathbb{C}})$ be the algebraic Clifford algebra of $(V_{\mathbb{C}}, q_{\mathbb{C}})$. It is an associative complex algebra with unit and contains $V_{\mathbb{C}}$ as a vector subspace. Its multiplication is called Clifford multiplication and denoted by "$\cdot$". It satisfies the Clifford relations

$$v \cdot w + w \cdot v = -2q_{\mathbb{C}}(v, w)1 \tag{15}$$

for all $v, w \in V_{\mathbb{C}}$. Define the $*$-operator on $\mathrm{Cl}_{\mathrm{alg}}(V_{\mathbb{C}}, q_{\mathbb{C}})$ to be the unique anti-multiplicative and anti-linear extension of the anti-linear map $V_{\mathbb{C}} \to V_{\mathbb{C}}$, $v_1 + i v_2 \mapsto -\overline{(v_1 + i v_2)} = -(v_1 - i v_2)$ for all $v_1, v_2 \in V$. In other words,

$$*\left(\sum_{i_1<\ldots<i_k} \alpha_{i_1,\ldots,i_k} z_{i_1} \cdot \ldots \cdot z_{i_k}\right) = (-1)^k \sum_{i_1<\ldots<i_k} \overline{\alpha_{i_1,\ldots,i_k}} \cdot \overline{z_{i_k}} \cdot \ldots \cdot \overline{z_{i_1}}$$

for all $k \in \mathbb{N}$ and $z_{i_1}, \ldots, z_{i_k} \in V_{\mathbb{C}}$. Let $\|\cdot\|_\infty$ be defined by

$$\|a\|_\infty := \sup_{\pi \in \mathrm{Rep}(V)} (\|\pi(a)\|)$$

for every $a \in \mathrm{Cl}_{\mathrm{alg}}(V_{\mathbb{C}}, q_{\mathbb{C}})$, where $\mathrm{Rep}(V)$ denotes the set of all (isomorphism classes of) $*$-homomorphisms from $\mathrm{Cl}_{\mathrm{alg}}(V_{\mathbb{C}}, q_{\mathbb{C}})$ to C*-algebras. Then $\|\cdot\|_\infty$ can be shown to be a well-defined C*-norm on $\mathrm{Cl}_{\mathrm{alg}}(V_{\mathbb{C}}, q_{\mathbb{C}})$, see e.g. [31, Sect. 1.2].

Definition A.2. The C*-Clifford algebra of a pre-Hilbert space $(V, (\cdot\,,\cdot))$ is the C*-completion of $\mathrm{Cl}_{\mathrm{alg}}(V_{\mathbb{C}}, q_{\mathbb{C}})$ with respect to the C*-norm $\|\cdot\|_\infty$ and the star operator defined above.

Theorem A.3. *For every complex pre-Hilbert space $(V, (\cdot\,,\cdot))$, the C*-Clifford algebra $\mathrm{Cl}(V_{\mathbb{C}}, q_{\mathbb{C}})$ provides a CAR-representation of $(V, (\cdot\,,\cdot))$ via $\mathbf{a}(v) = \frac{1}{2}(v + iJv)$, where J is the complex structure of V.*

Moreover, CAR-representations have the following universal property: Let $\widehat{A}$ be any unital C-algebra and $\widehat{\mathbf{a}} : V \to \widehat{A}$ be any anti-linear map satisfying Axioms (ii) and (iii) of Definition A.1. Then there exists a unique C*-morphism $\tilde{\alpha} : \mathrm{Cl}(V_{\mathbb{C}}, q_{\mathbb{C}}) \to \widehat{A}$ such that*

$$\begin{array}{ccc} V & \xrightarrow{\ \widehat{\mathbf{a}}\ } & \widehat{A} \\ {\scriptstyle \mathbf{a}}\downarrow & \nearrow{\scriptstyle \tilde{\alpha}} & \\ \mathrm{Cl}(V_{\mathbb{C}}, q_{\mathbb{C}}) & & \end{array}$$

commutes. Furthermore, $\tilde{\alpha}$ is injective.

Proof. Define $p_\mp : V \to \mathrm{Cl}(V_{\mathbb{C}}, q_{\mathbb{C}})$ by $p_-(v) := \frac{1}{2}(v + iJv)$ and $p_+(v) := \frac{1}{2}(v - iJv)$. Since $p_-(Jv) = -ip_-(v)$, the map $\mathbf{a} = p_-$ is anti-linear. Because of $\mathbf{a}(v) - \mathbf{a}(v)^* = p_-(v) + p_+(v) = v$, the C*-subalgebra of $\mathrm{Cl}(V_{\mathbb{C}}, q_{\mathbb{C}})$ generated by the image of $\mathbf{a}$ contains V. Hence $\mathbf{a}(V)$ generates $\mathrm{Cl}(V_{\mathbb{C}}, q_{\mathbb{C}})$ as a C*-algebra. Axiom (A.1) in Definition A.1 is proved.

Let $v_1, v_2 \in V$, then

$$\begin{aligned} \{\mathbf{a}(v_1), \mathbf{a}(v_2)\} &= p_-(v_1) \cdot p_-(v_2) + p_-(v_2) \cdot p_-(v_1) \\ &= -2q_{\mathbb{C}}(p_-(v_1), p_-(v_2)) \cdot 1 \\ &= 0, \end{aligned}$$

which is Axiom (iii) in Definition A.1. Furthermore,

$$\begin{aligned} \{\mathbf{a}(v_1)^*, \mathbf{a}(v_2)\} &= -p_+(v_1) \cdot p_-(v_2) - p_-(v_2) \cdot p_+(v_1) \\ &= 2q_{\mathbb{C}}(p_+(v_1), p_-(v_2)) \cdot 1 \\ &= \mathfrak{Re}(v_1, v_2) \cdot 1 + i\mathfrak{Re}(v_1, Jv_2) \cdot 1 \\ &= (v_1, v_2) \cdot 1, \end{aligned}$$

which shows Axiom (iii). in Definition A.1. Therefore $(\mathbf{a}, \mathrm{Cl}(V_{\mathbb{C}}, q_{\mathbb{C}}))$ is a CAR-representation of $(V, (\cdot\,,\cdot))$.

The second part of the theorem follows from $\mathrm{Cl}(V_{\mathbb{C}}, q_{\mathbb{C}})$ being simple, i.e., from the non-existence of non-trivial closed two-sided $*$-invariant ideals, see [31, Thm. 1.2.2]. Let $\widehat{\mathbf{a}} : V \to \widehat{A}$ be any other anti-linear map satisfying (ii) and (iii) in Definition A.1. Since $\mathbf{a}$ and $\widehat{\mathbf{a}}$ are injective (which is clear by Axiom (iii)) one may set $\alpha(\mathbf{a}(v)) := \widehat{\mathbf{a}}(v)$ for all $v \in V$. Axioms (ii) and (iii) allow us to extend α to a C*-morphism $\tilde{\alpha} : C^*(\mathbf{a}(V)) = \mathrm{Cl}(V_{\mathbb{C}}, q_{\mathbb{C}}) \to \widehat{A}$. The injectivity of $\widehat{\mathbf{a}}$ implies the non-triviality of $\tilde{\alpha}$ which, together with the simplicity of $\mathrm{Cl}(V_{\mathbb{C}}, q_{\mathbb{C}})$, provides the injectivity of $\tilde{\alpha}$. Therefore we found an injective C*-morphism $\tilde{\alpha} : \mathrm{Cl}(V_{\mathbb{C}}, q_{\mathbb{C}}) \to \widehat{A}$ with $\tilde{\alpha} \circ \mathbf{a} = \widehat{\mathbf{a}}$. It is unique since it is determined by $\mathbf{a}$ and $\widehat{\mathbf{a}}$ on a subset of generators. This concludes the proof of Theorem A.3. □

For an alternative description of the CAR-representation in terms of creation and annihilation operators on the fermionic Fock space we refer to [9, Prop. 5.2.2].

Corollary A.4. *For every complex pre-Hilbert space $(V, (\cdot\,,\cdot))$ there exists a* CAR-*representation of $(V, (\cdot\,,\cdot))$, unique up to C*-isomorphism.*

Proof. The existence has already been proved in Theorem A.3. Let $(\widehat{\mathbf{a}}, \widehat{A})$ be any CAR-representation of $(V, (\cdot\,,\cdot))$. Theorem A.3 states the existence of a unique injective C*-morphism $\tilde{\alpha} : \mathrm{Cl}(V_{\mathbb{C}}, q_{\mathbb{C}}) \to \widehat{A}$ such that $\tilde{\alpha} \circ \mathbf{a} = \widehat{\mathbf{a}}$. Now $\tilde{\alpha}$ has to be surjective since Axiom (A.1) holds for $(\widehat{\mathbf{a}}, \widehat{A})$. □

From now on, given a complex pre-Hilbert space $(V, (\cdot\,,\cdot))$, we denote the C*-algebra $\mathrm{Cl}(V_{\mathbb{C}}, q_{\mathbb{C}})$ associated with the CAR-representation $(\mathbf{a}, \mathrm{Cl}(V_{\mathbb{C}}, q_{\mathbb{C}}))$ of $(V, (\cdot\,,\cdot))$ by $\mathrm{CAR}(V, (\cdot\,,\cdot))$. We list the properties of CAR-representations which are relevant for quantization, see also [9, Vol. II, Thm. 5.2.5, p. 15].

Proposition A.5. *Let $(V, (\cdot\,,\cdot))$ be a complex pre-Hilbert space and $(\mathbf{a}, \mathrm{CAR}(V, (\cdot\,,\cdot)))$ its* CAR-*representation.*

(i) For every $v \in V$ one has $\|\mathbf{a}(v)\| = |v| = (v, v)^{\frac{1}{2}}$, where $\|\cdot\|$ denotes the C-norm on $\mathrm{CAR}(V, (\cdot\,,\cdot))$.*
(ii) The C-algebra $\mathrm{CAR}(V, (\cdot\,,\cdot))$ is simple, i.e., it has no closed two-sided $*$-ideals other than $\{0\}$ and the algebra itself.*
(iii) The algebra $\mathrm{CAR}(V, (\cdot\,,\cdot))$ is $\mathbb{Z}_2$-graded,

$$\mathrm{CAR}(V, (\cdot\,,\cdot)) = \mathrm{CAR}^{\mathrm{even}}(V, (\cdot\,,\cdot)) \oplus \mathrm{CAR}^{\mathrm{odd}}(V, (\cdot\,,\cdot)),$$

and $\mathbf{a}(V) \subset \mathrm{CAR}^{\mathrm{odd}}(V, (\cdot\,,\cdot))$.
(iv) Let $f : V \to V'$ be an isometric linear embedding, where $(V', (\cdot\,,\cdot)')$ is another complex pre-Hilbert space. Then there exists a unique injective C-morphism $\mathrm{CAR}(f) : \mathrm{CAR}(V, (\cdot\,,\cdot)) \to \mathrm{CAR}(V', (\cdot\,,\cdot)')$ such that*

$$
\begin{array}{ccc}
V & \xrightarrow{\quad f \quad} & V' \\
\Big\downarrow{\scriptstyle \mathbf{a}} & & \Big\downarrow{\scriptstyle \mathbf{a}'} \\
\mathrm{CAR}(V,(\cdot\,,\cdot)) & \xrightarrow{\ \mathrm{CAR}(f)\ } & \mathrm{CAR}(V',(\cdot\,,\cdot)')
\end{array}
$$

commutes.

Proof. We show assertion (A.5) . On the one hand, the C^*-property of the norm $\|\cdot\|$ implies

$$\begin{aligned}\|\mathbf{a}(v)\|^4 &= \|\mathbf{a}(v)\mathbf{a}(v)^*\|^2 \\ &= \|(\mathbf{a}(v)\mathbf{a}(v)^*)^2\|.\end{aligned}$$

On the other hand,

$$\begin{aligned}(\mathbf{a}(v)\mathbf{a}(v)^*)^2 &= \mathbf{a}(v)\{\mathbf{a}(v)^*,\mathbf{a}(v)\}\mathbf{a}(v)^* \\ &= |v|^2\mathbf{a}(v)\mathbf{a}(v)^*,\end{aligned}$$

where we used $\mathbf{a}(v)^2 = 0$ which follows from the second axiom. We deduce that

$$\begin{aligned}\|\mathbf{a}(v)\|^4 &= |v|^2\cdot\|\mathbf{a}(v)\mathbf{a}(v)^*\| \\ &= |v|^2\cdot\|\mathbf{a}(v)\|^2.\end{aligned}$$

Since $\mathbf{a}$ is injective, we obtain the result.

Assertion (ii) follows from $\mathrm{Cl}(V_{\mathbb{C}}, q_{\mathbb{C}})$ being simple, see [31, Thm. 1.2.2]. Alternatively, it can be deduced from the universal property formulated in Theorem A.3.

To see (iii) we recall that the Clifford algebra $\mathrm{Cl}(V_{\mathbb{C}}, q_{\mathbb{C}})$ has a $\mathbb{Z}_2$-grading where the even part is generated by products of an even number of vectors in $V_{\mathbb{C}}$ and, similarly, the odd part is the vector space span of products of an odd number of vectors in $V_{\mathbb{C}}$, see [31, p. 27]. This is compatible with the Clifford relations (15). Clearly, $\mathbf{a}(V) \subset \mathrm{CAR}^{\mathrm{odd}}(V,(\cdot\,,\cdot))$.

It remains to show (iv). It is straightforward to check that $\mathbf{a}' \circ f$ satisfies Axioms (ii) and (iii) in Definition A.1. The result follows from Theorem A.3. □

One easily sees that $\mathrm{CAR}(\mathrm{id}) = \mathrm{id}$ and that $\mathrm{CAR}(f' \circ f) = \mathrm{CAR}(f') \circ \mathrm{CAR}(f)$ for all isometric linear embeddings $V \xrightarrow{f} V' \xrightarrow{f'} V''$. Therefore we have constructed a covariant functor

$$\mathrm{CAR} : \mathsf{HILB} \longrightarrow \mathsf{C^*Alg},$$

where HILB denotes the category whose objects are the complex pre-Hilbert spaces and whose morphisms are the isometric linear embeddings.

For real pre-Hilbert spaces there is the concept of self-dual CAR-representations.

Definition A.6. A *self-dual* CAR-*representation* of a real pre-Hilbert space $(V, (\cdot\,,\cdot))$ is a pair $(\mathbf{b}, A)$, where A is a unital C*-algebra and $\mathbf{b} : V \to A$ is an $\mathbb{R}$-linear map satisfying:

(i) $A = C^*(\mathbf{b}(V))$,
(ii) $\mathbf{b}(v) = \mathbf{b}(v)^*$ and
(iii) $\{\mathbf{b}(v_1), \mathbf{b}(v_2)\} = (v_1, v_2) \cdot 1$,

for all $v, v_1, v_2 \in V$.

Given a self-dual CAR-representation, one can extend $\mathbf{b}$ to a $\mathbb{C}$-linear map from the complexification $V_{\mathbb{C}}$ to A. This extension $\mathbf{b} : V_{\mathbb{C}} \to A$ then satisfies $\mathbf{b}(\bar{v}) = \mathbf{b}(v)^*$ and $\{\mathbf{b}(v_1), \mathbf{b}(v_2)\} = (v_1, \bar{v}_2) \cdot 1$ for all $v, v_1, v_2 \in V_{\mathbb{C}}$. These are the axioms of a self-dual CAR-representation as in [1, p. 386].

Theorem A.7. *For every real pre-Hilbert space $(V, (\cdot\,,\cdot))$, the C*-Clifford algebra* $\mathrm{Cl}(V_{\mathbb{C}}, q_{\mathbb{C}})$ *provides a self-dual* CAR-*representation of $(V, (\cdot\,,\cdot))$ via $\mathbf{b}(v) = \frac{i}{\sqrt{2}} v$.*

Moreover, self-dual CAR-*representations have the following universal property: Let $\widehat{A}$ be any unital C*-algebra and $\widehat{\mathbf{b}} : V \to \widehat{A}$ be any $\mathbb{R}$-linear map satisfying Axioms (ii) and (iii) of Definition A.6. Then there exists a unique C*-morphism $\tilde{\beta} : \mathrm{Cl}(V_{\mathbb{C}}, q_{\mathbb{C}}) \to \widehat{A}$ such that*

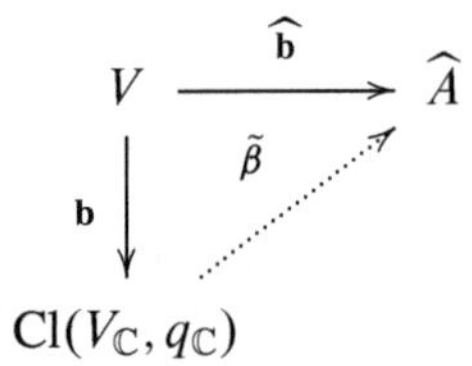

commutes. Furthermore, $\tilde{\beta}$ is injective.

Corollary A.8. *For every real pre-Hilbert space $(V, (\cdot\,,\cdot))$ there exists a* CAR-*representation of $(V, (\cdot\,,\cdot))$, unique up to C*-isomorphism.*

From now on, given a real pre-Hilbert space $(V, (\cdot\,,\cdot))$, we denote the C*-algebra $\mathrm{Cl}(V_{\mathbb{C}}, q_{\mathbb{C}})$ associated with the self-dual CAR-representation $(\mathbf{b}, \mathrm{Cl}(V_{\mathbb{C}}, q_{\mathbb{C}}))$ of $(V, (\cdot\,,\cdot))$ by $\mathrm{CAR}_{\mathrm{sd}}(V, (\cdot\,,\cdot))$.

Proposition A.9. *Let $(V, (\cdot\,,\cdot))$ be a real pre-Hilbert space and $(\mathbf{b}, \mathrm{CAR}_{\mathrm{sd}}(V, (\cdot\,,\cdot)))$ its self-dual* CAR-*representation.*

(i) For every $v \in V$ one has $\|\mathbf{b}(v)\| = \frac{1}{\sqrt{2}}|v|$, where $\|\cdot\|$ denotes the C-norm on $\mathrm{CAR}_{\mathrm{sd}}(V, (\cdot\,,\cdot))$.*
(ii) The C-algebra $\mathrm{CAR}_{\mathrm{sd}}(V, (\cdot\,,\cdot))$ is simple.*

(iii) The algebra $\mathrm{CAR}_{\mathrm{sd}}(V,(\cdot\,,\cdot))$ *is* $\mathbb{Z}_2$*-graded,*

$$\mathrm{CAR}_{\mathrm{sd}}(V,(\cdot\,,\cdot)) = \mathrm{CAR}^{\mathrm{even}}_{\mathrm{sd}}(V,(\cdot\,,\cdot)) \oplus \mathrm{CAR}^{\mathrm{odd}}_{\mathrm{sd}}(V,(\cdot\,,\cdot)),$$

and $\mathbf{b}(V) \subset \mathrm{CAR}^{\mathrm{odd}}_{\mathrm{sd}}(V,(\cdot\,,\cdot))$.

(iv) Let $f : V \to V'$ *be an isometric linear embedding, where* $(V',(\cdot\,,\cdot)')$ *is another real pre-Hilbert space. Then there exists a unique injective* C^**-morphism* $\mathrm{CAR}_{\mathrm{sd}}(f) : \mathrm{CAR}_{\mathrm{sd}}(V,(\cdot\,,\cdot)) \to \mathrm{CAR}_{\mathrm{sd}}(V',(\cdot\,,\cdot)')$ *such that*

$$\begin{array}{ccc} V & \xrightarrow{\quad f \quad} & V' \\ \downarrow{\scriptstyle \mathbf{b}} & & \downarrow{\scriptstyle \mathbf{b}'} \\ \mathrm{CAR}_{\mathrm{sd}}(V,(\cdot\,,\cdot)) & \xrightarrow{\mathrm{CAR}_{\mathrm{sd}}(f)} & \mathrm{CAR}_{\mathrm{sd}}(V',(\cdot\,,\cdot)') \end{array}$$

commutes.

The proofs are similar to the ones for CAR-representations of complex pre-Hilbert spaces. We have constructed a functor

$$\mathrm{CAR}_{\mathrm{sd}} : \mathsf{HILB}_{\mathbb{R}} \longrightarrow \mathsf{C^*Alg},$$

where $\mathsf{HILB}_{\mathbb{R}}$ denotes the category whose objects are the real pre-Hilbert spaces and whose morphisms are the isometric linear embeddings.

Remark A.10. Let $(V,(\cdot\,,\cdot))$ be a complex pre-Hilbert space. If we consider V as a real vector space, then we have the real pre-Hilbert space $(V,\mathfrak{Re}(\cdot\,,\cdot))$. For the corresponding CAR-representations we have

$$\mathrm{CAR}(V,(\cdot\,,\cdot)) = \mathrm{CAR}_{\mathrm{sd}}(V,\mathfrak{Re}(\cdot\,,\cdot)) = \mathrm{Cl}(V_{\mathbb{C}},q_{\mathbb{C}})$$

and

$$\mathbf{b}(v) = \frac{i}{\sqrt{2}}(\mathbf{a}(v) - \mathbf{a}(v)^*).$$

A.2 CCR Algebras

In this section, we recall the construction of the representation of any (real) symplectic vector space by the so-called canonical commutation relations (CCR). Proofs can be found in [4, Sect. 4.2].

Definition A.11. A CCR-*representation* of a symplectic vector space (V,ω) is a pair (w,A), where A is a unital C^*-algebra and w is a map $V \to A$ satisfying:

(i) $A = C^*(w(V))$,
(ii) $w(0) = 1$,
(iii) $w(-\varphi) = w(\varphi)^*$,
(iv) $w(\varphi + \psi) = e^{i\omega(\varphi,\psi)/2} w(\varphi) \cdot w(\psi)$,

for all $\varphi, \psi \in V$.

The map w is in general neither linear, nor any kind of group homomorphism, nor continuous [4, Prop. 4.2.3].

Example A.12. Given any symplectic vector space (V, ω), consider the Hilbert space $H := L^2(V, \mathbb{C})$, where V is endowed with the counting measure. Define the map w from V into the space $\mathcal{L}(H)$ of bounded endomorphisms of H by

$$(w(\varphi)F)(\psi) := e^{i\omega(\varphi,\psi)/2} F(\varphi + \psi),$$

for all $\varphi, \psi \in V$ and $F \in H$. It is well-known that $\mathcal{L}(H)$ is a C*-algebra with the operator norm as C*-norm, and that the map w satisfies the Axioms (ii) and (iii) from Definition A.11, see e.g. [4, Ex. 4.2.2]. Hence setting $A := C^*(w(V))$, the pair (w, A) provides a CCR-representation of (V, ω).

This is essentially the only example of CCR-representation:

Theorem A.13. *Let (V, ω) be a symplectic vector space and $(\hat{w}, \widehat{A})$ be a pair satisfying the Axioms (ii)-(iv) of Definition A.11. Then there exists a unique C*-morphism $\Phi : A \to \widehat{A}$ such that $\Phi \circ w = \hat{w}$, where (w, A) is the* CCR*-representation from Example A.12. Moreover, Φ is injective.*

In particular, (V, ω) has a CCR*-representation, unique up to C*-isomorphism.*

We denote the C*-algebra associated to the CCR-representation of (V, ω) from Example A.12 by $\mathrm{CCR}(V, \omega)$. As a consequence of Theorem A.13, we obtain the following important corollary.

Corollary A.14. *Let (V, ω) be a symplectic vector space and $(w, \mathrm{CCR}(V, \omega))$ its* CCR*-representation.*

(i) The C-algebra $\mathrm{CCR}(V, \omega)$ is simple, i.e., it has no closed two-sided $*$-ideals other than $\{0\}$ and the algebra itself.*
(ii) Let (V', ω') be another symplectic vector space and $f : V \to V'$ a symplectic linear map. Then there exists a unique injective C-morphism $\mathrm{CCR}(f) : \mathrm{CCR}(V, \omega) \to \mathrm{CCR}(V', \omega')$ such that*

$$\begin{array}{ccc} V & \xrightarrow{\quad f \quad} & V' \\ \downarrow{\scriptstyle w} & & \downarrow{\scriptstyle w'} \\ \mathrm{CCR}(V,\omega) & \xrightarrow{\mathrm{CCR}(f)} & \mathrm{CCR}(V',\omega') \end{array}$$

commutes.

Obviously $\mathrm{CCR}(\mathrm{id}) = \mathrm{id}$ and $\mathrm{CCR}(f' \circ f) = \mathrm{CCR}(f') \circ \mathrm{CCR}(f)$ for all symplectic linear maps $V \xrightarrow{f} V' \xrightarrow{f'} V''$, so that we have constructed a covariant functor

$$\mathsf{CCR} : \mathsf{Sympl} \longrightarrow \mathsf{C^*Alg}.$$

References

1. Araki, H.: On quasifree states of CAR and Bogoliubov automorphisms. Publ. Res. Inst. Math. Sci. **6**, 385–442 (1970/71)
2. Bär, C., Becker, C.: C*-algebras. In: Bär,C., Fredenhagen, K. (eds.) Quantum Field Theory on Curved Spacetimes, Lecture Notes in Physics, vol. 786, pp. 1–37. Springer, Berlin (2009)
3. Bär, C., Gauduchon, P., Moroianu, A.: Generalized cylinders in semi-Riemannian and spin geometry. Math. Z. **249**, 545–580 (2005)
4. Bär, C., Ginoux, N., Pfäffle, F.: Wave Equations on Lorentzian Manifolds and Quantization. EMS, Zürich (2007)
5. Baum, H.: Spin-Strukturen und Dirac-Operatoren über pseudoriemannschen Mannigfaltigkeiten. Teubner, Leipzig (1981)
6. Bernal, A.N., Sánchez, M.: Smoothness of time functions and the metric splitting of globally hyperbolic spacetimes. Comm. Math. Phys. **257**, 43–50 (2005)
7. Bernal,A.N., Sánchez, M.: Further results on the smoothability of Cauchy hypersurfaces and Cauchy time functions. Lett. Math. Phys. **77**, 183–197 (2006)
8. Branson, T., Hijazi, O.: Bochner–Weitzenböck formulas associated with the Rarita–Schwinger operator. Int. J. Math. **13**, 137–182 (2002)
9. Bratteli, O., Robinson, D.W.: Operator Algebras and Quantum Statistical Mechanics, I–II. Texts and Monographs in Physics, 2nd edn. Springer, Berlin (1997)
10. Bröcker, T., tom Dieck, T.: Representations of Compact Lie Groups. Graduate Texts in Mathematics, vol. 98, Springer, New York (1995)
11. Brunetti, R., Fredenhagen, K., Verch, R.: The generally covariant locality principle—a new paradigm for local quantum field theory. Comm. Math. Phys. **237**, 31–68 (2003)
12. Buchdahl, H.A.: On the compatibility of relativistic wave equations in Riemann spaces. II. J. Phys. A **15** (1982), 1–5.
13. Buchdahl, H.A.: On the compatibility of relativistic wave equations in Riemann spaces. III. J. Phys. A **15**, 1057–1062 (1982)
14. Dappiaggi, C., Hack, T.-P., Pinamonti, N.: The extended algebra of observables for Dirac fields and the trace anomaly of their stress-energy tensors. Rev. Math. Phys. **21**, 1241–1312 (2009)
15. Dappiaggi, C., Moretti, V., Pinamonti, N.: Distinguished quantum states in a class of cosmological spacetimes and their Hadamard property. J. Math. Phys. **50**, 062304, 38 pp (2009)
16. Dappiaggi, C., Pinamonti, N., Porrmann, M.: Local causal structures, Hadamard states and the principle of local covariance in quantum field theory. Comm. Math. Phys. **304**(2), 459–498 (2011)
17. Dimock, J.: Algebras of local observables on a manifold. Comm. Math. Phys. **77**, 219–228 (1980)
18. Dimock, J.: Dirac quantum fields on a manifold. Trans. Am. Math. Soc. **269**, 133–147 (1982)
19. Fewster, C.J., Verch, R.: A quantum weak energy inequality for Dirac fields in curved spacetime. Comm. Math. Phys. **225** (2002), 331–359.
20. Furlani, E.: Quantization of massive vector fields in curved space-time. J. Math. Phys. **40**, 2611–2626 (1999)
21. Geroch, R.P.: Domain of dependence. J. Math. Phys. **11**, 437–449 (1970)
22. Gibbons, G.W.: A note on the Rarita–Schwinger equation in a gravitational background. J. Phys. A **9**, 145–148 (1976)

23. Hörmander, L.: The Analysis of Linear Partial Differential Operators. I. Distribution Theory and Fourier Analysis. 2nd edn. Grundlehren der Mathematischen Wissenschaften, vol. 256. Springer, Berlin (1990)
24. Hörmander, L.: The Analysis of Linear Partial Differential Operators. III. Pseudodifferential Operators. Grundlehren der Mathematischen Wissenschaften, vol. 274. Springer, Berlin (1985)
25. Hollands, S., Wald, R.M.: Axiomatic quantum field theory in curved spacetime. Comm. Math. Phys. **293**, 85–125 (2010)
26. Kay, B.S.: Linear spin-zero quantum fields in external gravitational and scalar fields. Comm. Math. Phys. **62**, 55–70 (1978)
27. Lawson, H.B., Michelsohn, M.-L.: Spin Geometry. Princeton University Press, Princeton (1989)
28. Mühlhoff, R.: Higher spin fields on curved spacetimes. Diplomarbeit, Universität Leipzig (2007)
29. Mühlhoff, R.: Cauchy problem and Green's functions for first order differential operators and algebraic quantization. J. Math. Phys. **52**, 022303, 7 pp (2011)
30. O'Neill, B.: Semi-Riemannian Geometry. Academic Press, San Diego (1983)
31. Plymen, R.J., Robinson, P.L.: Spinors in Hilbert Space. Cambridge Tracts in Mathematics, vol. 114. Cambridge University Press, Cambridge (1994)
32. Radzikowski, M.J.: Micro-local approach to the Hadamard condition in quantum field theory on curved space-time. Comm. Math. Phys. **179**, 529–553 (1996)
33. Rarita, W., Schwinger, J.: On a theory of particles with half-integral spin. Phys. Rev. **60**, 61 (1941)
34. Reed, M., Simon, B.: Methods of Modern Mathematical Physics I: Functional Analysis. Academic Press, Orlando (1980)
35. Reed, M., Simon, B.: Methods of Modern Mathematical Physics II: Fourier Analysis, Self-Adjointness. Academic Press, Orlando (1975)
36. Sahlmann, H., Verch, R.: Passivity and microlocal spectrum condition. Comm. Math. Phys. **214**, 705–731 (2000)
37. Sahlmann, H., Verch, R.: Microlocal spectrum condition and Hadamard form for vector-valued quantum fields in curved spacetime. Rev. Math. Phys. **13**, 1203–1246 (2001)
38. Sanders, K.: The locally covariant Dirac field. Rev. Math. Phys. 22, 381–430 (2010)
39. Strohmaier, A.: The Reeh–Schlieder property for quantum fields on stationary spacetimes. Comm. Math. Phys. **215**, 105–118 (2000)
40. Taylor, M.E.: Partial Differential Equations I—Basic Theory. Springer, New York, Berlin, Heidelberg (1996)
41. Verch, R.: A spin-statistics theorem for quantum fields on curved spacetime manifolds in a generally covariant framework. Comm. Math. Phys. **223**, 261–288 (2001)
42. Wald, R.M.: Quantum Field Theory in Curved Spacetime and Black Hole Thermodynamics. University of Chicago Press, Chicago (1994)
43. Wang, McK.Y.: Preserving parallel spinors under metric deformations. Indiana Univ. Math. J. **40**, 815–844 (1991)
44. V. Wünsch: Cauchy's problem and Huygens' principle for relativistic higher spin wave equations in an arbitrary curved space-time. Gen. Relat. Gravit. **17**, 15–38 (1985)

Computations and Applications of η Invariants

Sebastian Goette

Abstract We give a survey on η-invariants including methods of computation and applications in differential topology.

Introduction

The η-invariant has been introduced by Atiyah, Patodi and Singer as a boundary contribution in an index theorem for elliptic operators in the series of papers [2–5]. There are several invariants of odd-dimensional manifolds M in differential topology that are originally defined by finding a compact manifold N with boundary $\partial N = M$ and evaluating certain characteristic numbers on N. The Atiyah–Patodi–Singer index Theorem 1.1 often allows to compute these invariants in terms of η-invariants and other magnitudes that can be defined directly on M without choosing N first. Sometimes this leads to generalisations of these invariants to manifolds that are not 0-cobordant. However, to determine such an invariant for a given manifold M, one needs ways to compute η-invariants of operators defined on M without using the Atiyah–Patodi–Singer index theorem.

In the present paper, we give a short survey on applications of η-invariants, with a focus on situations where the corresponding η-invariants can be computed. The η-invariant also appears in possible generalisations of the analytic torsion, in conformal geometry, and in the definition of certain smooth extensions of

Sebastian Goette (✉)
Mathematisches Institut
Universität Freiburg
Eckerstr. 1
79104 Freiburg
Germany
e-mail: sebastian.goette@math.uni-freiburg.de

C. Bär et al. (eds.), *Global Differential Geometry*, Springer Proceedings in Mathematics 17, DOI 10.1007/978-3-642-22842-1_13,

topological K-theory. To keep this article reasonably short, we will not touch upon these and several other issues.

We start by reviewing the definition of η-invariants as spectral invariants in Sect. 1. We also review the Atiyah–Patodi–Singer theorem and some of its immediate consequences. We list some examples where η-invariants have been computed directly. Sometimes it is easier to compute η-invariants of modified operators first and to determine their difference to the original η-invariants, see Sects. 2.3 and 4.2.

The Atiyah–Patodi–Singer theorem has generalisations in different directions. In Sect. 2, we consider families of manifolds, group actions and orbifolds. The corresponding generalisations of Theorem 1.1 involve generalisations of η-invariants that are sometimes easier to compute.

In Sect. 3, we discuss the behaviour of η-invariants for direct images under proper maps and under gluing constructions. These methods sometimes give rise to explicit computations, see Sect. 4.5.

Finally, in Sect. 4, we discuss some applications of η-invariants mainly to differential topology, but also to questions ranging from algebraic K-theory to Riemannian manifolds of positive scalar curvature.

The author wishes to thank M. Braverman, U. Bunke, D. Crowley, P. Piazza and A. Ranicki for some helpful comments and explanations.

1 The Atiyah–Patodi–Singer η-Invariant and Related Invariants

The η-invariant of a selfadjoint elliptic differential operator on an odd-dimensional manifold M first appeared in the Atiyah–Patodi–Singer index theorem for manifolds N with boundary M, which was announced in [2] and proved in [3]. They already noted that the η-invariant was related to other topological invariants known at the time.

1.1 The Index Theorem for Manifolds with Boundary

Let N be a compact Riemannian manifold with boundary $M = \partial N$, and let $A\colon \Gamma(E^+) \to \Gamma(E^-)$ be an elliptic differential operator on N. Assume that a neighbourhood U of M in N is isometric to a product $M \times [0,\varepsilon)$, that $\nu\colon E^+|_U \to E^-|_U$ is a vector bundle isomorphism, and that $E^\pm|_U$ are identified with $E^\pm|_M \times [0,\varepsilon)$ in such a way that on U,

$$A|_{\Gamma(E^+|_U)} = \nu \circ \left(\frac{\partial}{\partial t} + B\right), \tag{1}$$

where $\frac{\partial}{\partial t}$ denotes differentiation in the direction of $[0, \varepsilon)$ and B is a selfadjoint elliptic differential operator acting on smooth sections of $E^+|_{\partial N} \to M$.

A typical example consists of a Dirac operator $A = D_N^+$ on an even-dimensional manifold. With

$$\nu = c_N\Big(\frac{\partial}{\partial t}\Big) \colon E^+|_U \to E^-|_U ,$$

one can construct a Clifford multiplication c_M of TM on $E^+|_M$, such that

$$c_N(v) = \nu \circ c_M(v) \qquad \text{for all} \qquad v \in TM .$$

Then (1) holds with $B = D_M$ a Dirac operator on the odd-dimensional boundary M.

There are topological obstructions against local elliptic boundary conditions for the operator A on N of (1). However, there are elliptic spectral boundary conditions on each connected component of M, inspired by replacing U by an infinite cylinder $M \times (-\infty, \varepsilon)$ and imposing L^2-boundary conditions in the special case that B is invertible. Concretely, one restricts A and its adjoint A^* to

$$\begin{aligned} \Gamma_<(E^+) &= \big\{ \sigma \in \Gamma(E^+) \bigm| P_\geq(\sigma|_M) = 0 \big\} , \\ \Gamma_\leq(E^-) &= \big\{ \tau \in \Gamma(E^-) \bigm| P_<(\nu^{-1}\tau|_M) = 0 \big\} , \end{aligned} \tag{2}$$

where $P_\geq$, $P_< \colon \Gamma(E^+|_M) \to \Gamma(E^+|_M)$ denote spectral projections onto the nonnegative and the negative eigenspaces of B, respectively. These are the so-called *APS boundary conditions*, and one defines

$$\operatorname{ind}_{\mathrm{APS}}(A) = \ker\big(A|_{\Gamma_<(E^+)}\big) - \ker\big(A^*|_{\Gamma_\leq(E^-)}\big) .$$

The index of a suitable double of A is given as the integral of a local index density α_0 over the double of N. In the case of a Dirac operator, we know from the Atiyah–Singer index theorem that

$$\alpha_0 = \big(\hat{A}(TN, \nabla^{TN}) \wedge \operatorname{ch}(E/S, \nabla^E)\big)^{\max} ,$$

where $\operatorname{ch}(E/S, \nabla^E)$ denotes the twist Chern character form, see [9].

The η-invariant $\eta(B)$ of B is defined as the value at $s = 0$ of the meromorphic continuation of the η-function, that is for $\operatorname{Re} s \gg 0$ given by

$$\eta_B(s) = \sum_{\lambda \in \operatorname{Spec}(B)\setminus\{0\}} \operatorname{sign}(\lambda) \cdot |\lambda|^{-s} = \frac{1}{\Gamma\big(\frac{s+1}{2}\big)} \int_0^\infty t^{\frac{s-1}{2}} \operatorname{tr}(Be^{-tB^2})\, dt . \tag{3}$$

It is proved in [5] that the η-function indeed has a meromorphic continuation and that $\eta(B) = \eta_B(0)$ is finite. For a Dirac operator $B = D_M$, one can show directly that the integral expression in (3) converges for $\operatorname{Re} s > -1$, see [15]. One also defines

$$h(B) = \dim \ker B .$$

Theorem 1.1 (Atiyah–Patodi–Singer [2, 3]). *Let A be an elliptic differential operator on a compact manifold N with boundary $M = \partial N$ as in* (1). *Then the Fredholm index of A under the APS boundary conditions* (2) *is given as*

$$\operatorname{ind}_{\mathrm{APS}}(A) = \int_N \alpha_0 - \frac{\eta + h}{2}(B) .$$

The signature operator of a $4k$-dimensional compact oriented manifold with boundary is an important special case. Here, one considers the symmetric bilinear form on

$$\operatorname{im}\bigl(H^{2k}(N, \partial N; \mathbb{R}) \to H^{2k}(N; \mathbb{R})\bigr)$$

given by the evaluation of the cup product of two such classes on the relative fundamental cycle $[N, \partial N]$. The signature of this form is denoted by $\operatorname{sign}(N)$.

On the odd-dimensional manifold $M = \partial N$, the bundle $\Lambda^{\mathrm{ev}} T^*M$ of even differential forms constitutes a Dirac bundle. The Hodge star operator $*$ interchanges even and odd forms. The Dirac operator $B = \pm(*d - d*)$ on $\Lambda^{\mathrm{ev}} T^*M$ is usually named the *odd signature operator* on M.

Theorem 1.2 (Atiyah–Patodi–Singer [3]). *Let N be a $4k$-dimensional manifold with totally geodesic boundary $M = \partial N$, and let B denote the odd signature operator on M, then*

$$\operatorname{sign}(N) = \int_N L(TN) - \eta(B) .$$

Comparing with Theorem 1.1, one notes that the boundary operator consists of two copies of the odd signature operator. Also, the index of the signature operator on N under APS boundary conditions is not precisely $\operatorname{sign}(N)$ due to the asymmetric treatment of $\ker(B)$. In fact, if $h(B)$ was present on the right hand side in Theorem 1.2, the equation would not be compatible with a change of orientation. The most prominent feature for applications is the fact that the signature of N is a topological invariant, in contrast to most other APS indices, which depend on the geometry of N near its boundary.

Some elementary properties of η-invariants can be deduced directly from Theorems 1.1 and 1.2. For simplicity, we will stick to Dirac operators, and we let B denote the odd signature operator.

If $P(V, \nabla^V) \in \Omega^\bullet(M)$ denotes a Chern–Weil form associated to a vector bundle $V \to M$ with connection ∇^V and an invariant polynomial P, we let $\tilde{P}(V, \nabla^{V,0} \nabla^{V,1}) \in \Omega^\bullet(M)/\operatorname{im} d$ denote the Chern–Simons class satisfying

$$d\tilde{P}\bigl(V, \nabla^{V,0}, \nabla^{V,1}\bigr) = P\bigl(V, \nabla^{V,1}\bigr) - P\bigl(V, \nabla^{V,0}\bigr) . \tag{4}$$

The Dirac operator D on a Dirac bundle $E \to M$ depends on smoothly on the Riemannian metric g on M and on a Clifford multiplication and a connection ∇^E on E that is compatible with the Levi–Civita connection. Applying Theorems 1.1 and 1.2 to an adapted Dirac operator D_N on the cylinder $N = M \times [0,1]$ gives a variation formula.

Corollary 1.3 (Atiyah–Patodi–Singer [3]). *Let $(g_s)_{s\in[0,1]}$ be a family of Riemannian metrics on M with Levi–Civita connections $\nabla^{TM,s}$, and let $(E, c_s, \nabla^{E,s})_{s\in[0,1]}$ be compatible bundles with Dirac operators D^s_M. Then*

$$\frac{\eta+h}{2}(D^1_M) - \frac{\eta+h}{2}(D^0_M) = \int_M \Big(\tilde{\hat{A}}\big(TM, \nabla^{TM,0}, \nabla^{TM,1}\big)\,\mathrm{ch}\big(E/S, \nabla^{E,0}\big) - \hat{A}\big(TM, \nabla^{TM,1}\big)\,\tilde{\mathrm{ch}}\big(E/S, \nabla^{E,0}, \nabla^{E,1}\big)\Big) \in \mathbb{R}/\mathbb{Z}\,. \tag{5}$$

For the odd signature operator $(B^s)_{s\in[0,1]}$, one has

$$\eta(B^1) - \eta(B^0) = \int_M \tilde{L}\big(TM, \nabla^{TM,0}, \nabla^{TM,1}\big) \in \mathbb{R}\,. \tag{6}$$

Thus, η-invariants have similar variation formulas as Cheeger–Simons numbers, by which we mean the evaluation of (products of) Cheeger–Simons classes on the fundamental cycle of an odd-dimensional compact oriented manifold. We may think of Cheeger–Simons numbers as geometric $\mathbb{R}/\mathbb{Z}$-valued refinements of integral characteristic classes of vector bundles, whereas η-invariants are geometric $\mathbb{R}/\mathbb{Z}$-valued refinements of indices of Dirac operators. In general, these numbers are difficult to compare.

Example 1.4. Let M be a oriented three-manifold. Then the variation formula for $\eta(B)$ of Corollary 1.3 becomes

$$\eta(B^1) - \eta(B^0) = \frac{1}{3}\int_M \tilde{p}_1\big(TM, \nabla^{TM,0}, \nabla^{TM,1}\big)\,,$$

where $\tilde{p}_1$ is the Chern–Simons class associated to the first Pontrijagin class. The variation formula for the associated Cheeger–Simons character $\hat{p}_1$ of a general vector bundle is

$$\Big(\hat{p}_1\big(E, \nabla^{E,1}\big) - \hat{p}_1\big(E, \nabla^{E,0}\big)\Big)[M] = \int_M \tilde{p}_1\big(E, \nabla^{E,0}, \nabla^{E,1}\big)\,.$$

This implies that $3\eta(B)$ is an $\mathbb{R}$-valued refinement of $\hat{p}_1(TM, \nabla^{TM})[M]$. For other vector bundles, we do not get a natural $\mathbb{R}$-valued refinement of $\hat{p}_1(E, \nabla^E)[M] \in \mathbb{R}/\mathbb{Z}$ due to the presence of a nontrivial gauge group.

For higher dimensional manifolds, the situation is more complicated due to the formulas for the multiplicative sequences,

$$\hat{A} = 1 - \frac{p_1}{24} + \frac{7p_1^2 - 4p_2}{2^7 \cdot 3^2 \cdot 5} - \frac{31p_1^3 - 44p_1p_2 + 16p_3}{2^{10} \cdot 3^3 \cdot 5 \cdot 7} \pm \dots, \tag{7}$$

$$L = 1 + \frac{p_1}{3} - \frac{p_1^2 - 7p_2}{45} + \frac{2p_1^3 - 13p_1p_2 + 62p_3}{3^3 \cdot 5 \cdot 7} \pm \dots \tag{8}$$

Finally, one can use Corollary 1.3 to count sign changes of eigenvalues of D_M^s for $s \in [0,1]$ by comparing the actual difference of η-invariants with the value predicted by the local Chern–Simons variation terms. The so-called *spectral flow* of the family $(D_M^s)_{s\in[0,1]}$ is given by

$$\begin{aligned} \mathrm{sf}\big((D_M^s)_{s\in[0,1]}\big) &= \frac{\eta+h}{2}(D_M^1) - \frac{\eta+h}{2}(D_M^0) \\ &\quad - \int_M \Big(\tilde{\hat{A}}\big(TM, \nabla^{TM,0}, \nabla^{TM,1}\big)\,\mathrm{ch}\big(E/S, \nabla^{E,0}\big) \\ &\quad + \hat{A}\big(TM, \nabla^{TM,1}\big)\tilde{\mathrm{ch}}\big(E/S, \nabla^{E,0}, \nabla^{E,1}\big)\Big) \in \mathbb{Z}\,. \end{aligned} \tag{9}$$

If the Dirac bundles for $s = 0$ and $s = 1$ are isomorphic, then the spectral flow defines an odd index sf: $K^1(M) \to \mathbb{Z}$, see [5].

1.2 Direct Computation of η-Invariants

For generic Riemannian manifolds, it seems impossible to determine the spectrum of a given differential operator B. And even if one succeeds, one often needs techniques from analytic number theory in order to describe the function $\eta_B(s)$ explicitly and compute its special value $\eta(B) = \eta_B(0)$ at $s = 0$. This section is devoted to a few examples where this has been done. All examples are locally homogeneous spaces, and representation theory plays a prominent role in the determination of the relevant spectrum.

For the operator $B_\lambda = i\frac{d}{dt} + \lambda$ on a circle of length 2π, the η-invariant is computed in [2] as

$$\eta(B_\lambda) = \begin{cases} 0 & \lambda \in \mathbb{Z}\,, \\ 1 - 2(\lambda - n) & \lambda \in (n, n+1)\,. \end{cases}$$

Next, consider three-dimensional Berger spheres. Thus, one rescales the fibres of the Hopf fibration $S^3 \to S^2$ by $\lambda > 0$, while the metric orthogonal to the fibres is unchanged. This metric is still $U(2)$-invariant. Let D_λ denote the untwisted Dirac operator on S^3_λ. Using $U(2)$-invariance and a suitable Hilbert basis of sections of the spinor bundle, Hitchin determined the eigenvalues of D_λ in [51] as follows,

$$\frac{\lambda}{2} + \frac{p}{\lambda} \qquad \text{with multiplicity } 2p\,,$$

$$\frac{\lambda}{2} \pm \frac{\sqrt{4pq\lambda^2 + (p-q)^2}}{\lambda} \qquad \text{with multiplicity } p+q\,,$$

for p, $q > 0$. From these values, Hitchin computes the η-invariant explicitly and obtains

$$\eta(D_\lambda) = -\frac{(\lambda^2-1)^2}{6}$$

for $0 < \lambda < 4$. For larger values of λ, the formula holds only up to spectral flow, see (9). In his diploma thesis [49], Habel does analogous computations for a few higher dimensional Berger spheres. Bechtluft–Sachs gets the same result in low dimensions by applying Theorem 1.1 to disk bundles over $\mathbb{C}P^n$ [8].

Let $\Gamma \subset PSL(2,\mathbb{R})$ be a cocompact subgroup, then $M = PSL(2,\mathbb{R})/\Gamma$ is a compact three-manifold, Seifert fibred over a hyperbolic surface. Seade and Steer compute $\eta(D_\lambda)$ when Γ is a Fuchsian group [77]. The parameter λ refers to the length of the generic fibre as in the case of the Berger sphere. As in Hitchin's computations, representation theory plays a prominent role in these computations. The results of Seade and Steer are generalised to noncompact quotients of finite volume by Loya, Moroianu and Park in [63].

The spectrum of an untwisted Dirac D operator on a flat torus M depends on the spin structure and is always symmetric, so $\eta(D) = 0$. However, among the Bieberbach manifolds M/Γ, where $\Gamma \subset SO(n)$ is a finite subgroup that acts freely on M, there exist examples with asymmetric spectra and non-vanishing η-invariants. Pfäffle computes the spectra and the η-invariants of all three-dimensional examples in [72]. Higher-dimensional examples are studied by Sadowski and Szczepánski in [75], by Miatello and Podestá in [66], and by Gilkey, Miatello and Podestá in [40]. In all these cases, the η-invariant of an untwisted Dirac operator can be expressed in number theoretic terms.

Similar computations are also possible for spherical space forms S^n/Γ with $\Gamma \subset SO(n)$ a finite subgroup. In [24], Cisneros–Molina gives a general formula for the spectra of Dirac operators on $M = S^3/\Gamma$ twisted by flat vector bundles and computes the corresponding η-invariants. These η-invariants are closely related to the ξ-invariant of M and the Γ-equivariant η-invariants of S^3, see Sect. 4.1. Seade [76] and Tsuboi [79] compute η-invariants for certain spherical space forms as average over equivariant η-invariants, see also [7]. Degeratu extends these computations to orbifold quotients in [33] and exhibits a relation with the Molien series.

In [68], Millson expresses the η-invariant of the odd signature operator on a compact hyperbolic manifold as a special value of a ζ-function associated to the closed geodesics on M and their holonomy representations and Poincaré maps. This result is extended to Dirac operators on locally symmetric spaces M of noncompact type by Moscovici–Stanton [69]. A generalisation to the finite-volume case is given by Park [71].

The η-invariant of a Dirac operator on an interval $[0,1]$, twisted by a symplectic vector space (V,ω), with different Lagrangian boundary conditions L_0, $L_1 \subset \ker(D_N)$ is computed by Cappell, Lee and Miller in [22]. It is related to the Maslov index of Lagrangians in (V,ω). Indeed, Maslov indices naturally occur when considering η-invariants on manifolds with boundary, for example in generalisations of Theorem 3.4.

2 Families, Group Actions, and Orbifolds

Instead of regarding Dirac operators on a single manifold N, one may consider families of manifolds, or manifolds with the action of some Lie-group, or even orbifolds with boundary. Under certain conditions, Theorems 1.1 and 1.2 extend to these situations. We state a few of these generalisations below and indicate relations between them. We also explain how ordinary η-invariants can be computed using equivariant methods.

2.1 Families of Manifolds with Boundary

Assume that $p\colon W \to B$ is a proper submersion with typical even-dimensional fibre N, such that the fibrewise boundaries form another submersion $V \to B$. Let g^{TN} be a fibrewise Riemannian metric and let $T^HW \to W$ be a horizontal complement for the fibrewise tangent bundle $TN = \ker dp \subset TW$. These data uniquely define a generalised Levi–Civita connection ∇^{TN} on $TN \to W$. Let $E = E^+ \oplus E^- \to W$ be a fibrewise Dirac bundle, i.e., TN acts on E by Clifford multiplication, and there is a compatible metric g^E and a compatible connection ∇^E on E. Then one can define a family of Dirac operators D_N on the fibres of p. We assume that condition (1) is satisfied on each fibre, and we also assume that the kernels of the boundary operators D_M form a family over B. Then let us assume for simplicity that the kernels of the family D_N under APS boundary conditions also form a family over B.

In this situation, there exist natural families of Bismut–Levi–Civita superconnections $(\mathbb{A}_t)_{t\in(0,\infty)}$ and $(\mathbb{B}_t)_{t\in(0,\infty)}$ on the infinite dimensional vector bundles $p_*E \to B$ and $p_*(E^+|_V) \to B$. These superconnections define ordinary connections ∇^H, $\nabla^{K^\pm}$ on $H = \ker(D_M) \to B$ and on $K^\pm = \ker(D_N^\pm) \to B$. The η-invariant generalises to a natural η*-form*

$$\tilde\eta(\mathbb{B}) = \frac{1}{\sqrt{\pi}} \int_0^\infty \operatorname{tr}\left(\frac{\partial \mathbb{B}_t}{\partial t}\, e^{-\mathbb{B}_t^2}\right) dt \in \Omega^\bullet(B)\,. \tag{10}$$

Note that the component of degree 0 is $\tilde\eta(\mathbb{B})^{[0]} = \frac{\eta}{2}(B)$.

Theorem 2.1 (Bismut–Cheeger [12–14]). *Under the assumptions above,*

$$\begin{aligned}\operatorname{ch}(K^+,\nabla^{K^+}) - \operatorname{ch}(K^-,\nabla^{K^-}) &= \int_{W/B} \hat{A}(TN,\nabla^{TN})\,\operatorname{ch}(E/S,\nabla^E)\\ &\quad - \tilde{\eta}(\mathbb{B}) - \frac{1}{2}\operatorname{ch}(H,\nabla^H) \in H^\bullet(B;\mathbb{R})\,.\end{aligned}$$

There exists a similar generalisation of Theorem 1.2. Note that the kernels of the signature operator and the odd signature operator on the boundaries automatically form bundles over B by Hodge theory.

Melrose and Piazza relax the condition that the kernels of the boundary operator D_M form a bundle over B. For the definition of boundary conditions, it is sufficient to have a spectral section, see [65]. It is also not necessary to demand that $\ker D_N$ forms a bundle over B, since the virtual index bundle always exists.

2.2 *Group Actions on Manifolds with Boundary*

Theorems 1.1 and 1.2 generalise to manifolds with group actions in the same way that the Atiyah–Singer index theorem becomes the Atiyah–Segal fixpoint theorem. In particular, from invariants on the boundary one can conclude the existence of fixpoints in the interior.

Let N and D_N be as in Sect. 1.1. Let G be a group that acts on N by isometries. Assume that this action also lifts to the Dirac bundle E, and that the induced action on $\Gamma(E)$ commutes with D_N. Then G also acts on M and $E^+|_M$ such that the induced action on sections commutes with D_M. One can define an *equivariant index* and an *equivariant η-invariant* for all $g \in G$ by

$$\begin{aligned}\operatorname{ind}_{\mathrm{APS},g}(D_N) &= \operatorname{tr}(g|_{\ker D_N^+}) - \operatorname{tr}(g|_{\ker D_N^-})\\ \eta_{D_M,g}(s) &= \sum_{\lambda\in\operatorname{Spec}(D_M)\setminus\{0\}} \operatorname{sign}(\lambda)\cdot|\lambda|^{-s}\cdot\operatorname{tr}(g|_{E_\lambda})\\ &= \frac{1}{\Gamma\left(\frac{s+1}{2}\right)}\int_0^\infty t^{\frac{s-1}{2}}\operatorname{tr}\left(g\,D_M\,e^{-tD_M^2}\right)dt\,.\end{aligned} \tag{11}$$

The equivariant η-function has a meromorphic continuation to $\mathbb{C}$, and 0 is a regular value. Again, we put $\eta_g(D_M) = \eta_{D_M,g}(0)$. If one generalises the proof of Theorem 1.1 to this new setting, then the index density α_0 localises to the fix-point set N_g of g. For a Dirac operator, we will write the equivariant index density as

$$\hat{A}_g(TN,\nabla^{TN})\,\operatorname{ch}_g(E/S,\nabla^E) \in \Omega^\bullet(N_g;o(N_g))\,,$$

where $o(N_g)$ denotes the orientation line bundle. Note that $\hat{A}_g(TN, \nabla^{TN})$ itself is a product of $\hat{A}(TN_g, \nabla^{TN_g})$ and a contribution from the action of g on the normal bundle of N_g in N. Both forms are unique only up to a sign that depends on the choice of a lift of g to the spin group of N_g, but their product is well-defined, see the discussion in [9]. We also put $h_g(D_M) = \operatorname{tr}(g|_{\ker D_M})$.

Theorem 2.2 (Donnelly [36]). *The G-equivariant index is given by*

$$\operatorname{ind}_{\mathrm{APS},g}(D_N) = \int_{N_g} \hat{A}_g(TN, \nabla^{TN}) \operatorname{ch}_g(E/S, \nabla^E) - \frac{\eta_g + h_g}{2}(D_M) .$$

Remark 2.3. The integral vanishes if g acts freely on N, and the equivariant index is always a virtual character of G. This has two consequences:

1. There is an analogue of Corollary 1.3 with values in functions on G modulo virtual characters. For each $g \in G$, the local contribution is an integral over M_g. Hence, equivariant η-invariants are rigid modulo virtual characters for $g \in G$ that act freely on M.
2. Let D_M be a G-equivariant operator and let $G_0 \subset G$ denote a subset of elements that act freely on M. If there is no virtual character χ of G that extends $\frac{\eta+h}{2}|_{G_0}$, and there is a compact manifold N with $\partial N = M$ and D_N as in Donnelly's theorem, then some elements of $g \in G$ will have fixpoints on N.

If G is a compact connected Lie group, then the equivariant index theorem can be stated in a different way. Let $\mathfrak{g}$ denote the Lie algebra of G. Then consider Cartan's complex of *equivariant differential forms,*

$$(\Omega_G^\bullet(N), d_{\mathfrak{g}}) = \left((\Omega^\bullet(N)[\![\mathfrak{g}^*]\!])^G, d - \frac{\iota_X}{2\pi i} \right) .$$

Here, a monomial in $\mathfrak{g}^*$ of degree ℓ with values in the k-forms has total degree $k + 2\ell$, and ι_X denotes the inner product of a differential form with a variable Killing field X, which is of total degree $1 = -1 + 2$. Classical Chern–Weil theory generalises to G-equivariant vector bundles with invariant connections, giving classes $\hat{A}_X, \operatorname{ch}_X$ with values in the equivariant cohomology

$$H_G^\bullet(N; \mathbb{R}) = H^\bullet(\Omega_G^\bullet(N), d_{\mathfrak{g}}) .$$

The classical equivariant index theorem can be stated in terms of these equivariant characteristic classes as explained by Berline, Getzler and Vergne [9].

Following Bismut's proof of the equivariant index theorem in [10], put

$$D_{X,t} = \sqrt{t}\, D_M + \frac{1}{4\sqrt{t}} c_X ,$$

where c_X denotes Clifford multiplication with the Killing field associated to $X \in \mathfrak{g}$. The *infinitesimally equivariant η-invariant* of D_M is defined as

$$\eta_X(D_M) = \frac{2}{\sqrt{\pi}} \int_0^\infty \operatorname{tr}\Big(\frac{\partial D_{X,t}}{\partial t} e^{-D_{X,t}^2 - \mathcal{L}_X}\Big)\, dt \in \mathbb{C}[\![\mathfrak{g}^*]\!] \,, \tag{12}$$

where $\mathcal{L}_X$ denotes the Lie derivative. We can now state another version of the equivariant index theorem for manifolds with boundary.

Theorem 2.4 ([42]). *The equivariant index for $g = e^{-X}$ is given by the formal power series*

$$\operatorname{ind}_{\mathrm{APS},e^{-X}}(D_N) = \int_N \hat{A}_X(TN, \nabla^{TN})\, \operatorname{ch}_X(E/S, \nabla^E) - \frac{\eta_X + h_{e^{-X}}}{2}(D_M) \in \mathbb{C}[\![\mathfrak{g}^*]\!] \,.$$

This theorem can be deduced from the Bismut–Cheeger Theorem 2.1 by regarding fibre bundles with structure group G and applying the general Chern–Weil principle.

Another possible proof uses Donnelly's Theorem 2.2 and Bott's localisation formula. Let $\vartheta = \frac{1}{2\pi i} g^{TN}(\,\cdot\,, X)$ denote the dual of a variable Killing field, then

$$d_X\Big(\frac{\vartheta_X}{d_X \vartheta_X} \alpha_X\Big) = \alpha_X - \frac{\alpha_X}{e_X(\nu)} \cdot \delta \,,$$

where δ denotes the distribution of integration over the zero-set N_X of X, and $e_X(\nu)$ denotes the equivariant Euler class of the normal bundle $\nu \to N_X$. Note that though the single terms are not defined on all of N, the equation above still makes sense in an L^1-sense, i.e., after integration over N. Theorem 2.4 follows from Theorem 2.2 and the following result by an application of the localisation formula. Both proofs are explained in [9] in the case $\partial N = \emptyset$.

Theorem 2.5 ([42]). *Assume that the Killing field X has no zeros on M. Then*

$$\eta_X(D_M) = \eta_{e^{-X}}(D_M) + 2 \int_M \frac{\vartheta_X}{d_X \vartheta_X} \hat{A}_X(TM, \nabla^{TM})\, \operatorname{ch}_X(E/S, \nabla^E) \in \mathbb{C}[\![\mathfrak{g}^*]\!] \,.$$

One expects that for sufficiently small $X \in \mathfrak{g}$, the formal power series in Theorems 2.4 and 2.5 converge, and that a similar formula also holds if X vanishes somewhere on M. One also expects that one can apply both theorems to ge^{-X}, where $g \in G$ and $X \in \mathfrak{g}$ with $Ad_g X = X$, and the local contributions are integrated over N_g.

Remark 2.6. In general, the equivariant η-invariant $\eta_g(D_M)$ is only continuous in g as long as the fixpoint set M_g varies continuously in g. In particular, it is usually singular at $g = e$. The singularity near $g = e$ is encoded in the integral in Theorem 2.5. Arguing as in Remark 2.3 (2), we see that the singularity of the integral above at $X = 0$ contains information about fixpoints of elements of G on compact G-manifolds N with $\partial N = M$.

2.3 Homogeneous Spaces

It seems that the introduction of families and group actions is an unnecessary complication if one is mainly interested in the η-invariants of Sect. 1.1. In the presence of a Lie group action, Theorem 2.5 allows to split the infinitesimal η-invariant $\eta_X(D)$ into a rigid global object $\eta_{e^{-X}}(D)$ and a locally computable correction term if $X \neq 0$ everywhere on M. If both terms can computed, then their sum extends continuously to $X = 0$ and gives the ordinary η-invariant.

Assume that $H \subset G$ are compact Lie groups, and let D be the geometric Dirac operator on the homogeneous space $M = G/H$ with a normal metric. In [41], we consider the *reductive Dirac operator* $\tilde{D}$. It is a selfadjoint differential operator with the same principal symbol as D, but $\tilde{D}$ is better adapted to homogeneous spaces. This operator was independently discovered by Kostant [55]. If G and H are not of the same rank, then most elements $g \in G$ act freely on G/H. By Remark 2.3 (1),

$$\frac{\eta_g + h_g}{2}(\tilde{D}) - \frac{\eta_g + h_g}{2}(D) = \chi(g) \tag{13}$$

for all $g \in G$ that act freely on M, where the *equivariant spectral flow* χ is a virtual character of G. Moreover $\chi = 0$ for the untwisted Dirac operator. On the other hand, the kernel of the reductive odd signature operator $\tilde{B}$ has no topological significance, and hence the spectral flow does not vanish in general for $D = B$.

Given three compact Lie groups $H \subset K \subset G$, one considers the fibration $G/H \to G/K$ with fibre K/H. The equivariant η-invariant $\eta_G(\tilde{D})$ for G/H can be computed from the equivariant η-invariant of a reductive Dirac operator either on the base G/K or on the fibre K/H, whichever is odd-dimensional. The formula is similar to the adiabatic limit formula in Theorem 2.1, but an equivariant η-invariant appears instead of an η-form, and no limit has to be taken. Suppose that $S \subset T$ are maximal tori of H and G, we consider the fibrations $G/S \to G/H$ and $G/S \to G/T$. This way, the computation of $\eta_G(\tilde{D})$ is reduced in two steps to the computation of an equivariant η-invariant of a twisted Dirac operator on the flat torus T/S, which vanishes unless $\operatorname{rk} G - \operatorname{rk} H = \dim T - \dim S = 1$. Although the formula for $\eta_G(\tilde{D})$ in [41] contains representation theoretic expressions, explicit knowledge of the representations of G is not needed. In particular, the spectrum of $\tilde{D}$ on M is not computed, in contrast to the examples in Sect. 1.2.

In [43], a formula for the correction term in Theorem 2.5 is given, again using the fibrations $G/S \to G/H$ and $G/S \to G/T$ considered above. Combining this with (13), one obtains a formula for $\frac{\eta_X + h_{e^{-X}}}{2}(D)$ up to a virtual character of G. Evaluating at $X = 0$ gives $\frac{\eta + h}{2}(D) \in \mathbb{R}/\mathbb{Z}$. By estimation of sufficiently many small eigenvalues of D and $\tilde{D}$, one can even determine the equivariant spectral flow. This method is applied to compute the Eells–Kuiper invariant of the Berger space $SO(5)/SO(3)$ in [45], see Sect. 4.5.

2.4 Orbifolds with Boundary

Theorem 1.1 is generalised to orbifolds by Farsi [38]. For simplicity, we state the version for Dirac operators. Farsi's original theorem holds in the generality of Theorem 1.1.

Let M be an n-dimensional orbifold. In particular, for each $p \in M$ there exists a local parametrisation of the form $\psi: V \to \Gamma_p \backslash V \cong U \subset M$. Here, the *isotropy group* $\Gamma_p \subset O(n)$ of p is a finite subgroup acting linearly on $V \subset \mathbb{R}^n$. If $\gamma \in \Gamma$, let (γ) denote its conjugacy class, and let $C_\Gamma(\gamma)$ denote its centraliser in Γ. Then the *inertia orbifold* ΛM consists of all pairs $(p,(\gamma))$ with (γ) a conjugacy class in Γ_p. A parametrisation of ΛM around $(p,(\gamma))$ is given by

$$\psi_{(\gamma)}: C_\Gamma(\gamma)\backslash V^\gamma \to \psi(V^\gamma) \times \{(\gamma)\} \subset \Lambda M \ .$$

In general, the inertia orbifold is not effective. The multiplicity $m(p,(\gamma))$ defines a locally constant function on M that says how many elements of the isotropy group $C_\Gamma(\gamma)$ act trivially on the fixpoint set V^γ.

An orbifold vector bundle E over M is given by trivialisations of $\psi^* E \to V$ for all parametrisations ψ, together with an action of the isotropy group Γ on $\psi^* E$ and compatible gluing data. A smooth section is represented locally by a Γ-equivariant section of $\psi^* E$. There are natural notions of Dirac bundles and Dirac operators. Because $\Gamma(E)$ is a vector space, one can define the index and the η-invariant of a Dirac operator as before.

On ΛM, one defines characteristic differential forms $\hat{A}_{\Lambda M}$ and $\mathrm{ch}_{\Lambda M}$ such that

$$\psi_{(\gamma)}^* \hat{A}_{\Lambda M}(TM,\nabla^{TM}) = \frac{1}{m(\gamma)} \hat{A}_\gamma(TM,\nabla^{TM}) \in \Omega^\bullet(V^\gamma, o(V^\gamma))$$

$$\text{and} \qquad \psi_{(\gamma)}^* \mathrm{ch}_{\Lambda M}(E/S,\nabla^E) = \mathrm{ch}_\gamma(E/S,\nabla^E) \in \Omega^\bullet(V^\gamma) \ .$$

Apart from the multiplicity, these forms are the same as in Theorem 2.2. In particular, the signs of both forms depend on the choice of a lift of γ, but their product is well-defined. The integrand $\hat{A}_{\Lambda M}(TM,\nabla^{TM}) \wedge \mathrm{ch}_{\Lambda M}(E/S,\nabla^E)$ on ΛM is the same as in Kawasaki's index theorem, and on the regular part of $M \cong M \times \{\mathrm{id}\} \subset \Lambda M$, it agrees with the classical index density on a manifold.

We now assume that N is an orbifold with boundary M, and that D_N, D_M are Dirac operators satisfying (1).

Theorem 2.7 (Farsi [38]). *The orbifold index under APS boundary conditions is given by*

$$\operatorname{ind}_{\mathrm{APS}}(D_N) = \int_{\Lambda N} \hat{A}_{\Lambda N}\big(TN, \nabla^{TN}\big)\, \mathrm{ch}_{\Lambda N}\big(E/S, \nabla^E\big) - \frac{\eta + h}{2}(D_M)\,.$$

If N is a quotient of a compact manifold by a finite group of isometries, then Theorem 2.7 can be deduced from Theorem 2.2. In general, one combines the proof of Kawasaki's index theorem with the proof of Theorem 1.1.

3 Properties of η-Invariants

We state some formulas that do not directly follow from the Atiyah–Patodi–Singer index theorem and its generalisations in the previous section. The formulas are useful to understand properties of secondary invariants derived from η-invariants as in Sect. 4, and sometimes even to compute them.

3.1 *The Adiabatic Limit*

Let $p\colon M \to B$ be proper Riemannian submersion with fibre F and $TM = T^H M \oplus TF$ with $T^H M = TF^\perp \cong p^*TB$. Write $g^{TM} = g^{TF} \oplus p^* g^{TB}$ and define

$$g_\varepsilon^{TM} = g^{TF} \oplus \frac{1}{\varepsilon^2} g^{TB}\,.$$

The limit $\varepsilon \to 0$ is called the *adiabatic limit*. As the distance between different fibres becomes arbitrarily large in the adiabatic limit, heat kernels of adapted Laplacians localise to a fibrewise operators as $\varepsilon \to 0$ for bounded times. This allows to localise a large part of the integral (3) to the fibres of p.

Let $(D_{M,\varepsilon})_{\varepsilon>0}$ be a family of Dirac operators on a bundle $E \to M$ that are compatible with the metrics g_ε^{TM}. We assume that the connections $\nabla^{E,\varepsilon}$ converge to a limit connection $\nabla^{E,0}$. Associated to the limit $\varepsilon \to 0$, there exists a family of superconnections $(\mathbb{A}_t)_{t>0}$ as in Sect. 2.1. The vertical Dirac operator D_F appears as the degree zero component of $\mathbb{A}_1$. We assume that $H = \ker D_F$ forms a vector bundle over B. Then we can define the η-form $\eta(\mathbb{A}) \in \Omega^\bullet(B)$ as in (10). More precisely, if $SB \to B$ is a local spinor bundle on B, then there exists a fibrewise Dirac bundle $W \to M$ such that $E = p^*SB \otimes W$, and we consider the η-form of a superconnection $\mathbb{A}$ on $p_* W$.

The bundle $H \to B$ with the connection induced by $\nabla^{E,0}$ becomes a Dirac bundle on (B, g^{TB}), and one can construct a limit Dirac operator D_B^0 acting on H. We assume that $D_{M,\varepsilon}$ can be continued analytically in ε to $\varepsilon = 0$. Then $\ker D_{M,\varepsilon}$ has constant dimension for all $\varepsilon \in (0, \varepsilon_0)$ if $\varepsilon_0 > 0$ is sufficiently small. There are finitely many *very small eigenvalues* $\lambda = \lambda_\nu(\varepsilon)$ of $D_{M,\varepsilon}$ such that

$$\lambda_\nu(\varepsilon) = O(\varepsilon^2) \qquad \text{and} \qquad 0 \neq \lambda_\nu(\varepsilon) \text{ for } \varepsilon \in (0, \varepsilon_0) .$$

Theorem 3.1 (Bismut–Cheeger [11]; Dai [30]). *Under the assumptions above and for* $\varepsilon \in (0, \varepsilon_0)$*, one has*

$$\lim_{\varepsilon \to 0} \eta(D_{M,\varepsilon}) = \int_B \hat{A}(TB, \nabla^{TB}) 2\eta(\mathbb{A}) + \eta(D_B^0) + \sum_\nu \operatorname{sign}(\lambda_\nu(\varepsilon)) .$$

Both the Levi–Civita connection $\nabla^{TM,\varepsilon}$ and the connection $\nabla^{E,\varepsilon}$ converge as $\varepsilon \to 0$, so one can still define Chern–Simons classes as in (4). Moreover, the spectral flow of (9) eventually becomes constant by our assumptions above, so one can recover $\eta(D_M)$ from

$$\begin{aligned}\frac{\eta + h}{2}(D_M) &= \lim_{\varepsilon \to 0} \Big(\frac{\eta + h}{2}(D_{M,\varepsilon}) - \operatorname{sf}\big((D_{M,s})_{s \in (\varepsilon,1]}\big)\Big) \\ &\quad + \int_M \Big(\tilde{\hat{A}}(TM, \nabla^{TM,0}, \nabla^{TM}) \operatorname{ch}\big(E/S, \nabla^{E,0}\big) \\ &\quad + \hat{A}\big(TM, \nabla^{TM}\big) \tilde{\operatorname{ch}}\big(E/S, \nabla^{E,0}, \nabla^{E}\big)\Big) .\end{aligned}$$

Theorem 3.1 can be generalised to Seifert fibrations. Here, a *Seifert fibration* is a map $p \colon M \to B$, where M is a manifold and B an orbifold, such that locally for a parametrisation ψ as in Sect. 2.4, p pulls back to

$$\psi^* p \colon \psi^* M \cong V \times F \to V .$$

Then we call F the generic fibre of p. Equivalently, a Seifert fibration is a Riemannian foliation of M with compact leaves. We define metrics g_ε^{TM} as above.

Over the inertia orbifold ΛB, we define an equivariant η-form $\eta_{\Lambda B}(\mathbb{A})$ such that

$$\psi_{(\tilde{\gamma})}^* \eta_{\Lambda B}(\mathbb{A}) = \frac{1}{\sqrt{\pi}} \int_0^\infty \operatorname{tr}\Big(\tilde{\gamma} \frac{\partial \mathbb{A}_t}{\partial t} e^{-\mathbb{A}_t^2}\Big)\, dt \in \Omega^\bullet(V^\gamma) .$$

Again, the sign of $\eta_{\Lambda B}(\mathbb{A})$ depends on the choice of a certain lift $\tilde{\gamma}$ of γ, but the integrand $\hat{A}_{\Lambda B}\big(TB, \nabla^{TB}\big)\, 2\eta_{\Lambda B}(\mathbb{A})$ in the theorem below is well-defined.

We assume that $\ker D_F$ forms an orbifold vector bundle over B and define $\eta(D_B^H)$ as in Sect. 2.4.

Theorem 3.2 ([44]). *Under the assumptions above and for $\varepsilon \in (0, \varepsilon_0)$, one has*

$$\lim_{\varepsilon \to 0} \eta(D_{M,\varepsilon}) = \int_{\Lambda B} \hat{A}_{\Lambda B}(TB, \nabla^{TB})\, 2\eta_{\Lambda B}(\mathbb{A}) + \eta(D_B^H) + \sum_\nu \operatorname{sign}(\lambda_\nu(\varepsilon)) .$$

It is likely that this result still holds if M is an orbifold, provided the generic fibres are still compact manifolds.

Remark 3.3. In principle, Theorems 3.1 and 3.2 simplify the computations of η-invariants and other invariants derived from them as in Sect. 4. However, the η-forms needed are at least as difficult to compute as the η-invariants of the fibres. There are explicit formulas for circle bundles in [84] and three-sphere bundles in [42]. In [31], the Kreck–Stolz invariants of Sect. 4.4 are computed this way for circle bundles. And in Sect. 2.3, we have exhibited a method of computation if the structure group is compact and the fibre a quotient of compact Lie groups.

If the family $M \to B$ bounds a family $N \to B$ as in Sect. 2.1, one can use the original Atiyah–Patodi–Singer Theorem 1.1 in place of Theorem 3.1, see [8] for the case of circle bundles. Similarly, one can use Theorem 2.7 in place of Theorem 3.2 if the generic fibre F bounds a compact manifold and one can construct an orbifold fibre bundle $N \to B$ that bounds M. Nevertheless, the computation of the local index density still requires some work.

3.2 Gluing Formulas

The η-invariants of connected sums can be computed by applying the APS Index Theorem 1.1 to the boundary connected sum $M_1 \times [0,1] \natural M_2 \times [0,1]$ with boundary $-(M_1 \# M_2) \sqcup M_1 \sqcup M_2$. Because Pontrijagin forms are conformally invariant, one can choose the geometry in such a way that the index density vanishes completely. Hence,

$$\frac{\eta + h}{2}(D_1 \# D_2) = \frac{\eta + h}{2}(D_1) + \frac{\eta + h}{2}(D_2) \in \mathbb{R}/\mathbb{Z}$$

under suitable geometric assumptions. As a consequence, many of the invariants introduced in Sect. 4 are additive under connected sums. We will now describe the behaviour of η-invariants under gluing along more complicated hypersurfaces.

We assume that M can be cut along a hypersurface N in two pieces M_1 and M_2. We also assume that N has a neighbourhood U isometric to $N \times (-\varepsilon, \varepsilon)$. Let $A = D_M$ be a Dirac operator on M that is of a form similar to (1) on U, with $B = D_N$ a Dirac operator on N and $\nu = c_M\left(\frac{\partial}{\partial t}\right)$. If D_N is invertible, one can define η-invariants for the operators $D_{M_i} = D_M|_{M_i}$ under APS boundary conditions similar to (2).

Theorem 3.4 (Wojciechowski [82]; Bunke [20]). *If D_N is invertible, then*

$$\eta(D_M) = \eta(D_{M_1}) + \eta(D_{M_2}) \qquad \in \mathbb{R}/\mathbb{Z} .$$

This formula holds in $\mathbb{R}$ up to an integer correction term that is also described in [20] and [82]. If D_N is not invertible, one chooses Lagrangian subspaces L_1, $L_2 \subset \ker(D_N)$, with respect to a symplectic structure on $\ker(N)$ defined in terms of the Clifford volume element on N. The APS boundary conditions modified by the projections onto these subspaces give rise to selfadjoint operators D_{M_i,L_i}. Their η-invariants are described by Lesch and Wojciechowski in [62]. Bunke and Wojciechowski generalise Theorem 3.4 to this setting in [20] and [83]. Their formula involves the Maslov index of the Lagrangians L_1, L_2.

Gluing results for η-invariants as in Theorem 3.4 allow to understand the behaviour of the secondary invariants of Sect. 4 under operations like surgery. But since manifolds with boundary appear only as intermediate steps in these constructions, it would be nice to have a general gluing formula where no manifolds with boundary occur. Bunke states such a result in [20].

3.3 Embeddings

In this section, let $\iota\colon M \to N$ be a smooth embedding of compact spin manifolds. If D_M is a Dirac operator on M, one constructs a K-theoretic direct image D_N on N and compares the associated η-invariants. Hence, the main result of this section is similar in spirit to Theorem 3.1 of Bismut–Cheeger and Dai.

More precisely, let $SM \to M$ and $SN \to N$ denote spinor bundles on M and N. Then the normal bundle $\nu \to M$ of the embedding has a spinor bundle $S\nu \to M$ such that $SN|_M \cong SM \otimes S\nu$. A *direct image* of a complex vector bundle $V \to M$ consists of a complex vector bundle $W = W^+ \oplus W^- \to N$ and a selfadjoint endomorphism $A = a + a^*$ of W with $a\colon W^+ \to W^-$, such that A is invertible on $N \setminus M$ and degenerates linearly along M, and one has an isomorphism $\ker A \cong S\nu \otimes V$ that relates the compression of $dA|_\nu$ to $\ker A$ to Clifford multiplication by normal vectors. We also assume that V, W and ν carry compatible metrics and connections, see [17] for details.

Let δ_M denote the current of integration on $M \subset N$. Then there exists a natural current $\gamma(W, \nabla^W, A)$ on N such that

$$d\gamma(W,\nabla^W,A) = \operatorname{ch}(W^+,\nabla^{W^+}) - \operatorname{ch}(W^-,\nabla^{W^-}) - \hat{A}^{-1}(\nu,\nabla^\nu)\operatorname{ch}(V,\nabla^V)\delta_M\ .$$

Theorem 3.5 (Bismut–Zhang [17]). *Under the assumptions above,*

$$\frac{\eta+h}{2}(D_N^{W+}) - \frac{\eta+h}{2}(D_N^{W-}) = \frac{\eta+h}{2}(D_M^V) + \int_N \hat{A}(TN,\nabla^{TN})\gamma(W,\nabla^W,A)$$
$$+ \int_M \tilde{\hat{A}}(TN|_M,\nabla^{TM\oplus\nu},\nabla^{TN})\hat{A}^{-1}(\nu,\nabla^\nu)\operatorname{ch}(V,\nabla^V) \in \mathbb{R}/\mathbb{Z}\,.$$

One can get rid of the last term on the right hand side by assuming that M is totally geodesic in N. Moreover, if A is ∇^W-parallel outside a small neighbourhood of M in N, then $\gamma(W,\nabla^W,A)$ is supported near M. In this case, the difference of the η-invariants of D_N^{W+} and D_N^{W-} localises near M. On the other hand, let $M=\emptyset$, so $a\colon W^+\to W^-$ is an isomorphism of vector bundles. Then Theorem 3.5 reduces to Corollary 1.3 (5) with g_s constant.

Remark 3.6. Theorem 3.5 is formally similar to Theorem 3.1. In fact, since every proper map $F\colon M\to N$ can be decomposed into the embedding $M\to M\times N$ given by the graph of F, followed by projection onto N, both theorem can be combined to compute η-invariants of direct images under proper maps. These direct images carry additional geometric information (like a connection), so one would like to have a generalisation of topological K-theory that takes care of the relevant additional data. In principle, some kind of smooth K-theory should be the right choice for this, but it seems difficult to construct a smooth K-theory that covers both proper submersions and embeddings.

4 Differential Topological Invariants

We have seen that η-invariants have local variation formulas with respect to variations of the geometric structure, but they still contain global differential-topological information. The common theme of the following sections will be the construction of invariants that do not depend on the geometric structure of the manifold.[1]

4.1 Invariants of Flat Vector Bundles

Let $\alpha\colon\pi_1(M)\to U(n)$ be a unitary representation of the fundamental group, then $F_\alpha=\tilde M\times_\alpha\mathbb{C}^n\to M$ is a flat Hermitian vector bundle with holonomy α. In particular, we may regard the twisted odd signature operator B_α acting on even F_α-valued smooth forms on M. Because $\ker(B_\alpha)=H^{\mathrm{ev}}(M;F_\alpha)$ is independent of the Riemannian metric on M, the variation formula in Corollary 1.3 becomes

$$\eta(B_\alpha^1)-\eta(B_\alpha^0)=\int_M n\tilde L\big(TM,\nabla^{TM,0},\nabla^{TM,1}\big)\in\mathbb{R}\,.$$

Note that α enters on the right hand side only through its rank $n=\mathrm{ch}(F_\alpha)$.

Theorem 4.1 (Atiyah–Patodi–Singer [4]). *The* ρ-invariant

$$\rho_\alpha(M)=\eta(B_\alpha)-n\,\eta(B)\in\mathbb{R}$$

is a diffeomorphism invariant of M *and* α.

---

[1]Bunke gives a systematic construction of such invariants in: Bunke, U.: On the topological contents of eta invariants. Preprint, `arXiv: 1103.4217`

If α factors through a finite group G, one can consider the compact manifold $\bar{M} = \tilde{M}/\ker\alpha$. Then G acts on $\bar{M}$ with quotient M, and one can compute $\rho_\alpha(M)$ from the equivariant signature η-invariants $\eta_g(\bar{M})$ of M. This proves in particular that $\rho_\alpha(M)$ is rational in this case. The equivariant η-invariants here are related to the invariants $\sigma_g(M)$ considered in [6].

If $\pi_1(M)$ is torsion free and a certain Baum–Connes assembly map is an isomorphism, then $\rho_\alpha(M)$ is a homotopy invariant. This is proved by Keswani in [53] and generalised by Piazza and Schick [73], earlier similar results are due to Neumann [70], Mathai [64] and Weinberger [81]. Hence for such fundamental groups, $\rho_\alpha(M)$ behaves almost as a primary invariant.

If one replaces the odd signature operator in the construction of $\rho_\alpha(M)$ by a different Dirac operator D on M, one gets similar invariants with values in $\mathbb{R}/\mathbb{Z}$ due to the possible spectral flow. However, instead of diffeomorphism invariants, one now obtains cobordism invariants.

Theorem 4.2 (Atiyah–Patodi–Singer [4,5]). *The ξ-invariant*

$$\xi_\alpha(D) = \frac{\eta+h}{2}(D_\alpha) - n\cdot\frac{\eta+h}{2}(D) \in \mathbb{R}/\mathbb{Z}$$

is a cobordism invariant in the sense that $\xi_\alpha(D) = 0$ if there exists a compact manifold N with $M = \partial N$ such that D extends to an operator on N in the sense of (1) *and α extends to a representation of $\pi_1(N)$.*

The representation α defines a class $[\alpha] \in K^{-1}(M;\mathbb{R}/\mathbb{Z})$ and the symbol of D gives $\sigma \in K^1(TM)$. Then there exists a topological index $\operatorname{Ind}_{[\alpha]}(\sigma) \in \mathbb{R}/\mathbb{Z}$, *and*

$$\xi_\alpha(D_M) = \operatorname{Ind}_{[\alpha]}(\sigma)\,.$$

Example 4.3. If M is spin and SM is a fixed spinor bundle on M, then all other spinor bundles arise by twisting SM with real line bundles. In particular, the difference of $\frac{\eta+h}{2}(D)$ for different spin structures is a ξ-invariant. Real line bundle are classified by $H^1(M;\mathbb{Z}/2\mathbb{Z})$. Dahl investigates these ξ-invariants for spin structures induced by the mod 2 reduction of integer classes, and also their dependence on the initial spin structure in [29].

It is possible to define $\xi_\alpha(D) \in \mathbb{C}/\mathbb{Z}$ for flat vector bundles associated to non-unitary representations $\alpha\colon \pi_1(M) \to Gl(n,\mathbb{C})$, see [5, 52]. In this case, the imaginary part of $\xi_\alpha(D)$ is related to the Kamber–Tondeur classes (also known as Borel classes) of α. Choose a Hermitian metric on F_α and a unitary connection ∇^u on F_α and let D_u be the Dirac operator twisted by (F_α, ∇^u). Arguing as in [16],

$$\begin{aligned}\operatorname{Im}\xi_\alpha(D) &= \operatorname{Im}\left(\frac{\eta+h}{2}(D_\alpha) - \frac{\eta+h}{2}(D_u)\right)\\ &= \Big(\hat{A}(TM)\,\operatorname{Im}\tilde{\operatorname{ch}}\big(F_\alpha,\nabla^u,\nabla^\alpha\big)\Big)[M]\,,\end{aligned}$$

and $\operatorname{Im}\tilde{\operatorname{ch}}\big(F_\alpha,\nabla^u,\nabla^\alpha\big) \in H^{\mathrm{odd}}(M;\mathbb{R})$ represents the Kamber–Tondeur class.

Assume that M is an m-dimensional homology sphere, then the fundamental group $\Gamma = \pi_1(M)$ satisfies $\Gamma = [\Gamma, \Gamma]$. Let $F_\alpha \to M$ be associated to a representation $\alpha\colon \Gamma \to Gl(n, \mathbb{C})$, classified by a map $M \to BGL(n, \mathbb{C})$. Then Quillen's plus construction by functoriality gives an element $[M, \alpha]$ of the algebraic K-group $K_m(\mathbb{C}) = \pi_m(BGL(n, \mathbb{C})^+)$ by

$$S^m = M^+ \longrightarrow BGL(n, \mathbb{C})^+ \,.$$

If m is odd, clearly $\hat{A}(TM) = 1 \in H^{\mathrm{ev}}(M; \mathbb{Q})$ because M is a homology sphere, in particular, $\operatorname{Im} \xi_\alpha(D) = \operatorname{Im} \tilde{\mathrm{ch}}\big(F_\alpha, \nabla^u, \nabla^\alpha\big)[M]$ then gives the Borel regulator of $[M, \alpha] \in K_m(\mathbb{C})$, see [52]. Jones and Westbury prove that the map $(M, \alpha) \mapsto \xi_\alpha(D)$ induces an isomorphism $K_1(\mathbb{C}) \cong \mathbb{C}/\mathbb{Z}$ for $m = 1$, and an isomorphism of the torsion subgroup of $K_m(\mathbb{C})$ with $\mathbb{Q}/\mathbb{Z}$ for $m > 1$ odd.

Example 4.4. Let $M = \Gamma \backslash SL(2, \mathbb{C})/SU(2)$ be a hyperbolic homology three-sphere, and let $\alpha\colon \Gamma \to SL(2, \mathbb{C})$ be the representation corresponding to the embedding of Γ as a cocompact subgroup. By [52],

$$\operatorname{Im} \xi_\alpha(D) = -\frac{1}{4\pi^2} \operatorname{vol}(M) \,,$$

which proves that $[M, \alpha]$ is never torsion.

Jones and Westbury also show that all torsion elements of $K_3(\mathbb{C})$ can be realised as $[M, \alpha]$ where M now is a Seifert fibred three-manifold. A similar analysis of $K_3(\mathbb{R})$ is done in [25].

The classical Lichnerowicz theorem asserts that a spin Dirac operator on a closed spin manifold N has vanishing index if N carries a metric of positive scalar curvature $\kappa > 0$. It is shown in [4] that Lichnerowicz' theorem extends to compact spin manifolds N with totally geodesic boundary $M = \partial N$. If N has $\kappa > 0$, then so does M, so $\operatorname{ind}(D_N) = h(D_M) = 0$ for a spin Dirac operator D_N on N, and Theorem 1.1 becomes

$$\int_N \hat{A}(TN, \nabla^{TN}) = \frac{1}{2} \eta(D_M) \,.$$

The analogous statement also holds for the Dirac operator D_α twisted by a flat vector bundle on N associated to $\alpha\colon \pi_1(N) \to U(n)$.

Let us call two closed Riemannian spin manifolds (M_0, g_0), (M_1, g_1) *spin*$^+$-*cobordant* if there exists a compact Riemannian spin manifold (N, g) with totally geodesic boundary $(M_1, g_1) - (M_0, g_0)$. If $M_0 = M = M_1$ and there exists a family $(g_t)_{t \in [0,1]}$ of positive scalar curvature metrics, then the metric $g_{\varphi(t/a)} \oplus dt^2$ on $N = M \times [0, a]$ has $\kappa > 0$ for a sufficiently large, where $\varphi\colon [0, 1] \to [0, 1]$ is smooth and locally constant 0 (1) near 0 (1). Thus metrics in the same connected component of the moduli space of positive scalar curvature metrics are spin$^+$-cobordant.

Theorem 4.5 (Atiyah–Patodi–Singer [4]; Botvinnik–Gilkey [18]). *The number*

$$\bar{\xi}_\alpha(M, g) = \eta(D_\alpha) - n \cdot \eta(D) \in \mathbb{R}$$

is a spin$^+$*-cobordism invariant in the sense that* $\bar{\xi}_{\alpha_0}(M_0, g_0) = \bar{\xi}_{\alpha_1}(M_1, g_1)$ *if* (N, g) *is a spin*$^+$*-cobordism of* (M_0, g_0) *and* (M_1, g_1) *and* α_0, α_1 *extend to a representation of* $\pi_1(N)$.

This result is used by Botvinnik and Gilkey to construct and detect Riemannian metrics with $\kappa > 0$ lying in countably many different connected components in the moduli space of positive scalar curvature metrics on M, whenever M has a non trivial finite fundamental group and admits at least one metric of positive scalar curvature [18, 19]. Apart from the computation of $\bar{\xi}_\alpha$ for sufficiently many examples, the proof relies on the surgery techniques for positive scalar curvature metrics introduced by Gromov and Lawson in [46]. Using various generalisations of η-invariants, the results of Botvinnik–Gilkey are extended to manifolds M whose fundamental group contains torsion by Leichtnam and Piazza [61] and Piazza and Schick [74]. Other (and in fact earlier) results in this direction will be discussed at the end of Sect. 4.3.

On the other hand, if $\pi_1(M)$ is torsion free and a certain Baum–Connes assembly map is an isomorphism, then $\bar{\xi}_\alpha(D) = 0$ for the untwisted Dirac operator D by a result of Piazza and Schick [73]. Hence, for such fundamental groups, the invariant $\bar{\xi}_\alpha(D)$ behaves similar as the index of the untwisted Dirac operator in Lichnerowicz' theorem.

4.2 The Adams *e*-Invariant

In this subsection, we regard a framed bordism invariant. Recall that a closed manifold M is framed by an embedding $M \hookrightarrow \mathbb{R}^n$ for n sufficiently large together with a trivialisation of the normal bundle $\nu \to M$ of the embedding. This defines a stable parallelism of TM, i.e., a trivialisation of $TM \oplus \mathbb{R}^{n-m}$, because

$$TM \oplus \mathbb{R}^{n-m} \cong TM \oplus \nu \cong M \times \mathbb{R}^n .$$

In fact, framings and stable parallelisms are equivalent notions. By the Pontrijagin–Thom construction, the framed bordism classes of manifolds of dimension m are in bijection with the m-th stable homotopy group π_m^s of spheres.

Let M be a framed closed manifold of dimension $4k - 1$ with parallelism π. Then M carries a preferred spin structure. Because the spin cobordism group in dimension $4k - 1$ is trivial, there exists a compact spin manifold N with $\partial N = M$. Because TM is stably trivial, there exists a well-defined relative class $\hat{A}(TN) \in H^\bullet(N, M; \mathbb{Q})$ and one defines

$$e(M,\pi) = \begin{cases} \hat{A}(TN)[N] & \text{if } k \text{ is even, and} \\ \frac{1}{2}\hat{A}(TN)[N] & \text{if } k \text{ is odd .} \end{cases} \tag{14}$$

On the other hand, since TM is stably trivial, we can consider the Chern–Simons class $\tilde{\hat{A}}(TM, \nabla^{TM}, \nabla^{\pi})$, where ∇^{TM} denotes the connection induced on $TM \oplus \mathbb{R}^{n-m}$ by the Levi–Civita connection with respect to a Riemannian metric, and ∇^{π} the connection induced by the trivialisation. It follows from Corollary 1.3 that

$$\frac{\eta + h}{2}(D_M) + \int_M \tilde{\hat{A}}(TM, \nabla^{TM}, \nabla^{\pi})$$

is invariant under variation of g modulo $\mathbb{Z}$ if k is even, and modulo $2\mathbb{Z}$ if k is odd due to a quaternionic structure on the spinor bundle of M.

Theorem 4.6 (Atiyah–Patodi–Singer [4]). *The e-invariant of a framed $4k - 1$-dimensional manifold (M, π) is given by*

$$e(M,\pi) = \varepsilon(k)\Big(\frac{\eta + h}{2}(D_M) + \int_M \tilde{\hat{A}}(TM, \nabla^{TM}, \nabla^{\pi})\Big) \in \mathbb{Q}/\mathbb{Z}$$

with $\varepsilon(k) = 1$ if k is even and $\varepsilon(k) = \frac{1}{2}$ if k is odd.

Seade uses this formula to determine the e-invariants of quotients of S^3 in [76].

Example 4.7. Let $H(n) \subset Gl_{n+2}(\mathbb{R})$ denote the $2n + 1$-dimensional Heisenberg group and let $\Gamma(n) = H(n) \cap Gl_{n+2}(\mathbb{Z})$ be the subgroup with integer entries. Then $TH(n)$ is trivialised by right translation, and this descends to a trivialisation π of $TH(n)/\Gamma(n)$. For odd $n = 2k - 1$, the e-invariant is calculated by Deninger and Singhof in [34],

$$e(H(n)/\Gamma(n), \pi) = -(-1)^k \varepsilon(k) \zeta(-n) + \delta(n) ,$$

where $\delta(1) = \frac{1}{2}$ and $\delta(n) = 0$ otherwise. Here ζ is the Riemann zeta function. Comparing with the possible values of $e(M,\pi)$, one sees that $e(H(n)/\Gamma(n))$ is a generator of $\operatorname{im}(e\colon \pi^s_{4k-1} \to \mathbb{Q}/\mathbb{Z})$ for odd k and twice a generator for even k.

For the proof, the Dirac operator D is replaced by $\tilde{D}$, where $\tilde{D} - D$ is an operator of order 0. The spectrum and the η-invariant of $\tilde{D}$ are computed explicitly. Since $e(H(n)/\Gamma(n), \pi) - \varepsilon(k) \cdot \frac{\eta+h}{2}(\tilde{D}) \in \mathbb{R}/\mathbb{Z}$ is given as the integral of a locally defined invariant density on $H(n)/\Gamma(n)$, an argument involving finite covering spaces allows to reconstruct the e-invariant from the η-invariant of the modified operator $\tilde{D}$.

Bunke and Naumann give a similar description of the f-invariant [21]. Their formula uses η-invariants on manifolds with boundary that are related to a certain elliptic genus.

4.3 The Eells–Kuiper Invariant

In this section, we consider closed oriented spin manifolds M of dimension $m = 4k - 1$ such that

$$H^{4l}(M;\mathbb{R}) = 0 \qquad \text{for all } l \geq 1 . \tag{15}$$

If M bounds a compact spin manifold N, this conditions allows to define relative Pontrijagin classes $p_j(TN) \in H^{4j}(N, M;\mathbb{Q})$ for $1 \leq j < k$. We express the universal characteristic classes $\hat{A}$ and L in terms of Pontrijagin classes as in (7), (8). Then there exists a unique constant $t_k \in \mathbb{Q}$ such that the homogeneous component $(\hat{A} - t_k L)^{[4k]}$ in degree $4k$ does not involve p_k. With $\varepsilon(k)$ as in Theorem 4.6, the Eells–Kuiper invariant of M is defined in [37] as

$$\mu(M) = \varepsilon(k) \left(t_k \operatorname{sign}(N) + (\hat{A} - t_k L)(TN)[N, M]\right) \in \mathbb{Q}/Z . \tag{16}$$

Condition (15) allows one to express the Pontrijagin forms $p_j(TM, \nabla^{TM})$ with respect to some Riemannian metric g on M as

$$p_j(TM, \nabla^{TM}) = d\hat{p}_j(TM, \nabla^{TM}) \tag{17}$$

for $1 \leq j < k$. Moreover, $\hat{p}_j(TM, \nabla^{TM}) \in \Omega^{4j-1}(M)/\operatorname{im} d$ is unique because $H^{4j-1}(M;\mathbb{R}) = 0$ by Poincaré duality. Replacing one factor p_j in each monomial of $(\hat{A} - t_k L)^{[4k]}(TM)$ by $\hat{p}_j$, we obtain a natural class $\alpha(TM, \nabla^{TM}) \in H^{4k-1}(M;\mathbb{R}) = \Omega^{4k-1}(M)/\operatorname{im} d$ such that

$$\alpha(TM, \nabla^{TM,1}) - \alpha(TM, \nabla^{TM,0}) = \left(\tilde{\hat{A}} - t_k \tilde{L}\right)\left(TM, \nabla^{TM,0}, \nabla^{TM,1}\right)$$

for any two connections $\nabla^{TM,0}$, $\nabla^{TM,1}$ on TM. Note that α does not depend on the choice of the factors p_j above, because

$$(p_i\,\hat{p}_j - \hat{p}_i\,p_j)(TM, \nabla^{TM}) = d\left((\hat{p}_i\,\hat{p}_j)(TM, \nabla^{TM})\right) .$$

Let D again be the spin Dirac operator and B the odd signature operator on M. The following result is a consequence of Theorems 1.1 and 1.2.

Theorem 4.8 (Donnelly [35]; Kreck–Stolz [56]). *The Eells–Kuiper invariant of M equals*

$$\mu(M) = \varepsilon(k) \cdot \left(\frac{\eta + h}{2}(D) - t_k\,\eta(B) - \alpha(TM, \nabla^{TM})[M]\right) \in \mathbb{Q}/\mathbb{Z} .$$

Remark 4.9. Other interesting invariants have expressions similar to (16):

1. The Eells–Kuiper invariant distinguishes all diffeomorphism types of exotic spheres that bound parallelisable manifolds in dimension $4k - 1$ for $k = 1$,

2, 3, see [37]. Stolz constructs a similar invariant that detects all exotic spheres bounding parallelisable manifolds in all dimensions $4k - 1$ in [78]. Stolz' invariant also has a presentation in terms of η-invariants and Cheeger–Simons correction terms.
2. Rokhlin's theorem says that the signature of a spin manifold in dimension $8k+4$ is divisible by 16. One can define a secondary Rokhlin number in $\mathbb{R}/16\mathbb{Z}$ for spin structures on $8k + 3$-dimensional manifolds. Lee and Miller express the Rokhlin number as a linear combination of η-invariants as above and without local correction terms [67]. In particular, condition (15) is not needed. If spin structures differ only by the mod 2 reduction of an integer cohomology class, then Dahl proved that the Rokhlin number mod 8 remains unchanged [29].

Note that $\mu(M)$ changes sign if the orientation of M is reversed, and that μ is additive under connected sums. Hence, given any closed spin $4k - 1$-manifold M that satisfies assumption (15) and an exotic sphere Σ, $\mu(M\#\Sigma^{\#r})$ takes as many different values in $\mathbb{Q}/\mathbb{Z}$ as $\mu(\Sigma^{\#r})$ does for $r \in \mathbb{Z}$. This way, one can construct and detect a certain number of exotic smooth structures on manifolds M for which $\mu(M)$ is defined.

As an example, for $k = 2$ the Eells–Kuiper invariant becomes

$$\mu(M) = \frac{\eta + h}{2}(D) + \frac{\eta}{2^5 \cdot 7}(B) - \frac{1}{2^7 \cdot 7}(p_1\, \hat{p}_1)\big(TM, \nabla^{TM}\big)[M] \,.$$

This invariant is one of the main ingredients in the diffeomorphism classification of S^3-bundles over S^4 by Crowley–Escher [27], and also in the examples discussed in Sect. 4.5.

We now come back to manifolds of positive scalar curvature. Note that $\mu(M)$ is not a spin-cobordism invariant because $\mu(M_1) - \mu(M_0)$ depends on $t_k \cdot \operatorname{sign}(N)$ if $\partial N = M_1 - M_0$ by (16). We thus cannot expect to refine $\mu(M)$ to a spin$^+$-cobordism invariant. Thus, we call two positive scalar curvature metrics g_0, g_1 on M *concordant* if there exists a positive scalar curvature metric g on $M \times [0, T]$ for some $T > 0$ that is isometric to $g_0 \times dt^2$ on $M \times [0, \varepsilon)$ and to $g_1 \times dt^2$ on $M \times (T - \varepsilon, T]$. From the discussion preceding Theorem 4.5, we see that metrics in the same connected component of the space of scalar curvature metrics are concordant.

Theorem 4.10 (Kreck–Stolz [58]). *Let M be a closed spin $4k - 1$-manifold satisfying* (15). *Then the refined Eells–Kuiper invariant*

$$\bar{\mu}(M, [g]) = \varepsilon(k)\Big(\frac{\eta}{2}(D) - t_k\, \eta(B) - \alpha\big(TM, \nabla^{TM}\big)[M]\Big) \in \mathbb{R}$$

is well-defined on concordance classes $[g]$ of positive scalar curvature metrics on M. Moreover, if $[g_0], [g_1]$ are two such concordance classes, then

$$\bar{\mu}(M, [g_1]) - \bar{\mu}(M, [g_0]) \in \mathbb{Z} \,.$$

All members of the family of Aloff–Wallach spaces $M \cong SU(3)/S^1$ allow metrics of positive sectional curvature, and the numbers $\bar{\mu}(M)$ are computed in [57]. For Witten's family of Ricci-positive homogeneous Einstein manifolds $M \cong SU(3) \times SU(2) \times S^1/SU(2) \times S^1 \times S^1$, the invariants $\bar{\mu}(M)$ are computed in [56].

Theorem 4.11 (Kreck–Stolz [58]).

1. *There exist closed manifolds with a non-connected moduli space of positive sectional curvature metrics.*
2. *There exist closed manifolds for which the moduli space of Ricci positive metrics has infinitely many connected components.*

4.4 Kreck–Stolz Invariants of Complex and Quaternionic Line Bundles

In [56], Kreck and Stolz define three invariants that determine the diffeomorphism type of certain 7-manifolds completely. For spin manifolds, their first invariant is precisely the Eells–Kuiper invariant of the previous section. The other two invariants use Dirac operators twisted by complex line bundles. For non-spin-manifold, similar invariants are defined that use a spin^c-Dirac operator in place of the Dirac operator. We will restrict attention to the spin case for simplicity.

Thus assume that M is a closed simply connected spin 7-manifold with

$$H^1(M) = H^3(M) = 0, \qquad H^2(M) \cong \mathbb{Z}, \qquad \text{and} \qquad H^4(M) \cong \mathbb{Z}/\ell\mathbb{Z}\,, \tag{18}$$

where $H^4(M)$ is generated by the square of a generator of $H^2(M)$. In particular, condition (15) holds. Since $H^2(M)$ classifies complex line bundles, for each class $a \in H^2(M)$, there exists a complex line bundle $L \to M$ with Chern class $c_1(L) = a$, which is unique up to isomorphism. Let ∇^L be a connection, then as in (17) above, there exists a unique class $v(L, \nabla^l) \in \Omega^3(M)/\operatorname{im} d$ such that

$$dv(L, \nabla^L) = c_1(L, \nabla^L)^2\,.$$

We define a universal formal power series ch' in c_1 such that

$$\mathrm{ch}(L) - 1 - c_1(L) = c_1^2(L)\,\mathrm{ch}'(L)\,.$$

Now, let D^L denote the Dirac operator twisted by (L, ∇^L), and put

$$\begin{aligned} s_M(a) = {}& \frac{\eta + h}{2}(D^L) - \frac{\eta + h}{2}(D) \\ & - \Big(\hat{A}\big(TM, \nabla^{TM}\big)\,\big(v\,\mathrm{ch}'\big)\big(L, \nabla^L\big)\Big)[M] \in \mathbb{Q}/\mathbb{Z}\,. \end{aligned} \tag{19}$$

Let $u \in H^2(M)$ be a generator, then the remaining two Kreck–Stolz invariants are given by

$$s_2(M) = s_M(u) \qquad \text{and} \qquad s_3(M) = s_M(2u)\,.$$

Clearly, s_2 and s_3 determine s_M completely. Also, one can recover the linking form on $H^4(M)$ and the half Pontrijagin class $\frac{p_1}{2}(TM) \in H^4(M)$ from s_2 and s_3. Indeed, M bounds a compact spin manifold N, and by Theorem 1.1,

$$s_M(a) = \big(\hat{A}(TM)(\mathrm{ch}(L) - 1)\big)[N, M] = \left(\frac{a^2}{24}\left(a^2 - \frac{p_1}{2}(TM)\right)\right)[N, M]\,. \quad (20)$$

In particular,

$$24 s_M(a) = \mathrm{lk}_M\left(a^2, a^2 - \frac{p_1}{2}(TM)\right) \in \mathbb{Q}/\mathbb{Z}\,.$$

Hepworth generalises the Kreck–Stolz classification in his thesis [50] to simply connected closed spin 7-manifolds with

$$H^1(M) = H^3(M) = 0, \qquad H^2(M) \cong \mathbb{Z}^r \qquad \text{and} \qquad \#H^4(M) < \infty\,,$$

such that $H^4(M)$ is generated by products of elements of $H^2(M)$ and $\frac{p_1}{2}(TM)$.

A 7-manifold M is called *highly connected* if $\pi_1(M) = \pi_2(M) = 0$. If $\pi_3(M)$ is finite, then

$$H^1(M) = H^2(M) = H^3(M) = 0 \qquad \text{and} \qquad \#H^4(M) < \infty\,.$$

Since $H^4(M)$ is not necessarily generated by $\frac{p_1}{2}(TM)$, the results of Hepworth do not apply. Crowley has shown in [26] that a highly connected 7-manifolds is determined up to diffeomorphism by its Eells–Kuiper invariant and a quadratic form $q_M\colon H^4(M) \times H^4(M) \to \mathbb{Q}/\mathbb{Z}$ satisfying

$$q_M(a + b) = q_M(a) + q_M(b) + \mathrm{lk}_M(a, b)$$

$$\text{and} \qquad q_M(-a) = q_M(a) + \mathrm{lk}_M\left(a, \frac{p_1}{2}(TM)\right).$$

Note that these properties do not define q_M uniquely if $H^4(M)$ has 2-torsion. An extrinsic definition of q_M using a handlebody N with $\partial N = M$ can be found in [26]. The quadratic form q_M can also be recovered from a Kreck–Stolz type invariant t that we now describe.

Assume that M is a closed $4k - 1$-dimensional spin manifold satisfying

$$H^3(M; \mathbb{R}) = H^4(M; \mathbb{R}) = 0\,. \quad (21)$$

Let $H \to M$ be a quaternionic Hermitian line bundle. Equivalently, H is a complex rank 2 vector with structure group $SU(2)$. In particular, the determinant line bundle det H is trivialised. Then the Chern character of H is a formal power series in c_2, and there exists a formal power series ch$'$ in c_2 such that

$$2 - \mathrm{ch}(H) = c_2(H) \cdot \mathrm{ch}'(H)$$

for all quaternionic line bundles H.

We fix a compatible connection ∇^H on $H \to M$. By assumption (21) and as in (17), there exists a unique class $\hat{c}_2(H, \nabla^H) \in \Omega^3(M)/\operatorname{im} d$ such that

$$d\hat{c}_2(H, \nabla^H) = c_2(H, \nabla^H) .$$

Let D^H denote the Dirac operator twisted by H and note that $S \otimes H$ carries a quaternionic structure if and only if S carries a real structure and vice versa. Let $\varepsilon(k)$ be as in Theorem 4.6. In [28], we define the *t-invariant* of H in analogy with (19) by

$$t_M(H) = \varepsilon(k+1)\Bigg(\frac{\eta + h}{2}(D^H) - (\eta + h)(D)$$

$$+ \Big(\hat{A}(TM, \nabla^{TM})\big(\hat{c}_2 \,\mathrm{ch}'\big)\big(H, \nabla^H\big)\Big)[M]\Bigg) \in \mathbb{Q}/\mathbb{Z} . \qquad (22)$$

If M is a highly connected closed 7-manifold with $H^4(M)$ finite, then for each $a \in H^4(M)$, there exists a quaternionic line bundle $H \to M$ with $c_2(H) = a$, and similar as in (20), we find that

$$q_M(a) = 12\, t_M(H) .$$

Note that the invariants t_M and s_M are related. Let $L \to M$ be a complex line bundle with $c_1(L) = a$. Then $H = L \oplus \bar{L}$ carries a natural quaternionic structure. It follows from (19) and (22) that $t_M(H) = 2\varepsilon(k+1)\, s_M(a)$, so t_M generalises s_M in dimension $8\ell - 1$.

Example 4.12. Let $\pi \colon M \to S^4$ be the unit sphere bundle of a real vector bundle $W \to S^4$ of rank 4, and pick a quaternionic line bundle $H \to S^4$, such that

$$n = e(W), \qquad p = \frac{p_1}{2}(W), \qquad \text{and} \qquad a = c_2(H) \in \mathbb{Z} \cong H^4(S^4) .$$

Note that n and p are of the same parity. Then $c_2(\pi^* H) = a \in \mathbb{Z}/n\mathbb{Z} \cong H^4(M)$ by the Gysin sequence. As shown in [28],

$$t_M(\pi^* H) = \frac{a(p-a)}{24n} \qquad \text{and} \qquad q_M(a) = \frac{a(p-a)}{2n} \in \mathbb{Q}/\mathbb{Z} .$$

Together with the computation of the Eells–Kuiper invariant

$$\mu(M) = \frac{p^2 - n}{2^5 \cdot 7 \cdot n} \in \mathbb{Q}/\mathbb{Z}$$

in [27], one can recover the Crowley–Escher diffeomorphism classification of S^3-bundles over S^4.

The above example already shows that different quaternionic line bundles can have the same second Chern class, but different t-invariants. In fact, the classifying space $BSU(2) \cong \mathbb{H}P^\infty$ for quaternionic line bundles is not a $K(\pi, 4)$ because

$$\pi_{\ell+1}(BSU(2)) \cong \pi_\ell(SU(2)) \cong \pi_\ell(S^3)$$

by the exact sequence of the fibre bundle $ESU(2) \to BSU(2)$. Hence, c_2 alone does not classify quaternionic line bundles. Take $M = S^7$ as an example, then quaternionic line bundles are classified by elements of

$$\pi_7(BSU(2)) \cong \pi_6(S^3) \cong \mathbb{Z}/12\mathbb{Z}\,.$$

We prove in [28] that on highly connected 7-manifold M with $\pi_3(M)$ finite as above, $\pi_6(S^3)$ acts simply transitively on the set of isomorphism classes of quaternionic line bundles with a fixed second Chern class. The group $\pi_6(S^3)$ acts freely by a clutching construction over a small $S^6 \subset M$, and this action is detected by the t-invariant.

The t-invariant also distinguishes all quaternionic line bundles on S^{11}, but not on S^{15}. Regard the sequence of Hopf fibrations and inclusions

$$\begin{array}{ccccccccc} \cdots & & & & S^{4k-1} & & & & S^{4k+3} & & \\ & \searrow & & \swarrow & & \searrow & & \swarrow & & \searrow & \\ & & \mathbb{H}P^{k-1} & & & & \mathbb{H}P^{k} & & & & \cdots \end{array}$$

Here, $\mathbb{H}P^k$ decomposes along S^{4k-1} into a $4k$-disk and a 4-disk bundle over $\mathbb{H}P^{k-1}$. A quaternionic line bundle on $\mathbb{H}P^{k-1}$ can be extended to $\mathbb{H}P^k$ if and only if its pullback to S^{4k-1} is trivial. The t-invariant on S^{4k-1} is thus an obstruction against such an extension. By cellular approximation, quaternionic line bundles on $\mathbb{H}P^k$ are classified by homotopy classes of maps $\mathbb{H}P^k \to \mathbb{H}P^k \subset \mathbb{H}P^\infty \cong BSU(2)$. If we compute the t-invariants on $S^7, S^{11}, \dots, S^{4k-1}$ for a quaternionic line bundle with a given second Chern class, we recover precisely the obstructions against self maps of $\mathbb{H}P^k$ found by Feder and Gitler in [39].

Finally, Crowley also defines an analogous quadratic form q_M on $H^8(M)$ for highly connected 15-manifolds M in [26]. An intrinsic formula for q_M will probably involve the unique string structure on M in the same way that (22) above uses the unique spin structure.

4.5 Seven-Manifolds of Positive Curvature

Riemannian metrics of positive sectional curvature on closed manifolds are a rare phenomenon, and sharp conditions for their existence are far from being understood. Apart from the obvious symmetric examples, few other manifolds are known. Many of these other examples are seven-dimensional manifolds that are either of Kreck–Stolz type (18) or highly connected. The homogeneous Aloff–Wallach spaces $SU(3)/U(1)$ and their biquotient analogues, the Eschenburg space, have been classified using Kreck–Stolz invariants in [1, 57] and [59]. Kruggel uses a cobordism with lens spaces, whose η-invariants have already been given in [4].

The Berger space $SO(5)/SO(3)$ is diffeomorphic to a particular S^3-bundle over S^4. For the proof in [45], one needs to know that it is homeomorphic to such a bundle by [54]. Then the Eells–Kuiper invariant of $SO(5)/SO(3)$ together with the classification of all S^3-bundles over S^4 in [27] suffices to determine the diffeomorphism type.

One is still interested in finding new examples of positive curvature metrics. Grove, Wilking and Ziller [48] give two families (P_k), (Q_k) of 7-manifolds and one exceptional space R, which possibly allow such metrics and contain new examples. The spaces P_k are highly connected, whereas Q_k and R are of Kreck–Stolz type. In [47], Grove, Verdiani and Ziller constructed a positive sectional curvature metric on P_2 (note that $P_1 = S^7$); another construction is due to Dearricott [32]. On the other hand, the space R does not carry a metric of cohomogeneity one with positive sectional curvature by a result of Verdiani and Ziller [80].

The spaces P_k form Seifert fibrations with generic fibre S^3 over some base orbifold B_k as indicated in [48]. We apply Theorem 3.2 to determine the η-invariants in (15) and (22) in the adiabatic limit and compute $\mu(P_k)$ and t_{P_k} for all P_k.

Theorem 4.13 ([44]). *The Eells–Kuiper invariant of P_k is given by*

$$\mu(P_k) = -\frac{4k^3 - 7k + 3}{2^5 \cdot 3 \cdot 7} \quad \in \quad \mathbb{Q}/\mathbb{Z}\,. \tag{23}$$

Crowley's quadratic form q on $H^4(P_k) \cong \mathbb{Z}/k\mathbb{Z}$ is given by

$$q(\ell) = \frac{\ell(\ell - k)}{2k} \quad \in \quad \mathbb{Q}/\mathbb{Z}\,. \tag{24}$$

By comparing these values with the corresponding values for S^3-bundles over S^4 in [27] and [28], see Example 4.12, one can construct manifolds that are diffeomorphic to P_k.

Theorem 4.14 ([44]). *Let $E_{k,k} \to S^4$ denote the principal S^3-bundle with Euler class $k \in H^4(S^4) \cong \mathbb{Z}$, and let Σ_7 denote the exotic seven sphere with $\mu(\Sigma_7) = \frac{1}{28}$. Then there exists an orientation preserving diffeomorphism*

$$P_k \cong E_{k,k} \# \Sigma_7^{\# \frac{k-k^3}{6}} \, .$$

In particular, P_k and $E_{k,k}$ are homeomorphic.

This result also implies that P_2 with reversed orientation is diffeomorphic to some S^3-bundle over S^4, and to $T_1 S^4 \# \Sigma_7$, where $T_1 S^4$ denotes the unit tangent bundle of S^4.

Acknowledgements Supported in part by DFG special programme "Global Differential Geometry"

References

1. Astey, L., Micha, E., Pastor, G.: Homeomorphism and diffeomorphism types of Eschenburg spaces. Diff. Geom. Appl. **7**, 41–50 (1997)
2. Atiyah, M.F., Patodi, V.K., Singer, I.M.: Spectral asymmetry and Riemannian geometry. Bull. London Math. Soc. **5**, 229–234 (1973)
3. Atiyah, M.F., Patodi, V.K., Singer, I.M.: Spectral asymmetry and Riemannian geometry, I. Math. Proc. Cambridge Philos. Soc. **77**, 43–69 (1975)
4. Atiyah, M.F., Patodi, V.K., Singer, I.M.: Spectral asymmetry and Riemannian geometry, II. Math. Proc. Cambridge Philos. Soc. **78**, 405–432 (1975)
5. Atiyah, M.F., Patodi, V.K., Singer, I.M.: Spectral asymmetry and Riemannian geometry, III. Math. Proc. Cambridge Philos. Soc. **79**, 71–99 (1976)
6. Atiyah, M.F., Singer, I.M.: The index of elliptic operators, III. Ann. of Math. **87**, 546–604 (1968)
7. Bär, C.: Dependence of the Dirac spectrum on the Spin structure. in Global analysis and harmonic analysis (Marseille-Luminy, 1999), pp. 17–33, Sémin. Congr., 4, Soc. Math. France, Paris (2000)
8. Bechtluft-Sachs, S.: The computation of η-invariants on manifolds with free circle action. J. Funct. Anal. **174**, 251–263 (2000)
9. Berline, N., Getzler, E., Vergne, M.: Heat kernels and Dirac operators. Springer, Berlin (1992)
10. Bismut, J.-M.: The infinitesimal Lefschetz formulas: A heat equation proof. J. Funct. Anal. **62**, 435–457 (1985)
11. Bismut, J.-M., Cheeger, J.: η-invariants and their adiabatic limits. J. Amer. Math. Soc. 2 33–70 (1989)
12. Bismut, J.-M., Cheeger, J.: Families index for manifolds with boundary, superconnections, and cones, I, Families of manifolds with boundary and Dirac operators. J. Funct. Anal. **89**, 313–363 (1990)
13. Bismut, J.-M., Cheeger, J.: Families index for manifolds with boundary, superconnections and cones, II, The Chern character. J. Funct. Anal. **90**, 306–354 (1990)
14. Bismut, J.-M., Cheeger, J.: Remarks on the index theorem for families of Dirac operators on manifolds with boundary. In: Lawson, H.B., Tenenblat, K. (eds.) Differential geometry, pp. 59–83, Pitman Monogr. Surveys Pure Appl. Math., 52, Longman Sci. Tech., Harlow (1991)
15. Bismut, J.M., Freed, D.S.: The Analysis of Elliptic Families II, Dirac Operators, Êta Invariants, and the Holonomy Theorem. Comm. Math. Phys. **107**, 103–163 (1986)
16. Bismut, J.-M., Lott, J.: Flat vector bundles, direct images and higher real analytic torsion. J. Amer. Math. Soc. **8**, 291–363 (1995)
17. Bismut, J.-M., Zhang, W.: Real embeddings and eta invariants. Math. Ann. **295**, 661–684 (1993)

18. Botvinnik, B., Gilkey, P.B.: The eta invariant and metrics of positive scalar curvature. Math. Ann. **302**, 507–517 (1995)
19. Botvinnik, B., Gilkey, P.B.: Metrics of positive scalar curvature on spherical space forms. Canad. J. Math **48**, 64–80 (1996)
20. Bunke, U.: On the gluing problem for the η-invariant. J. Diff. Geom. **41**, 397–448 (1995)
21. Bunke, U., Naumann, N.: The f-invariant and index theory. Manuscr. Math. **132**, 365–397 (2010)
22. Cappell, S.E., Lee, R., Miller, E.Y.: On the Maslov index. Comm. Pure Appl. Math. **47**, 121–186 (1994)
23. Cheeger, J., Simons, J.: Differential characters and geometric invariants. In: Geometry and topology (College Pack, 1983/84), 50–80, Lecture Notes in Math. 1167, Springer, Berlin (1985)
24. Cisneros-Molina, J.L.: The η-invariant of twisted Dirac operators of S^3/Γ. Geom. Dedicata **84**, 207–228 (2001)
25. Cisneros-Molina, J.L.: A note on torsion in K_3 of the real numbers. Bol. Soc. Mat. Mexicana (3) **10**, 117–128 (2004)
26. Crowley, D.: The classification of highly connected manifolds in dimensions 7 and 15. preprint, `arXiv:math/0203253`.
27. Crowley, D., Escher, C.: Classification of S^3-bundles over S^4. Diff. Geom. Appl. **18**, 363–380 (2003)
28. Crowley, D., Goette, S.: Kreck-Stolz invariants for quaternionic line bundles. preprint, `arXiv:1012.5237`
29. Dahl, M.: Dependence on the Spin structure of the eta and Rokhlin invariants. Topology Appl. **118**, 345–355 (2002)
30. Dai, X.: Adiabatic limits, nonmultiplicativity of signature, and Leray spectral sequence. J. Amer. Math. Soc. **4**, 265–321 (1991); Ann. Global Anal. Geom. **27**, 333–340 (2005)
31. Dai, X., Zhang, W.: Circle bundles and the Kreck-Stolz invariant. Trans. Amer. Math. Soc. **347**, 3587–3593 (1995)
32. Dearricott, O.: A 7-manifold with positive curvature. Duke Math. J. **158**, 307–346 (2011)
33. Degeratu, A.: Eta-invariants from Molien series. Q. J. Math. **60**, 303–311 (2009)
34. Deninger, C., Singhof, W.: The e-invariant and the spectrum of the Laplacian for compact nilmanifolds covered by Heisenberg groups. Invent. math. **78**, 101–112 (1984)
35. Donnelly, H.: Spectral geometry and invariants from differential topology. Bull. London Math. Soc. **7**, 147–150 (1975)
36. Donnelly, H.: Eta invariants for G-spaces. Indiana Univ. Math. J. **27**, 889–918 (1978)
37. Eells, J. Jr., Kuiper, N.: An invariant for certain smooth manifolds. Ann. Mat. Pura Appl. (4) **60**, 93–110 (1962)
38. Farsi, C.: Orbifold η-invariants. Indiana Univ. Math. J. **56**, 501–521 (2007)
39. Feder, S., Gitler, S.: Mappings of quaternionic projective spaces. Bol. Soc. Mat. Mexicana (2) **18**, 33–37 (1973)
40. Gilkey, P.B., Miatello, R.J., Podestá, R.A.: The eta invariant and equivariant bordism of flat manifolds with cyclic holonomy group of odd prime order. Ann. Glob. Anal. Geom. **37**, 275–306 (2010)
41. Goette, S.: Equivariant η-invariants on homogeneous spaces. Math. Z. **232**, 1–42 (1999)
42. Goette, S.: Equivariant eta-Invariants and eta-Forms. J. reine angew. Math. **526**, 181–236 (2000)
43. Goette, S.: Eta invariants of homogeneous spaces. Pure Appl. Math. Q. **5**, 915–946 (2009)
44. Goette, S.: Adiabatic limits of Seifert fibrations, Dedekind sums, and the diffeomorphism types of certain 7-manifolds, Preprint, `arXiv:1108.5614`
45. Goette, S., Kitchloo, N., Shankar, K.: Diffeomorphism type of the Berger space SO(5)/SO(3). Am. J. Math. **126**, 395–416 (2004)
46. Gromov, M., Lawson, H.B.: The classification of simply connected manifolds of positive scalar curvature and the Dirac operator on complete Riemannian manifolds. Publ. Math. IHES **58**, 83–196 (1983)

47. Grove, K., Verdiani, L., Ziller, W.: An exotic $T_1\mathbb{S}^4$ with positive curvature. Geom. Funct. Anal. **21**, 499–524 (2011)
48. Grove, K., Wilking, B., Ziller, W.: Positively curved cohomogeneity one manifolds and 3-sasakian geometry. J. Diff. Geo. **78**, 33–111 (2008); `arXiv:math/0511464`
49. Habel, M.: Die η-Invariante der Berger-Sphäre. Diplomarbeit Universität Hamburg (2000)
50. Hepworth, R.: Generalized Kreck-Stolz invariants and the topology of certain 3-Sasakian 7-manifolds. thesis, University of Edinburgh, Scotland, UK (2005)
51. Hitchin, N.: Harmonic spinors. Adv. Math. **14**, 1–55 (1974)
52. Jones, J.D., Westbury, B.W.: Algebraic K-theory, homology spheres, and the η-invariant. Topology **34**, 929–957 (1995)
53. Keswani, N.: Relative eta-invariants and C^*-algebra K-theory. Topology **39**, 957–983 (2000)
54. Kitchloo, N., Shankar, K.: On complexes equivalent to S^3-bundles over S^4. Int. Math. Research Notices **8**, 381–394 (2001)
55. Kostant, B.: A cubic Dirac operator and the emergence of Euler number multiplets of representations for equal rank subgroups. Duke Math. J. **100**, 447–501 (1999)
56. Kreck, M., Stolz, S.: A diffeomorphism classification of 7-dimensional homogeneous Einstein manifolds with $SU(3) \times SU(2) \times U(1)$-symmetry. Ann. of Math. **127**, 373–388 (1988)
57. Kreck, M., Stolz, S.: Some nondiffeomorphic homeomorphic homogeneous 7-manifolds with positive sectional curvature. J. Differ. Geom. **33**, 465–486 (1991)
58. Kreck, M., Stolz, S.: Nonconnected moduli spaces of positive sectional curvature metrics. J. Amer. Math. Soc. **6**, 825–850 (1993)
59. Kruggel, B.: Homeomorphism and diffeomorphism classification of Eschenburg spaces. Q. J. Math. **56**, 553–577 (2005)
60. Lawson, H.B., Michelsohn, M.-L.: Spin geometry. p. xii + 427. Princeton University Press, Princeton, NJ (1989)
61. Leichtnam, E., Piazza, P.: On higher eta-invariants and metrics of positive scalar curvature. K-Theory **24**, 341–359 (2001)
62. Lesch, M., Wojciechowski, K.P.: On the η-invariant of generalized Atiyah-Patodi-Singer boundary value problems. Illinois J. Math. **40**, 30–46 (1996)
63. Loya, P., Moroianu, S., Park, J.: Adiabatic limit of the eta invariant over cofinite quotients of $PSL(2,\mathbb{R})$. Compos. Math. **144**, 1593–1616 (2008)
64. Mathai, V.: Spectral flow, eta invariants, and von Neumann algebras. J. Funct. Anal. **109**, 442–456 (1992)
65. Melrose, R.B., Piazza, P.: Families of Dirac operators, boundaries and the b-calculus. J. Differ. Geom. **46**, 99–180 (1997)
66. Miatello, R.J., Podestá, R.A.: Eta invariants and class numbers. Pure Appl. Math. Q. **5**, 729–753 (2009)
67. Miller, E.Y., Lee, R.: Some invariants of spin manifolds. Topology Appl. **25**, 301–311 (1987)
68. Millson, J.J.: Closed geodesics and the eta-invariant. Ann. Math. **108**, 1–39 (1978)
69. Moscovici, H., Stanton, R.J.: Eta invariants of Dirac operators on locally symmetric spaces. Invent. Math. **95**, 629–666 (1989)
70. Neumann, W.: Signature related invariants of manifolds I: monodromy and γ-invariants. Topology **18**, 147–172 (1979)
71. Park, J.: Eta invariants and regularized determinants for odd dimensional hyperbolic manifolds with cusps. Amer. J. Math. **127**, 493–534 (2005)
72. Pfäffle, F.: The Dirac spectrum of Bieberbach manifolds. J. Geom. Phys. **35**, 367–385 (2000)
73. Piazza, P., Schick, T.: Bordism, rho-invariants and the Baum-Connes conjecture. J. Noncommut. Geom. **1** 27–111 (2007)
74. Piazza, P., Schick, T.: Groups with torsion, bordism and rho invariants. Pacific J. Math. **232**, 355–378 (2007)
75. Sadowski, M., Szczepánski, A.: Flat manifolds, harmonic spinors and eta invariants. Adv. Geom. **6**, 287–300 (2006)
76. Seade, J.A.: On the η-function of the Dirac operator on $\Gamma\backslash S^3$. An. Inst. Mat. Univ. Nac. Autónoma México **21**, 129–147 (1981)

77. Seade, J., Steer, B.: A note on the eta function for quotients of $PSL_2(R)$ by co-compact Fuchsian groups. Topology **26**, 79–91 (1987)
78. Stolz, S.: A note on the bP-component of $(4n-1)$-dimensional homotopy spheres. Proc. Amer. Math. Soc. **99**, 581–584 (1987)
79. Tsuboi, K.: Eta invariants and conformal immersions. Publ. Res. Inst. Math. Sci. **17**, 201–214 (1981)
80. Verdiani, L., Ziller, W.: Obstructions to positive curvature. preprint (2010); `arXiv:1012.2265`
81. Weinberger, S.: Homotopy invariance of eta invariants. Proc. Nat. Acad. Sci. **85**, 5362–5365 (1988)
82. Wojciechowski, K.P.: The additivity of the η-invariant: The case of an invertible tangential operator. Houston J. Math. **20**, 603–621 (1994)
83. Wojciechowski, K.P.: The additivity of the η-invariant: The case of a singular tangential operator. Comm. Math. Phys. **169**, 315–327 (1995)
84. Zhang, W.: Circle bundles, η-invariants and Rokhlin congruences. Ann. Inst. Fourier (Grenoble) **44**, 249–270 (1994)

Part III
Symplectic Geometry

Rabinowitz Floer Homology: A Survey

Peter Albers and Urs Frauenfelder

Abstract Rabinowitz Floer homology is the semi-infinite dimensional Morse homology associated to the Rabinowitz action functional used in the pioneering work of Rabinowitz. Gradient flow lines are solutions of a vortex-like equation. In this survey article we describe the construction of Rabinowitz Floer homology and its applications to symplectic and contact topology, global Hamiltonian perturbations and the study of magnetic fields.

2000 Mathematics Subject Classification. 53D70, 37J10, 58J05

1 Rabinowitz Floer Homology

1.1 The Rabinowitz Action Functional

In his pioneering work [36, 37] Rabinowitz used the Rabinowitz action functional to prove existence of periodic orbits on star-shaped hypersurfaces in $\mathbb{R}^{2n}$. This fundamental work was one of the motivations for Weinstein to state his famous conjecture on the existence of Reeb orbits, [42].

P. Albers (✉)
Department of Mathematics, Purdue University, West Lafayette, IN 47907, USA
e-mail: palbers@math.purdue.edu

U. Frauenfelder
Department of Mathematics and Research Institute of Mathematics, Seoul National University, Gwanak-gu, Seoul, Korea 151-742
e-mail: frauenf@snu.ac.kr

C. Bär et al. (eds.), *Global Differential Geometry*, Springer Proceedings in Mathematics 17, DOI 10.1007/978-3-642-22842-1_14,

1.1.1 The Unperturbed Rabinowitz Action Functional and Reeb Dynamics

1.1.2 Exact Symplectic Manifolds

Let $(M, \omega = d\lambda)$ be an exact symplectic manifold, for example $(\mathbb{R}^{2n}, \omega_0)$ or a cotangent bundle (T^*B, ω_{std}) each with its canonical symplectic form. We fix an autonomous Hamiltonian, i.e. a smooth time-independent function $F : M \to \mathbb{R}$. The Hamiltonian vector field X_F of F is defined implicitly by

$$\iota_{X_F}\omega = dF \ . \tag{1.1}$$

Since F is autonomous the Hamiltonian vector field X_F is tangent to level sets of F and therefore its flow $\phi_F^t : M \longrightarrow M$ leaves level sets invariant. This means that the energy F is preserved under the flow ϕ_F^t.

Let $\mathscr{L} := C^\infty(S^1, M)$, $S^1 = \mathbb{R}/\mathbb{Z}$, be the loop space of M. The Rabinowitz action functional is defined as follows:

$$\begin{aligned} \mathcal{A}^F : \mathscr{L} \times \mathbb{R} &\longrightarrow \mathbb{R} \\ (v, \eta) \mapsto \mathcal{A}^F(v, \eta) &:= -\int_{S^1} v^*\lambda - \eta \int_0^1 F\big(v(t)\big)dt \ . \end{aligned} \tag{1.2}$$

The real number η can be thought of as a Lagrange multiplier. Hence critical points $(v, \eta) \in \mathrm{Crit}\mathcal{A}^F$ are critical points of the area functional restricted to the space of loops with F-mean value zero. They are solutions of

$$\begin{cases} \dot v(t) = \eta X_F\big(v(t)\big), \ \forall t \in S^1 \\ \displaystyle\int_0^1 F\big(v(t)\big)dt = 0 \ . \end{cases} \tag{1.3}$$

The first equation can be integrated to $v(t) = \phi_F^{\eta t}(v(0))$ and thus, by preservation of energy, the critical point equation is equivalent to

$$\begin{cases} \dot v(t) = \eta X_F\big(v(t)\big), & \forall t \in S^1 \\ v(t) \in F^{-1}(0), & \forall t \in S^1 \ . \end{cases} \tag{1.4}$$

Hence, critical points of $\mathcal{A}^F$ correspond to periodic orbits of X_F with period η and lie on the energy hypersurface $\Sigma := F^{-1}(0)$. Here, the period η is understood in a generalized sense; it is allowed to be negative in which case the periodic orbit is traversed backwards. Moreover, if the period is zero then v is constant and corresponds to a point on the energy hypersurface $F^{-1}(0)$.

If Σ is a regular hypersurface for two functions $F, \widetilde{F} : M \to \mathbb{R}$, $\Sigma = \widetilde{F}^{-1}(0) = F^{-1}(0)$, then critical points of $\mathcal{A}^{\widetilde{F}}$ agree up to reparametrization with critical points of $\mathcal{A}^F$. In fact, they are closed characteristics of the canonical line bundle $\ker \omega|_\Sigma \to \Sigma$ or constant.

It is interesting to compare the critical points of the Rabinowitz action functional $\mathcal{A}^F$ to critical points of the action functional $\mathcal{A}_F$ of classical mechanics

$$\begin{aligned} \mathcal{A}_F : \mathscr{L} &\longrightarrow \mathbb{R} \\ v \mapsto \mathcal{A}_F(v) &:= -\int_{S^1} v^*\lambda - \int_0^1 F\big(v(t)\big)dt \,. \end{aligned} \tag{1.5}$$

Critical points $v \in \mathrm{Crit}\mathcal{A}_F$ are 1-periodic solutions of

$$\dot{v}(t) = X_F\big(v(t)\big) \,. \tag{1.6}$$

In this case the period of v is fixed but the energy is arbitrary.

1.1.3 Symplectically Aspherical Manifolds

A symplectic manifold (M,ω) is called symplectically aspherical if the homomorphism $I_\omega : \pi_2(M) \to \mathbb{R}$ obtained by integrating ω over a smooth representative vanishes identically. This holds for example if $(M,\omega = d\lambda)$ is an exact symplectic manifold. In this situation the Rabinowitz action functional can still be defined on the component $\mathscr{L}_0 \subset \mathscr{L}$ of the loop space of contractible loops. For $v \in \mathscr{L}_0$ there exists a filling disk $\bar{v} : \mathbb{D}^2 \to M$ where $\mathbb{D}^2 = \{z \in \mathbb{C} \mid |z| \le 1\}$ and $\bar{v}(e^{2\pi i t}) = v(t)$. In this case we set

$$\begin{aligned} \mathcal{A}^F : \mathscr{L}_0 \times \mathbb{R} &\longrightarrow \mathbb{R} \\ (v,\eta) \mapsto \mathcal{A}^F(v,\eta) &:= -\int_{\mathbb{D}^2} \bar{v}^*\omega - \eta\int_0^1 F\big(v(t)\big)dt \,. \end{aligned} \tag{1.7}$$

Due to the symplectic asphericity the definition of $\mathcal{A}^F$ does not depend on the choice of the filling disk. If in addition M is symplectically atoroidal, i.e. $\int_{T^2} f^*\omega = 0$, $\forall f : T^2 \to M$, the Rabinowitz action functional $\mathcal{A}^F$ can be extended to the whole loop space $\mathscr{L}$. An interesting class of symplectically atoroidal manifolds are certain twisted cotangent bundles. A twisted cotangent bundle is $(T^*B, \omega_{std} + \tau^*\sigma)$, where $\tau : T^*B \to B$ is the projection and such that $\sigma \in \Omega^2(B)$ is closed. If the pull-back of σ to the universal cover $\widetilde{B}$ has a bounded primitive then $(T^*B, \omega_{std} + \tau^*\sigma)$ is symplectically atoroidal. This fact was used by Merry in [33].

1.1.4 The Perturbed Rabinowitz Action Functional and Global Hamiltonian Perturbations

Since $F : M \to \mathbb{R}$ is autonomous the energy hypersurface $\Sigma = F^{-1}(0)$ is preserved under ϕ_F^t. Therefore, Σ is foliated by leaves $L_x := \{\phi_F^t(x) \mid t \in \mathbb{R}\}$, $x \in \Sigma$. It is a challenging problem to compare the system F before and after a global perturbation occurring in the time interval [0, 1]. Such a perturbation is described by a function $H : M \times [0,1] \to \mathbb{R}$. Moser observed in [34] that it is not possible to destroy all trajectories of the unperturbed system if the perturbation is sufficiently small, that is, there exists $x \in \Sigma$

$$\phi_H^1(x) \in L_x \,. \tag{1.8}$$

Such a point x is referred to as a leaf-wise intersection point. Equivalently, there exists $(x, \eta) \in \Sigma \times \mathbb{R}$ such that

$$\phi_F^\eta(x) = \phi_H^1(x) \,. \tag{1.9}$$

We point out that the time shift η is uniquely defined by the above equation unless the leaf L_x is closed. If the time shift is negative then the perturbation moves the system back into its own past. Likewise, if the time shift is positive the perturbation moves the system forward into its own future.

Already the existence problem for leaf-wise intersection points is highly non-trivial. The search for leaf-wise intersection points was initiated by Moser in [34] and pursued further in [3–7,9,19,20,24,26,27,29,33,43]. We refer to [3] for a brief history.

In [4] Albers – Frauenfelder developed a variational approach to the study of leaf-wise intersection points.

Definition 1.1. A pair $\mathfrak{M} = (F, H)$ of Hamiltonians $F, H : M \times S^1 \longrightarrow R$ is called a Moser pair if it satisfies

$$F(\cdot, t) = 0 \quad \forall t \in [\tfrac{1}{2}, 1] \qquad \text{and} \qquad H(\cdot, t) = 0 \quad \forall t \in [0, \tfrac{1}{2}] \,, \tag{1.10}$$

and F is of the form $F(x,t) = \rho(t) f(x)$ for some smooth map $\rho : S^1 \to [0,1]$ with $\int_0^1 \rho(t)dt = 1$ and $f : M \longrightarrow \mathbb{R}$.

If we start with an autonomous Hamiltonian $\widehat{F} : M \to \mathbb{R}$ and an arbitrary $\widehat{H} : M \times S^1 \to \mathbb{R}$ we can find $F, H : M \times S^1 \to \mathbb{R}$ such that the Hamiltonian flows ϕ_F^t, ϕ_H^t are time reparametrizations of the flows $\phi_{\widehat{F}}^t, \phi_{\widehat{H}}^t$ and such that (F, H) is a Moser pair.

For simplicity we assume that $(M, \omega = d\lambda)$ is an exact symplectic manifold. For a Moser pair $\mathfrak{M} = (F, H)$ the perturbed Rabinowitz action functional is defined by

$$\begin{aligned} \mathcal{A}^{(F,H)} \equiv \mathcal{A}^{\mathfrak{M}} &: \mathscr{L} \times \mathbb{R} \longrightarrow \mathbb{R} \\ (v, \eta) &\mapsto -\int_0^1 v^*\lambda - \int_0^1 H(v,t)dt - \eta \int_0^1 F(v,t)dt \,. \end{aligned} \tag{1.11}$$

Critical points (v, η) of $\mathcal{A}^{\mathfrak{M}}$ are solutions of

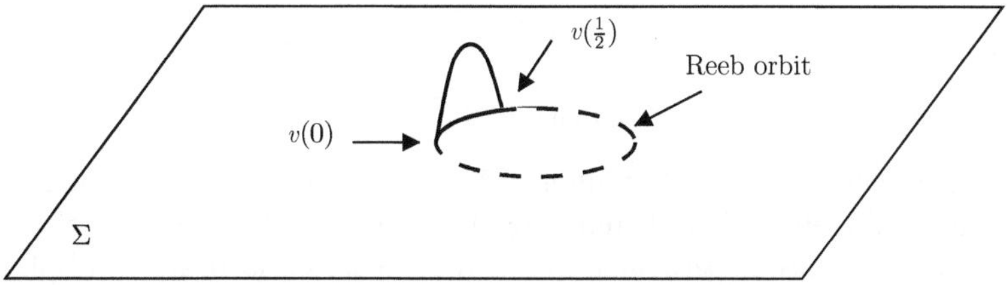

Fig. 1

$$\begin{cases} \partial_t v = \eta X_F(v,t) + X_H(v,t), \ \forall t \in S^1 \\ \int_0^1 F(v,t)dt = 0 \end{cases} \tag{1.12}$$

In [4] we observed that critical points of the perturbed Rabinowitz action functional $\mathcal{A}^{\mathfrak{M}}$ give rise to leaf-wise intersection points.

Proposition 1.2 ([4]). *Let (v,η) be a critical point of $\mathcal{A}^{\mathfrak{M}}$ then $x := v(\frac{1}{2}) \in f^{-1}(0)$ and*

$$\phi_H^1(x) \in L_x \tag{1.13}$$

thus, x is a leaf-wise intersection. Moreover, the map $\mathrm{Crit}\mathcal{A}^{\mathfrak{M}} \to$ *{leaf-wise intersections} is injective unless there exists a leaf-wise intersection point on a closed leaf, see Fig. 1.*

1.1.5 First Properties

In this paragraph we make the **wrong** assumption that Rabinowitz Floer homology

$$\text{“}\mathbf{RFH}(\Sigma, M)\text{”} \equiv \text{“HM}(\mathcal{A}^F)\text{”}, \tag{1.14}$$

i.e. the semi-infinite dimensional Morse homology (in the sense of Floer) for the Rabinowitz action functional, can always be constructed and is independent of the defining function F for a regular hypersurface Σ and invariant under Hamiltonian perturbations as described above. Thus, $\mathbf{RFH}(\Sigma, M)$ should have the following properties:

(1) $\mathbf{RFH}(\Sigma, M) \cong \mathrm{H}_*(\Sigma)$ if there are no closed characteristics on Σ. Indeed, then the only critical points of $\mathcal{A}^F$ correspond to constant loops in Σ, see above. Then $\mathcal{A}^F$ is Morse-Bott with $\mathrm{Crit}\mathcal{A}^F \cong \Sigma$ and $\mathcal{A}^F|_{\mathrm{Crit}\mathcal{A}^F} = 0$ since Σ is a regular hypersurface.
(2) If Σ is displaceable, that is, $\exists H : M \times S^1 \to \mathbb{R}$ such that $\phi_H^1(\Sigma) \cap \Sigma = \emptyset$, then $\mathbf{RFH}(\Sigma, M) \cong 0$. This follows from invariance under Hamiltonian perturbations

$$\text{“}\mathbf{RFH}(\Sigma, M)\text{”} \equiv \text{“HM}(\mathcal{A}^F)\text{”} \cong \text{“HM}(\mathcal{A}^{(F,H)})\text{”} \cong 0 \qquad (1.15)$$

together with the observation that $\text{Crit}\mathcal{A}^{(F,H)} = \emptyset$.

The counterexamples to the Hamiltonian Seifert conjecture, see Ginzburg – Gürel [22] and the literature cited therein, are closed hypersurfaces $\Sigma \subset \mathbb{R}^{2n}$ with no closed characteristics. In particular, since any compact subset of $\mathbb{R}^{2n}$ is displaceable we arrive at the contradiction

$$\text{H}_*(\Sigma) \cong \text{“}\mathbf{RFH}(\Sigma, M)\text{”} \cong 0\,. \qquad (1.16)$$

The reason behind the fact that Rabinowitz Floer homology cannot be defined in full generality is that the moduli spaces of gradient flow lines do not have the necessary compactness properties. However, if certain topological/dynamical assumptions1 on (Σ, M) are made the desired compactness properties can be established. This is described in the next section.

1.2 Gradient Flow Lines

1.2.1 Gradient Flow Equation

To compute the gradient of the Rabinowitz action functional we need to specify a metric on $\mathscr{L} \times \mathbb{R}$. We take the product of an L^2-metric on $\mathscr{L}$ and the standard metric on $\mathbb{R}$. In order to specify the L^2-metric on $\mathscr{L}$ we choose an S^1-family $J \equiv \{J_t\}_{t\in S^1}$ of ω-compatible almost complex structures on M. For $(\hat{v}_1, \hat{\eta}_1), (\hat{v}_2, \hat{\eta}_2) \in T_{(v,\eta)}\big(\mathscr{L} \times \mathbb{R}\big) = \Gamma(v^*TM) \times \mathbb{R}$ we set

$$\mathfrak{m}\big((\hat{v}_1, \hat{\eta}_1), (\hat{v}_2, \hat{\eta}_2)\big) \equiv \mathfrak{m}_J\big((\hat{v}_1, \hat{\eta}_1), (\hat{v}_2, \hat{\eta}_2)\big) := \int_0^1 \omega\big(\hat{v}_1, J_t(v)\hat{v}_2\big)dt + \hat{\eta}_1\hat{\eta}_2\,. \qquad (1.17)$$

Then the gradient of $\mathcal{A}^F$ is

$$\nabla^{\mathfrak{m}}\mathcal{A}^F(v, \eta) = \begin{pmatrix} J_t(v)\big(\partial_t v - \eta X_F(v)\big) \\ -\int_0^1 F(v)dt \end{pmatrix}. \qquad (1.18)$$

A gradient flow line is formally a map $w = (v, \eta) \in C^\infty(\mathbb{R}, \mathscr{L} \times \mathbb{R})$ satisfying

$$\partial_s w(s) + \nabla^{\mathfrak{m}}\mathcal{A}^F\big(w(s)\big) = 0 \qquad (1.19)$$

in the sense of Floer, that is $v : \mathbb{R} \times S^1 \to M$, $\eta : \mathbb{R} \to \mathbb{R}$ satisfy

$$\begin{cases} \partial_s v + J_t(v)\big(\partial_t v - \eta X_F(v)\big) = 0 \\ \partial_s \eta - \displaystyle\int_0^1 F(v)dt = 0\,. \end{cases} \tag{1.20}$$

If we replace $\mathcal{A}^F$ by the perturbed Rabinowitz action functional $\mathcal{A}^{(F,H)}$ then the corresponding gradient flow equation changes to

$$\begin{cases} \partial_s v + J_t(v)\big(\partial_t v - \eta X_F(t,v) - X_H(t,v)\big) = 0 \\ \partial_s \eta - \displaystyle\int_0^1 F(t,v)dt = 0\,. \end{cases} \tag{1.21}$$

1.2.2 Compactness

1.2.3 The Case of Restricted Contact Type

Let $(W,\omega = d\lambda)$ be a compact, exact symplectic manifold with contact type boundary $\Sigma = \partial W$, that is, the Liouville vector field L, defined by $\iota_L\omega = \lambda$, points outward along Σ. In particular, $(\Sigma, \alpha := \lambda|_\Sigma)$ is a contact manifold.[1] We denote by M the completion of W obtained by attaching the positive half of the symplectization of Σ, that is, $(M = W \cup_\Sigma (\Sigma\times\mathbb{R}_+), \omega = d\lambda)$ where λ is extended over $\Sigma\times\mathbb{R}_+$ by $e^r\alpha$, $r \in \mathbb{R}_+$. We assume that $F : M \to \mathbb{R}$ is a defining function for Σ such that dF has compact support. The main compactness theorem is as follows:

Theorem 1.3. *In the above situation let $w_n = (v_n, \eta_n)$ be a sequence of gradient flow lines of $\mathcal{A}^{\mathfrak{M}}$ for which there exists $a < b$ such that*

$$a \leq \mathcal{A}^{\mathfrak{M}}\big(w_n(s)\big) \leq b \qquad \forall s \in \mathbb{R}\,. \tag{1.22}$$

Then for every reparametrisation sequence $\sigma_n \in \mathbb{R}$ the sequence $w_n(\cdot + \sigma_n)$ has a subsequence which converges in $C^\infty_{\text{loc}}(\mathbb{R}, \mathscr{L}\times\mathbb{R})$.

The proof from standard arguments in Floer theory as soon as we establish

(1) A uniform $C^0(\mathbb{R})$ bound on η_n,
(2) A uniform $C^0(\mathbb{R}\times S^1)$ bound on v_n,
(3) A uniform $C^0(\mathbb{R}\times S^1)$ bound on the derivatives of v_n.

Assertion (2) follows from a maximum principle since M is convex at infinity. Assertion (3) follows from standard bubbling-off analysis for holomorphic spheres

[1] In general, a closed hypersurface in a symplectic manifold (W,ω) is of *restricted contact type* if there exists a globally defined primitive λ of ω with Liouville vector field L satisfying $L \pitchfork \Sigma$. If λ only exists locally near Σ then Σ is of *contact type*.

in Floer theory together with the assumption that (M, ω) is exact. Indeed, a non-constant holomorphic sphere in M has to exist if the derivatives of v_n explode . This contradicts Stokes theorem since ω is exact. Obviously, symplectic asphericity of M would be sufficient.

An interesting feature of gradient flow lines is that they have an infinite amount of flow time but a finite amount of energy available. This leads to the paradoxical conclusion that it is favorable for a gradient flow line to run slowly in order to get far away. The following "Fundamental Lemma" prevents that for slowly running gradient flow lines $w = (v, \eta)$ the Lagrange multiplier η explodes, i.e. slowly running gradient flow lines cannot get too far.

Fundamental Lemma (restricted contact type case)**.** There exists a constant $C > 0$ such that for all $(v, \eta) \in \mathscr{L} \times \mathbb{R}$

$$||\nabla^{\mathrm{m}} \mathcal{A}^{\mathfrak{M}}(v, \eta)|| < \frac{1}{C} \quad \Longrightarrow \quad |\eta| \leq C(|\mathcal{A}^{\mathfrak{M}}(v, \eta)| + 1) \tag{1.23}$$

Remark 1.4. With help of the Fundamental Lemma it can be proved that along gradient flow lines the Lagrange multiplier is bounded in terms of asymptotic action values. This depends heavily on the fact that the action $\mathcal{A}^{\mathfrak{M}}$ is decreasing along gradient flow lines. For details we refer to Cieliebak – Frauenfelder, [11].

Remark 1.5. The assumption of restricted contact type enters the proof of the Fundamental Lemma through the following observation. If we normalize F such that $X_F|_\Sigma$ is the Reeb vector field of α then in the unperturbed case at a critical points $(v, \eta) \in \mathrm{Crit}\mathcal{A}^F$ we have

$$\mathcal{A}^F(v, \eta) = -\eta\,. \tag{1.24}$$

Since $v(t/\eta)$ is a η-periodic Reeb orbit equation (1.24) can be thought of as a period-action equality for Reeb orbits. Consequently, (1.23) can be interpreted as a period-action inequality for almost Reeb orbits.

Remark 1.6. For the unperturbed Rabinowitz action functional the Fundamental Lemma and Theorem 1.3 have been proved in [11] and for the perturbed Rabinowitz action functional in [4].

Remark 1.7. If the defining function F and the almost complex structure J are adapted to the symplectization structure near the hypersurface Σ then the constant C in the Fundamental Lemma can be chosen universally, i.e. independent of F, J, and the contact structure. This has been used in the definition of spectral invariants for Rabinowitz Floer homology in [6].

Remark 1.8. The assumption that the differential of the defining function $F : M \to \mathbb{R}$ has compact support can be replaced by an appropriate asymptotic growth condition without changing the Rabinowitz Floer homology. This property plays an important role in the proof relating Rabinowitz Floer homology to symplectic homology and loop space topology, see [2, 16, 33].

Remark 1.9. Let (M,ω) be an exact symplectic manifold which is convex at infinity. If $\Sigma \subset M$ is a compact, bounding hypersurface of restricted contact type then the symplectization of Σ embeds into M. It has been examined by Cieliebak – Frauenfelder – Oancea in [16] under which conditions gradient flow lines of the Rabinowitz action functional do not leave the positive part of the symplectization of Σ. The upshot is that M can be replaced by the completion of the region bounded by Σ without changing the Rabinowitz Floer homology.

1.2.4 The Case of Stable Hypersurfaces

Let (M,ω) be a symplectically aspherical symplectic manifold. We assume further that (M,ω) is convex at infinity or geometrically bounded to guarantee C^0-bounds for gradient flow lines. A closed hypersurface $\Sigma \subset M$ is called stable if there exists $\alpha \in \Omega^1(\Sigma)$ such that

$$\alpha \wedge \omega^{m-1}|_\Sigma > 0 \quad \text{and} \quad \ker\omega \subset \ker d\alpha \tag{1.25}$$

where $m = \frac{1}{2}\dim M$. The 1-form α is called a stabilizing 1-form. Σ is called of contact type if $\omega|_\Sigma = d\alpha$. Since $\ker\omega|_\Sigma \to \Sigma$ is a rank-1-distribution the Reeb vector field R on Σ is uniquely defined by

$$\iota_R\alpha = 1 \quad \text{and} \quad \iota_R\omega|_\Sigma = 0\,. \tag{1.26}$$

The Rabinowitz action functional $\mathcal{A}^F$ is defined as above. However, the Fundamental Lemma does not carry over to the stable case since already at critical points the period-action equality fails to hold for $\mathcal{A}^F$ in general. We define a modified Rabinowitz action functional by

$$\begin{aligned} &\widehat{\mathcal{A}}^F : \mathscr{L}\times\mathbb{R} \longrightarrow \mathbb{R} \\ &(v,\eta) \mapsto \widehat{\mathcal{A}}^F(v,\eta) := -\int_{S^1} v^*\lambda - \eta\int_0^1 F\big(v(t)\big)dt\,. \end{aligned} \tag{1.27}$$

where $\lambda \in \Omega^1(M)$ satisfies $\lambda|_\Sigma = \alpha$. Then the period-action equality holds for $\widehat{\mathcal{A}}^F$. If λ is chosen appropriately then the Fundamental Lemma in the stable case is as follows.

Fundamental Lemma (stable case)**.** There exists a constant $C > 0$ such that for all $(v,\eta) \in \mathscr{L}\times\mathbb{R}$

$$||\nabla^{\mathrm{m}}\mathcal{A}^F(v,\eta)|| < \frac{1}{C} \quad \Longrightarrow \quad |\eta| \le C(|\widehat{\mathcal{A}}^F(v,\eta)| + 1) \tag{1.28}$$

This was proved by Cieliebak – Frauenfelder – Paternain in [18]. Establishing uniform bounds on the Lagrange multiplier along gradient flow lines in terms of asymptotic action values as described in Remark 1.4 in the stable case involves additional arguments. This is due to the fact that $\widehat{\mathcal{A}}^F$ is not necessarily decreasing along gradient flow lines of $\mathcal{A}^F$. The crucial observation is that

$$\mathcal{A}_\alpha(v,\eta) := \mathcal{A}^F(v,\eta) - \widehat{\mathcal{A}}^F(v,\eta) = \int v^*\lambda - \int \bar{v}^*\omega : \mathscr{L}_0 \times \mathbb{R} \longrightarrow \mathbb{R} \quad (1.29)$$

is non-increasing along gradient flow lines of $\mathcal{A}^F$, i.e. $\mathcal{A}_\alpha$ serves as a Lyapunov function for the gradient flow of $\mathcal{A}^F$. This immediately implies that if w is a gradient flow line of $\mathcal{A}^F$ with $\lim_{s\to\pm\infty} w(s) =: w_\pm \in \mathrm{Crit}\mathcal{A}^F$ we can estimate

$$\mathcal{A}^F(w_+) - \mathcal{A}_\alpha(w_-) \le \widehat{\mathcal{A}}^F(w(s)) \le \mathcal{A}^F(w_-) - \mathcal{A}_\alpha(w_+)\,. \quad (1.30)$$

Proving that $\mathcal{A}_\alpha$ is non-increasing along gradient flow lines involves the bilinear form

$$\widehat{\mathfrak{m}}\big((\hat{v}_1,\hat{\eta}_1),(\hat{v}_2,\hat{\eta}_2)\big) := \int_0^1 d\lambda\big(\hat{v}_1, J_t(v)\hat{v}_2\big)dt + \hat{\eta}_1\hat{\eta}_2\,. \quad (1.31)$$

for $(\hat{v}_1,\hat{\eta}_1),(\hat{v}_2,\hat{\eta}_2) \in T_{(v,\eta)}\big(\mathscr{L}\times\mathbb{R}\big) = \Gamma(v^*TM)\times\mathbb{R}$. We note that in general $\widehat{\mathfrak{m}}$ is neither symmetric nor positive definite.

Proposition 1.10. *If F, J, and λ are chosen appropriately, see [18], the following two properties hold*

$$\begin{aligned} &(1) \quad \mathfrak{m} - \widehat{\mathfrak{m}} \ge 0 \\ &(2) \quad d\widehat{\mathcal{A}}^F(\cdot) = \widehat{\mathfrak{m}}(\nabla^{\mathfrak{m}}\mathcal{A}^F,\cdot)\,. \end{aligned} \quad (1.32)$$

Remark 1.11. Property (2) in Proposition 1.10 can be interpreted as $\nabla^{\widehat{\mathfrak{m}}}\widehat{\mathcal{A}^F} = \nabla^{\mathfrak{m}}\mathcal{A}^F$ although $\nabla^{\widehat{\mathfrak{m}}}\widehat{\mathcal{A}^F}$ is not well-defined since $\widehat{\mathfrak{m}}$ might be degenerate.

Remark 1.12. Proposition 1.10 is very sensitive to Hamiltonian perturbations. It is not clear if compactness continues to hold in the perturbed case. However, Kang proved in [29] an analogue of Proposition 1.10 in the case that the hypersurface is of contact type (but not necessarily restricted contact type) and that the Hamiltonian perturbation is localized in the symplectization.

Corollary 1.13. *If F, J, and λ are chosen as in Proposition 1.10 then $\mathcal{A}_\alpha$ is a Lyapunov function for the gradient flow of $\mathcal{A}^F$.*

Proof. Let w be a gradient flow line of $\mathcal{A}^F$, i.e. a solution of $\partial_s w(s) + \nabla^{\mathfrak{m}}\mathcal{A}^F\big(w(s)\big) = 0$. Then we estimate using Proposition 1.10

$$\begin{aligned} \frac{d}{ds}\mathcal{A}_\alpha(w(s)) &= d\mathcal{A}^F(w(s))[\partial_s w] - d\widehat{\mathcal{A}}^F(w(s))[\partial_s w] \\ &= d\widehat{\mathcal{A}}^F(w(s))[\nabla^{\mathfrak{m}}\mathcal{A}^F] - d\mathcal{A}^F(w(s))[\nabla^{\mathfrak{m}}\mathcal{A}^F] \\ &= \widehat{\mathfrak{m}}(\nabla^{\mathfrak{m}}\mathcal{A}^F,\nabla^{\mathfrak{m}}\mathcal{A}^F) - \mathfrak{m}(\nabla^{\mathfrak{m}}\mathcal{A}^F,\nabla^{\mathfrak{m}}\mathcal{A}^F) \\ &\le 0\,. \end{aligned} \quad (1.33)$$

□

1.2.5 Relation to Symplectic Vortex Equations

Symplectic vortex equations have been found independently by Mundet and Salamon and were known in the physics literature as gauged sigma models. They are usually studied for Hamiltonian group actions of compact groups. Symplectic vortex equation with domain the 2-dimensional cylinder can be derived from the gradient flow equation of the moment action functional. We explain in this section how the gradient flow equation for the Rabinowitz action functional can be thought of as the substitute for the (not well-defined) symplectic vortex equations for Hamiltonian $\mathbb{R}$-actions.

Let (M, ω) be a symplectically aspherical manifold and $F : M \to \mathbb{R}$ a compactly supported autonomous Hamiltonian function. Then the Hamiltonian flow $\phi_F^t : M \to M$ can be thought of as a Hamiltonian $\mathbb{R}$-action with moment map F. The moment action functional corresponding to this $\mathbb{R}$-action as introduced by Cieliebak – Gaio – Salamon in [15] is

$$\begin{aligned} \mathbb{A}^F &: \mathscr{L}_0 \times C^\infty(S^1, \mathbb{R}) \longrightarrow \mathbb{R} \\ &(v, \eta) \mapsto \mathbb{A}(v, \eta) := -\int \bar{v}^*\omega - \int \eta(t) F(v(t)) dt \,. \end{aligned} \tag{1.34}$$

This differs from the Rabinowitz action functional in so far that we have an entire loop $\eta(t)$ of Lagrange multiplies, i.e. there are infinitely many Lagrange multipliers. Critical points of $\mathbb{A}^F$ are solutions of

$$\begin{cases} \partial_t v = \eta(t) X_F(v), \ \forall t \in S^1 \\ F(v(t)) = 0, \ \forall t \in S^1 \,. \end{cases} \tag{1.35}$$

The gauge group $\mathscr{H} := C^\infty(S^1, \mathbb{R})$ acts on $\mathscr{L}_0 \times C^\infty(S^1, \mathbb{R})$ by

$$(v, \eta) \mapsto h_*(v, \eta) := (\phi_F^{h(t)}(v(t)), \eta + h') \,. \tag{1.36}$$

The differential of $\mathbb{A}^F$ is invariant under the action of the gauge group:

$$d\mathbb{A}^F(w)\big[\hat{w}\big] = d\mathbb{A}^F(h_* w)\Big[dh(w)\big[\hat{w}\big]\Big] \tag{1.37}$$

for all $w \in \mathscr{L}_0 \times C^\infty(S^1, \mathbb{R})$, $\hat{w} \in T_w\big(\mathscr{L}_0 \times C^\infty(S^1, \mathbb{R})\big)$, and $h \in \mathscr{H}$. Since $\mathscr{H}$ is connected we conclude that $\mathbb{A}^F$ is invariant under $\mathscr{H}$, too. We define the based gauge group

$$\mathscr{H}_0 := \{h \in \mathscr{H} \mid h(0) = 0\} \,. \tag{1.38}$$

The based gauge group acts freely on $\mathscr{L}_0 \times C^\infty(S^1, \mathbb{R})$. Since for any $\eta \in C^\infty(S^1, \mathbb{R})$ there exists a unique $h_\eta \in \mathscr{H}_0$ such that $\eta + h'_\eta = \int_0^1 \eta(t) dt$ we can identify

$$\mathscr{C} : \mathscr{L}_0 \times C^\infty(S^1, \mathbb{R}) / \mathscr{H}_0 \cong \mathscr{L}_0 \times \mathbb{R}$$
$$[(v, \eta)] \mapsto (h_\eta)_*(v, \eta) = \left(\phi_F^{h_\eta(t)}(v(t)), \int_0^1 \eta(t)dt\right). \tag{1.39}$$

Taking the mean-value of η corresponds to the Coulomb gauge on the trivial $\mathbb{R}$-bundle over S^1.

Remark 1.14. It is a remarkable fact that there exists a global slice for the gauge group action. This is related to the fact that $\mathbb{R}$ is abelian. For non-abelian gauge theories in general there exist no global slices. If the moment action functional $\mathbb{A}^F$ is restricted to this global Coulomb slice we obtain the Rabinowitz action functional $\mathcal{A}^F$. This gives another explanation why a single Lagrange multiplier in the Rabinowitz action functional $\mathcal{A}^F$ eventually leads to a point-wise constraint at critical points (as opposed to an integral constraint).

In order to proceed to the symplectic vortex equations we need to assume that there exists an $\mathbb{R}$-invariant compatible almost complex structure J on M, i.e.

$$(\phi_F^t)_* J = J \quad \forall t \in \mathbb{R}. \tag{1.40}$$

On compact Lie groups invariant almost complex structure exist by averaging. In the current situation existence is not at all clear but occurs for instance if the flow of F is periodic. Given such an invariant J we define an L^2 inner product on $\mathscr{L}_0 \times C^\infty(S^1, \mathbb{R})$ by

$$\mathfrak{g}\big((\hat{v}_1, \hat{\eta}_1), (\hat{v}_2, \hat{\eta}_2)\big) := \int_0^1 \omega\big(\hat{v}_1, J(v)\hat{v}_2\big)dt + \int_0^1 \hat{\eta}_1(t)\hat{\eta}_2(t)dt \tag{1.41}$$

for $(\hat{v}_i, \hat{\eta}_i) \in T_{(v,\eta)}(\mathscr{L}_0 \times C^\infty(S^1, \mathbb{R}))$. Then the gradient of $\mathbb{A}^F$ equals

$$\nabla^{\mathfrak{g}} \mathbb{A}^F(v, \eta) = \begin{pmatrix} J(v)\big(\partial_t v - \eta X_F(v)\big) \\ F(v)dt \end{pmatrix} \tag{1.42}$$

and the gradient flow equation is

$$\begin{cases} \partial_s v + J(v)\big(\partial_t v - \eta X_F(v)\big) = 0 \\ \partial_s \eta - F(v) = 0 \end{cases} \tag{$\star$}$$

where $v : \mathbb{R} \times S^1 \to M$ and $\eta : \mathbb{R} \times S^1 \to \mathbb{R}$ are smooth maps. Since the functional $\mathbb{A}$ and the metric $\mathfrak{g}$ are invariant under $\mathscr{H} = C^\infty(S^1, \mathbb{R})$ so is the gradient flow equation. These are the symplectic vortex equations on the cylinder in temporal gauge. The full symplectic vortex equations are

$$\begin{cases} \partial_s v - \zeta X_F(v) + J(v)\big(\partial_t v - \eta X_F(v)\big) = 0 \\ \partial_s \eta - \partial_t \zeta = F(v) \end{cases} \tag{$\star\star$}$$

where $v : \mathbb{R} \times S^1 \to M$ and $\eta, \zeta : \mathbb{R} \times S^1 \to \mathbb{R}$ are smooth maps. This is invariant under the enlarged gauge group $\mathscr{G} := C^\infty(\mathbb{R} \times S^1, \mathbb{R})$ acting by

$$g_*(v, \eta, \zeta) = (\phi_F^g(v), \eta + \partial_t g, \zeta + \partial_s g) \,. \tag{1.43}$$

Then we have

$$\{(v, \eta) \text{ solves } (\star)\}\big/\mathscr{H} \cong \{(v, \eta, \zeta) \text{ solves } (\star\star)\}\big/\mathscr{G} \tag{1.44}$$

since in each $\mathscr{G}$-orbit $[v, \eta, \zeta]$ there is a representative with $\zeta = 0$ obtained by regauging with $g \in \mathscr{G}$ satisfying $\partial_s g = -\zeta$. The second equation in $(\star\star)$ should be thought of as an equation for the curvature of the connection $\mathbf{A} := \eta dt + \zeta ds$ on the cylinder. Indeed $(\star\star)$ can be written as

$$\begin{cases} \bar\partial_{J,\mathbf{A}} v = 0 \\ * F_{\mathbf{A}}(v) = F(v) \end{cases} \tag{$\star\star'$}$$

where $F_{\mathbf{A}}$ is the curvature of $\mathbf{A}$ and $*$ is the Hodge star operator.

Remark 1.15. We point out that we have two inner products on $\mathscr{L}_0 \times \mathbb{R}$, see diagram (1.45). One is the natural L^2 inner product $\mathfrak{m}$, see (1.17). The other is $\mathscr{C}^*[\mathfrak{g}]$ where $\mathscr{C}$ is the Coulomb map, see (1.39). We recall that $\mathfrak{g}$ is $\mathscr{H}_0$-invariant and we denote by $[\mathfrak{g}]$ the induced metric on the quotient.

$$\begin{array}{ccc} & & (\mathscr{L}_0 \times C^\infty(S^1, \mathbb{R}), \mathfrak{g}) \\ & \overset{\iota}{\nearrow} & \big\downarrow \pi \\ (\mathscr{L}_0 \times \mathbb{R}, \mathfrak{m}) & \overset{\mathscr{C}}{\longleftarrow} & (\mathscr{L}_0 \times C^\infty(S^1, \mathbb{R})/\mathscr{H}_0, [\mathfrak{g}]) \end{array} \tag{1.45}$$

We point out that

$$\mathfrak{m} \neq \mathscr{C}^*[\mathfrak{g}] \tag{1.46}$$

which is due to the fact that the infinitesimal gauge action is not $\mathfrak{g}$-orthogonal to the Coulomb slice. Thus, gradient flow lines of $\nabla^{\mathfrak{m}}\mathcal{A}^F$ are different from gauge orbits of gradient flow lines of $\nabla^{\mathscr{C}^*[\mathfrak{g}]}\mathcal{A}^F$. The latter are the symplectic vortex equations. Therefore, the gradient flow equation for the Rabinowitz action functional is substantially different from the symplectic vortex equations. This also reflected in the respective compactness proofs for gradient flow lines.

1.2.6 Definition of Rabinowitz Floer Homology

For a generic Moser pair $\mathfrak{M} = (F, H)$ the perturbed Rabinowitz action functional $\mathcal{A}^{\mathfrak{M}}$ is Morse. Interestingly enough, if Σ is a regular hypersurface, it is enough to perturb H in order to make $\mathcal{A}^{\mathfrak{M}}$ Morse, see [4]. On the other hand, if $H = 0$ then $\mathcal{A}^F$ is never Morse. This is due to the fact that $\mathcal{A}^F$ is invariant under the S^1-action on the loop space $\mathscr{L}$. Consequently, S^1 acts on the critical set $\mathrm{Crit}\mathcal{A}^F$. Indeed, if $(v, \eta) \in \mathrm{Crit}\mathcal{A}^F$ is a Reeb orbit then $v(t + \tau, \eta)$, $\tau \in S^1$, is also a critical point. Moreover, $\mathrm{Crit}\mathcal{A}^F$ always contains the constant solutions $(x, 0)$, $x \in \Sigma$. These are the fixed points of the S^1-action. For generic choice of F the unperturbed Rabinowitz action functional $\mathcal{A}^F$ is Morse-Bott, see [11].

In the following we only give a definition of Rabinowitz Floer homology in the case where $\mathcal{A}^{\mathfrak{M}}$ is Morse. The Morse-Bott case can be defined by choosing an auxiliary Morse function on the critical manifolds and counting gradient flow lines with cascades, see Frauenfelder [11, 21]. Moreover, we restrict our attention to $\mathbb{Z}/2$-coefficients. There's no doubt that Rabinowitz Floer homology can also be defined with $\mathbb{Z}$-coefficients but so far there is no treatment of coherent orientations for Rabinowitz Floer homology in the literature.

1.2.7 The Case of Restricted Contact Type

Let Σ be a hypersurface of restricted contact type and $\mathfrak{M}$ a Moser pair such that $\mathcal{A}^{\mathfrak{M}}$ is Morse. The Rabinowitz Floer chain complex is defined as the Morse complex of $\mathcal{A}^{\mathfrak{M}}$ with help of Novikov sums, see Hofer – Salamon, [28], for Floer theory with Novikov rings:

$$\mathbf{RFH}(\mathfrak{M}) := \Big\{\xi = \sum_c \xi_c\, c \mid \#\{c \in \mathrm{Crit}\mathcal{A}^{\mathfrak{M}} \mid \xi_c \neq 0 \in \mathbb{Z}/2 \text{ and } \mathcal{A}^{\mathfrak{M}}(c) \geq \kappa\} < \infty\ \forall \kappa \in \mathbb{R}\Big\}. \tag{1.47}$$

The boundary operator

$$\partial : \mathbf{RFC}(\mathfrak{M}) \longrightarrow \mathbf{RFC}(\mathfrak{M}) \tag{1.48}$$

is defined by counting gradient flow lines by the standard procedure in Floer homology and satisfies $\partial \circ \partial = 0$ thanks to compactness. Then Rabinowitz Floer homology is

$$\mathbf{RFH}(\mathfrak{M}) := \mathrm{H}(\mathbf{RFC}(\mathfrak{M}), \partial))\ . \tag{1.49}$$

The usual theory of continuation homomorphisms in Floer theory implies that Rabinowitz Floer homology is independent of auxiliary data such as the almost complex structure J and the perturbation H. Moreover, if the hypersurface Σ is isotoped through restricted contact type hypersurfaces Rabinowitz Floer homology does not change either.

Remark 1.16. Since Rabinowitz action functional is defined on the full loop space and the differential in the Rabinowitz Floer complex counts topological cylinders we can split Rabinowitz Floer homology into factors labeled by free homotopy classes

$$\mathbf{RFH}(\mathfrak{M}) = \bigoplus_{\gamma \in [S^1, M]} \mathrm{RFH}(\mathfrak{M}, \gamma)\,. \tag{1.50}$$

Definition 1.17. We abbreviate by

$$\mathrm{RFH}(\Sigma, M) := \mathrm{RFH}(\mathfrak{M}, \mathrm{pt}) \tag{1.51}$$

where pt denotes the free homotopy class of a point and

$$\mathbf{RFH}(\Sigma, M) := \mathbf{RFH}(\mathfrak{M}) \tag{1.52}$$

where $\mathfrak{M}$ is a Moser pair such that $\mathcal{A}^{\mathfrak{M}}$ is Morse.

Definition 1.18. Let $(W, \omega = d\lambda)$ be an exact, compact symplectic manifold with contact type boundary $\Sigma := \partial W$. We denote by $\widehat{W}$ the completion of W as described at the beginning of Sect. 1.2.2 and set

$$\begin{aligned} \mathbf{RFH}(\partial W, W) &:= \mathbf{RFH}(\Sigma, \widehat{W})\,, \\ \mathrm{RFH}(\partial W, W) &:= \mathrm{RFH}(\Sigma, \widehat{W})\,. \end{aligned} \tag{1.53}$$

Theorem 1.19 ([16]). *Let $\Sigma \subset (M, \omega = d\lambda)$ be a restricted contact type closed hypersurface which bounds the compact region W, then Rabinowitz Floer homology does not depend on the exterior $M \setminus W$:*

$$\begin{aligned} \mathbf{RFH}(\Sigma, W) &= \mathbf{RFH}(\Sigma, M)\,, \\ \mathrm{RFH}(\Sigma, W) &= \mathrm{RFH}(\Sigma, M)\,. \end{aligned} \tag{1.54}$$

There are two other versions of Rabinowitz Floer homology. For this we fix numbers $a < b$ and define the $\mathbb{Z}/2$ vector space

$$\mathbf{RFC}^{(a,b)}(\mathfrak{M}) := \Big\{ \sum_c \xi_c c \mid c \in \mathrm{Crit}\mathfrak{M}, a < \mathcal{A}^{\mathfrak{M}}(c) < b, \xi_c \in \mathbb{Z}/2 \Big\}\,. \tag{1.55}$$

Counting gradient flow lines again defines a differential on $\mathbf{RFC}^{(a,b)}(\mathfrak{M})$. We denote the homology by

$$\mathbf{RFH}^{(a,b)}(\mathfrak{M})\,. \tag{1.56}$$

The natural inclusion and projection homomorphisms between $\mathbf{RFC}^{(a,b)}(\mathfrak{M})$ and $\mathbf{RFC}^{(a',b')}(\mathfrak{M})$ turn $\left\{\mathbf{RFH}^{(a,b)}(\mathfrak{M})\right\}$ into a bidirected system of vector spaces. Thus, we can take direct and inverse limits:

$$\begin{aligned} \overline{\mathbf{RFH}}(\mathfrak{M}) &:= \varinjlim_{b\to\infty} \varprojlim_{a\to-\infty} \mathbf{RFH}^{(a,b)}(\mathfrak{M}) \,, \\ \underline{\mathbf{RFH}}(\mathfrak{M}) &:= \varprojlim_{a\to-\infty} \varinjlim_{b\to\infty} \mathbf{RFH}^{(a,b)}(\mathfrak{M}) \,. \end{aligned} \tag{1.57}$$

The three vector spaces $\mathbf{RFH}(\mathfrak{M})$, $\overline{\mathbf{RFH}}(\mathfrak{M})$, and $\underline{\mathbf{RFH}}(\mathfrak{M})$ are related by a commutative diagram as follows

$$\begin{array}{ccc} \mathbf{RFH}(\mathfrak{M}) & \xrightarrow{\ \overline{\rho}\ } & \overline{\mathbf{RFH}}(\mathfrak{M}) \\ & {\scriptstyle \underline{\rho}}\searrow & \downarrow{\scriptstyle \kappa} \\ & & \underline{\mathbf{RFH}}(\mathfrak{M}) \end{array} \tag{1.58}$$

in which $\overline{\rho}$ is an isomorphism and κ is surjective. We point out that the canonical map $\overline{\rho}$ is not necessarily an isomorphism if $\mathbb{Z}$-coefficients instead of field coefficients are used. For details and proofs see [12].

Remark 1.20. A hypersurface $\Sigma \subset (M,\omega)$ is of virtual restricted contact type if there exists a cover $\pi : \widetilde{M} \longrightarrow M$ and a primitive $\widetilde{\lambda}$ of $\pi^*\omega$ which is bounded on $\pi^{-1}(\Sigma)$ with respect to π^*g where g is a Riemannian metric on Σ. The notion of virtual restricted contact type plays an important role in the study of twisted cotangent bundles since energy hypersurfaces above Mañé's critical value are of virtual restricted contact type but in general not of restricted contact type, see [18] for details. Rabinowitz Floer homology for virtual restricted contact type hypersurfaces can be defined as in the case of restricted contact type on the component of the loop space containing the contractible loops.

1.2.8 The Case of Stable Hypersurfaces

In this section the Rabinowitz action functional is only considered on the space of contractible loops. If restricted contact type is replaced by stability of the hypersurface two difficulties appear. The first is of technical nature namely stability is not an open condition, see [17]. Therefore, in general it cannot be expected that the unperturbed Rabinowitz action functional can be made Morse-Bott on a given stable hypersurface. It remains true that for a small Hamiltonian perturbation the perturbed Rabinowitz action function is Morse. However, Proposition 1.10 does not allow for Hamiltonian perturbations. Therefore, compactness for families of gradient flow lines might fail as follows: new gradient flow lines from infinity, i.e. with very large Lagrange multipliers, might appear. Nevertheless, for sufficiently small perturbation there is no interaction between gradient flow lines with small and large Lagrange multipliers. To implement this the filtration on the Rabinowitz Floer complex has to be modified to involve both action functionals $\mathcal{A}^F$ and $\mathcal{A}_\alpha$ defined above. Since both action functional are Lyapunov functions along gradient flow lines (see above) we obtain a doubly filtered complex $\mathrm{RFC}^{(a,b),(a',b')}$. Then using a Hamiltonian perturbation of size depending on a, b, a', b' a boundary operator ∂ can still be defined by ignoring gradient flow lines with very large Lagrange multipliers. Moreover, the resulting filtered homology is independent of the small Hamiltonian perturbation. Thus, $\overline{\mathrm{RFH}}$ can be defined as above by taking an inverse-direct limit over a, b, a', b'. We point out that there is no analogue of the definition with Novikov sums for stable hypersurfaces.

The second, more serious difficulty is that the just defined Rabinowitz Floer homology might depend on the stabilizing 1-form $\alpha \in \Omega^1(\Sigma)$ in a very subtle way. Indeed, even though the critical points and the gradient flows do not depend on α Rabinowitz Floer homology $\overline{\mathrm{RFH}}$ might depend on α through the double action filtration by $\mathcal{A}^F$ and $\mathcal{A}_\alpha$ in the inverse-direct limit. To guarantee independence of Rabinowitz Floer homology we impose the additional assumption of tameness.

Definition 1.21. A stable hypersurface $\Sigma \in (M, \omega)$ is tame if there exists a stabilizing 1-form $\alpha \in \Omega^1(\Sigma)$ and a constant $C > 0$ such that

$$\left| \int v^*\alpha \right| \leq C \left| \int \bar{v}^*\omega \right| \tag{1.59}$$

for all Reeb orbits v of α which are contractible in M. Here $\bar{v}$ again denotes a filling disk and the right hand side is independent of $\bar{v}$ since (M, ω) is symplectically aspherical.

There are examples of stable hypersurfaces which are not tame. In fact, even contact type hypersurface need not be tame. For more details we refer to [18]. More examples of non-tame hypersurfaces are provided in Macarini – Paternain [31] and Cieliebak – Volkov [13]. An isotopy of hypersurfaces is called a stable isotopy if the stabilizing 1-form depends smoothly on the isotopy parameter. It is called a tame stable isotopy if in addition all hypersurfaces are tame with a taming constant independent of the isotopy parameter.

Theorem 1.22 ([18]). *If Σ is tame then Rabinowitz Floer homology $\overline{\mathrm{RFH}}$ is invariant under tame stable isotopies. Since the space of stabilizing 1-forms is a cone, hence connected, $\overline{\mathrm{RFH}}$ is independent of α.*

Definition 1.23. We abbreviate

$$\mathrm{RFH}(\Sigma, M) := \overline{\mathrm{RFH}}\,. \tag{1.60}$$

This is justified since in the case of restricted contact type the natural map $\overline{\rho}$ in the commuting diagram 1.58 is an isomorphism.

1.2.9 Grading

If the homomorphism $I_{c_1(M)} : \pi_2(M) \to \mathbb{Z}$ obtained by integrating the first Chern class $c_1(M)$ over a smooth representative vanishes identically then Rabinowitz Floer homology RFH_* carries a $\mathbb{Z}$-grading in terms of the Conley-Zehnder index, see [11]. In general only a relative $\mathbb{Z}/N$ can be defined where N is the minimal Chern number. There are different normalization conventions. In [11] the grading takes values in the set $\frac{1}{2} + \mathbb{Z}$. Underlying this convention is the philosophy that the index in a semi-infinite homology is $\frac{1}{2}$ times some kind of signature index of the Hessian at critical points. We refer to the article [38] by Salamon – Robbin for research in this direction. The half signature index takes values in $\mathbb{Z}$ on even dimensional manifolds and in $\frac{1}{2} + \mathbb{Z}$ on odd dimensional manifolds. Since the loop space $\mathscr{L}$ of a symplectic manifolds is itself symplectic $\mathscr{L}$ is an "even" infinite-dimensional manifold and consequently $\mathscr{L} \times \mathbb{R}$ is an "odd" infinite-dimensional manifold. Therefore the convention used in [11] is consistent with this interpretation. Moreover, with this convention

$$\mathrm{RFH}^{-*} \cong \mathrm{RFH}_* \tag{1.61}$$

holds. This isomorphism is induced by the involution

$$\begin{aligned} I : \mathscr{L} \times \mathbb{R} &\longrightarrow \mathscr{L} \times \mathbb{R} \\ (v, \eta) &\mapsto (v^-, -\eta) \end{aligned} \tag{1.62}$$

where $v^-(t) := v(1-t)$ under which the unperturbed Rabinowitz action functional is anti-invariant

$$\mathcal{A}^F \circ I = -\mathcal{A}^F \; . \tag{1.63}$$

Another useful convention is to replace the above convention by adding $\frac{1}{2}$. This conventions fits better with the computational results comparing Rabinowitz Floer homology with symplectic homology and loop space topology as carried out in [2, 16, 33].

1.3 Computations

1.3.1 The Displaceable Case

Let (M, ω) be symplectically aspherical, and convex at infinity or geometrically bounded and $\Sigma \subset (M, \omega)$ of (virtual) restricted contact type or stable tame.

Theorem 1.24. *(1) If the hypersurface* Σ *is displaceable then Rabinowitz Floer homology* $\mathrm{RFH}_*(\Sigma, M)$ *vanishes.*

(2) If in addition Σ *is of restricted contact type then even the full Rabinowitz Floer homology* $\mathbf{RFH}_*(\Sigma, M)$ *vanishes.*

This is proved for restricted contact type in [11] and for the remaining cases in [18].

Theorem 1.25. *If the hypersurface* Σ *carries no Reeb orbits which are contractible in* M *then*

$$\mathrm{RFH}_*(\Sigma, M) \cong \mathrm{H}_*(\Sigma) . \tag{1.64}$$

This follows immediately from the definition of RFH_* in the Morse-Bott case, see above. The previous two Theorems imply the following Corollary.

Corollary 1.26. *If the hypersurface* Σ *is displaceable then there exists a Reeb orbit which is contractible in* M.

Remark 1.27. If the hypersurface Σ is stable then Corollary 1.26 has been proved by Schlenk in [40] using different methods. Schlenk does not need the tameness assumption. Using a local version of Rabinowitz Floer homology Corollary 1.26 can also be proved without the tameness assumption in the framework of Rabinowitz Floer homology, see [18].

1.3.2 Relations to Symplectic Homology and Loop Space Topology

For the next theorem we assume that we are in the same setup as at the beginning of Sect. 1.2.2, *The case of restricted contact type*. Namely, $(W, \omega = d\lambda)$ is a compact exact symplectic manifold with boundary $\partial W = \Sigma$ of restricted contact type and (M, ω) is the completion of W. In this situation symplectic homology $\mathrm{SH}_*(\Sigma, M)$ resp. cohomology $\mathrm{SH}^*(\Sigma, M)$ of M can be defined, see Cieliebak – Floer – Hofer [14] and Viterbo [41].

Theorem 1.28 ([16]). *There exists a long exact sequence*

$$\cdots \to \mathrm{SH}^{-*}(\Sigma, M) \to \mathrm{SH}_*(\Sigma, M) \to \mathbf{RFH}_*(\Sigma, M) \to \mathrm{SH}^{-*+1}(\Sigma, M) \to \cdots \tag{1.65}$$

From now on $M = T^*B$ is the cotangent bundle with its standard symplectic structure $\omega_{std} = d\lambda_{std}$ over a closed manifold B and $\Sigma = S_g^*B$ is the unit cotangent bundle with respect to some Riemannian metric g on B. Using that the symplectic (co-)homology of cotangent bundles has been computed before by Viterbo, Salamon – Weber, and Abbondandolo – Schwarz [1, 39, 41] the following Theorem has been proved via the long exact sequence in [16]. An independent and direct proof has been given in [2].

Theorem 1.29 ([2,16]).

$$\mathbf{RFH}_*(S_g^*B, T^*B) = \begin{cases} \mathrm{H}^{-*+1}(\mathscr{L}_B) & \text{if } * < 0 \\ \mathrm{H}_*(\mathscr{L}_B) & \text{if } * > 1 \end{cases} \tag{1.66}$$

where $\mathscr{L}_B := C^\infty(S^1, B)$ *is the free loop space of* B.

Remark 1.30. In the remaining degrees $* = 0, 1$ the answer is known and depends on the Euler class, see [2, 16].

Remark 1.31. In [33] Merry extends Theorem 1.29 to energy hypersurfaces above the Mañé critical value in twisted cotangent bundles.

2 Applications and Results

2.1 Symplectic and Contact Topology

As we pointed out in Theorem 1.19 Rabinowitz Floer homology does not depend on the exterior. It is an open question to what extent Rabinowitz Floer homology $\mathrm{RFH}(\Sigma, M)$ depends on the filling M. A partial independence result is the following Theorem.

Theorem 2.1. *Let* $\dim B \geq 4$ *and* $\pi_1(B) = \{1\}$ *and let* $(W, \omega = d\lambda)$ *be a compact exact symplectic manifold with* $\partial W \cong S_g^*B$. *If* $(\partial W, \lambda|_{\partial W})$ *is contactomorphic to* $(S_g^*B, \lambda_{std}|_{S_g^*B})$ *then*

$$\mathbf{RFH}_*(\partial W, W) \cong \mathbf{RFH}_*(S_g^*B, T^*B)\ . \tag{2.1}$$

This is a special case of a result proved in [16]. The crucial ingredient in the above Theorem is that there exist no rigid finite energy planes in the filling W. If B is a sphere then the above theorem can in fact be checked by a direct computation, see [11].

Corollary 2.2. *Under the assumptions of Theorem 2.1* S_g^*B *does not admit an exact contact embedding into* $\mathbb{R}^{2n}$ *or, more generally, into a subcritical Stein manifold.*

Proof. We assume by contradiction that there exists an exact contact embedding of S_g^*B into a subcritical Stein manifold $(M, \omega = d\lambda)$. We denote by Σ the image of S_g^*B in M. Since $\mathrm{H}_{2n-1}(M) = 0$, $2n = \dim M$, the hypersurface Σ bounds a compact region W in M. Because any compact subset of a subcritical Stein manifold is displaceable, see Biran – Cieliebak [10], $\mathbf{RFH}_*(\Sigma, M) \cong 0$. On the other hand by Theorem 1.19 and 2.1 we know $0 \cong \mathbf{RFH}_*(\Sigma, M) \cong \mathbf{RFH}_*(\Sigma, W) \cong \mathbf{RFH}_*(S_g^*B, T^*B)$. This contradicts Theorem 1.29. □

2.2 Global Perturbations of Hamiltonian Systems

We recall from Sect. 1.1.4 that critical points of $\mathcal{A}^{\mathfrak{M}}$, $\mathfrak{M} = (F, H)$, give rise to leaf-wise intersections.

Theorem 2.3. *Let $\Sigma \subset (M, \omega = d\lambda)$ be a closed, bounding, restricted contact type hypersurface and M convex at infinity or geometrically bounded. We denote by $\wp > 0$ the smallest period of a Reeb orbit on $(\Sigma, \lambda|_\Sigma)$ which is contractible in M. Let $\psi \in \mathrm{Ham}_c(M, \omega)$, that is, ψ is a Hamiltonian diffeomorphism generated by a Hamiltonian with compact support. If the Hofer norm $||\psi|| < \wp$ then there exists a leaf-wise intersection.*

This was proved in [4]. An alternative proof is given by Gürel in [26]. Kang [29] found an extension of Theorem 2.3 if Σ is only of contact type. In [8] a cup-length estimate for leaf-wise intersections is proved under the assumptions of Theorem 2.3.

Remark 2.4. The existence of displaceable hypersurfaces Σ shows that a smallness assumption $||\psi|| < \wp$ is necessary.

Remark 2.5. If $\psi = \mathrm{id}$ in Theorem 2.3 then each point in Σ is a leaf-wise intersection point via constant critical points. The leaf-wise intersection point found in Theorem 2.3 arises as a perturbation of a constant critical point by a stretching-of-the-neck argument for gradient flow lines. If this stretching-of-the-neck argument is replaced by local Floer homology around the constant critical points then for a generic Hamiltonian diffeomorphism ψ in Theorem 2.3 there exists $\sum_{i=0}^{\dim \Sigma} b_i(\Sigma)$ different leaf-wise intersection points, see [4,29].

We recall that $\mathscr{L}_B = C^\infty(S^1, B)$ denotes the free loop space of B.

Theorem 2.6. *Let $\dim \mathrm{H}_*(\mathscr{L}_B) = \infty$. If $\dim B \geq 2$ and $\Sigma \subset T^*B$ is a generic fiber-wise star-shaped hypersurface then for a generic $\psi \in \mathrm{Ham}_c(T^*B)$ there exist infinitely many leaf-wise intersection points.*

We point out that there is no assumption on the Hofer norm $||\psi||$ of ψ in Theorem 2.6. This was proved in [3] along the following lines. Since Σ is fiber-wise star-shaped it is of restricted contact type and isotopic to S_g^*B through restricted contact type hypersurfaces. Thus, by the assumption $\dim \mathrm{H}_*(\mathscr{L}_B) = \infty$, invariance of Rabinowitz Floer homology, and Theorem 1.29 we conclude that the Rabinowitz Floer homology $\mathrm{RFH}(\Sigma, T^*B)$ is infinite dimensional. In particular, since for a generic perturbation the corresponding Rabinowitz action functional $\mathcal{A}^{\mathfrak{M}}$ is Morse, $\mathcal{A}^{\mathfrak{M}}$ has infinitely many critical points. Then a transversality result using $\dim B \geq 2$ yields that generically critical points of $\mathcal{A}^{\mathfrak{M}}$ won't lie on a closed leaf. Thus, the assertion of Theorem 2.6 follows from Proposition 1.2.

Remark 2.7. Theorems 2.3 and 2.6 have been proved for energy hypersurfaces above Mañé's critical value in twisted cotangent bundles in [33].

Remark 2.8. As the sketch of the proof of Theorem 2.6 shows to obtain generically infinitely many leaf-wise intersection points it is enough to prove that Rabinowitz

Floer homology is infinite dimensional. A large class of examples with this property has been constructed by Albers – McLean in [7].

Using spectral invariants in Rabinowitz Floer homology Theorem 2.6 can be improved as follows, see [6].

Theorem 2.9. *Let* $\dim \mathrm{H}_*(\mathscr{L}_B) = \infty$. *If* $\dim B \geq 2$ *and* $\Sigma \subset T^*B$ *is a fiber-wise star-shaped hypersurface then for* $\psi \in \mathrm{Ham}_c(T^*B)$ *there exist infinitely many leaf-wise intersection points or a leaf-wise intersection point on a closed leaf.*

This Theorem is equivalent to the assertion that the Rabinowitz action functional always, even in degenerate situations, has infinitely many critical points. This does not follow from Theorem 2.6 since for degenerate functions the Morse inequalities might fail.

The new ingredient in Theorem 2.9 are spectral invariants. The idea behind spectral invariants is the following. To each homology class a critical value is associated. If the action functional is Morse this is done by a mini-max procedure. It can then be shown that this assignment is locally Lipschitz continuous in changes of the Moser pair. Thus, it extends to all Moser pairs. The crucial property of the extension is that it still assigns critical values to homology classes, even in degenerate situations.

To prove Theorem 2.9 it is shown that the set of spectral values is unbounded and therefore gives rise to infinitely many critical points which are distinguished by their critical values.

It is an interesting open question whether a Gromoll-Meyer type theorem, [25, 32], holds for leaf-wise intersection points, that is, if there exist infinitely many leaf-wise intersection points even in the case when there is a leaf-wise intersection point on a closed leaf. This problem is intimately related to the existence problem of (geometrically distinct) geodesics. Katok's example, [30] suggests that, as in the Gromoll-Meyer Theorem, a growth condition for the homology of the loop space is necessary. Interesting research in the direction can be found in [23].

Since $S^{2n-1} \subset \mathbb{R}^{2n}$ is displaceable for general Hamiltonian perturbations there need not exist leaf-wise intersection points. However, if the class of Hamiltonian perturbations is restricted to preserve symmetries this picture might change dramatically. Such a phenomenon was discovered by Ekeland – Hofer in [20]. They prove that if Σ is a centrally symmetric, restricted contact type hypersurface in $\mathbb{R}^{2n}$ and ϕ^t is an isotopy of centrally symmetric Hamiltonian perturbations there always exists a leaf-wise intersection point for ϕ^1. If restricted contact type is replaced by star-shaped it is proved in [5] that under the same symmetry assumptions there exist infinitely many leaf-wise intersection points or a leaf-wise intersection point on a closed leaf. The proof uses a computation of $\mathbb{Z}/2$-invariant Rabinowitz Floer homology.

2.3 Mañé's Critical Value

Let (B, g) be a closed Riemannian manifold and $\sigma \in \Omega^2(B)$ be a closed 2-form which vanishes on $\pi_2(B)$. Thus, on the universal cover $\pi : \widetilde{B} \to B$ the 2-form $\pi^*\sigma$ is exact. The twisted cotangent bundle is $(T^*B, \omega := \omega_{std} + \tau^*\sigma)$ where $\tau : T^*B \to B$ is the projection. We fix a potential $U : B \to \mathbb{R}$ and define $F : T^*B \to \mathbb{R}$ by $F(q, p) := \frac{1}{2}||p||_g^2 + U(q)$. Then the Mañé critical value is

$$c = c(g, \sigma, U) := \inf_{\theta} \sup_{q \in \widetilde{B}} \tilde{F}(q, \theta_q) \tag{2.2}$$

where $\widetilde{F} = F \circ \pi$ and $\theta \in \Omega^1(\widetilde{B})$ satisfies $d\theta = \pi^*\sigma$.

In physical terms the Hamiltonian dynamics of X_F taken with respect to the twisted symplectic form describes the motion of a particle on B subject to a conservative force $-\nabla U$ and a magnetic field σ. For energies above the Mañé critical value the magnetic field is interpreted as small compared to the energy.

The hypersurfaces $\Sigma_k := F^{-1}(k)$ above the Mañé critical value, i.e. $k > c$, are of virtual restricted contact type, see Remark 1.20. If there are no topological obstructions Σ_k is displaceable for sufficiently small k if $\sigma \neq 0$. This follows from the fact that the zero-section B in T^*B is displaceable, see Polterovich [35].

Theorem 2.10 ([33]). *The hypersurface Σ_k for $k > c$ is not displaceable.*

This follows from the fact that Rabinowitz Floer homology above the Mañé critical value does not vanish, see Remark 1.31 .

At the Mañé critical value the hypersurface Σ_c is in general not stable or of virtual restricted contact type. It might even be singular, e.g. if the magnetic field is zero. Moreover, there are examples where Σ_c carries no closed characteristics, e.g. the horocycle flow. The article [18] proposes the following *paradigms*:

($k > c$) Above the Mañé critical value, Σ_k is virtually contact. It may or may not be stable. Its Rabinowitz Floer homology $\mathrm{RFH}(\Sigma_k)$ is defined and nonzero, so Σ_k is non-displaceable (now proved by Merry [33]). The dynamics on Σ_k is like that of a geodesic flow; in particular, it has a periodic orbit in every nontrivial free homotopy class.

($k = c$) At the Mañé critical value, Σ_k is non-displaceable and can be expected to be non-stable (hence non-contact).

($k < c$) Below the Mañé critical value, Σ_k may or may not be of contact type. It is stable and displaceable (provided that $\chi(M) = 0$), so its Rabinowitz Floer homology $\mathrm{RFH}(\Sigma_k)$ is defined and vanishes. In particular, Σ_k has a contractible periodic orbit.

It is unknown whether Σ_k is always stable below the Mañé critical value. In [18] many examples can be found supporting these paradigms.

Acknowledgements The authors thank Will Merry and Gabriel Paternain for various helpful comments on an earlier version.

This article was written during visits of the authors at the Institute for Advanced Study, Princeton. The authors thank the Institute for Advanced Study for their stimulating working atmospheres.

This material is based upon work supported by the National Science Foundation under agreement No. DMS-0635607 and DMS-0903856. Any opinions, findings and conclusions or recommendations expressed in this material are those of the authors and do not necessarily reflect the views of the National Science Foundation.

References

1. Abbondandolo, A., Schwarz, M.: On the Floer homology of cotangent bundles. Comm. Pure Appl. Math. **59**(2), 254–316 (2006)
2. Abbondandolo, A., Schwarz, M.: Estimates and computations in Rabinowitz-Floer homology. J. Topol. Anal. **1**(4), 307–405 (2009)
3. Albers, P., Frauenfelder, U.: Infinitely many leaf-wise intersections on cotangent bundles, arXiv:0812.4426 (2008)
4. Albers, P., Frauenfelder, U.: Leaf-wise intersections and Rabinowitz Floer homology. J. Topol. Anal. **2**(1), 77–98 (2010)
5. Albers, P., Frauenfelder, U.: On a Theorem by Ekeland-Hofer, arXiv:1001.3386. Israel Journal of Mathematics (2010, to appear)
6. Albers, P., Frauenfelder, U.: Spectral invariants in Rabinowitz Floer homology and global Hamiltonian perturbations, arXiv:1001.2920 (2010)
7. Albers, P., McLean, M.: Non-displaceable contact embeddings and infinitely many leaf-wise intersections, J. Symplectic Geom. **9**(3), 271–284 (2011)
8. Albers, P., Momin, A.: Cup-length estimates for leaf-wise intersections, arXiv:1002.3283 (2010)
9. Banyaga, A.: On fixed points of symplectic maps. Invent. Math. **56**(3), 215–229 (1980)
10. Biran, P., Cieliebak, K.: Lagrangian embeddings into subcritical Stein manifolds, Israel J. Math. **127** (2002), 221–244.
11. Cieliebak, K., Frauenfelder, U.: A Floer homology for exact contact embeddings. Pacific J. Math. **293**(2), 251–316 (2009)
12. Cieliebak, K., Frauenfelder, U.: Morse homology on noncompact manifolds, arXiv:0911.1805 (2009)
13. Cieliebak, K., Volkov, E.: First steps in stable Hamiltonian topology, arXiv:1003.5084 (2010)
14. Cieliebak, K., Floer, A., Hofer, H.: Symplectic homology. II. A general construction. Math. Z. **218**(1), 103–122 (1995)
15. Cieliebak, K., Gaio, A.R., Salamon, D.A.: J-holomorphic curves, moment maps, and invariants of Hamiltonian group actions. Internat. Math. Res. Notices 16, 831–882 (2000)
16. Cieliebak, K., Frauenfelder, U., Oancea, A.: Rabinowitz Floer homology and symplectic homology, Ann. Sci. Éc. Norm. Supér. (4) **43**(6), 957–1015 (2010)
17. Cieliebak, K., Frauenfelder, U., Paternain, G.: Stability is not open, arXiv:0908.2540 (2009)
18. Cieliebak, K., Frauenfelder, U., Paternain, G.: Symplectic topology of Mañé's critical value, arXiv:0903.0700 (2009)
19. Dragnev, D.L.: Symplectic rigidity, symplectic fixed points, and global perturbations of Hamiltonian systems. Comm. Pure Appl. Math. **61**(3), 346–370 (2008)
20. Ekeland, I., Hofer, H.: Two symplectic fixed-point theorems with applications to Hamiltonian dynamics. J. Math. Pures Appl. (9) **68**(4), 467–489 (1989/1990)
21. Frauenfelder, U.: The Arnold-Givental conjecture and moment Floer homology. Int. Math. Res. Not. 42, 2179–2269 (2004)

22. Ginzburg, V.L., Gürel, B.: A C^2-smooth counterexample to the Hamiltonian Seifert conjecture in $\mathbb{R}^4$. Ann. Math. (2) **158**(3), 953–976 (2003)
23. Ginzburg, V.L., Gürel, B.: Local Floer Homology and the Action Gap, arXiv:0709.4077 (2007, to appear) Journal of Symplectic Geometry
24. Ginzburg, V.L.: Coisotropic intersections. Duke Math. J. **140**(1), 111–163 (2007)
25. Gromoll, D., Meyer, W.: Periodic geodesics on compact Riemannian manifolds. J. Diff. Geom. **3**, 493–510 (1969)
26. Gürel, B.: Leafwise Coisotropic Intersections. Int. Math. Res. Not. (2009), article ID rnp 164
27. Hofer, H.: On the topological properties of symplectic maps. Proc. Roy. Soc. Edinburgh Sect. A **115**(1–2), 25–38 (1990)
28. Hofer, H., Salamon, D.A.: Floer homology and Novikov rings. The Floer Memorial Volume, Progr. Math., vol. 133, pp. 483–524. Birkhäuser, Basel (1995)
29. Kang, J.: Existence of leafwise intersection points in the unrestricted case, arXiv:0910.2369 (2009)
30. Katok, A.B.: Ergodic perturbations of degenerate integrable Hamiltonian systems. Izv. Akad. Nauk SSSR Ser. Mat. **37**, 539–576 (1973)
31. Macarini, L., Paternain, G.: On the stability of Mañé critical hypersurfaces, arXiv:0910.5728 (2009)
32. Matthias, H.: Zwei Verallgemeinerungen eines Satzes von Gromoll und Meyer. Bonner Mathematische Schriften 126. Universität Bonn Mathematisches Institut (1980)
33. Merry, W.: On the Rabinowitz Floer homology of twisted cotangent bundles, arXiv:1002.0162 (2010)
34. Moser, J.: A fixed point theorem in symplectic geometry. Acta Math. **141**(1–2), 17–34 (1978)
35. Polterovich, L.: An obstacle to non-Lagrangian intersections. The Floer memorial volume, Progr. Math., vol. 133, pp. 575–586. Birkhäuser, Basel (1995)
36. Rabinowitz, P.H.: Periodic solutions of Hamiltonian systems. Comm. Pure Appl. Math. **31**(2), 157–184 (1978)
37. Rabinowitz, P.H.: Periodic solutions of a Hamiltonian system on a prescribed energy surface. J. Diff Eq. **33**(3), 336–352 (1979)
38. Robbin, J., Salamon, D.A.: Feynman path integrals on phase space and the metaplectic representation. Math. Z. **221**(2), 307–335 (1996)
39. Salamon, D.A., Weber, J.: Floer homology and the heat flow. Geom. Funct. Anal. **16**(5), 1050–1138 (2006)
40. Schlenk, F.: Applications of Hofer's geometry to Hamiltonian dynamics. Comment. Math. Helv. **81**(1), 105–121 (2006)
41. Viterbo, C.: Functors and computations in Floer homology with applications. I. Geom. Funct. Anal. **9**(5), 985–1033 (1999)
42. Weinstein, A.: On the hypotheses of Rabinowitz' periodic orbit theorems. J. Diff. Eq. **33**(3), 353–358 (1979)
43. Ziltener, F.: Coisotropic Submanifolds, Leafwise Fixed Points, and Presymplectic Embeddings, J. Symplectic Geom. **8**(1), 95–118 (2010)

Contact Structures and Geometric Topology

Hansjörg Geiges

1 Introduction

A *contact structure* on a manifold M of dimension $2n + 1$ is a tangent hyperplane field, i.e. a $2n$-dimensional sub-bundle ξ of the tangent bundle TM, satisfying the following *maximal non-integrability* condition: if ξ is written locally as the kernel of a differential 1-form α, then $\alpha \wedge (\mathrm{d}\alpha)^n$ is required to be nowhere zero on its domain of definition. Notice that ξ determines α up to multiplication by a smooth nowhere zero function f. So the contact condition is independent of the choice of 1-form defining ξ, since $(f\alpha) \wedge \mathrm{d}(f\alpha)^n = f^{n+1}\alpha \wedge (\mathrm{d}\alpha)^n$. I shall always assume our contact structures to be *coorientable*, which is equivalent to saying that we can write $\xi = \ker \alpha$ with a 1-form α defined on all of M; such an α is called a *contact form*. Then $\alpha \wedge (\mathrm{d}\alpha)^n$ is a volume form on M, so a *contact manifold* $(M, \xi = \ker \alpha)$ has to be orientable.

The classical Darboux theorem states that any contact form α can locally be written, in suitable coordinates, as $\alpha = \mathrm{d}z + \sum_{i=1}^{n} x_i \, \mathrm{d}y_i$. This is one of the reasons why the most interesting aspects of contact geometry are of global nature.

Contact structures provide the mathematical language for many phenomena in classical mechanics, geometric optics and thermodynamics. Equally important for the interest in these structures are their relations with symplectic, Riemannian and complex geometry. These aspects are surveyed in [28] and [31, Chap. 1].

In the last two decades it has become increasingly apparent that contact manifolds constitute a natural framework for many problems in low-dimensional geometric topology. As hypersurfaces in symplectic 4-manifolds, 3-dimensional

H. Geiges (✉)
Mathematisches Institut, Universität zu Köln, Weyertal 86–90, 50931 Köln, Germany
e-mail: geiges@math.uni-koeln.de

C. Bär et al. (eds.), *Global Differential Geometry*, Springer Proceedings in Mathematics 17, DOI 10.1007/978-3-642-22842-1_15,

contact manifolds build a bridge to 4-manifold topology. This interplay between dimensions three and four has helped solve some long-standing problems in knot theory. One salient example is the result of Kronheimer–Mrowka that all non-trivial knots in the 3-sphere S^3 have the so-called property P; see Sect. 4.2 below. Their proof is based on a result of Eliashberg and, independently, Etnyre that any symplectic filling of a 3-dimensional contact manifold can be capped off to a closed symplectic 4-manifold.

Moreover, contact topology has inspired new approaches to some known results. Pride of place has to be given to Eliashberg's proof [19] of Cerf's theorem that any diffeomorphism of S^3 extends to the 4-ball, based on the classification of contact structures on S^3; see [31] for an exposition of that proof.

Arguably the most influential contact topological result of the last decade is due to Giroux [35], cf. [24] and [8]. He established a correspondence between contact structures on a given manifold and open book decompositions of that manifold; in dimension three and subject to suitable equivalences on either set of structures, this correspondence is actually one-to-one.

In the present article I want to survey a selection of these recent developments in contact topology. In Sect. 2 a few basic contact geometric concepts will be reviewed. I then discuss some of the results highlighted above, and others besides, from a somewhat idiosyncratic point of view. As starting point I take a surgery presentation of contact 3-manifolds due to Fan Ding and yours truly; this is the content of Sect. 3.

In Sect. 4 we then turn to applications of this structure theorem. For instance, one can use it to derive an adapted open book decomposition (see Sect. 4.1), thus providing an alternative proof for one direction of the Giroux correspondence in dimension three. In Sect. 4.2 I shall also explain in outline how symplectic caps can be constructed directly from the surgery presentation theorem, without any appeal to open books. In Sect. 4.3 I offer the reader an *amuse gueule* illustrating the use of contact surgery in Heegaard Floer theory. Surgery diagrams also play a supporting role in a contact topological argument for computing the diffeotopy group of the 3-manifold $S^1 \times S^2$, as will be explained in Sect. 4.4. An example how contact surgery can be used to detect so-called non-loose (or exceptional) Legendrian knots will be given in Sect. 4.5. Finally, in Sect. 4.6 I allow the reader a glimpse of some recent results in collaboration with Fan Ding and Otto van Koert on the diagrammatic representation of 5-dimensional contact manifolds. The Giroux correspondence reduces the description of such manifolds to that of a page of an open book (here: a 4-dimensional Stein manifold) and the monodromy of the open book. The 4-dimensional Stein manifold, in turn, can be described by a surgery picture that describes the attachment of Stein handles; the attaching circles for the 2-handles are Legendrian knots, which can be visualised in terms of their front projection from S^3 (with a point removed) to a 2-plane. In conclusion, one obtains an essentially 2-dimensional representation of a contact 5-manifold. This has implications on the classification of subcritically Stein fillable contact 5-manifolds.

2 Basic Notions and Results in Contact Geometry

Here I want to recall some fundamental concepts of the subject. I also mention a few classification and structure theorems necessary for understanding or putting into perspective the more recent results described in the subsequent sections.

2.1 *Tight vs. Overtwisted*

We begin with a dichotomy of contact structures that is specific to dimension three. A smooth knot L in a contact 3-manifold (M, ξ) is called *Legendrian* if it is everywhere tangent to the contact structure. If L is homologically trivial in M, one can find an embedded surface $\Sigma \subset M$ with boundary $\partial\Sigma = L$, a so-called *Seifert surface* for L. Then L has two distinguished framings (i.e. trivialisations of its normal bundle, which can alternatively be described by a vector field along and transverse to L, or by a parallel curve obtained by pushing L in the direction of that vector field): the *surface framing*, given by a vector field tangent to the surface Σ, and the *contact framing*, given by a vector field tangent to the contact structure ξ. (The surface framing turns out to be independent of the choice of Seifert surface.)

An embedded 2-disc $\Delta \subset M$ in a contact 3-manifold (M, ξ) is called *overtwisted*, if the boundary $\partial\Delta$ is a Legendrian curve whose contact framing coincides with the surface framing. If one wishes, one may then arrange that $T_x\Delta = \xi_x$ for all $x \in \partial\Delta$.

A contact 3-manifold is called *overtwisted* if it contains an overtwisted disc; otherwise it is called *tight*. It was shown by Eliashberg [17] that the classification of overtwisted contact structures on closed 3-manifolds is a purely homotopical problem: each homotopy class of tangent 2-plane fields contains a unique overtwisted contact structure (up to isotopy). For a detailed exposition of Eliashberg's proof see [31, Chap. 4.7].

Example. Let (z, r, φ) be cylindrical coordinates on $\mathbb{R}^3$. The contact structure $\xi_{\mathrm{ot}} = \ker(\cos r\, \mathrm{d}z + r \sin r\, \mathrm{d}\varphi)$ is an overtwisted contact structure; each disc $\Delta_{z_0} = \{z = z_0, r \leq \pi\}$ is overtwisted.

The classification of tight contact structures, on the other hand, is a very intricate problem that has not yet been solved completely. It was shown by Bennequin [5], *avant la lettre*, that the standard contact structure $\xi_{\mathrm{st}} = \ker(\mathrm{d}z + x\, \mathrm{d}y)$ on $\mathbb{R}^3$ is tight. We shall return to the classification of tight structures in Sect. 2.3.

2.2 *Symplectic Fillings*

A contact manifold $(M^{2n-1}, \xi = \ker\alpha)$ with a *cooriented* contact structure is naturally oriented by the volume form $\alpha \wedge (\mathrm{d}\alpha)^{n-1}$. Likewise, a symplectic manifold

(W^{2n}, ω), i.e. with ω a closed non-degenerate 2-form, is naturally oriented by the volume form ω^n.

Definition. (a) A compact symplectic manifold (W^{2n}, ω) is called a *weak (symplectic) filling* of $(M^{2n-1}, \xi = \ker\alpha)$ if $\partial W = M$ as oriented manifolds and $\omega^{n-1}|_\xi > 0$. Here ∂W is oriented by the 'outward normal first' rule.

(b) A compact symplectic manifold (W^{2n}, ω) is called a *strong (symplectic) filling* of $(M^{2n-1}, \xi = \ker\alpha)$ if $\partial W = M$ and there is a Liouville vector field Y defined near ∂W, pointing outwards along ∂W, and satisfying $\xi = \ker(i_Y\omega|_{TM})$ (as cooriented contact structure). In this case we say that (M, ξ) is the *convex* boundary of (W, ω). Here *Liouville vector field* means that the Lie derivative $\mathcal{L}_Y\omega$ – which is the same as $\mathrm{d}(i_Y\omega)$ because of $\mathrm{d}\omega = 0$ and Cartan's formula $\mathcal{L}_Y = i_Y \circ \mathrm{d} + \mathrm{d} \circ i_Y$ – equals ω.

(c) A *Stein filling* (W, J) of (M, ξ) is a sublevel set of an exhausting strictly plurisubharmonic function on a Stein manifold such that $M = \partial W$ is the corresponding level set and ξ coincides with the complex tangencies $TM \cap J(TM)$.

For the details of (c) I refer the reader to [31, Chap. 5.4]. The following implications hold for contact structures:

$$\text{Stein fillable} \Longrightarrow \text{strongly fillable} \Longrightarrow \text{weakly fillable} \Longrightarrow \text{tight}$$

The first implication is fairly straightforward; the second one is obvious. For the third implication and references to examples that the converse implications fail, in general, see [31, Chap. 5].

2.3 Topology of the Space of Contact Structures

Let M be a closed (i.e. compact without boundary) odd-dimensional manifold. The space $\Xi(M)$ of contact structures on M is an open (possibly empty) subset of the space of all differential 1-forms on M. According to the Gray stability theorem [31, Theorem 2.2.2], any smooth homotopy of contact structures on M is induced by an isotopy of the manifold. So the isotopy classification of contact structures on M amounts to determining the set $\pi_0(\Xi(M))$ of path-components.

For $\dim M = 3$ it is opportune, thanks to Eliashberg's classification of overtwisted contact structures, to restrict attention to tight contact structures. Moreover, we observe that the sign of $\alpha \wedge \mathrm{d}\alpha$ is independent of the choice of contact form α defining a given contact structure ξ. If M is oriented, we can thus speak of *positive* and *negative* contact structures. In what follows a choice of orientation for M will be understood (or specified, if M does not admit an orientation-reversing diffeomorphism), and we only consider positive contact structures.

Here are examples of 3-manifolds with a unique tight contact structure (up to isotopy); we call this structure the standard contact structure on the respective manifold and denote it by ξ_{st}:

- $S^3 \subset \mathbb{R}^4$, $\xi_{st} = \ker(x\,dy - y\,dx + z\,dt - t\,dz)$,
- $S^1 \times S^2 \subset S^1 \times \mathbb{R}^3$, $\xi_{st} = \ker(z\,d\theta + x\,dy - y\,dx)$, and
- $\mathbb{R}^3$, $\xi_{st} = \ker(dz + x\,dy)$.

Remark. The standard contact structure on S^3, when restricted to the complement of a point, equals the standard contact structure on $\mathbb{R}^3$ [31, Proposition 2.1.8].

These results are due to Eliashberg, cf. [31, Chap. 4.10]. Etnyre and Honda [26] have shown that the Poincaré homology sphere P with the opposite of its natural orientation does not admit a tight contact structure. From a splitting theorem for tight contact structures due to Colin [7], cf. [11], it follows that the connected sum of two copies of P, one with its natural and one with the opposite orientation, does not admit any tight contact structure for either orientation.

On the 3-torus T^3 the contact structures $\xi_n = \ker(\sin(n\theta)\,dx + \cos(n\theta)\,dy)$, $n \in \mathbb{N}$, constitute a complete list (without repetition) of the tight contact structures *up to diffeomorphism*. The classification up to isotopy is a little more subtle, see [22]. As tangent 2-plane fields, however, the ξ_n are all homotopic to $\ker d\theta$. This is an instance of a general phenomenon for toroidal manifolds, i.e. manifolds admitting an embedding of a 2-torus that induces an injection on fundamental groups: all such manifolds admit infinitely many tight contact structures.

On the other hand, there are the following finiteness results, due to Colin–Honda–Giroux [9]:

- On each closed, oriented 3-manifold there are only finitely many homotopy classes of tangent 2-plane fields that contain a tight contact structure.
- Unless the 3-manifold is toroidal, there are only finitely many tight contact structures up to isotopy.

For $\dim M \geq 5$ there are some existence results for contact structures, cf. [29], but no complete classification on any contact manifold. An interesting result in this context is due to Seidel [55, Corollary 6.8]: the isomorphism problem for simply connected closed contact manifolds is algorithmically unsolvable – beware, though, that this does not rule out the practical solution of the problem for a given manifold.

A few things are known about the fundamental group $\pi_1(\Xi(M))$ with a chosen basepoint. For instance, for each $n \in \mathbb{N}$ the group $\pi_1(\Xi(T^3), \xi_n)$ contains an infinite cyclic subgroup [32]. Or, as shown in [13], the component of $\Xi(S^1 \times S^2)$ containing the unique tight contact structure ξ_{st} has fundamental group isomorphic to $\mathbb{Z}$. For results about higher homotopy groups of $\Xi(M)$ for higher-dimensional contact manifolds M see [6].

These results are intimately connected with the topology of the group $\mathrm{Diff}_0(M)$ of diffeomorphisms of M isotopic to the identity. Write $\Xi_0(M)$ for the component of $\Xi(M)$ containing a chosen contact structure ξ_0. Then the map

$$\begin{array}{ccc} \mathrm{Diff}_0(M) & \longrightarrow & \Xi_0(M) \\ f & \longmapsto & Tf(\xi_0) \end{array}$$

is a Serre fibration, the homotopy lifting property being a consequence of Gray stability. The fibre $\mathrm{Cont}_0(M)$, which need not be connected, consists of those contactomorphisms (i.e. diffeomorphisms that preserve ξ_0) that are are isotopic (as diffeomorphisms) to the identity. Thus, the homotopy exact sequence of this Serre fibration allows us to translate homotopical information about two of the three spaces Cont_0, Diff_0, and Ξ_0 into information about the third.

For the mentioned result $\pi_1(\Xi(S^1 \times S^2), \xi_{\mathrm{st}}) \cong \mathbb{Z}$, the homotopy type of the topological group $\mathrm{Diff}_0(S^1 \times S^2)$ is taken as a given. But there are also examples where contact topology can be used to extract information about the diffeomorphism group, scc Scct. 4.4 below.

2.4 Convex Hypersurfaces

The notion of a convex hypersurface has been introduced into contact geometry by Giroux [33].

Definition. A vector field X on a contact manifold (M, ξ) is called a *contact vector field* if its flow preserves the contact structure ξ. When ξ is written as $\xi = \ker \alpha$, the condition on X can be stated as $\mathcal{L}_X \alpha = \mu\alpha$ for some smooth function $\mu\colon M \to \mathbb{R}$.

A hypersurface $\Sigma \subset M$ is called *convex* if there is a contact vector field defined near and transverse to Σ.

Example. On $S^1 \times \mathbb{R}^2$ with contact structure $\xi = \ker(\cos\theta\, dx - \sin\theta\, dy)$, the circle $L = S^1 \times \{0\}$ is Legendrian, $X = x\,\partial_x + y\,\partial_y$ is a contact vector field, and $\Sigma = S^1 \times \partial D^2$ is a convex surface. This is actually the universal model for the neighbourhood of a Legendrian knot in a contact 3-manifold.

Convex hypersurfaces, notably in 3-dimensional contact manifolds, play an important role in the classification of contact structures and topological constructions such as surgery. The reason is the following.

Given a surface Σ in a contact 3-manifold (M, ξ), the intersection $T\Sigma \cap \xi$ defines a singular 1-dimensional foliation Σ_ξ on Σ, the so-called *characteristic foliation*. Singularities occur at points $x \in \Sigma$ where the tangent plane $T_x\Sigma$ coincides with the contact plane ξ_x. It can be shown that the characteristic foliation Σ_ξ determines the germ of ξ near Σ. This permits, for instance, the gluing of contact manifolds along surfaces with the same characteristic foliation.

In general, the characteristic foliation is difficult to control. For convex surfaces, however, it turns out that all the essential information is contained in the *dividing set*, which is defined as the set of points in Σ where the contact vector field is contained in the contact plane; in a closed surface this set is a collection of embedded circles. The characteristic foliations of two convex surfaces with the same dividing set can be made to coincide after a C^0-small perturbation.

2.5 Open Book Decompositions

Given a topological space W and a homeomorphism $\phi\colon W \to W$, the *mapping torus* $W(\phi)$ is the quotient space obtained from $W \times [0,2\pi]$ by identifying $(x,2\pi)$ with $(\phi(x),0)$ for each $x \in W$. If W is a differential manifold and ϕ a diffeomorphism equal to the identity near the boundary ∂W, then $W(\phi)$ is in a natural way a differential manifold with boundary $\partial W \times S^1$.

According to an old theorem of Alexander, cf. [24], any closed, connected, orientable 3-manifold can be written in the form

$$M_{(\Sigma,\phi)} := \Sigma(\phi) \cup_{\mathrm{id}} (\partial\Sigma \times D^2),$$

with Σ a compact, orientable surface with boundary; it can be arranged that the boundary $\partial\Sigma$ is connected (i.e. a single copy of S^1). Write $B \subset M$ for the link (i.e. collection of knots) corresponding to $\partial\Sigma \times \{0\}$ under this identification. Then we can define a smooth, locally trivial fibration $\mathfrak{p}\colon M \setminus B \to S^1 = \mathbb{R}/2\pi\mathbb{Z}$ by

$$\mathfrak{p}([x,\varphi]) = [\varphi] \quad \text{for} \quad [x,\varphi] \in \Sigma(\phi)$$

and

$$\mathfrak{p}(\theta, re^{\mathrm{i}\varphi}) = [\varphi] \quad \text{for} \quad (\theta, re^{\mathrm{i}\varphi}) \in \partial\Sigma \times D^2 \subset \partial\Sigma \times \mathbb{C}.$$

In other words $B \subset M$ has a tubular neighbourhood of the form $B \times D^2$, where the fibration $\mathfrak{p}$ is given by the projection onto the angular coordinate in the D^2-factor. Such a fibration is called an *open book decomposition* with *binding* B and *pages* the closures of the fibres $\mathfrak{p}^{-1}(\varphi)$. Notice that each page is a codimension 1 submanifold of M with boundary B, see Fig. 1.

A submanifold $B \subset M$ that arises as the binding of an open book decomposition is called a *fibred link*.

Conversely, from an open book decomposition of M one can derive a description of M in the form $M_{(\Sigma,\phi)}$, so we may think of an open book decomposition as a pair (Σ,ϕ). The diffeomorphism ϕ is called the *monodromy* of the open book.

In the following definition we call a contact (resp. symplectic) form on an oriented manifold *positive* if the volume form it defines on the manifold gives the positive orientation.

Definition. Let M be a manifold with an open book decomposition $(B,\mathfrak{p})$, where M and B are oriented. The pages of the open book are oriented consistently with their boundary B. A contact structure $\xi = \ker\alpha$ on M defined by a positive contact form is said to be *supported* by the open book decomposition $(B,\mathfrak{p})$ if

(i) The 2-form $d\alpha$ induces a positive symplectic form on each fibre of $\mathfrak{p}$, and
(ii) The 1-form α induces a positive contact form on B.

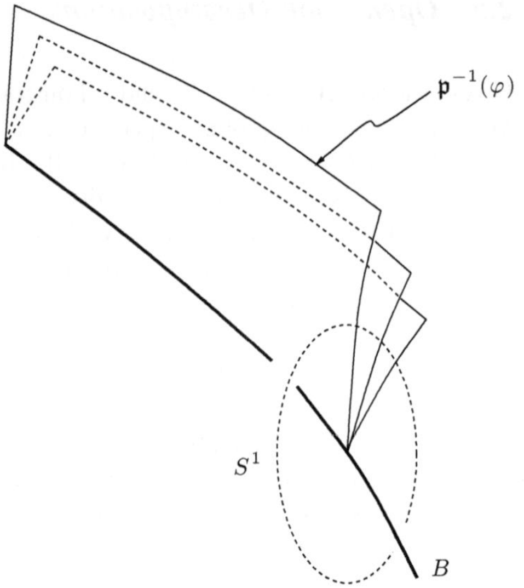

Fig. 1 An open book near the binding

Condition (i) is equivalent to the Reeb vector field R of α being positively transverse to the fibres of $\mathfrak{p}$. Recall that R is defined by the conditions $i_R d\alpha \equiv 0$ and $\alpha(R) \equiv 1$.

Examples. (1) The standard contact form on $S^3 \subset \mathbb{C}^2$ can be written in polar coordinates as $\alpha = r_1^2\, d\varphi_1 + r_2^2\, d\varphi_2$. Set $B = \{r_1 = 0\}$. Then $\mathfrak{p}\colon S^3 \setminus B \to S^1$, $\mathfrak{p}(r_1 e^{i\varphi_1}, r_2 e^{i\varphi_2}) = \varphi_1$ defines an open book whose pages are 2-discs, and whose monodromy is the identity map. This open book supports $\xi_{st} = \ker \alpha$, since α restricts to $d\varphi_2$ along the binding B, and $d\alpha$ to $r_2\, dr_2 \wedge d\varphi_2$ on the tangent spaces to the pages.

(1^+) Set $B_+ = \{r_1 r_2 = 0\}$, which is the Hopf link in S^3. Then $\mathfrak{p}_+\colon S^3 \setminus B_+ \to S^1$, $\mathfrak{p}_+(r_1 e^{i\varphi_1}, r_2 e^{i\varphi_2}) = \varphi_1 + \varphi_2$ is an open book whose pages are annuli, and whose monodromy is a right-handed Dehn twist along the core circle of the annulus. When oriented as the boundary of a single page, the binding is a *positive* Hopf link; the annulus is called a *positive Hopf band*. For details of these claims, and the fact that this open book also supports ξ_{st}, see [31, Example 4.4.8]. Notice that the linking number of an oriented core circle of the annulus with a push-off along that annulus equals -1.

(2) The 2-sphere S^2 admits an open book decomposition where the binding consist of the north and the south pole, the pages are half great circles between the poles, and the monodromy is the identity. When we cross this picture with S^1 we obtain an open book for $S^1 \times S^2$ with binding consisting of two circles, pages equal to annuli, and monodromy equal to the identity. The standard contact form $z\, d\theta + x\, dy - y\, dx$ restricts to $\pm d\theta$ along the binding, and the Reeb vector field $z\, \partial_\theta + x\, \partial_y - y\, \partial_x$ is transverse to the interior of the pages. So this open book supports the standard contact structure ξ_{st}.

It was shown by Thurston and Winkelnkemper [57] that any open book decomposition of a 3-manifold supports a contact structure. Giroux [35] observed that the construction carries over to higher dimensions, provided the page admits an exact symplectic form $\omega = \mathrm{d}\beta$ which makes it a strong symplectic filling of its boundary, and the monodromy is symplectic; for details see [31, Chap. 7.3].

Giroux has also shown the converse, which is a much deeper result:

Theorem 1 (Giroux). *Every contact structure on a closed manifold is supported by an open book decomposition whose fibres are Stein manifolds, and whose monodromy is a symplectomorphism.* □

Moreover, for 3-dimensional manifolds he has further refined this correspondence between contact structures and open books. Given an open book decomposition of a closed 3-manifold M with page Σ and monodromy ϕ, one can form a *positive stabilisation* by adding a band to Σ along $\partial\Sigma$ and composing ϕ with a right-handed Dehn twist along a simple closed curve running once over the band. This does not change the underlying 3-manifold M. Examples (1) and (1^+) above are an instance of this phenomenon. There is then a one-to-one correspondence between contact structures on M up to isotopy and open book decompositions of M up to positive stabilisations and isotopy.

An intrinsic view of this positive stabilisation is to say that the page Σ is replaced by the plumbing of Σ with a positive Hopf band; the plumbing is done in a neighbourhood of a proper arc in Σ and in the Hopf band, respectively; see [40].

Analogously, there is a *negative stabilisation*, corresponding to a left-handed Dehn twist or a plumbing with a negative Hopf band. This will play a role in Corollary 5. The corresponding open book of S^3 has the negative Hopf link B_- as binding (which equals B_+ as a point set, but one of the two link components gets the reverse orientation), and the open book decomposition is given by $\mathfrak{p}_-\colon S^3 \setminus B_- \to S^1$, $\mathfrak{p}_-(r_1\mathrm{e}^{\mathrm{i}\varphi_1}, r_2\mathrm{e}^{\mathrm{i}\varphi_2}) = \varphi_1 - \varphi_2$.

3 A Surgery Presentation of Contact 3-Manifolds

3.1 Dehn Surgery

Let K be a homologically trivial knot in a 3-manifold M. Write $\nu K \cong S^1 \times D^2$ for a (closed) tubular neighbourhood of K. On the boundary $\partial(\nu K) \cong T^2$ of this tubular neighbourhood there are two distinguished curves:

1. The meridian μ, defined as a simple closed curve that bounds a disc in νK.
2. The preferred longitude λ, defined as a simple closed curve parallel to K corresponding to the surface framing.

Given an orientation of M, orientations of μ and λ are chosen such that the tangent direction of μ followed by the tangent direction of λ at a transverse intersection point of μ and λ gives the orientation of T^2 (as boundary of νK).

Let p, q be coprime integers. The manifold $M_{p/q}(K)$ obtained from M by *Dehn surgery* along K with *surgery coefficient* $p/q \in \mathbb{Q} \cup \{\infty\}$ is defined as

$$M_{p/q}(K) := \overline{M \setminus \nu K} \cup_g S^1 \times D^2,$$

where the gluing map g sends the meridian $* \times \partial D^2$ to $p\mu + q\lambda$, i.e. a simple closed curve on T^2 in the class $p[\mu] + q[\lambda] \in H_1(T^2)$. The resulting manifold is determined up to diffeomorphism by the surgery coefficient (changing p, q to $-p, -q$ yields the same manifold).

For $p/q = \infty$ the surgery is trivial. If $p/q \in \mathbb{Z}$, there is a diffeomorphism $S^1 \times D^2 \to \nu K$ sending a standard longitude $\lambda_0 = S^1 \times \{*\}$ (with some point $* \in \partial D^2$) to $p\mu + q\lambda$. This implies that integer Dehn surgery can be described as cutting out $S^1 \times D^2$ and gluing in $D^2 \times S^1$ with the obvious identification of boundaries. If M is thought of as the boundary of some 4-manifold W, the surgered manifold will be the new boundary after attaching a 2-handle $D^2 \times D^2$ to W along M. For that reason, integer Dehn surgery is also called *handle surgery*.

3.2 Contact Dehn Surgery

Now suppose that K is a Legendrian knot with respect to some contact structure ξ on M. Then we may replace λ by the longitude corresponding to the contact framing of K. We now consider Dehn surgery along K with coefficient p/q as before, but we define the surgery coefficient with respect to the contact framing. Notice that the two surgery coefficients differ by an integer depending only on the Legendrian knot K. This integer, the difference between the contact framing and the surface framing, is called the *Thurston–Bennequin invariant* $\mathtt{tb}(K)$ of K. (Notice that the contact framing is defined for any Legendrian knot; the surface framing and $\mathtt{tb}$ are only defined for homologically trivial ones.)

It turns out that for $p \neq 0$ one can always extend the contact structure $\xi|_{M \setminus \nu K}$ to one on the surgered manifold in such a way that the extended contact structure is tight on the glued-in solid torus $S^1 \times D^2$. Moreover, subject to this tightness condition there are but finitely many choices for such an extension, and for $p/q = 1/k$ with $k \in \mathbb{Z}$ the extension is in fact unique. These observations hinge on the fact that $\partial(\nu K)$ is a convex surface in the sense of Sect. 2.4. On solid tori with convex boundary condition, tight contact structures have been classified by Giroux [34] and Honda [41,42].

We can therefore speak sensibly of *contact* $(1/k)$*-surgery*. So the contact surgeries that are well defined *and* correspond to handle surgeries are precisely the contact (± 1)-surgeries.

There is also an *ad hoc* definition for a contact 0-surgery, but here the extension over the glued-in solid torus is necessarily overtwisted, since the contact framing and the surface framing of a meridional disc coincide.

The notion of contact Dehn surgery was introduced in [10], and the following surgery presentation of contact 3-manifolds is the main result from that paper.

Theorem 2. *Let* (M,ξ) *be a closed, connected contact 3-manifold. Then* (M,ξ) *can be obtained from* (S^3,ξ_{st}) *by contact* (± 1)*-surgery along a Legendrian link.*

Sketch proof. According to a theorem of Lickorish and Wallace, M can be obtained from S^3 by surgery along some link. Since the reverse of a surgery is again a surgery, we may likewise obtain S^3 by surgery along a link in M.

It is possible to isotope that link in (M,ξ) to a Legendrian link. Then perform the surgeries as contact surgeries. This yields S^3 with some contact structure ξ'.

Now there is an algorithm for turning each contact surgery into a sequence of contact (± 1)-surgeries. Moreover, the contact structures on S^3 are known explicitly (the unique tight one ξ_{st}, and an overtwisted one in each homotopy class of tangent 2-plane fields). This allows one to find a further sequence of contact (± 1)-surgeries that turns (S^3,ξ') into (S^3,ξ_{st}).

In conclusion, we can obtain (S^3,ξ_{st}) from (M,ξ) by contact (± 1)-surgery along a Legendrian link. The theorem is then a consequence of the fact that the converse of a contact (± 1)-surgery is a contact (∓ 1)-surgery. This "cancellation lemma" is proved as follows, see Fig. 2.

Write the Cartesian coordinates on $\mathbb{R}^4$ as $(\mathbf{p},\mathbf{q}) = (p_1,p_2,q_1,q_2)$. The standard symplectic form on $\mathbb{R}^4$ can then be written as $\omega = \mathrm{d}\mathbf{p}\wedge\mathrm{d}\mathbf{q} := \mathrm{d}p_1\wedge\mathrm{d}q_1 + \mathrm{d}p_2\wedge\mathrm{d}q_2$. Consider the hypersurfaces $g^{-1}(\pm 1)$, where $g(\mathbf{p},\mathbf{q}) = \mathbf{p}^2 - \mathbf{q}^2/2$, and the Liouville vector field $Y = 2\mathbf{p}\,\partial_{\mathbf{p}} - \mathbf{q}\,\partial_{\mathbf{q}}$. Notice that Y is the gradient vector field of g with respect to the standard metric on $\mathbb{R}^4$. Figure 2 gives the local model for a contact

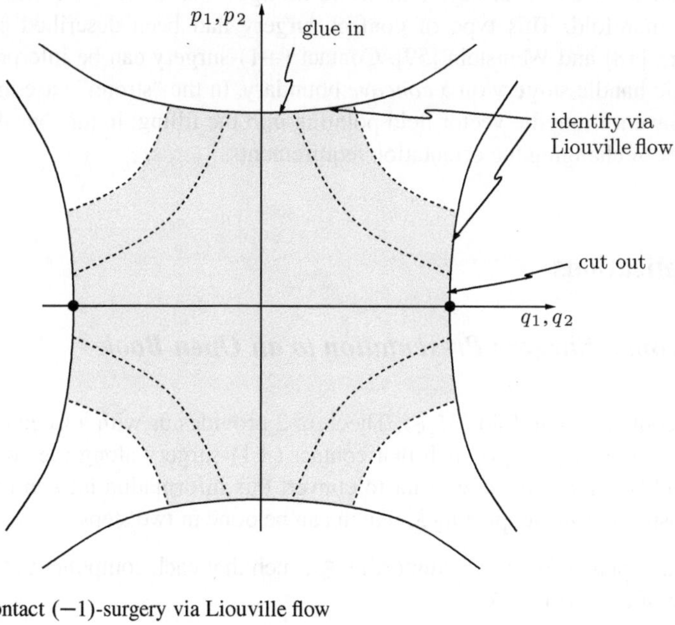

Fig. 2 Contact (-1)-surgery via Liouville flow

Fig. 3 Contact (−1)-surgery via handle attachment

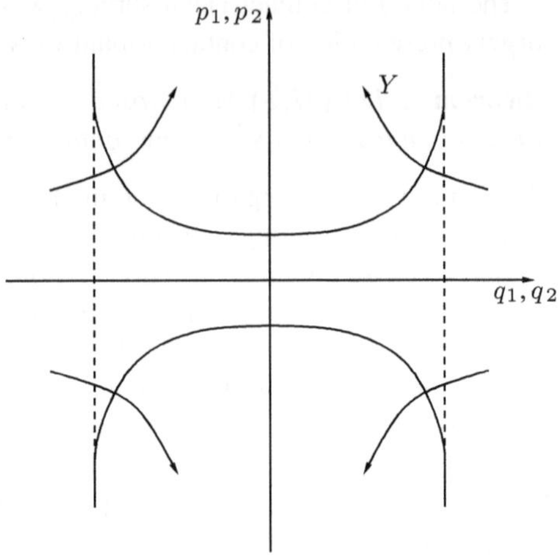

(−1)-surgery along the Legendrian circle $\{\mathbf{p} = 0,\ \mathbf{q}^2 = 1\} \subset g^{-1}(-1)$; this follows from the neighbourhood theorem for Legendrian knots, and a computation of framings in the local model.

It is clear that the reverse surgery is the one along $\{\mathbf{p}^2 = 1,\ \mathbf{q} = 0\} \subset g^{-1}(1)$ in this local model, and here a computation of framings shows this to be a contact (+1)-surgery. □

Remark. In Fig. 3 of Sect. 4.1 below we give an alternative description of contact (−1)-surgery that shows how to perform such a surgery as a symplectic handle surgery on a (weak or strong) symplectic filling, so as to obtain a filling of the surgered manifold. This type of contact surgery had been described earlier by Eliashberg [18] and Weinstein [59]. Contact (+1)-surgery can be interpreted as a symplectic handle surgery on a *concave* boundary. In the "strong" case this means that we have a Liouville vector field pointing *into* the filling; in the "weak" case it is a matter of changing the orientation requirements.

4 Applications

4.1 From a Surgery Presentation to an Open Book

Given a contact 3-manifold (M, ξ), Theorem 2 provides us with a Legendrian link $\mathbb{L} = \mathbb{L}^- \sqcup \mathbb{L}^+$ in (S^3, ξ_{st}) such that contact (±1)-surgery along the components of $\mathbb{L}^\pm$ yields (M, ξ). We now want to convert this information into an open book decomposition of M supporting ξ, which can be done in two steps:

1. Find an open book for S^3 supporting ξ_{st}, such that each component of $\mathbb{L}$ sits on a page of the open book.

2. Show that contact (-1)-surgery (resp. $(+1)$-surgery) along a Legendrian knot sitting on a page of a supporting open book amounts to changing the monodromy by a right-handed (resp. left-handed) Dehn twist.

The first step is carried out by Plamenevskaya in [54, Proposition 4], building on work of Akbulut–Özbağcı [2, 3]. The second step is done by Gay [27, Proposition 2.8] for contact (-1)-surgeries, and for contact surgeries of both signs by Stipsicz [56, pp. 78–79]. Their proofs rely on a result of Torisu about Heegaard splittings of contact 3-manifolds along a convex surface into two handlebodies with a tight contact structure. An alternative proof of the second step, using only local considerations, is given by Etnyre [24, Theorem 5.7]; here I give an independent and self-contained proof. Like Etnyre's, it is done in a local model, but the surgery is described by a smooth model rather than a cut-and-paste procedure. This proof arose in discussions with Otto van Koert. Together with Niederkrüger he has extended this argument to higher dimensions; see also [1, Proposition 6.2].

In the proposition, we use the following notation: given a surface Σ and a simple closed curve $L \subset \Sigma$, we write D_L^+ for the diffeomorphism of Σ given by a right-handed Dehn twist along L; a left-handed Dehn twist will be denoted by D_L^-.

Proposition 3. *Let $(M, \xi = \ker\alpha)$ be a contact 3-manifold with supporting open book (Σ, ϕ), and let L be a Legendrian knot sitting on a page of this open book. Then the contact manifold obtained from (M, ξ) by contact (± 1)-surgery along L has a supporting open book $(\Sigma, \phi \circ \mathrm{D}_L^{\mp})$.*

Proof. We prove this for a contact (-1)-surgery along L; the case of a contact $(+1)$-surgery is completely analogous. We begin with a modified local model for a contact (-1)-surgery, see Fig. 3. As in Fig. 2 we consider $\mathbb{R}^4$ with symplectic form $\omega = \mathbf{dp} \wedge \mathbf{dq}$ and Liouville vector field $Y = 2\mathbf{p}\,\partial_{\mathbf{p}} - \mathbf{q}\,\partial_{\mathbf{q}}$. But instead of the hypersurface $g^{-1}(-1)$, we now take the hypersurface $\{\mathbf{q}^2 = 1\}$ as a model for our contact manifold in a neighbourhood of the Legendrian knot L, which we identify with $\{\mathbf{p} = 0,\ \mathbf{q}^2 = 1\}$. Perform the surgery along L by attaching a handle as shown in Fig. 3, whose boundary is transverse to the Liouville vector field Y and hence inherits the contact form $i_Y\omega = 2\mathbf{p}\,\mathbf{dq} + \mathbf{q}\,\mathbf{dp}$.

Consider the map

$$\begin{aligned} \mathfrak{p}\colon\ \mathbb{R}^4 &\longrightarrow \mathbb{R} \\ (\mathbf{p}, \mathbf{q}) &\longmapsto \mathbf{pq}. \end{aligned}$$

On the hypersurface $\{\mathbf{q}^2 = 1\}$ the Reeb vector field R of $i_Y\omega$ takes the form $R = \mathbf{q}\,\partial_{\mathbf{p}}$, so we have $R(\mathfrak{p}) = \mathbf{q}^2 \equiv 1$ along that hypersurface, which implies that the Reeb vector field R is transverse to the fibres of the map $\mathfrak{p}$. (These fibres, inside the hypersurface $\{\mathbf{q}^2 = 1\}$, are annuli.) Therefore, by a standard argument involving Gray stability, cf. [31, Chap. 2], we may identify a neighbourhood of $L \subset M$ with a neighbourhood $\{\mathbf{p}^2 < \varepsilon,\ \mathbf{q}^2 = 1\}$ in such a way that α becomes identified with $i_Y\omega$ (restricted to the tangent spaces of the hypersurface $\{\mathbf{q}^2 = 1\}$), and such that the map $\mathfrak{p}$ describes the open book $M \setminus B \to S^1$ in that neighbourhood. Notice that the Legendrian knot L lies on the page $\mathfrak{p}^{-1}(0)$.

I claim that the map $\mathfrak{p}$, restricted to the surgered hypersurface in the local model, still describes an open book supporting the contact structure after the surgery. In order to prove this claim, we need to describe the handle in the model more explicitly. Following the approach in [59], we write the surgered manifold in the model as a hypersurface $\{F(\mathbf{p}^2, \mathbf{q}^2) = 0\}$, where $F\colon \mathbb{R}_0^+ \times \mathbb{R}_0^+ \to \mathbb{R}$ is a smooth function with the properties

$$\begin{cases} F(0,0) < 0, \\ \dfrac{\partial F}{\partial u} \geq 0, \quad \dfrac{\partial F}{\partial u} > 0 \text{ for } v = 0, \\ \dfrac{\partial F}{\partial v} \leq 0, \\ \left(\dfrac{\partial F}{\partial u}\right)^2 + \left(\dfrac{\partial F}{\partial v}\right)^2 > 0, \\ F(u,1) = 0 \text{ for } u > \varepsilon^2/4. \end{cases}$$

With $\widetilde{F}(\mathbf{p}, \mathbf{q}) := F(\mathbf{p}^2, \mathbf{q}^2)$ we have

$$\mathrm{d}\widetilde{F}(Y) = 4\mathbf{p}^2 \frac{\partial F}{\partial u} - 2\mathbf{q}^2 \frac{\partial F}{\partial v} > 0 \text{ along } \{\widetilde{F} = 0\},$$

so $\{\widetilde{F} = 0\}$ is indeed a hypersurface transverse to Y that coincides with $\{\mathbf{q}^2 = 1\}$ for $|\mathbf{p}| > \varepsilon/2$.

The Reeb vector field R of the contact form induced by $i_Y\omega$ on the hypersurface $\{\widetilde{F} = 0\}$ is determined, up to scale, by the condition that $i_R\mathrm{d}(i_Y\omega) = i_R\omega$ be proportional to $\mathrm{d}\widetilde{F}$. This implies that, up to a positive factor, the Reeb field is given by

$$R' := \frac{\partial F}{\partial u}\mathbf{p}\,\partial_{\mathbf{q}} - \frac{\partial F}{\partial v}\mathbf{q}\,\partial_{\mathbf{p}}.$$

From

$$R'(\mathfrak{p}) = -\frac{\partial F}{\partial v}\mathbf{q}^2 + \frac{\partial F}{\partial u}\mathbf{p}^2 > 0 \text{ along } \{\widetilde{F} = 0\}$$

it follows that $\mathfrak{p}$ does indeed define an open book supporting the contact structure on the surgered manifold.

It remains to verify that this surgery amounts to changing the monodromy by a right-handed Dehn twist D_L^+. In 3-manifold topology it is well known that a Dehn twist on a splitting surface is equivalent to a surgery along the corresponding curve; this observation forms the basis for deriving a surgery presentation of a 3-manifold from a Heegaard splitting. For completeness I shall presently provide the argument. *A priori*, this only shows that the surgered manifold admits *some* open book decomposition where the monodromy has changed as described, but not that this is in fact the open book decomposition given by the map $\mathfrak{p}$ in the local

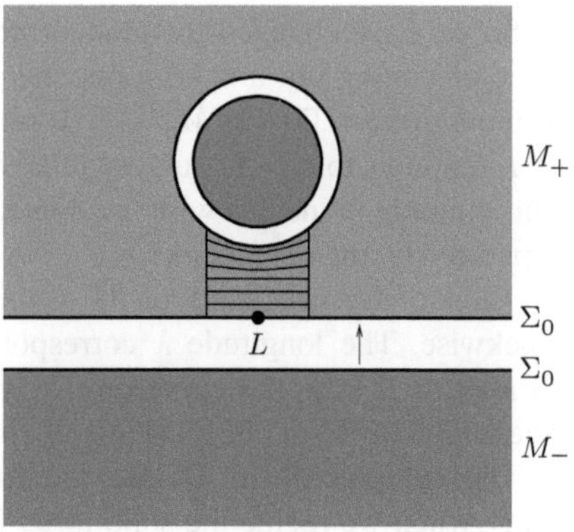

Fig. 4 The effect of surgery on the monodromy

model above. A result of Waldhausen [58, Lemma 3.5] comes to our rescue: any diffeomorphism of $\Sigma \times [0,1]$ equal to the identity near the boundary is isotopic rel boundary to a fibre-preserving diffeomorphism; this implies that the monodromy is determined by a single page and the global topology. Beware that this is a result specific to dimension 3. Moreover, since I promised a self-contained proof, I show in Example (1) following this proof how to give a direct argument.

Imagine that we make an incision in our manifold M along the page Σ_0 containing L. Figure 4 shows a cross-section of this incision, orthogonal to L. In other words, in the figure we see L as a black dot, and the incision is seen as a horizontal cut. We think of the positive coorientation to the pages as pointing up in the figure – this is the direction of the flow transverse to the pages that determines the monodromy. With $M_\pm$ we denote neighbourhoods of L on either side of Σ_0

If we want to recover the original M, we simply reglue using the identity map. In other words, in our local picture we form

$$(M_- + M_+)/(\partial M_- \ni x \sim x \in \partial M_+).$$

Now cut M_+ open along a 2-torus lying vertically over L, as shown in Fig. 4. The disc seen in that figure corresponds to a solid torus T. Then the right-handed Dehn twist D_L^+, which can be thought of as moving only points in a thin annulus around L, extends to a diffeomorphism D_+ of $M_+ \setminus T$ moving only points in the interior of the region indicated by (more or less) horizontal lines, which correspond to annuli, and acting as a right-handed Dehn twist on each of these annuli.

This diffeomorphism D_+ of $M_+ \setminus T$, and the identity map on M_-, induce a diffeomorphism from

$$\big(M_- + (M_+ \setminus T)\big)/\big(\partial M_- \ni x \sim x \in \partial M_+\big)$$

to

$$\big(M_- + (M_+ \setminus T)\big)/\big(\partial M_- \ni x \sim D_L^+(x) \in \partial M_+\big).$$

So we have changed the monodromy by a right-handed Dehn twist D_L^+, at the price of cutting out a solid torus and gluing it back after we have performed the diffeomorphism D_+ on $M_+ \setminus T$. It remains to show that this cutting and regluing of T amounts to a (-1)-surgery relative to the framing of L given by the page Σ_0 (this framing coincides with the contact framing in the case of a contact structure supported by the open book).

Think of the meridian μ on ∂T as the boundary of the disc seen in Fig. 4, oriented clockwise. The longitude λ corresponding to the mentioned framing is a curve parallel to L (e.g. the curve on ∂T lying vertically above L). With the standard orientation of $\mathbb{R}^3$ in our local model, this longitude points into the picture.

The diffeomorphism D_- has the effect of sending λ to itself and μ to $\mu + \lambda$. Thus, when we reglue the solid torus T, its meridian is glued to $\mu - \lambda$. So this is indeed a (-1)-surgery. □

We observed in the above proof that the fibres of $\mathfrak{p}$ in the local model are annuli, so it is clear that after the contact surgery the map $\mathfrak{p}$ describes an open book whose monodromy can only have changed by a multiple of a Dehn twist along the core curve of such a fibre. The same can be said about the open book obtained from the surgery illustrated in Fig. 4. Therefore, the first of the following two examples, where the monodromy directly affects the topology, implies that the change in monodromy is the same in both cases, i.e. a single Dehn twist. This argument allows us to do away with the reference to Waldhausen.

Examples. (1) Consider the open book for S^3 with binding the positive Hopf link, with pages diffeomorphic to annuli, and with monodromy a right-handed Dehn twist along the core circle of the annulus. For any $k \in \mathbb{Z}$ we now want to find the surgery necessary to turn this into an open book where the monodromy is a k-fold right-handed Dehn twist (for $k < 0$ this means a $|k|$-fold left-handed Dehn twist), i.e. we want to add $k - 1$ right-handed Dehn twists to the monodromy. According to the preceding proof, this surgery is given by regluing the solid torus T by sending its meridian to $\mu - (k-1)\lambda$. Beware, though, that λ does not give the surface framing (in S^3) of the core circle of T. By Example (1) in Sect. 2.5, the linking number of L with its push-off along the page is -1. So the surface framing of the core circle of T is given by $\lambda' = \mu + \lambda$. From

$$\mu - (k-1)\lambda = k\mu - (k-1)\lambda'$$

we deduce that the required surgery is a surgery along an unknot in S^3 with surgery coefficient $-k/(k-1)$. This is the well-known surgery description of the lens space $L(k, k-1)$, cf. [39, Example 5.3.2].

For an alternative proof that the open book with page an annulus and monodromy a k-fold right-handed Dehn twist is a lens space $L(k, k-1)$ see [45]. That proof uses Brieskorn manifolds and generalises to higher dimensions.

(2) The Legendrian unknot $L = \{(e^{i\varphi}, e^{-i\varphi})\colon \varphi \in [0, 2\pi]\}$ in (S^3, ξ_{st}) is the core circle in the annulus fibre $\mathfrak{p}_+^{-1}(0)$ of the open book $(B_+, \mathfrak{p}_+)$ supporting ξ_{st}, see Example (1^+) of Sect. 2.5. As mentioned there, the monodromy of that open book is a right-handed Dehn twist D_L^+ along L. Thus, when we perform a contact $(+1)$-surgery on L we obtain the contact structure supported by the open book with annulus fibres and monodromy equal to $D_L^+ \circ D_L^- = \mathrm{id}$, which by Example (2) of Sect. 2.5 is the standard contact structure on $S^1 \times S^2$.

Notice that L is a standard Legendrian unknot in (S^3, ξ_{st}) with Thurston–Bennequin invariant $\mathtt{tb}(L) = -1$ (this characterises L up to Legendrian isotopy). For alternative proofs that contact $(+1)$-surgery along L produces $(S^1 \times S^2, \xi_{st})$, see [15, Lemma 4.3], which uses a splitting along a convex torus, and [49, Lemma 2.5], which uses the contact invariant from Heegaard Floer theory (see Sect. 4.3 below). The proof in the present example is essentially equivalent to that of [56, Proposition 4.1].

The following corollary, in a slightly weaker form, was first proved by Loi–Piergallini [51]. In the form presented here, it is due to Giroux [35], cf. [24, Theorem 5.11].

Corollary 4 (Loi-Piergallini, Giroux). *A contact 3-manifold is Stein fillable if and only if it admits a supporting open book whose monodromy is a composition of right-handed Dehn twists.*

Sketch proof. Suppose the contact 3-manifold (M, ξ) is Stein fillable. According to a result of Eliashberg, cf. [38, Theorem 1.3], (M, ξ) can be obtained from a connected sum $\#S^1 \times S^2$ with its unique tight contact structure ξ_{st} by contact (-1)-surgery along a Legendrian link $\mathbb{L}$. There is an open book supporting ξ_{st} with trivial monodromy, just as in the preceding example. One can also construct an open book supporting ξ_{st} that contains $\mathbb{L}$ on its pages, but may have left-handed Dehn twists in its monodromy. When we pass to a common stabilisation of these two open books, we have an open book whose monodromy can be described by right-handed Dehn twists only *and* contains $\mathbb{L}$ on its pages. Now apply Proposition 3.

Conversely, suppose that ξ is supported by an open book (Σ, ϕ) with ϕ a composition of right-handed Dehn twists. The contact manifold described by (Σ, id) admits a Stein filling by the product $\Sigma \times D^2$; observe that

$$\partial(\Sigma \times D^2) = (\Sigma \times S^1) \cup (\partial\Sigma \times D^2) = M_{(\Sigma, \mathrm{id})}.$$

If the Dehn twists that make up ϕ are along homologically essential curves L_i, one can realise each of them as a Legendrian curve on a page of the open book. By Proposition 3, (M, ξ) is then Stein fillable as a manifold obtained by contact (-1)-surgery on a Stein fillable manifold. If an L_i is homologically trivial in Σ, one first writes $D_{L_i}^+$ as a composition of right-handed Dehn twists along non-separating curves, and then concludes as before. □

Remark. There is a related criterion for a contact structure to be tight. Honda–Kazez–Matić [44] introduce the notion of *right-veering* diffeomorphisms of a surface; the class of such diffeomorphisms contains those that can be written as a composition of right-handed Dehn twist. These authors show that a contact structure is tight if and only if *all* its supporting open books have right-veering monodromy.

The next topological application of contact open books, due to Giroux–Goodman [36], answers a question of Harer [40, Remark 5.1 (a)].

Corollary 5 (Giroux–Goodmann). *Any fibred link in S^3 can be obtained from the unknot by finitely many plumbings and "deplumbings" of Hopf bands.*

Sketch proof. Suppose $B \subset S^3$ is a fibred link, i.e. we have an open book $(B, \mathfrak{p})$. We formulate everything in the language of open books, where the plumbing of a Hopf band corresponds to a positive or negative stabilisation. Consider the negative stabilisation $(B_-, \mathfrak{p}_-)$ of $(B, \mathfrak{p})$.

In a negative Hopf band in S^3, the two boundary circles have linking number -1 when oriented as the boundary of the band. Thus, when we orient the core circle in this band and consider a push-off of this core circle along the band, with the induced orientation, their linking number will be $+1$. It follows that in the contact structure ξ_- on S^3 supported by $(B_-, \mathfrak{p}_-)$ one can find a Legendrian unknot with $\mathtt{tb} = +1$, which forces ξ_- to be overtwisted, since such a knot violates the Bennequin inequality [31, Theorem 4.6.36] that holds true in tight contact 3-manifolds.

Likewise, the unknot in S^3 is fibred, and after a negative stabilisation we obtain an open book $(B', \mathfrak{p}')$ supporting an overtwisted contact structure.

Once a trivialisation of the tangent bundle of S^3 has been chosen, tangent 2-plane fields on S^3 are in one-to-one correspondence with maps $S^3 \to S^2$, which are classified by the Hopf invariant, cf. [31, Chap. 4.2]. One can check that a positive stabilisation does not change the Hopf invariant of the contact structure supported by the respective open book; the examples (1) and (1^+) in Sect. 2.5 illustrate this claim. A negative stabilisation, on the other hand, leads to a contact structure whose Hopf invariant is one greater.

Thus, by negatively stabilising one of $(B_-, \mathfrak{p}_-)$ or $(B', \mathfrak{p}')$ sufficiently often, we obtain two open books supporting overtwisted contact structures ξ_1, ξ_2 with the same Hopf invariant. So ξ_1, ξ_2 are homotopic as tangent 2-plane fields. By Eliashberg's classification of overtwisted contact structures, ξ_1 and ξ_2 are in fact isotopic as contact structures. Then the Giroux correspondence guarantees that the underlying open books become isotopic after a suitable number of further *positive* stabilisations. □

4.2 Symplectic Caps

In this section I sketch how Theorem 2 can be used to give a proof of the following theorem, due to Eliashberg [20] and Etnyre [23], and then discuss some of its topological applications. Both Eliashberg and Etnyre base their proof on an open book decomposition supporting a given contact structure; the idea for the proof indicated here belongs to Özbağcı and Stipsicz [52], see [30] for details.

Theorem 6 (Eliashberg, Etnyre). *Any weak symplectic filling* (W, ω) *of a contact* 3*-manifold* (M, ξ) *embeds symplectically into a closed symplectic* 4*-manifold.*

Sketch proof. We need to show that the given convex filling can be "capped off", i.e. we need to find a concave filling of the contact 3-manifold that can be glued to the convex filling so as to produce a closed 4-manifold.

The desired cap is constructed in three stages. By Theorem 2 we know that (M, ξ) can be obtained by performing contact (± 1)-surgeries on some Legendrian link $\mathbb{L}$ in (S^3, ξ_{st}). For each component L_i of $\mathbb{L}$ choose a Legendrian knot K_i in (S^3, ξ_{st}) with linking number $\mathtt{lk}(K_i, L_i) = 1$, and $\mathtt{lk}(K_i, L_j) = 0$ for $i \neq j$. Moreover, we require that K_i have Thurston–Bennequin invariant $\mathtt{tb}(K_i) = 1$, which is the same as saying that contact (-1)-surgery along K_i is the same as a topological 0-surgery; such K_i can always be found. Now attach to (W, ω) the weak symplectic cobordism W_1 between (M, ξ) at the concave end and a new contact manifold (Σ^3, ξ') at the convex end, corresponding to attaching symplectic handles along the K_i. By our choices, Σ^3 will be a homology 3-sphere.

Thus, after the first step we have embedded (W, ω) symplectically into a weak filling $(W \cup_M W_1, \omega')$ of (Σ^3, ξ'). Since Σ^3 is a homology 3-sphere, the symplectic form is exact near $\Sigma^3 = \partial(W \cup_M W_1)$. This allows one to write down an explicit symplectic form on the cylinder $W_2 = \Sigma^3 \times [0, 1]$ that coincides with ω' near $\Sigma^3 \times \{0\}$ and makes $\Sigma^3 \times \{1\}$ a *strong* convex boundary (with the same induced contact structure ξ').

We now have a strong filling $(W \cup_M W_1 \cup_{\Sigma^3} (\Sigma^3 \times [0, 1]), \omega'')$ of (Σ^3, ξ'). One could then quote to a result of Gay [27] that strong fillings can be capped off; this result, however, is again based on open book decompositions. Alternatively, we appeal once more to Theorem 2, and argue as follows. Attach a (strong) symplectic cobordism corresponding to contact (-1)-surgeries that cancel the contact $(+1)$-surgeries in a surgery presentation of (Σ^3, ξ'). The new boundary has a surgery description involving only contact (-1)-surgeries on (S^3, ξ_{st}), which implies that it is Stein fillable. Symplectic caps for Stein fillings have been constructed by Akbulut–Özbağcı [4] and Lisca–Matić [47]. □

This theorem has a number of topological applications, which are nicely surveyed by Etnyre [25]. For instance, Kronheimer–Mrowka [46] used it to show that every non-trivial knot K in S^3 has the (unfortunately named) property P, which says that Dehn surgery along K with any surgery coefficient $p/q \neq \infty$ leads to a 3-manifold $S^3_{p/q}(K)$ with non-trivial fundamental group. A more palatable consequence of this fact is the Gordon–Luecke theorem, which states that knots in S^3 are determined by their complement, cf. [30]. Theorem 6 enters in the Kronheimer–Mrowka proof as follows. Given a purported counterexample, i.e. a non-trivial knot $K \subset S^3$ and some $p/q \neq \infty$ for which $\pi_1(S^3_{p/q}(K)) = \{1\}$, one constructs with the help of Theorem 6 a certain closed symplectic 4-manifold that contains, essentially,

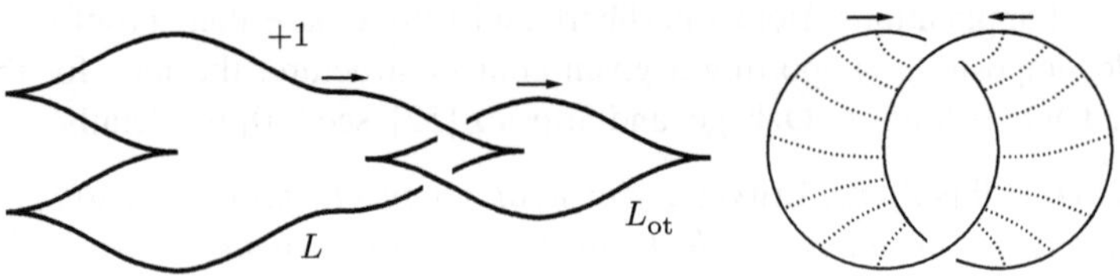

Fig. 5 The overtwisted contact manifold $(S^3, \xi_{st})_{+1}(L)$

$S^3_{p/q}(K)$ as a separating hypersurface. Deep gauge theoretic results show that such a 4-manifold cannot exist.

4.3 Heegaard Floer Theory

As remarked at the end of Sect. 3, any contact manifold obtained from a symplectically fillable contact manifold via contact (-1)-surgery will again be symplectically fillable, and hence in particular tight. It is not known, in general, whether contact (-1)-surgery on a tight contact 3-manifold will preserve tightness. For manifolds with boundary, Honda [43] has an example where tightness is destroyed by contact (-1)-surgery; for closed manifolds no such example is known.

Contact $(+1)$-surgery may well turn a fillable contact manifold into an overtwisted one. An example is shown in Fig. 5 (where the Legendrian knots in $(\mathbb{R}^3, \ker(dz + x\,dy)) \subset (S^3, \xi_{st})$ are represented in terms of their so-called front projection to the yz-plane; the missing x-coordinate can be recovered as the negative slope $x = -dz/dy$). The contact manifold $(S^3, \xi_{st})_{+1}(L)$ obtained from (S^3, ξ_{st}) via contact $(+1)$-surgery on the "shark" L is overtwisted. Indeed, the Legendrian knot L_{ot} bounds an overtwisted disc in $(S^3, \xi_{st})_{+1}(L)$, as is indicated on the right side of Fig. 5. The Seifert surface of the Hopf link $L \sqcup L_{ot}$ shown there glues with a new meridional disc in in $(S^3, \xi_{st})_{+1}(L)$ to form an embedded disc bounded by L_{ot} in the surgered manifold, and the contact framing of L_{ot} coincides with the disc framing.

On the other hand, a manifold obtained via contact $(+1)$-surgery may also be tight, as is shown by example (2) in Sect. 4.1: the tight contact manifold $(S^1 \times S^2, \xi_{st})$ is obtained from (S^3, ξ_{st}) by contact $(+1)$-surgery along a standard Legendrian unknot.

So far the most effective approach towards the question whether contact (-1)-surgery on closed contact 3-manifolds preserves tightness comes from the Heegaard Floer theory introduced by Ozsváth and Szabó [53]. Let (M, ξ) be a closed contact 3-manifold with orientation induced by the contact structure ξ. We write $-M$ for the manifold with the opposite orientation. The contact structure ξ determines a natural Spin^c structure $\mathbf{t}_\xi$ on M. Suffice it to say here that Ozsváth and Szabó define a contact invariant $c(M,\xi)$, which lives in the Heegaard Floer group $\widehat{HF}(-M,\mathbf{t}_\xi)$, with the following properties:

- If (M,ξ) is overtwisted, then $c(M,\xi)=0$.
- If (M,ξ) is Stein fillable, then $c(M,\xi)\neq 0$.

If (M',ξ') is obtained from (M,ξ) by a single contact $(+1)$-surgery (and hence (M,ξ) by contact (-1)-surgery on (M',ξ')), the cobordism W given by the contact $(+1)$-surgery induces a homomorphism

$$F_{-W}\colon \widehat{HF}(-M,\mathbf{t}_\xi)\longrightarrow \widehat{HF}(-M',\mathbf{t}_{\xi'}).$$

As shown by Lisca and Stipsicz [49, Theorem 2.3], this homomorphism maps one contact invariant to the other:

$$F_{-W}(c(M,\xi))=c(M',\xi').$$

This immediately implies the following result.

Theorem 7 (Lisca–Stipsicz). *If $c(M',\xi')\neq 0$, then $c(M,\xi)\neq 0$. In particular, (M,ξ) is tight.* □

In a masterly series of papers, Lisca and Stipsicz have refined this approach to obtain wide-ranging existence results for tight contact structures, culminating in their paper [50], where they give a complete solution to the existence problem for tight contact structures on Seifert fibred 3-manifolds.

4.4 Diffeotopy Groups

The *diffeotopy group* $\mathcal{D}(M)$ of a smooth manifold M is the quotient of the diffeomorphism group Diff(M) by its normal subgroup Diff$_0(M)$ of diffeomorphisms isotopic to the identity. Alternatively, one may think of the diffeotopy group as the group $\pi_0(\mathrm{Diff}(M))$ of path components of Diff(M), since any continuous path in Diff(M) can be approximated by a smooth one, i.e. an isotopy.

The theorem of Cerf (in its strong form) says that $\mathcal{D}(S^3)=\mathbb{Z}_2$, that is, up to isotopy there are only two diffeomorphisms of S^3, the identity and an orientation reversing one. The diffeotopy groups of a number of 3-manifolds are known, for instance those of all lens spaces.

The diffeotopy group $\mathcal{D}(S^1\times S^2)$ was shown to be isomorphic to $\mathbb{Z}_2\oplus\mathbb{Z}_2\oplus\mathbb{Z}_2$ by Gluck [37]. In [13] we give a contact geometric proof of Gluck's result. The starting point for this proof is the uniqueness of the tight contact structure ξ_{st} on $S^1\times S^2$. With Gray stability this easily translates into saying that any orientation preserving diffeomorphism of $S^1\times S^2$ is isotopic to a contactomorphism of ξ_{st}.

In order to find an isotopy of such a contactomorphism f to one in a certain standard form, and thus to derive Gluck's theorem, one observes the effect of the contactomorphism on some Legendrian knot L in $(S^1 \times S^2, \xi_{st})$ generating the homology of $S^1 \times S^2$. This can be done in a contact surgery diagram for $(S^1 \times S^2, \xi_{st})$. The general "Kirby moves" in such a diagram, as described in [12], then allow one to find a contact isotopy from $f(L)$ back to L. This translates into an isotopy from f to a contactomorphism fixing L. This gives one enough control over the contactomorphism to determine its isotopy type.

As an application of such methods, [13] contains examples of homologically trivial Legendrian knots in $(S^1 \times S^2, \xi_{st})\#(S^1 \times S^2, \xi_{st})$ that cannot be distinguished by their classical invariants (i.e. the Thurston–Bennequin invariant and the rotation number, which counts the rotations of the tangent vector of the Legendrian knot relative to a trivialisation of the contact structure over a Seifert surface) – but which may well be distinguished by performing contact (-1)-surgery on them.

4.5 Non-Loose Legendrian Knots

A Legendrian knot L in an overtwisted contact 3-manifold (M, ξ) is called *non-loose* or *exceptional* if the restriction of ξ to $M \setminus L$ is tight. In other words, L has to intersect each overtwisted disc Δ in (M, ξ) in such a manner that no Legendrian isotopy will allow one to separate L from Δ. This is quite a surprising phenomenon, since overtwisted discs always appear in infinite families, as in the example given in Sect. 2.1.

Exceptional knots were first described by Dymara [16]. For a classification of exceptional unknots in S^3 see [21, Theorem 4.7]; there is in fact a unique overtwisted contact structure on S^3 that admits exceptional unknots.

Here I want to exhibit an example, due to Lisca et al. [48, Lemma 6.1], which illustrates the use of contact surgery in detecting exceptional Legendrian knots. Figure 6 (courtesy of Paolo Lisca and András Stipsicz) shows a surgery link in (S^3, ξ_{st}) (in the front projection); the labels ± 1 indicate contact (± 1)-surgeries. The additional Legendrian knot $L(n)$, which is an unknot in S^3, then represents a Legendrian knot in the surgered contact manifold (M, ξ).

By Kirby moves on this surgery diagram one can show that M is simply another copy of S^3, and that $L(n)$ becomes the torus knot $T_{2,2n+1}$ in this 3-sphere. Moreover, with a formula given in [15, Corollary 3.6], one can easily compute the Hopf invariant of the contact structure ξ; it turns out that this differs from the Hopf invariant ξ_{st} on S^3. This implies that ξ and ξ_{st} are not homotopic as tangent 2-plane fields. Hence, by the uniqueness of the tight contact structure on S^3, the contact structure ξ must be overtwisted.

We now want to convince ourselves that $L(n)$ is an exceptional knot in (S^3, ξ). When we perform contact (-1)-surgery along $L(n)$, this cancels one of the previous $(+1)$-surgeries. So the resulting contact manifold is the same as the one obtained from the original diagram in (S^3, ξ_{st}), with one of the two $(+1)$-surgery knots

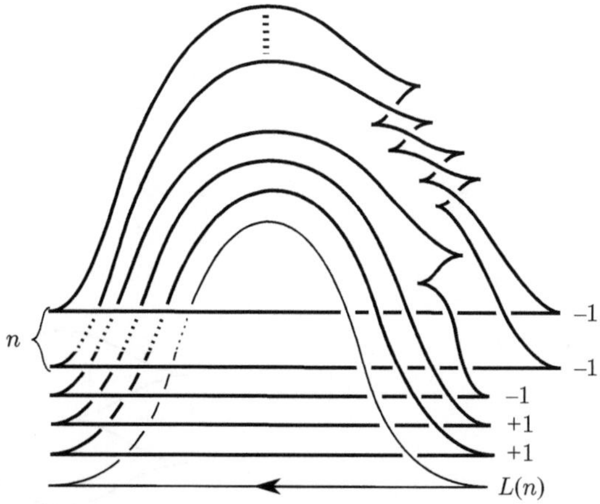

Fig. 6 Non-loose Legendrian torus knot $T_{2,2n+1}$ in S^3 ($n \geq 1$)

removed. As seen in Example (2) of Sect. 4.1, a single contact (+1)-surgery on a Legendrian unknot as in Fig. 6 results in $S^1 \times S^2$ with its unique tight (and Stein fillable) contact structure. Further contact (−1)-surgeries on this contact manifold preserve the fillability and hence tightness of the contact structure. This implies that $T_{2,2n+1}$ is exceptional, for if there were an overtwisted disc in $S^3 \setminus T_{2,2n+1}$, it would survive to the manifold obtained by surgery along $T_{2,2n+1}$.

4.6 *Diagrams for Contact 5-Manifolds*

In the proof of Corollary 4 we alluded to a result of Eliashberg about the surgery description of Stein fillable contact 3-manifolds. That theorem is in fact a statement about the fillings; in other words, the Stein filling is obtained by attaching 1-handles to the 4-ball (resulting in a boundary connected sum of copies of $S^1 \times D^3$), and then attaching 2-handles along Legendrian knots in the boundary with framing −1 relative to the contact framing.

According to Theorem 1, any 5-dimensional contact manifold (M, ξ) is supported by an open book whose fibres are Stein surfaces. By what we just said, those fibres can be described in terms of a Kirby diagram [39] containing the information how to attach the 1- and 2-handles to the 4-ball with its standard Stein structure along its boundary (S^3, ξ_{st}). As in Sect. 4.3, the pairs of attaching balls for the 1-handles and the Legendrian knots along which the 2-handles are attached can be drawn in the front projection of $(\mathbb{R}^3, \ker(dz + x\,dy))$ to the yz-plane.

It is not clear how to describe a general symplectic monodromy in such a diagram. Some monodromies can be encoded in the diagram, though. For instance, there are situations where one can "see" Lagrangian spheres in the diagram

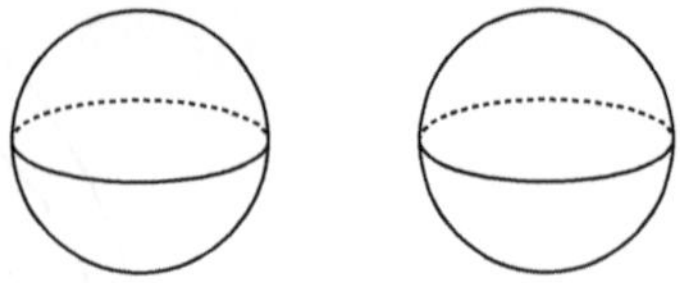

Fig. 7 A contact structure on $S^1 \times S^4$

Fig. 8 A contact structure on $S^2 \times S^3$

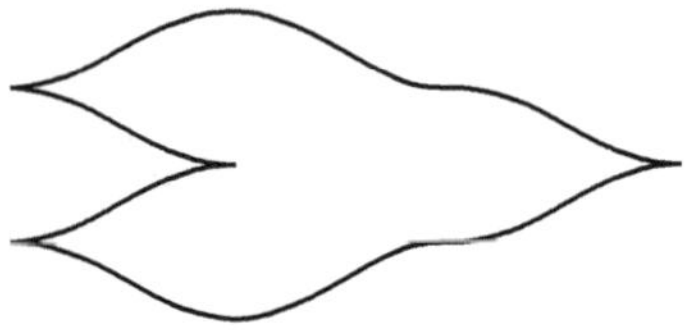

Fig. 9 A contact structure on $S^2 \tilde{\times} S^3$

(i.e. spheres of half the dimension of the page on whose tangent spaces the symplectic form of the page vanishes identically), and one can speak of Dehn twists along such spheres.

Here are some simple examples with trivial monodromy. Recall from the proof of Corollary 4 that the manifold M given by an open book with pages Σ and trivial monodromy is diffeomorphic to $\partial(\Sigma \times D^2)$.

Examples. (1) The diagram in Fig. 7 shows a single 1-handle; this describes the 4-manifold $S^1 \times D^3$. So this is a diagram for a contact structure on $\partial(S^1 \times D^3 \times D^2) = S^1 \times S^4$.

(2) The diagram in Fig. 8 shows an unknot with Thurston–Bennequin invariant $\mathtt{tb} = -1$. So this corresponds topologically to attaching a 2-handle with framing -2 relative to the surface framing (given by a spanning disc), which produces the D^2-bundle Σ_{-2} over S^2 with Euler number -2, see [39, Example 4.4.2]. Then $\partial(\Sigma_{-2} \times D^2)$ is the trivial S^3-bundle over S^2. (Observe that the S^3-bundles over S^2 are classified by $\pi_1(\mathrm{SO}_3) = \mathbb{Z}_2$; the non-trivial bundle is detected by the non-vanishing of the second Stiefel–Whitney class.)

(3) In Fig. 9 we have an unknot with $\mathtt{tb} = -2$. So the 4-manifold encoded by this diagram is the D^2-bundle Σ_{-3} over S^2 with Euler number -3. It follows that $\partial(\Sigma_{-3} \times D^2)$ is the unique non-trivial S^3-bundle over S^2, which we write as $S^2 \tilde{\times} S^3$.

In a forthcoming paper with Fan Ding and Otto van Koert [14] we exploit the information contained in such diagrams, and the handle moves introduced in [13], in order to derive a number of equivalences of 5-dimensional contact manifolds. For instance, one consequence of such moves is that the contact manifold described by a single Legendrian knot L (and trivial monodromy) will always be diffeomorphic to $S^2 \times S^3$ or $S^2 \tilde{\times} S^3$, and the contact structure is completely determined by the rotation number of L.

Here is one further observation about open books with trivial monodromy. From the Seifert–van Kampen theorem one sees that the fundamental group of

$$\partial(\Sigma \times D^2) = \partial\Sigma \times D^2 \cup_\partial \Sigma \times S^1$$

is isomorphic to $\pi_1(\Sigma)$, since any loop in $\partial\Sigma$ is in particular a loop in Σ, and $S^1 = \partial D^2$ becomes homotopically trivial in D^2. From a Kirby diagram for Σ one can easily read off a presentation of $\pi_1(\Sigma)$: each 1-handle gives a generator, and the attaching circles for the 2-handles provide the relations when read as words in the generators.

As observed by Cieliebak, subcritical Stein fillings (i.e. Stein fillings with no handles of maximal index) split off a D^2-factor. Thus, contact manifolds with subcritical Stein fillings are precisely those admitting an open book with trivial monodromy.

Combining these two observations, we show in our forthcoming paper that the contactomorphism type of a subcritically fillable contact 5-manifold is, up to a certain stable equivalence, determined by its fundamental group. This result is achieved by showing how handle moves in contact surgery diagrams can be used to effect the so-called Tietze moves on the corresponding presentation of the fundamental group; any two finite presentations of a given group are related by such Tietze moves.

Acknowledgements Some parts of the research described in this survey were supported by DFG grant GE 1245/1 within the framework of the Schwerpunktprogramm 1154 "Globale Differentialgeometrie".

References

1. Akbulut, S., Arikan, M.F.: Lefschetz fibrations on compact Stein manifolds. arXiv: 1003.2200v3
2. Akbulut, S., Özbağcı, B.: Lefschetz fibrations on compact Stein surfaces. Geom. Topol. **5**, 319–334 (2001)
3. Akbulut, S., Özbağcı, B.: Lefschetz fibrations on compact Stein surfaces (erratum). Geom. Topol. **5**, 939–945 (2001)
4. Akbulut, S., Özbağcı, B.: On the topology of compact Stein surfaces. Int. Math. Res. Not. **2002**, 769–782 (2002)
5. Bennequin, D.: Entrelacements et équations de Pfaff. In: IIIe Rencontre de Géométrie du Schnepfenried, vol. 1. Astérisque **107–108**, 87–161 (1983)
6. F. Bourgeois, Contact homology and homotopy groups of the space of contact structures. Math. Res. Lett. **13**, 71–85 (2006)
7. Colin, V.: Chirurgies d'indice un et isotopies de sphères dans les variétés de contact tendues. C. R. Acad. Sci. Paris Sér. I Math. **324**, 659–663 (1997)
8. Colin, V.: Livres ouverts en géométrie de contact (d'après Emmanuel Giroux), Séminaire Bourbaki, vol. 2006/2007. Astérisque **317**, 91–117 (2008)
9. Colin, V., Giroux, E., Honda, K.: Finitude homotopique et isotopique des structures de contact tendues. Publ. Math. Inst. Hautes Études Sci. **109**, 245–293 (2009)

10. Ding, F., Geiges, H.: A Legendrian surgery presentation of contact 3-manifolds. Math. Proc. Cambridge Philos. Soc. **136**, 583–598 (2004)
11. Ding, F., Geiges, H.: A unique decomposition theorem for tight contact 3-manifolds. Enseign. Math. (2) **53**, 333–345 (2007)
12. Ding, F., Geiges, H. Handle moves in contact surgery diagrams. J. Topol. **2**, 105–122 (2009)
13. Ding, F., Geiges, H.: The diffeotopy group of $S^1 \times S^2$ via contact topology. Compos. Math. **146**, 1096–1112 (2010)
14. Ding, F., Geiges, H., van Koert, O.: Diagrams for contact 5-manifolds. arXiv: 1012.4693
15. Ding, F., Geiges, H., Stipsicz, A.I.: Surgery diagrams for contact 3-manifolds. Turkish J. Math. **28**, 41–74 (2004); also in Proceedings of the 10th Gökova Geometry–Topology Conference 2003, pp. 41–74
16. Dymara, K.: Legendrian knots in overtwisted contact structures on S^3. Ann. Global. Anal. Geom. **19**, 293–305 (2001)
17. Eliashberg, Ya.: Classification of overtwisted contact structures on 3-manifolds. Invent. Math. **98**, 623–637 (1989)
18. Eliashberg, Ya.: Topological characterization of Stein manifolds of dimension > 2. Internat. J. Math. **1**, 29–46 (1990)
19. Eliashberg, Ya.: Contact 3-manifolds twenty years since J. Martinet's work. Ann. Inst. Fourier (Grenoble) **42**, 165–192 (1992)
20. Eliashberg, Ya.: A few remarks about symplectic filling. Geom. Topol. **8**, 277–293 (2004)
21. Eliashberg, Ya., Fraser, M.: Topologically trivial Legendrian knots. J. Symplectic Geom. **7**, 77–127 (2009)
22. Eliashberg, Ya., Polterovich, L.: New applications of Luttinger's surgery. Comment. Math. Helv. **69**, 512–522 (1994)
23. Etnyre, J.B.: On symplectic fillings. Algebr. Geom. Topol. **4**, 73–80 (2004)
24. Etnyre, J.B.: Lectures on open book decompositions and contact structures. In: Floer Homology, Gauge Theory, and Low-Dimensional Topology, Clay Math. Proc., vol. 5, pp. 103–141. American Mathematical Society, Providence, RI (2006)
25. Etnyre, J.B.: Contact geometry in low dimensional topology. Low Dimensional Topology, IAS/Park City Math. Ser., vol. 15, pp. 229–264. American Mathematical Society, Providence, RI (2009)
26. Etnyre, J.B., Honda, K.: On the nonexistence of tight contact structures. Ann. Math. (2) **153**, 749–766 (2001)
27. Gay, D.T.: Explicit concave fillings of contact three-manifolds. Math. Proc. Cambridge Philos. Soc. **133**, 431–441 (2002)
28. Geiges, H.: A brief history of contact geometry and topology. Expo. Math. **19**, 25–53 (2001)
29. Geiges, H.: Contact topology in dimension greater than three. In: European Congress of Mathematics (Barcelona, 2000), vol. II, pp. 535–545. Progr. Math., vol. 202. Birkhäuser, Basel (2001)
30. Geiges, H.: Contact Dehn surgery, symplectic fillings, and Property P for knots. Expo. Math. **24**, 273–280 (2006)
31. Geiges, H.: An Introduction to contact topology. In: Cambridge Stud. Adv. Math., vol. 109. Cambridge University Press, Cambridge (2008)
32. Geiges, H., Gonzalo Pérez, J.: On the topology of the space of contact structures on torus bundles. Bull. London Math. Soc. **36**, 640–646 (2004)
33. Giroux, E.: Convexité en topologie de contact. Comment. Math. Helv. **66**, 637–677 (1991)
34. Giroux, E.: Structures de contact en dimension trois et bifurcations des feuilletages de surfaces. Invent. Math. **141**, 615–689 (2000)
35. Giroux, E.: Géométrie de contact: de la dimension trois vers les dimensions supérieures. In: Proceedings of the International Congress of Mathematicians (Beijing, 2002), vol. II, pp. 405–414. Higher Education Press, Beijing (2002)
36. Giroux, E., Goodman, N.: On the stable equivalence of open books in three-manifolds. Geom. Topol. **10**, 97–114 (2006)

37. Gluck, H.: The embedding of two-spheres in the four-sphere. Trans. Am. Math. Soc. **104**, 308–333 (1962)
38. Gompf, R.E.: Handlebody construction of Stein surfaces. Ann. Math. (2) **148**, 619–693 (1998)
39. Gompf, R.E., Stipsicz, A.I.: 4-Manifolds and Kirby calculus. Grad. Stud. Math. vol. 20. American Mathematical Society, Providence, RI (1999)
40. Harer, J.: How to construct all fibered knots and links. Topology **21**, 263–280 (1982)
41. Honda, K.: On the classification of tight contact structures I. Geom. Topol. **4**, 309–368 (2000)
42. Honda, K.: Factoring nonrotative $T^2 \times I$ layers. Geom. Topol. **5**, 925–938 (2001)
43. Honda, K.: Gluing tight contact structures. Duke Math. J. **115**, 435–478 (2002)
44. Honda, K., Kazez, W.H., Matić, G.: Right-veering diffeomorphisms of compact surfaces with boundary. Invent. Math. **169**, 427–449 (2007)
45. van Koert, O., Niederkrüger, K.: Open book decompositions for contact structures on Brieskorn manifolds. Proc. Am. Math. Soc. **133**, 3679–3686 (2005)
46. Kronheimer, P.B., Mrowka, T.S.: Witten's conjecture and property P. Geom. Topol. **8**, 295–310 (2004)
47. Lisca, P., Matić, G.: Tight contact structures and Seiberg–Witten invariants. Invent. Math. **129**, 509–525 (1997)
48. Lisca, P., Ozsváth, P., Stipsicz, A.I., Szabó, Z.: Heegaard Floer invariants of Legendrian knots in contact three-manifolds. J. Eur. Math. Soc. (JEMS) **11**, 1307–1363 (2009)
49. Lisca, P., Stipsicz, A.I.: Seifert fibered contact three-manifolds via surgery. Algebr. Geom. Topol. **4**, 199–217 (2004)
50. Lisca, P., Stipsicz, A.I.: On the existence of tight contact structures on Seifert fibered 3-manifolds. Duke Math. J. **148**, 175–209 (2009)
51. Loi, A., Piergallini, R.: Compact Stein surfaces with boundary as branched covers of B^4. Invent. Math. **143**, 325–348 (2001)
52. Özbağcı B., Stipsicz, A.I.: Surgery on contact 3-manifolds and Stein surfaces. Bolyai Soc. Math. Stud. vol. 13. Springer, Berlin (2004)
53. Ozsváth, P., Szabó, Z.: Heegaard Floer homology and contact structures. Duke Math. J. **129**, 39–61 (2005)
54. Plamenevskaya, O.: Contact structures with distinct Heegaard Floer invariants. Math. Res. Lett. **11**, 547–561 (2004)
55. Seidel, P.: A biased view of symplectic cohomology. Current Developments in Mathematics 2006, pp. 211–253. International Press, Somerville, MA (2008)
56. Stipsicz, A.I.: Surgery diagrams and open book decompositions of contact 3-manifolds. Acta Math. Hungar. **108**, 71–86 (2005)
57. Thurston, W.P., Winkelnkemper, H.E.: On the existence of contact forms. Proc. Am. Math. Soc. **52**, 345–347 (1975)
58. Waldhausen, F.: On irreducible 3-manifolds which are sufficiently large. Ann. of Math. (2) **87**, 56–88 (1968)
59. Weinstein, A.: Contact surgery and symplectic handlebodies. Hokkaido Math. J. **20**, 241–251 (1991)

On Product Structures in Floer Homology of Cotangent Bundles

Alberto Abbondandolo and Matthias Schwarz

Abstract In an earlier paper we have shown that the pair-of-pants product on the Floer homology of the cotangent bundle of an oriented compact manifold Q corresponds to the Chas-Sullivan loop product on the singular homology of the free loop space of Q. We now give chain level constructions of further product structures in Floer homology, corresponding to the cup product on the homology of any path space, and to the Goresky-Hingston product on the relative cohomology of the free loop space modulo constant loops. Moreover, we give a explicit construction for the inverse isomorphism between Floer homology and loop space homology.

1 Introduction and Main Results

Let Q be a closed, smooth manifold, and let $H: \mathbb{T} \times T^*Q \to \mathbb{R}$ be a time-periodic smooth Hamiltonian on its cotangent bundle. The cotangent bundle is viewed as a symplectic manifold with the canonical symplectic structure $\omega = d\lambda$, where λ is the Liouville one-form, whose expression in local coordinates is $\lambda = \sum p_j \, dq_j$. The corresponding Liouville vector field Y, which is defined by $\omega(Y, \cdot) = \lambda$, has the local expression $Y = \sum p_j \frac{\partial}{\partial p_j}$.

We assume that H is of *quadratic type*, i.e., it satisfies the conditions

(H1) $dH(t,q,p)[Y] - H(t,q,p) \geq h_0 |p|^2 - h_1$,
(H2) $|\nabla_q H(t,q,p)| \leq h_2(1 + |p|^2)$, $|\nabla_p H(t,q,p)| \leq h_2(1 + |p|)$,

M. Schwarz (✉)
Universität Leipzig, Mathematisches Institut, Postfach 10 09 20, D-04009 Leipzig, Germany
e-mail: mschwarz@math.uni-leipzig.de

A. Abbondandolo
Università di Pisa, Dipartimento di Matematica, Largo B. Pontecorvo, 5, I-56127 Pisa, Italy
e-mail: abbondandolo@dm.unipi.it

C. Bär et al. (eds.), *Global Differential Geometry*, Springer Proceedings in Mathematics 17, DOI 10.1007/978-3-642-22842-1_16,

for every (t, q, p), for some positive constants h_0, h_1 and h_2. Here the norm $|\cdot|$ and the covariant derivative ∇ are induced by some fixed metric on Q, but the conditions are actually independent on the choice of this metric. Condition (H1) essentially says that H grows at least quadratically in p on each fiber of T^*Q, and that it is radially convex for $|p|$ large. Condition (H2) implies that H grows at most quadratically in p on each fiber. Such Hamiltonians include in particular physical Hamiltonians with magnetic fields,

$$H(t, q, p) = \frac{1}{2}|p - A(t, q)|^2 + V(t, q),$$

where $A(t, \cdot)$ is a one-form and $V(t, \cdot)$ is a smooth function on Q, both depending 1-periodically on $t \in \mathbb{R}$. Generically, the Hamiltonian system

$$\dot{x}(t) = X_H(t, x(t)), \tag{1}$$

for the Hamiltonian vector field X_H defined by $\omega(X_H, \cdot) = -dH$, has a discrete set $\mathcal{P}_1(H)$ of 1-periodic orbits. In fact, the following non-degeneracy condition holds for a generic set of H:

(H0) The time-1-map of the flow Φ_H^t generated by X_H has only non-degenerate fixed points, i.e. $D\Phi_H^1(x)$ has no eigenvalue 1 for any fixed point x of Φ_H^1.

The free abelian group $F_*(H)$ generated by the elements $x \in \mathcal{P}_1(H)$, which by $x \mapsto x(0)$ correspond exactly to the fixed points of Φ_H^1, graded by their Conley-Zehnder index $\mu_{cz}(x)$, supports a chain complex, the *Floer complex* $(F_*(H), \partial)$. The boundary operator ∂ is defined by an algebraic count of the maps u from the cylinder $\mathbb{R} \times \mathbb{T}$ to T^*Q, solving the Cauchy-Riemann type equation

$$\partial_s u(s,t) + J(t, u(s,t))\big(\partial_t u(s,t) - X_H(t, u(s,t))\big) = 0, \quad \text{for all } (s,t) \in \mathbb{R}\times\mathbb{T}, \tag{2}$$

in short $\overline{\partial}_{J,H} u = 0$, and converging to two 1-periodic orbits x, y for $s \to -\infty$ and $s \to \infty$. Here, J is an almost-complex structure on T^*Q calibrated by the symplectic structure in the sense that $\omega(J\cdot, \cdot)$ gives a positive definite and symmetric form.

The equation (2) can be seen as the negative L^2-gradient equation for the Hamiltonian action functional

$$\mathcal{A}_H : C^\infty(\mathbb{T}, T^*Q) \to \mathbb{R}, \quad \mathcal{A}(x) = \int_{\mathbb{T}} \big(x^*\lambda - H(t, x(t))\, dt\big). \tag{3}$$

The almost complex structure J is chosen in a generic way, so that for every pair (x, y) of 1-periodic orbits, the space of solutions of (2) with asymptotics x and y is the zero-set of a Fredholm section of a Banach bundle which is transverse to the zero-section, and in particular it is a finite dimensional manifold.

This construction is due to A. Floer (see e.g. [13–16]) in the case of a closed symplectic manifold (M,ω), in order to prove a conjecture of V. Arnold on the number of periodic Hamiltonian orbits. The extension to non-compact symplectic manifolds, such as the cotangent bundles we consider here, requires suitable conditions on the asymptotic behavior of both the Hamiltonian H and the almost complex structure J. A possibility is to assume that H satisfies the asymptotic quadratic growth conditions (H1) and (H2) and that J is C^0-close to the Levi-Civita almost complex structure on T^*Q which is induced by the Riemannian metric on Q (see [2]). Another possibility is to consider Hamiltonians which are superlinear functions of $|p|$ for $|p|$ large and almost complex structures which are of contact type with respect to λ (see e.g. [25]). Here we stick to the former set of conditions, although everything we say could also be adapted to the latter one.

The Floer complex obviously depends on the Hamiltonian H, but its homology often does not, so it makes sense to call this homology the *Floer homology* of the underlying symplectic manifold (M,ω), and to denote it by $HF_*(M)$. The Floer homology of a compact symplectic manifold M without boundary is isomorphic to the singular homology of M, as proved by A. Floer for special classes of symplectic manifolds, and later extended to larger and larger classes by several authors (the general case requiring special coefficient rings, see [18, 20, 21]).

Unlike the compact case, the Floer homology of a cotangent bundle T^*Q for Hamiltonians of quadratic type is a truly infinite-dimensional homology theory, being isomorphic to the singular homology of the free loop space ΛQ of Q. This fact was proved by C. Viterbo (see [26]) using a generating functions approach, later by D. Salamon and J. Weber using the heat flow for curves on a Riemannian manifold (see [23]) and then by the authors in [2].

In particular, our proof reduces the general case to the case of a Hamiltonian which is uniformly convex in the momenta, meaning that it satisfies the condition

(H3) $\nabla_{pp}H(t,q,p) \geq h_3 I$, for some $h_3 > 0$,

and for such a Hamiltonian it constructs an explicit isomorphism between the Floer complex $(F_*(H),\partial)$ and the *Morse complex* $(M_*(\mathbb{S}_L),\partial)$ of the action functional

$$\mathbb{S}_L(\gamma) = \int_{\mathbb{T}} L(t,\gamma(t),\dot\gamma(t))dt, \quad \gamma \in W^{1,2}(\mathbb{T},Q),$$

associated to the Lagrangian $L\colon \mathbb{T}\times TQ \to \mathbb{R}$ which is the Fenchel dual of H,

$$L(t,q,v) = \max_{p\in T_q^*Q} \left(\langle p,v\rangle - H(t,q,p)\right),$$

a Lagrangian of Tonelli type. The latter complex is the standard chain complex associated to the Lagrangian action functional $\mathbb{S}_L$. The domain of such a functional is the infinite dimensional Hilbert manifold $W^{1,2}(\mathbb{T},Q)$ consisting of closed loops of Sobolev class $W^{1,2}$ on Q. The functional $\mathbb{S}_L$ is bounded from below, it has non-degenerate critical points a with finite Morse index $i(a)$, it satisfies the Palais-Smale condition, and, although in general it is not of class C^2, it admits a smooth Morse-Smale pseudo-gradient flow (see [3]). The construction of the Morse complex in this infinite-dimensional setting and the proof that its homology is isomorphic to the singular homology of the ambient manifold are described in [1]. The isomorphism goes from the Morse to the Floer complex and is obtained by coupling the Cauchy-Riemann type equation on the half-cylinder $\mathbb{R}^+ \times \mathbb{T}$ with the pseudo-gradient flow equation for the Lagrangian action. We call this the *hybrid method.*

Since the space $W^{1,2}(\mathbb{T}, Q)$ is homotopy equivalent to ΛQ, we get the asserted isomorphism

$$\Phi^\Lambda : H_*(\Lambda Q) \xrightarrow{\cong} HF_*(T^*Q)\,.$$

This isomorphism result was generalized in [6] for more general path spaces than the free loop space. In fact, given a closed submanifold $R \subset Q \times Q$, we can consider the path space

$$\Omega_R Q = \big\{ c \in W^{1,2}([0,1], Q) \mid (c(0), c(1)) \in R \big\}.$$

In particular, the choice $R = \Delta$, the diagonal in $Q \times Q$, produces the free loop space ΛQ, while the based loop space $\Omega_{q_o} Q$ is given by the choice $R = \{(q_o, q_o)\}$.

Given a submanifold $S \subset Q$ we have its associated conormal bundle

$$N^*S = \big\{ (q,p) \in T^*Q \mid q \in S,\ p_{|T_q S} \equiv 0 \big\},$$

which is a Lagrangian submanifold of $(T^*Q, d\lambda)$ on which the Liouville one-form λ vanishes identically. The non-degeneracy assumption for a Hamiltonian $H: [0,1] \times T^*Q \to \mathbb{R}$ is now that the Lagrangian submanifold

$$G_H = \big\{ \big(\alpha, \mathsf{C}\phi_H^1(\alpha)\big) \mid \alpha \in T^*Q \big\} \subset T^*Q \times T^*Q = T^*(Q \times Q)$$

should have a transverse intersection with N^*R in $T^*(Q \times Q)$, where $\mathsf{C}: (q,p) \mapsto (q,-p)$ is the anti-symplectic conjugation on T^*Q.

In [6] it was shown that we have an associated Floer homology HF_*^R, with the chain complex $F_*^R(H)$ generated by the Hamiltonian paths

$$\mathcal{P}_R(H) = \big\{ x: [0,1] \to T^*Q \mid \dot{x}(t) = X_H(t, x(t)),\ \big(x(0), \mathsf{C}x(1)\big) \in N^*R \big\}, \tag{4}$$

and the boundary operator $\partial: F_*^R \to F_{*-1}^R$ defined by counting the Floer trajectories

$$u: \mathbb{R} \times [0,1] \to T^*Q, \quad \overline{\partial}_{J,H} u = 0, \quad \big(u(s,0), \mathsf{C}u(s,1)\big) \in N^*R \quad \forall s \in \mathbb{R},$$

converging to x and $y \in \mathcal{P}_R(H)$ as $s \to -\infty$ and $s \to \infty$. Note that this is a well-posed Fredholm problem because N^*R is a Lagrangian submanifold of $T^*(Q \times Q)$. Compactness and energy estimates hold because $(\lambda \oplus \lambda)_{|N^*R} \equiv 0$.

Theorem 1.1. *[6] We have $HF_*^R(T^*Q) \cong H_*(\Omega_R Q)$ via an explicit chain complex isomorphism*

$$\Phi^R\colon M_*(\mathbb{S}_L|_{\Omega_R Q}) \xrightarrow{\cong} F_*^R(H)$$

where $L : [0,1]\times TQ \to \mathbb{R}$ is the Lagrangian which is Fenchel dual to the quadratic type Hamiltonian H.

The first aim of this paper is to give an explicit chain level construction of a chain complex homomorphism

$$\Psi^R\colon F_*^R(H) \to M_*(\mathbb{S}_L|_{\Omega_R Q})$$

which might not be a chain complex isomorphism, but which induces an isomorphism

$$\Psi_*^R\colon HF_*^R(H) \xrightarrow{\cong} HM_*(\mathbb{S}_L|_{\Omega_R Q})$$

such that $\Psi_*^R = \left(\Phi_*^R\right)^{-1}$. Such a chain map brings methodical advantages when comparing the ring structures on the Floer and on the topological side, as we are going to show.

An important structure in Floer homology is its canonical ring structure, the so-called *pair-of-pants product* in the case of the free loop space (see [24]), or triangle product in the case of the path space with endpoints on Lagrangian submanifolds. Already in the case of a closed symplectic $2n$-dimensional manifold (M,ω), the pair-of-pants product

$$m_\Delta\colon HF_*(M) \otimes HF_*(M) \to HF_{*-n}(M)$$

encodes a truly symplectic invariant. While $HF_*(M)$ as an abelian group is isomorphic to the ordinary singular homology of M, the pair-of-pants product in general deviates from the expected intersection product (note that the grading of m_Δ becomes consistent with that of the intersection product by the grading shift in the isomorphism $HF_*(M) \cong H_{*+n}(M)$). In fact, as shown in [22], Floer homology with the pair-of-pants product is ring isomorphic to the quantum homology of $QH_*(M,\omega)$ of (M,ω), a deformation of the intersection ring structure due to the presence of pseudoholomorphic spheres.

In the context of cotangent bundles, such a deformation by pseudoholomorphic spheres cannot occur, since they simply cannot exist for the exact symplectic structure $\omega = d\lambda$. But the question remains, what the pair-of-pants ring structure corresponds to in view of the isomorphism $HF_*(T^*Q) = HF_*^\Delta(H) \cong H_*(\Lambda Q)$. In [5], we finally give the proof that the same isomorphism Φ^Δ intertwines m_Δ with the Chas-Sullivan loop product (see [9]), provided that we consider closed and oriented smooth manifolds Q.

For the definition of the pair-of-pants product on chain level

$$m_\Delta\colon F_*^\Delta(H) \otimes F_*^\Delta(H) \to F_{*-n}^\Delta(H^{(2)}),$$

in [5] we use as a model for the domain surface the branched 2:1-covering of the standard cylinder, a smooth pair-of-pants surface with two cylindrical entrances and one cylindrical exit and a conformal structure globally given in the cylindrical coordinates as $s+it$. Note that, for precise energy estimates, we use the Hamiltonian $H^{(2)}(t,q,p) = 2H(2t,q,p)$ whose 1-periodic orbits equal the 2-periodic ones for H. Equivalently, we define m_Δ by counting the solutions of the following problem

$$\begin{aligned} &u = (u_1,u_2)\colon \mathbb{R}\times[0,1] \to T^*(Q\times Q),\ \overline{\partial}_{J,H} u_i = 0,\ i = 1,2, \\ &\big(u_1(s,0), \mathrm{C}u_1(s,1), u_2(s,0), \mathrm{C}u_2(s,1)\big) \in \begin{cases} N^*(\Delta_{12}\times\Delta_{34}), & s \le 0, \\ N^*(\Delta_{14}\times\Delta_{23}), & s \ge 0, \end{cases} \end{aligned} \tag{5}$$

with asymptotics $(x,y) \in \mathcal{P}_1(H)\times\mathcal{P}_1(H)$ for $s \to -\infty$ and $z \in \mathcal{P}_2(H)$ for $s \to \infty$ (see Fig. 1). Here

$$\begin{aligned} \Delta_{12}\times\Delta_{34} &= \big\{(q,q,q',q') \,|\, q,q' \in Q\big\}, \\ \Delta_{14}\times\Delta_{23} &= \big\{(q,q',q',q) \,|\, q,q' \in Q\big\}. \end{aligned} \tag{6}$$

Similarly, when $R = \{(q_0,q_0)\}$ we have the triangle product

$$m_{\{(q_0,q_0)\}} : HF_*^{\{(q_0,q_0)\}}(H) \otimes HF_*^{\{(q_0,q_0)\}}(H) \to HF_*^{\{(q_0,q_0)\}}(H^{(2)}),$$

and [5] contains the proof of the following:

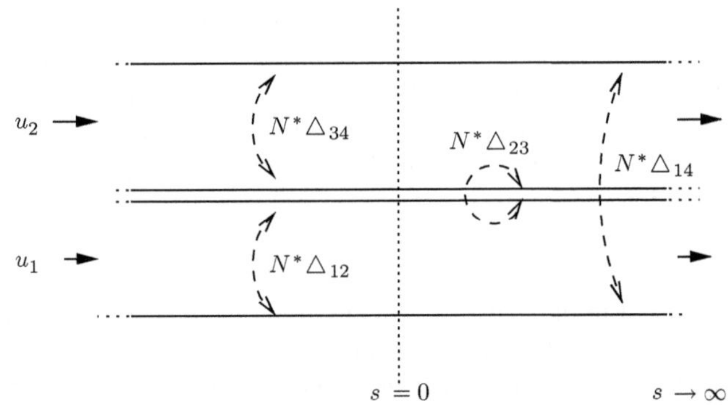

Fig. 1 The pair-of-pants product

Theorem 1.2. *The chain complex isomorphisms* $\Phi^R\colon M_*(\mathbb{S}_L|_{\Omega_R Q}) \to F_*^R(H)$, *for* $R = \Delta$ *or* $R = \{(q_0, q_0)\}$, *induces ring isomorphisms*

$$(H_*(\Lambda Q), \circ) \cong (HF_*^{\Delta}, m_\Delta), \quad (H_*(\Omega_{q_o} Q), \#) \cong (HF_*^{\{(q_0,q_0)\}}, m_{\{(q_0,q_0)\}}),$$

for the Chas-Sullivan product $\circ$ *on the singular homology of the free loop space and the Pontrjagin product # on the singular homology of the based loop space.*

If we view the submanifold $R \subset Q \times Q$ as a correspondence, these products have natural generalizations in terms of composition of correspondences. In fact, given two correspondences $R_1, R_2 \subset Q \times Q$, their composition is defined as $R_2 \circ R_1 = \pi_{13}\big((R_1 \times Q) \cap (Q \times R_2)\big)$, where $\pi_{13}\colon Q \times Q \times Q \to Q \times Q$ is the projection on the first and third coordinate. We actually have $R \circ R = R$ both for the free loop case $R = \Delta$ and for the based loop case $R = \{(q_o, q_o)\}$. When $R_1 \times Q$ and $Q \times R_2$ intersect cleanly in Q^3, and the restriction of π_{13} to such an intersection is regular, meaning that the kernel of its differential has constant dimension, then R_1 and R_2 are said to be smoothly composable. In this case, $R_2 \circ R_1$ is a closed submanifold of $Q \times Q$, so the Floer homology $HF_*^{R_2 \circ R_1}(H)$ is still defined.

One can show that the pair-of-pants product m_Δ on HF_*^Δ and the triangle product $m_{\{(q_0,q_0)\}}$ on $HF_*^{\{(q_0,q_0)\}}$ can be unified in terms of a binary operation

$$m_{R_1,R_2}\colon HF_*^{R_1} \otimes HF_*^{R_2} \to HF_{*-d(R_1,R_2)}^{R_2 \circ R_1}$$

for composable correspondences. In fact, in (5) we have to replace $\Delta_{12} \times \Delta_{34}$ for $s \le 0$ by $R_1 \times R_2$, and $\Delta_{14} \times \Delta_{23}$ for $s \ge 0$ by $(R_2 \circ R_1) \times \Delta_{23}$. Depending on the correspondences R_1 and R_2, there is a degree shift $d(R_1, R_2)$, which equals the codimension of the clean intersection $(R_1 \times R_2) \cap (Q \times \Delta \times Q)$ in $R_1 \times R_2$.

In general, m_{R_1,R_2} is isomorphic to a binary operator

$$H_*(\Omega_{R_1} Q) \otimes H_*(\Omega_{R_2} Q) \to H_{*-d(R_1,R_2)}(\Omega_{R_2 \circ R_1} Q),$$

generalizing the loop product. Such a binary operator is defined as the composition

$$\begin{aligned} H_j(\Omega_{R_1} Q) \otimes H_k(\Omega_{R_2} Q) &\xrightarrow{\times} H_{j+k}(\Omega_{R_1} Q \times \Omega_{R_2} Q) \\ &= H_{j+k}(\Omega_{R_1 \times R_2} Q \times Q) \to \\ &\xrightarrow{i_!} H_{j+k-d}(\Omega_{(R_1 \times R_2) \cap (Q \times \Delta \times Q)} Q \times Q) \longrightarrow H_{j+k-d}(\Omega_{R_2 \circ R_1} Q), \end{aligned}$$

where $\times$ is the exterior product, $i_!$ is the Umkehr morphism induced by the d-co-dimensional and co-oriented inclusion

$$i : \Omega_{(R_1 \times R_2) \cap (Q \times \Delta \times Q)} Q \times Q \hookrightarrow \Omega_{R_1 \times R_2} Q \times Q,$$

and the last homomorphism is induced by the concatenation map.

In this paper, we want to emphasize the general rule that Floer homology on cotangent bundles should be able to remodel any known algebro-topological structure in classical (co-)homology of loop spaces of closed, oriented manifolds. In fact, there should always be an independent chain level construction which, under the isomorphism Φ, is isomorphic to a corresponding structure on the classical side. This has been carried out successfully with the loop product and the Pontrjagin product in [5], where in fact, for the loop product, it was the pair-of-pants product which had been considered first, whereas the loop product had for whatever reason essentially eluded the topologists' attention until [9].

In the present paper we want to address in the same light two more product structures on the classical side. One is the cup-product on cohomology, which can be equivalently seen as a coproduct on the homology of $\Omega_R Q$,

$$\cup\colon H_*(\Omega_R Q) \to H_*(\Omega_R Q) \otimes H_*(\Omega_R Q)\,.$$

We give a Floer-theoretical construction of such a product, and we prove the following:

Theorem 1.3. *Given a generic triple of quadratic type Hamiltonians, we have a chain level operation* $u\colon F_*^R(H_1) \to F_*^R(H_2)\otimes F_*^R(H_3)$ *which induces a coproduct* $u_*\colon HF_*^R \to HF_*^R \otimes HF_*^R$ *isomorphic to the cup-coproduct on* $H_*(\Omega_R Q)$ *via the isomorphism* Φ_*^R.

An interesting question is whether the coalgebra structure u_* on HF_*^R can be seen to be an algebra homomorphism $(HF_*^R, m^R) \to (HF_*^R \otimes HF_*^R, m_R \otimes m_R)$, or equivalently, whether m_R is a coalgebra morphism for u_*. In other words, this is the question of whether (HF_*^R, m_R, u_*) carries a Hopf algebra structure, which for the based loop space homology $(H_*(\Omega Q), \#, \cup)$ is classically known to hold. Clearly, the fact that the isomorphism Φ^Ω intertwines # with $m_{\{(q_0,q_0)\}}$ and $\cup$ with u_* (Theorems 1.2 and 1.3) implies that the Hopf algebra structure also exists on the Floer side for $R = (q_o, q_o)$. In fact, this Hopf algebra property can be verified directly on chain level on the Floer side for the based loop space version. For general R with $R \circ R = R$, this Hopf algebra property cannot hold already for dimensional reasons, e.g. for the free loop space version $R = \Delta$.

The other structure we are interested in is a coproduct derived from the obvious pair-of-pants type coproduct with one entrance and two exits (see [11]). This coproduct, however, is essentially trivial, but it gives rise to a secondary coproduct on homology of loop space relative to the constant loops,

$$\square\colon H_*(\Lambda Q, Q) \to \big(H_*(\Lambda Q, Q) \otimes H_*(\Lambda Q, Q)\big)_{*-n+1}\,.$$

This coproduct has been constructed by M. Goresky and N. Hingston in [19], and computed for interesting examples such as spheres.

Given the special Hamiltonian $\frac{1}{2}|p|^2$ with a generic and small potential perturbations $V(t,q)$ we can consider Floer cohomology filtered by the action, $F^*_{\geq a}(H)$.

On the level of cohomology we can perform a limit for the perturbation $V \to 0$, and we have the following:

Theorem 1.4. *For every action values $a, b > 0$, Floer cohomology comes equipped with a product operation*

$$\tilde{w}\colon HF^*_{\geq a}(\tfrac{1}{2}|p|^2) \otimes HF^*_{\geq b}(\tfrac{1}{2}|p|^2) \to HF^{*+n-1}_{\geq a+b}(\tfrac{1}{2}|p|^2).$$

When the positive numbers a, b are small enough, the isomorphism Φ^* induces a ring isomorphism from $(HF^*_{>0}, \tilde{w})$ to $(H^*(\Lambda Q, Q), \Box)$.

In fact, it is possible to replace $\frac{1}{2}|p|^2$ by any superlinear $c|p|^{1+\delta}$, $\delta > 0$. This is not of quadratic type and requires a somewhat different argument for the C^0-estimates of the moduli spaces involved. In this paper, we give an explicit construction of $\tilde{w}$. The proof of the equivalence with $\Box$ will be given elsewhere.

2 The Inverse Isomorphism

Let us recall the construction of the isomorphism from $H_*(\Omega_R Q)$ to $HF^R_*(T^*Q)$ from [2] and [6]. When the Hamiltonian $H \in C^\infty([0,1] \times T^*Q)$ satisfies (H1), (H2) and (H3), its Fenchel dual Lagrangian $L \in C^\infty([0,1] \times TQ)$ is well-defined and satisfies the analogous quadratic growth and strict convexity assumptions. We denote by $\mathbb{S}^R_L$ the restriction of the Lagrangian action functional

$$\mathbb{S}_L(\gamma) = \int_0^1 L(t, \gamma, \dot{\gamma})\, dt,$$

to the path space $\Omega_R Q$. Here $\Omega_R Q$ carries a $W^{1,2}$-Hilbert manifold structure, $\mathbb{S}^R_L$ is of class $C^{1,1}$ on $\Omega_R Q$ and it is twice Gateaux-differentiable. The fact that the Hamiltonian H is non-degenerate with respect to the correspondence R implies also the non-degeneracy of all critical points of $\mathbb{S}^R_L$. This fact allows to construct a smooth negative pseudo-gradient Morse vector field for $\mathbb{S}^R_L$, see [3]. We denote by $M_*(\mathbb{S}^R_L)$ the chain complex generated by the critical points $a \in \operatorname{Crit} \mathbb{S}^R_L$, graded by the non-negative Morse index $i(a)$, with boundary operator $\partial\colon M_*(\mathbb{S}^R_L) \to M_{*-1}(\mathbb{S}^R_L)$ defined by algebraically counting the unparametrized connecting trajectories for the generically chosen negative pseudo-gradient vector field for $\mathbb{S}^R_L$. A result from [1] shows that $H_*(M_*(\mathbb{S}^R_L), \partial) \cong H_*(\Omega_R Q)$ in a natural way, i.e. compatible with the continuation isomorphism $H_*(M_*(\mathbb{S}^R_L), \partial) \cong H_*(M_*(\mathbb{S}^R_{L'}), \partial)$ for homotopies of the Lagrangian.

In [2] and generalized for the path spaces $\Omega_R Q$ in [6], a chain complex isomorphism

$$\Phi^R\colon \big(M_*(\mathbb{S}^R_L), \partial\big) \xrightarrow{\cong} \big(F^R_*(H), \partial\big)$$

was constructed explicitly building on the Legendre-Fenchel duality of H and L. Given generators $x \in \mathcal{P}_R(H)$, $a \in \mathrm{Crit}(\mathbb{S}_L^R)$, we have the moduli space of hybrid type trajectories

$$\begin{aligned}\mathcal{M}_{a;x} &= \{u\colon [0,\infty)\times[0,1] \to T^*Q \mid \bar{\partial}_{J,H} u = 0,\ u(+\infty) = x,\\ &\quad \big(u(s,0), \mathrm{C}u(s,1)\big) \in N^*R, (\pi\circ u)(0,\cdot) \in W^u(\mathbb{S}_L^R; a)\},\end{aligned} \tag{7}$$

where $W^u(\mathbb{S}_L^R; a)$ denotes the unstable manifold of a for the negative pseudo-gradient flow of $\mathbb{S}_L^R$. For generic choices of J and of the pseudo-gradient vector field, $\mathcal{M}_{a;x}$ is a manifold of dimension $i(a) - \mu^R(x)$, where $\mu^R(x)$ is the Maslov-type index of x as a solution of the non-local Lagrangian boundary value problem (4) (see [6] for the precise definition). Assuming arbitrary orientations for all unstable manifolds $W^u(\mathbb{S}_L^R; a)$ and using the concept of coherent orientation for Floer homology according to [17], we show in [2] that all $\mathcal{M}_{a;x}$ are orientable in a coherent way, that is, compatible with the splitting-off of boundary trajectories on either side. The compactness proof for this moduli space follows from the energy estimate for $u \in \mathcal{M}_{a;x}$

$$\mathbb{S}_L(a) \geq \mathbb{S}_L\big((\pi\circ u)(0)\big) \geq \mathcal{A}_H(u(0,\cdot)) \geq \mathcal{A}_H(x),$$

with equality if and only if $\pi\circ x = a$ and u is constant in s with $\pi(u(s,\cdot)) = a$, in particular $\#\mathcal{M}_{\pi(x);x} = 1$. The central estimate is an immediate consequence of the Fenchel-Legendre duality between L and H.

As a consequence from the identification of the generating sets, consistent even with index and critical value

$$\pi\colon \mathcal{P}_R(H) \xrightarrow{\cong} \mathrm{Crit}\,\mathbb{S}_L^R, \quad i(\pi(x)) = \mu^R(x),\ \mathbb{S}_L(\pi(x)) = \mathcal{A}_H(x),$$

the chain morphism

$$\Phi^R a = \sum_{\substack{x\in\mathcal{P}_R(H)\\ \mathcal{A}_H(x)\leq\mathbb{S}_L(a)}} \big(\#_{\mathrm{alg}}\mathcal{M}_{a;x}\big)\cdot x,$$

gives a chain complex isomorphism, as it is representable by a semi-infinite triangular matrix with ± 1 on the diagonal.

We now give an equally explicit chain level construction of a chain morphism

$$\Psi^R\colon F_*^R(H) \to M_*(\mathbb{S}_L^R)$$

such that at the homology level $\Psi_*^R = (\Phi_*^R)^{-1}$. Here, we cannot give an argument why the given Ψ^R should already be a chain complex isomorphism, certainly not necessarily equal to $(\Phi^R)^{-1}$. However, the concrete form of Ψ^R allows for simpler proofs of ring isomorphism properties of Φ^R_*, compared with the construction from [5].

Let us consider the moduli space for $x \in \mathcal{P}_R(H)$,

$$\begin{aligned} \mathcal{M}_x^- = \big\{ u\colon (-\infty,0]\times[0,1] \to T^*Q \,\big|\, \bar\partial_{J,H} u = 0,\ u(-\infty) = x, \\ u(0,\cdot) \in 0_Q,\ \big(u(s,0), \mathsf{C}u(s,1)\big) \in N^*R \big\}, \end{aligned} \tag{8}$$

where 0_Q denotes the zero-section of T^*Q. For generic J, this is a smooth manifold of dimension $\mu^R(x)$, compact modulo splitting-off Floer trajectories at $-\infty$, in particular C^∞_{loc}-compact. Hence, we find an upper bound $c = c(x)$ depending on x for the Lagrangian action of the path $(\pi \circ u)(0,\cdot) \in \Omega_R Q$,

$$\mathbb{S}_L\big((\pi \circ u)(0,\cdot)\big) \le c(x) \quad \text{for all } u \in \mathcal{M}_x^- .$$

Given $x \in \mathcal{P}_R(H)$, $a \in \mathrm{Crit}\,\mathbb{S}_L^R$, we now set

$$\mathcal{M}_{x;a} = \big\{ u \in \mathcal{M}_x^- \,\big|\, (\pi \circ u)(0) \in W^s(\mathbb{S}_L^R; a) \big\},$$

where $W^s(\mathbb{S}_L^R; a)$ denotes the stable manifold of a. Provided that $x \not\subset 0_Q$ or $\pi \circ x \neq a$ if $x \subset 0_Q$ (Fig. 2), we find for generic J and pseudo-gradient vector field for $\mathbb{S}_L^R$ that $\mathcal{M}_{x;a}$ is a smooth manifold of dimension $\mu^R(x) - i(a)$, compact up to splitting-off boundary trajectories, and oriented via coherent orientation. We set

$$\Psi^R\colon F_*^R(H) \to M_*(\mathbb{S}_L), \quad \Psi^R x = \sum_{\substack{a \in \mathrm{Crit}\,\mathbb{S}_L^R \\ \mathbb{S}_L(a) \le c(x)}} \big(\#_{\mathsf{alg}}\mathcal{M}_{x;a}\big)\cdot a,$$

and we obtain a chain complex morphism.

However, in general $c(x) > \mathcal{A}_H(x)$ is possible, in fact necessary if $\mathcal{M}_{x;\pi(x)} \neq \emptyset$, so that we cannot expect Ψ^R to be of triangular shape similarly to Φ^R. In fact, Ψ^R

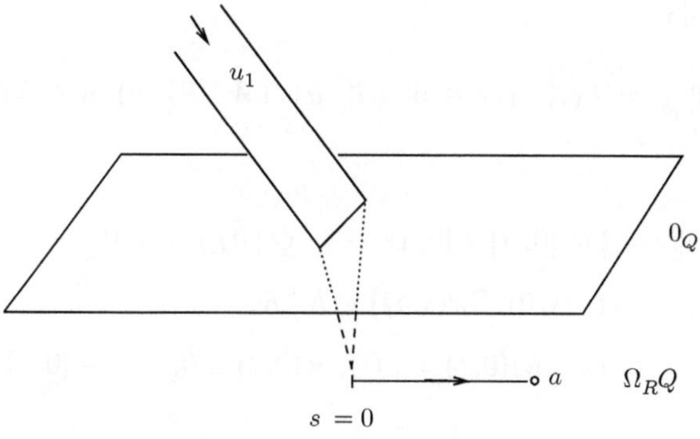

Fig. 2 The inverse construction, $\mathcal{M}_{x;a}$

can easily be defined for any pair (H, L) of a quadratic type Hamiltonian and a Lagrangian which does not need to be Fenchel dual to H.

The idea of using half-cylinders with boundary on the zero section of the cotangent bundle in order to provide cycles in the path space from cycles in the Floer chain complex via the evaluation at the zero section has been known for a while. In [10] this technique is used towards an isomorphism for linearized contact homology instead of Floer homology. The same idea is also used in [7].

Let us now give the proof that $\Psi^R \circ \Phi^R$ is chain homotopy equivalent to $\mathrm{id}_{M_*(\mathbb{S}_L)}$, which already implies that $\Psi^R_* = (\Phi^R_*)^{-1}$ since we know Φ^R_* to be an isomorphism.

Proposition 2.1. *Given H of quadratic type we have $\Psi^R \circ \Phi^R \simeq \mathrm{id}$ on $M_*(\mathbb{S}^R_L)$.*

Proof. Via the usual gluing result for Floer theory we clearly have that $\Psi^R \circ \Phi^R$ is chain homotopy equivalent to the chain morphism $M_*(\mathbb{S}^R_L) \to M_*(\mathbb{S}^R_L)$ defined by counting

$$\begin{aligned} \mathcal{M}^\sigma_{a,b} = \big\{ & w\colon [0,\sigma] \times [0,1] \to T^*Q \mid \bar\partial_{J,H} w = 0, \\ & \big(w(s,0), \mathsf{C}w(s,1)\big) \in N^*R,\ w(\sigma,\cdot) \subset 0_Q, \\ & (\pi \circ w)(0,\cdot) \in W^u(\mathbb{S}^R_L; a),\ (\pi \circ w)(\sigma,\cdot) \in W^s(\mathbb{S}^R_L; b) \big\} \end{aligned} \tag{9}$$

for $a, b \in \mathrm{Crit}\, \mathbb{S}^R_L$ with equal Morse index, and for $\sigma > 0$ fixed. The chain homotopy to $\mathrm{id}_{M_*(\mathbb{S}^R_L)}$ then follows from letting σ shrink to 0.

In order to simplify this argument, let us insert a further cobordism step. Namely, we clearly obtain a chain homotopy equivalence to the chain morphism on $M_*(\mathbb{S}^R_L)$ defined by counting

$$\begin{aligned} \widetilde{\mathcal{M}}^{\sigma,\lambda}_{a,b} = \big\{ & w\colon [0,\sigma] \times [0,1] \to T^*Q \mid \bar\partial_{J,H} w = 0, \\ & \big(w(s,0), \mathsf{C}w(s,1)\big) \in N^*R,\ w(\sigma,\cdot) \subset 0_Q, \\ & (\pi \circ w)(0,\cdot) \in W^u(\mathbb{S}^R_L; a),\ (\pi \circ w)(\lambda,\cdot) \in W^s(\mathbb{S}^R_L; b) \big\} \end{aligned} \tag{10}$$

for $\sigma > 0$ fixed and $\lambda \in [0,\sigma]$ given. For $\lambda = \sigma$ we have exactly $\mathcal{M}^\sigma_{a,b}$, and for $\lambda = 0$ we obtain

$$\widetilde{\mathcal{M}}^\sigma_{a,b} = \big\{ (c,w) \,\big|\, c \in W^u(\mathbb{S}^R_L; a) \cap W^s(\mathbb{S}^R_L; b),\ w \in \mathcal{M}^\sigma_c \big\}$$

with

$$\begin{aligned} \mathcal{M}^\sigma_c = \big\{ & w\colon [0,\sigma] \times [0,1] \to T^*Q \mid \bar\partial_{J,H} w = 0, \\ & \big(w(s,0), \mathsf{C}w(s,1)\big) \in N^*R, \\ & (\pi \circ w)(0,t) = c(t),\ w(\sigma,t) \in 0_Q\ \forall\, t \in [0,1] \big\}. \end{aligned} \tag{11}$$

If $i(a) = i(b)$ we have for $(c,w) \in \widetilde{\mathcal{M}}^\sigma_{a,b}$ that $a = b = c$, $w \in \mathcal{M}^\sigma_a$. The proof of the Proposition then follows from the following:

Lemma 2.2. *Given $c \in \Omega_R Q$ there exists a $\sigma_o = \sigma_o(c) > 0$ such that for each $\sigma \in (0, \sigma_o]$ the solution space $\mathcal{M}_c^\sigma$ contains a unique solution, compatible with coherent orientation.*

In fact, for $\sigma_n \to 0$, the solution sequence w_n converges uniformly with all derivatives to the path $(c, 0) \in \Omega_{N^*R} T^*Q$. Compatibility with coherent orientation implies that

$$\#_{\mathsf{alg}}\mathcal{M}_a^\sigma = \#\mathcal{M}_a^\sigma = 1 \quad \text{for } \sigma \in (0, \sigma_o],\ a \in \mathrm{Crit}\mathbb{S}_L\,.$$

Hence, counting $\widetilde{\mathcal{M}}_{a,b}^\sigma$ for $\sigma \in (0, \sigma_o]$ defines exactly the identity operator on $M_*(\mathbb{S}_L^R)$. This concludes the proof of Proposition 2.1 □

For the proof of Lemma 2.2 we refer to Proposition 4.10 in [5]. It follows from a uniform convergence analysis of solutions $w_n \in \mathcal{M}_c^{\sigma_n}$ as $\sigma_n \to 0$ together with a Newton type method to prove the unique existence of solutions for σ small enough. Note that, for example for $H_o = \frac{1}{2}|p|^2$, a first order approximation of solutions $w \in \mathcal{M}_c^\sigma$ is given by $w_{\text{approx}}^\sigma(s,t) = \big(c(t), (\sigma - s)\dot{c}(t)\big)$, where we identify $TQ \cong T^*Q$ via the Legendre transformation from H_o.

Moreover, there is also a parametric version of Lemma 2.2, where we allow c to vary in a relatively compact family $K \subset \Omega_R Q$, for example an unstable manifold $W^u(\mathbb{S}_L; a)$. This would be the version to use in order to show $\Psi^R \circ \Phi^R \simeq \mathrm{id}$ directly by considering $\mathcal{M}_{a,b}^\sigma$ above for σ running from ∞ to 0.

3 Cup Product

We now show that also the cup-coproduct structure on path space homology

$$\cup\colon H_*(\Omega_R Q) \to H_*(\Omega_R Q) \otimes H_*(\Omega_R Q)$$

has a Floer theoretic counterpart given by a chain level construction, isomorphic to $\cup$ via Φ^R.

Given three R-nondegenerate Hamiltonians H_i, $i = 0, 1, 2$, we define a chain operation

$$u\colon F_*^R(H_0) \to F_*^R(H_1) \otimes F_*^R(H_2)$$

as follows. Given generators $x_i \in \mathcal{P}_R(H_i)$, $i = 0, 1, 2$, we consider three-fold Floer half-strips coupled by a conormal boundary condition

$$\begin{aligned} \mathcal{M}_{x_0;x_1,x_2}^{\cup,R} = \big\{ & u = (u_0, \bar{u}_1, \bar{u}_2)\colon (-\infty, 0] \times [0,1] \to T^*Q^3 \;\big| \\ & \bar{\partial}_{J,H_i} u_i = 0,\ i = 0,1,2,\ u_i(\pm\infty, \cdot) = x_i, \\ & \big(u_i(s,0), \mathsf{C}u_i(s,1)\big) \in N^*R,\ 0 \le |s| < \infty, \\ & u(0,t) \in N^*\triangle^{(3)} \big\}, \end{aligned} \tag{12}$$

where $\bar{u}_i(s,t) = \mathsf{C}u_i(-s,t)$ and $\Delta^{(3)} = \{(q,q,q) \mid q \in Q\} \subset Q^3$. Note that the conormal condition $u(0,\cdot) \in N^*\Delta^{(3)}$ means that

$$\begin{aligned} \pi \circ u_0(0,\cdot) &= \pi \circ u_1(0,\cdot) = \pi \circ u_2(0,\cdot) =: q(\cdot) \quad \text{and} \\ u_0(0,\cdot) &= u_1(0,\cdot) + u_2(0,\cdot) \quad \text{in } T^*_{q(\cdot)}Q\,. \end{aligned} \tag{13}$$

Hence, we have a well-posed Fredholm problem for $\mathcal{M}^{\cup,R}_{x_0;x_1,x_2}$ with

$$\dim \mathcal{M}^{\cup,R}_{x_0;x_1,x_2} = \mu^R(x_0) - \mu^R(x_1) - \mu^R(x_2)\,.$$

For the index formula for half-strips with piecewise conormal boundary condition see [5], Theorems 5.24 and 5.25. It remains to provide an energy estimate in order to obtain the usual compactness result. We compute with $u_i(0,\cdot) = \big(q(\cdot), p_i(\cdot)\big)$ and (13)

$$\begin{aligned} \mathcal{A}_{H_0}(x_0) \geq \mathcal{A}_{H_0}(u_0(0,\cdot)) &= \int_0^1 \big(\langle p_0, \dot{q}\rangle - H_0(t,q,p_0)\big)dt \\ &\overset{(13)}{=} \int_0^1 \big(\langle p_1 + p_2, \dot{q}\rangle - H_0(t,q,p_0)\big)dt \\ &= \mathcal{A}_{H_1}(u_1(0,\cdot)) + \mathcal{A}_{H_2}(u_2(0,\cdot)) \\ &\quad + \int_0^1 \big(H_1(t,q,p_1) + H_2(t,q,p_2) - H_0(t,q,p_0)\big)dt\,. \end{aligned} \tag{14}$$

Thus, we obtain the required action monotonicity provided that the Hamiltonians satisfy

$$\begin{gathered} H_0(t,q,p+p') \leq H_1(t,q,p) + H_2(t,q,p') \\ \text{for all } t \in [0,1],\, q \in Q,\, p,p' \in T_q^*Q\,. \end{gathered}$$

For example, this is satisfied for geodesic type Hamiltonians with time-dependent potential perturbation,

$$H_0(t,q,p) = \frac{1}{2}|p|^2 + V(t,q), \quad H_1(t,q,p) = H_2(t,q,p) = |p|^2 + \frac{1}{2}V(t,q)\,.$$

Note that we have canonical isomorphisms $HF_*^R(H_0) \cong HF_*^R(H_i)$, $i = 1,2$ from the standard continuation argument. We define u by counting $\mathcal{M}^{\cup,R}_{x_0;x_1,x_2}$ with the usual orientation procedure,

$$u\colon F_*(H_0) \to F_*(H_1) \otimes F_*(H_2), \quad u(x) = \sum_{\substack{(y,z) \in \mathcal{P}_R(H_1) \times \mathcal{P}_R(H_2) \\ \mu^R(y) + \mu^R(z) = \mu^R(x)}} \big(\#_{\mathrm{alg}}\mathcal{M}^{\cup,R}_{x;y,z}\big)\, y \otimes z\,. \tag{15}$$

We shall prove the following:

Theorem 3.1. *The chain level operation* $u\colon F_*^R(H_0) \to F_*^R(H_1) \otimes F_*^R(H_2)$ *induces a coproduct* $u_*\colon HF_*^R \to HF_*^R \otimes HF_*^R$ *which is isomorphic to the cup coproduct on* $H_*(\Omega_R Q)$ *via the isomorphism* Φ_*^R.

Before proving the ring isomorphism property, let us remark that we have a variety of homotopically equivalent definitions for the cup coproduct in Floer homology. In fact, given $x_i \in \mathcal{P}_R(H_i)$, $i = 0,1,2$, we can consider the problem for $\lambda \in [0,1]$,

$$u_0\colon (-\infty,0]\times[0,1] \to T^*Q,\ u_i\colon [0,\infty)\times[0,1]\to T^*Q,\ i=1,2,$$

$$\overline{\partial}_{J_i,H_i} u_i = 0;\ u_1(-\infty) = x_1,\ u_i(+\infty) = x_i,\ i = 1,2,$$

$$\begin{aligned} &\big(u_i(s,0), \mathrm{C}u_i(s,1)\big) \in N^*R,\ \text{f.a. } 0 \le |s| < \infty,\ i = 0,1,2,\\ &(\pi\circ u_0)(0,\cdot) = (\pi\circ u_1)(0,\cdot) = (\pi\circ u_2)(0,\cdot) =: q,\\ &\text{i.e. } u_i(0,\cdot) = (q,p_i),\ i=0,1,2,\ p_0 = \lambda p_1 + (1-\lambda)p_2\,. \end{aligned} \tag{16}$$

This is a well-posed Fredholm problem for all $\lambda \in [0,1]$, and for $\lambda = 1/2$ we obtain a problem which is essentially equivalent to (12) (Fig. 3). In order to get compactness for the above problem, it is convenient to assume that the Hamiltonians H_0, H_1 and H_2 are physical Hamiltonians with the same kinetic part,

$$H_j(t,q,p) = \frac{1}{2}|p|^2 + V_j(t,q), \qquad \forall j = 0,1,2,$$

and that J is C^0-close enough to the Levi-Civita almost complex structure J_0. Under these assumptions we have the following compactness result, where as usual on the space of maps we consider the C^∞_{loc} topology:

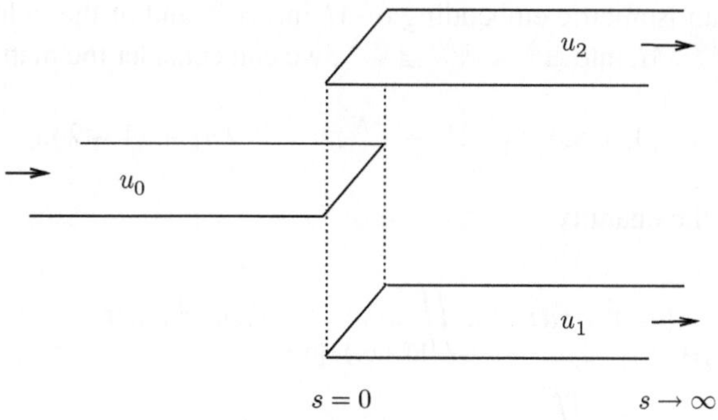

Fig. 3 The cup coproduct

Lemma 3.2. *For every triple $x_j \in \mathcal{P}_R(H_j)$, the space of solutions (λ, u_0, u_1, u_2) of (16) is pre-compact. Moreover, the existence of a solution (λ, u_0, u_1, u_2) of (16) gives rise to the estimate*

$$\lambda \mathcal{A}_{H_1}(x_1) + (1-\lambda)\mathcal{A}_{H_2}(x_2) \leq \mathcal{A}_{H_0}(x_0) + \|V_0\|_\infty + \max\{\|V_1\|_\infty, \|V_2\|_\infty\}. \tag{17}$$

Proof. By the special form of the Hamiltonians, we have

$$\begin{aligned} H_0\big(t,q,\lambda p_1 + (1-\lambda)p_2\big) - \lambda H_1(t,q,p_1) - (1-\lambda)H_2(t,q,p_2) \\ \leq \|V_0\|_\infty + \max\{\|V_1\|_\infty, \|V_2\|_\infty\}. \end{aligned}$$

Therefore,

$$\begin{aligned} \mathcal{A}_{H_0}(u_0(0,\cdot)) &= \int p_0 dq - H_0(t,q,p_0)\,dt \\ &= \int \big(\lambda p_1 + (1-\lambda)p_2\big)dq - H_0(t,q,p_0)\,dt \\ &= \lambda \mathcal{A}_{H_1}(u_1(0,\cdot)) + (1-\lambda)\mathcal{A}_{H_2}(u_2(0,\cdot)) \\ &\quad - \int \big(H_0(t,q,p_0) - \lambda H_1(t,q,p_1) - (1-\lambda)H_2(t,q,p_2)\big)\,dt \\ &\geq \lambda \mathcal{A}_{H_1}(u_1(0,\cdot)) + (1-\lambda)\mathcal{A}_{H_2}(u_2(0,\cdot)) \\ &\quad - \|V_0\|_\infty - \max\{\|V_1\|_\infty, \|V_2\|_\infty\}, \end{aligned} \tag{18}$$

and the estimate (17) follows from the bounds

$$\begin{aligned} \mathcal{A}_{H_0}(u_0(0,\cdot)) \leq \mathcal{A}_{H_0}(x_0), \quad \mathcal{A}_{H_1}(u_1(0,\cdot)) \geq \mathcal{A}_{H_1}(x_1), \\ \mathcal{A}_{H_2}(u_2(0,\cdot)) \geq \mathcal{A}_{H_2}(x_2). \end{aligned} \tag{19}$$

By means of an isometric embedding of M into $\mathbb{R}^N$ and of the induced isometric embedding of T^*M into $\mathbb{R}^N \times \mathbb{R}^N \cong \mathbb{C}^N$, we can consider the map

$$v : [0,+\infty) \times [0,1] \to \mathbb{C}^N, \quad v = \lambda u_1 + (1-\lambda)u_2.$$

Then by (18), the quantity

$$\begin{aligned} \iint_{[0,+\infty)\times[0,1]} |\partial_s v|^2\,ds dt \leq \lambda \iint_{[0,+\infty)\times[0,1]} |\partial_s u_1|^2\,ds dt \\ + (1-\lambda) \iint_{[0,+\infty)\times[0,1]} |\partial_s u_2|^2\,ds dt \end{aligned}$$

$$
\begin{aligned}
&= \lambda\big(\mathcal{A}_{H_1}(u_1(0,\cdot)) - \mathcal{A}_{H_1}(x_1)\big) + (1-\lambda)\big(\mathcal{A}_{H_2}(u_2(0,\cdot)) - \mathcal{A}_{H_2}(x_2)\big) \\
&\le \mathcal{A}_{H_0}(x_0) + \|V_0\|_\infty + \max\big\{\|V_1\|_\infty, \|V_2\|_\infty\big\} \\
&\qquad + \big|\mathcal{A}_{H_1}(x_1)\big| + \big|\mathcal{A}_{H_2}(x_2)\big|
\end{aligned}
$$

has a uniform bound. Since also $\|\partial_s u_0\|_2$ is uniformly bounded, because of (18) and (19), the L^2 norm of the s-derivative of the map

$$
w : [0,+\infty) \times [0,1] \to \mathbb{C}^N \times \mathbb{C}^N, \quad w(s,t) = \big(\overline{u_0(-s,t)}, v(s,t)\big),
$$

has a uniform bound. Since $\|J - J_0\|_\infty$ is small, w solves a Cauchy-Riemann type equation, and $w(0,t)$ belongs to the totally real space given by the conormal of the diagonal in $\mathbb{R}^N \times \mathbb{R}^N$, the argument of [2, Sect. 1.5] shows that w is uniformly bounded in C^∞. In particular, u_0 and

$$
q(t) := \pi \circ u_0(0,t) = \pi \circ u_1(0,t) = \pi \circ u_2(0,t)
$$

are uniformly bounded in C^∞, and we get uniform upper bounds for

$$
\mathcal{A}_{H_1}(u_1(0,\cdot)) \le \mathbb{S}_{L_1}(q) \quad \text{and} \quad \mathcal{A}_{H_2}(u_2(0,\cdot)) \le \mathbb{S}_{L_2}(q).
$$

Together with the lower bounds of (19), we conclude that $\|\partial_s u_1\|_2$ and $\|\partial_s u_2\|_2$ are both uniformly bounded. By [2, Theorem 1.14 (iii)] and the usual elliptic bootstrap argument, we conclude that also u_1 and u_2 have uniform C^∞ bounds. □

Let now, for given $\lambda \in [0,1]$, $W^\lambda_{x_0;x_1,x_2}$ denote the set of solutions of (16) with generically chosen J_i for each u_i, $i = 0,1,2$, as well as generically chosen triple (V_0, V_1, V_2) of perturbing potentials. Then, we can define a chain level operation

$$
u_\lambda : F^R_*(H_0) \to \bigoplus_{i+j=*} F^R_i(H_1) \otimes F^R_j(H_2),
$$

from counting $\#_{\mathrm{alg}} W^\lambda_{x_0;x_1,x_2}$. Using the full solution space $W_{x_0;x_1,x_2}$ of (16) with variable $\lambda \in [0,1]$ and accordingly generically chosen structures J_i and V_i and index relation $\mu^R(x_0) = \mu^R(x_1) + \mu^R(x_2) - 1$ we obtain easily the following:

Proposition 3.3. *The induced coproducts* $(u_\lambda)_* : HF^R_*(H_0) \to HF^R_*(H_1) \otimes HF^R_*(H_2)$ *do not depend on* $\lambda \in [0,1]$, *and they are equal to the cup-coproduct* u.

In fact, the cup-coproduct (15) is essentially given by $u_{\frac{1}{2}}$.

As a consequence, in dual cohomological formulation, we can apply the above action estimates to the notion of cohomologically critical values

$$
c^*(\alpha, H) := \sup\big\{a \in \mathbb{R} \mid \alpha \in \operatorname{Im}\big(HF^*_{\ge a}(H) \to HF^*(H)\big)\big\}
$$

for given $\alpha \in HF^*(H)$, where $HF^*_{\geq a}$ is the cohomology of the subcochain complex $F^*_{\geq a}(H) = \mathbb{Z}^{\{x \in \mathcal{P}_R(H) \mid \mathcal{A}_H(x) \geq a\}}$, and we are omitting the superscript R.

We have in this cohomological formulation, with $\cup$ dual to u:

Corollary 3.4 *Given $H_i(t,q,p) = \frac{1}{2}|p|^2 + V_i(t,q)$ as above, we have for $\alpha_i \in HF^*(H_i)$, $i = 1,2$ with $\alpha_1 \cup \alpha_2 \in HF^*(H_0)$*

$$c * (\alpha_1 \cup \alpha_2, H_0) \geq \max\{c^*(\alpha_1, H_1), c^*(\alpha_2, H_2)\} - \|V_0\|_\infty - \max\{\|V_1\|_\infty, \|V_2\|_\infty\}.$$

We now complete the proof of Theorem 3.1. At first, we give a Morse-homological definition of the cup-product.

Suppose we have three non-degenerate Lagrangians L_i, $i = 0,1,2$, such that $\mathbb{S}^R_{L_1}$ and $\mathbb{S}^R_{L_2}$ have no common critical points. Then we define

$$\begin{aligned} \cup\colon M_*(\mathbb{S}^R_{L_0}) &\to M_*(\mathbb{S}^R_{L_1}) \otimes M_*(\mathbb{S}^R_{L_2}), \\ \cup a &= \sum_{(b,c) \in \operatorname{Crit} \mathbb{S}^R_{L_1} \times \operatorname{Crit} \mathbb{S}^R_{L_2}} \langle a; b, c\rangle\, b \otimes c\,, \end{aligned} \tag{20}$$

where $\langle a; b, c\rangle$ is the oriented count of

$$W^u(\mathbb{S}^R_{L_0}; a) \cap W^s(\mathbb{S}^R_{L_1}; b) \cap W^s(\mathbb{S}^R_{L_2}; c),$$

provided that we have chosen three generic pseudogradient fields so that the triple intersection is transverse. The dimensions of this intersection is $i(a) - i(b) - i(c)$, and the intersection is oriented if the unstable manifolds (which are all finite-dimensional) are oriented.

The usual splitting-off argument for boundary trajectories proves the Leibniz rule for $\cup$, and it is well-known see e.g. [8] that $\cup_*$ defines the cup-coproduct. One can also show Morse homologically that the cohomological product $\cup^*$ satisfies $\cup^* = \Delta^* \circ \times$, where $\times$ is the exterior product and Δ^* the pull-back by the diagonal embedding $\Delta\colon \Omega_R Q \hookrightarrow \Omega_R Q \times \Omega_R Q$, for which we also have Morse homological functoriality.

We now want to show that the isomorphism

$$\Psi^R_*\colon HF^R_*(H) \to HM_*(\mathbb{S}^R_L)$$

intertwines the coproducts u and $\cup_*$, i.e.

$$\cup \circ \Psi^R \simeq (\Psi^R \otimes \Psi^R) \circ u$$

are chain homotopic on F^R_*.

Clearly, $\cup \circ \Psi^R$ is chain homotopic to the operation

$$w_1 \colon F_*^R(H) \to M_*(\mathbb{S}_{L_1}^R) \otimes M_*(\mathbb{S}_{L_2}^R),$$
$$w_1(x) = \sum_{\substack{(b,c)\\ i(b)+i(c)=\mu^R(x)}} \left(\#_{\mathsf{alg}} \widetilde{\mathcal{M}}^{(1)}(x;b,c)\right) \cdot b \otimes c\,, \tag{21}$$

with

$$\widetilde{\mathcal{M}}^{(1)}(x;b,c) = \left\{u \in \mathcal{M}_x^- \,\middle|\, (\pi \circ u)(0,\cdot) \in W^s(\mathbb{S}_{L_1}^R;b) \cap W^s(\mathbb{S}_{L_2}^R;c)\right\}.$$

Then we find generic J for $\mathcal{M}_x^-$ and pseudo-gradient vector fields for $\mathbb{S}_{L_i}^R$, $i = 1,2$, such that $\widetilde{\mathcal{M}}^{(1)}(x;b,c)$ satisfies transversality for all x,b,c.

Next, we use Proposition 3.3, which allows us to replace u by u_λ for $\lambda = 1$. We obtain

$$\left(\Psi^R \otimes \Psi^R\right) \circ u_1 \simeq w_2,$$

with w_2 given by the oriented count of

$$\begin{aligned}\widetilde{\mathcal{M}}_\sigma^{(2)}(x;b,c) = \Big\{ (u,v) \,\Big|\, & u \in \mathcal{M}_x^-,\ (\pi \circ u)(0,\cdot) \in W^s(\mathbb{S}_{L_1}^R;c),\\ & v \colon [0,\sigma] \times [0,1] \to T^*Q,\ \left(v(s,0), \mathsf{C}v(s,1)\right) \in N^*R,\\ & (\pi \circ v)(0,\cdot) = (\pi \circ u)(-\sigma,\cdot),\\ & v(\sigma,\cdot) \subset 0_Q,\ (\pi \circ v)(\sigma,\cdot) \in W^s(\mathbb{S}_{L_2}^R;b) \Big\}\end{aligned} \tag{22}$$

for any fixed $\sigma > 0$ (Figs. 4 and 5). Moreover, w_2 is clearly chain homotopic to $w_3 \colon F_*^R \to M_* \otimes M_*$ given by

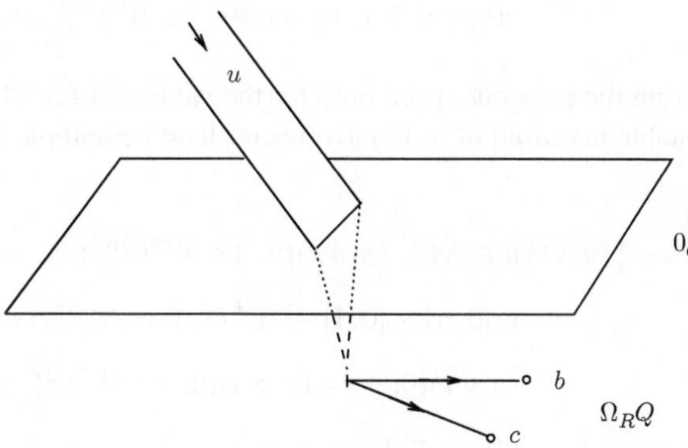

Fig. 4 $\widetilde{\mathcal{M}}_\sigma^{(1)}(x;b,c)$

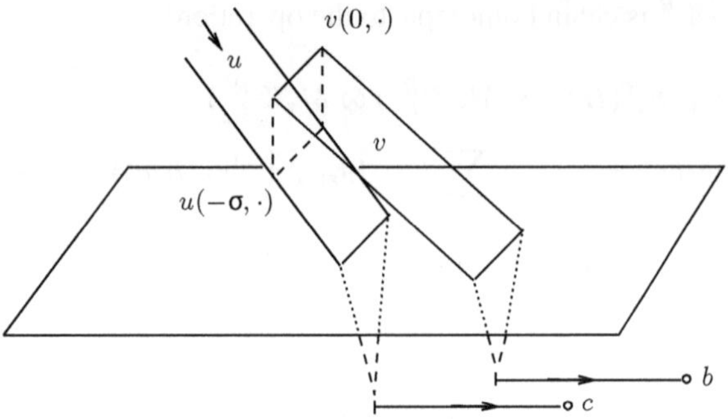

Fig. 5 $\widetilde{\mathcal{M}}^{(2)}_\sigma(x;b,c)$

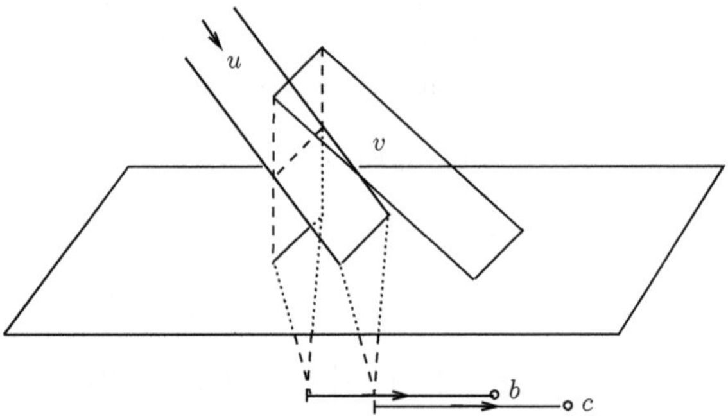

Fig. 6 $\widetilde{\mathcal{M}}^{(3)}_\sigma(x;b,c)$

$$\begin{aligned}\widetilde{\mathcal{M}}^{(3)}_\sigma(x;b,c) = \Big\{ (u,v) \,\Big|\, & u \in \mathcal{M}^-_x,\ (\pi\circ u)(0,\cdot) \in W^s(\mathbb{S}^R_{L_1};c),\\ & v\colon [0,\sigma]\times[0,1] \to T^*Q,\ \big(v(s,0),\mathrm{C}v(s,1)\big) \in N^*R,\\ & (\pi\circ v)(0,\cdot) = (\pi\circ u)(-\sigma,\cdot),\\ & v(\sigma,\cdot) \subset 0_Q,\ (\pi\circ v)(0,\cdot) \in W^s(\mathbb{S}^R_{L_2};b) \Big\}, \end{aligned} \tag{23}$$

which differs from the previous space only for the value of s for which $\pi \circ v(s,\cdot)$ belongs to the stable manifold of b. Finally, w_3 is chain homotopic to w_4 given by (Fig. 6)

$$\begin{aligned}\widetilde{\mathcal{M}}^{(4)}_\sigma(x;b,c) = \Big\{ (u,v) \,\Big|\, & u \in \mathcal{M}^-_x,\ (\pi\circ u)(0,\cdot) \in W^s(\mathbb{S}^R_{L_1};c),\\ & v\colon [0,\sigma]\times[0,1] \to T^*Q,\ \big(v(s,0),\mathrm{C}v(s,1)\big) \in N^*R,\\ & (\pi\circ v)(0,\cdot) = (\pi\circ u)(0,\cdot) \in W^s(\mathbb{S}^R_{L_2};b),\\ & v(\sigma,\cdot) \subset 0_Q \Big\}\end{aligned}$$

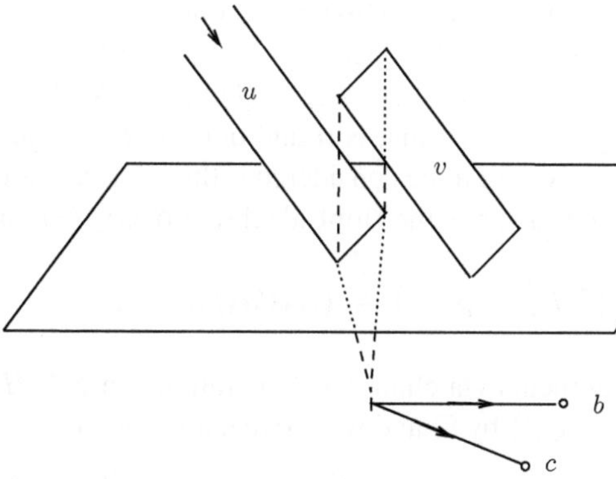

Fig. 7 $\widetilde{\mathcal{M}}_\sigma^{(4)}(x;b,c)$

$$\begin{aligned}
&= \Big\{ (u,v) \,\Big|\, u \in \widetilde{\mathcal{M}}^{(1)}(x;b,c), \\
&\qquad v\colon [0,\sigma]\times[0,1] \to T^*Q,\ \big(v(s,0), \mathrm{C}v(s,1)\big) \in N^*R, \\
&\qquad (\pi\circ v)(0,\cdot) = (\pi\circ u)(0,\cdot), \\
&\qquad v(\sigma,\cdot) \subset 0_Q \Big\}, \\
&= \Big\{ (u,v) \,\Big|\, u \in \widetilde{\mathcal{M}}^{(1)}(x;b,c),\ v \in \mathcal{M}^\sigma_{\pi\circ u(0)} \Big\}, \qquad (24)
\end{aligned}$$

with $\mathcal{M}^\sigma_{\pi\circ u(0)}$ as in (11). The chain homotopy $w_4 \simeq w_1$ then follows from Lemma 2.2 if we choose $\sigma > 0$ small enough. This finishes the proof of Theorem 3.1 (Fig. 7). □

4 The Proof of the Hopf Algebra Property

Let us now analyze the compatibility of the coproduct $u_*\colon HF_*^R(H_0) \to HF_*^R(H_1) \otimes HF_*^R(H_2)$ with the product $m_R\colon HF_*^R(H) \otimes HF_*^R(H) \to HF^R_{*-d(R,R)}(H^{(2)})$ where $R\circ R = R$. In fact, we are interested in the relation

$$u\circ m = (m\otimes m)\circ\tau\circ(u\otimes u), \qquad (25)$$

where

$$\begin{aligned}
\tau\colon & HF^R(H_1)\otimes HF^R(H_2)\otimes HF^R(H_1)\otimes HF^R(H_2) \\
&\to HF^R(H_1)\otimes HF^R(H_1)\otimes HF^R(H_2)\otimes HF^R(H_2)
\end{aligned}$$

commutes the second and third factor. Obviously, a necessary condition for (8) to hold besides $R \circ R = R$ is that the degree $d(R,R)$ vanishes. Since this degree equals the codimension of $(R \times R) \cap (Q \times \Delta \times Q)$ in $R \times R$, $d(R,R)$ vanishes if and only if $R \times R \subset Q \times \Delta \times Q$, that is if and only if $R = \{(q_0, q_0)\}$ for some $q_0 \in Q$. Therefore, the only case to be considered is the classical case of based loop homology, that is we want to verify the Hopf-algebra property (8) for

$$\big(HF_*^R, m_R, u_*\big) \cong \big(H_*(\Omega Q), \#, \cup\big)$$

by Floer-theoretical arguments via chain level operations on $F_*^R(H_i)$. We replace the superscript $R = \{(q_o, q_o)\}$ by Ω and we prove the following:

Theorem 4.1. *The chain maps* $u^i \colon F_*^\Omega(H_0^{(i)}) \to F_*^\Omega(H_1^{(i)}) \otimes F_*^\Omega(H_2^{(i)})$, $i = 1, 2$, *and* $m_j \colon F_*^\Omega(H_j) \otimes F_*^\Omega(H_j) \to F_*^\Omega(H_j^{(2)})$, $j = 0, 1, 2$, *satisfy the chain homotopy property*

$$u^2 \circ m_0 \simeq (m_1 \otimes m_2) \circ \tau \circ (u^1 \otimes u^1).$$

Proof. We recall from [5] that $m_j \colon F_*^\Omega(H_j) \otimes F_*^\Omega(H_j) \to F_*^\Omega(H_j^{(2)})$ is defined by counting

$$(u_1, u_2) \colon \mathbb{R} \times [0,1] \to (T^*Q)^2,\ \overline{\partial}_{J,H_j} u_i = 0,\ i = 1, 2,$$

$$\big(u_1(s,0), \mathrm{C}u_1(s,1), u_2(s,0), \mathrm{C}u_2(s,1)\big) \in \begin{cases} (T_{q_o}^* Q)^4 = N^*(\{q_o\}^4), & s \le 0, \\ T_{q_o}^* Q \times N^*\Delta_{23} \times T_{q_o}^* Q, & s \ge 0. \end{cases}$$

Hence, using the definition of u^i via (12) and $H_j = \frac{1}{2}|p|^2 + V_j(t,q)$, with V_j 1-periodic in time, for $j = 0, 1, 2$, we obtain via the usual gluing argument that for every $\rho > 0$ the chain map $u^2 \circ m_0$ is chain homotopic to the operator

$$A_\rho \colon F_*^\Omega(H_0) \otimes F_*^\Omega(H_0) \to F_*^\Omega(H_1^{(2)}) \otimes F_*^\Omega(H_2^{(2)}),$$

which is defined by counting

$$\begin{aligned}
& w \colon (-\infty, 0] \times [0,1] \to (T^*Q)^6 \quad \text{with}, \\
& w(s,t) = \big(u_{10}(s,t), \mathrm{C}u_{20}(s,1-t), \mathrm{C}u_{11}(-s,t), u_{21}(-s,1-t), \\
& \qquad\qquad \mathrm{C}u_{12}(-s,t), u_{22}(-s,1-t)\big) \\
& \overline{\partial}_{J,H_j} u_{ij} = 0,\ i = 1, 2,\ j = 0, 1, 2, \\
& w(s,0) \in (T_{q_o}^* Q)^6, \ \text{for} \ -\infty < s \le 0, \\
& w(0,t) \in N^*\big(\Delta^{(3)} \times \Delta^{(3)}\big),\ 0 \le t \le 1,
\end{aligned} \tag{26}$$

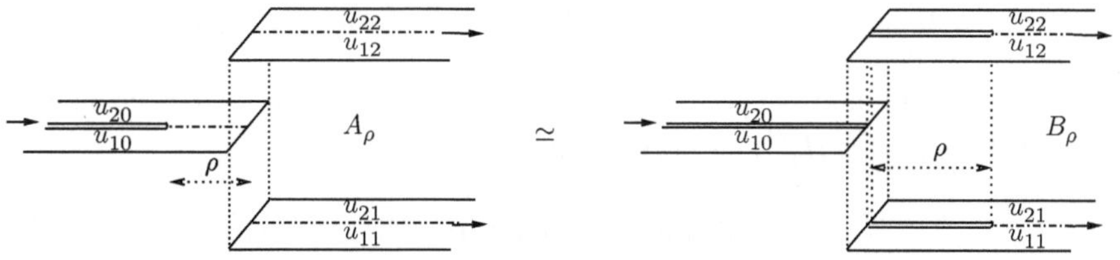

Fig. 8 The Hopf algebra argument

$$w(s,1) \in N^*\big(\{q_o\}^2 \times \Delta \times \Delta\big),\ -\infty < s \le -\rho, \quad \text{and}$$

$$w(s,1) \in N^*\big(\Delta \times \Delta \times \Delta\big),\ -\rho \le s \le 0. \tag{27}$$

Likewise, we have that for every $\rho > 0$ the chain map $(m_1 \otimes m_2) \circ \tau \circ (u^1 \otimes u^1)$ is chain homotopic to the operator

$$B_\rho \colon F_*^\Omega(H_0) \otimes F_*^\Omega(H_0) \;\to\; F_*^\Omega(H_1^{(2)}) \otimes F_*^\Omega(H_2^{(2)}),$$

which is defined by counting $w \colon (-\infty, 0] \times [0,1] \to (T^*Q)^6$ as above satisfying instead of the last equation of (26) the condition

$$w(s,1) \in N^*(\{q_o\}^6) \quad \text{for } -\rho \le s \le 0\,. \tag{28}$$

We now need to show that A_ρ and B_ρ are chain homotopic. Instead of identifying the limit operators A_0 and B_0 as $\rho \to 0$, we apply a different argument from [5] (Fig. 8).

Note that it is not possible to homotope the last line of (26) into (28) through conormal boundary conditions N^*R for $s \in [-\rho, 0]$ since $\{q_o\}^6$ and $\Delta \times \Delta \times \Delta$ are not isotopic in Q^6 already by dimensional reasons. However, we have the following:

Lemma 4.2. *The chain maps $A_\rho \otimes B_\rho$ and $B_\rho \otimes A_\rho$ are chain homotopic.*

The proof is completely analogous to that of Proposition 4.7 in [5]. In order to deduce from the above lemma that A_ρ is chain homotopic to B_ρ, we can use the following algebraic fact, which is proven as Lemma 4.6 in [5]:

Lemma 4.3. *Let (C, ∂) and (C', ∂) be chain complexes, bounded from below. Let $\varphi, \psi : C \to C'$ be chain maps. Assume that there is an element $\epsilon \in C_0$ with $\partial\epsilon = 0$ and a chain map $\delta : C' \to (\mathbb{Z}, 0)$ such that*

$$\delta(\varphi(\epsilon)) = \delta(\psi(\epsilon)) = 1.$$

If $\varphi \otimes \psi$ is homotopic to $\psi \otimes \varphi$ then φ is homotopic to ψ.

Here $(\mathbb{Z}, 0)$ denotes the trivial chain complex all of whose groups vanish, except for the one in degree zero which is $\mathbb{Z}$.

It remains to find a cycle $\epsilon \in (F^{\Omega}(H_0) \otimes F^{\Omega}(H_0))_0$ and a chain map $\delta \colon F^{\Omega}(H_1^{(2)}) \otimes F^{\Omega}(H_2^{(2)}) \to (\mathbb{Z}, 0)$ such that

$$\delta(A_\rho(\epsilon)) = \delta(A_\rho(\epsilon)) = 1. \tag{29}$$

Without loss of generality, we can assume that the potentials V_j are time-independent Morse functions and that they have a common unique maximum at q_0. The constant orbit $x_0 = (q_0, 0)$ is an element of $F_0^{\Omega}(H_j^{(i)})$ for every $i = 1, 2$ and $j = 0, 1, 2$ and it defines a cycle

$$\epsilon = (x_0, x_0) \in (F^{\Omega}(H_0) \otimes F^{\Omega}(H_0))_0.$$

Since x_0 is the critical point with minimal action $\mathcal{A}_{H_j^{(i)}}$, for every $i = 1, 2$ and $j = 0, 1, 2$, we have

$$A_\rho(\epsilon) = B_\rho(\epsilon) = \epsilon. \tag{30}$$

Let $L_1^{(2)}$ and $L_2^{(2)}$ be the Lagrangians on TQ which are Fenchel dual to $H_1^{(2)}$ and $H_2^{(2)}$. Let

$$\tilde{\delta} : M\left(\mathbb{S}_{L_1^{(2)}}\right) \otimes M\left(\mathbb{S}_{L_2^{(2)}}\right) = M\left(\mathbb{S}_{L_1^{(2)}} \oplus \mathbb{S}_{L_2^{(2)}}\right) \to (\mathbb{Z}, 0)$$

be the standard augmentation on the Morse complex of the functional

$$\mathbb{S}_{L_1^{(2)}} \oplus \mathbb{S}_{L_2^{(2)}} : \Omega_{q_0} Q \times \Omega_{q_0} Q \to \mathbb{R}, \quad (\gamma_1, \gamma_2) \mapsto \mathbb{S}_{L_1^{(2)}}(\gamma_1) + \mathbb{S}_{L_2^{(2)}}(\gamma_2),$$

that is the homomorphism which maps each critical point of Morse index zero into 1. The homomorphism $\tilde{\delta}$ is a chain map because the boundary of every critical point γ of Morse index one has the form $\gamma_1 - \gamma_2$, where γ_1 and γ_2 are the critical points of Morse index zero to which the two sides of the one-dimensional unstable manifold of γ converge. We can now use the isomorphism Φ^{Ω} between the Morse complex and the Floer complex to read the chain map $\tilde{\delta}$ on the Floer complex, thus defining the chain map

$$\delta : F^{\Omega}(H_1^{(2)}) \otimes F^{\Omega}(H_2^{(2)}) \to (\mathbb{Z}, 0).$$

Since Φ^{Ω} is the identity mapping on global minimizers, we have

$$\delta(\epsilon) = 1.$$

Together with (30), this proves (29) and concludes the proof of the theorem. □

5 The Goresky-Hingston Coproduct

Throughout this section, we deal only with periodic boundary conditions, i.e. to the case $R = \Delta$. In order to simplify the notation, we omit the superscript Δ from all the objects which would require it (such as F_*^{Δ}, $\mathbb{S}_L^{\Delta}$, μ^{Δ}, and so on).

Let us consider the coproduct of degree $-n$, $w\colon F_*(H_0) \to \big(F_*(H_1) \otimes F_*(H_2)\big)_{*-n}$ defined by counting

$$
\begin{aligned}
& u = (u_1, u_2)\colon \mathbb{R} \times [0,1] \to T^*(Q \times Q), \quad \text{solving} \\
& \overline{\partial}_{J,H_0} u_i = 0 \text{ for } s \le 0,\ i = 1,2, \\
& \overline{\partial}_{J,H_1} u_1 = \overline{\partial}_{J,H_2} u_2 = 0 \text{ for } s \ge 0, \\
& \big(u_1(s,0), \mathsf{C}u_1(s,1), u_2(s,0), \mathsf{C}u_2(s,1)\big) \in \begin{cases} N^*(\Delta_{14} \times \Delta_{23}), & s \le 0, \\ N^*(\Delta_{12} \times \Delta_{34}), & s \ge 0, \end{cases}
\end{aligned}
\tag{30}
$$

with asymptotics $x \in \mathcal{P}_2(H_0)$ for $s \to -\infty$ and $(y,z) \in \mathcal{P}_1(H_1) \times \mathcal{P}_1(H_2)$ for $s \to \infty$.

Then, completely analogous to the ring isomorphism $\Phi_*\colon \big(H_*(\Lambda Q), \circ\big) \xrightarrow{\cong} \big(HF_*(T^*Q), m\big)$, one can show that Φ_* identifies the coproduct w on HF_* with the comultiplication

$$
\mu := \mu_{0,3}^{\text{top}}\colon H_*(\Lambda Q) \to \big(H_*(\Lambda Q) \otimes H_*(\Lambda Q)\big)_{*-n},
$$

of degree $-n$ from [11] (see Theorem 3).

We now give a short argument which explains why this coproduct is essentially trivial, i.e. 0 to large extents. Let us assume for simplicity that Q is simply connected and hence $H_0(\Lambda Q) \cong \mathbb{Z}$ generated by 1, where this class 1 is represented by any constant loop $q_o \in Q \subset \Lambda Q$ as a 0-cycle. Moreover, we denote by $e = [Q] \in H_n(\Lambda Q)$ the neutral element for the Chas-Sullivan loop product, which is given by the fundamental class of Q, as an n-cycle of constant loops. In Floer homology, $\Phi_*(e)$ is given by the Floer cycle

$$
\sum_{\mu(x)=n} \big(\#_{\text{alg}} \mathcal{M}_x^+\big) \cdot x \in F_n(H),
$$

$$
\mathcal{M}_x^+ = \big\{ u\colon [0,\infty) \times \mathbb{T} \to T^*Q \;\big|\; \overline{\partial}_{J,H} u = 0,\ u(+\infty) = x,\ \frac{\partial}{\partial t}(\pi \circ u)(0,\cdot) = 0 \big\}
\tag{31}
$$

for a generic J. We have $e \circ a = a \circ e = a$ for all $a \in H_*(\Lambda Q)$ and $\mu(e) = \alpha \cdot 1 \otimes 1$ for some $\alpha \in \mathbb{Z}$ by dimensional reasons. In fact, it is not hard to show that

$$
\mu(e) = \chi(Q) \cdot 1 \otimes 1\,. \tag{32}
$$

Lemma 5.1. *For any $a \in H_k(\Lambda Q)$, we have*

$$\mu(a) = \begin{cases} 0, & \text{if } k \neq n, \\ \beta \cdot 1 \otimes 1, & \text{if } k = n, \end{cases}$$

with $\beta \cdot 1 = \chi(Q) \cdot (a \circ 1) \in H_0(\Lambda Q)$.

Proof. From [11] or the property of HF_* to be a (noncompact) 2-dimensional topological field theory (see also [12]) it follows that

$$\begin{aligned}(\mathrm{id} \otimes m) \circ (\mu \otimes \mathrm{id}) &= (m \otimes \mathrm{id}) \circ (\mathrm{id} \otimes \mu) \\ &= \mu \circ m \colon H_*(\Lambda Q) \otimes H_*(\Lambda Q) \to \big(H_*(\Lambda Q) \otimes H_*(\Lambda Q)\big)_{*-2n} \end{aligned} \tag{33}$$

where for notational clarity we write m for the loop product $\circ$. Applying this identity on $a \otimes e$ and $e \otimes a$ for the given $a \in H_k(\Lambda Q)$ gives

$$\begin{aligned}\mu(a) = (\mu \circ m)(a \otimes e) &= (m \times \mathrm{id}) \circ (\mathrm{id} \otimes \mu)(a \otimes e) \\ &= \chi(Q) \cdot (m \otimes \mathrm{id})(a \otimes 1 \otimes 1), \\ &= \chi(Q) \cdot m(a, 1) \otimes 1, \quad \text{as well as} \\ &= \chi(Q) \cdot 1 \otimes m(a, 1), \end{aligned} \tag{34}$$

which leaves only the possibility $\chi(Q) \cdot m(a, 1) = 0$ in the case $k \neq n$ and $\mu(a) = \beta \cdot 1 \otimes 1, b \cdot 1 = \chi(Q) \cdot m(a, 1)$ if $k = n$. □

Hence, apart from degree n classes, the coproduct has to be trivial. This, however, can be seen as a possibility to define a secondary structure, namely a coproduct on relative homology $H_*(\Lambda Q, Q)$, or equivalently a cohomological product

$$\square \colon H^*(\Lambda Q, Q) \otimes H^*(\Lambda Q, Q) \to H^{*+n-1}(\Lambda Q, Q)\,.$$

This cohomological product has been explicitly constructed and carefully analyzed in [19]. It gives an interesting nontrivial operation in particular for spheres $Q = S^n$.

Here, we now want to give an explicit chain-level construction for the Floer-homological counterpart of □. Let us consider a special Hamiltonian of physical type $H = \frac{1}{2}|p|^2 + V(t,q)$, where $V(t,q)$ is only a small potential perturbation in order to achieve Morse-nondegeneracy for the action $\mathcal{A}_H$. Let us pick $V(t,q)$ generically with $\|V\|_\infty$ small enough compared to the smallest length of a closed geodesic, so that the orbits $x \in \mathcal{P}_1(H)$ with $\mathcal{A}_H(x) > \epsilon$ for some $\epsilon > \|V\|_\infty$ can be seen as the generators of the quotient chain complex $F_*(H)/F_*^{\le\epsilon}(H)$ which defines the homology $HF_*^{\epsilon>0}(H)$. Then, $HF_*(H)^{\epsilon>0}$ becomes isomorphic to $H_*(\Lambda Q, Q)$ under Φ_* for $\epsilon > 0$ small enough. Let us denote

$$HF_*^{>0}(T^*Q) := \lim_{\epsilon>\|V\|_\infty\to 0} HF_*^{\epsilon>0}(H).$$

We will now construct a coproduct

$$\tilde{w}: HF_*^{>0}(T^*Q) \to \big(HF_*^{>0}(T^*Q)\otimes HF_*^{>0}(T^*Q)\big)_{*-n+1}. \tag{35}$$

Given $0<\lambda<1$ we consider the disjoint union of strips

$$\Sigma_\lambda = (-\infty,0]\times[0,\lambda]\;\dot\cup\;(-\infty,0]\times[\lambda,1].$$

Given 1-periodic solutions $x_i\in\mathcal{P}_1(H_i)$, $i=0,1,2$ with $H_i=\frac{1}{2}|p|^2+V_i(t,q)$ for a generic triple of small perturbations as above (V_0,V_1,V_2), we consider $\widetilde{\mathcal{M}}_{x_0;x_1,x_2}$ as the space of solutions (u,v,w,λ) of

$$\begin{aligned}
&\lambda\in(0,1),\ u:\Sigma_\lambda\to T^*Q,\ (v,w):[0,\infty)\times\mathbb{T}\to T^*Q\times T^*Q,\\
&\big(v(+\infty),w(+\infty)\big)=(x_1,x_2),\ u(-\infty,t)=x_1(t)\text{ for }0\le t\le 1,\\
&\overline{\partial}_{J,H_1}v=\overline{\partial}_{J,H_2}w=0,\\
&\overline{\partial}_{J,H_0}u(s,t)=0\text{ for all }0\le t\le 1,\ s\le -1,\\
&\overline{\partial}_{J,\frac{1}{2}|p|^2}u(s,t)=0\text{ for all }0\le t\le 1,\ -1\le s\le 0,\\
&\big(u(s,0),u(s,\lambda+)\big)=\begin{cases}\big(u(s,\lambda-),u(s,1)\big), & -1\le s\le 0,\\ \big(u(s,1),u(s,\lambda-)\big), & s\le -1,\end{cases}\\
&v(0,t)=u(0,\lambda t),\quad w(0,t)=u(0,\lambda+(1-\lambda)t)\quad\text{for all }0\le t\le 1.
\end{aligned}\tag{36}$$

Note that the variation of $\lambda\in(0,1)$ can be equivalently regarded as a particular variation of the conformal structure on a pair-of-pants surface $\bar\Sigma$ with boundary (Fig. 9), given by $\Sigma_{\frac{1}{2}}$ sewed along $(s,0)=(s,\frac{1}{2}-)$ and $(s,\frac{1}{2}+)=(s,1)$ for $-1\le s\le 0$ and $(s,0)=(s,1)$ and $(s,\frac{1}{2}-)=(s,\frac{1}{2}+)$ for $s\le -1$. In fact, $\bar\Sigma$ relative to $\partial\bar\Sigma$ has a topologically nontrivial Riemann moduli space and in order to define $\tilde w$ we are using a particular 1-cycle in its homology relative to its Deligne-Mumford compactification (Fig. 9).

Again, it is not hard to show that for generic choices of J and (V_0,V_1,V_2), $\widetilde{\mathcal{M}}_{x_0;x_1,x_2}$ is a smooth manifold of dimension

$$\dim\widetilde{\mathcal{M}}_{x_0;x_1,x_2} = \mu(x_0)-\mu(x_1)-\mu(x_2)-n+1. \tag{37}$$

In order to obtain the important compactness modulo splitting-off of Floer trajectories, let us compute the energy estimate. We clearly have $\lambda|p|^2,\,(1-\lambda)|p|^2\le|p|^2$

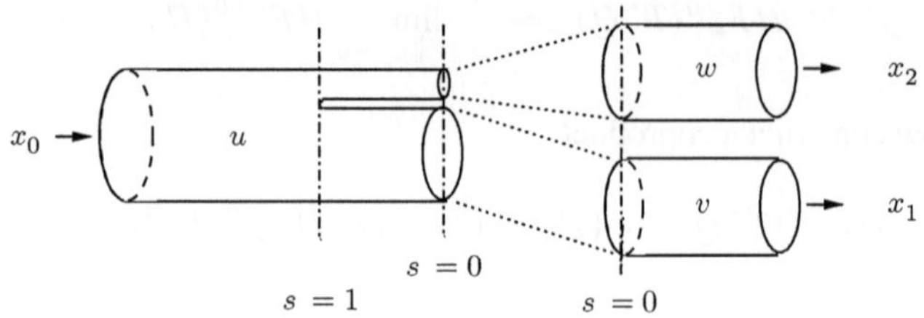

Fig. 9 The coproduct construction. $\widetilde{\mathcal{M}}_{x_1;x_2,x_3}$

for all $\lambda \in [0,1]$. Hence we have for any solution $(u,v,w,\lambda) \in \widetilde{\mathcal{M}}_{x_0;x_1,x_2}$

$$\mathcal{A}_{\frac{1}{2}|p|^2}(v(0)) \le \mathcal{A}_{\lambda\frac{1}{2}|p|^2}(v(0)) = \mathcal{A}_{\frac{1}{2}|p|^2}\big(u(0,\cdot)_{|[0,\lambda]}\big) \quad \text{and}$$

$$\mathcal{A}_{\frac{1}{2}|p|^2}(w(0)) \le \mathcal{A}_{(1-\lambda)\frac{1}{2}|p|^2}(w(0)) = \mathcal{A}_{\frac{1}{2}|p|^2}\big(u(0,\cdot)_{|[\lambda,1]}\big)\,.$$

Using $\epsilon \ge \|V_i\|_\infty$, we have

$$\begin{aligned}\mathcal{A}_{\frac{1}{2}|p|^2}(u(-1,\cdot)) &= \mathcal{A}_{H_1}(u(-1,\cdot)) + \int_0^1 V_0(t,(\pi\circ u)(-1,t))dt\\ &\le \mathcal{A}_{H_1}(x_1) - \int_{-\infty}^{-1}\int_0^1 |\partial_s u|^2 ds dt + \epsilon\end{aligned}$$

and

$$\begin{aligned}\mathcal{A}_{\frac{1}{2}|p|^2}(u(0,\cdot)) &= \mathcal{A}_{\frac{1}{2}|p|^2}\big(u(0,\cdot)_{|[0,\lambda]}\big) + \mathcal{A}_{\frac{1}{2}|p|^2}\big(u(0,\cdot)_{|[\lambda,1]}\big)\\ &\le \mathcal{A}_{\frac{1}{2}|p|^2}(u(-1,\cdot)) - \int_{-1}^{0}\int_0^1 |\partial_s u|^2 ds dt\,.\end{aligned}$$

Assembling all this gives

$$\mathcal{A}_{H_1}(x_1) \le \mathcal{A}_{\frac{1}{2}|p|^2}(v(0,\cdot)) - \iint_0^\infty |\partial_s v|^2 ds dt + \epsilon$$

$$\mathcal{A}_{H_2}(x_2) \le \mathcal{A}_{\frac{1}{2}|p|^2}(w(0,\cdot)) - \iint_0^\infty |\partial_s w|^2 ds dt + \epsilon$$

and

$$\begin{aligned}\mathcal{A}_{H_1}(x_1) + \mathcal{A}_{H_2}(x_2) &\le \mathcal{A}_{\frac{1}{2}|p|^2}(u(0,\cdot)) \\ &\quad - \iint |\partial_s v|^2 ds dt - \iint |\partial_s w|^2 ds dt + 2\epsilon \\ &\le \mathcal{A}_{H_0}(x_0) - E(u,v,w) + 3\epsilon,\end{aligned}$$

that is

$$0 \le E(u,v,w) \le \mathcal{A}_{H_0}(x_0) - \mathcal{A}_{H_1}(x_1) - \mathcal{A}_{H_2}(x_2) + 3\epsilon, \tag{38}$$

with

$$E(u,v,w) = \int_{-\infty}^{0}\int_0^1 |\partial_s u|^2 ds dt + \int_0^{\infty}\int_0^1 \left(|\partial_s v|^2 + |\partial_s w|^2\right) ds dt .$$

With the usual arguments from the compactness theory for Floer trajectories in T^*Q for Hamiltonians of quadratic type, we see that $\widetilde{\mathcal{M}}_{x_0;x_1,x_2}$ is C^∞_{loc}-precompact.

The only new case here concerns sequences $(u_n, v_n, w_n, \lambda_n) \in \mathcal{M}_{x_0;x_1,x_2}$ with $\lambda_n \to 0$ or $\lambda_n \to 1$. Assume without loss of generality $\lambda_n \to 0$. After choosing a C^∞_{loc}-convergent subsequence we view the restriction $u_n|[-1,0] \times [0,\lambda_n]$ as

$$u_n\colon [-1,0]\times\mathbb{R} \to T^*Q, \quad \overline{\partial}_{J,\frac{1}{2}|p|^2} u_n = 0 \quad \text{and}$$

$$u_n(s, t+\lambda_n) = u_n(s,t) \quad \text{for all } (s,t) \in [-1,0]\times\mathbb{R} .$$

We have $u_n \to u_\infty$ in C^∞_{loc},

$$u_\infty\colon [-1,0]\times\mathbb{R} \to T^*Q, \quad \partial_t u_\infty \equiv 0,$$

that is, $u_\infty(0) \in T^*Q$ is a point. On the other side

$$v_n(0,t) = u_n(0,\lambda_n t) \text{ f.a. } t \in \mathbb{R},\ n \in \mathbb{N}, \quad \text{and} \quad v_n \overset{C^\infty_{\text{loc}}}{\to} v_\infty .$$

It follows that $v_\infty(0,t) = u_\infty(0)$ for all $t \in \mathbb{R}$. Hence

$$\mathcal{A}_{\frac{1}{2}|p|^2}\big(v_n(0)\big) \to \mathcal{A}_{\frac{1}{2}|p|^2}\big(u_\infty(0)\big) = -\frac{1}{2}|w_\infty(0)|^2 \le 0,$$

and thus

$$\mathcal{A}_{H_1}(x_1) \le \epsilon - \frac{1}{2}|u_\infty(0)|^2 \le \epsilon .$$

This proves

Proposition 5.2. *If* $\mathcal{A}_{H_1}(x_1), \mathcal{A}_{H_2}(x_2) > \epsilon \geq \max\{\|V_1\|_\infty, \|V_2\|_\infty\}$, *then for all* $x_0 \in \mathcal{P}_1(H_0)$, *the solution space* $\widetilde{\mathcal{M}}_{x_0;x_1,x_2}$ *is compact modulo splitting of Floer trajectories.*

By counting the 0-dimensional solutions of $\widetilde{\mathcal{M}}_{x_0;x_1,x_2}$ we obtain a well-defined cochain operation on the Floer cochain complexes from the ascending $\mathcal{A}_H$-flow,

$$\tilde{w}^\bullet\colon F^k_{\geq a}(H_1) \otimes F^l_{\geq b}(H_2) \rightarrow F^{k+l+n-1}_{\geq a+b-3\epsilon}(H_0)$$
$$\tilde{w}^\bullet(x,y) = \sum_z \#_{\mathsf{alg}} \widetilde{\mathcal{M}}_{z;x,y}\, z\,,$$

for all $a, b > \epsilon$.

After using the usual continuation isomorphism of Floer theory in order to eliminate the perturbation V_i of $H = \frac{1}{2}|p|^2$, we obtain the product

$$\tilde{w}\colon HF^k_{\geq a}(H) \otimes HF^l_{\geq b}(H) \rightarrow HF^{k+l+n-1}_{\geq a+b}(H)$$

for all positive $a, b > 0$ and a ring $\left(HF^*_{>0}(H), \tilde{w}\right)$.

The proof that this product on cohomology is isomorphic to $\Box$ from [19] will appear elsewhere.

Acknowledgements We wish to thank the Max Planck Institute for Mathematics in the Sciences at Leipzig for its kind hospitality. The first author thanks the Humboldt Foundation for financial support in form of a *Humboldt Research Fellowship for Experienced Researchers*. The second author thanks the Deutsche Forschungsgemeinschaft for the support by the grant DFG SCHW 892/2-3. Both authors thank for the support by the DAAD-Vigoni exchange programm, D/08/15177.

References

1. Abbondandolo, A., Majer, P.: Lectures on the Morse complex for infinite dimensional manifolds. In: Biran, P., Cornea, O., Lalonde, F. (eds.) Morse Theoretic Methods in Nonlinear Analysis and in Symplectic Topology (Montreal), pp. 1–74. Springer, Berlin (2006)
2. Abbondandolo, A., Schwarz, M.: On the Floer homology of cotangent bundles. Comm. Pure Appl. Math. **59**, 254–316 (2006)
3. Abbondandolo, A., Schwarz, M.: A smooth pseudo-gradient for the Lagrangian action functional. Adv. Nonlinear Stud. **9**, 597–624 (2009)
4. Abbondandolo, A., Schwarz, M.: Estimates and computations in Rabinowitz-Floer homology. J. Topol. Anal. **1**(4), 307–405 (2009)
5. Abbondandolo, A., Schwarz, M.: Floer homology of cotangent bundles and the loop product. Geom. Topol. **14**(3), 1569–1722 (2010)
6. Abbondandolo, A., Portaluri, A., Schwarz, M.: The homology of path spaces and Floer homology with conormal boundary conditions. J. Fixed Point Theory Appl. **4**, 263–293 (2008)

7. Abouzaid, M.: A cotangent fibre generates the Fukaya category. Advances in Mathematics, **228**, 894–939 (2011)
8. Betz, M., Cohen, R.L.: Moduli spaces of graphs and cohomology operations. Turkish J. Math. **18**, 23–41 (1994)
9. Chas, M., Sullivan, D.: String topology. arXiv:math.GT/9911159 (1999)
10. Cieliebak, K., Latschev, J.: The role of string topology in symplectic field theory. In: Polterovich, L., Abreu, M., Lalonde, F. (eds.) New Perspectives and Challenges in Symplectic Field Theory CRM Proceedings and Lecture Notes, vol. 49, pp. 113–146. American Mathematical Society, Providence, RI (2009)
11. Cohen, R.L., Schwarz, M.: A Morse theoretic description of string topology. In: Polterovich, L., Abreu, M., Lalonde, F. (eds.) New Perspectives and Challenges in Symplectic Field Theory, CRM Proceedings and Lecture Notes, vol. 49, pp. 147–172. American Mathematical Society, Providence, RI (2009)
12. Cohen, R.L., Hess, K., Voronov, A.A.: String topology and cyclic homology. Lectures from the Summer School held in Almería, September 1620, 2003. Advanced Courses in Mathematics. CRM Barcelona, Birkhäuser Verlag, Basel, 2006
13. Floer, A.: A relative Morse index for the symplectic action. Comm. Pure Appl. Math. **41**, 393–407 (1988)
14. Floer, A.: The unregularized gradient flow of the symplectic action. Comm. Pure Appl. Math. **41**, 775–813 (1988)
15. Floer, A.: Symplectic fixed points and holomorphic spheres. Comm. Math. Phys. **120**, 575–611 (1989)
16. Floer, A.: Witten's complex and infinite dimensional Morse theory. J. Diff. Geom. **30**, 207–221 (1989)
17. Floer, A., Hofer, H.: Coherent orientations for periodic orbit problems in symplectic geometry. Math. Zeit. **212**, 13–38 (1993)
18. Fukaya, K., Ono, K.: Arnold conjecture and Gromov-Witten invariant. Topology **38**(5), 933–1048 (1999)
19. Goresky, M., Hingston, N.: Loop products and closed geodesics. Duke Math. J. **150**(1), 117–209 (2009)
20. Hofer, H. Salamon, D.: Floer homology and Novikov rings. In: Hofer, H., Taubes, C., Weinstein, A., Zehnder, E. (eds.) The Floer Memorial Volume, Progress in Mathematics, vol. 133, pp. 483–524. Birkhäuser, Berlin (1995)
21. Liu, G. Tian, G.: Floer homology and Arnold conjecture. J. Differential Geom. **49**(1), 1–74 (1998)
22. Piunikhin, S., Salamon, D. Schwarz, M.: Symplectic Floer-Donaldson theory and quantum cohomology. In: Thomas, C.B. (ed.) Contact and Symplectic Geometry (Cambridge 1994) Publ. Newton Inst., vol. 8, pp. 171–200. Cambridge University Press, London (1996)
23. Salamon, D., Weber, J.: Floer homology and the heat flow. Geom. Funct. Anal. **16**, 1050–1138 (2006)
24. Schwarz, M.: Cohomology operations from S^1-cobordisms in Floer homology, Ph.D. thesis, Swiss Federal Inst. of Techn. Zurich, Diss. ETH No. 11182 (1995)
25. Seidel, P.: A biased view of symplectic cohomology. in *Current Developments in Mathematics, 2006*. Edited by D. Jenison, B. Mazur, T. Mrowka, W. Schmid, R. Stanley and S.T. Yau. International Press, Somerville, MA, 2008, pp. 211–253
26. Viterbo, C.: Functors and computations in Floer homology with applications, II, preprint (revised in 2003) (1996)

MIX
Papier aus verantwortungsvollen Quellen
Paper from responsible sources
FSC® C105338

If you have any concerns about our products,
you can contact us on
ProductSafety@springernature.com

In case Publisher is established outside the EU,
the EU authorized representative is:
Springer Nature Customer Service Center GmbH
Europaplatz 3, 69115 Heidelberg, Germany

Printed by Libri Plureos GmbH
in Hamburg, Germany